Fundamentals of Recovery, Regeneration, and Adaptation to Exercise Stress: An Integrated Approach

Nikos C. Apostolopoulos • Gregory C. Bogdanis
Loren R. Seagrave • Michael J. Plyley
Editors

Fundamentals of Recovery, Regeneration, and Adaptation to Exercise Stress: An Integrated Approach

Springer

Editors
Nikos C. Apostolopoulos
Founder microStretching®
and Stretch Therapy™
Toronto, ON, Canada

Loren R. Seagrave
Speed Dynamics
Marietta, GA, USA

Gregory C. Bogdanis
Physical Education and Sport Science
National and Kapodistrian University
of Athens
Athens, Greece

Michael J. Plyley
Department of Kinesiology
Brock University
St. Catharines, ON, Canada

ISBN 978-3-031-44269-8 ISBN 978-3-031-44270-4 (eBook)
https://doi.org/10.1007/978-3-031-44270-4

This Springer imprint is published by the registered company Springer Nature Switzerland AG
The registered company address is: Gewerbestrasse 11, 6330 Cham, Switzerland

To the continued beating of my heart and my breath, I dedicate this book to my granddaughter Juniper Lea.

—Nikos

To my daughters, Nafsika and Christina, in whose talent, curiosity, and determination I see the future of science. You, and all young promising scientists, will contribute your best to advancing health and well-being through the transformative power of exercise.

—Greg

I would like to dedicate this book to my families for understanding my passion in coaching to help athletes realize their goals and dreams, and the athletes who have chosen me to be their coach, believing in my abilities to guide them to their pinnacle and creating long-lasting friendships.

—Loren

To my family, who have always provided me with support and encouragement, and instilled inspiration for the future.

—Mike

Preface

During the pandemic, regular video-conference discussions between colleagues (see below) concerning muscle and connective tissue adaptation from training, and recovery and regeneration from injury/over-use, led to the realization that there was a considerable gap between research and practice in terms of coaching and therapy, that the speed with which new knowledge was emerging made keeping up with advances difficult to maintain, and worse yet, that there was a lack of communication and "translation" of the advances from current research into actual practice, and in turn, how current practice might voice issues and questions to fuel further research.

While research focused on explicit, systematic, controlled work aimed at the growth of theoretical knowledge, practice (therapy and coaching) is concerned with the application of research and practical wisdom in terms of planning, assessment, and treatment. Combining both research and practice legitimizes or refutes anecdotal evidence, itself born out of pragmatism. Together, research and practice would best serve each other by borrowing from, informing, and supporting each other. Within this paradigm, research provides practitioners with facts, evidence, and new ideas that in turn become instrumental in informing practice. On the flip side, the practitioner's perspective in terms of issues, questions, and observations, provides a potential framework to formulate ideas and hypotheses concerning the response of the body to, during, and post, therapy. Within this approach, the human body becomes the common ground between research and practice, providing an interface to develop new research questions, and for incorporating new knowledge and ideas from various fields, such as biology, biochemistry, biomechanics, medicine, physics, and physiology, into practice.

This book will be of interest to researchers who investigate muscle and muscle function, including (in alphabetical order): biologists, biomechanists, exercise biochemists, neuroscientists, physiologists, zoologists, and healthcare professionals who focus on treating and rebuilding muscle function and mobility following musculoskeletal injury, including (in alphabetical order): athletic therapists, biomechanists, chiropractors, kinesiologists, massage therapists, occupational therapists, osteopaths, physiotherapists, physicians, and others interested in the latest developments in research and practice in terms of adaptation, recovery, and regeneration.

The editors of *Fundamentals of Recovery, Regeneration, and Adaptation to Exercise Stress: An Integrated Approach* include Nikos C. Apostolopoulos, PhD, an internationally acclaimed therapist, researcher, and developer of the recovery and regeneration technique, microStretching®, and the founder of Stretch Therapy™; Gregory C. Bogdanis, PhD, Professor, sport and exercise training and physiology (muscle metabolism, fatigue, recovery); Loren R. Seagrave, world-class coach specializing in speed development for Track and Field and American Football; and Michael J. Plyley, PhD, Professor Emeritus, basic and applied researcher (microcirculation, muscle adaptation, training responses).

Toronto, ON, Canada — Nikos C. Apostolopoulos
Athens, Greece — Gregory C. Bogdanis
Marietta, GA, USA — Loren R. Seagrave
St. Catharines, ON, Canada — Michael J. Plyley

Acknowledgments

We gratefully acknowledge the participation of everyone involved in the making of this textbook. Our special thanks go to the contributing authors, whose dedicated time, expertise, and enthusiastic commitments made this volume a reality. We especially appreciate the energy and excitement of the Springer Publishing team (Merry Stuber, Kate Lazaro, Sofia Valsendur, and Anthineth Iroudayanadin), who guided us through this whole process. Their unwavering patience and tenacity ensured that our project went the distance. Finally, our deepest thanks go to our families and all those who encouraged us with their support through these several years of continuous intensity and concentration that is required for this effort to become a reality.

Nikos C. Apostolopoulos
Gregory C. Bogdanis
Loren R. Seagrave
Michael J. Plyley

Contents

1 **Introduction** . . . 1
Nikos C. Apostolopoulos, Gregory C. Bogdanis, Loren R. Seagrave, and Michael J. Plyley

Part I Basic Sciences

2 **Fundamentals of Muscle Damage and Repair of Skeletal Muscle** . . . 13
Michael Kamal, Amanda Johnson, Mai Wageh, Tim Snijders, and Gianni Parise

3 **Central and Neuromuscular Structural and Functional Influences on the Initiation and Control of Movement** . . . 41
David G. Behm

4 **Proprioception—The Sixth Sense: Understanding Our Relationship with the Internal and External Environment** . . . 75
Nikos C. Apostolopoulos

5 **Hormonal and Metabolic Responses During Exercise and Recovery** . . . 105
Anthony C. Hackney

6 **Microvascular Blood Flow in Skeletal Muscle in Health and Injury** . . . 123
Coral L. Murrant

7 **Mechanotransduction in Skeletal Muscle** . . . 163
Eirini Chatzinikita, Maria Maridaki, and Anastassios Philippou

8 **Muscle Damage Induced by Eccentric Exercise, Recovery and Adaptations** . . . 185
Kazunori Nosaka and Trevor C. Chen

9 **Inflammatory Response to Different Types of Exercise** 213
Konstantinos Papanikolaou, Dimitrios Draganidis,
Ioannis G. Fatouros, and Athanasios Z. Jamurtas

10 **Addition and Loss of Serial Sarcomeres—Implications for Muscle Mechanical Function** 235
Avery Hinks and Geoffrey A. Power

Part II Applied Sciences

11 **Training Adaptation and Maladaptation in Athletes** 283
Jonathon Weakley, Shona L. Halson, and Iñigo Mujika

12 **Heart Rate Variability as a Means of Assessing Workload Effects and Recovery During Exercise Training** 303
Fabio Y. Nakamura, Petrus Gantois, Christoforos Giannaki,
Angelos Vlahoyiannis, and Gregory C. Bogdanis

13 **Key Considerations in the Preparation of the Elite Athlete: Planning, Performance, and Recovery** 317
Andrew N. Guard

14 **Interactions Between Sleep and Exercise Performance** 347
Angelos Vlahoyiannis, Gregory C. Bogdanis, Fabio Y. Nakamura,
Petrus Gantois, and Christoforos D. Giannaki

15 **Neurophysiology of Mindfulness in Sports—From Performance to Rehabilitation** 365
Ann Friedman, Robin Jump, and Alejandro Chaoul

16 **Application of Sport Nutrition and Sports Psychology to Recovery** 391
Ian Rollo, Clyde Williams, and Leslie Podlog

17 **Vitamins, Minerals, and Antioxidants** 407
Anatoli Petridou and Vassilis Mougios

Part III Manual Therapies

18 **Effects of Anti-Inflammatory and Analgesic Drugs on Recovery and Performance in Sports** 427
Glyn Howatson and Tommy Lundberg

19 **Stretching to Improve Muscle Performance, Recovery, and Health** 445
Olyvia Donti and Gregory C. Bogdanis

20 **The Genesis and Evolution of Low-Intensity Gentle Stretching—microStretching®** 479
Nikos C. Apostolopoulos

21 **Healing Champions: The Power of Physical Therapy in Athlete Recovery and Performance** ... 523
Tugce Kalaycioglu and Gul Baltaci

22 **Acupuncture: It's Relationship to Recovery, Regeneration and Function** ... 555
Colin Harris and Richard Hazel

23 **Neuromusculoskeletal Complaints Related to the Spine and Peripheral Joints** ... 571
Sean Abdulla, Simon Wang, Eric St-Onge, and Karin Hammerich

24 **Manual Soft-Tissue Manipulation and Its Use in Recovery, Regeneration, and Adaptation to Exercise Stress** ... 597
Seth Forman and Kika Mela

Part IV Machine Based Therapies

25 **In the Pursuit of Excellence—The Importance of Electrical Stimulation Modalities for Athletics** ... 635
Amir Pakravan

26 **Pulsed Electromagnetic Field Therapy** ... 647
Siddharth M. Agrawal

27 **Photobiomodulation: The Use of Specific Light Wavelengths to Accelerate Recovery, Repair, and Regeneration** ... 675
Jennifer Novak

28 **Cryotherapy: Extreme Cold-Based Physiological Stimulation to Improve Recovery and Enhance Adaptation to Exercise** ... 713
Giovanni Lombardi, Paolo Capodaglio, and Ewa Ziemann

29 **Musculoskeletal Extracorporeal Shockwave Therapy in Sports Medicine** ... 737
Nikolaos Malliaropoulos and Amir Pakravan

30 **Hyperbaric Oxygen Therapy** ... 745
Navid Moghadam, Ali Ahmadzadeh Amiri, and Amir Ahmadzadeh Amiri

Last Word: Challenges and Opportunities Ahead ... 771

Index ... 775

About the Editors

Nikos C. Apostolopoulos, PhD founder of microStretching® and Stretch Therapy™, is a recovery and regeneration specialist focused on the use and efficacy of low-intensity passive and active stretching; his research concerns the adaptive responses of muscle and connective tissue, and the role played by rest and recovery in the adaptation process. Nikos has worked both in Canada and abroad, with high performance professional and amateur athletes, including Olympic and Paralympic competitors, and provided care for numerous individuals following injury and various musculoskeletal disorders. Both amateur athletic and professional sports organizations (National Basketball Association (NBA), National Football League (NFL), National Hockey League (NHL), Major League Soccer (MLS), Canadian Football League (CFL), English Premier League (EPL)) have sought Nikos to treat athletes after injury during competition, and as part of long-term prevention programs.

In a recent publication in Springer Nature ("Stretch Intensity and the Inflammatory Response: A Paradigm Shift"), Nikos reported on his groundbreaking research on the importance of the magnitude of stretch intensity and how it influences the adaptation response of muscle and connective tissue. Nikos has developed several manuals on microStretching® techniques for both clinicians and coaches. He is a member of the International Association for the Study of Pain and the International Society of Exercise and Immunology, and has served as Chief Scientific Officer for several biotechnology companies researching endocannabinoids and psychedelics and their effects on the body's inflammatory responses. Nikos has lectured internationally and has also served as an editor and invited reviewer for numerous scholarly journals. Lastly, Nikos has explored philosophical issues in sport and the adaptation of athletes in a recent chapter ("Selling the Dream: The Darker Side of the 'Superman' Athlete") in the Springer publication, *Handbook of Theory and Research in Cultural Studies and Education*.

Email: nikos.apostolopoulos@microstretching.com

Gregory C. Bogdanis, PhD, FACSM, FECSS is a Professor of Sport and Exercise Training at the School of P.E. and Sport Science of the National and Kapodistrian University of Athens. He is also the General Secretary of the Hellenic Society of Biochemistry and Physiology of Exercise (http://www.eevfa.gr) and Fellow of the American College of Sports Medicine (FACSM) and the European College of Sport Science (FECSS). Prof. Bogdanis has studied exercise physiology at Loughborough University, UK (MSc and PhD) and has conducted post-doctoral research at the Department of Histochemistry and Metabolism of the Hellenic Sports Research Institute.

His research work focuses on exercise physiology and muscle biochemistry applied to sport and exercise, with seminal work on muscle fatigue and recovery. His recent work is on muscle functional and architectural adaptations following strength and power training, with emphasis on power and eccentric training, stretching training, as well as on oxidative stress and inflammation during high intensity exercise and training. Over the last 30 years, Prof. Bogdanis has published more than 170 papers in national and international peer-reviewed journals (>8200 citations, h-index: 48, i10 index: 118), 13 books or book chapters, and >450 abstracts in National and International Congresses. In terms of research achievements, Prof. Bogdanis is listed in the "Top 2%" scientists in the Sports Science field worldwide for the last 5 years (https://topresearcherslist.com/Home/Profile/828122). He serves as Section Editor of *European Journal of Sport Science*, *Journal of Science and Medicine in Sport*, and *Journal of Human Kinetics*, and is a member of the Editorial Board of several high-impact journals.

Email: gbogdanis@phe.d.uoa.gr, gregbogdanis@gmail.com

Loren R. Seagrave, BSc is currently serving as the Head Coach and Director of Sport Performance at the Shanghai Elite Athletes Center in China. As an Athletics Coach (Track and Field Specialist), Seagrave has guided numerous Olympic and World Champion Medalists in the sprints, hurdles, and jumping events for the past 50 years. As a Coach Educator, Seagrave has been the original editor for the USA Track & Field, Central America and Caribbean, and the World Athletics, formerly the International Amateur Athletic Federation, Coach Education Program. In 1990, Seagrave founded "Speed Dynamics," a company that promoted coach education through instructional videos and seminars. Seagrave launched the unique concept of Speed and Performance Training for Developmental Youth Athletes in all sports by founding Velocity Sports Performance, where young athletes learned how to be faster so they could get off the bench, get into the game, play more, and enjoy physical activity and sport the rest of their life

Email: lorenseagrave@aol.com

Michael J. Plyley, PhD, FCSEP is Professor Emeritus (2019), Department of Kinesiology, Faculty of Applied Health Sciences, Brock University, Ontario, Canada. He is a former President of the Canadian Society for Exercise Physiology (1989–1990) and the Sports Medicine & Science Council of Canada (1994–1995). Mike studied biophysics and exercise physiology at Western University (BSc

(Hon.), Biophysics, 1968–1972; PhD, Biophysics, 1972–1977; Post-doc Fellow, Kinesiology, 1977–1979). Initially appointed to the Department of Physical Education, Ball State University (1979–1980), Mike moved to the Faculty of Kinesiology & Physical Education, University of Toronto (1980–2001), and later to the Faculty of Applied Health Sciences, Brock University (2002–2019). His career has included positions of Associate Dean, Research and Graduate Studies (University of Toronto and Brock University), and Dean of Graduate Studies (Brock University, 2011–2016). Mike's research has concerned both basic and applied research, involving human subjects and animal models. The focus of the research has aimed at developing a better understanding of how the number and geometrical arrangement of the capillaries supplying muscle fibers affect the supply of oxygen, nutrients, and other physiologically active materials to, and the removal of waste by-products from, skeletal muscle in health and disease. The research includes the etiology of angiogenesis in skeletal muscle, and the relationship of capillary gains (and losses) to various fiber attributes, such as fiber size, metabolism, mitochondrial supply, and trans-sarcolemmal transport activity. Mike has published over 100 scientific papers/reports, 15 books/book chapters, and over 200 conference presentations. Mike was elected as a Fellow, Canadian Society of Exercise Physiology in 2020.
Email: mplyley@brocku.ca

Chapter 1
Introduction

Nikos C. Apostolopoulos, Gregory C. Bogdanis, Loren R. Seagrave, and Michael J. Plyley

> Declare the past, diagnose the present, foretell the future; practice these acts.
> As to diseases, make a habit of two things to help, or at least to do no harm.
>
> –Hippocrates, *Epidemics, Book 1, Section XI*

Skeletal muscle is a remarkable example of structure matching function. Through its contractile and metabolic arrangement, supported by a well-developed vascular and neural supply, and effective adaptive system, muscle functions to support our overall health by preserving skeletal integrity, providing force development for various body movements, and generating heat [1]. The approximately 600 or so, named skeletal muscles of the human body account for 40–45% of the body mass and collectively, muscle comprises the largest organ system in the body [2]. Therefore, an understanding of the structure and function of muscle is of both practical and theoretical importance to understanding muscle responses to acute exercise and adaptation to chronic training or injury [3–7] **(see Chapter 3. Central and Neuromuscular Structural and Functional Influences on the Initiation and Control of Movement;**

N. C. Apostolopoulos (✉)
Founder microStretching® and Stretch Therapy™, Toronto, ON, Canada
e-mail: nikos.apostolopoulos@microstretching.com

G. C. Bogdanis
Physical Education and Sport Science, National and Kapodistrian University of Athens, Athens, Greece

L. R. Seagrave
Speed Dynamics, Marietta, GA, USA

M. J. Plyley
Department of Kinesiology, Brock University, St. Catharines, ON, Canada

N. C. Apostolopoulos et al. (eds.), *Fundamentals of Recovery, Regeneration, and Adaptation to Exercise Stress: An Integrated Approach*,
https://doi.org/10.1007/978-3-031-44270-4_1

Chapter 4. Proprioception—The Sixth Sense: Understanding Our Relationship with the Internal and External Environment; Chapter 5. Hormonal and Metabolic Responses During Exercise and Recovery; Chapter 6. Microvascular Blood Flow in Skeletal Muscle in Health and Injury; Chapter 11. Training Adaptation and Maladaptation in Athletes).

Skeletal muscle is quite capable of adapting, repairing, or even regenerating itself, depending on the nature and extent of the cellular disruption and damage. In response to exercise training or injury, the regular disruption of cell homeostasis initiates structural and functional adaptations through a series of adaptive pathways within cellular and sub-cellular structures [1, 8–10]. Crucial components in shaping recovery and regeneration from exercise, and promoting the intended adaptive responses to training are proper nutritional and fluid intake, appropriate rest and sleep, and downloading of extraneous stress **(see Chapter 12. Heart Rate Variability as a Means of Assessing Workload Effects and Recovery During Exercise Training; Chapter 13. Key Considerations in the Preparation of the Elite Athlete: Planning, Performance, and Recovery; Chapter 14. Interactions Between Sleep and Exercise Performance; Chapter 15. Neurophysiology of Mindfulness in Sports—From Performance to Rehabilitation; Chapter 16. Application of Sport Nutrition and Sports Psychology to Recovery; Chapter 17. Vitamins, Minerals, and Antioxidants).**

The adaptive response to training includes all components of the muscle, including cell proteins, metabolic pathways, neuromuscular function, vascular network, and support system. The changes in cell structure and function result from alterations in gene expression, and occur via direct cooperation between the fibres and the central nervous system [3, 4, 11]. For instance, Williams et al. [12] have recently shown that 6 weeks of cycle ergometer training (1 hour, 5 days/week, for 6 weeks) resulted in up-regulation of 641 genes and down-regulation of 176 genes (of the 13,108 genes found in the biopsies).

Unfortunately, although musculoskeletal tissues are built to withstand the rigors of most activities of daily life, significant alterations in the mechanical stress brought about by unexpected, violent, and high strain movements, falls and collisions, and chronic overuse, often lead to injury. This damage and the accompanying soreness results in a dysfunction of the muscle proprioceptors [7, 13], and the resulting disruption of cellular homeostasis triggers an integrated series of inflammatory and mechanosensitive events [9] **(see Chapter 2. Fundamentals of Muscle Damage and Repair of Skeletal Muscle; Chapter 7. Mechanotransduction in Skeletal Muscle; Chapter 8. Muscle Damage Induced by Eccentric Exercise, Recovery and Adaptations; Chapter 9. Inflammatory Response to Different Types of Exercise; Chapter 10. Addition and Loss of Serial Sarcomeres—Implications for Muscle Mechanical Function).**

It has been estimated that approximately "1.7 billion people worldwide" suffer from adverse musculoskeletal conditions yearly [14], and roughly "10–55% are categorized as acute injuries" [15, 16]. While many of these injuries (90%) are bruises or strains the more severe types of injury produce chronic pain, dysfunction, increased recurrence, and compartment syndrome [15–17].

In the workplace, 126.6 million Americans (1 in 2 adults) report work-related musculoskeletal injuries annually, at a cost estimated at $213 billion in treatment, care, and lost wages [18], and "a median of 8 days away from work" [19].

Young individuals are not immune to injury as it has been estimated that "more than 3.5 million American children, 14 years of age and younger", get hurt annually playing sports or participating in recreational activities [20]. About "775,000 of these young individuals are treated in hospital emergency rooms for sports-related injuries resulting from falls, being struck by an object, collision, or over exertion during organized or informal sports activities" [20]. The estimated annual cost of managing sports injury is estimated to be $2 billion [21].

Of course, these data do not include the vast numbers of contusions, strains, sprains, and other "nicks" that occur on a regular basis in daily life that are treated through self-care. Recovery in these instances often only requires following one of the current self-care protocols, with or without pain management using analgesic or anti-inflammatory medication [22, 23] **(see Chapter 18. Effects of Anti-Inflammatory and Analgesic Drugs on Recovery and Performance in Sports).**

Following injury, repair and regeneration of muscle proceeds through a series of three recognized phases: degeneration, inflammation, regeneration and remodelling [24–26]. Both the time and the effectiveness of the repair varies according to the nature of the injury, i.e., injured muscle regenerates slowly, and sometimes produces incomplete functional recovery, potential scar formation, and/or a tendency for recurrent injuries [22, 25, 27–29], which presents a challenge for both researchers and practitioners. Therefore, with the primary goal of the treatment plan being to minimize further damage, while promoting healing, a thorough knowledge of the repair and regeneration mechanisms of adaptation, including inflammation, may help the practitioner reduce or avoid imminent hazards and accelerate the return to activity [17].

While some of these injuries mend on their own, many need specialized treatment to increase recovery responses. A number of approaches have been studied in terms of their efficacy in improving muscle repair and regeneration after injury, including ultrasound, mechanical stimulation, eccentric contractile activity, etc. [28, 30]. Interestingly, research would suggest that muscle undergoing repeated damage and repair exhibits a diminished response [10, 31].

The Two Worlds of Basic Science Researchers and Physical Therapists

On the surface, basic research and clinical practice would seem to be two solitudes, as there is a number of fundamental and cultural differences between the two pursuits. Scientists and practitioners work in different settings, perform their craft along disparate lines of inquiry, and most critically, speak different languages. These dissimilarities can hinder communication, leading to the loss of critical information that might inform treatment, or provide important practical questions for developing significant research directions, with the result that opportunities are lost.

Basic research occurs in university, hospital, or industry-based labs, using a strict set of procedures for knowledge generation. The steps involve data collection, analysis, and hypothesis testing to produce knowledge that is mainly theoretical in

Table 1.1 Therapy modalities in therapeutic use[a]

Acupuncture: It's Relationship to Recovery, Regeneration, and Function—Chapter 22
Cryotherapy: Extreme Cold-Based Physiological Stimulation to Improve Recovery and Enhance Adaptation to Exercise—Chapter 28
Healing Champions: The Power of Physical Therapy in Athlete Recovery and Performance—Chapter 21
Hyperbaric Oxygen Therapy—Chapter 30
In the Pursuit of Excellence—The Importance of Electrical Stimulation Modalities for Athletics—Chapter 25
Musculoskeletal Extracorporeal Shockwave Therapy in Sports Medicine—Chapter 29
Neuromusculoskeletal Complaints Related to the Spine and Peripheral Joints—Chapter 23
Photobiomodulation: The Use of Specific Light Wavelengths to Accelerate Recovery, Repair, and Regeneration—Chapter 27
Pulsed Electromagnetic Field Therapy—Chapter 26
Manual Soft-Tissue Manipulation and Its Use in Recovery, Regeneration, and Adaptation to Exercise Stress—Chapter 24
Stretching to Improve Muscle Performance, Recovery, and Health—Chapter 19
The Genesis and Evolution of Low-Intensity Gentle Stretching—microStretching®—Chapter 20

[a]Listed alphabetically

nature, and aimed at broad application to future problems. In practice, biomedical researchers use categorical definitions of injury, and develop animal or computer models, with standardized conditions and controlled environments, to ensure reliable and reproducible models to simulate real-life injuries and body responses [32, 33].

The main objectives of a therapist and the treatment plan include enhancing healing, while managing pain and controlling swelling, and getting the individual back to "normal" activity, with a plan of self-care activities to help in remaining healthy and avoiding re-injury. Practitioners are involved in the delivery of care to individuals with specific problems in settings within the community or health centres. The focus of their work is on resolving one-off injuries/physical issues through the administration of a specific treatment plan involving various skills and modalities. A key research element for therapists in developing the treatment plan is obtaining information concerning the nature of the injury/issue, the medical and injury history of the individual, and the results of patient assessment and various diagnostic tools. From this information, the practitioner develops the treatment plan, including what is the intended end point to be achieved by the treatment plan, what needs to be done to achieve the established goals, what are the various options in terms for treatment techniques and modalities (Table 1.1), and what is the anticipated duration of treatment? Of course, injuries and injury history being what they are, there are often specific issues of concern that develop with any given case, which requires tailoring, or even changing, the plan to meet the situation presented.

Research and practice can be seen as two sides of the same coin, with researchers directed at generating knowledge concerning the "why, how, and when" of phenomena, and practitioners looking for the latest utilitarian information to understand "what" does this material mean in terms of helping us right now with our current issue [34, 35], which in the case of therapists means "how" can we refine existing procedures, or develop future applications, to aid in injury management and promote the health and well-being of individuals.

Fundamental to the ongoing development of best practice in therapeutic techniques and the evolution of new modalities is a thorough understanding of the latest research in recovery and regeneration from the perspectives of both training and injury, and integration of this knowledge into current practice [36].

Practitioners have considerable need for up-to-date information in order to follow the latest research findings, to acquire information on advances in therapeutic modalities, and to learn about new developments in technology and techniques. Such information is important to the development of new techniques and optimal treatment plans. However, the speed with which research and technology progress today, the different directions that inquiries take, and the infusion of new terms in the language of discovery, makes being and staying informed both time consuming and challenging. It is easy to fall behind in the latest developments in the field, often leading to significant 'knowledge gaps'.

New studies in basic science are published each day on both the cellular and molecular components of adaptation, recovery, and regeneration, and on the implications of this knowledge in terms of health and well-being. While many of these discoveries are important in their own right as they represent critical knowledge for the researchers to continue to unravel biological processes, they are also of potential importance in improving therapeutic treatment—if the information is adapted into language and formats useful to practitioners.

Communication and the Exchange of Ideas and Thoughts

As described above, there are several operational differences that impede the development of a regular dialogue between researchers and health practitioners in terms of exchanging essential information and ideas, including (a) different work settings and environments (i.e., university or private research lab vs. community clinic or hospital); (b) few opportunities or forums to exchange viewpoints and thoughts (i.e., few regular joint conferences, existing publications are in purpose-specific, divergent journals, with mismatched messages); (c) a lack of a common language to translate the research into useable material for application by practitioners, or for practitioners to be able to voice what they see as the next critical pieces of knowledge they need to continue the development of best practice (i.e., each group has a vernacular specific to the area in which they work, and there are no "translators" of information); (d) culture—each group focuses on different problems, asks unique questions, and has specialized goals/objectives (i.e., the creation of new theory and increasing understanding of general topics vs. helping individuals through effective treatment of specific physical issues and development of new applied techniques) [37].

Added to these systematic differences, scientists more often than not, publish their research for other scientists, and have little interest in translating their findings for practical application. And, practitioners often do not have the time within busy treatment days to search for the latest basic research studies, and often lack the background to extract the essence of the research for application in their practice. What happens then is that important research that could be used to improve current practice or develop new practice, languishes in journals.

Lastly, it is important to acknowledge that many clinical decisions are based to a great degree on judgements made by practitioners based on chief complaint, medical/injury history, and physical assessment, which often results in any number of questions as to etiology of injury and suggests options for the treatment plan. In this process, there are often pieces of "anecdotal information" that are gleaned by practitioners, which can provide vital data that can be used as a framework for asking clinical questions aimed at improving therapeutic techniques or modalities, or in the formulation of new research questions concerning mechanisms of injury, recovery, and regeneration.

Clearly, practitioners would greatly benefit from ongoing, regular interaction with research scientists to discuss questions and issues and share their observations, and in turn scientists would benefit from the insights and ideas provided by the practitioners [37]. This approach, in which knowledge and ideas are actively shared, represents the best opportunity for researchers and therapists to find a common footing to develop knowledge, advance research, improve treatment, and help individuals perform better, work without pain, and lead a rich and rewarding lifestyle [38].

Thus, the issues here are relatively clear, the solution less so. First, we need to understand that similar to the two-headed Janus, the transfer of knowledge is a two-way process, which can be best understood as being circular in nature (Fig. 1.1). This "Knowledge-to-Action" and complementary "Action-to-Research" scheme is an important concept that provides a potential plan for developing an "integrative systems approach" to connecting the latest advances in injury and muscle regeneration research to practical application, and in turn, identifies gaps and questions in current practice to highlight and focus on questions that need to be addressed in future health research [38, 39].

Fundamentals of Recovery, Regeneration, and Adaptation to Exercise Stress: An Integrated Approach

Our interest in formulating this book began as an exercise amongst ourselves concerning how best to 'translate" and communicate the latest knowledge developed from basic research to use effectively in current practice in therapy and coaching, either to improve existing techniques or to develop new methods of practice. In turn, we recognized the need for those involved in current practice to voice relevant issues and pertinent questions that could fuel future research. Thus, a key objective for this book was to provide an opportunity for crosstalk between researchers and practitioners in terms of how muscle works, what can happen adversely with exercise, and what can be done to promote recovery, regeneration, and adaptation (Fig. 1.1).

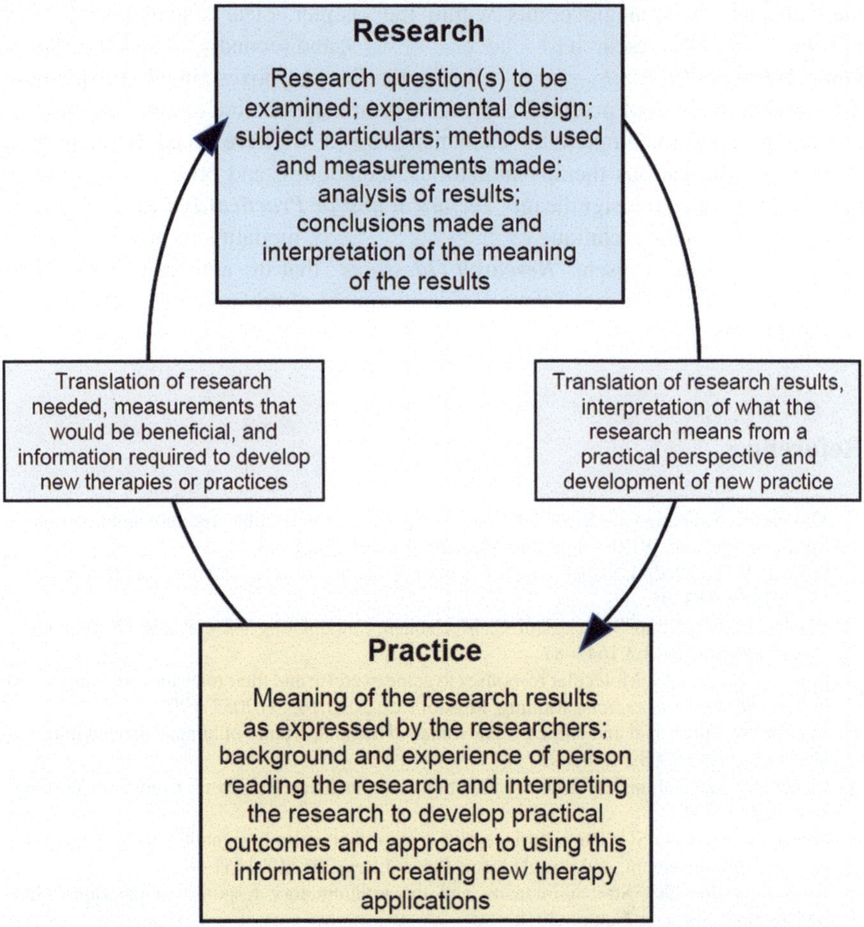

Fig. 1.1 Recommended interactive pattern between basic scientists and physical therapists modelled after the Knowledge to Action Framework of Graham et al. [38]

Clearly, given the diverse obstacles that confront basic scientists and health practitioners in having a meaningful exchange of information and ideas, it is important to work across existing boundaries to build effective working relationships to explore common problems [40].

Therefore, the aims of this book were to explore the latest research regarding muscle function, injury, and recovery/rehabilitation from a multi-disciplinary perspective, to highlight critical knowledge concerning research and application issues and questions, and to integrate the latest information to improve our knowledge, understanding, and best practice on how muscle reacts, recovers, regenerates, and adapts to a physical stimulus and changes in its environment.

To actualize the process of bridging the gap between researchers and practitioners, and to begin the exchange of information and ideas, we have requested that authors of chapters on "basic" and "applied" muscle/exercise research include in their chapters at strategic points within the chapter, clear statements of "***Key Findings***" from the research (in light blue colour), and secondly, to develop "***Take-home Messages***" (in light green colour) in the form of boxes aimed at explaining the research in terms of potential current, or future, application to clinical practice.

Similarly, and to complete the exchange process, we have asked the authors of chapters on the various therapy modalities, techniques, and protocols to provide boxes that outlined the significant "***Technical and/or Practical Issues***" that practitioners have with the technique (in light blue colour), modality, or protocol, and in addition, boxes that present "***Research Questions***" that the authors feel would be beneficial to them in terms of their use/development of the modality, technique, or protocol (in light green colour).

References

1. Mukund K, Subramaniam S. Skeletal muscle: a review of molecular structure and function, in health and disease. WIREs Syst Biol Med. 2020;12(e1462):1–46.
2. Frontera WR, Ochala J. Skeletal muscle: a brief review of structure and function. Calcif Tissue Int. 2015;96:183–95.
3. Blaauw B, Schiaffino S, Reggiani C. Mechanisms modulating skeletal muscle phenotype. Compr Physiol. 2013;3:1645–87.
4. Egan B, Sharples AP. Molecular responses to acute exercise and their relevance for adaptations in skeletal muscle to exercise training. Physiol Rev. 2023;103(3):2057–170.
5. Koryak YA. Functional and clinical significance of the architecture of human skeletal muscles. Human Physiol. 2008;34:482–92.
6. Lieber RL. Skeletal muscle structure, function, and plasticity. Baltimore: Lippincott Williams & Wilkins; 2009.
7. Proske U, Gandevia SC. The proprioceptive senses: their roles in signaling body shape, body position and movement, and muscle force. Physiol Rev. 2012;92:1651–97.
8. Apostolopoulos NC. Stretch intensity and the inflammatory response: a paradigm shift. Switzerland: Springer-Nature; 2018.
9. Smith JA, Murach KA, Dyar KA, Zierath JR. Exercise metabolism and adaptation in skeletal muscle. Nat Rev Mol Cell Biol. 2023;24:607–32.
10. Tidball JG. Mechanisms of muscle injury, repair, and regeneration. Compr Physiol. 2011;1(4):2029–62.
11. Kapreli E, Athanasopoulos S. The anterior cruciate ligament deficiency as a model of brain plasticity. Med Hypotheses. 2006;67:645–50.
12. Williams K, Carrasquilla GD, Ingerslev LR, Hochreuter MY, Hansson S, Pillon NJ, Donkin I, Versteyhe S, Zierath JR, Kilpeläinen TO, Barrès R. Epigenetic rewiring of skeletal muscle enhancers after exercise training supports a role of whole-body function and human health. Mol Metab. 2021;53:101290.
13. Weerakkody NS, Percival P, Canny BJ, Morgan DL, Proske U. Force matching at the elbow joint is disturbed by muscle soreness. Somatosens Mot Res. 2003;20:27–32.
14. Cieza A, Causey K, Kamenov K, Hanson SW, Chatterji S, Vos T. Global estimates of the need for rehabilitation based on the Global Burden of Disease Study 2019: a systematic analysis for the Global Burden of Disease Study 2019. Lancet. 2020;396(10267):2006–17.
15. Delos D, Maak T, Rodeo SA. Muscle injuries in athletes: enhancing recovery through scientific understanding and novel therapies. Sports Health. 2013;5(4):346–52.
16. Maffulli N, Del Buono A, Oliva F, Via AG, Frizziero A, Barazzuol M, Brancaccio P, Freschi M, Galletti S, Lisitano G, Melegati G, Nanni G, Pasta G, Ramponi C, Rizzo D, Testa V,

Valent A. Muscle injuries: a brief guide to classification and management. Transl Med UniSa. 2015;12:14–8.
17. Fernandes TL, Pedrinelli A, Hernandez A. Muscle injury – physiopathology, diagnosis, treatment and clinical presentation. Rev Bras Ortop. 2011;46(3):247–55.
18. U.S. Bone and Joint Initiative. The impact of musculoskeletal disorders on Americans – opportunities for action. 2016. Retrieved from: https://www.boneandjointburden.org/docs/BMUSExecutiveSummary2016.pdf
19. NIOSH. NIOSH Workers Health Chartbook 2004. Washington, DC: NIOSH Publication No. 2004-146; 2004.
20. John Hopkins Medicine. Sports injury statistics. 2023. Retrieved from www.hopkinsmedicine.org/health/conditions-and-diseases/sports-injuries/sports-injury-statistics
21. Merkel DL. Youth sport: positive and negative impact on young athletes. J Sports Med. 2013;4:151–60.
22. Mackey A, Mikkelsen U, Magnusson S, Kjaer M. Rehabilitation of muscle after injury–the role of anti-inflammatory drugs. Scand J Med Sports. 2012;22(4):e8–e14.
23. Rahusen FT, Weinhold PS, Almekinders LC. Nonsteroidal anti-inflammatory drugs and acetaminophen in the treatment of an acute muscle injury. Am J Sports Med. 2004;32(8):1856–9.
24. Forcina L, Cosentino M, Musarò A. Mechanisms regulating muscle regeneration: insights into the interrelated and time-dependent phases of tissue healing. Cells. 2020;9(5):1297–325.
25. Huard J, Li Y, Fu FH. Muscle injuries and repair: current trends in research. J Bone Joint Surg. 2002;84(5):822–32.
26. Jarvinen TAH, Jarvinen TLN, Kaariainen M, Kalimo H, Jarvinen M. Muscle injuries: biology and treatment. Am J Sports Med. 2005;33:745–64.
27. Barroso GC, Thiele ES. Muscle injuries in athletes. Rev Bras Ortop. 2011;46:354–8.
28. Järvinen IAH, Järvinen M, Kalimo H. Regeneration of injured skeletal muscle after the injury. Muscles Ligaments Tendons J. 2013;3(4):337–45.
29. Stožer A, Vodopivc P, Bombek LK. Pathophysiology of exercise-induced muscle damage and its structural, functional, metabolic, and clinical consequences. Physiol Res. 2020;69(4):565–98.
30. Laumonier T, Menetrey J. Muscle injuries and strategies for improving their repair. J Exp Orthop. 2016;3:15–24.
31. Yang W, Hu P. Skeletal muscle regeneration is modulated by inflammation. J Orthop Translat. 2018;13:25–32.
32. Hausken-Sutter SE, af Gennäs KB, Schubring A, Grau S, Jungmalm J, Barker-Ruchti N. Interdisciplinary sport injury research and the integration of qualitative and quantitative research. BMC Med Res Methodol. 2023;23:110.
33. Weber B, Lackner I, Haffer-Luntzer M, Palmer A, Pressmar, Scharffetter-Kochanek K, Knoll B, Schrezenemeier H, Relja B, Kalbitz M. Modeling trauma in rats: similarities to humans and potential pitfalls to consider. J Transl Med. 2019;17:305.
34. Anandarajan M, Lippert SK. Competing mistresses? Academic vs. practitioner perceptions of systems analysis. J Comput Inf Syst. 2016;46:114–26.
35. Chang Y-W. A comparison of researcher – practitioner collaborations in library and information science, education, and sociology. J Librariansh Inf Sci. 2013;51:208–17.
36. Chen J, Zhou R, Feng Y, Cheng L. Molecular mechanisms of exercise contributing to tissue regeneration. Signal Transduct Target Ther. 2022;7:383–407.
37. McClintock C. Spanning boundaries of knowledge and organization: collaborations for mind and management in higher education. Organization. 2001;8(2):349–57.
38. Graham I, Logan J, Harrison M, Straus S, Tetroe J, Caswell W, Robinson N. Lost in knowledge translation: time for a map. J Contin Educ Heal Prof. 2006;26:13–24.
39. Affret A, Prigent O, Porcherie M, Aromatario O, Cambon L. Development of a knowledge translation taxonomy in the field of health prevention: a participative study between researchers, decision-makers and field professionals. Health Res Policy Syst. 2020;18:91–112.
40. Rhoten D, Parker A. Education. Risks and rewards of an interdisciplinary research path. Science. 2004;306(5704):2046.

Part I
Basic Sciences

Chapter 2
Fundamentals of Muscle Damage and Repair of Skeletal Muscle

Michael Kamal, Amanda Johnson, Mai Wageh, Tim Snijders, and Gianni Parise

Introduction

Exercise is known to improve skeletal muscle mass, function, and metabolism, and provide protection against obesity, insulin resistance, atherosclerosis and other metabolic diseases [1–3]. As the largest organ in the body, it is not surprising that skeletal muscle remodeling in response to stress or damage is highly implicated in overall health. Ultrastructural muscle fibre damage is part of a healthy growth and remodeling response, such as after resistance exercise training. However, repetitive exercise that elongates the muscle during contraction (known as an eccentric motion), like with downhill running, can produce extensive damage and painful side effects. For the purposes of this chapter, we will limit the discussion to damage caused by exercise, also known as exercise-induced muscle damage.

Skeletal muscle is responsible for providing the contractile forces needed for locomotion (i.e., movement), respiration, and posture, which it does through the action of thousands of muscle fibres, also known as myofibres [4, 5]. The contractile units of skeletal muscle are sarcomeres, which are composed of thick filaments (myosin) and thin filaments (actin) that interact with each other to contract the muscle fibre and produce force (Fig. 2.1a). The lateral boundaries of each sarcomere are the Z-bands, which are responsible for anchoring the thin filaments together [4, 5]. Following a bout of damaging exercise, under a high-resolution electron microscope, the Z-bands appear in disarray and are often disconnected (Fig. 2.1b). While

M. Kamal · A. Johnson · M. Wageh · G. Parise (✉)
Department of Kinesiology, Faculty of Science, McMaster University, Hamilton, ON, Canada
e-mail: pariseg@mcmaster.ca

T. Snijders
Department of Human Biology, NUTRIM School for Nutrition and Translational Research in Metabolism, Maastricht University Medical Centre, Maastricht, The Netherlands

N. C. Apostolopoulos et al. (eds.), *Fundamentals of Recovery, Regeneration, and Adaptation to Exercise Stress: An Integrated Approach*,
https://doi.org/10.1007/978-3-031-44270-4_2

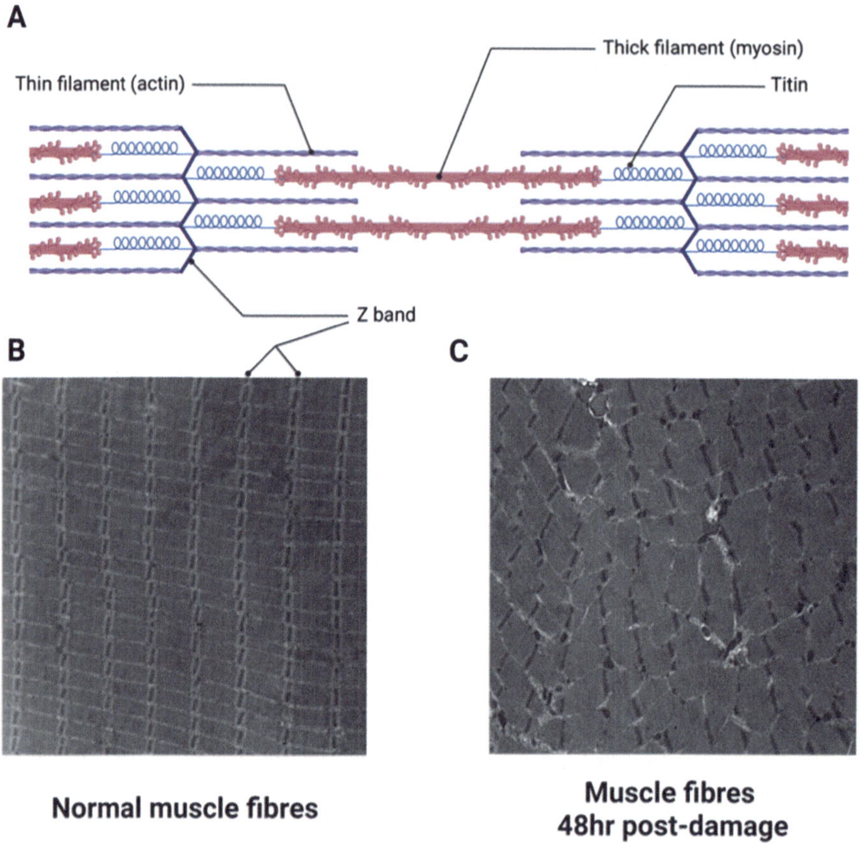

Fig. 2.1 The ultrastructural components of skeletal muscle fibres. (**a**) A diagram of the sarcomere—the contractile unit of skeletal muscle. High-magnification TEM images of skeletal muscle fibres (**b**) at rest or (**c**) 48 h following a single bout of eccentric contractions (M. Wageh, unpublished images)

the predominant cell type within skeletal muscle are the myofibres, there are several other cell types present that are essential for the function and maintenance of the tissue. Satellite cells (discussed in further detail in section "The Role of Satellite Cells in Muscle Regeneration") are the primary drivers of muscle fibre regeneration and repair caused by damage, injury, or disease [2, 4, 6–8]. These cells reside in a niche between the sarcolemma and the basal lamina which is composed of structural proteins such as laminin, collagen IV, and glycoproteins [4]. Satellite cells are implicated in the adaptive response to exercise and are an integral component in the exercise-induced muscle damage recovery process. Other cell types, such as fibroblasts, fibro/adipogenic progenitors, immune cells, and neurons also contribute to this process, and will be discussed in the following sections.

This chapter will provide a thorough characterization of exercise-induced muscle damage, including: (i) the etiology of exercise-induced muscle damage, (ii) the inflammatory response to muscle damage, (iii) the role of satellite cells in skeletal muscle remodeling, (iv) the role of supporting cells and structures in muscle repair and regeneration, (v) muscle recovery following damage, and (vi) sex differences in the response to exercise-induced muscle damage.

> Eccentric exercise produces the greatest amount of damage to muscle tissues compared to other exercise modes. The lengthening of the muscle during contraction stretches the sarcomeres beyond their optimal length, resulting in the tearing of myofibrils, exposing membranes (particularly T-tubules), and disrupting calcium homeostasis.

Etiology of Exercise-Induced Muscle Damage

The severity of exercise-induced muscle damage can vary significantly based on several factors. To best understand the relationship between exercise and muscle damage, it is important to understand the process of skeletal muscle contraction. To initiate a muscle contraction, an action potential is sent to the muscle via an alpha motor neuron terminating at the neuromuscular junction. The result is depolarization of the muscle membrane (or sarcolemma), triggering another action potential to travel along the sarcolemma and down into invaginations within the fibre known as t-tubules. This triggers a release of calcium ions from the sarcoplasmic reticulum, which bind to troponin. Troponin is a small protein that is bound to tropomyosin, which prevents the thick filament, myosin, from binding to the thin filament, actin. Once calcium ions bind to troponin, there is a conformational shift in the protein tropomyosin, exposing the myosin binding sites on actin. The myosin heads then bind to actin and pull the filaments towards the centre of the sarcomere, causing shortening of the myofibre, i.e., a muscle contraction. This entire process is known as excitation-contraction coupling, and while the steps described here result in a shortening of muscle length (concentric contraction), it is also possible to contract a muscle with no change in length (isometric contraction) or with increasing length (eccentric contraction) (Fig. 2.2).

During exercise, repetitive muscle contractions cause breaks in the structure of the myofibres, primarily at the level of the sarcomere. This damage is evidenced by streaming of the Z-band and results in structural disorganization of myofibrillar proteins and the sarcoplasmic reticulum (Fig. 2.1b) [9]. Damage to the sarcoplasmic reticulum dysregulates calcium release and disrupts calcium homeostasis, causing excitation-contraction coupling dysfunction and subsequently reduces muscle force output [10]. Elevated calcium levels increase calcium-dependent protease activity, which attenuates sarcolemmal integrity by cleaving cytoskeletal proteins [11, 12]. Furthermore, damage to myofibres includes disruption of the sarcolemma and release of intramuscular factors into circulation, including creatine kinase, myoglobin and certain metabolites [10, 13]. These circulating factors are often used as

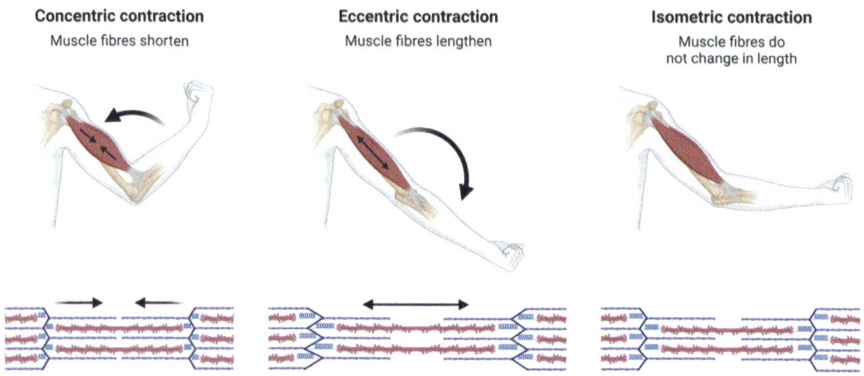

Fig. 2.2 The three types of muscle contractions. (**a**) Concentric contractions occur when muscle fibres shorten as the muscle contracts. (**b**) Eccentric contractions occur when muscle fibres lengthen as the muscle contracts. (**c**) Isometric contractions occur when muscle fibres remain the same length as the muscle contracts. Representative sarcomere diagrams display how the thick and thin myofilaments interact during each contraction type

proxy measures of muscle damage when evaluating post-exercise recovery. In addition to the morphological changes in myofibres following exercise-induced muscle damage, there are also metabolic consequences. These include elevations in temperature, free radical production, and lipid peroxidation, as well as decreased local pH, all of which can trigger an autogenic degradation of the damaged fibres [14].

> Exercise-induced muscle damage leads to a significant decrease in the muscle's ability to generate force—it tends to preferentially affect high-force generating muscle fibres and range of motion. Repair and regeneration may take 5–6 days or longer to return to baseline.

There are several consequences related to extensive damage of muscle fibres following exercise; however, the most recognizable hallmark of exercise-induced muscle damage is the reduced ability to generate force [15]. In the immediate hours following muscle damage, diminished force production is most likely attributed to fatigue [16]. Over the next day or two, an even greater decrease in force occurs as the excitation-contraction coupling dysfunction is compounded with damage to the sarcolemma, myofilaments, and sarcoplasmic reticulum (Fig. 2.3). Moreover, the displacement of sarcomeres beyond their normal length, as typically occurs with eccentric exercise, further renders it difficult to maintain the linear formation that brings about peak muscle force [17, 18].

As discussed earlier, eccentric contractions result in the greatest amount of muscle damage, although concentric or isometric exercises are capable of producing damage as well. There are three types of damage caused by muscle contraction: Type I (delayed onset muscle soreness; Type II (muscle tears ranging from the level

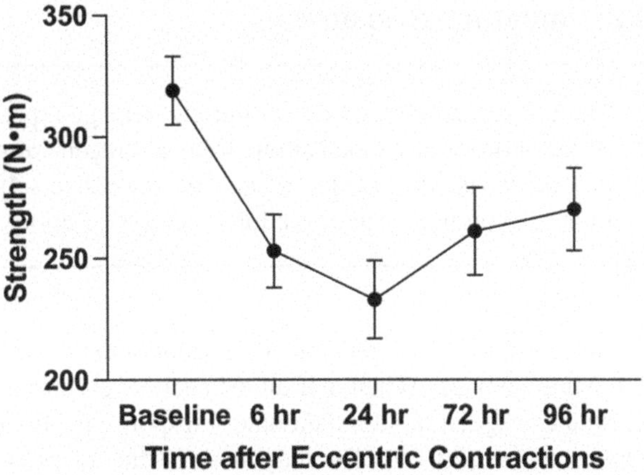

Fig. 2.3 Force Production following Eccentric Exercise. Graph depicting the change in strength after a single session of eccentric exercise. (Data adapted from Nederveen et al. [58])

of the fibre to the fascia), and Type III (muscle soreness or cramps occurring during or immediately after exercise) [15]. Eccentric exercise produces the greatest amount of damage because as the muscle lengthens while attempting to contract, the sarcomeres are stretched beyond their optimal length, resulting in the shearing of myofibrils, exposed membranes (particularly T-tubules), and the disruption of calcium homeostasis [18–20]. However, exercise-induced muscle damage does not affect all muscles equally, i.e., it tends to preferentially affect high-force generating muscle fibres.

Human skeletal myofibres are classified by the isoform expression of myosin heavy chain and are labelled type I (slow oxidative), type IIA (fast oxidative), or type IIX (fast glycolytic) fibres [4, 5]. Type I muscle fibres are fatigue-resistant, utilize mainly oxygen-mediated metabolism, and are more active during aerobic exercise, such as long-distance running [4, 5]. Type IIA/X fibres are quicker to fatigue, utilize aerobic and anaerobic glycolysis, and are commonly recruited for strength-based exercises like weightlifting [4, 5]. These faster type II fibres are more susceptible to eccentric damage compared to the slower type I fibres, most likely due to mechanisms related to fatigability, the ultrastructural composition of the fibre types, and differential tension during exercise [21–24]. For example, the protein titin is the main contributor to elasticity in muscle fibres and exists in less elastic isoforms in fast twitch fibres, making them less compliant and thus more susceptible to damage during eccentric contractions [25, 26].

In addition to the genetic factors that determine the magnitude of muscle damage, there are a number of factors that can influence exercise-induced muscle damage, including the type, intensity, duration, frequency, and volume of exercise [15, 27]. An individual's previous experience with similar stimuli may also attenuate the magnitude of damage incurred due to the repeated bout effect, whereby multiple exposures to the same exercise adds structural integrity through connective tissue deposition, increases motor unit efficiency and synchronization, and allows for more efficient recruitment of synergistic muscles in an effort to prevent damage following subsequent bouts of exercise [14, 28].

Inflammatory Response

The inflammatory response to muscle damage represents the beginning of a series of processes aimed at repairing and regenerating the muscle fibres; this process starts with: (a) the release of vasoactive substances by the damaged fibres, (b) immune cell migration to the site of injury, and (c) repair initiation.

Inflammation is an essential component of the repair response in skeletal muscle following damage. When a stimulus such as eccentric exercise induces extensive muscle damage, a rapid inflammatory response is initiated to signal the start of the repair process [29–35]. The magnitude of this response is dependent on the type, duration, and intensity of exercise performed; however, the fundamental steps are relatively consistent: (a) the release of vasoactive substances by the damaged fibres, (b) immune cell migration to the site of injury, and (c) repair initiation [36]. Within skeletal muscle, the inflammatory response to damage is driven by white blood cells, also known as leukocytes [35, 37, 38]. These cells circulate throughout the vasculature or reside within the muscle, essentially ready and waiting to be triggered by a stressor [35]. Often, these triggers are in the form of small signalling molecules, known as cytokines, which are secreted by other immune cells or the muscle fibre directly [37, 39]. Cytokines can mediate effects on tissues and cells via autocrine signalling (act on the originating cell), endocrine signalling (travel through circulation to act on another cell), or paracrine signalling (act on adjacent cells) [37, 39]. Certain subtypes of cytokines operate as chemoattractants by recruiting immune cells to migrate to the site of damage within the myofibre. Following exercise-induced muscle damage, there is both a local inflammatory response within the injured muscle and a systemic inflammatory response that occurs throughout the entire body.

Local Inflammation After Exercise-Induced Muscle Damage

Exercise-induced muscle damage elicits a significant immune reaction at the site of injury—the result of the damage to myofibres and as part of the timely initiation of repair mechanisms. The local muscle response involves the recruitment and activation of certain leukocytes, primarily neutrophils, macrophages, and T-lymphocytes [29–35]. These inflammatory cells work in concert with myogenic precursor cells to initiate proper muscle repairs following damage. This subsection will discuss the role of these leukocytes on the local inflammatory response to muscle damage after exercise.

Neutrophils make up 50–60% of circulating leukocytes and are the first responders to exercise-induced muscle damage. These immune cells can react as early as 1 h after exercise and peak between 6 and 24 h post-damage before rapidly declining [30, 32, 40, 41]. Neutrophils are primarily responsible for removing debris and secreting chemoattractants, such as the pro-inflammatory interleukin cytokines (IL-1 and IL-8) and tumour necrosis factor alpha, that recruit macrophages to the damaged site [31, 35, 42]. Lysosomal proteases and free O_2 radicals are also released by neutrophils to help break down damaged fibres that need to be removed for the proper repair of myofibres [40, 41]. As a result, in the early post-exercise period, additional damage accumulates before the repair process begins.

Within the pool of leukocytes that respond to exercise-induced muscle damage, macrophages are considered to be the most important to the muscle repair process. Neutrophil infiltration and activity are closely followed by macrophage migration to the damaged myofibres, beginning between 12 and 24 h post-damage [6, 7, 36, 43]. Macrophages phagocytose (i.e., ingest) debris while secreting cytokines and growth factors, which can have either beneficial or detrimental effects on muscle repair depending on their polarization [44–46]. The first set of macrophages that appear in damaged muscle are pro-inflammatory and act to remove cellular debris via phagocytosis. Pro-inflammatory macrophages also work in conjunction with neutrophils by secreting reactive oxygen species, an important signalling molecule that is integral for the proper response to exercise-induced muscle damage [40, 41]. One effect of an elevated level of reactive oxygen species is the degradation of the sarcolemma of injured muscle fibres, which in turn causes membrane leakage of more pro-inflammatory cytokines that further attracts neutrophils and macrophages to the site of injury [31, 32, 43].

Following the phagocytosis of muscle debris by pro-inflammatory macrophages, these cells begin to transition to an anti-inflammatory state to mark the initiation of the repair process. Both the anti-inflammatory cytokine IL-10 and T-lymphocytes (T-cells) regulate macrophage polarization and promote the transition towards an anti-inflammatory profile [6, 7, 36, 46, 47]. Macrophages in this state help down-regulate the immune response by secreting anti-inflammatory cytokines, including IL-4 and IL-10, to avoid excessive damage due to an extended pro-inflammatory environment [45, 46]. These cells also promote the activity of resident muscle stem cells, also known as satellite cells, which are responsible for skeletal muscle repair following damage. Pro-inflammatory macrophages secrete cytokines and growth factors that stimulate satellite cells to begin proliferating, while anti-inflammatory macrophages secrete factors that promote the fusion of the differentiated muscle cell with the myofibres in order to repair the tissue [31, 32, 45, 48]. A more detailed discussion of the role of muscle satellite cells can be found in section "The Role of Satellite Cells in Muscle Regeneration".

Systemic Response to Exercise-Induced Muscle Damage

In addition to the local inflammatory response, a damaging bout of exercise also induces a systemic inflammatory response. Several of the pro- and anti-inflammatory cytokines that are found locally at the site of damage are also found in the circulation, including IL-6, tumour necrosis factor alpha, IL-1β, and IL-10 [27, 35, 49]. Most of these cytokines exert similar effects systemically as they do locally, but some have varying roles that are dependent on timing, concentration, and location. One such example is IL-6, a "myokine" that derives its classification owing to the fact that it can be secreted by skeletal muscle. Despite its pro-inflammatory profile, elevated IL-6 levels following exercise are needed to restore metabolic homeostasis, regulate antioxidant defenses, and stimulate the increase of other pro-inflammatory cytokines [30, 49–51]. When bound to its receptor IL-6R on satellite cells, IL-6 triggers important regulatory events including cell activation and proliferation [33, 49–52]. However, chronically elevated IL-6 levels can cause more muscle damage and atrophy [53]. The balance between pro- and anti-inflammatory cytokines is important for the regulation of the systemic inflammatory response and for reducing excess inflammation following exercise-induced muscle damage. In addition to the systemic cytokine reaction, several proteins, such as creatine kinase and myoglobin, leak into circulation due to breaks in the sarcolemma [41, 52, 54]. As a result, creatine kinase and myoglobin are often highly elevated following extensive eccentric exercise and are commonly used as proxy measures of muscle damage, although inter-individual differences exist in serum concentrations which should be considered when interpreting data [55, 56].

The Role of Satellite Cells in Muscle Regeneration

The remarkable regenerative capacity of skeletal muscle relies on resident muscle stem cells known as satellite cells. Despite having several thousand nuclei within a single skeletal muscle fibre, these cells are terminally differentiated and do not proliferate to repair damaged muscle fibres. Satellite cells are the only appreciable source of new myonuclei when they undergo differentiation. Under resting conditions, satellite cells exist in a quiescent, or dormant, state and are commonly labelled with the biomarker paired box 7 (Pax7). These cells reside on the periphery of the muscle fibre, between the sarcolemma and basal lamina (referred to as the satellite cell niche) and exhibit a remarkable capacity for division to support repair, adaptation or regeneration of skeletal muscle.

The process through which satellite cells are activated, proliferate, and differentiate is termed myogenesis (Fig. 2.4), which is predominantly regulated through the action of myogenic regulatory factors (MRFs). Once activated, satellite cells begin expressing the various MRFs, including myogenic factor 5 (Myf5) and myoblast determination protein (MyoD), and migrate to the site of muscle damage to start

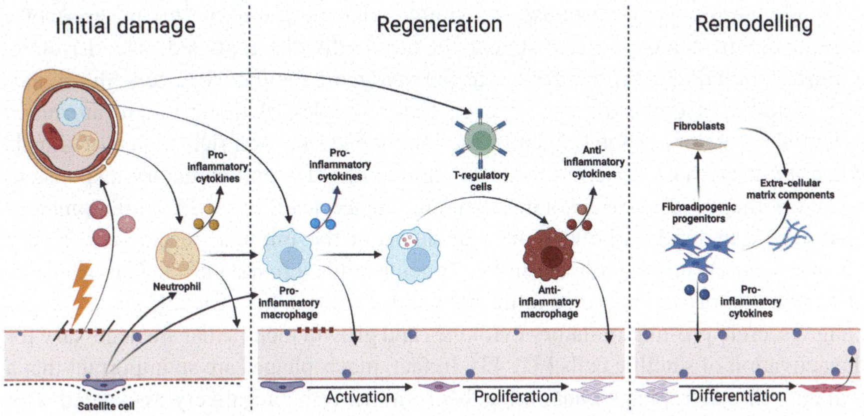

Fig. 2.4 Schematic representation of the muscle repair process following exercise-induced muscle damage. Following a damaging stimulus, such as eccentric exercise, muscle fibres begin releasing intramuscular components like myoglobin and creatine kinase into circulation. Neutrophils begin the muscle repair process by recruiting pro-inflammatory macrophages. These macrophages, along with pericytes and fibro/adipogenic progenitors, release factors to promote the activation of satellite cells. This process occurs over several days, during which time macrophages transition into an anti-inflammatory phase which influences myoblasts to undergo differentiation. The remodelling phase includes the addition of extracellular matrix proteins in order to repair the satellite cell niche

proliferating. At this stage, these cells are termed myoblasts and have two possible fates: they can self-renew and return to the quiescent state to repopulate the basal satellite cell pool or undergo terminal differentiation by fusing to an existing muscle fibre or by forming nascent myotubes.

Satellite cells can be activated as early as 3 h following a single bout of exercise, and the population of Pax7+ / MyoD+ cells continues to rapidly increase for 24–48 h [57]. To study muscle damage in humans, a single bout of exercise (generally focused on eccentric muscle fibre contraction) is utilized to induce ultrastructural muscle damage. In response to a damaging eccentric exercise protocol, the total number of satellite cells increases after 24 h, and peaks at 3 days post-exercise before returning to baseline levels by 96–120 h post-exercise [58, 59]. When stratified by muscle fibre type, it becomes evident that the increase in satellite cell content following muscle damage is primarily the result of an expansion of satellite cells associated with type II fibres [58, 60–63]. These findings suggest that severe muscle damage following eccentric contractions is largely due to the selective recruitment of high force-generating type II muscle fibres.

> In response to a damaging eccentric exercise protocol, the total number of satellite cells increases after 24 h, peaks at 3 post-exercise and returns to baseline levels by 5–6 days post-exercise.

While the myogenic response is a simple concept describing the progress of a satellite cell from the quiescent state to the terminally differentiated state, driven by a transcriptional network referred to as the myogenic regulatory factors, the signalling factors that orchestrate these events are complex. Although there are many potential regulators of satellite cells, the immune system is a rich source of signals when it comes to exercise-induced muscle damage. The inflammatory response to muscle damage was previously discussed in greater detail in section "Inflammatory Response". In brief, following muscle damage, the immune response is nearly instantaneous, with a significant influx of neutrophils into the muscle that promotes pro-inflammatory macrophage infiltration and activation [31, 40, 42]. These macrophages secrete pro-inflammatory cytokines and growth factors that are necessary for the activation of satellite cells [31, 43]. In fact, macrophages are so important that a reduction in macrophage content has been shown to be negatively associated with impairments in muscle regeneration [64]. After the initial inflammatory response, macrophages transition to an anti-inflammatory phenotype and begin secreting anti-inflammatory cytokines [46]. This transition is essential for the next stage of myogenesis to occur, as these macrophages promote myoblast differentiation [46].

There are many regulators of satellite cell activity. Circulating growth factors, such as insulin-like growth factor-1 or hepatocyte growth factor, are released by skeletal muscle in response to muscle damage and play a role in muscle regeneration. Insulin-like growth factor-1, which can also be released by the liver, is thought to play a beneficial role in muscle regeneration by promoting satellite cell proliferation and differentiation [65, 66]. Following damage, hepatocyte growth factor levels rise in skeletal muscle as the growth factor binds to its receptor c-Met on the satellite cell membrane and consequently induces satellite cell activation and proliferation [67, 68]. Various cytokines are also involved in myogenic regulation, such as IL-6. In response to extensive damage, IL-6 is produced and released by skeletal muscle, establishing this cytokine as a myokine. IL-6 has a complicated relationship with skeletal muscle regulation. Whereas a chronic systemic elevation of IL-6 is associated with an increase in muscle catabolism and atrophy, the acute increase in local IL-6 levels following damage is required for normal satellite cell function [49, 50, 53, 69]. Further information regarding the various growth factors and cytokines that regulate satellite cell activity can be found in the following reviews [50, 59, 70–72].

Until this point, the models of exercise-induced muscle damage discussed have involved single bouts of eccentric muscle damage. This is primarily due to a paucity of research utilizing multiple eccentric exercise sessions. However, in some circumstances such as professional athletes with intense exercise regimes, skeletal muscle is subjected to repeated bouts of exercise-induced muscle damage. Research shows that a second bout of eccentric exercise 5.5 weeks after the first results in a greater elevation in neutrophil levels in young females [73]. However, the subsequent eccentric exercise did not result in a larger macrophage response compared to the first bout [73]. A second study demonstrated that six consecutive days of eccentric exercise training did not elicit a response in terms of the myogenic regulatory factors—MyoD and Myf5 [74]. These findings suggest that repetitive eccentric damage with minimal recovery time is not accompanied by the expected transcriptional response of myogenic regulatory factors. This could be due to a number of reasons. One possibility is that there is insufficient recovery time between exercise sessions to allow for satellite cells to adequately respond to the stimulus. Another is that there is diminished muscle damage following subsequent bouts of exercise, thereby reducing the need for satellite cell activation and repair. More work is needed to directly investigate the effect of multiple bouts of eccentric exercise on satellite cell activity.

Supporting Cells and Structures in Muscle Repair

> Muscle repair after extensive damage is a highly coordinated process involving a number of cell types, but the most critical of these are the satellite cells located on the periphery of a fibre. Satellite cells act as stem cells to support muscle repair and regeneration, and provide new myonuclei for muscle growth and adaptation. The other critical components of repair and regeneration is an intact and functional microvasculature and supporting mononuclear cells found between muscle fibres.

Satellite cells are the primary driver for repair in response to muscle tissue damage, but there are several other components of the muscle microenvironment that contribute to this process. As a highly metabolic tissue, skeletal muscle vascularization is a significant factor during recovery from exercise-induced muscle damage. At the myofibre level, a complex network of capillaries surrounds each fibre carrying oxygen, carbon dioxide, nutrients, inflammatory factors, and growth factors in response to various stimuli. Following muscle contraction, blood flow increases rapidly to match metabolic demand, which can stimulate capillary angiogenesis and vascular remodelling [75, 76]. A dense microvascular network has a beneficial effect on satellite cell activity as it tightly regulates endocrine and paracrine signalling. Christov et al. [77] reported the first evidence of a spatial relationship between muscle fibre capillaries and satellite cells, observing that activated satellite cells were located closer to capillaries than their quiescent counterparts. These findings triggered further investigations into the relationship between satellite cells and capillaries. Later work confirmed these observations and determined that muscle fibre capillarization was related to satellite cell activation status [78]. Following a bout of damaging eccentric exercise, young males with a denser muscle fibre capillary network had significantly greater satellite cell activation [58]. These more vascularized myofibres were also associated with a greater expansion of the satellite cell pool following exercise-induced muscle damage, which accelerated the recovery of muscle function [58]. These findings suggest that satellite cell content and activation are regulated by their proximity to the nearest capillary, which can influence the regenerative response to muscle damage.

Skeletal muscle contains several different mononuclear cell types that play a role in muscle repair following exercise-induced muscle damage, including (but not limited to) vascular progenitor cells, immune cells, and fibroblasts. Over the past 15 years, mesenchymal stem cells have emerged as a key contributor to skeletal muscle repair and regeneration, either through direct or indirect impacts on satellite cells and muscle fibres. Pericytes, a subpopulation of mesenchymal stem cells, facilitate angiogenesis and blood flow regulation by secreting several growth factors and structural proteins, which help form new blood vessels, or by manipulating vasoconstriction and vasodilation [79]. There have been mixed reports regarding the pericyte response to muscle damage. Whereas some studies have shown an increase in pericyte content following eccentric damage [80, 81], others have not [82]. Yet another study has reported a substantial decline in pericyte number during recovery from exercise-induced muscle damage, which could potentially be explained by a transition from a pericyte phenotype to a more mesenchymal progenitor cell type [83]. The discrepancy within the literature may be explained by a difference in pericyte identification, as they have been historically labelled with varying markers based on their location or lineage [79]. Within skeletal muscle, pericytes can adopt a myogenic fate in conditions of extreme muscle damage (e.g., following toxin injection or with muscular dystrophy) by either assuming a satellite cell identity or directly fusing with muscle fibres [84, 85]. Pericytes can perform as both vessel stabilizers and myogenic support cells to actively promote muscle regeneration.

Another mesenchymal stem cell subpopulation highly implicated in muscle regeneration is the fibro/adipogenic progenitor cell. Fibro/adipogenic progenitor cells, as their name suggests, are progenitor cells that reside in the interstitial space between muscle fibres and give rise to adipocytes and fibroblasts, which can be detrimental to skeletal muscle structure and function [86, 87]. In healthy muscles under normal conditions, fibro/adipogenic progenitor cells function to maintain homeostasis by secreting growth factors and cytokines (e.g., insulin-like growth factor-1, IL-6 and IL-33) that support muscle repair [86, 88]. Furthermore, fibro/adipogenic progenitor cells cannot adopt a myogenic fate and instead support satellite cell activity by providing extracellular matrix proteins, such as collagen, laminin, and fibronectin, which are components of the satellite cell niche [89, 90]. Fibro/adipogenic progenitor cells are capable of producing a large variety of cytokines and proteins which can have a beneficial impact on regeneration in conditions of extensive muscle damage. Much of what we know about the fibro/adipogenic progenitor response to muscle damage is from pre-clinical models, usually in mice which have been subjected to either exercise or an injectable toxin (e.g., cardiotoxin or notexin) to produce muscle damage. While these models are not perfect, they allow researchers to thoroughly evaluate the stem cell and myofibre response to extensive muscle damage. After toxin injection, the fibro/adipogenic progenitor cell population begins to expand rapidly in response to IL-4 release by type 2 immune cells termed eosinophils [91]. Activation in this manner signals the fibro/adipogenic progenitor cells to support myogenesis and inhibit their adipogenic and fibrogenic tendencies [91]. Following muscle damage, assistance for fibro/adipogenic progenitor cells comes through the secretion of various factors, such as follistatin, insulin-like growth factor-1, and IL-6, in order to stimulate satellite cell proliferation and differentiation [86, 92]. Fibro/adipogenic progenitor cells also produce IL-33 that recruits immune regulatory T-cells to support muscle regeneration [93]. After muscle repair is complete, macrophages release tumour necrosis factor alpha which promotes fibro/adipogenic progenitor cell apoptosis and a return to pre-damage levels [94].

A thorough discussion on the complex nature of supporting cells and structures involved in muscle damage and repair is beyond the scope of this chapter and readers are encouraged to consult the following cited work for a more complete review of these topics [75, 79, 88, 95–97].

Muscle Recovery from Exercise-Induced Damage

Muscle recovery from exercise-induced muscle damage can be determined by evaluating functional indicators such as delayed onset muscle soreness or muscle force production. Delayed onset muscle soreness is the stiffening and swelling of injured muscle, which results in painful contractions and soreness following exercise-induced muscle damage [98]. This discomfort can begin as early as 8 h post-damage and usually peaks after 1–2 days (depending on damage severity) before returning to normal up to 5 days later [99–101]. Several explanations have been hypothesized as the cause of delayed onset muscle soreness, such as lactic acid buildup, connective or muscle tissue damage, excessive inflammation, or enzyme efflux, but the primary cause is not certain [101, 102]. A reduction in maximal voluntary contraction is routinely associated with delayed onset muscle soreness. These criteria are considered hallmark indicators of (extensive) muscle damage and reflect the decrease in muscle function that occurs after exercise-induced muscle damage [98, 103]. The decrease in muscle force production following muscle damage occurs immediately after exercise and can take up to 120 h to fully recover, depending on the severity of the damage [100–102, 104].

Athletic trainers and physiotherapists rely on range of motion as an indicator of recovery following exercise-induced muscle damage. Range of motion, or the degree of joint movement, can be reduced immediately following a damaging bout of exercise and may take 5 days to return to baseline [99, 105–107]. In extreme cases of exercise-induced muscle damage, this period may be longer, and evaluating range of motion has been found to be a reliable predictor of functional recovery [108]. While delayed onset muscle soreness, reduced force production, and limited range of motion are commonly found following exercise-induced muscle damage, they are not accurate proxy measures of muscle damage per se.

Influence of Female Sex Hormones on Muscle Damage

There are significant differences in the damage level and response to extensive muscle damage between males and females. It has been suggested that estrogen may have a protective role by providing membrane stabilization, antioxidant protection, and satellite cell regulation.

Over several decades, the potential effects of estrogen and other female sex steroids on skeletal muscle have been explored. Emerging literature has found differences in muscle structure, inflammation, and myogenic regulation between human males and females; however, the underlying mechanisms remain poorly understood [109–119]. Much of our current understanding comes from preclinical investigations, usually through rodent or cell culture models. These experiments provide a path to examining sex-based differences in muscle following damage in human participants, using gene editing or following surgical procedures. However, while informative, preclinical investigations are limited by their relevance to humans. Nonetheless, these models provide important insight into the impact of female sex hormones on muscle damage.

In the late 1990s, several researchers sought to identify if there was a differential response to muscle damage in male and female rodents. Interestingly, after a damaging stimulus, male rodents displayed greater myofibre swelling, decreased sarcolemmal integrity, and impaired inflammatory response relative to their female counterparts [109, 110, 120]. More recently, investigators observed that female rats subjected to eccentric exercise had less physical damage and less disruption of calcium homeostasis, as evidenced by significantly lower Ca^{2+} concentrations [121]. These sex-based differences may be due to the presence of estrogen in female rats, as it has been suggested that estrogen may have a protective effect against damage [119]. However, the mechanism through which estrogen protects skeletal muscle is not well known but may be related to its properties as an antioxidant and membrane stabilizer [119]. Overall, it appears that female rodent skeletal muscles are partially protected from muscle damage, potentially due to estrogen.

There is less known about the differences between males and females in response to exercise-induced muscle damage in humans. This is due, in part, to the very few studies that have examined both males and females following exercise-induced muscle damage, but also because of the large inter-individual variability that is inherent to human investigations. Early studies indicate no sex differences in measures of muscle damage except for the inflammatory response. Plasma granulocyte counts increased after damaging unilateral leg extensions and leg press in men, but decreased in women, despite both sexes having similar increases in creatine kinase post-exercise [122]. Similarly, others observed no differences in circulating markers of muscle damage or force production between men and women following exercise-induced muscle damage [123–125]. While these findings suggest that sex-based differences in the response to exercise-induced muscle damage in humans are not as pronounced as in rodent models, it is worth noting that most human studies used proxy measures of damage. More high-quality human studies are needed to comprehensively understand the post-damage response between males and females.

Female Sex Hormones and Recovery from Exercise-Induced Muscle Damage

There is some evidence that female sex hormones contribute to muscle recovery following exercise-induced muscle damage. Female mice subjected to repeated eccentric exercise bouts displayed faster force recovery compared to males 3 days post-exercise [126]. Additionally, female mice that underwent ovariectomies to deplete estrogen levels, followed by subsequent estrogen replacement, showed similar recovery patterns compared to sham-treated female mice, as well as promoting greater mitochondrial activity which alleviates the excessive elevation of cytoplasmic calcium content [126]. It appears that, in mice at least, estrogen is a mediator of skeletal muscle recovery. In humans, females reported lower pain symptoms relative to males after a bout of eccentric exercise, indicating a sex-based difference in the degree of delayed onset muscle soreness response [123]. Females have also been found to experience greater initial strength loss after exercise, but a faster recovery compared to males, suggesting that the protective effects of estrogen might be occurring in the recovery phase and not during exercise-induced muscle damage [127, 128]. While there is only minimal evidence to support this notion, the pattern of recovery following muscle damage in females warrants further investigation.

Estrogen as a Key Factor Underlying Sex-Based Differences

The limited evidence available suggests that estrogen could be a significant variable in the observed differences in damage characteristics and recovery between females and males following exercise-induced muscle damage. Estrogen is best known as the hormone that promotes the growth and development of female sex characteristics, but it has also been found to influence several other functions throughout the body [112, 113, 119, 129, 130]. Its molecular structure is similar to cholesterol, and thus contributes to decreasing sarcolemma fluidity and increasing membrane stability [111, 131]. Estrogen does this by inserting itself into the phospholipid bilayer of the sarcolemma and altering not only the fluidity but the function of the membrane. This intercalation of estrogen into the sarcolemma stabilizes it, meaning the fibre could be more resistant to damage when greater forces are imposed on it, such as during eccentric contractions. It is likely that muscle ultrastructural composition would be protected with greater estrogen levels [111, 119, 131].

In addition to its membrane stabilizing effects, estrogens have prominent antioxidant properties due to a carbon-ring structure that originates from a phenol species [132]. With their additional OH group, phenol species reduce electrons, acting as a buffer for the increase in free radical production after damaging exercise [132]. The phenolic hydroxyl group on estrogen may donate hydrogen atoms to pair with these unpaired electrons, thus terminating peroxidation chain reactions and reducing susceptibility to exercise-induced oxidative damage [132]. Estrogen's antioxidant capacity might also be linked to its regulation of nuclear factor kappa B, a protein that regulates gene expression of multiple cytokines and adhesion molecules (e.g., vascular cell adhesion protein 1 and leukocyte-endothelial cell adhesion). The inhibition of nuclear factor kappa B by estrogen has been shown to prevent leukocyte-endothelial cell adhesion, which is a key factor in suppressing inflammation [133]. In summary, estrogen appears to offer membrane stabilizing and antioxidant effects, which may provide females with a biological advantage in their response to exercise-induced muscle damage.

Estrogen and the Satellite Cell Response

As previously discussed, satellite cells are a primary driver of muscle repair following exercise-induced muscle damage. Given the known effects of estrogen on skeletal muscle structure and function, the relationship between estrogen and satellite cells requires some consideration. Ovariectomized rats that underwent a damaging downhill running session had greater total, activated, and proliferating satellite cell levels if supplemented with estrogen compared to the control exercising rats [134]. Furthermore, satellite cell number drops approximately 40% in ovariectomized mice relative to both sham and ovariectomized mice re-supplemented with estrogen [114]. The latter is most likely because estrogen-deficient mice have significantly compromised satellite cell expansion, differentiation, and self-renewal in addition to inducing a phenotypic switch to a faster myosin heavy chain fibre isoform [113]. In older women, estrogen supplementation via hormone replacement therapy increases myoblast determination protein, myogenin, myogenic factor 5, and myogenic regulatory factor 4 gene expression in response to eccentric exercise [112]. These findings suggest an evolutionarily conserved necessity of estrogen in maintaining satellite cell number and function, especially with advancing age.

There are several proposed mechanisms underlying estrogen's effect on satellite cells. One theory is that estrogen's anti-inflammatory properties can promote satellite cell activity. At greater estrogen levels, there is a reduction in macrophage infiltration and an increase in the M1/M2 macrophage ratio, which is known to promote satellite cell activation and proliferation [31, 43, 135]. Another possibility is that the binding of estrogen to its receptor on satellite cells elicits the observed beneficial effects. Ovariectomized female rats supplemented with estrogen were subjected to eccentric exercise and treated with an estrogen receptor antagonist to evaluate the impact of this receptor on satellite cell dynamics [136]. The elevated post-exercise increases in satellite cell content, activation, and proliferation in the estrogen-supplemented rats were significantly reduced when estrogen receptors were inhibited [136]. Additionally, the presence of estrogen receptor beta in satellite cells has been found to be essential for muscle regeneration in female mice [137]. While further studies are needed to determine the exact mechanism, it seems likely that satellite cells are regulated directly by estrogen and indirectly by estrogen's anti-inflammatory properties.

Other Hormonal Factors Underlying Sex-Based Differences in Exercise-Induced Muscle Damage

In addition to the study of estrogen, other sex hormones have been investigated in the context of exercise-induced muscle damage. More specifically, there is evidence of a negative impact of the sex hormone progesterone on the inflammatory response to exercise. Progesterone treatment in rats results in a significant attenuation of muscle leukocyte infiltration post-exercise [138]. However, it appears that this observation also required the presence of estrogen. A study investigating leukocyte infiltration post-injury found no independent effect of progesterone, but a complete loss of leukocyte infiltration when combined with estrogen [115]. These findings suggest that progesterone is another important hormone to consider in the skeletal muscle response to exercise-induced muscle damage, potentially through its antagonistic effects of estrogen's inflammatory regulation.

Considering the interplay between estrogen and progesterone, it is important to consider the effects of the hormonal cycle in women. Menstrual cycle phase has become a significant area of focus for training, exercise performance, and response to exercise-induced muscle damage. Generally speaking, the menstrual cycle is comprised of the early follicular phase (low estrogen, low progesterone), late follicular phase (high estrogen, low progesterone), and mid-luteal phase (high estrogen, high progesterone). One study investigating the impact of the menstrual cycle on muscle damage and inflammation found slightly higher levels of IL-6 in the mid-luteal phase, which may be due to the antagonistic effects of progesterone on estrogen in reducing muscle inflammation [116]. Furthermore, pre-exercise progesterone concentrations are negatively correlated to the relative change in neutrophil content 4 h post-exercise during the mid-luteal phase [117]. While these findings have implications for all females, a significant proportion of the population use hormonal contraceptives that contain a synthetic version of progesterone. Recent studies on the effect of contraceptive use on the structural and inflammatory response to exercise-induced muscle damage have reported equivocal findings [139–141]. Satellite cells also respond differently to natural vs. synthetic hormones, highlighting the need for further work on the effects of hormonal contraceptives on muscle physiology [142].

In summary, there appear to be sex-based differences with respect to both the extent of muscle damage and the regenerative/repair response to exercise. Presently, with much of the evidence originating from rodent models, there is a lack of understanding concerning the mechanisms underlying these differences in terms of human responses. Human studies to date are largely equivocal, likely due to the significant variability in hormone levels between individuals. Furthermore, a full consideration of hormones involved in the menstrual and hormonal contraceptive cycles should be given when interpreting outcomes of muscle damage, regeneration, and inflammation in females.

Conclusion

In this chapter, the etiology of exercise-induced muscle damage was discussed including: the role of the immune system, the influence of satellite cells, the contribution of supporting cells and structures, the recovery of skeletal muscle following damage, and sex-based differences in the response to exercise-induced muscle damage.

Skeletal muscle can be damaged through various means, but how effectively muscle is repaired is directly influenced by several key factors discussed in this chapter. In the lens of exercise, the most severe form of damage occurs following eccentric contractions. In terms of recovery, it would be especially detrimental if skeletal muscle did not exhibit such a remarkable capacity for regeneration as it does.

Muscle repair after extensive damage is a highly coordinated process between many cell types and structures within the muscle microenvironment. This includes the action of the resident muscle satellite cells, several mesenchymal stem cell populations, resident and infiltrating immune cells, circulating growth factors and hormones, and the vast capillary system interspersed throughout muscle tissues.

Exercise-induced muscle damage and subsequent recovery occur in several phases, beginning with the initial damage phase where myofibres are injured by excessive muscle contraction, resulting in ultrastructural damage. Following this, immune cells, known as neutrophils, begin the process of muscle regeneration by recruiting another immune cell type, pro-inflammatory macrophages. The macrophages, along with mesenchymal stem cells, such as fibro/adipogenic progenitors and pericytes, support muscle repair by releasing factors involved in the activation of satellite cells. This process occurs over several days, during which time macrophages transition into an anti-inflammatory phase, which influences satellite cells to undergo differentiation in order to aid in the repair of muscle damage.

Sex-specific differences in muscle damage and recovery are important to consider when discussing exercise-induced muscle damage. Despite a paucity of research, preliminary findings suggest there is a difference in how males and females respond to exercise-induced muscle damage and how efficiently muscle repair occurs. Furthermore, the process may be influenced by hormonal contraceptives, which a significant portion of the female population uses regularly. More high-quality investigations on human subjects are needed to fully elucidate the influence of sex and sex-related hormones on exercise-induced muscle damage and recovery.

References

1. Nelke C, Dziewas R, Minnerup J, Meuth SG, Ruck T. Skeletal muscle as potential central link between sarcopenia and immune senescence. EBioMedicine. 2019;49:381–8. https://doi.org/10.1016/J.EBIOM.2019.10.034.
2. Mikkelsen UR, Langberg H, Helmark IC, Skovgaard D, Andersen LL, Kjær M, et al. Local NSAID infusion inhibits satellite cell proliferation in human skeletal muscle after eccentric exercise. J Appl Physiol. 2009;107:1600–11. https://doi.org/10.1152/JAPPLPHYSIOL.00707.2009/ASSET/IMAGES/LARGE/ZDG0110987830005.JPEG.
3. Paulsen G, Mikkelsen UR, Raastad T, Peake JM. Leucocytes, cytokines and satellite cells: what role do they play in muscle damage and regeneration following eccentric exercise? Exerc Immunol Rev. 2012;18:42–97.
4. Wang J, Khodabukus A, Rao L, Vandusen K, Abutaleb N, Bursac N. Engineered skeletal muscles for disease modeling and drug discovery. Biomaterials. 2019;221:119416. https://doi.org/10.1016/j.biomaterials.2019.119416.
5. Bottinelli R, Reggiani C. Human skeletal muscle fibres: molecular and functional diversity. Prog Biophys Mol Biol. 2000;73:195–262.
6. Boppart MD, De Lisio M, Zou K, Huntsman HD. Defining a role for non-satellite stem cells in the regulation of muscle repair following exercise. Front Physiol. 2013;4:310. https://doi.org/10.3389/FPHYS.2013.00310/BIBTEX.
7. Perandini LA, Chimin P, Lutkemeyer DD, Câmara NOS. Chronic inflammation in skeletal muscle impairs satellite cells function during regeneration: can physical exercise restore the satellite cell niche? FEBS J. 2018;285:1973–84. https://doi.org/10.1111/FEBS.14417.
8. Cornelison DDW. Context matters: in vivo and in vitro influences on muscle satellite cell activity. J Cell Biochem. 2008;105:663–9. https://doi.org/10.1002/jcb.21892.
9. Clarkson PM, Hubal MJ. Exercise-induced muscle damage in humans. Am J Phys Med Rehabil. 2002;81:S52–69. https://doi.org/10.1097/00002060-200211001-00007.
10. Armstrong RB. Initial events in exercise-induced muscular injury. Med Sci Sports Exerc. 1990;22:429–35. https://doi.org/10.1249/00005768-199008000-00002.
11. Hyatt HW, Powers SK. The role of calpains in skeletal muscle remodeling with exercise and inactivity-induced atrophy. Int J Sports Med. 2020;41:994–1008. https://doi.org/10.1055/A-1199-7662.
12. Zhang BT, Yeung SS, Allen DG, Qin L, Yeung EW. Role of the calcium-calpain pathway in cytoskeletal damage after eccentric contractions. J Appl Physiol. 1985;2008(105):352–7. https://doi.org/10.1152/JAPPLPHYSIOL.90320.2008.
13. Kanda K, Sugama K, Hayashida H, Sakuma J, Kawakami Y, Miura S, et al. Eccentric exercise-induced delayed-onset muscle soreness and changes in markers of muscle damage and inflammation. Exerc Immunol Rev. 2013;19:72–85.
14. Brentano MA, Martins Kruel LF. A review on strength exercise-induced muscle damage: applications, adaptation mechanisms and limitations. J Sports Med Phys Fitness. 2011;51:1–10.
15. Nosaka K, Lavender A, Newton M, Sacco P. Muscle damage in resistance training—is muscle damage necessary for strength gain and muscle hypertrophy? Int J Sport Health Sci. 2003;1:1–8. https://doi.org/10.5432/IJSHS.1.1.

16. Edwards RH, Hill DK, Jones DA, Merton PA. Fatigue of long duration in human skeletal muscle after exercise. J Physiol. 1977;272:769–78. https://doi.org/10.1113/JPHYSIOL.1977.SP012072.
17. Gibala MJ, MacDougall JD, Tarnopolsky MA, Stauber WT, Elorriaga A. Changes in human skeletal muscle ultrastructure and force production after acute resistance exercise. J Appl Physiol. 1985;1995(78):702–8. https://doi.org/10.1152/JAPPL.1995.78.2.702.
18. Morgan DL. New insights into the behavior of muscle during active lengthening. Biophys J. 1990;57:209–21. https://doi.org/10.1016/S0006-3495(90)82524-8.
19. Allen DG, Whitehead NP, Yeung EW. Mechanisms of stretch-induced muscle damage in normal and dystrophic muscle: role of ionic changes. J Physiol. 2005;567:723–35. https://doi.org/10.1113/JPHYSIOL.2005.091694.
20. Schoenfeld BJ. Does exercise-induced muscle damage play a role in skeletal muscle hypertrophy? J Strength Cond Res. 2012;26:1441–53. https://doi.org/10.1519/JSC.0B013E31824F207E.
21. Friden J, Sjostrom M, Ekblom B. Myofibrillar damage following intense eccentric exercise in man. Int J Sports Med. 1983;4:170–6. https://doi.org/10.1055/S-2008-1026030.
22. Yamaguchi M, Izumimoto M, Robson RM, Stromer MH. Fine structure of wide and narrow vertebrate muscle Z-lines. A proposed model and computer simulation of Z-line architecture. J Mol Biol. 1985;184:621–43. https://doi.org/10.1016/0022-2836(85)90308-0.
23. Proske U, Morgan DL. Muscle damage from eccentric exercise: mechanism, mechanical signs, adaptation and clinical applications. J Physiol. 2001;537:333–45. https://doi.org/10.1111/J.1469-7793.2001.00333.X.
24. Nosaka K, Clarkson PM. Influence of previous concentric exercise on eccentric exercise-induced muscle damage. J Sports Sci. 1997;15:477–83. https://doi.org/10.1080/026404197367119.
25. Prado LG, Makarenko I, Andresen C, Krüger M, Opitz CA, Linke WA. Isoform diversity of giant proteins in relation to passive and active contractile properties of rabbit skeletal muscles. J Gen Physiol. 2005;126:461–80. https://doi.org/10.1085/JGP.200509364.
26. Horowits R. Passive force generation and titin isoforms in mammalian skeletal muscle. Biophys J. 1992;61:392–8. https://doi.org/10.1016/S0006-3495(92)81845-3.
27. Malm C. Exercise-induced muscle damage and inflammation: fact or fiction? Acta Physiol Scand. 2001;171:233–9. https://doi.org/10.1046/j.1365-201X.2001.00825.x.
28. Tee JC, Bosch AN, Lambert MI. Metabolic consequences of exercise-induced muscle damage. Sports Med. 2007;37:827–36. https://doi.org/10.2165/00007256-200737100-00001.
29. Leinroth AP, Mirando AJ, Rouse D, Kobayahsi Y, Tata PR, Rueckert HE, et al. Identification of distinct non-myogenic skeletal-muscle-resident mesenchymal cell populations. Cell Rep. 2022;39:110785. https://doi.org/10.1016/J.CELREP.2022.110785.
30. Pedersen BK, Hoffman-Goetz L. Exercise and the immune system: regulation, integration, and adaptation. Physiol Rev. 2000;80:1055–81. https://doi.org/10.1152/PHYSREV.2000.80.3.1055.
31. Otis JS, Niccoli S, Hawdon N, Sarvas JL, Frye MA. Pro-inflammatory mediation of myoblast proliferation. PLoS One. 2014;9:e92363. https://doi.org/10.1371/journal.pone.0092363.
32. Tidball JG, Villalta SA. Regulatory interactions between muscle and the immune system during muscle regeneration. Am J Physiol Regul Integr Comp Physiol. 2010;298:1173–87. https://doi.org/10.1152/AJPREGU.00735.2009/ASSET/IMAGES/LARGE/ZH60051071990005.JPEG.
33. Bay ML, Heywood S, Wedell-Neergaard AS, Schauer T, Lehrskov LL, Christensen RH, et al. Human immune cell mobilization during exercise: effect of IL-6 receptor blockade. Exp Physiol. 2020;105:2086–98. https://doi.org/10.1113/EP088864.
34. Ho CSK, López JA, Vuckovic S, Pyke CM, Hockey RL, Hart DNJ. Surgical and physical stress increases circulating blood dendritic cell counts independently of monocyte counts. Blood. 2001;98:140–5. https://doi.org/10.1182/BLOOD.V98.1.140.
35. Geissmann F, Manz MG, Jung S, Sieweke MH, Merad M, Ley K. Development of monocytes, macrophages and dendritic cells. Science. 2010;327:656–61. https://doi.org/10.1126/SCIENCE.1178331.

36. Malm C, Sjödin B, Sjöberg B, Lenkei R, Renström P, Lundberg IE, et al. Leukocytes, cytokines, growth factors and hormones in human skeletal muscle and blood after uphill or downhill running. J Physiol. 2004;556:983–1000. https://doi.org/10.1113/JPHYSIOL.2003.056598.
37. Parkin J, Cohen B. An overview of the immune system. Lancet. 2001;357:1777–89. https://doi.org/10.1016/S0140-6736(00)04904-7.
38. Gray JI, Farber DL. Tissue-resident immune cells in humans. Annu Rev Immunol. 2022;40:195–220. https://doi.org/10.1146/ANNUREV-IMMUNOL-093019-112809.
39. Zoico E, Roubenoff R. The role of cytokines in regulatingprotein metabolism and muscle function. Nutr Rev. 2002;60:39–51. https://doi.org/10.1301/00296640260085949.
40. Fielding RA, Manfredi TJ, Ding W, Fiatarone MA, Evans WJ, Cannon JG. Acute phase response in exercise. III. Neutrophil and IL-1 beta accumulation in skeletal muscle. Am J Phys. 1993;265:R166–72. https://doi.org/10.1152/AJPREGU.1993.265.1.R166.
41. Toft AD, Jensen LB, Bruunsgaard H, Ibfelt T, Halkjær-Kristensen J, Febbraio M, et al. Cytokine response to eccentric exercise in young and elderly humans. Am J Physiol Cell Physiol. 2002;283:289–95. https://doi.org/10.1152/AJPCELL.00583.2001/ASSET/IMAGES/LARGE/H00721023004.JPEG.
42. Bosurgi L, Manfredi AA, Rovere-Querini P. Macrophages in injured skeletal muscle: a perpetuum mobile causing and limiting fibrosis, prompting or restricting resolution and regeneration. Front Immunol. 2011;2:2. https://doi.org/10.3389/FIMMU.2011.00062.
43. Tidball JG. Regulation of muscle growth and regeneration by the immune system. Nat Rev Immunol. 2017;17:165–78. https://doi.org/10.1038/NRI.2016.150.
44. Bernard C, Zavoriti A, Pucelle Q, Chazaud B, Gondin J. Role of macrophages during skeletal muscle regeneration and hypertrophy—implications for immunomodulatory strategies. Physiol Rep. 2022;10:e15480. https://doi.org/10.14814/PHY2.15480.
45. Saclier M, Yacoub-Youssef H, Mackey AL, Arnold L, Ardjoune H, Magnan M, et al. Differentially activated macrophages orchestrate myogenic precursor cell fate during human skeletal muscle regeneration. Stem Cells. 2013;31:384–96. https://doi.org/10.1002/STEM.1288.
46. Arnold L, Henry A, Poron F, Baba-Amer Y, Van Rooijen N, Plonquet A, et al. Inflammatory monocytes recruited after skeletal muscle injury switch into antiinflammatory macrophages to support myogenesis. J Exp Med. 2007;204:1057–69. https://doi.org/10.1084/JEM.20070075.
47. Marklund P, Mattsson CM, Wåhlin-Larsson B, Ponsot E, Lindval B, Lindval L, et al. Extensive inflammatory cell infiltration in human skeletal muscle in response to an ultraendurance exercise bout in experienced athletes. J Appl Physiol. 2013;114:66–72. https://doi.org/10.1152/JAPPLPHYSIOL.01538.2011/ASSET/IMAGES/LARGE/ZDG0241204120004.JPEG.
48. Tidball JG, Wehling-Henricks M. Macrophages promote muscle membrane repair and muscle fibre growth and regeneration during modified muscle loading in mice in vivo. J Physiol. 2007;578:327–36. https://doi.org/10.1113/jphysiol.2006.118265.
49. Serrano AL, Baeza-Raja B, Perdiguero E, Jardí M, Muñoz-Cánoves P. Interleukin-6 is an essential regulator of satellite cell-mediated skeletal muscle hypertrophy. Cell Metab. 2008;7:33–44. https://doi.org/10.1016/J.CMET.2007.11.011.
50. Belizário JE, Fontes-Oliveira CC, Borges JP, Kashiabara JA, Vannier E. Skeletal muscle wasting and renewal: a pivotal role of myokine IL-6. Springerplus. 2016;5:619. https://doi.org/10.1186/S40064-016-2197-2.
51. Cantini M, Massimino ML, Rapizzi E, Rossini K, Catani C, Libera LD, et al. Human satellite cell-proliferation in vitro is regulated by autocrine secretion of IL-6 stimulated by a soluble factor(s) released by activated monocytes. Biochem Biophys Res Commun. 1995;216:49–53. https://doi.org/10.1006/BBRC.1995.2590.
52. Steensberg A, Van Hall G, Osada T, Sacchetti M, Saltin B, Pedersen BK. Production of interleukin-6 in contracting human skeletal muscles can account for the exercise-induced increase in plasma interleukin-6. J Physiol. 2000;529:237–42. https://doi.org/10.1111/J.1469-7793.2000.00237.X.

53. Haddad F, Zaldivar F, Cooper DM, Adams GR. IL-6-induced skeletal muscle atrophy. J Appl Physiol. 2005;98:911–7. https://doi.org/10.1152/JAPPLPHYSIOL.01026.2004/ASSET/IMAGES/LARGE/ZDG0030536820007.JPEG.
54. Kawanishi N, Mizokami T, Niihara H, Yada K, Suzuki K. Neutrophil depletion attenuates muscle injury after exhaustive exercise. Med Sci Sports Exerc. 2016;48:1917–24. https://doi.org/10.1249/MSS.0000000000000980.
55. Sherwood RA, Lambert A, Newhaml DJ, Wassif WS, Peters TJ. The effect of eccentric exercise on serum creatine kinase activity in different ethnic group. Ann Clin Biochem. 1996;33:324–9.
56. Newham DJ, Jones DA, Edwards RHT. Plasma creatine kinase changes after eccentric and concentric contractions. Muscle Nerve. 1986;9:59–63. https://doi.org/10.1002/MUS.880090109.
57. McKay BR, Ogborn DI, Bellamy LM, Tarnopolsky MA, Parise G. Myostatin is associated with age-related human muscle stem cell dysfunction. FASEB J. 2012;26:2509–21. https://doi.org/10.1096/FJ.11-198663.
58. Nederveen JP, Joanisse S, Snijders T, Thomas ACQ, Kumbhare D, Parise G. The influence of capillarization on satellite cell pool expansion and activation following exercise-induced muscle damage in healthy young men. J Physiol. 2018;596:1063–78. https://doi.org/10.1113/JP275155.
59. Snijders T, Nederveen JP, McKay BR, Joanisse S, Verdijk LB, van Loon LJC, et al. Satellite cells in human skeletal muscle plasticity. Front Physiol. 2015;6:283. https://doi.org/10.3389/FPHYS.2015.00283/ABSTRACT.
60. Fortino SA, Wageh M, Pontello R, McGlory C, Kumbhare D, Phillips SM, et al. Sex-based differences in the myogenic response and inflammatory gene expression following eccentric contractions in humans. Front Physiol. 2022;13:13. https://doi.org/10.3389/FPHYS.2022.880625.
61. Brown A, Thomas ACQ, Hatt AA, McGlory C, Phillips SM, Kumbhare D, et al. Aerobic conditioning augments the satellite cell and ribosome response to acute eccentric contractions in young men and women. Am J Physiol Cell Physiol. 2022;323:C1577–85. https://doi.org/10.1152/AJPCELL.00418.2022/ASSET/IMAGES/LARGE/AJPCELL.00418.2022_F003.JPEG.
62. Imaoka Y, Kawai M, Mori F, Miyata H. Effect of eccentric contraction on satellite cell activation in human vastus lateralis muscle. J Physiol Sci. 2015;65:461–9. https://doi.org/10.1007/S12576-015-0385-4/FIGURES/6.
63. Cermak NM, Snijders T, McKay BR, Parise G, Verdijk LB, Tarnopolsky MA, et al. Eccentric exercise increases satellite cell content in type II muscle fibers. Med Sci Sports Exerc. 2013;45:230–7. https://doi.org/10.1249/MSS.0B013E318272CF47.
64. Wang H, Melton DW, Porter L, Sarwar ZU, McManus LM, Shireman PK. Altered macrophage phenotype transition impairs skeletal muscle regeneration. Am J Pathol. 2014;184:1167–84. https://doi.org/10.1016/J.AJPATH.2013.12.020.
65. Barton-Davis ER, Shoturma DI, Sweeney HL. Contribution of satellite cells to IGF-I induced hypertrophy of skeletal muscle. Acta Physiol Scand. 1999;167:301–5. https://doi.org/10.1046/J.1365-201X.1999.00618.X.
66. Engert JC, Berglund EB, Rosenthal N. Proliferation precedes differentiation in IGF-I-stimulated myogenesis. J Cell Biol. 1996;135:431–40. https://doi.org/10.1083/JCB.135.2.431.
67. McKay BR, O'Reilly CE, Kawka S, Phillips SM, Tarnopolsky MA, Parise G. Hepatocyte growth factor signaling in mediating human muscle satellite cell activation and proliferation following eccentric exercise. FASEB J. 2008;22:962.23. https://doi.org/10.1096/FASEBJ.22.1_SUPPLEMENT.962.23.
68. Tatsumi R, Anderson JE, Nevoret CJ, Halevy O, Allen RE. HGF/SF is present in normal adult skeletal muscle and is capable of activating satellite cells. Dev Biol. 1998;194:114–28. https://doi.org/10.1006/DBIO.1997.8803.

69. Toth KG, McKay BR, de Lisio M, Little JP, Tarnopolsky MA, Parise G. IL-6 Induced STAT3 signalling is associated with the proliferation of human muscle satellite cells following acute muscle damage. PLoS One. 2011;6:e17392. https://doi.org/10.1371/JOURNAL.PONE.0017392.
70. Schiaffino S, Dyar KA, Ciciliot S, Blaauw B, Sandri M. Mechanisms regulating skeletal muscle growth and atrophy. FEBS J. 2013;280:4294–314. https://doi.org/10.1111/FEBS.12253.
71. Anderson JE. Hepatocyte growth factor and satellite cell activation. Adv Exp Med Biol. 2016;900:1–25. https://doi.org/10.1007/978-3-319-27511-6_1/FIGURES/4.
72. Nielsen AR, Pedersen BK. The biological roles of exercise-induced cytokines: IL-6, IL-8, and IL-15. Appl Physiol Nutr Metab. 2007;32:833–9. https://doi.org/10.1139/H07-054.
73. Stupka N, Tarnopolsky MA, Yardley NJ, Phillips SM. Cellular adaptation to repeated eccentric exercise-induced muscle damage. J Appl Physiol. 2001;91:1669–78. https://doi.org/10.1152/JAPPL.2001.91.4.1669/ASSET/IMAGES/LARGE/DG1011023007.JPEG.
74. Costa A, Dalloul H, Hegyesi H, Apor P, Csende Z, Racz L, et al. Impact of repeated bouts of eccentric exercise on myogenic gene expression. Eur J Appl Physiol. 2007;101:427–36. https://doi.org/10.1007/S00421-007-0510-Z/FIGURES/6.
75. Parise G, Murrant CL, Cocks M, Snijders T, Baum O, Plyley MJ. Capillary facilitation of skeletal muscle function in health and disease. Appl Physiol Nutr Metab. 2020;45:453–62. https://doi.org/10.1139/APNM-2019-0416/ASSET/IMAGES/LARGE/APNM-2019-0416F3.JPEG.
76. Hardy D, Besnard A, Latil M, Jouvion G, Briand D, Thépenier C, et al. Comparative study of injury models for studying muscle regeneration in mice. PLoS One. 2016;11:e0147198. https://doi.org/10.1371/JOURNAL.PONE.0147198.
77. Christov C, Chrétien F, Abou-Khalil R, Bassez G, Vallet G, Authier FJ, Bassaglia Y, Shinin V, Tajbakhsh S, Chazaud B, Gherardi RK. Muscle satellite cells and endothelial cells: close neighbors and privileged partners. Mol Biol Cell. 2017;18(4):1397–409. https://doi.org/10.1091/mbc.e06-08-0693.
78. Nederveen JP, Joanisse S, Snijders T, Ivankovic V, Baker SK, Phillips SM, et al. Skeletal muscle satellite cells are located at a closer proximity to capillaries in healthy young compared with older men. J Cachexia Sarcopenia Muscle. 2016;7:547–54. https://doi.org/10.1002/JCSM.12105.
79. Munroe M, Dvoretskiy S, Boppart MD. Skeletal muscle-resident pericyte responses to conditions of use and disuse. In: Stem cell biology and regenerative medicine, vol. 68. Springer International Publishing; 2021. p. 203–17. https://doi.org/10.1007/978-3-030-62129-2_9.
80. Hyldahl RD, Xin L, Hubal MJ, Moeckel-Cole S, Chipkin S, Clarkson PM. Activation of nuclear factor-κPB following muscle eccentric contractions in humans is localized primarily to skeletal muscle-residing pericytes. FASEB J. 2011;25:2956–66. https://doi.org/10.1096/FJ.10-177105.
81. Valero MC, Huntsman HD, Liu J, Zou K, Boppart MD. Eccentric exercise facilitates mesenchymal stem cell appearance in skeletal muscle. PLoS One. 2012;7:e29760. https://doi.org/10.1371/JOURNAL.PONE.0029760.
82. De LM, Farup J, Sukiennik RA, Clevenger N, Nallabelli J, Nelson B, et al. The acute response of pericytes to muscle-damaging eccentric contraction and protein supplementation in human skeletal muscle. J Appl Physiol. 2015;119:900–7. https://doi.org/10.1152/JAPPLPHYSIOL.01112.2014/ASSET/IMAGES/LARGE/ZDG0181515640004.JPEG.
83. Farup J, De Lisio M, Rahbek SK, Bjerre J, Vendelbo MH, Boppart MD, et al. Pericyte response to contraction mode-specific resistance exercise training in human skeletal muscle. J Appl Physiol. 2015;119:1053–63. https://doi.org/10.1152/JAPPLPHYSIOL.01108.2014/ASSET/IMAGES/LARGE/ZDG0221516300005.JPEG.
84. Dellavalle A, Sampaolesi M, Tonlorenzi R, Tagliafico E, Sacchetti B, Perani L, et al. Pericytes of human skeletal muscle are myogenic precursors distinct from satellite cells. Nat Cell Biol. 2007;9:255–67. https://doi.org/10.1038/NCB1542.

85. Dellavalle A, Maroli G, Covarello D, Azzoni E, Innocenzi A, Perani L, et al. Pericytes resident in postnatal skeletal muscle differentiate into muscle fibres and generate satellite cells. Nat Commun. 2011;2:1–11. https://doi.org/10.1038/ncomms1508.
86. Joe AWB, Yi L, Natarajan A, Le Grand F, So L, Wang J, et al. Muscle injury activates resident fibro/adipogenic progenitors that facilitate myogenesis. Nat Cell Biol. 2010;12:153–63. https://doi.org/10.1038/NCB2015.
87. Uezumi A, Ito T, Morikawa D, Shimizu N, Yoneda T, Segawa M, et al. Fibrosis and adipogenesis originate from a common mesenchymal progenitor in skeletal muscle. J Cell Sci. 2011;124:3654–64. https://doi.org/10.1242/jcs.086629.
88. Biferali B, Proietti D, Mozzetta C, Madaro L. Fibro-adipogenic progenitors cross-talk in skeletal muscle: the social network. Front Physiol. 2019;10:1074. https://doi.org/10.3389/fphys.2019.01074.
89. Chapman MA, Mukund K, Subramaniam S, Brenner D, Richard X, Lieber L. Three distinct cell populations express extracellular matrix proteins and increase in number during skeletal muscle fibrosis. Am J Physiol Cell Physiol. 2017;312:131–43. https://doi.org/10.1152/ajpcell.00226.2016.-Tissue.
90. Lukjanenko L, Jung MJ, Hegde N, Perruisseau-Carrier C, Migliavacca E, Rozo M, et al. Loss of fibronectin from the aged stem cell niche affects the regenerative capacity of skeletal muscle in mice. Nat Med. 2016;22:897–905. https://doi.org/10.1038/nm.4126.
91. Heredia JE, Mukundan L, Chen FM, Mueller AA, Deo RC, Locksley RM, et al. Type 2 innate signals stimulate fibro/adipogenic progenitors to facilitate muscle regeneration. Cell. 2013;153:376–88. https://doi.org/10.1016/J.CELL.2013.02.053.
92. Mozzetta C, Consalvi S, Saccone V, Tierney M, Diamantini A, Mitchell KJ, et al. Fibroadipogenic progenitors mediate the ability of HDAC inhibitors to promote regeneration in dystrophic muscles of young, but not old Mdx mice. EMBO Mol Med. 2013;5:626–39. https://doi.org/10.1002/EMMM.201202096.
93. Kuswanto W, Burzyn D, Panduro M, Wang KK, Jang YC, Wagers AJ, et al. Poor repair of skeletal muscle in aging mice reflects a defect in local, interleukin-33-dependent accumulation of regulatory T cells. Immunity. 2016;44:355–67. https://doi.org/10.1016/j.immuni.2016.01.009.
94. Lemos DR, Babaeijandaghi F, Low M, Chang CK, Lee ST, Fiore D, et al. Nilotinib reduces muscle fibrosis in chronic muscle injury by promoting TNF-mediated apoptosis of fibro/adipogenic progenitors. Nat Med. 2015;21:786–94. https://doi.org/10.1038/nm.3869.
95. Nederveen JP, Betz MW, Snijders T, Parise G. The importance of muscle capillarization for optimizing satellite cell plasticity. Exerc Sport Sci Rev. 2021;49:284–90. https://doi.org/10.1249/JES.0000000000000270.
96. Giuliani G, Rosina M, Reggio A. Signaling pathways regulating the fate of fibro/adipogenic progenitors (FAPs) in skeletal muscle regeneration and disease. FEBS J. 2022;289:6484–517. https://doi.org/10.1111/FEBS.16080.
97. Molina T, Fabre P, Dumont NA. Fibro-adipogenic progenitors in skeletal muscle homeostasis, regeneration and diseases. Open Biol. 2021;11:210110. https://doi.org/10.1098/rsob.210110.
98. Huang CC, Lee MC, Ho CS, Hsu YJ, Ho CC, Kan NW. Protective and recovery effects of resveratrol supplementation on exercise performance and muscle damage following acute plyometric exercise. Nutrients. 2021;13:3217. https://doi.org/10.3390/NU13093217/S1.
99. Chen TC, Chen HL, Lin MJ, Wu CJ, Nosaka K. Muscle damage responses of the elbow flexors to four maximal eccentric exercise bouts performed every 4 weeks. Eur J Appl Physiol. 2009;106:267–75. https://doi.org/10.1007/S00421-009-1016-7/FIGURES/7.
100. Newham DJ, Mills KR, Quigley BM, Edwards RHT. Pain and fatigue after concentric and eccentric muscle contractions. Clin Sci. 1983;64:55–62. https://doi.org/10.1042/CS0640055.
101. Ranchordas MK, Rogerson D, Soltani H, Costello JT. Antioxidants for preventing and reducing muscle soreness after exercise. Cochrane Database Syst Rev. 2017;12:CD009789. https://doi.org/10.1002/14651858.CD009789.PUB2.

102. Cheung K, Hume PA, Maxwell L. Delayed onset muscle soreness: treatment strategies and performance factors. Sports Med. 2003;33:145–64. https://doi.org/10.2165/00007256-200333020-00005/METRICS.
103. Hunter AM, Galloway SDR, Smith IJ, Tallent J, Ditroilo M, Fairweather MM, et al. Assessment of eccentric exercise-induced muscle damage of the elbow flexors by tensiomyography. J Electromyogr Kinesiol. 2012;22:334–41. https://doi.org/10.1016/J.JELEKIN.2012.01.009.
104. Macgregor LJ, Hunter AM. High-threshold motor unit firing reflects force recovery following a bout of damaging eccentric exercise. PLoS One. 2018;13:e0195051. https://doi.org/10.1371/JOURNAL.PONE.0195051.
105. Schoenfeld BJ, Grgic J. Effects of range of motion on muscle development during resistance training interventions: a systematic review. SAGE Open Med. 2020;8:205031212090155. https://doi.org/10.1177/2050312120901559.
106. Clarkson P, Nosaka K, Braun B. Muscle function after exercise-induced muscle damage and rapid adaptation. Med Sci Sports Exerc. 1991;24:512–20.
107. Clarkson PM, Nosaka K, Braun B. Muscle function after exercise-induced muscle damage and rapid adaptation. Med Sci Sports Exerc. 1992;24:512–20. https://doi.org/10.1249/00005768-199205000-00004.
108. Larkin-Kaiser KA, Parr JJ, Borsa PA, George SZ. Range of motion as a predictor of clinical shoulder pain during recovery from delayed-onset muscle soreness. J Athl Train. 2015;50:289–94. https://doi.org/10.4085/1062-6050-49.5.05.
109. Komulainen J, Koskinen SOA, Kalliokoski R, Takala TES, Vihko V. Gender differences in skeletal muscle fibre damage after eccentrically biased downhill running in rats. Acta Physiol Scand. 1999;165:57–63. https://doi.org/10.1046/j.1365-201x.1999.00481.x.
110. Tiidus PM, Bombardier E. Oestrogen attenuates post-exercise myeloperoxidase activity in skeletal muscle of male rats. Acta Physiol Scand. 1999;166:85–90. https://doi.org/10.1046/j.1365-201x.1999.00550.x.
111. Wiseman H, Quinn P. The antioxidant action of synthetic oestrogens involves decreased membrane fluidity: relevance to their potential use as anticancer and cardioprotective agents compared to tamoxifen? Free Radic Res. 1994;21:187–94. https://doi.org/10.3109/10715769409056569.
112. Dieli-Conwright CM, Spektor TM, Rice JC, Sattler FR, Schroeder ET. Influence of hormone replacement therapy on eccentric exercise induced myogenic gene expression in postmenopausal women. J Appl Physiol. 2009;107:1381–8. https://doi.org/10.1152/japplphysiol.00590.2009.
113. Kitajima Y, Ono Y. Estrogens maintain skeletal muscle and satellite cell functions. J Endocrinol. 2016;229:267–75. https://doi.org/10.1530/JOE-15-0476.
114. Larson AA, Baumann CW, Kyba M, Lowe DA. Oestradiol affects skeletal muscle mass, strength and satellite cells following repeated injuries. Exp Physiol. 2020;105:1700–7. https://doi.org/10.1113/EP088827.
115. Xing D, Miller A, Novak L, Rocha R, Chen YF, Oparil S. Estradiol and progestins differentially modulate leukocyte infiltration after vascular injury. Circulation. 2004;109:234–41. https://doi.org/10.1161/01.CIR.0000105700.95607.49.
116. Romero-parra N, Barba-moreno L, Rael B, Alfaro-magallanes VM, Cupeiro R, Díaz ÁE, et al. Influence of the menstrual cycle on blood markers of muscle damage and inflammation following eccentric exercise. Int J Environ Res Public Health. 2020;17:1618. https://doi.org/10.3390/IJERPH17051618.
117. Funaki A, Gam H, Matsuda T, Ishikawa A, Yamada M, Ikegami N, et al. Influence of menstrual cycle on leukocyte response following exercise-induced muscle damage. Int J Environ Res Public Health. 2022;19:9201. https://doi.org/10.3390/IJERPH19159201.
118. Oxfeldt M, Dalgaard LB, Farup J, Hansen M. Sex hormones and satellite cell regulation in women. Transl Sports Med. 2022;2022:1–12. https://doi.org/10.1155/2022/9065923.

119. Kendall B, Eston R. Exercise-induced muscle damage and the potential protective role of estrogen. Sports Med. 2002;32:103–23. https://doi.org/10.2165/00007256-200232020-00003.
120. St. Pierre Schneider B, Correia LA, Cannon JG. Sex differences in leukocyte invasion in injured murine skeletal muscle. Res Nurs Health. 1999;22:243–50. https://doi.org/10.1002/(sici)1098-240x(199906)22:3<243::aid-nur6>3.0.co;2-x.
121. Sonobe T, Inagaki T, Sudo M, Poole DC, Kano Y. Sex differences in intracellular Ca(2+) accumulation following eccentric contractions of rat skeletal muscle in vivo. Am J Physiol Regul Integr Comp Physiol. 2010;299:R1012. https://doi.org/10.1152/AJPREGU.00623.2009.
122. Stupka N, Lowther S, Chorneyko K, Bourgeois JM, Hogben C, Tarnopolsky MA. Gender differences in muscle inflammation after eccentric exercise. J Appl Physiol. 2000;89:2325–32. https://doi.org/10.1152/jappl.2000.89.6.2325.
123. Dannecker EA, Koltyn KF, Riley JL, Robinson E. Sex differences in delayed onset muscle soreness. J Sports Med Phys Fitness. 2003;43:78–84.
124. Dannecker EA, Liu Y, Rector RS, Thomas TR, Fillingim RB, Robinson ME. Sex differences in exercise-induced muscle pain and muscle damage. J Pain. 2012;13:1242–9. https://doi.org/10.1016/j.jpain.2012.09.014.
125. Bruce CD, Ruggiero L, Dix GU, Cotton PD, McNeil CJ. Females and males do not differ for fatigability, muscle damage and magnitude of the repeated bout effect following maximal eccentric contractions. Appl Physiol Nutr Metab. 2021;46:238–46. https://doi.org/10.1139/APNM-2020-0516/ASSET/IMAGES/LARGE/APNM-2020-0516F5.JPEG.
126. Watanabe D, Ikegami R, Kano Y. Predominant cause of faster force recovery in females than males after intense eccentric contractions in mouse fast-twitch muscle. J Physiol. 2021;599:4337–56. https://doi.org/10.1113/JP281927.
127. Flores DF, Gentil P, Brown LE, Pinto RS, Carregaro RL, Bottar M. Dissociated time course of recovery between genders after resistance exercise. J Strength Cond Res. 2011;25:3039–44. https://doi.org/10.1519/JSC.0B013E318212DEA4.
128. Sewright KA, Hubal MJ, Kearns A, Holbrook MT, Clarkson PM. Sex differences in response to maximal eccentric exercise. Med Sci Sports Exerc. 2008;40:242–51. https://doi.org/10.1249/MSS.0B013E31815AEDDA.
129. Hansen M. Female hormones: do they influence muscle and tendon protein metabolism? Proc Nutr Soc. 2018;77:32–41. https://doi.org/10.1017/S0029665117001951.
130. Hansen M, Miller BF, Holm L, Doessing S, Petersen SG, Skovgaard D, et al. Effect of administration of oral contraceptives in vivo on collagen synthesis in tendon and muscle connective tissue in young women. J Appl Physiol. 2009;106:1435–43. https://doi.org/10.1152/JAPPLPHYSIOL.90933.2008.
131. Wiseman H, Quinn P, Halliwell B. Tamoxifen and related compounds decrease membrane fluidity in liposomes. Mechanism for the antioxidant action of tamoxifen and relevance to its anticancer and cardioprotective actions? FEBS Lett. 1993;330:53–6. https://doi.org/10.1016/0014-5793(93)80918-K.
132. Karlsson J. Antioxidants and exercise. Champagne: Human Kinetics; 1997.
133. Yoshikawa T, Yoshida N. Vitamin E and leukocyte-endothelial cell interactions. Antioxid Redox Signal. 2000;2:821–5. https://doi.org/10.1089/ARS.2000.2.4-821.
134. Enns DL, Tiidus PM. Estrogen influences satellite cell activation and proliferation following downhill running in rats. J Appl Physiol. 2008;104:347–53. https://doi.org/10.1152/japplphysiol.00128.2007.
135. Le G, Novotny SA, Mader TL, Greising SM, Chan SSK, Kyba M, et al. A moderate oestradiol level enhances neutrophil number and activity in muscle after traumatic injury but strength recovery is accelerated. J Physiol. 2018;596:4665–80. https://doi.org/10.1113/JP276432.
136. Enns DL, Iqbal S, Tiidus PM. Oestrogen receptors mediate oestrogen-induced increases in post-exercise rat skeletal muscle satellite cells. Acta Physiol. 2008;194:81–93. https://doi.org/10.1111/J.1748-1716.2008.01861.X.

137. Seko D, Fujita R, Kitajima Y, Nakamura K, Imai Y, Ono Y. Estrogen receptor β controls muscle growth and regeneration in young female mice. Stem Cell Rep. 2020;15:577–86. https://doi.org/10.1016/J.STEMCR.2020.07.017.
138. Iqbal S, Thomas A, Bunyan K, Tiidus PM. Progesterone and estrogen influence postexercise leukocyte infiltration in overiectomized female rats. Appl Physiol Nutr Metab. 2008;33:1207–12. https://doi.org/10.1139/H08-108.
139. Shenouda N, Priest SE, Rizzuto VI, MacDonald MJ. Brachial artery endothelial function is stable across a menstrual and oral contraceptive pill cycle but lower in premenopausal women than in age-matched men. Am J Physiol Heart Circ Physiol. 2018;315:H366–74. https://doi.org/10.1152/ajpheart.00102.2018.
140. Torgrimson BN, Meendering JR, Kaplan PF, Minson CT. Endothelial function across an oral contraceptive cycle in women using levonorgestrel and ethinyl estradiol. Am J Physiol Heart Circ Physiol. 2007;292:H2874–80. https://doi.org/10.1152/ajpheart.00762.2006.
141. Dragoman MV. The combined oral contraceptive pill- recent developments, risks and benefits. Best Pract Res Clin Obstet Gynaecol. 2014;28:825–34. https://doi.org/10.1016/j.bpobgyn.2014.06.003.
142. Wallner C, Rausch A, Drysch M, Dadras M, Wagner JM, Becerikli M, et al. Regulatory aspects of myogenic factors GDF-8 and Follistatin on the intake of combined oral contraceptives. Gynecol Endocrinol. 2020;36:406–12. https://doi.org/10.1080/09513590.2019.1666816.

Chapter 3
Central and Neuromuscular Structural and Functional Influences on the Initiation and Control of Movement

David G. Behm

> *Lack of activity destroys the good condition of every human being, while movement and methodical physical exercise save it and preserve it.*
>
> *–Plato*

Introduction

When writing and attempting to describe corticospinal and neuromuscular components of any kind in a text, it is difficult to reflect the integrated functional and temporal complementarity of these components. Likewise, during the reading of some portions of a chapter, it may be perceived that the functions of one area occur before, and act separately, from the actions of another area. For example, in the following text, you may perceive that the pre-frontal cortex formulates the idea to move and then initially activates the basal ganglia, followed by the cerebellum, and subsequently the primary motor cortex, and other components. However, activation of all these areas is synchronized and influence each other continuously in a united effort both temporally and functionally. Anatomical cortical differences are better described as smoothly varying continuities than by discrete borders, patterns, or timing, and therefore, they need correspondingly smooth models of functional organization. It is impossible to provide the full complexity of the corticospinal and neuromuscular structures and functions in one brief chapter. Hence, please keep in

D. G. Behm (✉)
School of Human Kinetics and Recreation, Memorial University of Newfoundland, St. John's, NL, Canada
e-mail: dbehm@mun.ca

N. C. Apostolopoulos et al. (eds.), *Fundamentals of Recovery, Regeneration, and Adaptation to Exercise Stress: An Integrated Approach*,
https://doi.org/10.1007/978-3-031-44270-4_3

mind that the following explanations may unfortunately seem simplistic at times, but they hopefully provide a conceptual idea of how the corticospinal and neuromuscular system work to initiate, coordinate, and execute not only the expected movement, but also the motivation to move or train or perform rehabilitation exercises. Hopefully, this information will be helpful in formulating exercise programs that are successful in terms of physiological adaptations and the individual's persistence in participating in a rehabilitation program.

Movement Executive Functions and Decisions

Although our autonomic nervous system does not need much input in terms of conscious control to ensure our heart keeps pumping at regular intervals, our blood pressure is maintained at healthy levels, our digestive muscles push nutrients through the alimentary system, or perform other autonomic functions, our somatic or voluntary motor control system is initiated and driven by our cortex, with executive decisions greatly influenced by our prefrontal cortex.

Our executive decisions, strategies, and motivations for undertaking exercise training or attending and performing rehabilitation exercises depends heavily on the input to, and interactions with, components of the pre-frontal cortex, as well as the limbic system, basal ganglia, cerebellum, and other areas [1]. The dorsolateral prefrontal cortex has extensive connections with visual and proprioceptive somatosensory regions accessing and implementing working memory for planning and movement [2]. Hence, to move an individual (patient) from the sofa to the gym or clinic, it is important to know the structures and factors affecting how that decision is made.

One might argue that many of our decisions to move are based on sensory feedback. We change positions in bed because a joint may be uncomfortable, we stand up and go to the kitchen because we sense hunger, we move our head when we see a baseball in a trajectory to cause damage, a persistent injury makes us decide to get up and start rehabilitating it. Our primary visual cortex, frontal eye field, and associated cortical areas cannot interpret that a cup of coffee in front of us is actually a cup of coffee. Visually, it can only provide information about light, colours, form, and movement [1, 3, 4]. Similarly, the acoustic system only informs us of changes in the amplitude and frequency of the sound signals being received [3, 4]. However, loud acoustic stimulation can affect the startle system, leading to inhibition of corticospinal excitability [5]. Touch, pressure, vibration, and proprioceptive signals are also only a series of low to high frequency signals sent to specific areas of our brain. All these signals are processed through related cortical areas and then sent primarily to the hippocampus, amygdala (limbic area) and prefrontal cortex for interpretation [1, 3, 4].

The hippocampus is a primary component for memory [6], while the amygdala is a principal area for attaching the type (e.g., pleasure, fear, anger) and intensity of emotions to incoming information [1, 7]. Every sensory input has emotions attached through this limbic system [7], and thus, we may fear a nearby spider or snake, we may have positive feelings about seeing a hot cup of coffee, we may be excited to see a tennis ball hit by us flying by our opponent, and we may enjoy seeing greater weights lifted at the gym or clinic. The amygdala has direct connections to the brainstem affecting arousal and response systems [8] as evidenced by the increases in heart rate, blood pressure, muscle tone among other physiological responses when fear or anger is experienced. Whether we want to respond to these sensory inputs is decided upon chiefly by the pre-frontal cortex. The limbic system not only influences voluntary decisions, but also connects to autonomic motor control to affect muscle tension, breathing patterns, blood pressure, and other functions [1]. Three major pre-frontal areas involved in making executive decisions about movement are the orbitofrontal cortex, anterior cingulate cortex and ventral striatum [1].

The orbitofrontal cortex evaluates sensory information (e.g., touch, taste, olfactory, vision and others) which has transited through the limbic system (memory, emotion and meaning from the hippocampus and amygdala) [9] and the uncinate fasciculus to be interpreted by the orbitofrontal cortex [1]. The orbitofrontal cortex decodes and represents some primary reinforcers to control and modify the emotional and social context of reward-related and punishment-related behaviour (attractor networks) [10]. In simpler terms, the orbitofrontal cortex attempts to determine whether reality is friendly and safe, thus an approach (reward) behaviour is adopted, or perhaps the incoming information is threatening and uncertain (punishment-related), in which case, withdrawal is a more appropriate strategy. Orbitofrontal cortex activation is reported to be related to the sensitivity of winning [11]. In the context of emotional and social changes, the lateral orbitofrontal cortex can make very rapid changes in the decision to act when a reward is not present [12].

The anterior cingulate cortex receives input from the orbitofrontal cortex about reward and non-reward outcomes [13] and then compares competing ideas and strategies to aid in the execution of a corporate strategy. The posterior cingulate cortex adds spatial and movement-related information from parietal cortical areas contributing to action-outcome learning [13]. The orbitofrontal cortex, anterior cingulate cortex, and ventromedial prefrontal cortex learn actions in order to obtain rewards, thus furnishing movement goals, and ingraining reward-related effects on memory [14]. For example, risk taking individuals present greater interconnections of the medial orbitofrontal cortex, the anterior cingulate cortex, the ventromedial prefrontal cortex, and the parahippocampal areas [15]. Since the anterior cingulate cortex couples rewards to actions, it influences emotion; and as the posterior cingulate cortex connects to the hippocampal system, it is involved in memory. These strategies are directly affected by the ventral striatum, which is a reward system of the brain and highly affected by dopamine neurotransmitter release. Compared to the greater focus on primary rewards activating the orbitofrontal cortex and the anterior cingulate cortex (e.g., tastes, smells, sounds, touch, and others), the ventral striatum provides a greater emphasis on secondary rewards (e.g., money, power, competition) to help motivate or initiate movement [16]. The ventral striatum is densely and extensively interconnected with the amygdala and basal ganglia [16]. Thus, the orbitostriatal circuits link thoughts with action, as attention is focused on specific goals and motivations are activated and energized [1].

Why would this basic information be of any importance to professionals and individuals interested in recovery, rehabilitation, and training? Although the preceding paragraph was a simplified version of the decision process, it is evident that how the proposed activities, exercises, or training programs are presented will determine whether the individual will actually positively process that input and act upon it. If the environment is threatening as a resistance training facility may be to an untrained individual, especially females in a predominately male environment, the orbitofrontal cortex will sense this as a threatening input and will influence the anterior cingulate cortex to adopt a strategy to remove the individual from this space. To overcome this reticence, the anterior cingulate cortex responsibility for developing movement strategies and decisions to escape from this uncertain environment must be overridden by a strong reward in concert with the ventral striatum. If the rehabilitation or training facility is considered dirty, malodorous (sweaty smelling), or slimy, then the primary rewards related to smells, sounds, and touch that activate the orbitofrontal cortex and anterior cingulate cortex will influence the prefrontal cortex to forego initiating movement or exercise. Therefore, the training environment must be made safe, friendly, and inviting, and the secondary positive rewards from training or rehabilitation must be made both intellectually apparent, and positively magnified with some rapid physical improvements or positive social reward (e.g., group exercises). Progressive introduction to the training environment, without negative consequences (such as delayed onset muscle soreness from exercise that is too intense, or loads too high), and continual positive feedback, is essential for fostering executive decisions to start, and eventually continue with, a training or rehabilitation program.

The prefrontal cortex is highly involved in making executive decisions to initiate and plan movement, and is particularly associated with intrinsic and extrinsic rewards. As such, the rehabilitation and training environment, as well as the perceived personal benefit, greatly determine the motivation to begin, and/or continue, exercise or rehabilitation, and therefore dictates the ultimate success.

Corticospinal Structure and Function

Once the executive decision from the input of the pre-frontal cortex is made, action must be taken. While the primary motor cortex possesses specific motor neurons corresponding to parallel spinal motor neurons, they are insufficient to actually perform a coordinated movement [1, 3, 4]. While the proposed motor homunculus (topographic representation of the body parts based on the concentration of cortical motoneurons) has the ability to activate all cortical motoneurons, and hence, all spinal motor units (motoneurons and their corresponding muscle fibres), a coordinated movement also necessitates the involvement of the basal ganglia, the cerebellum, and the supplementary motor cortex [1, 3, 4]. The integration of these areas is described as "motion adjustment loops" since it achieves movement regulation by adjusting cortical motoneurons activity [17].

Basal Ganglia

The basal ganglia's major role is the control of conscious and proprioceptive movements (refer to Fig. 3.1). Transmissions from the cortex are weighted in order to determine what movements should be disinhibited [18]. The basal ganglia nuclei (components: caudate, putamen, globus pallidus nuclei, and associated components: subthalamic nuclei, substantia nigra [18]) do not have direct afferent or efferent connections with spinal cord motoneurons [1, 3, 4]. While the basal ganglia also receives afferent sensory information, its main output is to the thalamus [18]. There are two primary connections to the thalamus. The direct pathway disinhibits thalamic neurons, thereby activating the thalamus, whereas the indirect pathway inhibits unwanted movements [18]. Working in combination, they contribute to the precision of the movement [19]. The activation or inhibition from the thalamus then modulates the supplementary motor area, premotor, and primary motor cortices (motor loop) [17, 18]. Based on the sensory input, the basal ganglia can filter movement signals that may be deemed unnecessary or incorrect, enhancing specific muscle control. Various other loops provide input from the limbic system (limbic loop: memory, emotion, and motivation), the pre-frontal cortex (integration of cognitive processes, such as planning, memory, attention and reward-based motivation: prefrontal loop) as well as the frontal eye field and supplementary eye field (oculomotor loop), which contribute to learning, planning, and performing various activities and movements [16, 17].

Cerebellum

The basal ganglia and the cerebellum are reciprocally interconnected. The cerebellum has been labelled as the "Grand Comparator" [1]. It compares the movement intent to the actual action or performance. Any disparities between the intent and the motor output are compared and alternative strategies are employed to minimize the differences between intent and action. The cerebrocerebellum, the largest functional subdivision of the cerebellum receives information regarding movement strategies or plans from the pre-frontal cortex and basal ganglia among other structures (refer to Fig. 3.2). Improvements in proprioception, which are essential for optimal movement strategies, are also accounted for within the cerebro-cerebellar loop [20]. Based on practice and continual feedback, the cerebellum predicts limb position [20]. This exchange of movement planning and execution information is termed corollary discharge or internal feedback [21]. The vestibulocerebellum and spinocerebellum receive sensory feedback for balance (vestibular system and vision), as

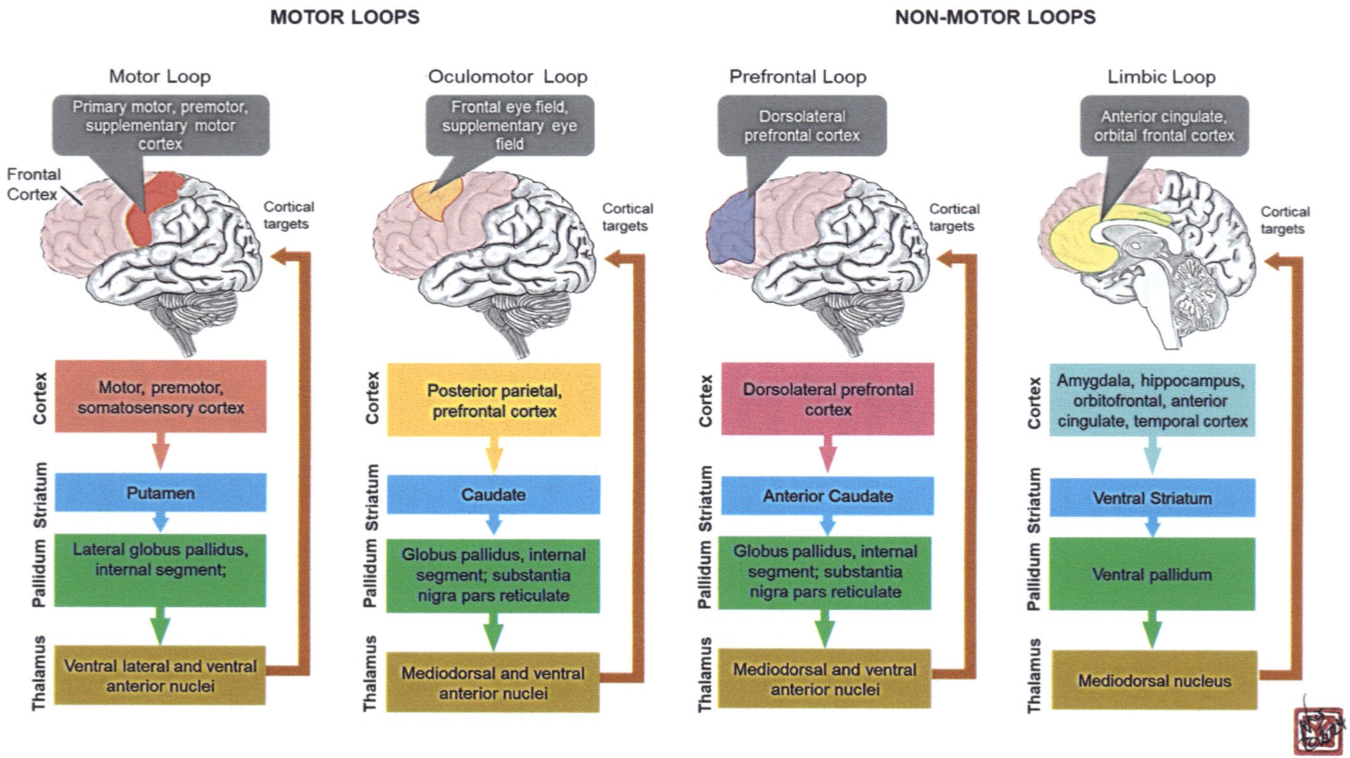

Fig. 3.1 Basal Ganglia. (Reprinted with permission from Nikos C. Apostolopoulos)

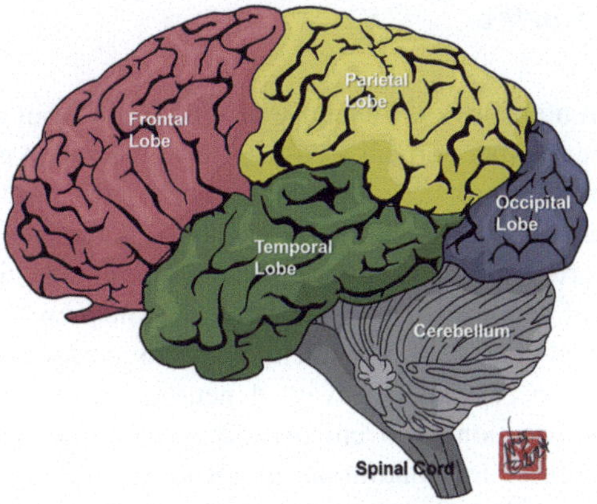

Fig. 3.2 The Cerebrocerebellum. (Reprinted with permission from Nikos C. Apostolopoulos)

well as the status of the periphery (proprioception), and is termed re-afference or external feedback [21].

Sensory input is extensive with afferents from the olivary nucleus (pons: olivocerebellar), vestibular cortex (vestibulocerebellar), reticular formation (reticulocerebellar), pre-motor areas (corticopontocerebellar) and proprioception area (trigeminocerebellar). The efferent pathways (tracts) are equally extensive, with connections to the axial extensor muscles (cerebellovestibular tract), brain stem (cerebelloreticular), cerebral cortex (corticonuclear), thalamus (cerebellothalamic), and rubrospinal system (cerebellorubral) [1]. The comparison of corollary discharge or internal feedback (intent) with re-afference or external feedback (actual action or performance), results in a subsequent modification of movement strategy for future attempts [21]. Hence, through these extensive connections, the cerebellum plays a primary role in developing coordinated movement programs [17]. For example, an experienced tennis player does not hit a fluid forehand by sequentially activating each component of the movement separately, but implements a command from the pre-frontal cortex to activate a coordinated, integrated forehand tennis movement strategy that has been continually modified (intent vs. action) by the basal ganglia and cerebellum over the years to fully execute and integrate all contributing neuromuscular factors (e.g., appropriate activation or inhibition of the agonists, synergists, antagonists, core and joint stabilizer muscles).

> In terms of training and rehabilitation, the optimization of the cerebellum and basal ganglia would involve sufficient repetition of a movement until the intent and action are adequately aligned (corollary discharge or internal feedback orients with reafference or external feedback). Providing a variety of related movements with different velocities, ranges of motion, planes and axes, and other variables would challenge the system to adapt to a range of related coordinated movements. Finally, a gradual progression helping to ensure success would result in intrinsic rewards (e.g., increase dopamine release) that would promote motivation to continue with the activity.

Primary Motor Cortex

Although the pre-frontal cortex is generally assigned the role of executive decision-making for movement, other cortical areas are also involved in a decision concerning movement intent. The primary motor cortex also contributes to movement intent and action, and includes the premotor cortex, the supplementary motor area, the posterior parietal cortex, and several subcortical brain regions, that aid in the planning and execution of voluntary movements [1, 3, 4]. The premotor cortex is divided into lateral and medial areas [22]. The lateral premotor cortex responses are associated with movement occurrence by way of neurons activated more strongly in response to specific directional movements [22, 23]. Lateral premotor cortex neurons have responsibilities for conditional motor tasks, and are activated with the onset of an afferent stimuli, before the intent to perform a movement. When a stimulus or cue is learned to be associated with a subsequent movement, the lateral premotor cortex neurons increase their discharge rate in the interval between the stimuli and the signal for movement. Thus, the lateral premotor cortex seems to be specifically involved in movement selection initiated by external (afferents) events [22, 23]. On the other hand, the medial premotor cortex is more specialized for movements initiated by internal rather external stimuli [24, 25]. Thus, the medial premotor cortex has greater responsibility for self-initiated or "spontaneous" movements, firing shortly (<2 s) prior to the onset of a self-initiated movement. Premotor cortex motor neurons have extensive reciprocal interconnections with the primary motor cortex, as well as projecting through the corticobulbar and corticospinal pathways [24, 25].

The supplementary motor complex consists of the supplementary motor area, the supplementary eye field, and the pre-supplementary motor area [26, 27]. Compared to the primary motor cortex, the supplementary motor complex is more sensitive to broader coordinated movements. The supplementary motor area responsibilities are postulated to include postural stabilization, bilateral coordination, control of internally initiated movements (versus external afferent stimuli), and sequences of movements [26, 27]. For example, the premotor cortex is more activated with a single finger tapping movement, whereas the supplementary motor area would be activated to a greater degree with sequential movements of multiple fingers. Similar to the premotor cortex and cerebellum, the supplementary motor area projects directly to the spinal cord [1, 3, 4].

It is proposed that primary motor cortex motoneuron activation is controlled by inhibition and facilitation systems. The inhibition system involves afferent sensory information, which decreases motor output (supraspinal fatigue). The facilitation system increases the primary motor output to combat supraspinal fatigue by connecting the limbic system, basal ganglia, orbitofrontal cortex, anterior cingulate cortex, and other components of the pre-frontal cortex to increase the motivational input to increase the motor output to the peripheral system [28]. It is quite apparent from the aforementioned descriptions that primary and secondary rewards play a very important role in initiating movement and continuing to move and exercise, and thus are vital considerations to ensure success in training and rehabilitation.

Descending Motor Tracts

Activation of cortical motoneurons transmit to spinal motoneurons from the Brodmann areas 4 (primary motor cortex) and 6 (supplementary motor area) via the lateral corticospinal and ventral corticospinal tracts [1, 3, 4] (refer to Fig. 3.3). While both tracts help in the control of voluntary movement, the lateral corticospinal has a greater responsibility for the contralateral limbs whereas, the ventral corticospinal tract provides greater input to proximal muscles (i.e., core or trunk muscles) [1, 3, 4]. The lateral corticospinal tract crosses over at the pyramidal decussation (the border between the spinal cord and the medulla oblongata). The ventral tract crosses over at the vertebral level of the target muscle [1, 3, 4]. Damage to one tract can be partially taken over by the other tract. Although full motor function cannot be achieved, over time there may be some ability to perform cruder limb movements through the ventral corticospinal tract for example. Another movement tract is the rubrospinal tract projecting from the red nucleus (labelled as such due to the iron in haemoglobin and ferritin) in the rostral midbrain beside the substantia nigra (nuclei of the basal ganglia) [29, 30]. The rubrospinal tract does not project to the spinal cord. The parvocellular component transfers information from the motor cortex to the cerebellum through the inferior olivary complex (medulla) [29, 30]. The magnocellular component of the red nucleus and rubrospinal tract aids in the control of shoulder and upper arm muscles [31–34]. All these tracts innervate particular spinal motoneurons contributing to the control of limb (lateral corticospinal), shoulder, and arms (additional input from rubrospinal and axial) and axial (trunk) (ventral corticospinal tract) muscles [1, 3, 4].

Not all movement is totally directed by supraspinal / cortical control. Spinal neuronal networks, such as the central pattern generator, can produce rhythmic

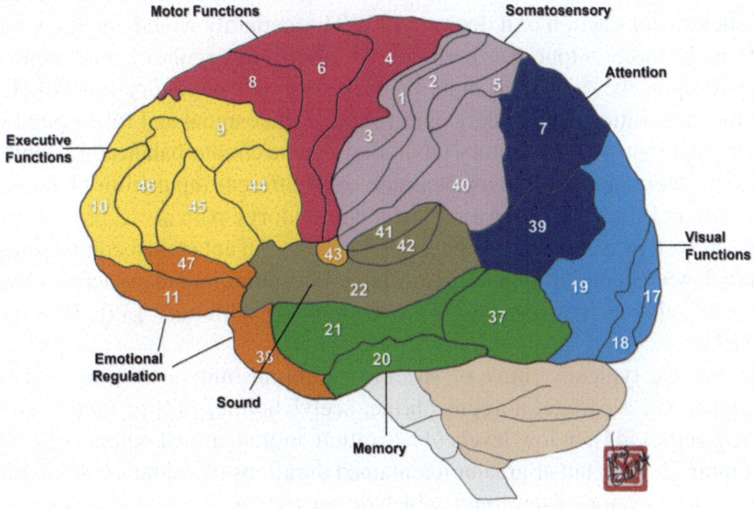

Fig. 3.3 Brodmann areas. (Reprinted with permission from Nikos C. Apostolopoulos)

muscle activation to control the limbs without modulation from the brain or sensory feedback [35].

Stereotyped locomotor movement patterns were first discovered over a century ago by Sherrington [36]. However, sole control of locomotor movement would only be apparent with spinal cord injuries [35]. In healthy individuals, the central pattern generator is integrated with supraspinal input. Spontaneous locomotor type muscle activity (i.e., forceful and patterned phasic bursts of alternating bilateral muscle activity) occurs more often with incomplete spinal cord injuries [37, 38], which provides evidence for supraspinal modulatory input. Furthermore, with healthy locomotion, central pattern generators need sensory afferent information to modulate and modify the rhythmic output. Without afferent feedback, reacting to the external environment (e.g., hills, curbs, holes, etc.) would not be possible [35]. While not all limb movements are under full voluntary control, there is an integration of spinal reflexes and supraspinal modulation.

Control of Muscle Force

Once we have determined that there are sufficient internal and external rewards motivating us to plan a movement, the plan must be executed. Activation of specific cortical and spinal motoneurons and their associated motor units (corticomotoneuronal pathway) [39] in order to initiate movement is not a simple one to one relationship. First of all, when a spinal motoneuron is depolarized, an "all or none phenomenon" is initiated whereby all the muscle fibres innervated by that nerve, i.e., the complete motor unit, will contract synchronously if sufficient excitatory stimuli trigger a motoneuron's action potential [40]. The single action potential of a single motor unit cannot modulate between strong or weak contractions. An impulse either elicits a contraction or it does not [40]. This certainly would not allow for fine gradations in force output. Furthermore, the ability to depolarize motoneurons is dependent upon the combined input of excitatory and inhibitory inputs [41, 42]. While the innervation of the lateral and ventral corticospinal and rubrospinal tracts would typically provide excitatory input, they can be counterbalanced by inhibitory input from interneurons, such as Renshaw cells, afferent input from Golgi tendon organs (Group Ib: can be both excitatory and inhibitory), pain nociceptors (Group II and IV), as well as reciprocal inhibition associated with antagonist contractions that counterbalance excitation inputs [41, 42]. Corticospinal, monoaminergic, propriospinal, and other drives can be both excitatory or inhibitory [39]. How is that possible?

Whereas we typically think of specific neurotransmitters as either excitatory (e.g., glutamate, aspartate, norepinephrine, acetylcholine) [43] or inhibitory (e.g., GABA, glycine) [44], at low levels of activation, motoneuronal output is facilitated by serotonin (5-HT), but at greater maintained durations of voluntary drive, there is a higher concentration of serotonin, which disperses to activate inhibitory receptors, which reduces the motoneuron firing frequency [39]. Thus, the same neurotransmitter can either be inhibitory or excitatory, depending on the concentration.

Extensive inhibitory input would seem to be an adverse strategy, diminishing the possibility of fully activating a muscle, which would undermine the capacity to exert maximal forces. However, full muscle activation is not a common occurrence or necessity in daily life, and maximal force output can result in musculotendinous injuries if maximal force output is not accustomed. Thus, neural inhibition can serve as a protective function, reducing input of unwanted stimuli to produce smoother, purposeful contractions.

Other peripheral input can be excitatory and augment the excitation of the descending tracts. For example, the activation of nuclear bags and chains from the muscle spindles (Group Ia) typically provide excitatory input [45–47]. Dynamic exercise that increases the extent of muscle elongation (activate nuclear chains) and rate and extent of muscle elongation (activate nuclear bags) augment excitatory input to the motoneurons [48]. Rapid stretch-shortening cycle actions, such as sprinting, bounding, and jumping, would increase the likelihood of activating the higher threshold motor units (e.g., type II or fast twitch motoneurons). Hence, the ability to depolarize a motoneuron is dependent upon the cumulative effects of cortical and peripheral excitatory and inhibitory input to the motoneurons. However as mentioned, if depolarization is achieved, the "all or none" phenomenon does not permit any modulation of force for that single depolarization of the motor unit.

Factors Affecting the Ability to Fully Activate Motoneurons

Not all individuals can fully activate all muscles consistently with some muscles and actions more difficult to activate than others. Allen et al. [49] reported that their volunteers could at times fully activate the elbow flexors, but failed in 75% of their attempts. There are conflicting reports revealing muscle activation differences between trained and untrained individuals [41, 50]. Muscles with higher type II motor unit fibre compositions and larger volumes, such as the quadriceps, are reported to have a lower capacity for full muscle activation (~85%) compared to the elbow flexors (~95%), plantar flexors (~95%), and dorsiflexors (~99%) [51]. Greater muscle activation was evident with bilateral, multi-articular knee extensions (~87–92%) and squats (~96–98%) versus the same unilateral exercises (~82%) [50]. A fatigued muscle typically will show a decreased capacity to fully activate [52–54]; however the muscle activation levels of trained individuals seem to be less affected by fatigue [41]. When placed under extremely unstable conditions, the ability to fully activate a muscle can decrease by up to 40% [55, 56]. Individuals recovering from ankle fractures demonstrated significantly lower plantar flexors' muscle activation (78.3 vs 98.7%) than controls, which was attributed to the inhibitory effects of the injured muscle and inflammation [57]. Decreases in muscle activation (~11%) are also reported with delayed onset muscle soreness [58], prolonged stretching [59], and when subjected to environmental stresses, such as hyperbaric pressures (~3% inactivation) [60]. Hence, a greater extent of muscle inactivation is evident with injuries, pain, fatigue, and other factors.

To ensure the ability to fully activate muscles, or at least increase the extent of activation during training or rehabilitation, it is important to minimize inhibitory inputs. Strategies might include decreasing inflammation and pain beforehand, improving balance and stability, enhancing excitatory inputs with resistance training that progresses from isometric (safe, low chance of injury) to more dynamic isoinertial actions. The progression from isometric to dynamic should expose the individual to ballistic (explosive—higher velocity) intent contractions [61, 62]. Although ballistic intent resistance training does not necessitate high speed ballistic movements (e.g., high velocity intent contractions against an immovable resistance or higher loads with greater inertia), the motor units will adapt with higher activation and firing frequencies to enable a transition to higher velocity and explosive type contractions [61, 62]. Although, ballistic intent training might be perceived as only important for explosive athletes, all individuals need to be able to react quickly, forcefully, and powerfully against threats to balance, such as trips, slips, and falls, which can induce grievous injuries.

To promote improved muscle activation when evaluating muscle, or during training and rehabilitation, it is advisable to reduce or minimize the various sources of neural inhibition which reduce muscle activation by relieving pain, decreasing inflammation, ensuring positioning to provide a stability via a solid balance, or using an isometric to dynamic progression in contraction effort.

Motor Unit Recruitment

Force modulation can first be achieved by the recruitment of additional motor units. Recruitment of more motor units will add additional force increments. Furthermore, recruitment is based on Henneman's size principle [63–65]. The motoneurons with smaller soma (cell bodies) correspond with type I (slow twitch) motor units, which possess lower activation thresholds and thus are the first to be activated. Type I motor units typically exert less force (3–5 g) than the higher threshold type II motor units (30–50 g) [66–69]. Hence, the activation of type I motor units is more appropriate for more delicate low force, slower actions whereby type II unit activation would be involved in more gross and higher force motor control [70–72]. Moreover, type I motor units receive more afferents than type II units, which promotes more sensitive proprioception and control (i.e., smaller force outputs and the ability to detect small changes in force). Additionally, type I motor units have smaller innervation ratios than type II motor units. For example, ocular muscles, innervated by predominately type I motor units, may only activate less than 10 muscle fibres, whereas a type II motor unit for the quadriceps may innervate hundreds of muscle fibres [1, 3, 4]. Thus, more complex and delicate actions necessitate lower innervation ratios.

Rate Coding

Once recruited the motoneuron can respond with firing frequencies ranging from low to high (rate coding) [73, 74]. An S-shape has been described for the rate coding curve, with force gradations primarily controlled by rate coding at low and high levels of force production, whereas in the mid-region of force output, recruitment would play a more substantial role [41, 75]. Hence, with the initiation of force, there would be a steep increase in firing rates followed by a plateau (recruitment emphasis) until force starts approaching maximum levels, and then, another steeper increase in firing rates occurs [41, 75–77]. However, this strategy is muscle dependent. Muscles, such as the soleus, reach a particular firing rate, which is sustained as force increases [78], whereas the flexor digitorum firing frequency increases linearly as the force increases [67]. With the biceps brachii, recruitment seems to be the primary force strategy from 0% to 80% of maximal voluntary isometric contraction force, which contrasts with the abductor pollicis which ceases recruitment beyond 50% of maximal voluntary isometric contraction [79]. Therefore, muscles with more delicate changes in force output, such as the intrinsic hand muscles, would rely more on rate coding, whereas larger muscles emphasizing more gross control would be more reliant upon recruitment.

Reflex Inhibition and Potentiation

While some of the aforementioned excitatory reflex actions, such as the myotatic reflex (muscle spindle excitation), can contribute to motoneuron activation, the differential control of reflexes might aid in movement control [80]. Reflexes have been reported to be inhibited in some explosive-jump trained athletes (i.e., sprinters and volleyball players) [81], but in another study, reflex potentiation was enhanced with sprinters [82]. Additional force inputs from reflex activity may place more complexity on fine motor control, and hence, suppression of reflexes may contribute to a more simplified input/output control.

Motoneuron Synchronization

Although increased synchronization of motor unit firing (coincident timing of impulses) is found with strength training and acute bouts of fatigue [83, 84], asynchronous firing is more advantageous for smooth motor control [85]. Additionally, greater force outputs are achieved at submaximal stimulation frequencies with asynchronous stimulation [86, 87], as well as a more rapid rate of force development [88]. This increased force output was attributed to a mechanical coupling that affected the series elastic element of the accompanying muscle fibres [85]. Recurrent inhibition by Renshaw interneurons help to maintain asynchronous stimulation [89–91]. When fatigued, recurrent inhibition can be overcome by increasing voluntary drive, so stimulation becomes more synchronous and there is a loss of smooth movement coordination.

Antagonist Co-contractions

Antagonist co-contractions also play a role in muscle force control. The extent of antagonist co-contractions is influenced by load [92–94], movement velocity, acceleration and deceleration [92, 95–97], range of motion [96, 98], contraction type [99], and movement precision [100]. Antagonist activity has been reported to be lower when training for certain types of work [101], but was also reported to be increased when learning a new task [102]. Greater co-contractions occur if the novice skill involves higher velocities or alternating movements [41]. Antagonist co-activation can provide a defence mechanism (increased resistance to the movement) against higher velocity, angular momentum, and alternating movements [41]. The inclusion of antagonist activity as well as synergists can apply an additional layer of control to the intended movement. However, fatigue, alters the load sharing patterns among antagonist and synergistic muscles, which will adversely affect movement control [103]. Trained individuals, who typically would exhibit more efficient movement control at particular tasks, demonstrate lower levels of co-contractions [53].

> Fine motor control, as well as the ability to fully activate motor units, involves an interaction of excitatory and inhibitory influences that modulate changes in recruitment, rate coding, synchronization of motor unit firing, reflex inhibition, and potentiation, and co-contractions, as well as optimal balance and stability, to provide a solid foundation for the force output.

Balance and Joint Stability

Optimal control of balance is essential to perform coordinated athletic actions and activities of daily living [104, 105]. A lack of balance is associated with an increased risk of injury [105]. The ability to exert forces whether they be maximal or executing smooth, coordinated, submaximal intensity contractions is adversely affected by balance impairments and joint instability. There are reports of 20–70% reductions in maximal isometric force when performed on an unstable platform (e.g., exercise balls, wobble boards) [55, 106, 107]. However, in contrast to this instability-induced force decrement, muscle activation of the limbs and trunk tend either to remain similar to, or greater than, that seen under stable conditions [106, 108]. This maintenance or increase of muscle activation is an attempt by the neuromuscular system to stabilize the trunk (core) and associated joints [104]. Related to the prior paragraph, antagonist co-contractions also increase with unstable and uncertain environments, again in order to stabilize the target joints [104]. Hence, a training or rehabilitation program that improves stability or balance could subsequently improve force output, coordination, and confidence. Balance training can improve balance within 1–3 sessions, independent of age [109].

On the other hand, if this instability-induced co-activation is too high, then a stiffening strategy is adopted, which hampers coordination, force, and performance [110, 111]. Similarly, postural threats that promote a stiffening strategy negatively impact the magnitude and rate of voluntary movements [112]. Fatigue can also induce a stiffening strategy or exaggerated postural rigidity, resulting in increased antagonist co-contractions, articular stiffness, reduced joint range of motion, and decreased movement velocity [113]. Without an adequately stable foundation (balance), higher forces or torques cannot be generated nor maintained, and any excessive co-contractions will lead to stiffness, impairing motor control. With training-induced improvements in balance though, there are decreases in antagonist co-activity [114].

Optimal balance necessitates continuous passive (stiffness versus compliance of connective and muscular tissue around the joints), as well as active, afferent and efferent control, with feedback from the somatosensory, vestibular, and visual systems. Sensory information from these systems is integrated, and their priority is dependent on the movement expectations, goals, environmental context, and prior experiences [115].

Anticipatory postural adjustments (efferent control) activate stabilizing muscles, such as lower abdominal (e.g., transversus abdominus, internal obliques), lower back (e.g., erector spinae), and diaphragm muscles, prior to movement and predicted balance disturbances in order to counteract the disruptive perturbations [116–118]. There is greater anticipatory corticospinal excitability with intended dynamic posture adjustments [119], and thus, anticipation and timing of motor output may be as, or more, important for training and rehabilitation considerations [115, 120] than strength training programs of the trunk and lower body for balance. Inefficient and improper timing of trunk muscular stabilization can contribute to chronic low back pain [121].

Balance training to improve postural control provided moderate to large magnitude improvements for static and dynamic balance in young athletes [122], which was in accord with balance training reviews on young (16–40 years) [123], and older adults (>65 years) [124], although youth tend to obtain greater training-induced improvements in balance measures. As mentioned earlier, balance deficits can adversely affect force and power, and thus, there is evidence that balance training not only improves postural control but also other physical fitness measures, such as muscle strength and jump performance [122, 125]. For greater improvements in balance, it is recommended that a training period of 11–12 weeks, with three training sessions per week, i.e., a total of 36–40 training sessions, with single training session durations of 31–45 min per week (a total weekly duration of 91–120 min), is effective [122, 126].

The use of unstable surfaces when strength training has been suggested to provide an effective action specific combination of metastability [127] and resistance training [56, 128–130]. However, while the force, power, velocity, and range of motion deficits associated with unstable surfaces may not be recommended for athletes or others seeking maximal muscle strength and hypertrophy gains, the similar or higher core and limb muscle activation levels as when stable and lifting heavy loads may be of high value to individuals undergoing recovery and rehabilitation (inability to lift heavier loads) or non-athletic individuals who are not interested in the physical stress of greater resistive loads [129]. Furthermore, ground based free weight resistance training exercises, such as squats, deadlifts, cleans, snatches, and other movements, possess an inherent degree of instability that precludes the need for additional instability (e.g., wobble boards, balls) [129, 131]. It has been demonstrated with youth athletes that performing 4 weeks of balance training prior to 4 weeks of plyometric training augments balance and muscular fitness training responses [125]. In addition, a training program of combined balance and plyometric training improved sprint and shuttle run performance in youth to a greater degree than a plyometric only program [132]. While chronic stretch training programs improve balance, an acute session of balance has no substantial effect [133].

The ability to produce maximal or near maximal forces, power, movement velocity, or sustain coordinated muscular efforts can be improved with balance training or combinations of balance and strength or power training.

> Improving stability and/or balance is important in programming to improve force output, coordination, and confidence, and therefore, employing 1–3 stability/balance sessions prior to initiating other training might be advantageous to initiating. For more substantial improvements in balance, 11–12 weeks, with 3 sessions of 30–45 per week is recommended.

Proprioception

Proprioception is a component of the afferent somatosensory system that senses, produces, predicts, and simulates joint positions, motions, and forces [134, 135]. Proprioception works in combination with the visual and vestibular systems in detecting motion and spatial orientation to monitor and correct balance [134]. Muscle spindles (intrafusal muscle fibres) are primary proprioceptive sensors that detect changes in muscle length and velocity [136]. The direction of motion and joint position can also be perceived by assessing the shortening or lengthening activity of the agonists and antagonists [134, 136]. The sensitivity of the muscle spindle is altered by gamma fusimotor control [137]. The muscle spindle has contractile elements, which when activated by gamma efferents, modifies spindle length

[134, 137]. Spindle sensitivity is also influenced by its immediate contractile history [134, 136, 138]. Golgi tendon organs also play a role as they contribute afferent information regarding changes in tensile force [139, 140]. Sensory information is transmitted by ascending tracts to supraspinal centres via the dorsal column, medial lemniscus system (primarily proprioception), spinothalamic / anterolateral system (primarily pain) and spinocerebellar system (unconscious proprioception) [1, 3, 4].

As mentioned earlier, inhibition of peripheral reflexes, such as the Hoffman reflex (H-reflex) (represents the afferent excitability of the alpha motoneurons [141, 142]), and increased co-contractions may be used to enhance balance control as reported by Alizadehsaravi et al. [143], who had younger and older participants perform unipedal balancing on a compliant surface (robot controlled balance platform). Reduced presynaptic inhibition (H-reflexes) minimizes spinal reflex contributions to balance increasing supraspinal control. Hence, this spinal reflex inhibition could reduce joint oscillations as control is moved more to the cortex [109]. However, a meta-analysis by Bakker et al. [109] indicated there is a lack of temporal correspondence between balance training improvements and H-reflex inhibition. They suggest that the H-reflex changes are more of a general training effect (remember the aforementioned inhibition of H-reflexes with volleyball players) rather than a balance training specific change. Bakker's review [109] reports reduced corticospinal excitability, increased levels of intracortical inhibition, and increased active motor thresholds during balance tasks, which suggests that there is a shift in movement control from cerebral cortical to more subcortical and cerebellar structures. Hence, balance control becomes more automatized and pre-programmed with training.

The accuracy or sensitivity of these receptors may be adversely affected by pain and fatigue [144]. The group III and IV afferents which aid in the transmission of pain impulses may interfere with proprioceptive sensitivity [145, 146]. The impact of fatigue, and the associated metabolite accumulation resulting from fatigue, which are transmitted by group III and IV afferents, impair position sense [144, 147]. Accumulated fatigue, while diminishing proprioceptive sensitivity, will shift position sense responsibilities more to the perception of effort, which is influenced and constantly modulating based on the extent and duration of fatigue; in this way, there is no consistent or sensitive information provided in terms of position sense feedback [148]. Furthermore, fear and anxiety have been shown to modulate the vestibulo-ocular reflexes [149], which in turn, negatively influences positional feedback.

Since balance is controlled by a number of components, deficits in one component may elevate the responsibilities of another component. For example, acute obstruction of vision impairs balance, whereas individuals with chronic visual impairments develop a more advanced vestibular system and more sensitive proprioceptive senses [150]. Decreased plantar sole sensitivity (proprioception) increased the triceps surae muscular activity (somatosensory response) to limit centre of pressure excursions [151]. When vision and proprioception are limited, the evoked sway reflex of the vestibular system can increase by approximately an order of magnitude greater than normal [152].

In summary, the control of muscle force is a result of highly interdependent actions. The activation (depolarization) of motor units and their associated muscle fibres are attributed to an interplay of excitatory and inhibitory influences. While a single depolarization of a target motor unit cannot control force, the recruitment of additional motor units with variable firing frequencies (rate coding) in concert with antagonist co-contractions, and optimal or at least adequate balance and stability from which to project the forces is necessary. Reflex inhibition can aid in decreasing the complexity of afferent inputs to the central nervous system, simplifying the role of cortical control to predict and proactively assist movement and balance with anticipatory postural adjustments and continuing asynchronous motor output. Feedback on the successful coupling of movement intent and action is provided by a number of afferent receptors including the vestibular system, vision (ocular) and proprioception, which is transmitted to the cerebellum, and somatosensory cortex among many other cortical structures.

Neuromuscular Fatigue

Neuromuscular fatigue can adversely affect motor output and movement control. There are a variety of fatigue definitions that emphasize the external manifestations of fatigue, such as a transient decrease in working capacity [153], a loss of force output leading to reduced performance [154], and a decline in the force generating capacity of the muscle [155]. But fatigue can also present as internal sensations, and as such has been defined as an increase in the perceived effort needed to exert a desired force and an eventual inability to produce this force [148, 156]. Fatigue-induced changes can occur centrally throughout the central nervous system as well as peripherally at the muscle level.

Neural Responses to Fatigue

The initial studies of Mosso (lecture-induced excitement) and Setchenov (active sawing recovery) proposed that the central nervous system was the primary focus of fatigue [157]. Gandevia's review [158] proposed that muscle afferent signals, reflecting the muscle's biochemical and force-generating status, can impair cortical activation. Cortical motoneuron activity has been demonstrated to decrease with sustained muscle fatigue [159], which has been attributed to increased cortical tryptophan levels [160]. Tryptophan promotes the synthesis rate of the neurotransmitter serotonin, which enhances the perceptions for sleep and relaxation [160]. With prolonged activity, serotonin spreads to the initial segments of the cortical motoneuron axons where it inhibits firing [161]. In addition, higher concentrations of ammonia can induce supraspinal dysfunction by inhibiting specific cortical centres, or by altering the membrane permeability of certain amino acids [162].

Increases in g-Aminobutyric acid (GABA) have also been posited to decrease motor cortex excitability, which contributes to central fatigue [163]. The neurotransmitter, dopamine, is a neuromodulator whose effects on fatigue follows a U-shaped pattern [164]. While appropriate levels of dopamine or dopamine agonists (e.g., methylphenidate) can reduce fatigue, too little or too much dopamine can increase fatigue effects. The dopamine response is dose dependent and individual, as greater levels of dopamine for one individual may reduce fatigue, while in another individual, the same dose amplifies fatigue effects [164]. Hence, dopamine agonists must be carefully titrated based on individual needs and responses.

Transcranial magnetic stimulation-induced muscle motor evoked potentials are commonly assumed to reflect corticospinal excitability. A number of studies have reported reductions in transcranial magnetic stimulation-induced motor evoked potentials with various fatigue protocols [165–170], which typically peak within 5 min after exercise and recover within 10 min [171, 172]. However not all of these studies normalized the motor evoked potential to an evoked maximum muscle action potential (Mmax). Hence, when examining the MEP/Mmax responses to fatigue, Gruet et al. [173] suggested that the evidence for diminished corticospinal excitability lacks consensus. For example, there are reports of increased corticospinal and spinal excitability after fatiguing exercise of the elbow flexors [174] and hand grip [175]. To differentiate between cortical and spinal motoneuron excitability, corticomedullary evoked potentials elicited with transmastoid stimulation are used to directly stimulate the spinal cord. Research in this area suggests that reduced spinal excitability may contribute to MEP/Mmax deficits in the first 2 min after exercise. However, Gandevia et al. [176] found reduced MEP/Mmax ratios for 10 min following exercise, and suggested that the later depression could be more related to cortical inhibition.

When single-pulse transcranial magnetic stimulation is elicited during a voluntary contraction, there is a silent period in the EMG signal, which is purported to be mediated by the activation of long-lasting $GABA_B$ receptors [44], with spinal motoneuron inhibition contributing to the initial component of the silent period, and intracortical (supraspinal) inhibitory mechanisms mediating the latter segment [177, 178]. While some studies report an increased duration of the silent period with fatigue [179], there is considerable variability in the responses, with the inconsistencies being dependent on subject instructions, inter-examiner variability, and other protocol issues [173]. Paired transcranial magnetic stimulation pulses can reveal changes in short interval and long interval cortical inhibition (SICI and LICI, respectively). Long interval cortical inhibition is modulated by $GABA_B$ inhibitory networks [44]. The difference between the silent period and long interval cortical inhibition is that the silent period represents the inhibition duration whilst the long interval cortical inhibition estimates the magnitude of inhibition. Short interval cortical inhibition is representative of $GABA_A$ intracortical inhibition [173, 180]. There are few studies examining changes in short interval cortical inhibition with fatigue. Using 10 min of intermittent maximal contractions of the first dorsal interosseus muscle, short interval cortical inhibition was observed to increase at exercise onset and then decreased progressively in parallel with the force [181]. While there is difficulty interpreting long interval cortical inhibition, short interval cortical inhibition and the silent period interactions with fatigue, the few studies seem to reveal an increased long interval cortical inhibition mediated primarily by a spinal component [182].

Fatigue-induced depression of corticospinal excitability may be identified peripherally as a disruption of movement control that is concordant with decreases in muscle activation [42, 53, 57, 183], and is observed through reductions in the amplitude of the electromyographic signal [184–187]. Decreases in muscle activation, measured via the interpolated twitch technique or superimposed twitch technique, have also been reported [52, 53, 57, 59]. A decrease in the EMG amplitude or interpolated twitch with fatigue can represent a number of changes, including decreased motor unit recruitment and firing frequency. The term "muscle wisdom" is used to describe decreases in firing frequency during maximal and submaximal contractions held to maximal intensity, which may result from a muscle-afferent reflex inhibition of motor unit firing rates [188, 189]. Muscle wisdom may serve as a defence mechanism to prevent membrane excitation deficits impacting the activation and release of Ca^{++} from the sarcoplasmic reticulum, since a high motoneuron firing frequency is unnecessary with the fatigue-induced slowing of contractile speed [189, 190]. However, a decreased firing frequency may also be an indication of an intrinsic motoneuron property termed "late adaptation," [191], where motoneuron rate coding is diminished with sustained constant-intensity excitation [192]. The mechanism for late adaptation has been accredited to an increased outward current mediated by a calcium-activated potassium conductance [191].

The complexity (temporal structure) of muscle force production is indicative of the ability of motor output to adapt accurately and rapidly to the demands of an activity [193–197], and represents the functional capacity of the neuromuscular system [193, 194]. Fatigue-induced loss of torque complexity was demonstrated during intermittent isometric and sustained fatiguing contractions [195].

Decreases in the afferent excitability of the motoneurons with fatigue as observed with the Hoffman reflex [141], indicates that inhibitory muscle afferents may downregulate motoneuron excitability of the fatigued motor unit [198], as well as increases in long latency inhibitory reflexes to neighbouring muscles [199]. Duchateau and Hainaut [199] proposed that an increase in neural drive with fatigue, associated with prolonged submaximal intensity contractions, occurred with Group Ia afferents presynaptic inhibition, as well as inhibition of interneurons in the oligosynaptic pathways. Further peripheral afferent inhibition can arise from stimulation of Group III afferents from fatigue-induced metabolic products, such as bradykinin [200, 201], arachidonic acid, and prostaglandin E2 [202], potassium [203], and lactate [204–206], as well as Group III and Group IV-induced inhibition with pain [207] and inflammation [57, 208, 209]. Thus, afferent inhibition of the motoneuron may be an attempt to diminish the possibility of further fatigue-induced muscle damage [42]. A global feedback loop characterized as a critical threshold of peripheral fatigue or sensory tolerance limit, is postulated to process all afferent feedback and feedforward signals in order to regulate exercise intensity with the goal of ensuring that voluntary activity remains within a tolerable range [210].

Peripheral (Muscle) Responses to Fatigue

As mentioned earlier, muscular fatigue can alter the force output and its control. According to a review by Collins et al., peripheral (muscle) impairments are more predominant with the initial stages of fatigue, which is then followed later by a combination of central and peripheral fatigue-related deficits [211]. The primary mechanisms underlying peripheral fatigue are changes in action potential synaptic transmission, sarcolemmal action potential propagation properties, and sarcoplasmic reticulum Ca2+ release, as well as excitation–contraction coupling, cross-bridge kinetics, and Ca2+ re-uptake. Peripheral fatigue can also influence force transmission by modifying the musculotendinous mechanical properties, as well as the tissue viscoelasticity and stiffness. These peripheral alterations do not occur in isolation as there is extrafusal and intrafusal afferent feedback to the central nervous system influencing force output and movement control [103]. Thus, alterations and/or impairments at the neural and peripheral levels can adversely affect movement control by limiting the timing and array of muscle forces available, and the ability to coordinate the activation and inhibition signals at many levels of the corticospinal system.

Training Recommendations to Optimize Neuromuscular Activation and Movement Control

It has been highlighted here, that there are a number of threats to movement initiation and control, such as negative emotions or lack of intrinsic or extrinsic rewards linked with training or rehabilitation (prefrontal cortex), lack of movement intent and practice (basal ganglia and cerebellum), impaired balance (afferent vestibular, visual and proprioceptive systems linked to efferent motor systems), fatigue (central and peripheral components), as well as injuries, and pain among others. To counterbalance some of these issues, it was recommended to ensure that the training or rehabilitation facility provides a positive rewarding environment. The training program and location can be an intimidating environment and experience for many individuals who have little interest or enthusiasm for exercise.

Behm et al. [212], in their narrative review, proposed that although higher volumes of resistance training can elicit greater strength training adaptations, minimalist resistance training can still improve physical fitness for at least the first 12 weeks involving one weekly session at intensities below 50% 1RM, with <3 sets per multi-joint exercise. Strength and muscle hypertrophy can even be accrued with 10–60 min of daily passive static stretching [213–220]. Although the volume of work would be extensive for multiple muscle groups, the work intensity (passive stretch) is almost non-existent. Similarly, a lack of physical exercise intensity is assumed with mental imagery exercises that activate and coordinate a number of cortical areas, such as the prefrontal cortex, limbic area, basal ganglia, cerebellum, and supplementary motor cortex [221]. The effectiveness of mental imagery to enhance movement control has been cited in many studies [221, 222].

It was mentioned that an emphasis on neural adaptations with resistance training should progress to and include ballistic intent contractions. Balance training and combinations of balance and resistance training should ensure more optimal neuromuscular efficiency and capacity. Muscle strength, power, and endurance resistance training would combat the early onset of fatigue-induced physical impairments. There are innumerable books, chapters, and articles in the literature that capture the expansiveness of this topic. For those individuals attempting to recover from an injury, resistance training programs may be problematic. With limb immobilization, isometric contraction training has been reported to be relatively ineffective in attenuating atrophy, and thus a progression to mobility exercises is important [222]. However adding nerve or muscle electrical stimulation to the isometric contractions, even at 5% of a maximal voluntary isometric contraction, can attenuate the immobility-induced decrements in protein synthesis [222].

Although periodization of training is most often associated with athletic training [223–225], it would also be an important consideration with a rehabilitation program. An injured and/or previously immobilized muscle would not have the same capacity for exercise training. Thus, modulating the volume and intensity of exercises through microcycles and mesocycles (days and weeks) would ensure the individual's neuromuscular system would have sufficient time to recover, adapt, and overcompensate to become stronger and more resilient [41, 222]. Rehabilitation training is not just a matter of becoming stronger, more powerful or attaining greater endurance. Optimal recovery of movement requires the coordination of these higher force outputs, and thus, functional movement assessments and corrective exercises would offer valuable insights for health and training professionals aiming to address neuromuscular deficits and prevent re-injury.

The decisions and subsequent actions involved in performing movement are complicated and integrated. Executive decisions to initiate and plan movement are highly connected to intrinsic and extrinsic rewards and motivation that are evaluated by the prefrontal cortex (i.e., orbitofrontal cortex, anterior cingulate cortex, ventral striatum) based on information received from the limbic system (amygdala and hippocampus). In concert with extensive interconnections of the basal ganglia, cerebellum, and supplementary motor cortex among other components, the primary motor cortex transmits impulses down the descending tracts using various strategies, such as changes in recruitment, rate coding, synchronization of motor unit firing, reflex inhibition and potentiation, and co-contractions to finely tune or attempt to fully activate the motoneurons. Optimal balance and stability are needed to provide a solid foundation for expected muscle force production. Matching of the movement intent and action is transmitted back to the spinal cord and cortex (e.g., cerebellum) via proprioceptive afferents (e.g., muscle spindles, Golgi tendon organs, and others). Sustained contractions can lead to fatigue, which disrupts optimal force control. A wide variety of training and rehabilitation protocols and strategies can be used to motivate individuals to exercise and optimize their training adaptations.

Suggestions for Future Exploration

There are manifold number of studies elucidating the most effective training programs to achieve maximum performance. Unfortunately, the motivation to exercise is quite low in much of the population. Therefore, future research should explore alternative strategies that will facilitate executive decisions (i.e., internal and external rewards affecting executive decisions within the prefrontal cortex) to exercise.

Future work should explore, in a manner similar to the review of resistance training by Behm et al. [212], the minimal requirements of time, volume, intensity, optimal schedule planning, etc. for developing efficient training programs that activate specific motor units while also providing training stress on balance, anticipatory postural adjustments, joint stability, and proprioception. Finally, all these interventions need to be explored in diverse populations.

Conclusion

The decisions and subsequent actions involved in planning and performing movement are complex and integrated. Executive decisions to initiate and plan movement are highly connected to intrinsic and extrinsic rewards and motivation that are evaluated by the prefrontal cortex (i.e., orbitofrontal cortex, anterior cingulate cortex, ventral striatum) based on information received from the limbic system (amygdala and hippocampus). In terms of training and rehabilitation, the efficacy of the cerebellum and basal ganglia would involve sufficient movement repetitions involving a variety of velocities, ranges of motion, planes and axes, as well as training progressions (i.e., volume, intensity, complexity) until the movement intent and action are adequately aligned (i.e., grand comparator) providing success (intrinsic rewards increase dopamine release) and continued motivation. The related connections between the basal ganglia, cerebellum, supplementary motor cortex, and primary motor cortex activate the spinal motoneurons which blend recruitment, rate coding, synchronization of motor unit firing, reflex inhibition and potentiation, and co-contractions to execute smooth movement control. Co-ordinated submaximal and maximal contractions can only be performed with adequate balance and stability to provide a solid foundation as well as minimizing the disruptive inhibitory effects of fatigue and pain or discomfort. Movement intent and action are compared through proprioceptive, vestibular, visual, and other afferents. A wide variety of training and rehabilitation protocols and strategies can be used to motivate individuals to exercise and optimize their training adaptations.

> Nothing happens until something moves.
>
> – Albert Einstein.

References

1. Kandel ER, Koester J, Mack S, Siegelbaum S. Principles of neural science. New York: McGraw Hill; 2021. p. 1. online resource (1, 1646 pages)
2. Rolls ET, Deco G, Huang CC, Feng J. Prefrontal and somatosensory-motor cortex effective connectivity in humans. Cereb Cortex. 2023;33:4939–63.
3. Berne RML, M.N. Physiology. Toronto: C.V. Mosby Publishers; 1983.
4. Martini FH, Nath JL. Fundamentals of anatomy and physiology. Upper Saddle River: Pearson Benjamin Cummings; 2008.
5. Kuhn AA, Sharott A, Trottenberg T, Kupsch A, Brown P. Motor cortex inhibition induced by acoustic stimulation. Exp Brain Res. 2004;158:120–4.
6. Anand KS, Dhikav V. Hippocampus in health and disease: an overview. Ann Indian Acad Neurol. 2012;15:239–46.
7. AbuHasan Q, Reddy V, Siddiqui W. Neuroanatomy, Amygdala, in StatPearls. Treasure Island: StatPearls Publishing; 2024.
8. Cardinal RN, Parkinson JA, Hall J, Everitt BJ. Emotion and motivation: the role of the amygdala, ventral striatum, and prefrontal cortex. Neurosci Biobehav Rev. 2002;26:321–52.
9. Rolls ET. The function of the orbitofrontal cortex. Brain Cogn. 2004;55:11–29.
10. Rolls ET, Cheng W, Feng J. The orbitofrontal cortex: reward, emotion and depression. Brain Commun. 2020;2:fcaa196.
11. Xie C, Jia T, Rolls ET, Robbins TW, Sahakian BJ, Zhang J, Liu Z, Cheng W, Luo Q, Zac Lo CY, Wang H, Banaschewski T, Barker GJ, ALW B, Buchel C, Quinlan EB, Desrivieres S, Flor H, Grigis A, Garavan H, Gowland P, Heinz A, Hohmann S, Ittermann B, Martinot JL, Paillere Martinot ML, Nees F, Orfanos DP, Paus T, Poustka L, Frohner JH, Smolka MN, Walter H, Whelan R, Schumann G, Feng J, Consortium I. Reward versus nonreward sensitivity of the medial versus lateral orbitofrontal cortex relates to the severity of depressive symptoms. Biol Psychiatry Cogn Neurosci Neuroimaging. 2021;6:259–69.
12. Rolls ET, Vatansever D, Li Y, Cheng W, Feng J. Rapid rule-based reward reversal and the lateral orbitofrontal cortex. Cereb Cortex Commun. 2020;1:tgaa087.
13. Rolls ET. The cingulate cortex and limbic systems for emotion, action, and memory. Brain Struct Funct. 2019;224:3001–18.
14. Rolls ET. Emotion, motivation, decision-making, the orbitofrontal cortex, anterior cingulate cortex, and the amygdala. Brain Struct Funct. 2023;228:1201–57.
15. Rolls ET, Wan Z, Cheng W, Feng J. Risk-taking in humans and the medial orbitofrontal cortex reward system. NeuroImage. 2022;249:118893.
16. Haber SN. Neuroanatomy of reward: a view from the ventral striatum. Boca Raton: CRC Press/Taylor and Francis; 2011.
17. Purves D, Augustine GJ, Fitzpatrick D. Neuroscience. New York: Sinauer Associates; 2001.
18. Young CB, Reddy V, Sonne J. Neuroanatomy, Basal Ganglia. Treasure Island: StatPearls Publishing; 2023.
19. Brooks VB. Roles of cerebellum and basal ganglia in initiation and control of movements. Can J Neurol Sci. 1975;2:265–77.
20. Bhanpuri NH, Okamura AM, Bastian AJ. Predictive modeling by the cerebellum improves proprioception. J Neurosci. 2013;33:14301–6.
21. Person AL. Corollary discharge signals in the cerebellum. Biol Psychiatry Cogn Neurosci Neuroimaging. 2019;4:813–9.
22. Alexander WH, Womelsdorf T. Interactions of medial and lateral prefrontal cortex in hierarchical predictive coding. Front Comput Neurosci. 2021;15:605271.
23. Schubotz RI, von Cramon DY. Functional organization of the lateral premotor cortex: fMRI reveals different regions activated by anticipation of object properties, location and speed. Brain Res Cogn Brain Res. 2001;11:97–112.
24. Alexander WH, Brown JW. Medial prefrontal cortex as an action-outcome predictor. Nat Neurosci. 2011;14:1338–44.

25. Alexander WH, Brown JW. A general role for medial prefrontal cortex in event prediction. Front Comput Neurosci. 2014;8:69.
26. Nachev P, Kennard C, Husain M. Functional role of the supplementary and pre-supplementary motor areas. Nat Rev Neurosci. 2008;9:856–69.
27. Nachev P, Wydell H, O'Neill K, Husain M, Kennard C. The role of the pre-supplementary motor area in the control of action. NeuroImage. 2007;36(Suppl 2):T155–63.
28. Tanaka M, Watanabe Y. Supraspinal regulation of physical fatigue. Neurosci Biobehav Rev. 2012;36:727–34.
29. Basile GA, Quartu M, Bertino S, Serra MP, Boi M, Bramanti A, Anastasi GP, Milardi D, Cacciola A. Red nucleus structure and function: from anatomy to clinical neurosciences. Brain Struct Funct. 2021;226:69–91.
30. Cacciola A, Milardi D, Basile GA, Bertino S, Calamuneri A, Chillemi G, Paladina G, Impellizzeri F, Trimarchi F, Anastasi G, Bramanti A, Rizzo G. The cortico-rubral and cerebello-rubral pathways are topographically organized within the human red nucleus. Sci Rep. 2019;9:12117.
31. Gibson AR, Houk JC, Kohlerman NJ. Magnocellular red nucleus activity during different types of limb movement in the macaque monkey. J Physiol. 1985;358:527–49.
32. Gibson AR, Houk JC, Kohlerman NJ. Relation between red nucleus discharge and movement parameters in trained macaque monkeys. J Physiol. 1985;358:551–70.
33. Van Kan PL, McCurdy ML. Contribution of primate magnocellular red nucleus to timing of hand preshaping during reaching to grasp. J Neurophysiol. 2002;87:1473–87.
34. van Kan PL, McCurdy ML. Discharge of primate magnocellular red nucleus neurons during reaching to grasp in different spatial locations. Exp Brain Res. 2002;142:151–7.
35. Klarner T, Zehr EP. Sherlock Holmes and the curious case of the human locomotor central pattern generator. J Neurophysiol. 2018;120:53–77.
36. Sherrington CS. Flexion-reflex of the limb, crossed extension reflex stepping and standing. J Physiol. 1910;40:28–121.
37. Harkema SJ. Neural plasticity after human spinal cord injury: application of locomotor training to the rehabilitation of walking. Neuroscientist. 2001;7:455–68.
38. Harkema SJ. Plasticity of interneuronal networks of the functionally isolated human spinal cord. Brain Res Rev. 2008;57:255–64.
39. Amann M, Sidhu SK, McNeil CJ, Gandevia SC. Critical considerations of the contribution of the corticomotoneuronal pathway to central fatigue. J Physiol. 2022;600:5203–14.
40. Pareti G. The "all-or-none" law in skeletal muscle and nerve fibres. Arch Ital Biol. 2007;145:39–54.
41. Behm DG. Neuromuscular implications and applications of resistance training. J Strength Cond Res. 1995;9:264–74.
42. Behm DG. Force maintenance with submaximal fatiguing contractions. Can J Appl Physiol. 2004;29:274–90.
43. Edwards C. The effects of innervation on the properties of acetylcholine receptors in muscle. Neuroscience. 1979;4:565–84.
44. McDonnell MN, Orekhov Y, Ziemann U. The role of GABA(B) receptors in intracortical inhibition in the human motor cortex. Exp Brain Res. 2006;173:86–93.
45. Partridge LD. Motor control and the myotatic reflex. Am J Phys Med Rehabil. 1961;40:96–103.
46. Poppele RE, Terzuolo CA. Myotatic reflex: its input – output relation. Science. 1968;159:743–5.
47. Smith LE. Facilitatory effects of myotatic stretch training upon leg strength and contralateral transfer. Am J Phys Med. 1970;49:132–8.
48. Gottlieb GL, Agarwal GC. Response to sudden torques about ankle in man: myotatic reflex. J Neurophysiol. 1979;42:91–106.
49. Allen GM, Gandevia SC, McKenzi DK. Reliability of measurements of muscle strength and voluntary activation using twitch interpolation. Muscle Nerve. 1995;18:593–600.

50. Behm DG, Power KE, Drinkwater EJ. Muscle activation is enhanced with multi- and uni-articular bilateral versus unilateral contractions. Can J Appl Physiol. 2003;28:38–52.
51. Behm DG, Whittle J, Button D, Power K. Intermuscle differences in activation. Muscle Nerve. 2002;25:236–43.
52. Behm DG, St-Pierre DM. Effects of fatigue duration and muscle type on voluntary and evoked contractile properties. J Appl Physiol. 1997;82:1654–61.
53. Behm DG, St-Pierre DMM. Fatigue mechanisms in trained and untrained plantar flexors. J Strength Cond Res. 1998;12:166–72.
54. Newham DJ, McCarthy T, Turner J. Voluntary activation of human quadriceps during and after isokinetic exercise. J Appl Physiol. 1991;71:2122–6.
55. Behm DG, Anderson K, Curnew RS. Muscle force and activation under stable and unstable conditions. J Strength Cond Res. 2002;16:416–22.
56. Behm DG, Anderson KG. The role of instability with resistance training. J Strength Cond Res. 2006;20:716–22.
57. Behm DG, St-Pierre DM. Fatigue characteristics following ankle fractures. Med Sci Sports Exerc. 1997;29:1115–23.
58. Behm DG, Baker KM, Kelland R, Lomond J. The effect of muscle damage on strength and fatigue deficits. J Strength Cond Res. 2001;15:255–63.
59. Behm DG, Button DC, Butt JC. Factors affecting force loss with prolonged stretching. Can J Appl Physiol. 2001;26:261–72.
60. Behm DG, Power K, White MD, LeDez K, Decker D, Drinkwater E. Effects of hyperbaric (6 ATA) pressure on voluntary and evoked skeletal muscle contractile properties. Undersea Hyperb Med Soc J. 2003;30:103–15.
61. Behm DG, Sale DG. Intended rather than actual movement velocity determines velocity-specific training response. J Appl Physiol. 1993;74:359–68.
62. Behm DG, Sale DG. Velocity specificity of resistance training. Sports Med. 1993;15:374–88.
63. Henneman E, Clamann HP, Gillies JD, Skinner RD. Rank order of motoneurons within a pool: law of combination. J Neurophysiol. 1974;37:1338–48.
64. Henneman E, Mendell LM. Functional organization of motoneuron pool and its inputs. In: Handbook of physiology – the nervous system II; 1980. p. 423–507.
65. Henneman E, Somjen G, Carpenter DO. Functional significance of cell size in spinal motoneurones. Neurophysiology. 1965;28:560–80.
66. Harrison PJ. The relationship between the distribution of motor unit mechanical properties and the forces due to recruitment and to rate coding for the generation of muscle force. Brain Res. 1983;264:311–5.
67. Milner-Brown HS, Stein RB, Yemm R. The orderly recruitment of human motor units during voluntary isometric contractions. J Physiol. 1973;230:359–70.
68. Secher N. Motor unit recruitment. Med Sci Sports. 1987;26:152–62. S. Karger AG
69. Zajac FE. Coupling of recruitment order to the force produced by motor units: the "size principle hypothesis" revisited. In: The orderly recruitment of motor units; 1989. p. 96–111.
70. Broman H, De Luca CJ, Mambrito B. Motor unit recruitment and firing rates interaction in the control of human muscles. Brain Res. 1985;337:311–9.
71. Burke RE. The control of muscle force: motor unit recruitment and firing patterns. In: Komi PV, editor. Strength and power in sport; 1992. p. 97–106.
72. Freund HJ. The relationship between the rate of rise of isometric tension and motor unit recruitment in a human foreman muscle. PflÅgers Arch Eur J Physiol. 1976;362:61–7.
73. Deschenes M. Short review: rate coding and motor unit recruitment patterns. J Appl Sports Sci Res. 1989;3:33–9.
74. Wani AM, Guha SK. A model for gradation of tension-recruitment and rate coding. Med Biol Eng. 1975;13:870–5.
75. Bigland B, Lippold OCJ. Motor unit activity in the voluntary contraction of human muscle. J Physiol. 1954;125:322–35.

76. Freund HJ. Motor unit and muscle activity in voluntary motor control. Physiol Rev. 1983;63:387–436.
77. Person RS, Kudina LP. Discharge frequency and discharge pattern of human motor units during voluntary contraction of muscle. Electroencephalogr Clin Neurophysiol. 1972;32:471–83.
78. Mori S. Discharge patterns of soleus motor units with associated changes in force exerted by foot during quiet stance in man. J Neurophysiol. 1973;36:458–71.
79. Kukulka CG, Clamann HP. Comparison of the recruitment and discharge properties of motor units in human brachial biceps and adductor pollicis during isometric contractions. Brain Res. 1981;219:45–55.
80. Evatt ML, Wolf SL, Segal RL. Modification of human stretch reflexes: preliminary studies. Neurosci Lett. 1989;105:350–5.
81. Casabona A, Polizzi MC, Perciavalle V. Differences in H-reflex between athletes trained for explosive contractions and non-trained subjects. Eur J Appl Physiol Occup Physiol. 1990;61:26–32.
82. Upton ARM, Radford PF. Motoneuron excitability in elite sprinters. Baltimore: University Park Press; 1975.
83. Buchtal F, Madsen A. Synchronous activity in normal and atrophic muscle. Electroencephalogr Clin Neurophysiol. 1950;2:425–44.
84. Milner-Brown HS, Stein RB, Lee RG. Synchronization of human motor units: possible roles of exercise and supraspinal reflexes. Electroencephalogr Clin Neurophysiol. 1975;38:245–54.
85. Clamann HP, Schelhorn TB. Nonlinear force addition of newly recruited motor units in the cat hindlimb. Muscle Nerve. 1988;11:1079–89.
86. Lind AR, Petrofsky JS. Isometric tension from rotary stimulation of fast and slow cat muscles. Muscle Nerve. 1978;1:213–8.
87. Rack PMH, Westbury DR. The effects of length and stimulus rate on tension in the isometric cat soleus muscle. J Physiol. 1969;204:443–60.
88. Miller RG, Mirka A, Maxfield M. Rate of tension development in isometric contractions of a human hand muscle. Exp Neurol. 1981;73:267–85.
89. Eccles JC, Eccles RM, Iggo A, Lundberg A. Electrophysiological investigations on Renshaw cells. J Physiol. 1961;159:461–78.
90. Hultborn HLS, Wigstrom H. On the function of recurrent inhibition in the spinal cord. Exp Brain Res. 1979;37:399–403.
91. Katz R, Pierrot-Deseilligny E. Recurrent inhibition in humans. Prog Neurobiol. 1999;57:325–55.
92. Lestienne F. Effects of inertial load and velocity on the braking process of voluntary limb movements. Exp Brain Res. 1979;35:407–18.
93. Mustard BE, Lee RG. Relationship between EMG patterns and kinematic properties for flexion movements at the human wrist. Exp Brain Res. 1987;66:247–56.
94. Wierzbicka MM, Wiegner AW, Shahani BT. Role of agonist and antagonist muscles in fast arm movements in man. Exp Brain Res. 1986;63:331–40.
95. Cooke JD, Brown SH. Movement-related phasic muscle activation II. Generation and functional role of the triphasic pattern. J Neurophysiol. 1990;63:465–72.
96. Karst GM, Hasan Z. Antagonist muscle activity during human forearm movements under varying kinematic and loading conditions. Exp Brain Res. 1987;67:391–401.
97. Marsden CD, Obeso JA, Rothwell JC. The function of the antagonist muscle during fast limb movements in man. J Physiol. 1983;335:1–13.
98. Hannaford B, Stark L. Roles of the elements of the triphasic control signal. Exp Neurol. 1985;90:619–34.
99. Ostering LR, Hamill J, Corcos DM, Lander J. Electromyographic patterns accompanying isokinetic exercise under varying speed and sequencing conditions. Am J Phys Med. 1984;63:289–96.
100. Gordon J, Ghez C. EMG patterns in antagonist muscles during isometric contractions in man: relations to response dynamics. Exp Brain Res. 1984;55:167–71.

101. Person RS. EMG study of co-ordination of activity of human antagonist muscles in the proces of developing motor habits. J Vysceit Nerveun Dejat. 1958;8:17–27.
102. Engelhorn R. Agonist and antagonist muscle EMG activity pattern changes with skill acquisition. Res Q Exerc Sport. 1983;54:315–23.
103. Ce E, Longo S, Limonta E, Coratella G, Rampichini S, Esposito F. Peripheral fatigue: new mechanistic insights from recent technologies. Eur J Appl Physiol. 2020;120:17–39.
104. Anderson K, Behm DG. The impact of instability resistance training on balance and stability. Sports Med. 2005;35:43–53.
105. Han J, Anson J, Waddington G, Adams R, Liu Y. The role of ankle proprioception for balance control in relation to sports performance and injury. Biomed Res Int. 2015;2015:842804.
106. Anderson KG, Behm DG. Maintenance of EMG activity and loss of force output with instability. J Strength Cond Res. 2004;18:637–40.
107. Kornecki S, Zschorlich V. The nature of stabilizing functions of skeletal muscles. J Biomech. 1994;27:215–25.
108. Anderson K, Behm DG. Trunk muscle activity increases with unstable squat movements. Can J Appl Physiol. 2005;30:33–45.
109. Bakker LBM, Lamoth CJC, Vetrovsky T, Gruber M, Caljouw SR, Nieboer W, Taube W, van Dieen JH, Granacher U, Hortobagyi T. Neural correlates of balance skill learning in Young and older individuals: a systematic review and meta-analysis. Sports Med Open. 2024;10:3.
110. Carpenter MG, Frank JS, Silcher CP. Surface height effects on postural control: a hypothesis for a stiffness strategy for stance. J Vestib Res. 1999;9:277–86.
111. Carpenter MG, Frank JS, Silcher CP, Peysar GW. The influence of postural threat on the control of upright stance. Exp Brain Res. 2001;138:210–8.
112. Adkin AL, Frank JS, Carpenter MG, Peysar GW. Fear of falling modifies anticipatory postural control. Exp Brain Res. 2002;143:160–70.
113. Ritzmann R, Freyler K, Werkhausen A, Gollhofer A. Changes in balance strategy and neuromuscular control during a fatiguing balance task-a study in perturbed unilateral stance. Front Hum Neurosci. 2016;10:289.
114. Behrens M, Mau-Moeller A, Wassermann F, Bader R, Bruhn S. Effect of balance training on neuromuscular function at rest and during isometric maximum voluntary contraction. Eur J Appl Physiol. 2015;115:1075–85.
115. Horak FB. Postural orientation and equilibrium: what do we need to know about neural control of balance to prevent falls? Age Ageing. 2006;35:7–11.
116. Callegari B, Saunier G, Duarte MB, Almeida G, Amorim CF, Mourey F, Pozzo T, Souza GDS. Anticipatory postural adjustments and kinematic arm features when postural stability is manipulated. PeerJ. 2018;6:e4309.
117. Hodges PW, Cresswell AG, Thorstensson A. Preparatory trunk motion accompanies rapid upper limb movement. Exp Brain Res. 1999;124:69–79.
118. Hodges PW, Gandevia SC. Activation of the human diaphragm during a repetitive postural task. Physiol J. 2000;522:165–75.
119. Chiou SY, Gottardi SE, Hodges PW, Strutton PH. Corticospinal excitability of trunk muscles during different postural tasks. PLoS One. 2016;11:e0147650.
120. Krause A, Freyler K, Gollhofer A, Stocker T, Bruderlin U, Colin R, Topfer H, Ritzmann R. Neuromuscular and kinematic adaptation in response to reactive balance training – a randomized controlled study regarding fall prevention. Front Physiol. 2018;9:1075.
121. Hodges PW, Richardson CA. Inefficient muscular stabilization of the lumbar spine associated with low back pain. Spine. 1996;21:2640–50.
122. Gebel A, Prieske O, Behm DG, Granacher U. Effects of balance training on physical fitness in youth and young athletes: a narrative review. Strength Cond J. 2020;42:35–44.
123. Lesinski M, Hortobagyi T, Muehlbauer T, Gollhofer A, Granacher U. Dose-response relationships of balance training in healthy young adults: a systematic review and meta-analysis. Sports Med. 2015;45:557–76.

124. Lesinski M, Hortobagyi T, Muehlbauer T, Gollhofer A, Granacher U. Effects of balance training on balance performance in healthy older adults: a systematic review and meta-analysis. Sports Med. 2015;45:1721–38.
125. Granacher U, Behm DG. Relevance and effectiveness of combined resistance and balance training to improve balance and muscular fitness in healthy youth and youth athletes: a scoping review. Sports Med. 2023;53:349–70.
126. Gebel A, Lesinski M, Behm DG, Granacher U. Effects and dose-response relationship of balance training on balance performance in youth: a systematic review and meta-analysis. Sports Med. 2018;48:2067–89.
127. Kibele A, Granacher U, Muehlbauer T, Behm DG. Stable, unstable and metastable states of equilibrium: definitions and applications to human movement. J Sports Sci Med. 2015;14:885–7.
128. Behm DG, Drinkwater EJ, Willardson JM, Cowley PM. The use of instability to train the core musculature. Appl Physiol Nutr Metab. 2010;35:91–108.
129. Behm DG, Drinkwater EJ, Willardson JM, Cowley PM, Canadian Society for Exercise Physiology. Canadian Society for Exercise Physiology position stand: the use of instability to train the core in athletic and nonathletic conditioning. Appl Physiol Nutr Metab. 2010;35:109–12.
130. Behm DG, Drinkwater EJ, Willardson JM, Cowley PM. The role of instability rehabilitative resistance training for the core musculature. Strength Cond J. 2011;33:72–81.
131. Hamlyn N, Behm DG, Young WB. Trunk muscle activation during dynamic weight-training exercises and isometric instability activities. J Strength Cond Res. 2007;21:1108–12.
132. Chaouachi A, Ben Othman A, Hammami R, Drinkwater EJ, Behm DG. The combination of plyometric and balance training improves sprint and shuttle run performances more often than plyometric-only training with children. J Strength Cond Res. 2013;28:401–12.
133. Behm DG, Kay AD, Trajano GS, Alizadeh S, Blazevich AJ. Effects of stretching on injury risk reduction and balance. J Clin Exerc Physiol. 2021;10:106–16.
134. Hillier S, Immink M, Thewlis D. Assessing proprioception: a systematic review of possibilities. Neurorehabil Neural Repair. 2015;29:933–49.
135. Riemann BL, Lephart SM. The sensorimotor system, part II: the role of proprioception in motor control and functional joint stability. J Athl Train. 2002;37:80–4.
136. Proske U. What is the role of muscle receptors in proprioception? Muscle Nerve. 2005;31:780–7.
137. Dubo HI, Darling RC. Gamma nervous system and muscle spindles. p. 49–63
138. Proske U, Gandevia SC. The proprioceptive senses: their roles in signaling body shape, body position and movement, and muscle force. Physiol Rev. 2012;92:1651–97.
139. Chalmers G. Re-examination of the possible role of Golgi tendon organ and muscle spindle reflexes in proprioceptive neuromuscular facilitation muscle stretching. Sports Biomech. 2004;3:159–83.
140. Houk JC, Crago PE, Rymer WZ. Functional properties of the Golgi tendon organs. In: Desmedt JE, editor. Spinal and supraspinal mechanisms of voluntary motor control and locomotion; 1980. p. 33–43.
141. Enoka RM, Hutton RS, Eldred E. Changes in excitability of tendon tap and Hoffmann reflexes following voluntary contractions. Electroencephalogr Clin Neurophysiol. 1980;48:664–72.
142. Zehr EP. Considerations for use of the Hoffmann reflex in exercise studies. Eur J Appl Physiol. 2002;86:455–68.
143. Alizadehsaravi L, Bruijn SM, Maas H, van Dieen JH. Modulation of soleus muscle H-reflexes and ankle muscle co-contraction with surface compliance during unipedal balancing in young and older adults. Exp Brain Res. 2020;238:1371–83.
144. Fortier S, Basset FA. The effects of exercise on limb proprioceptive signals. J Electromyogr Kinesiol. 2012;22:795–802.
145. Proske U, Gregory JE, Morgan DL, Percival P, Weerakkody NS, Canny BJ. Force matching errors following eccentric exercise. Hum Mov Sci. 2004;23:365–78.

146. Proske U, Weerakkody NS, Percival P, Morgan DL, Gregory JE, Canny BJ. Force-matching errors after eccentric exercise attributed to muscle soreness. Clin Exp Pharmacol Physiol. 2003;30:576–9.
147. Fortier S, Basset FA, Billaut F, Behm D, Teasdale N. Which type of repetitive muscle contractions induces a greater acute impairment of position sense? J Electromyogr Kinesiol. 2010;20:298–304.
148. Steele J. What is (perception of) effort? Objective and subjective effort during task performance. PsyArχiv; 2020.
149. Yardley L, Watson S, Britton J, Lear S, Bird J. Effects of anxiety arousal and mental stress on the vestibulo-ocular reflex. Acta Otolaryngol. 1995;115:597–602.
150. Aras D, Guler O, Akca F, Arslan E, Akalan C. Comparison of balance skills of visually impaired and non-impaired judo athletes and goalball/futsal players. Phys Educ Stud. 2018;22:292–7.
151. Billot M, Handrigan GA, Simoneau M, Teasdale N. Reduced plantar sole sensitivity induces balance control modifications to compensate ankle tendon vibration and vision deprivation. J Electromyogr Kinesiol. 2015;25:155–60.
152. Day BL, Cole J. Vestibular-evoked postural responses in the absence of somatosensory information. Brain. 2002;125:2081–8.
153. Asmussen E, Mazin B. A central nervous component in local muscular fatigue. Eur J Appl Physiol Occup Physiol. 1978;38:9–15.
154. Fitts RH, Metzger JM. Mechanisms of muscular fatigue. Med Sport Sci. 1993;38:248–68.
155. Degens H, Veerkamp JH. Changes in oxidative capacity and fatigue resistance in skeletal muscle. Int J Biochem. 1994;26:871–8.
156. Enoka RM, Stuart DG. Neurobiology of muscle fatigue. J Appl Physiol. 1992;72:1631–48.
157. Asmussen E. Muscle fatigue. Med Sci Sports. 1979;11:313–21.
158. Gandevia SC. Spinal and supraspinal actors in human muscle fatigue. Physiol Rev. 2001;81:1725–89.
159. Pappas GP, Olcott EW, Drace JE. Imaging of skeletal muscle function using 18 FDG PET: force production, activation and metabolism. J Appl Phys. 2001;90:329–37.
160. Newsholme EA, Blomstrand E. Tryptophan, 5-hydroxytryptamine and a possible explanation for central fatigue. In: Gandevia SC, et al., editors. Fatigue. New York: Plenum Press; 1995. p. 315–8.
161. Cotel F, Exley R, Cragg SJ, Perrier JF. Serotonin spillover onto the axon initial segment of motoneurons induces central fatigue by inhibiting action potential initiation. Proc Natl Acad Sci USA. 2013;110:4774–9.
162. Green HJ. Neuromuscular aspects of fatigue. Can J Sport Sci. 1987;12(suppl 1):7S–19S.
163. Tergau F, Geese R, Bauer A, Baur S, Paulus W, Detlev RC. Motor cortex fatigue in sports measured by transcranial magnetic double stimulation. Med Sci Sports Exerc. 2000;32:1942–8.
164. Dobryakova E, Genova HM, DeLuca J, Wylie GR. The dopamine imbalance hypothesis of fatigue in multiple sclerosis and other neurological disorders. Front Neurol. 2015;6:52.
165. Brasil-Neto JP, Pascual-Leone A, Valls-Sole J, Cammarota A, Cohen LG, Hallett M. Postexercise depression of motor evoked potentials: a measure of central nervous system fatigue. Exp Brain Res. 1993;93:181–4.
166. Fulton RC, Strutton PH, McGregor AH, Davey NJ. Fatigue-induced change in corticospinal drive to back muscles in elite rowers. Exp Physiol. 2002;87:593–600.
167. Hollge J, Kunkel M, Ziemann U, Tergau F, Geese R, Reimers CD. Central fatigue in sports and daily exercises. A magnetic stimulation study. Int J Sports Med. 1997;18:614–7.
168. Khedr EM, Galal O, Said A, Abd-elsameea M, Rothwell JC. Lack of post-exercise depression of corticospinal excitability in patients with Parkinson's disease. Eur J Neurol. 2007;14:793–6.
169. Milanovic S, Filipovic SR, Blesic S, Ilic TV, Dhanasekaran S, Ljubisavljevic M. Paired-associative stimulation can modulate muscle fatigue induced motor cortex excitability changes. Behav Brain Res. 2011;223:30–5.

170. Pitcher JB, Robertson AL, Clover EC, Jaberzadeh S. Facilitation of cortically evoked potentials with motor imagery during post-exercise depression of corticospinal excitability. Exp Brain Res. 2005;160:409–17.
171. McKay WB, Tuel SM, Sherwood AM, Stokic DS, Dimitrijevic MR. Focal depression of cortical excitability induced by fatiguing muscle contraction: a transcranial magnetic stimulation study. Exp Brain Res. 1995;105:276–82.
172. Zanette G, Bonato C, Polo A, Tinazzi M, Manganotti P, Fiaschi A. Long-lasting depression of motor-evoked potentials to transcranial magnetic stimulation following exercise. Exp Brain Res. 1995;107:80–6.
173. Gruet M, Temesi J, Rupp T, Levy P, Millet GY, Verges S. Stimulation of the motor cortex and corticospinal tract to assess human muscle fatigue. Neuroscience. 2013;231:384–99.
174. Aboodarda SJ, Fan S, Coates K, Millet GY. The short-term recovery of corticomotor responses in elbow flexors. BMC Neurosci. 2019;20:9.
175. Forman DA, Forman GN, Murphy BA, Holmes MWR. Sustained isometric wrist flexion and extension maximal voluntary contractions on corticospinal excitability to forearm muscles during low-intensity hand-gripping. Brain Sci. 2020;10:445.
176. Gandevia SC, Petersen N, Butler JE, Taylor JL. Impaired response of human motoneurones to corticospinal stimulation after voluntary exercise. J Physiol. 1999;521(Pt 3):749–59.
177. Inghilleri M, Berardelli A, Cruccu G, Manfredi M. Silent period evoked by transcranial stimulation of the human cortex and cervicomedullary junction. J Physiol. 1993;466:521–34.
178. Levenez M, Garland SJ, Klass M, Duchateau J. Cortical and spinal modulation of antagonist coactivation during a submaximal fatiguing contraction in humans. J Neurophysiol. 2008;99:554–63.
179. Hilty L, Langer N, Pascual-Marqui R, Boutellier U, Lutz K. Fatigue-induced increase in intracortical communication between mid/anterior insular and motor cortex during cycling exercise. Eur J Neurosci. 2011;34:2035–42.
180. Di Lazzaro V, Oliviero A, Saturno E, Dileone M, Pilato F, Nardone R, Ranieri F, Musumeci G, Fiorilla T, Tonali P. Effects of lorazepam on short latency afferent inhibition and short latency intracortical inhibition in humans. J Physiol. 2005;564:661–8.
181. Benwell NM, Sacco P, Hammond GR, Byrnes ML, Mastaglia FL, Thickbroom GW. Short-interval cortical inhibition and corticomotor excitability with fatiguing hand exercise: a central adaptation to fatigue? Exp Brain Res. 2006;170:191–8.
182. McNeil CJ, Martin PG, Gandevia SC, Taylor JL. The response to paired motor cortical stimuli is abolished at a spinal level during human muscle fatigue. J Physiol. 2009;587:5601–12.
183. Behm DG, Button DC, Barbour G, Butt JC, Young WB. Conflicting effects of fatigue and potentiation on voluntary force. J Strength Cond Res. 2004;18:365–72.
184. Dolmage T, Cafarelli E. Rate of fatigue during repeated submaximal contractions of human quadriceps muscle 1. Can J Physiol Pharmacol. 1991;69:1410–5.
185. Edwards RG, Lippold OCJ. The relation between force and integrated electrical activity in fatigued muscle. J Physiol. 1956;132:677–81.
186. Hakkinen K. Neuromuscular fatigue and recovery in women at different ages during heavy resistance loading. Electromyogr Clin Neurophysiol. 1995;35:403–13.
187. Moritani T, Nagata A, Muro M. Electromyographic manifestations of muscular fatigue. Med Sci Sports Exerc. 1982;14:198–202.
188. Garland SJ, Enoka RM, Serrano LP, Robinson GA. Behavior of motor units in human biceps brachii during a submaximal fatiguing contraction. J Appl Physiol. 1994;76:2411–9.
189. Woods JJ, Furbush F, Bigland-Ritchie B. Evidence for a fatigue-induced reflex inhibition of Motoneuron firing rates. J Neurophysiol. 1987;58:125–37.
190. Marsden CD, Meadows JC, Merton PA. "Muscular wisdom" that minimizes fatigue during prolonged effort in man: peak rates of motoneuron discharge and slowing of discharge during fatigue. In: Motor control mechanisms in health and disease. New York: Raven Press; 1983. p. 169–211.

191. Kernell D, Monster AW. Motoneurone properties and motor fatigue. Exp Brain Res. 1982;46:197–204.
192. Gardiner PF. Neuromuscular aspects of physical activity. Windsor: Human Kinetics; 2001.
193. Pethick J, Winter SL, Burnley M. Loss of knee extensor torque complexity during fatiguing isometric muscle contractions occurs exclusively above the critical torque. Am J Physiol Regul Integr Comp Physiol. 2016;310:R1144–53.
194. Pethick J, Winter SL, Burnley M. Caffeine ingestion attenuates fatigue-induced loss of muscle torque complexity. Med Sci Sports Exerc. 2018;50:236–45.
195. Pethick J, Winter SL, Burnley M. Fatigue reduces the complexity of knee extensor torque during fatiguing sustained isometric contractions. Eur J Sport Sci. 2019;19:1349–58.
196. Pethick J, Winter SL, Burnley M. Fatigue-induced changes in knee-extensor torque complexity and muscle metabolic rate are dependent on joint angle. Eur J Appl Physiol. 2021;121:3117–31.
197. Pethick J, Winter SL, Burnley M. Physiological complexity: influence of ageing, disease and neuromuscular fatigue on muscle force and torque fluctuations. Exp Physiol. 2021;106:2046–59.
198. Garland SJ, McComas AJ. Reflex inhibition of human soleus muscle during fatigue. J Physiol Lond. 1990;129:17–27.
199. Duchateau J, Hainaut K. Behaviour of short and long latency reflexes in fatigued human muscles. J Physiol. 1993;471:787–99.
200. Mense S. Nervous outflow from skeletal muscle following chemical noxious stimulation. J Physiol Lond. 1977;267:75–88.
201. Mense S, Schmidt RF. Activation of group IV afferent units from muscle by algesic agents. Brain Res. 1974;72:305–10.
202. Mense S. Sensitization of group IV muscle receptors to bradykinin by 5-hydroxytryptamime and prostaglandin E-2. Brain Res. 1981;225:95–105.
203. Kniffki KD, Mense S, Schmidt RF. Response of group IV afferent units from skeletal muscle to stretch, contraction and chemical stimuli. Exp Brain Res. 1978;31:511–22.
204. Rotto DM, Kaufman MP. Effects of metabolic products of muscular contraction on the discharge of group III and IV afferents. J Appl Physiol. 1988;64:2306–13.
205. Sinoway LI, Hill JM, Pickar JG, Kaufman MP. Effects of contraction and lactic acid on the discharge of group III afferents in cats. J Neurophysiol. 1993;69:1053–9.
206. Thimm F, Baum K. Response of chemosensitive nerve fibres of group III and IV to metabolic changes in rat muscles. Pflugers Arch. 2003;410:143–52.
207. Rutherford OM, White PD. Human quadriceps strength and fatiguability in patients with post viral fatigue. J Neurol Neurosurg Psychiatry. 1991;54:961–4.
208. Spencer JD, Hayes KC, Alexander IJ. Knee joint effusion and quadriceps reflex inhibition in man. Arch Phys Med Rehabil. 1984;65:171–7.
209. Wood L, Ferrell WR, Baxendale RH. Pressures in normal and acutely distended human knee joints and effects on quadriceps maximal voluntary contractions. J Exp Physiol. 1988;73:305–14.
210. Hureau TJ, Romer LM, Amann M. The 'sensory tolerance limit': a hypothetical construct determining exercise performance? Eur J Sport Sci. 2018;18:13–24.
211. Collins BW, Pearcey GEP, Buckle NCM, Power KE, Button DC. Neuromuscular fatigue during repeated sprint exercise: underlying physiology and methodological considerations. Appl Physiol Nutr Metab. 2018;43:1166–75.
212. Behm DG, Granacher U, Warneke K, Aragao-Santos JC, Da Silva-Grigoletto ME, Konrad A. Minimalist training: is lower dosage or intensity resistance training effective to improve physical fitness? A narrative review. Sports Med. 2023;54:289–302.
213. Arntz F, Markov A, Behm DG, Behrens M, Negra Y, Nakamura M, Moran J, Chaabene H. Chronic effects of static stretching exercises on muscle strength and power in healthy individuals across the lifespan: a systematic review with multi-level meta-analysis. Sports Med. 2023;53:723–45.

214. Warneke K, Brinkmann A, Hillebrecht M, Schiemann S. Influence of long-lasting static stretching on maximal strength, muscle thickness and flexibility. Front Physiol. 2022;13:878955.
215. Warneke K, Konrad A, Keiner M, Zech A, Nakamura M, Hillebrecht M, Behm DG. Using daily stretching to counteract performance decreases as a result of reduced physical activity-a controlled trial. Int J Environ Res Public Health. 2022;19:15571.
216. Warneke K, Lohmann LH, Keiner M, Wagner CM, Schmidt T, Wirth K, Zech A, Schiemann S, Behm D. Using long-duration static stretch training to counteract strength and flexibility deficits in moderately trained participants. Int J Environ Res Public Health. 2022;19:13254.
217. Warneke K, Lohmann LH, Lima CD, Hollander K, Konrad A, Zech A, Nakamura M, Wirth K, Keiner M. and Behm DG, vol. 53. Physiology of stretch-mediated hypertrophy and strength increases: a narrative review: Sports Med; 2023. p. 2055–75.
218. Warneke K, Wirth K, Keiner M, Lohmann LH, Hillebrecht M, Brinkmann A, Wohlann T, Schiemann S. Comparison of the effects of long-lasting static stretching and hypertrophy training on maximal strength, muscle thickness and flexibility in the plantar flexors. Eur J Appl Physiol. 2023;123:1773–87.
219. Warneke K, Zech A, Wagner CM, Konrad A, Nakamura M, Keiner M, Schoenfeld BJ, Behm DG. Sex differences in stretch-induced hypertrophy, maximal strength and flexibility gains. Front Physiol. 2022;13:1078301.
220. Warneke K, Hillebrecht M, Claasen-Helmers E, Wohlann T, Keiner M, Behm DG. Effects of a home-based stretching program on bench press maximum strength and shoulder flexibility. J Sports Sci Med. 2023;22:597–604.
221. Slimani M, Tod D, Chaabene H, Miarka B, Chamari K. Effects of mental imagery on muscular strength in healthy and patient participants: a systematic review. J Sports Sci Med. 2016;15:434–50.
222. Behm DG. Debilitation to adaptation. J Strength Cond Res. 1993;7:65–75.
223. Bompa T. Periodization-theory and methodology of training. Adelaide: Human Kinetics; 2000.
224. Bompa TO. Periodization of strength: the most effective methodology of strength training. Nat Strength Cond Assoc J. 1990;12:49–52.
225. Stone MH, Pierce KC, Haff GG, Stone M. Periodization: effects of manipulating volume and intensity. Part 2. Nat Strength Cond Assoc. 1999;21:54–60.

Chapter 4
Proprioception—The Sixth Sense: Understanding Our Relationship with the Internal and External Environment

Nikos C. Apostolopoulos

Introduction

> To move things is all that mankind can do … whether in whispering a syllable or in felling a forest.
>
> *Sir Charles S. Sherrington, 1920*

Skeletal muscle represents the largest metabolically active tissue in the body, accounting for approximately 40–50% of total body mass and containing 50–75% of all body proteins [1–3]. It functions to provide locomotion [4, 5], serves as an endocrine organ [6], and acts as a sensory organ detecting the position of the limbs and modulating posture [7]. Muscle mass depends on the balance between protein synthesis and degradation, processes that are sensitive to nutritional status, hormonal balance, level of physical activity/exercise and disease, or injury status [2]. Skeletal muscle has an intrinsic capacity to sense mechanical stimuli

N. C. Apostolopoulos (✉)
Founder microStretching® and Stretch Therapy™, Toronto, ON, Canada
e-mail: nikos.apostolopoulos@microstretching.com

N. C. Apostolopoulos et al. (eds.), *Fundamentals of Recovery, Regeneration, and Adaptation to Exercise Stress: An Integrated Approach*,
https://doi.org/10.1007/978-3-031-44270-4_4

(mechano-sensing) and to convert these into biochemical events (mechanotransduction) influencing its muscle growth, metabolism, and homeostasis [8].

As a complex and intricate organ, the activation, coordination, and execution of changes in muscle fibre length is reliant on constant sensory feedback to fine tune the envisaged contractile effort and to protect the system from critical over extension [9–11]. To accomplish this, evolution has provided multiple intersecting pathways serving to provide an optimal functional interaction of the various musculoskeletal tissues (skeletal bones, skeletal muscles, tendons, ligaments, cartilage, joints, and other connective tissue). This interconnected network is responsible for maintaining (within "normal" limits) the structural integrity and function of the body [12]. Research has shown that the interactive system provides a dynamic reciprocity of mechanochemical interactions of the musculoskeletal tissue(s), including components at the cellular (muscle fibre), tissue (muscle), and organismic level [13, 14].

Animals have evolved sensory cells to detect most, if not all of the Earth's physical stimuli, including mechanical force through various *mechanosensors* [15]. Adaptation to the myriad of mechanical force(s) (i.e., compression, tensile, shearing, hydrostatic, gravity, etc.) is important in shaping, developing, and maintaining the integrity and functionality of the tissues and organs [16–19]. The information generated, received, and integrated coordinates the behavior of the cells, tissues, organs of the body, locally and systemically. Force-dependent conformational changes often alter biochemical and cellular functions, such as gene expression, protein synthesis, cell proliferation and differentiation [20–25]. These changes are key for the body's homeostasis, development, adaptation, regeneration, and recovery.

When considering the role of the mechanosensors, the mechanical force variables include: magnitude/intensity, rate/frequency, and duration. These variables, both singly and collectively, are essential in terms of adaptability through the initiation of the necessary selectivity and handling of a biologically relevant stimulus pattern(s) leading to physiological change [26, 27]. In other words, mechanosensors evolve and adapt to match a particular biological need of the organism(s). In so doing, a resilience is developed by the body, in response to either a weak or a strong stimulus, which acts to maintain function, while adapting to future perturbations.

Mechanosensors are present in muscles, joints, ligaments, tendons, and skin; collectively, they provide input to the central nervous system regarding structural challenges/ deformations of tissue [28]. While subject to long-standing debate, it is now accepted that the mechanosensory signal includes both a peripheral afferent input, with a subsequent efferent motor response from the central nervous system [29].

In this chapter, the role and functionality of the various proprioceptive mechanosensors found within the musculoskeletal system will be examined, in particular the muscular (Table 4.1). Demonstrating different adaptive characteristics in response to a specific stimulus, each mechanosensor is responsible for transmitting precise mechanical deformation information via electrical signals to the central nervous system. The mechanosensors associated with muscle, in addition to those related to articular areas, are categorised as being either quick adapting or slow adapting.

Table 4.1 Characteristics of proprioceptive mechanosensors

Mechanosensor	Location	Function	Adaptation
Muscles[a]			
Golgi tendon organ	Joint capsule, ligaments, tendons	Reflex	Slow
Muscle spindle	Muscle (extrafusal)	Stretch reflex	Slow
Articular regions[b]			
Pacinian corpuscles	Skin, joint capsule, ligaments	High frequency vibration	Quick
Ruffini endings	Skin, joint capsule, ligaments	Joint pressure	Slow
Unmyelinated free nerve endings	Ligaments, related muscles	Joint pain	Slow

[a]Discussed in detail below

[b]Fleming, M.S. and Luo, W. (2013). The anatomy, function, and development of mammalian Aβ low-threshold mechanoreceptors. *Front Biol (Beijing)*; 8(4): 1–20

Efstathiou, M.A.; Giannaki, C.D.; Roupa, Z.; Hadjisavvas, S. and Stefanakis, M. (2022). Evidence of distorted proprioception and postural control in studies of experimentally induced pain: a critical review of the literature. *Scan J Pain*; 22(3): 445–456

These terms are qualitative and are used to describe a wide range of time dependence. A quick adapting response generally infers a cessation of activity soon after the start of a constant stimulus. These mechanosensors are very sensitive to stimulation and are able to fire action potentials as long as the stimulus is being generated, and are considered to mediate sensation related to joint motion [30, 31]. Slow adapting mechanosensors remain discharging in response to a continuous stimulus; they mediate the sensation of joint position, and are maximally stimulated at certain joint angles [30, 31]. Slow adapting mechanosensors are involved in the stretch reflex [32].

Knowledge of how the stimulation of the mechanosensors, in particular the proprioceptors, affect the body and their subsequent response(s) is important for therapists. Understanding the how, why, and when at which the magnitude of a mechanical force created during manual therapy (i.e., therapeutic stretching, massage, physiotherapy etc.) influences the response(s) of the body allows for the creation of novel concepts shaping both therapeutic actions and future somatic practices.

The rest of this chapter will explore proprioception in terms of a special group of non-nociceptive, low-threshold mechanosensors, the proprioceptors.

In 1906, the British neurophysiologist Sir Charles Scott Sherrington coined the term *proprio-ception from the Latin word, "proprius", meaning "one's own". Sherrington wrote, "in muscular receptivity, we see the body itself acting as a stimulus to its own receptors – the proprioceptors"* [33].

A "*Sensitive*", yet "*Receptive*", Brief History of Proprioception Research

In the nineteenth and twentieth centuries, the concept of the sensate body was advanced through discovery of the functioning of the nervous system. In 1826, the Scottish anatomist, Sir Charles Bell, first characterised "*muscle sense*". He referred to it as the "sixth sense", adding it to the five traditional senses: vison, audition, olfaction, gustation, and somatosensation [34, 35]. Bell's earlier anatomical discoveries of distinct specialised sensory and motor nerves were instrumental in investigating the mechanisms of the nervous system that governed and regulated muscular activity [35]. Using various weights, Bell estimated the "load" lifted by the hand and developed the concept of "*muscle sense*". Bell suggested that this "*sense*" was a muscular coordination heightened by an awareness of the exertion and spasm of the muscle during the lifting of the weight. He stated, *"I shall first enquire, if it is necessary to the governance of the muscular frame, that there be a consciousness of the state or degree of action of the muscles? This can be asked since "[w]e are sensible of the most minute changes in muscular exertion, by which we know the position of the body and limbs, when there is no other means of knowledge open to us"*[1] *(1826: p. 167).*

Using the example of a blind man balancing while moving, Bell provided a discussion of the interaction of the sensory and motor nerves, *"Between the brain and muscles there is a circle of nerves; one nerve conveys the influence from the brain to the muscle, another gives the sense of the condition of the muscle to the brain"*[2] *(1826: p. 170).* He further described the bidirectional nature of the nervous system by stating, *"there is a nerve of sensibility to convey a sensation of the coordination of the muscles to the sensorium, as well as a nerve of motion for conveying the mandate of the will to the muscles"*[3] *(1833: p. 196).*

In 1906, the British neurophysiologist Sir Charles Scott Sherrington published, "*The Integrative Action of The Nervous System*" [33]. Based on a collection of 10 lectures delivered in 1904 at the Silliman Memorial Lectures of Yale University, Sherrington summarised two decades of intensive research and thinking on the functionality of the neural system [36]. His research was inspired by the work of the Spaniard neuroanatomist, Santiago Y. Cajal, who shared the 1906 Nobel Prize in Physiology or Medicine with Italian neuroanatomist, Camillo Golgi (as in Golgi tendon organ) [37].

Sherrington's work changed the subsequent course of neurophysiology by setting forth in detail the manner by which the nervous system serves to bring together in a united action the various parts of the animal organism [38]. Sherrington wrote about the uniqueness of the function of the nervous system, *"it works through living*

---

[1] Bell, C. (1826), 'On the Nervous Circle Which Connects the Voluntary Muscles with the Brain,' *Philosophical Transactions of the Royal Society of London* 116: pp. 163–73.

[2] Ibid.

[3] Bell, C. (1833). *The Hand. Its Mechanism and Vital Endowments as Evincing Design*. London: W. Pickering.

lines of stationary cells along which it dispatches waves of physiochemical disturbance, and these act as releasing forces in distant organs where they finally impinge"[4] *(Lecture I: p. 3).* He later suggested that, *"A large part of co-ordination [of the nervous system and organism] consists in the orderly combining of reflexes. In studying this co-ordination we have to deal with and discriminate between simultaneous combinations and successive combinations of reflexes"*[5] *(Lecture V: 150).*

Based on his observations, Sherrington introduced three classifications of reception—*"the extero-ceptive, intero-ceptive, and proprio-ceptive"* [33]. Sherrington described extero-ceptive as, *"The surface field [the outer body] lies freely open to the numberless vicissitudes of the environment. It has felt for countless ages the full stream of the varied agencies forever pouring upon it from the outside world. The field, extero-ceptive as it may be called, is rich in the number and variety of receptors which adaptation has evolved in it"*[6] *(Lecture IV: p. 130).*

The extero-ceptive signals the information about the factors external to the body. Comparing the extero-ceptive to the intero-ceptive, Sherrington referred to the former as being "*rich*" consisting of a variety of receptors in constant interaction with the environment. However, the intero-ceptive was poor, and barren of receptors since it was internal and usually of alimentary function. Sherrington wrote, *"…[intero-ceptive] lies however less freely open to it [the environment]. It is partly screened by the organism itself. In this recess a fraction of the environment is more or less surrounded by the organism itself. Into that sequestered nook the organism by appropriate reactions gathers morsels of environmental material whence by chemical action and by its absorption it draws nutriment"*[7] *(Lecture IX: p. 317).*

With the extero-ceptive and intero-ceptive receptors signaling information about the external factors and the viscera of the body, Sherrington defined a third receptive field, the proprio-ceptive. Coined from the Latin word *proprius* (meaning "one's own"), this field signals information concerning the status of the body's internal and external environment [39]. Sherrington writes, *"…throughout a vast range of animal types the bulk formed by the organism presents to the environment a surface sheet of cells, and, beneath that, a mass of cells more less screened from the environment by the surface sheet…" "The deep tissues underlying the surface sheet are not devoid of receptors. They have receptors specific to themselves"… "adapted for excitation consonantly with changes going on in the organism itself, particularly in its muscles and their accessory organs (tendon, joints, blood-vessels). Since in this field the stimuli to the receptors are given by the organism itself, their field may be called the proprio-ceptive field"*[8] *(Lecture IV: pp. 129–130).*

The receptors that Sherrington considered to be proprioceptors were those found within muscles, tendons, joints, ligaments, and the skin.

---

[4] Sherrington [33].

[5] Ibid.

[6] Ibid.

[7] Ibid.

[8] Ibid.

Neural Function: The Kubernetes of Proprioception

The central nervous system consists of the spinal cord, and the brain (pons, cerebellum, midbrain, diencephalon and the two cerebral hemispheres) [40]. Although the 12 cranial nerves are generally considered part of the peripheral nervous system, recent evidence suggests otherwise. Since the first and second cranial nerves (olfactory and optic nerves, respectively) are myelinated by oligodendrocytes, glial cells of the central nervous system, as opposed to Schwann cells, the glial cells of the peripheral nervous system, they are now considered as extensions of the central nervous system [40–42]. Glial cells are the "glue" of the nervous system [43]. Besides regulating the formation, maintenance, and function of synapses, the key functional unit of the nervous system [44], the glial cells regulate inflammation and the inflammatory response within the nervous system [43, 44].

The peripheral nervous system consists of the majority of the cranial nerves, the spinal nerves with their roots and rami, the peripheral nerves, and the peripheral components of the autonomic nervous system, the sympathetic, the parasympathetic, and the enteric divisions [45]. Compared to the central nervous system, the peripheral nervous system, while extensive, is considered relatively straightforward [46]. Birch et al. stated that in no other system of the body does there exist a large functional and relay capacity that is concentrated in so small a volume of tissue as that of the peripheral nervous system [47].

All sensory input from the interior areas of the body, or exterior to the body, is transmitted to the central nervous system via the peripheral nerve [48]. This nerve is the "cable" used by the axons of the motor (efferent), sensory (afferent), and vegetative neurons to circulate information within the peripheral nervous system [49]. Information is transduced between these neurons and their effectors in both directions (sensitive receptors, skeletal muscles, and viscera) [49]. The afferent neurons carry information from the sensory receptors or organs (muscle, tendon, periosteum, bone, joint, and skin) to the central nervous system (i.e., brain and spinal cord) for further processing/analysis [47]. In turn, the efferent neurons transmit motor information away from the central nervous system to the organs (end plates in skeletal muscle, glands, etc.) for implementing the proper reaction [47]. Both the afferent and efferent neurons work together to sense and mediate relationships between the environment and the self. Proper integration of information between both neurons is relayed to the central nervous system via the interneurons (see [50, 51] for review). The interneurons are the central nodes of the neural circuits, the most common neurons of the spinal cord as well as an important modulating cell type of the spinal cord [52]. Interposed in pathways to α- and γ- motor neurons (more below), to autonomic efferent neurons, to ascending pathways, and to primary afferent terminals, they function as the relay station, enabling the communication between the sensory and motor neurons with the central nervous system [53].

A major task of the central nervous system is to extract and integrate information from both the internal and external environments. This is accomplished by merging information from the different sensory systems through the peripheral nervous system to create an adaptive response [54], i.e., a cross-modality integration [55]. The information collated by the brain depends on the state of the sensory systems and the characteristic of the stimulus (e.g., pain, discomfort, heat, cold, etc.) [48]. This "perception" is further processed [55] at sites in the midbrain, thalamus, and cortex, where individual multisensory neurons are dedicated to reviewing different combinations of sensory inputs and directing different functions [55–57]. In short, the brain evaluates stimuli with respect to their relevance for potential action(s).

The processing of any stimulus from the environment, and its categorisation requires the cooperation of the primary senses—vision, olfaction, vestibular, audition, and tactile [1]. The human body relies on the information provided by these multimodal sensory systems to relate to, and to function with, its immediate environment. In addition, a "sixth sense", *somatosensation*, is an overarching sense which includes the sub-modalities of thermoreception (sensing temperature), nociception (feeling pain), mechanoreception (sensing pressure), and proprioception (sensing body spatial position) [1, 58, 59] (See Fig. 4.1).

> The processing of any stimulus from the environment, and its categorisation requires the cooperation of the primary senses, vision, olfaction, vestibular, audition, tactile, and the sixth sense, "somatosensation". All sensory input from the interior areas of the body, or exterior to the body, is transmitted to the central nervous system via the peripheral nerve. Afferent neurons carry information from sensory receptors to the central nervous system for further processing/analysis. Efferent neurons transmit motor information away from central nervous system to the organs for implementing proper reaction. Proper information between both neurons is relayed to the central nervous system via the interneurons.

What Is Proprioception?

> *Proprioception is achieved through a summation of peripheral sensory input describing the degree of, and changes in, muscle length and tension, joint angle, and stretch of skin.*
>
> *Proske and Gandevia* [60]

Proprioception is central to the daily survival of animals [61]. It is phylogenetically old, and its development suggests the co-evolution of the nervous system in terms of the processing of and of making the optimal use of environmental information available to it. Proprioception defines the awareness of the internal and external status of the body in space and time [62–64]. It is a continuous feedback loop between the sensory mechanosensors throughout the body and the nervous system [58]. The afferent feedback from the somatic sensory system is essential for an effective and efficient outcome for both automatic and voluntary movement [65]. The control of movement is highly dependent on the quality of the afferent

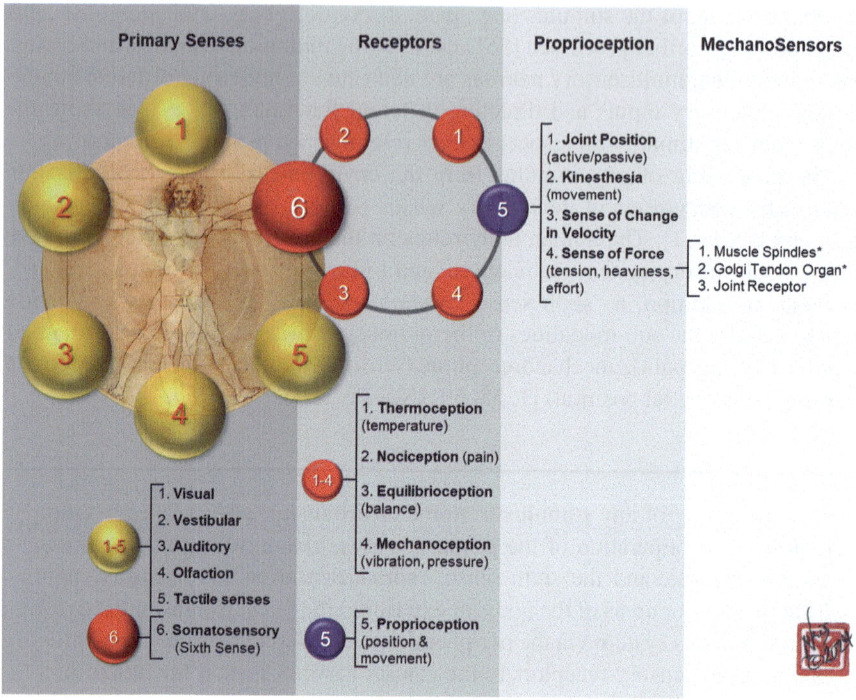

Fig. 4.1 Modalities. (Published with kind permission of copyright © Dr. Nikos C. Apostolopoulos 2024. All rights reserved)

information from the various somatosensory systems (i.e., interoceptors and mechanosensors) involved in proprioception [66]. Proprioception influences coordinated movements, body posture, postural control, balance, as well as motor learning and relearning [62].

Two levels of proprioception exist, the conscious (voluntary) and the unconscious (reflex initiated). Whereas the former enables a person to maintain proper joint function in sports, daily activities, and occupational tasks, the latter modulates muscle function by initiating a reflex stabilisation by way of the muscle receptors [67].

Proprioception encompasses three aspects of awareness—collectively referred to as the 'ABC of proprioception' [68, 69]. These attributes include: *agility*, *balance*, and *coordination*. *Agility* is the capacity of the body or body part to adjust position rapidly during movements. *Balance* is the ability to maintain equilibrium by keeping the line of the body within its base of support. *Coordination* is the smoothness of an activity produced by a combination of muscles acting together with appropriate intensity and timing [68, 69]. In the words of Kröger and Watkins "*Proprioceptive information informs us about the contractile state and movement of muscles, about muscle force, heaviness, stiffness, viscosity, and effort and, thus, is required for any coordinated movement, normal gait, and for the maintenance of a stable posture*" [70, p1].

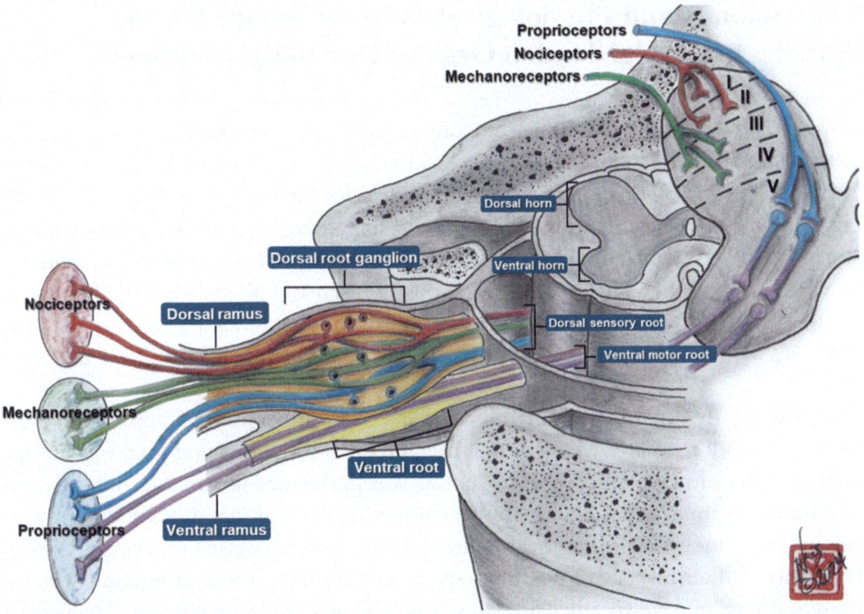

Fig. 4.2 Dorsal root ganglion. (Published with kind permission of copyright © Dr. Nikos C. Apostolopoulos 2024. All rights reserved)

The sensory inputs received from the mechanosensors are transduced by sensory neurons to the brain. These neurons have cell bodies within the dorsal root ganglia of the spinal nerves [71] (See Fig. 4.2). Previously thought to be metabolic storage "helpers" to aid in peripheral processes, the dorsal root ganglia play a direct role in various signaling processes [72]. Through sensory neurons found within their cell bodies, information about a variety of intrinsic and environmental cues, such as temperature, touch, muscle length, organ volume, regulation of blood supply, actual or potential harm to the body, as well as the changing of neuronal sensitivity is conveyed to the central nervous system [59, 73] with the goal of normalising proper function of the body. The entire neuron from the nerve ending in muscle to the presynaptic terminals in the spinal cord, including the cell body in the dorsal root ganglion, is a primary afferent neuron [74].

The sense of proprioception is not just limited to higher mammals. All motile animals rely on proprioceptive feedback to control and coordinate the movement of their bodies [63]. The signals forwarded by the proprioceptors vary throughout the action of the body [75]. Grill and Hallet [76] suggest that during discrete movements the muscles acting about a joint will be subject to different length changes and velocities of stretch. Furthermore, muscles spanning more than one joint are subject to length changes resulting from quite different actions, with the muscles at one joint influencing the pattern and activation of muscles acting about other joints. Consequently, the afferent information arising from the various sensors demands the coordination of many muscles to produce the required action [77].

The Anatomy and Physiology of Proprioception: Muscle Spindles and Golgi Tendon Organs, Our Body's Sentinels

Humans and terrestrial vertebrates embody several types of proprioceptive mechanosensors found within muscles, tendons, joint capsules, ligaments, and skin (See Table 4.1, p. 3) [78]. Mechanosensors are critical for the normal execution of voluntary motor tasks determined by the degree of the change in muscle length and tension, as well as joint angle [60, 64, 79]. Although joint receptors were initially viewed as important contributors to joint position sense at all angles [80], their input to kinesthesia only extends to that of being limit detectors, identifying only extreme positions within the normal range of a joint in both humans [81, 82] and cats [83, 84].

There are two main types of muscle-related proprioceptors—muscle spindles, and Golgi tendon organs. Much of the research on the anatomy and functional properties of these slow-adapting proprioceptors was performed in rats, cats, rabbits, and non-human primates, with comparative studies suggesting that the human proprioceptors are somewhat similar in structure, number, and response properties [85–87]. By virtue of their respective position—lying in parallel to force-producing skeletal muscle fibres, muscle spindles are stretched or shortened in response to changes in overall muscle length, as well as disturbances created by the nearby skeletal muscles, even if the contraction does not cause joint movement or a change in overall muscle length [78, 88]. The Golgi tendon organs are in series with the skeletal muscle fibres, and actively sense muscle tension in response to the force generated by the muscle [78, 89, 90].

> Proprioception defines the awareness of the internal and external status of the body in space and time. All motile animals, not just the higher animals, rely on proprioceptive feedback to control and coordinate the movement of their bodies in order to communicate with their external environment. Proprioception, can be either conscious (voluntary) or unconscious (reflex), encompassing three aspects of awareness—agility, balance, and coordination. Terrestrial vertebrates and humans embody several types of proprioceptive mechano-sensors found within muscles, tendons, joint capsules, ligaments, and the skin. The two most important, located in muscles and embedded within the musculotendinous junction, are the muscle spindle and the Golgi tendon organ, respectively. The former lies in parallel to the force generating skeletal muscle, with the Golgi tendon organ lying in series with the skeletal muscle.

Muscle Spindle: The Biomechanical Stretch Mechanosensor

After the special senses of the eye and ear, the muscle spindles are the most complex sensory receptor organ [64, 91]. Studies in both humans and other terrestrial vertebrates have observed that muscle spindles are present in the vast majority of striated muscles [92, 93], except for the facial and jaw opening muscles [94]. It is the only sensory receptor to have its own motor supply.

The human muscle spindle is a pleomorphic fusiform (spindle shape) structure of some 7–10 mm in length (slightly larger than a cat spindle) [95–97]. Found in tandem within muscles, the spindle acts as a major source of afferent feedback. It has multiple functions, such as generating antigravity thrust during a quiet upright stance, the timing of locomotor phases, correcting for muscle nonlinearities, compensation for muscle fatigue, determining synergy formation, modulating plasticity and motor learning, adjustment and maintenance of muscle tensile strength and tension, the reflex regulation of movement, and for partially augmenting and linearising muscle stiffness [98–102]. In addition, besides regulating motor control, spindles play a role in the appropriate realignment of fractured bones, successful regeneration of spinal cord axons after injury, and spinal alignment [103]. Rough estimates suggest that muscle spindles range from 1 to 100 per gram of muscle with approximately 30,000–50,000 contained within the entire human body [87, 104, 105]. An extrapolation from animal studies suggests that nearly as many Golgi tendon organs are present [87].

All muscle spindles are found to lie in parallel with the skeletal muscle fibres surrounding them, allowing the spindles to experience length changes representative of a whole muscle length [106, 107] (See Fig. 4.3a). In humans, the absolute number of muscle spindles varies from fewer than 50 for the intrinsic muscles of the hand (lumbricals) to 1350 for the quadriceps femoris muscle [88]. Voss [108] found that 70% of the spindles of the lumbrical muscles measured 1–3 mm compared to the average length of the human muscle spindles (7–10 mm) [95–97], suggesting that small muscles tend to have shorter spindles [108]. Given the small size of the individual intrinsic muscles of the hand, compared to the larger thigh muscles, the density (number of spindles per gram of muscle) is significantly higher for the intrinsic muscles of the hand [109].

The number and distribution of muscle spindles differs between skeletal muscle with different functions [110]. For instance, besides an overlap existing between the physiological flexors and physiological extensors, a comparison between the two reveals that the flexor muscle group exhibited a mean residual value of muscle spindles greater than that of the extensor muscle group [98]. In turn, research suggests that a non-linearity of muscle spindle abundance exists with increases in their number being a fractional power function of muscle mass [111]. This is observed across many species that progressively differ in body size by about an order of magnitude [111].

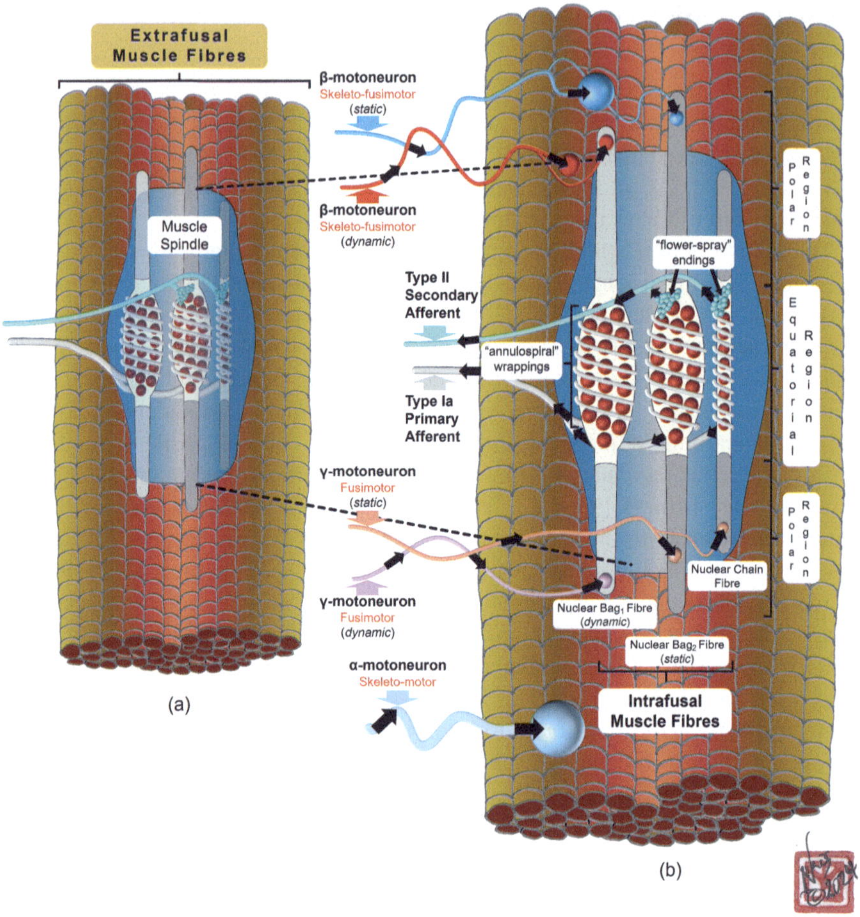

Fig. 4.3 Muscle spindle. (Published with kind permission of copyright © Dr. Nikos C. Apostolopoulos 2024. All rights reserved)

Muscle spindles consist of intrafusal [intra- + fusal (pertaining to a spindle)] muscle fibres, sensory axons, and motor axons (See Fig. 4.3b). They lie deep in the belly of the muscle near branches of the muscle nerve or blood vessels with some being close to the junction of the muscle and tendon [112]. Consisting of 8–20 intrafusal muscle fibres with a diameter of 8–25 μm [110], these muscle fibres are surrounded equatorially by a fusiform capsule of connective tissue, giving the muscle spindle its distinguished shape. The small longitudinally oriented specialised striated intrafusal muscle fibres run in parallel with, and are surrounded by the fascicles of the force producing extrafusal "skeleto-motor" muscle fibres [113]. The location of the muscle spindle complex within the muscle belly mimics its early development, paralleling the progression seen with the myogenesis of the extrafusal muscle fibres [100]. Interestingly, the number of intrafusal muscle fibres and the maximal diameter of the fusiform capsule depends on the intramuscular depth of the spindle location. The deeper a spindle is within a muscle belly, the more intrafusal muscle fibres it contains and the wider is its capsule [95, 114].

The fusiform capsule consists of an inner and an outer spindle sheath. Both sheaths are formed of a perineural-like covering of flattened fibroblasts and collagen, with the internal sheath forming delicate tubes surrounding the individual intrafusal muscle fibres, with thin collagen fibres running in spirals through the periaxial space between the two layers [115–117]. Histochemical studies have demonstrated that the capsular space is filled with a gelatinous fluid rich in glycosaminoglycan—*hyaluronic acid* [118, 119].

Based on physiological (myosin ATPase activity) and anatomical findings (number and location of myonuclei), mammalian muscle spindles contain three types of intrafusal muscle fibres [90, 112, 120–123]. The intensity of myofibrillar ATPase reactivity observed by an immunohistochemical staining profile at the level of the equatorial region of the fibres identified: (1) a dynamic nuclear bag_1 fibre (primarily sensitive to velocity), (2) a long static nuclear bag_2 fibre (primarily sensitive to length and position), and (3) a nuclear chain fibre (sensitive to both static and length) [112, 124–126] (See Fig. 4.3b). Anatomically, the arrangement of the myonuclei in the equatorial region of the fibres identified two intrafusal muscle fibres. The longer and larger in diameter nuclear bag (bag_1 and bag_2) muscle fibres containing a cluster of approximately 100 myonuclei, 5–6 abreast, at the equatorial and juxtaequatorial regions, and the smaller and thin nuclear chain muscle fibres. The nuclear chain muscle fibres are half the length and diameter of the longer bag muscle fibres with their myonuclei forming a single row (20–50) in the equatorial region [93, 112, 127] (See Fig. 4.3b).

The most plentiful type of intrafusal muscle fibres are the 'nuclear chain muscle fibre', commonly 3–10 per spindle [88]. A study of the cat muscle spindle observed approximately 5 nuclear chain muscle fibres compared to 2–3 nuclear bag muscle fibres [112]. This difference was also confirmed in human studies. Soukup et al. reported 3–11 nuclear chain muscle fibres, compared to 1–6 nuclear bag_1 (*dynamic*) muscle fibres, and 1–4 nuclear bag_2 (*static*) muscle fibres in 22 muscle spindles of human lumbricals [123]. A second study by Cazzato and Walton reported 2–11 nuclear chain muscle fibres compared to 1–5 nuclear bag muscle fibres based on 56 biopsies taken from various cases of neuromuscular diseases (i.e., denervation apathy, benign spinal muscular apathy, congenital muscular dystrophy etc.) [128].

The Golgi Tendon Organ: The Sensor of Whole Muscle Tension

Golgi tendon organs are encapsulated fusiform-like structures found at both the proximal and distal ends of the muscle, totaling 10–86 per muscle [114, 131, 132]. Golgi tendon organs consist of bundles of collagen fibres, and are distributed unevenly within the muscle, with a tendency to lie deep in the central zone of the muscle, i.e., in the "muscle core" [133, 134]. In contrast to muscle spindles, which are situated in parallel with the extrafusal muscle fibres, Golgi tendon organs are placed in series with muscle fibres and the tendon-aponeurosis connection to the bone [134]. Originating from 2 to 15 different motor units, the number of muscle fibres inserting into a single tendon organ varies, ranging from 3 to 50, with only 5% attached to more than 25 fibres [135]. At the "tendon end", the collagen bundles are attached obliquely to the whole muscle tendon or aponeurosis at an angle of about 5° [135–138]. Because of their size (1.6 mm long × 0.1 mm wide in humans), Golgi tendon organs can be seen by the naked eye [138, 139].

Motor Units

This is the basic functional unit of the skeletal muscle; it is the elementary unit of motor control. It is composed of the motor neuron (located in the spinal cord), its axon branching within a muscle, and the innervated fibres. When stimulated, all the fibres associated with the motor unit are activated to produce force. The amount of force generated depends on the size of the motor unit. Small motor units produce a smaller amount of force versus large motor units which generate greater force. Small delicate muscles, such as the eye muscle, have smaller number of motor units (≈ a dozen fibres), while the powerful leg muscles possess large motor units containing over 1000 fibres [129]. There are three types of motor units, based on their speed of contraction and their resistance to fatigue: (a) slow, fatigue resistant, (b) fast resistant to fatigue, and (c) fast fatigable. Differences between the three types involve a number of morphology and biochemical characteristics—for further information and discussion of motor units, please refer to Celichowski and Krutki [130].

As indicated above, Golgi tendon organs are exclusively located at the muscle tendon or muscle-aponeurosis junctions and not within tendons [140, 141]. A study looking at various thigh, shank, foot, and intercostal muscles of the cat reported that 92.4% of the 1337 receptors were located at the musculotendinous junction versus only 7.6% within the tendon proper [120]. In addition, it was found that 63% had a fusiform shape, with the others being either bifid or trifid at one or both ends [120].

The widespread distribution of the Golgi tendon organ allows them to monitor muscle contractions in every portion of the muscle. This ensures that the activity of virtually every motor unit in a muscle can be sensed by at least one tendon organ. Being highly sensitive to the contraction force of the motor unit, a Golgi tendon organ responds more strongly to the contraction of the muscle fibres of the motor unit to which it is attached, than to the contraction of the whole muscle [89, 142]. It has been observed that slow twitch motor units activate fewer Golgi tendon organs than fast twitch motor units [143].

A Golgi tendon organ is innervated by a single large, myelinated afferent axon. This afferent nerve fibre enters the capsule of the Golgi tendon organ via the *neural*

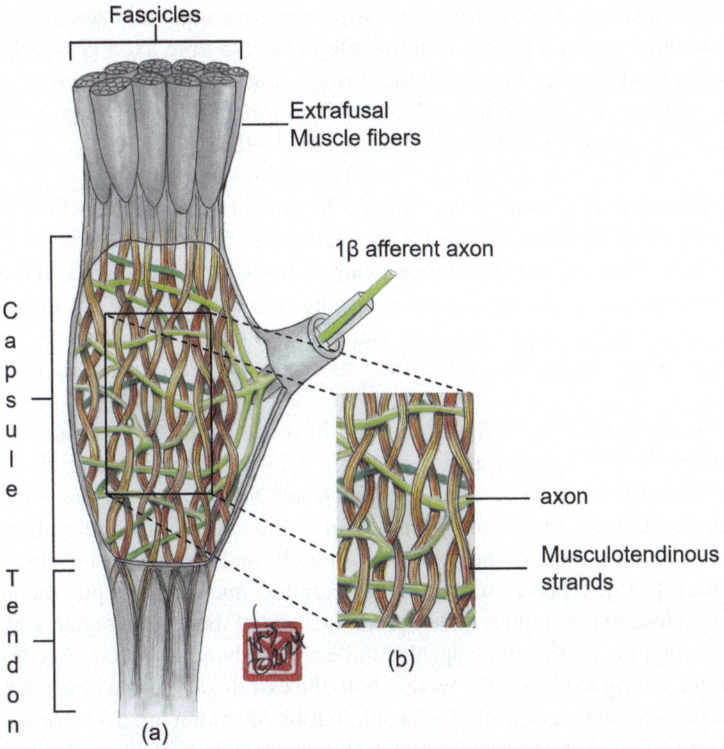

Fig. 4.4 Golgi tendon organ. (Published with kind permission of copyright © Dr. Nikos C. Apostolopoulos 2024. All rights reserved)

compartment near its equator (See Fig. 4.4a). This compartment contains the myelinated axon and axonal branches that have penetrated into the lumen [135]. Typically traveling into the center of the capsule, the myelinated afferent splits into two or more daughter branches, and repeatedly divides into smaller branches, eventually shedding the Schwann cells and hence, their myelin sheath [138, 144]. Within the core of the tendon, the thinner collagen fibrils belonging to different muscle fibres, which continuously divide, mix, and fuse with one another forming a collagenous mesh (See Fig. 4.4b). In rats, this mesh is composed of collagen bundles representing tendons of 5–10 muscle fibres. The loosely packed collagen mesh is densely innervated by the afferent sensory nerve fibre endings. The axonal branches wind back and forth between and around the collagen braids of the mesh, both longitudinally and transversely, to form the *neurotendinous core* [134, 145]. The close proximity of the unmyelinated sensory afferent nerve fibres to neighbouring collagen bundles exposes large areas of their cell surface [146]. In a relaxed muscle, the collagen mesh is loose and wavy, which allows for the unmyelinated nerve fibre axon to pass easily between them. However, activation of a motor unit, with at least one muscle fibre inserting into the tendon organ, straightens and brings tension to some of the loosely packed collagen strands. The straightened, elongated, and taut strands compress and depolarise the pressure-sensitive unmyelinated afferent endings, initiating an action potential in the afferent nerve fibre axon [134, 147]. This mechanism works rather like a "Chinese finger trap". It should be noted that the sensory afferents respond to increases in active force in discrete "steps", reflecting the recruitment of additional muscle fibres [148, 149].

Surrounding the neurotendinous core, is the third compartment of the tendon organ, the *tendinous compartment,* which is located in the lumen or within the capsular wall and has no contact with the axon terminals. The decreased level of innervation of this compartment constitutes a built-in basis for the so-called "in-parallel" effects observed for Golgi tendon organ responses to certain stimuli [145]. The collagen fibrils of this compartment run parallel with one another and are larger in diameter and more densely packed than those of the *neurotendinous core* [134, 145, 146].

In addition to the muscle fibres attached in series to the Golgi tendon organ, a large number of fibres run parallel to the Golgi tendon organ, inserting directly into the tendinous or aponeurotic end. The functional significance of this anatomical arrangement suggests that during contraction of the extrafusal muscle fibres, there is a reduction in the strain of the collagenous bundle and the sensory terminals of the tendon itself [135]. When contracting, the "parallel" muscle fibres pull on an insertion that is close to the aponeurotic or tendinous end of the tendon organ, which will limit deformation of the intracapsular collagenous bundles [135]. According to Fukami, the ability to unload the tendon with the extrafusal muscle fibres in parallel and in series accounts for the nonlinear summation of tendon organ responses [150]. In this way, if two motor units activate the same tendon at the same time, this arrangement will reduce each other's action on the receptor, and as a result, the amount of force actually sensed by an individual tendon organ represents a fraction of the force developed by the whole muscle force [135, 150]. The overall effect of this arrangement serves to provide an increased sensitivity of the system to respond to muscle force production.

Proprioception: Nerves Coming, Nerves Going

The philosophy of balance, where two opposites co-exist in harmony and are able to transmute into each other, thereby becoming the individual whole.

Tao philosophy 14th century BCE

Clearly, the central nervous system receives vast amounts of information from various receptors via the peripheral nervous system, concerning the status of both the internal and external environments, which must be quickly analysed, and the necessary outgoing responses must be coordinated, and transmitted simultaneously. To ensure a seamless response to any external stimulus, afferent feedback from the somatic sensory system is essential for an effective outcome for automatic and voluntary responses. The monitoring of muscle length by muscle spindles, and muscle tension by Golgi tendon organs, is a major part of this response. In the end, the change in muscle length and tension are essentially physically inseparable parameters, with the two proprioceptors functioning as a coupled system working together as an integrated whole.

Active control of the body depends on a complex interaction between the joints, the skeletal muscles, and the peripheral and central nervous systems. Functionally, the overall response can be subdivided into four components: (1) collection of the external stimuli by the sensory receptors (e.g., muscle spindles, and Golgi tendon organs), (2) afferent projection to the central nervous system via the sensory neurons, (3) processing, decision making, and development of a coordinated response by the central nervous system, (4) and efferent signaling from the central nervous system to the skeletal muscle via the motor neurons [138, 145, 146].

To facilitate these actions, the nervous system has evolved specific functional and morphological characteristics within its afferent and efferent neural axons. In addition to being either myelinated or unmyelinated, the size of the action potential and the velocity of transmission along each nerve fibre is dependent on the diameter size of the nerve fibres. Based on these characteristics, muscle afferent fibres are usually classified into four types (I–IV) by size, with the classification, "Type I", being used as the designation for sensory neural axons having the largest diameter and fastest impulse conduction [151] (Table 4.2).

As previously mentioned, the muscle spindle consists of three morphologically and functionally distinct intrafusal muscle fibres that are all innervated by afferent axons of dorsal root ganglion neurons [70]. Each has a unique structure, with a typical sarcomeric structure for most of its length, but in the equatorial region, just under the muscle spindle's juxtaequatorial bulge, the sarcomeres are almost completely replaced with clusters of myonuclei, which are found nowhere else in the body [32, 100, 152] (Table 4.3). This anatomical feature effectively divides the intrafusal fibres into two functional separate contractile regions (polar regions) on either side of the non-contractile equatorial region [152] (See Fig. 4.3b). The two nuclear bag muscle fibres extend beyond the capsule of the muscle spindle, terminating with the adjacent extrafusal muscle fibres or muscle connective tissue, while the nuclear chain muscle fibre terminates within the capsule [153]. These differences in attachment points, in conjunction with the lack of contractility at the equatorial region, suggest that contraction of select groups of intrafusal muscle fibres will result in deformation of the muscle spindle in different patterns [100].

The multinucleated, non-contractile region of the intrafusal muscle fibres have two kinds of sensory nerve endings which monitor deformation of the spindle—Type Ia (aka primary) and Type II (aka secondary) afferent nerve fibres [154–156].

Table 4.2 Classification of afferent, efferent, and autonomic nerve fibres

Group[a]	Group[b]	Diameter (μm)	Conduction velocity (m/s)	Function		Common term
Myelinated somatic afferent and efferent						
Aα	**Type I**	12–20	70–120	Largest afferent	Muscle spindle	Type Ia primary afferent
					Golgi tendon organ	Type Ib afferent
	[c]	12–20	70–120	Lower motor neuron [efferent]		α skeleto-motor
Aβ	**Type II**	5–12	30–70	Touch, pressure, muscle spindle (nuclear chain fibre) [afferent]		Type II secondary
Aγ	[c]	3–6	15–30	Motor for muscle spindle (fusimotor) [efferent]		γ fusimotor
Aδ	**Type III**	3–5	12–30	Some pain and visceral receptors, cold, touch [afferent]		δ
B	[c]	< 3	3–15	Myelinated preganglionic autonomic [efferent]		B
Unmyelinated somatic afferent (Dorsal horns)						
C	**Type IV**	0.4–1.2	0.5–2	Pain, temperature, mechanoreception reflex response [afferent]		C
Unmyelinated autonomic postganglionic (Sympathetic)						
C	[c]	0.3–1.3	0.7–2.3	Pilomotor, sudomotor, and vasomotor [efferent]		C

[a]Erlinger and Gasser classification includes afferent, efferent and autonomic nerves
[b]Lloyd and Hunt classification—afferent (sensory) nerves (only)
[c]Not classified by Lloyd and Hunt

Table 4.3 Intrafusal muscle fibres of the muscle spindle

Intrafusal muscle fibres	Myonuclei position and distribution	Group afferents	Velocity (ms) Diameter (μm)	Function/measures
Nuclear bag fibres				
[a]Bag_1 (*dynamic*)	Equatorial Cluster	Ia[a] (primary)	70–120 12–20	Rate of length change in muscle
[b]Bag_2 (*static*)	Equatorial Cluster	Ia[a] (primary)	70–120 12–20	Signal information about static length of muscle, not rate of change
		II[b] (secondary)	30–70 5–12	
Chain	Equatorial Row	Ia[a] (primary)	70–120 12–20	Signal information about static length of muscle, not rate of change
		II[b] (secondary)	30–70 5–12	

[a]"Annulospiral" wrappings (see Fig. 4.3b)
[b]"Flower-spray" endings (see Fig. 4.3)

The Type Ia nerve fibre coils around the equatorial regions of all three intrafusal muscle fibres, forming a synapsis with a characteristic "annulospiral" wrapping (See Fig. 4.3b) [157]. Because it innervates all three intrafusal muscle fibres, it exhibits a dynamic response, firing a burst at the beginning of a stretch and signals velocity and acceleration, as well as muscle length [155]. The length stimulus is transmitted to the sensory wrappings through the intermediary of the intrafusal muscle fibres. Alterations in the contractile states of the polar regions influences the form of the stimulus as seen by the "annulospiral" wrappings [152].

The Type II nerve fibres connect to the nuclear chain and nuclear bag$_2$ (*static*) muscle fibres. Compared to the Type Ia "annulospiral" wrappings", the Type II nerve fibres have "flower-spray" endings that are either "spray-shaped" or "annular" in shape [158] (See Fig. 4.3b). These endings only transmit information concerning muscle length, which is used for the perception of the body in space [39]. Both the "annulospiral" wrappings and the "flower-spray" endings form a mechanosensitive complex capable of relaying muscle stretch information to the spinal cord motor neurons [159].

Unlike the muscle spindle, the Golgi tendon organ is a simpler muscle proprioceptor, with an encapsulated afferent nerve ending (Type Ib), and no efferent innervation (refer to Table 4.2). Similar to the Type Ia spindle nerve fibre, the Type Ib nerve fibre of the Golgi tendon organ is a large nerve fibre capable of delivering information with great speed to the central nervous system (Table 4.2). For an in-depth discussion of the afferent nerve fibre in relation to the Golgi tendon organ, refer to the section above on the "Golgi tendon organ—the sensor of whole muscle tension" (pp. 11–12).

To coordinate the responses to the information provided from the internal and external environments, the intrafusal and extrafusal muscle fibres are both innervated by motor neurons. This relationship was formed during the development of the muscle spindle, in close parallel to the myogenesis of extrafusal muscle fibres [160]. The motor neuron (α-motor neuron) and the sensory afferent (Type Ia) arrive at the same time in the primordial muscle [100, 160]. The myotubes receiving motor axonal terminals evolve into the extrafusal muscle fibres, while those receiving the sensory afferent terminals become intrafusal muscle fibres [160].

According to Burke, the motor neurons are arguably the most important class of neurons in the central nervous system because without them, we would neither move nor breathe [161].The motor neurons are neuronal cells located in the central nervous system and control a variety of downstream targets [162]. The two main types of motor neurons are: (1) the "upper motor neurons", which originate from the cerebral cortex, and (2) the "lower motor neurons", which are located in the brainstem and spinal cord. The lower motor neurons are responsible for transmitting the signal from the upper motor neurons to the effector muscle(s) to signal the performance of a movement [163].

Three broad types of lower motor neurons exist: (a) somatic; (b) special visceral efferent (brachial); and (c) general visceral (for further information on these types of lower motor neurons (see [161, 163, 164] for review).

The somatic motor neurons are divided into three categories: (a) alpha (α); (b) beta (β); and (c) gamma (γ). These motor neurons are responsible for integrating and transmitting information from many sources from within and outside the central nervous system to the muscle fibres [161, 165, 166]. The alpha (skeleto-motor) motor neuron comprises one-third of the motor neurons in a given motor pool. A motor pool consists of all of the motor neurons that innervate a single muscle. The alpha motor neuron exclusively innervates the large *extrafusal* muscle fibres, the primary means of skeletal muscle contraction underlying active movement. The conduction velocity of an alpha motor neuron is similar to both the Type Ia (muscle spindle) and Type Ib (Golgi tendon organ) afferent nerve fibres (Table 4.2). The beta (skeleto-fusimotor) motor neuron possesses intermediate properties between alpha and gamma motor neurons. The name "beta" does not refer to conduction velocities, but is used to differentiate this motor neuron from the alpha and gamma motor neurons [167]. This motor neuron comprises about one-third of all motor units and innervates three-fourths of all muscle spindles [104, 167, 168]. This large motor neuron innervates both the extrafusal and the intrafusal muscle fibres, and regulates both muscle contraction and sensory information from the muscle spindle [164, 168]. Beta motor neurons are subdivided into two subtypes: static and dynamic. The static motor neurons innervate fatigue resistant extrafusal muscle fibre types and nuclear bag_2 (*static*) intrafusal muscle fibres, and the dynamic motor neurons innervate slow extrafusal muscle fibre types and nuclear bag_1 (*dynamic*) intrafusal muscle fibres [169]. The gamma (γ-axon—fusimotor) motor neuron represents one-third of the motor neurons in a given spinal motor pool. They innervate the smaller, highly specialised intrafusal muscle fibres within the muscle spindle and receive presynaptic input from sensory neurons within the muscle spindles. Gamma motor neurons respond to steady or changing length of the extrafusal muscle fibres that contain them. The gamma motor neurons also consist of two functional types—"dynamic" and "static". It is generally accepted that the dynamic type innervates nuclear bag_1 (*dynamic*) intrafusal muscle fibres, while the static type innervates both nuclear bag_2 (*static*) intrafusal muscle fibres and the chain intrafusal muscle fibres [98]. Unlike the large α-motor neuron, the gamma motor neuron does not elicit any force on the tendon [170]. Instead, it is believed that the gamma motor neuron is activated along with the alpha motor neuron and functions to "fine-tune" the muscle contraction [162]. In short, the gamma motor neuron enables continuous flow of sensory information about muscle length thereby ensuing fluid muscular action during muscular contractions [171, 172]—which is vital for proprioception.

The alpha motor neuron is an integral component of the motor unit. There are three main subtypes, with each innervating one of the three different physiological types of muscle fibres. Alpha motor neurons arise in parallel with the two subtypes of gamma motor neurons.

In total, there are seven types of motor neurons (namely 3 types of α, 2 types of β (static and dynamic), and 2 types of γ (also static and dynamic)), which innervate subgroups of the extrafusal and intrafusal muscle fibres. As more detailed information on the motor unit, their motor neurons, and the muscle fibre types (Type

I—slow; Type IIA—fatigue resistant; and Type IIB—fatigable) is beyond the scope of this chapter, the reader is referred to [163, 167, 173], for more information.

Closing: Putting It All Together (See Fig. 4.5)

Evolution has provided skeletal muscle with several specialised tissues termed "mechanosensors" (e.g., muscle spindles, Golgi tendon organs, joint receptors, and cutaneous receptors), which function to continuously provide the central nervous system with real-time updates on the status of working muscles in terms of the physical characteristics of force development by the muscle, the internal forces being applied to the muscle-tendon complex, and up-to-date information on limb and body position and movement pattern in time and space.

Both muscle spindles and Golgi tendon organs continuously provide the central nervous system with real-time updates on the position and movement of body and limbs, and the status of active muscles. When movement occurs, whether the movement is self-generated or the result of passive movement [174], the various specialised mechano-sensory receptors transduce the mechanical events into neural signals to the central nervous system through afferent sensory pathways [175].

Sending information about muscle contraction status to the spinal cord enables the motor centres of the brain to compute an internal body plan serving as a reference for motor planning [176]. Proprioceptive feedback modulates motor neuron activity output during unanticipated environmental circumstances through the neuromuscular reflex arc (stretch reflex arc). The muscle spindles embedded within the extrafusal muscle fibres sense changes in muscle length of the specific muscle they innervate through the intrafusal muscle fibres. The intrafusal muscle fibres possess a mechano-sensory capacity, forming part of the neuromuscular reflex arc. The γ- and β-motor neurons innervate the muscle spindle on the polar region of the intrafusal muscle fibres forming the fusimotor system. The γ-motor neuron plays a vital role in regulating the intrafusal muscle fibre's stretch, enabling the retention of proper tension and sensitivity during dynamic muscular activity. At the spinal cord, premotor interneurons provide direct access to environmental information to the central nervous system in order to control voluntary and automatic movement patterns, balance, and posture. To coordinate the muscle activity involved with these patterns, the three motor neurons, the α (skeleto-motor), β (skeleto-fusimotor) and γ (fusimotor) innervate the extrafusal muscle fibres and the specialised intrafusal muscle fibres of the muscle spindle, respectively. This motor signaling of the intrafusal and extrafusal muscle fibres results in their relaxation or contraction in response to the afferent sensory input. This simplified description of the coordinate response with the sensory afferents and the motor efferents characterises the fundamental features discussed in this chapter about the structural and functional organisation of proprioception intended for the body's ability to communicate with the environment.

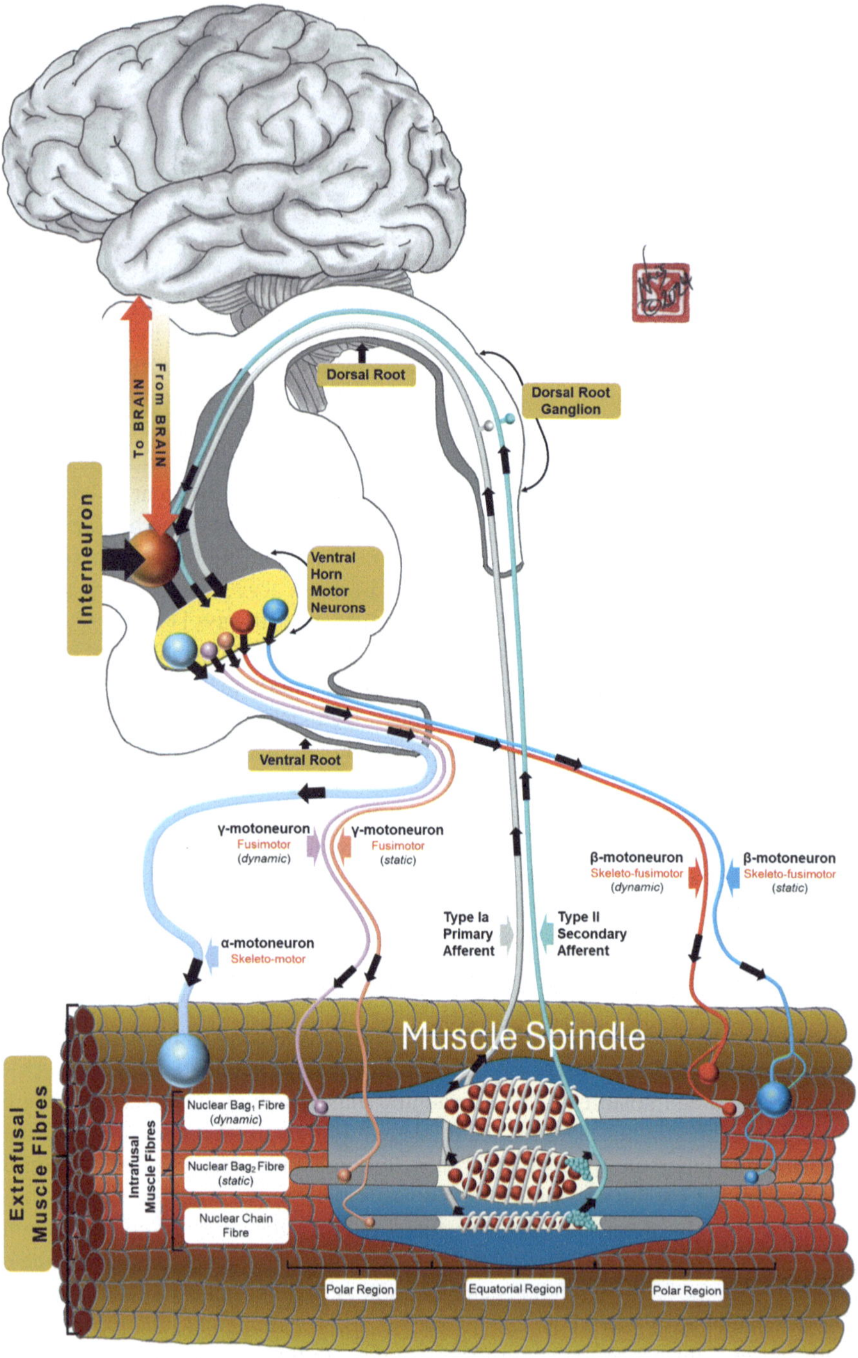

Fig. 4.5 Putting it altogether. (Published with kind permission of copyright © Dr. Nikos C. Apostolopoulos 2024. All rights reserved)

The neuromuscular reflex arc (stretch reflex arc)
Proprioception is integrally involved in the neuromuscular reflex arc, i.e., the stretch reflex. In the stretch reflex, information from the proprioceptors (muscle spindles and Golgi tendon organs) modify motor neuron output to achieve an appropriate graded muscle contraction [177]. The reflex can be broken down into two primary components: the efferent and afferent domains, with the former regulating mechanical actuation, muscle tension, and relaxation through the extrafusal muscle fibres, with the latter, signaling the intrafusal muscle fibres to relax or contract in response to sensory input [163]. The α-motor neurons regulate the tension and contraction of the extrafusal muscle fibres, with the sensory and gamma motor neurons regulating muscle function through the innervation of the intrafusal muscle fibres. The primary role of the intrafusal muscle fibres are the detection of the magnitude and speed of stretch or flexion of the muscle and position of the limbs, or proprioception [173]. With the change in muscle tension, the afferent sensory neurons of the muscle spindles send signals, which are relayed to neurons within the spinal cord via the interneurons. Upon receiving this afferent information (Type Ia, Type Ib, and Type II), the motor neurons (α and γ) signal the intrafusal or extrafusal muscle fibres to relax or contract in response to the sensory input. With the afferent sensory feedback, the reflex arc acts as an automated closed loop so that voluntary movement can be achieved as expected and accurately [177].

1. How effective is proprioception in injury prevention?
2. How does one assess/evaluate/measure proprioceptor function?
3. Should proprioception function be assessed for all athletes to have a baseline measure of proprioceptor function in order to know if recovery has occurred?
4. Does proprioception recover fully after muscle injury, i.e., is proprioception as effective post-injury as it was pre-injury?
5. What is the mechanism(s) by which the muscle spindle and Golgi tendon organ recover following injury?
6. In the time course of recovery from injury, where does the recovery of the muscle spindle and Golgi tendon organ fit?
7. Are there specific actions that therapists might/should use to enhance recovery of the muscle spindle and Golgi tendon organ?
8. At what point in the therapy process should therapists institute specific activity to promote recovery of proprioception?
9. Are there specific activities that coaches or athletes can do to enhance or maintain proprioceptor function as part of the training or pre-game activity?

References

1. Ager AL, Borms D, Descepper L, Dhooghe R, Dijkhuis J, Roy J-S, Cools A. Proprioception: how is it affected by shoulder pain? A systematic review. J Hand Ther. 2020;33:507–16.
2. Frontera WR, Ochala J. Skeletal muscle: a brief review of structure and function. Calcif Tissue Int. 2015;96:183–95.
3. Pedersen BK. Muscle as a secretory organ. Compr Physiol. 2013;3:1337–62.
4. Herrmann M, Engelke K, Ebert R, Muller-Deubert S, Rudert M, Ziouti F, Jundt F, Felsenberg D, Jakob F. Interactions between muscle and bone – where physics meets biology. MDPI-Biomolecules. 2020;10:1–30.
5. Liu J-X, Thornell L-E, Pedrosa-Domellof F. Muscle spindles in the deep muscles of the human neck: a morphological and immunocytochemical study. J Histochem Cytochem. 2003;51:175–86.
6. Karstoft K, Pedersen BK. Skeletal muscle as a gene regulatory endocrine organ. Curr Opin Clin Nutr Metab Care. 2016;19:270–5.
7. Blum KP, Campbell KS, Horslen BC, Nardelli P, Housley SP, Cope TC, Ting LH. Diverse and complex muscle spindle afferent firing properties emerge from multiscale muscle mechanics. elife. 2020;9:1–9.
8. Mathes S, Vanmunster M, Bloch W, Suhr F. Evidence for skeletal muscle fiber type-specific expressions of mechanosensors. Cell Mol Life Sci. 2019;76:2987–3004.
9. Gordon JC, Holt NC, Biewener A, Daley MA. Tuning of feedforward control enables stable muscle force-length dynamics after loss of autogenic proprioceptive feedback. Elife. 2020;9:1–9.
10. Mukund K, Subramaniam S. Skeletal muscle: a review of molecular structures and function, in health and disease. Wiley Interdiscip Rev Syst Biol Med. 2020;12(1):1–46.
11. Rohrle O, Davidson JB, Pullan AJ. A physiologically based, multi-scale model of skeletal muscle structure and function. Front Phys. 2012;3:1–14.
12. Dunlop JWC, Weinkamer R, Fratzl P. Artful interfaces within biological materials. Mater Today. 2011;14:70–8.
13. Felsenthal N, Zelzer E. Mechanical regulation of musculoskeletal system development. Development. 2017;144:4271–83.
14. Vogel V. Unraveling the mechanobiology of extracellular matrix. Annu Rev Physiol. 2018;80:353–87.
15. Katta S, Krieg M, Goodman MB. Feeling force: physical and physiological principles enabling sensory mechanotransduction. Annu Rev Cell Dev Biol. 2015;31:347–71.
16. Discher DE, Smith L, Choe S, Coloasurdo M, Garcia AJ, Safran S. Matrix Mechanosensing: from scaling concepts in 'omics data to mechanisms in the nucleus, regeneration, and cancer. Annu Rev Biophys. 2017;46:295–315.
17. Humphrey JD, Dufresnse ER, Schwartz MA. Mechanotransduction and extracellualr matrix homeostasis. Nat Rev Mol Cell Biol. 2014;15:820–12.
18. Iskratsch T, Wolfenson H, Sheetz MP. Appreciating force and shape-the rise of mechanotransduction in cell biology. Nat Rev Mol Cell Biol. 2014;15:825–33.
19. Janmey PA, Fletcher DA, Reinhart-King CA. Stiffness sensing by cells. Physiol Rev. 2020;100:695–724.
20. Alegre-Cebollada J. Protein nanomechanics in biological context. Biophys Rev. 2021;13:435–54.
21. Gomez D, Pena Ccoa WJ, Singh Y, Rojas E, Hocky GM. Molecular paradigms for biological mechanosensing. J Phys Chem B. 2021;125:12115–24.
22. Lim C-G, Jang J, Kim C. Cellular machinery for sensing mechanical force. BMB Rep. 2018;51:623–9.
23. Salzer I, Boehm S. Regulation of nociceptor signaling by serotonin. In: Serotonin. Elsevier; 2019.

24. Swaminathan V, Gloerich M. Decoding mechanical cues by molecular mechanotransduction. Curr Opin Cell Biol. 2021;72:72–80.
25. Wang J, Thampatty BP. An introductory review of cell mechanobiology. Biomech Model Mechanobiol. 2006;5:1–16.
26. Barth FG. Mechanics to pre-process information for the fine tuning of mechanoreceptors. J Comp Physiol A. 2019;205:661–86.
27. Wolfenson H, Yang B, Sheetz MP. Steps in mechanotransduction pathways that control cell morphology. Annu Rev Physiol. 2019;10:585–605.
28. Frontera WR. Clinical sports medicine: medical management and rehabilitation. Saunders/Elsevier; 2007.
29. Heroux ME, Butler AA, Robertson LS, Fisher G, Gandevia SC. Proprioception: a new look at an old concept. J Appl Physiol. 2022;132:811–4.
30. Ergen E, Ulkar B. Proprioception and coordination. In: Frontera WR, Herring SA, Micheli LJ, Silver JK, Young TP, editors. Clinical sports medicine: medical management and rehabilitation. Elsevier Health Sciences; 2007.
31. Wang H, Ji Z, Jiang G, Liu W, Jiao X. Correlation among proprioception, muscle strength, and balance. J Phys Ther Sci. 2016;28:3468–72.
32. Reschechtko S, Pruszynski A. Stretch reflexes. Curr Biol. 2020;30:1025–30.
33. Sherrington CS. The integrative action of the nervous system. New Haven: Yale University Press; 1906.
34. Moon KM, Kim J, Seong Y, Suh B-C, Kang KJ, Choe HK, Kim K. Proprioception, the regulator of motor function. BMB Rep. 2021;54:393–402.
35. Paterson M. Movement for movement's sake? On the relationship between kinesthesia and aesthetics. Essays Philos. 2012;13:471–97.
36. Burke RE. Sir Charles Sherrington's: The integrative action of the nervous system: a centenary appreciation. Brain. 2007;130:887–94.
37. Liddel EGT. The discovery of reflexes. London: Clarendon Press; 1960.
38. Freeman FN. The integrative action of the nervous system. J Philos Pssychol Sci Methods. 1907;4:301–4.
39. Taylor JL. Kinesthetic inputs. In: Pfaff DW, Volkow ND, Rubenstein JL, editors. Neuroscience in the 21st century: from basic to clinical. Springer Nature; 2022.
40. Romano N, Federici M, Castaldi A. Imaging of cranial nerves: a pictorial overview. Insights Imaging. 2019;10:1–21.
41. Bosch-Queralt M, Fledrich R, Stassart RM. Schwann cell functions in peripheral nerve development and repair. Neurobiol Dis. 2023;176:1–19.
42. Rueda-Lopes F. The cranial nerves: extensions of the central nervous system or components of the peripheral nervous system – how should we evaluate them? Radiol Bras. 2021;54:V–Vi.
43. Von Bernhardi R, Eugenin-Von Bernhardi J, Flores B, Eugenin Leon J. Glial cells and integrity of the nervous system. In: Bernhardi V, editor. Glial cells in health and disease of the CNS. Cham: Springer; 2016.
44. Jessen KR. Glial cells. Int J Biochem Cell Biol. 2004;36:1861–7.
45. Gardner E, Bunge RP. Gross anatomy of the peripheral nervous system. In: Dyck PJ, Thomas PK, editors. Peripheral neuropathy. 4th ed. Philadelphia: Saunders Elsevier; 2005.
46. Catala M, Kubis N. Gross anatomy and development of the peripheral nervous system. In: Said G, Krarup C, editors. Handbook of clinical neurology. Elsevier; 2013.
47. Birch R. The peripheral nervous system: anatomy and function. In: Birch R, editor. Peripheral nerve injuries: a clinical guide. London: Springer; 2013.
48. Pleger B, Villringer A. The human somatosensory system: from perception to decision making. Prog Neurobiol. 2013;103:76–97.
49. Rigoard P. The normal nerve. In: Rigoard P, editor. Atlas of anatomy of the peripheral nerves. Cham: Springer; 2017.
50. Jankowska E. Basic principles of processing of afferent information by spinal interneurons. J Neurophysiol. 2022;128:689095.

51. Jankowska E, Hammar I. Spinal interneurons; how can studies in animals contribute to the understanding of spinal interneuronal systems in man? Brain Res Rev. 2002;40:19–28.
52. Nógrádi A, Nógrádi A, Vrbová G. Anatomy and physiology of the spinal cord. In: Transplantation of neural tissue into the spinal cord. Austin: Landes Bioscience; 2006. p. 1–23.
53. Jankowska E, Lundberg A. Interneurones in the spinal cord. Trends Neurosci. 1981;4:230–3.
54. Stein BE. Neural mechanisms for synthesizing sensory information and producing adaptive behaviors. Exp Brain Res. 1998;123:124–37.
55. Mesulam MM. From sensation to cognition. Brain. 1998;121:1013–52.
56. Alex Meredith M. On the neuronal basis for multisensory convergence: a brief overview. Cog Brain Res. 2002;14:31–40.
57. Calvert GA, Thesen T. Multisensory integration: methodological approaches and emerging principles in the human brain. J Physiol-Paris. 2004;98:191–205.
58. Mukhopadyhyay K. Proprioception and kinesthesia: the sixth sense organ. Adv Health Exerc. 2021;1:12–7.
59. Nascimento AI, Mar FM, Sousa MM. The intriguing nature of dorsal root ganglion neurons: linking structure with polarity and function. Prog Neurobiol. 2018;168:86–103.
60. Proske U, Gandevia SC. The kinesthetic senses. J Physiol. 2009;587:4139–46.
61. Scott SH, Loeb GE. The computation of position sense from spindles in mono- and multiarticular muscles. J Neurosci. 1994;14:7529–40.
62. Ribeiro F, Oliveir J. Factors influencing proprioception: what do they reveal? In: Klika V, editor. Biomechanics in applications. Croatia, Balkans: InTech; 2011.
63. Tuthill JC, Azim E. Proprioception. Curr Biol. 2018;28:R194–203.
64. Vega JA, Cobo J. Structural and biological basis for proprioception. In: Vega JA, Cobo J, editors. Proprioception. IntechOpen; 2021.
65. Ellaway PH, Taylor A, Durbaba R. Muscle spindle and fusimotor activity in locomotion. J Anat. 2015;227:157–66.
66. Fortier S, Basset FA. The effects of exercise on limb proprioceptive signals. J Electromyogr Kinesiol. 2012;22:795–802.
67. Lephart S. Reestablishing proprioception, kinesthesia, joint position sense, and neuromuscular control in rehabilitation. In: Prentice WE, editor. Rehabilitation techniques in sports medicine. St. Louis: Mosby; 1994.
68. Houglum PA. Therapeutic exercise for athletic injuries. Human Kinetics: Champaign; 2001.
69. Norris CM. Exercise therapy. In: Norris CM, editor. Managing sports injuries a guide for students and clinicians. Churchill Livingston Elsevier; 2011.
70. Kroger S, Watkins B. Muscle spindle function in healthy and diseased muscle. Skelet Muscle. 2021;11:1–13.
71. Berta T, Qadri Y, Yan PH, Ji RR. Targeting dorsal root ganglia and primary sensory neurons for the treatment of chronic pain. Expert Opin Ther Targets. 2017;21:695–703.
72. Devor M. Unexplained peculiarities of the dorsal root ganglion. Pain. 1999;6:S27–35.
73. Herberger RV, Barry C, Dominguez N, Matusica D. Human dorsal root ganglia. Front Cell Neurosci. 2019;13:1–17.
74. Mense S. Functional anatomy of muscle: muscle, nocicpetors and afferent fibers. In: Mense S, Gerwin RD, editors. Muscle pain: understanding the mechanisms. Berlin Heidelberg: Springer; 2010.
75. Kerr GK, Worrington CJ. Velocity perception and proprioception. In: Gandevia SC, Proske U, Stuart DG, editors. Sensorimotor control of movement and posture. Boston: Springer; 2002.
76. Grill SE, Hallet M. Velocity sensitivity of human muscle spindle afferents and slowly adapting type II cutaneous mechanoreceptors. J Physiol. 1995;489(2):593–602.
77. Kasai T, Kawanishi M, Yahagi S. The effects of wrist-0muscle vibration on human voluntary elbow flexion-extension movements. Exp Brain Res. 1992;90:217–20.

78. Blecher R, Heinemann-Yerushalmi L, Assaraf E, Konstantin N, Chapman JR, Cope TC, Bewick GS, Banks RW, Zelzer E. New functions for the proprioceptive system in skeletal biology. Philos Trans R Soc B. 2018;373:1–7.
79. Proske U, Gandevia SC. The proprioceptive senses: their roles in signaling body shape, body position and movement, and muscle force. Physiol Rev. 2012;92:1651–97.
80. Mountcastle VB, Powell TP. Central nervous mechanisms subserving position sense and kinesthesis. Bull Johns Hopkins Hosp. 1959;105:173–200.
81. Burke D, Gandevia SC, Macefield G. Responses to passive movement of receptors in joint, skin and muscle of the human hand. J Physiol. 1988;402:347–61.
82. Macefield G, Gandevia SC, Burke D. Perceptual responses to microstimulaiton of single afferents innervating joints, muscles and skin of human hand. J Physiol. 1990;429:113–29.
83. Burgess PR, Clark FJ. Characteristic of knee joint receptors in the cat. J Physiol. 1969;203:317–35.
84. Clark FJ, Burgess PR. Slowly adapting receptors in cat knee joint: can they signal joint angle? J Neurophysiol. 1975;38:1448–63.
85. Newsom DJ. The response to stretch of human intercostal muscle spindles studied *in vitro*. J Physiol. 1975;249:561–79.
86. Poppele RE, Kennedy WR. Comparison between behavior of human and cat muscle spindles recorded *in vitro*. Brain Res. 1974;75:316–9.
87. Prochazka A. Proprioception: clinical relevance and neurophysiology. Curr Opin Physiol. 2021;23:100440.
88. Burke D. Fusimotor system. In: Reference module in biomedical sciences. United States: Elsevier; 2014.
89. Houk JC, Henneman E. Responses of Golgi tendon organs to active contractions of the soleus muscle of the cat. J Neurophysiol. 1967;30:466–81.
90. Macefield VG, Knellwolf TP. Functional properties of human muscle spindles. J Neurophysiol. 2018;120:452–67.
91. Bewick GS, Banks RW. Mechanotransduction in the muscle spindle. Pflugers Arch Eur J Physiol. 2015;467:175–90.
92. Dimitriou M. Human muscle spindles are wired to function as controllable signal-processing devices. Elife. 2022;11:1–14.
93. Santuz A, Akay T. Muscle spindles and their role in maintaining robust locomotion. J Physiol. 2023;601:275–85.
94. Cobo JL, Abbate F, De Vicente JC, Cobo J, Vega JA. Searching for proprioceptors in human facial muscles. Neurosci Lett. 2017;640:1–5.
95. Cooper S, Daniel PM. Muscle spindles in man; their morphology in the lumbricals and the deep muscles of the neck. Brain. 1963;86:563–86.
96. Loeb GE. The control and responses of mammalian muscle spindles during normally executed motor tasks. Exerc Sport Sci Rev. 1984;12:157–204.
97. Prochazka A, Hulliger M. Muscle afferent function and its significance for motor control mechanisms during voluntary movements in cat, monkey, and man. Adv Neurol. 1983;39:93–132.
98. Banks RW. The innervation of the muscle spindle: a personal history. J Anat. 2015;227:115–35.
99. Houk JC. Regulation of stiffness by skeletomotor reflexes. Annu Rev Physiol. 1979;41:99–114.
100. Kokkorogiannis T. Somatic and intramuscular distribution of muscle spindles and their relation to muscular angiotypes. J Theor Biol. 2004;229:263–80.
101. Matthews PBC. Evolving views on the internal operation and functional role of the muscle spindle. J Physiol (London). 1981a;320:1–30.
102. Windhorst U. Muscle spindles are multi-functional. Brain Res Bull. 2008;75:507–8.
103. Kröger S. Proprioception 2.0: novel functions for muscle spindles. Curr Opin Neurol. 2018;31:592–8.
104. Hulliger M. The mammalian muscle spindle and its central control. Rev Physiol Biochem Pharmacol. 1984;101:1–110.

105. Matthews PBC. The structure of receptors. In: Society MOTP, editor. Mammalian muscle receptors and their central actions. London: Edward Arnold; 1972.
106. Loeb GE, Mileusnic M. Proprioceptors and models of transduction. In: Prescott TJ, Ahissar E, Izhikevich E, editors. Scholarpedia of touch. Atlantis Press; 2016.
107. Voss VH. Tabelle der absoluten und relativen muskelspindelzahlen der menschlichen skeletmuskulatur. Anat Anz. 1971;129:562–72.
108. Voss VH. Untersuchungen uber zahl, anordnung and lange der muskelspindeln in den lumbrikalmuskeln des menschen und einiher tiere. Z Mikrosk-Anat Forsch. 1937;42:509–24.
109. Gandevia SC, Burke D. The peripheral motor system. In: The human nervous system. Elsevier; 2004.
110. Osterlund C, Liu J-X, Thornell L-E, Erikson P-O. Muscle spindle composition and distribution in human young masseter and biceps brachii muscles reveal early growth and maturation. Anat Rec. 2011;294:683–93.
111. Banks RW. An allometric analysis of the number of muscle spindles in mammalian skeletal muscles. J Anat. 2006;208:753–68.
112. Boyd IA. The isolated mammalian muscle spindle. Trends Neurosci. 1980;3:258–65.
113. Banks RW. The motor innervation of mammalian muscle spindles. Prog Neurobiol. 1994;43:323–62.
114. Eldred E, Bridgman CF, Sweet JE, Eldred B. Quantitative comparisons of muscle receptors of the cat medial gastrocnemius, soleus and extensor digitorum muscles. In: Barker D, editor. Symposium on muscle receptors. Hong Kong: Hong Kong University Press; 1962.
115. Mancall EL, Brock DG. Overview of the microstructure of the nervous system. In: Mancall EL, Brock DG, editors. Gray's clinical neuroanatomy: the anatomic basis for clinical neuroscience. Philadelphia: Elsevier Health Sciences; 2011.
116. Robertson DJ. Electron microscopy of the motor end-plate and the neuromuscular spindle. Am J Phys Med Rehabil. 1960;39:1–44.
117. Zeller-Plumhoff B, Roose T, Katsamenis OL, Mavrogordato MN, Torrens C, Schneider P, Clough GF. Phase contrast synchrotoron radiation computed tomography of muscle spindles in the mouse soleus muscle. J Anat. 2017;230:859–65.
118. Fukami Y. Studies of capsule and capsular space of cat muscle spindles. J Physiol. 1986;376:281–97.
119. Koike S, Mukudai S, Hisa Y. Muscle spindles and intramuscular ganglia. In: Koike S, Mukudai S, Hisa Y, editors. Neuroanatomy and neurophysiology of the larynx. Springer Japan: Japan; 2016.
120. Barker D. The morphology of muscle receptors. In: Hunt CC, editor. Handbook of sensory physiology. Muscle receptors. New York: Springer; 1974.
121. Boyd I. The mechanical properties of dynamic nuclear bag fibers, static nuclear bag fibers and nuclear chain fibers in isolated cat spindles. Prog Brain Res. 1976;44:33–50.
122. Matthews PBC. Muscle spindles: their messages and their fusimotor supply. In: Brookhart JM, Mountcastice WB, Brooks VB, editors. Handbook of physiology: the nervous system. Bethesda: American Physiological Society; 1981b.
123. Soukup T, Pedrosa-Domellof F, Thornell L-E. Intrafusal fiber type composition of muscle spindles in the first human lumbrical muscle. Acta Neuropathol. 2003;105:18–24.
124. Boyd IA, Gladden MH, Mcwilliam PN, Ward J. Control of dynamic and static nuclear bag fibres and nuclear chain fibres by gamma and beta axons in isolated cat muscle spindles. J Physiol. 1977;265:133–62.
125. Boyd IA, Ward J. Motor control of nuclear bag and nuclear chain intrafusal fibres in isolated living muscle spindles from the cat. J Physiol. 1975;244:83–112.
126. Te Kronnie G, Donselaar Y, Soukup T, Van Raamsdonk W. Immunohistochemical differences in myosin composition among intrafusal muscle fibres. Histochemistry. 1981;73:65–74.
127. Halata Z, Baumann KI. Anatomy of receptors. In: Grunwald M, editor. Human haptic perception: basic and applications. Basel: Birkhauser; 2008.

128. Cazzato G, Walton JN. The pathology of the muscle spindle. A study of biopsy material in various muscular and neuromuscular diseases. J Neurol Sci. 1968;7:15–70.
129. Gollapudi SK, Michel JJ, Chandra M. Striated muscle dynamics. In: Reference module in biomedical sciences. Elsevier; 2014.
130. Celichowski J, Krutki P. Motor units and muscle receptors. In: Zoladz JA, editor. Muscle and exercise physiology. Academic; 2019.
131. Barker D. The structure and distribution of muscle receptors. In: Barker D, editor. Symposium on muscle receptors. Hong Kong: Hong Kong University Press; 1962.
132. Scott JJA, Davies P, Petiti J. The number and distribution of muscle spindles and tendon organs in the peroneal muscles of the cat. J Anat. 1987;151:143.
133. Lund JP, Richmond FJR, Touloumis C, Patry Y, Lamarre Y. The distribution of Golgi tendon organs and muscle spindles in masseter and temporalis muscles of the cat. Neuroscience. 1978;3:259–70.
134. Mileusnic MP, Loeb GE. Mathematical models of proprioceptors. II. Structure and function of Golgi tendon organ. J Neurophysiol. 2006;96:1789–802.
135. Jami L. Golgi tendon organs in mammalian skeletal muscle: functional properties and central actions. Physiol Rev. 1992;72:623–66.
136. Binder MD, Osborn CE. Interactions between motor units and Golgi tendon organs in the tibialis posterior muscle of the cat. J Physiol. 1985;364:199–215.
137. Davies P, Petit J, Scott JA. The dynamic response of Golgi tendon organs to tetanic contraction of in-series motor units. Brain Res. 1995;690:82–91.
138. Nitatori T. The fine structure of human Golgi tendon organs as studied by three-dimensional reconstruction. J Neurocytol. 1988;17:27–41.
139. Bridgman CF. Sampling of total muscle force by tendon organs. J Neurophysiol. 1970;47:1069–83.
140. Golgi C. Intorno alla distribuzione e terminazione dei nervi nei tendini dell'uomo e di altri vertebrati. In: Golgi C, editor. Opera Omnia. Milano: Hoepli; 1903a.
141. Golgi C. Sui nervi dei tendini dell'Uomo et di altri vertebrati e di nuovo organo nervoso terminale musculo-tendineo. In: Golgi C, editor. Opera Omnia. Milano: Hoepli; 1903b.
142. Moore M. Golgi tendon organs-neuroscience update with relevance to stretching and proprioception in dancers. J Dance Med Sci. 2007;11:85–92.
143. Horcholle-Bossavit G, Jami L, Petit J, Vejsada R, Zytnicki D. Ensemble discharge from Golgi tendon organs of cat peroneus tertius muscle. J Neurophysiol. 1990;64:813–21.
144. Schoultz TW, Swett JE. Ultrastructural organization of the sensory fibers innervating the Golgi tendon organ. Anat Rec. 1973;179:147–62.
145. Zelena J, Soukup T. The in-series and in-parallel components in rat hindlimb tendon organs. Neuroscience. 1983;9:899–910.
146. Schoultz TW, Swett JE. The fine structure of the Golgi tendon organ. J Neurocytol. 1972;1:1–25.
147. Fukami Y, Wilkinson RS. Responses of isolated Golgi tendon organs of the cat. J Physiol. 1977;265:673–89.
148. Appentreng K, Prochazka A. Tendon organ firing during active muscle lengthening in awake, normally behaving cats. J Physiol. 1984;353:81–92.
149. Edin BB, Valbo AB. Muscle afferent responses to isometric contractions and relaxations in humans. J Neurophysiol. 1990;63:1307–13.
150. Fukami Y. Responses of isolated Golgi tendon organs of the cat to muscle contraction and electrical stimulation. J Physiol (London). 1981;318:429–43.
151. Lloyd DPC, Chang HT. Afferent fibers in muscle nerves. J Neurophysiol. 1948;11:199–207.
152. Banks R. Muscle spindles and tendon organs. In: Reference module in biomedical sciences. Elsevier; 2018.
153. Bridgman CF, Shumpert EE, Eldred E. Insertions of intrafusal fibers in muscle spindles of the cat and other mammals. Anat Rec. 1969;164:391–402.

154. Hunt CC. Relation of function to diameter in afferent fibers of muscle nerves. J Gen Physiol. 1954;38:117–31.
155. Taylor JL. Proprioception. In: Squire LR, editor. Encyclopedia of neuroscience. Academic; 2009.
156. Wei JY, Kripke BR, Burgess PR. Classification of muscle spindle receptors. Brain Res. 1986;370:119–26.
157. Cao R, Chen P, Wang H, Jing H, Zhang H, Xing G, Luo B, Pan J, Yu Z, Xiong W-C, Mei L. Intrafusal-fiberlrp4 for muscle spindle formation and maintenance in adult and aged animals. Nat Commun. 2023;14:1–17.
158. Clark BC, Thomas JS, Walkowski SA, Howell JN. The biology of manual therapies. J Am Osteopath Assoc. 2012;112:617–29.
159. Rumsey JW, Das M, Bhalkikar A, Stancescu M, Hickman JJ. Tissue engineering the mechanosensory circuit of the stretch reflex arc: sensory neuron innervation of intrafusal muscle fibers. Biomaterials. 2010;31:8218–27.
160. Milburn A. Stages in the development of cat muscle spindle. J Embryol Exp Morphol. 1984;82:177–216.
161. Burke R. Spinal motoneurons. In: Pfaff DW, Volkow ND, Rubenstein JL, editors. Neuroscience in the 21st century. Springer; 2022.
162. Zayia LC, Tadi P. Neuroanatomy, motor neuron. Treasure Island: StatPearls Publishing; 2024.
163. Stifani N. Motor neurons and the generation of spinal motor neuron diversity. Front Cell Neurosci. 2014;8:1–22.
164. Hultborn H, Fedirchuk B. Spinal motor neurons: properties. In: Squire LR, editor. Encyclopedia of neuroscience. Academic; 2009.
165. Laporte Y, Emonet-Denand F, Jami L. The skeletofusimotor or β-innervation of mammalian muscle spindles. Trends Neurosci. 1981;4:97–9.
166. Ten Donkelaar HJ. Motor systems. In: Ten Donkelaar HJ, editor. Clinical neuroanatomy: brain circuitry and its disorders. Cham: Springer; 2020.
167. Manuel M, Zytnicki D. Alpha, beta and gamma motoneurons: functional diversity in the motor system's final pathway. J Integr Neurosci. 2011;10:243–76.
168. Bessou P, Emonet-Denand F, Laporte Y. Motor fibres innervating extrafusal and intrafusal muscle fibres in the cat. J Physiol. 1965;180:649–72.
169. Khan MN, Marquardt T. Motor neurons– new insights. In: Motor neuron specification to function: filling in the gaps. IntechOpen; 2024.
170. Kuffler SW, Hunt CC, Quilliam JP. Function of medullated small-nerve fibers in mammalian ventral roots; efferent muscle spindle innervation. J Neurophysiol. 1951;14:29–54.
171. Murthy KS. Vertebrate fusimotor neurones and their influences on motor behavior. Prog Neurobiol. 1978;11:249–307.
172. Taylor A. Muscle receptors in the control of voluntary movement. Spinal Cord. 1972;9:167–72.
173. Kanning KC, Kaplan A, Henderson CE. Motor neuron diversity in development and disease. Ann Rev Neurosci. 2010;33:409–40.
174. Proske U, Gandevia SC. Kinesthetic senses. Compr Physiol. 2018;8:1157–83.
175. Grigg P. Peripheral neural mechanism in proprioception. J Sport Rehabil. 1994;3:2–17.
176. Zampieri N, De Nooij JC. Regulating muscle spindle and Golgi tendon organ proprioceptor phenotypes. Curr Opin Physio. 2021;19:204–19.
177. Colon A, Guo X, Akanda N, Cai Y, Hickman JJ. Functional analysis of human intrafusal fiber innervation by human γ-motoneurons. Sci Rep. 2017;7:1–11.

Chapter 5
Hormonal and Metabolic Responses During Exercise and Recovery

Anthony C. Hackney

Introduction

Numerous hormone-like substances exhibit endocrine, paracrine, and autocrine actions on the tissues in the human body. Biologically these substances engage in a multitude of regulatory roles for processes such as growth and development, reproduction, metabolism, hydration, cardiovascular, immune, and stress responses. Scientific investigations indicate that exercise has powerful effects on the production of these hormone-like substances and thus the processes they regulate [1–4]. The effects of exercise on the endocrine aspects of the body in healthy individuals are of a positive nature and improve the functional aspects of tissues and organs ultimately leading to improved health and/or performance. Unfortunately, however, when exercise is executed poorly or developed incorrectly counter-productive endocrine responses happen, and negative physiological consequences can occur (e.g., Overtraining Syndrome).

People doing exercise with the intent of improving health may typically conduct a training "exercise prescription" comprised of 3–5 days a week of activity, for 15–60 minutes in duration at a low-moderate to moderately vigorous intensity (of a continuous or intermittent nature) [5]. This ~1–5 hours commitment per week is challenging for many ordinary people to achieve. In the case of athletes who participate in sports, this volume of activity might represent what they typically do in a single day of their exercise training program. Such large dosages of exercise in athletes, when administered chronically bring about large and substantial physiological adjustment and biological adaptation, which is necessary to enhance sporting performance to the highest levels [6, 7]. Nonetheless whether an individual is a

A. C. Hackney (✉)
University of North Carolina, Chapel Hill, NC, USA
e-mail: ach@email.unc.edu

N. C. Apostolopoulos et al. (eds.), *Fundamentals of Recovery, Regeneration, and Adaptation to Exercise Stress: An Integrated Approach*,
https://doi.org/10.1007/978-3-031-44270-4_5

recreational exerciser or a striving athlete involved in intensive exercise training, there are critical changes in the endocrine system that allow the body to accommodate and adjust to the physiological needs of exercise.

This chapter presents an overview of the responses and adjustments as well as the adaptation of the hormonal components of the endocrine system concerning exercise and exercise training. There are many forms of exercise activity and in part, the type of hormonal response is influenced by the energy system demands and predominating physiological systems necessary to perform the activity; e.g., cardiovascular versus muscular force. Therefore, as a means to organize the topic dialog, the hormonal responses are presented in a dichotomous approach dealing with activities of cardiovascular–aerobic type (i.e., endurance-based) versus activities of muscular power–anaerobic type (resistance–strength-based).

Hormonal Exercise Responses

Acute Exercise Responses

Acute responses of the major physiological systems (e.g., metabolism, cardiovascular-respiratory [CVR]) to a single bout of exercise are usually proportional to the intensity of the exercise, although it is important to recognize that the relationship of this proportional response(s) is not always entirely linear [2]. These exercise hormonal changes occur for several physiological reasons: (*a*) to induce CVR adjustments, (*b*) to activate energy production pathways and mobilize energy substrates, (*c*) to facilitate maintenance of adequate hydration, and (*d*) to some extent as part of stress reactivity response. Table 5.1 illustrates the general changes in the major hormones of the endocrine system in response to endurance versus resistance exercise activities.

Many of these hormonal responses are not independent actions of a single endocrine gland but are highly interrelated and interdependent among several glands. As a means to illustrate this point, the "Hormonal Exercise Response Model" (HERM) explanatory paradigm is presented herein. This proposed model explains that the crux of the hormonal responses to an acute exercise session can be categorized into a series of three interactive phases [2, 8–15]. The HERM divides the hormonal sequence of events into three phases of responses, with the magnitude of the hormonal response being influenced by the intensity and duration of the exercise (i.e., "exercise load").

The first phase of this model deals with the hormonal response immediately at the onset of exercise, with responses taking just seconds to occur. These responses revolve around the increased sympathetic nervous system activation that occurs with the onset of body movement (*N.B.*, typically these are referred to as "neuroendocrine responses"). This increased sympathetic nervous system activity can also be a result of anticipation of the stress of the subsequent exercise activity, which is the

Table 5.1 Changes in the circulating levels of select hormones in response to an acute exercise session involving either endurance-based or resistance-based exercise activities

	Exercise—activity response	
Hormone	Endurance	Resistance
Adrenocorticotropic hormone (ACTH)	↑	↑
Aldosterone	↑	↑
Angiotensin	↑	↑
Antidiuretic Hormone (ADH; Vasopressin)	↑	↑↓
Cortisol	↑	↑
Dehydroepiandrosterone (DHEA)	↑	↑
β—Endorphin	↑	↑
Epinephrine (adrenaline)	↑↑	↑↑
Estrogen (estradiol-β-17)	↑	↑
Follicle-Stimulating Hormone	↑↓ nc	↑↓ nc
Glucagon	↑	↑
Growth Hormone (GH)	↑	↑
Insulin	↓	↑↓ nc
Insulin-Like Growth Factor-1 (IGF-1)	↑ nc	↑ nc
Leptin	↑↓ nc	↑↓ nc
Luteinizing Hormone (LH)	↑↓ nc	↑↓ nc
Norepinephrine (noradrenaline)	↑↑	↑↑
Progesterone (P4)	↑	↑
Prolactin (PRL)	↑	↑
Testosterone	↑↓	↑
Thyroxine (T4)	↑	↑
Triiodothyronine (T3)	↑	↑
Vitamin D	↑	?

The hormonal changes depicted here are relative to before versus immediately after the specific exercise activity. ↑ increase, ↓ decrease, *nc* no change, ? limited data exist for the determination

case typically in sporting competition situations. This increased sympathetic nervous system activity results in catecholamine (norepinephrine, neurotransmitter) release for post-ganglionic adrenergic neuronal fibers directly at target tissue organs. This response also results in blood elevations in circulating catecholamine from the sympathetic "spillover" effect (more neurotransmitter is released than the tissue can uptake, and the resulting excess diffuses into the blood) [11, 15, 16]. This effect is further amplified by the sympathetic neural connection to the adrenal medullary gland which in turn adds to the circulating catecholamines (epinephrine + norepinephrine) response [15, 17]. Concurrent with these sympathetic-adrenal actions, pancreatic insulin secretion begins to be inhibited, while glucagon secretion becomes stimulated [15]. This entire process involves a feed-forward mechanism of the central nervous system to drive these initial responses, although the events are also modified by peripheral afferent neural input from sensory receptors, in particular, those of skeletal muscle depending upon the rapidity and force of the muscular contractions during movement [13, 17].

The intermediate or secondary phase of HERM takes slightly longer to develop but is still typically very fast, beginning usually within the first minute of the onset of exercise. In this stage, the hypothalamus begins the process of releasing signaling hormones such as thyrotropin-releasing factor, corticotropin-releasing factor (CRF), and growth hormone-releasing factor in an attempt to provoke changes at the anterior pituitary gland; that is, to stimulate the anterior pituitary to release its specific hormones. As the anterior pituitary begins to respond to the hypothalamic stimulus, the "trophic hormones" (see Table 5.1) from this gland are secreted to the circulation, and these hormones in turn begin to act upon their specific peripheral target glands to stimulate additional hormonal release or have direct effects on non-glandular target tissues.

One of the most rapidly acting players at this stage in this cascade of events is the hypothalamic–pituitary–adrenocortical interaction where CRF brings about adrenocorticotrophic hormone (ACTH) release and this results in cortisol and dehydroepiandrosterone (DHEA) release from the adrenal cortex glands [12, 13, 18].

If the duration of an exercise session continues, there is a transition beyond the intermediate (second) phase into a third phase of response which is a more prolonged series of changes. In this third phase, the responses of the sympathetic-adrenal (medullary + cortical) axis are being augmented by other hormones from both the anterior (e.g., growth hormone, prolactin) and posterior pituitary (e.g., antidiuretic hormone [vasopressin]) and the peripheral endocrine glands subordinated to pituitary regulatory actions, such as testosterone, thyroxine, triiodothyronine, insulin-like growth factor-1 [2, 12, 15].

Furthermore, as more prolonged exercise exposure occurs, fluids shift from the vascular space, and total body water stores are utilized to allow sweating for heat dissipation, the renin–angiotensin–aldosterone system (RAAS) is activated (inducing vasoconstrictive actions to induce blood shunting and renal water resorption) [8, 11]. Additionally, during this phase, the skeletal muscle begins to release selected cytokines (e.g., interleukin-6 [IL-6]), hormonal-like agents, into the circulation which affect other hormones which are being released concurrently (e.g., IL-6 → cortisol) which can have actions to signal energy substrate mobilization and aspects of immune responses [19]. These additive effects of increasing hormonal release demonstrate the endocrine principle of amplification.

Phases one and two of the HERM model propose that neural factors are the primary stimuli are the driving force in the hormonal responses to exercise; however, in the third phase of the model, there is an ever-increasing influence of the humoral and hormonal factors regulating the overall responses due to the changes in the "internal milieu" influencing both target tissue as well as endocrine glands [15]. This shifting of primary regulatory factors allows an increasing reliance upon feedback rather than a feed-forward control mechanism to determine the magnitude of the hormonal response. The influence of humoral and hormonal stimuli in modulating the hormonal levels is magnified as the exercise duration is extended and energy substrate availability issues cause shifts in energy fuel usage (i.e., decreased carbohydrate availability leading to increased lipid reliance), or hydration issues (i.e., blood hemoconcentration and/or dehydration) begin compromising the thermoregulatory ability and leading to greater heat storage within the body. The latter, increased heat storage, leads to elevations in body core temperature which in turn augments hormonal responses to exercise (e.g., ↑ heat storage → ↑ norepinephrine, epinephrine, growth hormone, etc.) [8, 13, 17, 20]. Importantly, such core temperature effects are exacerbated if ambient environment conditions (or athlete behavior) do not allow for adequate cooling and result in an exceedingly greater hormonal response in several hormones [20, 21].

This HERM model [22], while not perfectly inclusive in capturing all nuances of hormonal changes to exercise, does provide an organized framework for the endocrine responses to exercise activities and illustrates the highly interactive, and complexity of these responses [2, 11, 12, 15]. Finally, it is important to recognize intermittent exercise (i.e., involving rest periods) can influence the phase transition in the HERM model.

Exercise hormonal changes occur in an attempt to:

(a) induce cardiovascular-respiratory adjustments,
(b) activate energy production pathways and mobilize energy substrates
(c) facilitate maintenance of adequate hydration—thermoregulation
(d) and, to some extent, as part of stress reactivity response.

Exercise Training Adaptations: Chronic Exercise Responses

In general, the responses of the physiological systems to an exercise session after performing a successful exercise training program are similar to those before such training. That is, an acute bout of exercise after chronic exercise training is still a stimulus to the physiology of the body, and the endocrine system responds accordingly.

That said, however, at all exercise intensities, less than a maximal effort, such responses are typically diminished to some extent. The more robust the exercise training adaption incurred due to the program, typically the greater the attenuation of the hormonal response (see later section).

An exception to this attenuated response is maximal or supramaximal exercise bouts. In these situations, training adaptations result in a greater level and ability of the athlete to perform higher workloads at maximal/supramaximal efforts. As such, the greater absolute maximal workload achieved produces a greater overall stimulus to physiological systems (including endocrine)—in other words, typically a greater maximal response in all measures/parameters and not an attenuation of physiological responses. A noted non-endocrine exception to this generalization is maximal heart rate, which is commonly slightly lower or unchanged after a training program even though there are substantial CVR adaptations (*N.B.*, in healthy individuals the most critical factor affecting maximal heart rate is aging) [8, 9, 13].

Undoubtedly, after training adaptations develop, a single exercise bout (either endurance or resistance-based activities) invokes hormonal responses and as noted above these tend to be attenuated, but a critical question is why are responses mitigated? These changed responses come about through a variety of training-related alterations in both the endocrine gland as well as the hormonal target tissue of the glands.

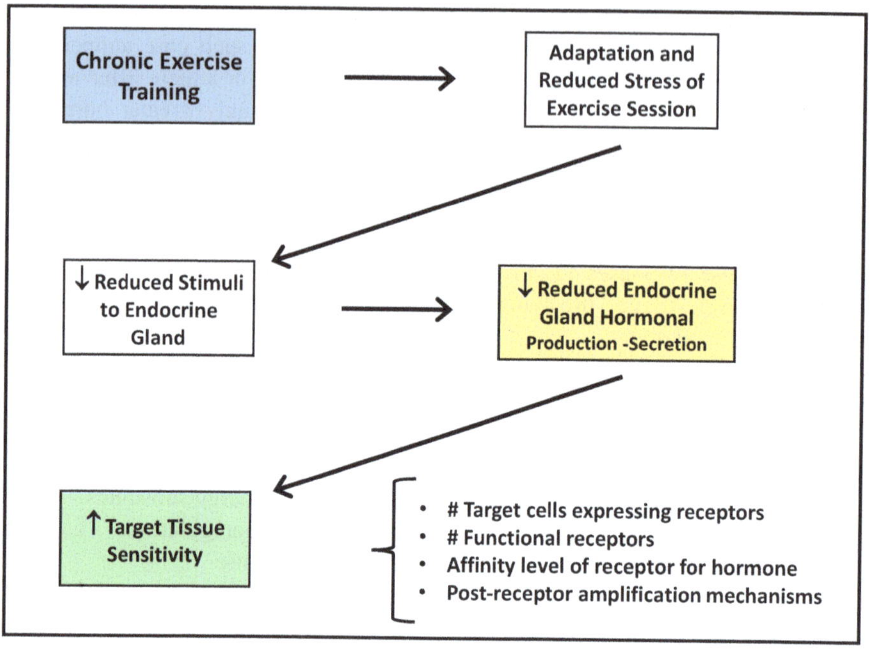

Fig. 5.1 Mechanisms by which chronic exercise training induces alterations in circulating hormonal levels and target tissue responsiveness

In brief, the reduced hormonal responses are due to improved sensitivity of the target tissue to the hormonal stimulus, and the level of neural, humoral, and hormonal stimuli disturbances in the blood that influence the various endocrine glands is greatly diminished [13]. In relation to the former point of sensitivity, after a training program, many target tissues; (*a*) increase the expression of functional hormone receptors, (*b*) receptor affinity for hormones becomes increased, and (*c*) the potential for post-receptor amplification mechanisms in target tissue cells are improved (See Fig. 5.1 for a summarization of endocrine gland–hormone–target tissue alterations with exercise training [22]). Effectively, these changes result in a target tissue needing a reduced amount of a hormone to bring about the physiological outcome/changes desired to accommodate exercise.

Basic Exercise Endocrinology of Hormonal Mechanisms

Hormonal Concentration Changes with Exercise

The HERM concept focuses on hormonal changes occurring to bring about physiological events which allow an individual to adjust to and mediate the homeostatic disruptions of an exercise session. Interestingly, not all these hormonal responses are brought about by endocrine glandular-tissue production.

In fact, the basic endocrinological mechanisms by which blood hormone concentrations change during exercise are associated with alterations in; (*a*) vascular fluid content (*b*) the rate of hormonal metabolic clearance (MCR), and (*c*) the glandular tissue secretion responses.

Relative to the first point, vascular fluid content alterations are a result of plasma fluid moving in or out of the circulatory vascular spaces. During exercise, plasma volume "shifts" do occur, specifically, a loss of plasma fluid, resulting in a hemoconcentration (increased levels) of any substance trapped within the blood vessel. This applies to large molecular weight hormones, or hormones in a bound state to their carrier protein(s), if relevant. In a 30–90-minute exercise session, a 10–20% hemoconcentration effect is not uncommon [21]. This effect is going to be intensified if the ambient environment condition creates a greater thermal load and/or an individual does not consume fluids during exercise. On the other hand, during the recovery from exercise (when ample fluids are usually consumed), there tends to be a rebound effect and fluid moves into the vascular space to cause a hemodilution of hormones (i.e., reduce concentrations) [14, 21].

Secondly, the rate of MCR of a hormone is in reference to its removal from the blood. Hormones are "taken up" by receptor mechanisms at their target tissues to implement physiological actions at the tissue. Additionally, some clearance is through the degradation—deactivation removal process at organs such as the liver or kidneys (i.e., renal filtration) [14, 23, 24]. Increases in the MCR can reduce circulating hormonal concentrations while decreases can have the opposite effect.

The third factor involving hormonal change during exercise deals with glandular secretion. That is the endocrine gland produces and, or secretes more hormones in response to a stimulus. This can usually result in increases in blood hormone levels, but it is possible for the increased rate of entry into the blood and the rate of MCR removal to match one another. In this case, the level of the hormone in the blood does not seem to change (e.g., increase), but actually, the turnover rate for the hormone has, and the target tissues are being presented with greater amounts of the hormone, as their uptake can increase to match the level of production/secretion from the gland [23, 24].

Exercise Hormonal Action on Tissues: Critical Factors

There is a misconception by some exercise scientists that if blood hormonal levels increase, then the physiological actions of that hormone on target tissue are activated. This, however, is a gross oversimplification of the interaction of endocrinological and physiological events. Blood hormonal concentrations are a key fundamental determinant of activation of physiological actions; i.e., a critical first step, but not the only required step for actions to occur.

For hormones to impact on target tissue physiology, the following are needed:

(a) activate number of target tissue cells expressing receptors for the hormone,
(b) adequate numbers of functional receptors on the specific cells,
(c) a high affinity level between the receptors on the cells and the hormone, and
(d) sufficient post-receptor amplification mechanisms within the target tissue cells to induce changes.

Specifically, for a hormone to initiate and activate a physiological process there must be several vital aspects of the target tissue functionally for desired actions to occur. These aspects are; (*a*) adequate numbers of target tissue cells expressing functional receptors for the hormone, (*b*) adequate numbers of functional receptors on the specific cells, (*c*) a high-affinity level of the receptors of the cells for the hormone, and (*d*) sufficient post-receptor amplification mechanisms within the cells to induce changes [22–24]. This last point is highly critical and can be influenced by the behavior and actions of the athlete. As an illustration, it might be assumed that increased testosterone yields increased tissue protein anabolism. This is only true if adequate amino acid quantity/types have been consumed in the diet and are available for cellular mechanisms (biochemical pathways, enzymes) to use in the protein synthesis process. The post-receptor amplification mechanisms are essential for a positive change to occur in skeletal muscle, arguably the most critical tissue, organ affecting exercise capacity, and a key focus of exercise training adaptation [5–7, 14].

Hence, in short, during exercise or in the period following (recovery) it is erroneous to automatically assume that if increased blood hormone concentrations occur then there is a ubiquitous occurrence of desired physiological actions and modifications going to occur at the target tissue (i.e., appropriate receptor actions are necessary too).

Endurance Versus Resistance Exercise Differentiation

Table 5.1 provides a summary of the hormonal change in response to an acute exercise session of either endurance or resistance-based activity. Evidence therein supports that the differentiation of response is not dramatically dissimilar. That said though, it is important to recognize that the magnitude of the response can be highly variable and influenced to some extent by exercise mode. Furthermore, as noted previously, critical to the issue of hormonal magnitude of change is the exercise load (intensity, duration, and volume of work performed) and muscle fiber type recruitment [22]. The latter in particular has critical impacts on an autonomic neural response which in turn impacts many aspects of endocrine gland responses. That is, endocrine glands are responsive to humoral, hormonal, and neural stimuli in the context of regulation of hormone production/secretion [23, 24].

Recovery

During the recovery period from an exercise session perhaps the most impactful factor to influence hormonal concentrations is the dietary practices of an individual. Consumption of food or beverages containing macro-nutrients (e.g., carbohydrates) has the ability to alter blood glucose levels. Changes in the glycemic status of an individual are a powerful stimulus to the hypothalamic-pituitary (H-P) axis as well as related axes, such as the H-P adrenocortical, adrenomedullary, thyroidal, and gonadal. This obviously can affect a multitude of circulating hormones (e.g., CRF, ACTH, cortisol, epinephrine, etc.). In a similar vein, fluid consumption in recovery can alter hormones that regulate the hydration levels of an individual. Exercise fluid loss (e.g., sweating) and plasma shifts (e.g., vascular efflux) can result in post-exercise elevations of antidiuretic hormone (vasopressin), angiotensin, and aldosterone throughout recovery. These hormones are critical in regulating bodily fluids by influencing fluid loss via renal mechanisms. Hence fluid consumption practices by an athlete can resultingly augment or attenuate the magnitude of these hormones in the circulation.

Exercise Metabolism: Energy Energetics Basics

Adenosine triphosphate (ATP) is the "energy currency" of muscle and while essential, it is not stored in large quantities intramuscularly. Due to its limited availability, ATP must be synthesized at a robust rate to meet the energy demands placed on muscle fibers (cells), this is especially true during exercise [25]. As bodily movements evolve from the "Sliding Filament" process within sarcomeres of contracting muscle fibers, ATP hydrolysis and resynthesis occur at a rate to match the metabolic demand (energy expenditure). To meet this demand, muscle fibers have a variety of biochemical ATP production pathways at their disposal [26]. These pathways are generally categorized into two broad classifications, being either anaerobic or aerobic-dependent pathways. Although, the use of this terminology which is very prevalent, tends to be a gross oversimplification, and more accurate phrasing would be substrate phosphorylation and oxidative phosphorylation, respectively.

Anaerobically, besides the limited stored ATP in muscle fibers (a few seconds worth of all-out effort), the most immediately available energy substrate source for ATP resynthesis is phosphocreatine (PCr). However, this substrate is also in extremely limited supply as skeletal muscle typically has sufficient PCr to sustain a maximum ATP turnover rate for only about an additional ~10–30 seconds [25].

Skeletal muscle also utilizes carbohydrates (CHO) as an energy substrate for ATP resynthesis in the pathway of anaerobic glycolysis as well as in the aerobic biochemical pathways (aerobic glycolysis, tricarboxylic acid cycle [TCA], electron transport system [ETS]). Aerobic energy metabolism is further supported by fat oxidation involving the β-oxidation, TCA, and ETS pathways. The rates at which

CHO and fat substrates, in their respective pathways, can resynthesize ATP are significantly slower than that of PCr or anaerobic glycolysis. However, the capacity (yield) for ATP resynthesis in these aerobic pathways is significantly and substantially greater [25].

Relative to this last point, metabolically in the ATP resynthesis during exercise, there is a constant ongoing trade-off within muscle fibers between the need for the *speed* (rate per unit time [power]) of ATP production and the need for *capacity* (yield in amount) of ATP resynthesis for specific activities. The trade-off and the give-and-take between speed vs. capacity of ATP resynthesis are apparent during maximal exercise session(s) as illustrated in Fig. 5.2 [25, 27].

> With ATP resynthesis during exercise, there is a constant ongoing trade-off within muscle fibers between the needed *speed* (rate per unit time) of ATP production and the needed *capacity* (yield in amount) of ATP production. This trade-off influences which energy pathway predominates (provided substrate is available and fatigue inducing mechanisms are mitigated).

Furthermore, the balance between substrate-level phosphorylation (glycolysis) and oxidative phosphorylation (TCA, β-Oxidation; i.e., utilization of pyruvate vs. free fatty acid utilization in acetyl-CoA production) is in part determined by the match between glycolytic flux and mitochondrial respiration (oxygen [O_2] consumption). These latter two factors are again not only influenced by exercise specifics (i.e.,

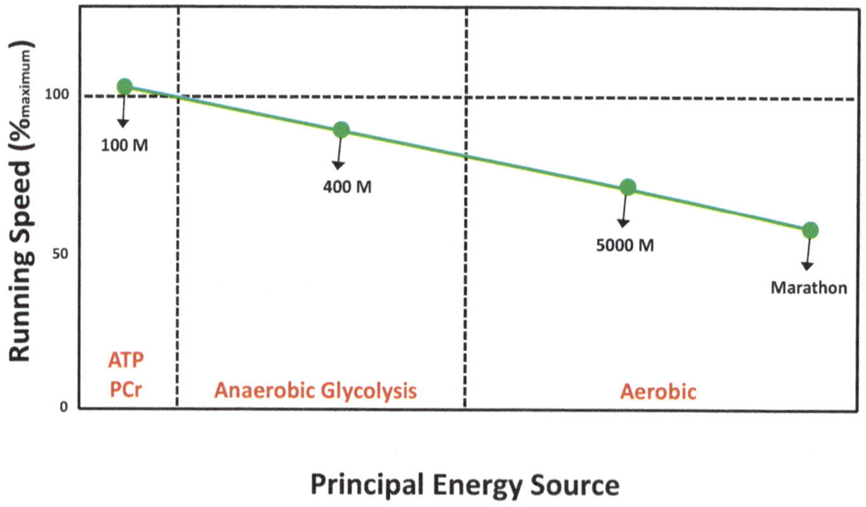

Fig. 5.2 Relationship between the predominant, principal energy source and maximal exercise effort (i.e., maximal running speed) of an individual

intensity, modality, duration), but also by the relative overall physical fitness level (training status) of an individual, and substrate availability via either the body energy stores or aspects of dietary consumption during an exercise session [25, 27].

The following is an illustration of the point raised just prior. Increases in exercise duration are achieved at the expense of the rate per unit time of ATP turnover (↑ duration, ↓ rate), and have more of a reliance on the capacity ATP being generated as a driving determinant of pathway utilization. As such, the lower rate of ATP turnover but greater capacity can be matched by oxidative phosphorylation pathways that utilize a combination of glucose/glycogen and fatty acids as the energy substrates which ultimately provides reducing equivalents for the electron transport chain production of ATP at the mitochondria [25, 27].

It is also critically important to recognize that preferential energy pathway utilized and hence ATP rate/capacity resynthesis varies within muscle fiber type (Type I vs. Type II and subtypes) across the continuum of energy substrates available. Additionally, such fiber-type recruitment patterns (high force—power activities vs. low force—dynamic activities) can influence greatly the energy substrate and ATP resynthesis pathway predominating during the increased energy utilization of exercise (e.g., this point is highly germane to the present discussion—resistance vs. endurance exercise) [10, 13, 28–32].

The mobilization of non-intra-muscular energy substrates hepatic glucose/glycogen and adipose tissue fatty acids are principally under endocrine control (i.e., substrate mobilization. Conventionally this mobilization has been attributed to increased sympathetic activity as well as H-P-adrenomedullary-cortical activity, and supportive anterior pituitary actions (in response to exercise stimulus), thereby increasing hepatic glycogenolysis and adipose lipolysis [25, 31–37].

Metabolic Energetics Regulation: Evolving Views

As implied previously, skeletal muscle fibers have a complex system to modulate their metabolism based on nutrient availability and ATP needs. Evidence points out that a key regulatory player in this system is the AMP-activated protein kinase (AMPK) protein system. That is, AMPK has the capacity to monitor the amount of energy available in a cell by actions involving directly binding to adenine nucleotides. In brief, through changes in energy availability, and the ATP to ADP or ATP to AMP ratio, AMPK can be activated by an allosteric mechanism that stimulates its kinase actions/activity. Subsequently, AMPK redirects metabolism towards decreased anabolism and increased catabolism through the phosphorylation of essential proteins in a multitude of pathways, including mTOR complex 1 (mTORC1; *N.B.*, also known as mammalian target of rapamycin complex 1—integrates signals from multiple growth factors, nutrients, and energy supply to promote cell growth when energy is sufficient), lipid homeostasis, glycolytic, and mitochondrial respiration-related homeostasis. Additionally, AMPK directly regulates major enzymes involved in energy-related pathways. These critical regulatory actions of

AMPK to metabolic demands are also coupled with its impact on an ability to alter and adapt a muscle fiber's metabolic profile.

Relative to this last point, AMPK can target transcriptional regulators. That is the genes involved specifically in the regulation of gene expression through direct binding to cis-regulatory elements. These, in turn, can bring about adaptive changes in the metabolic status associated with exercise training program adaptations [25, 35–37].

Endocrine: Metabolism Interaction

Conclusive evidence supports the endocrine system (more correctly, the neuroendocrine) and metabolism are inseparably linked. To that end, Fig. 5.3 gives a generalized illustrative perspective on that linkage [38].

> The interaction of neuroendocrine (autonomic nervous system + endocrine glands) and the metabolic system is a key and critical element in the physiological adjustments to an exercise session.

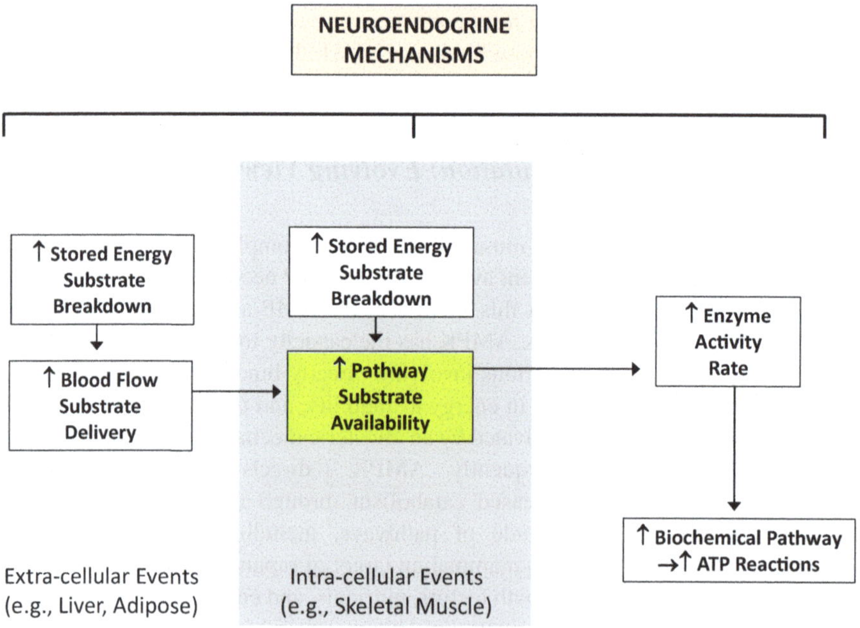

Fig. 5.3 The interaction of neuroendocrine (autonomic nervous system + hormonal glands) actions on metabolism during an exercise session

Table 5.2 Energy metabolism roles of selected hormones of the endocrine system on tissues and organs essential to exercise (*N.B.*, note only major actions a depicted herein)

Hormone (Source)	Target tissue	Principle processes influenced
Glucagon (Pancreas Alpha Cells)	Muscle Adipose Liver	↑ Glycogenolysis, Proteolysis (protein degraded) — ↑ Glycogenolysis, Gluconeogenesis, Proteolysis
Insulin (Pancreas Beta Cells)	Muscle Adipose Liver	↑ Glycogenesis, Protein synthesis, ↓ Lipolysis ↑ Lipogenesis, ↓ Lipolysis ↑ Glycogenesis
Epinephrine (Adrenal Medulla)	Muscle Adipose Liver	↑ Gluconeogenesis, Lipolysis ↑ Lipolysis ↑ Glycogenolysis
Norepinephrine (Adrenal Medulla + ANS)	Muscle Adipose Liver	↑ Glycogenolysis, Lipolysis ↑ Lipolysis ↑ Glycogenolysis
Growth Hormone (Anterior Pituitary)	Muscle Adipose Liver	↑ Protein synthesis, Lipolysis ↑ Lipolysis ↑ Gluconeogenesis
Cortisol (Adrenal Cortex)	Muscle Adipose Liver	↑ Lipolysis, Proteolysis ↑ Lipolysis ↑ Gluconeogenesis
Testosterone (Testes ♂, Adrenal Cortex ♀)	Muscle Adipose Liver	↑ Protein synthesis ↑ Lipolysis ↑ Protein synthesis, Glycogenesis
Estrogens (Ovaries ♀, Adrenal Cortex ♂)	Muscle Adipose Liver	↑ Lipolysis, Glycogenesis ↑ Lipolysis ↑ Protein synthesis
Progesterone (Ovaries ♀, Adrenal Cortex ♂)	Muscle Adipose Liver	↑ Glycogenesis ↑ Lipolysis ↑ Gluconeogenesis, Protein synthesis

ANS autonomic nervous system, ♀ female, ♂ male (principal hormonal source noted for each sex)

To aid in the linkage of these two systems, Table 5.2 lists the specific actions of a variety of hormones to influence and upregulate the actions of a variety of metabolic pathways which are associated with ATP production, substrate availability, and the other biochemical processes/pathways (e.g., protein synthesis) [39].

Conclusions

Exercise of any type is in the most classic perspective of physiology a "stress", and as such, provokes homeostatic disruptions in the *internal milieu* of the human body. Disruption herein, however, is not a negative consequence or outcome, but a biological signaling to the various physiological systems to induce accommodating changes to deal with the disruption. The endocrine and metabolic systems undergo such accommodating changes proportional to the exercise load experienced and display tremendous flexibility and capacity to respond appropriately to demanding exercise loads of various degrees and modes of activity.

In this chapter, the intent was to provide the reader with the basic fundamentals of exercise endocrinology as related to energy metabolism; i.e., in response to two highly diverse exercise modalities/activities, endurance versus resistance exercise. Hopefully, the reader recognizes that while invoking differing physiological demands, these exercise activities have a great deal of commonality in their endocrine-metabolic profile responses. That said, the nature of the training differences in select physiological systems, muscle fiber recruitment patterns, as well as the plasticity of select aspects of body composition changes (…etc.) result in dramatically different physique and performance characteristics (typically) between endurance and resistance-focused athletes undergoing training (see Table 5.3). It must be noted, however, that due to space limitations, one potentially key element not addressed herein is the sex differentiation of responses. In women (females) of reproductive age who are eumenorrheic or on hormonal contraceptives, the changes in circulating estrogens and progesterone do potentially influence many factors and aspects of the hormonal and metabolic responses to exercise; although, relative to the last point consensus does not exist in the scientific community [40–42].

Table 5.3 Endurance versus resistance exercise activities are classified based upon the level of intensity of effort and the predominating exercise pathway for adenosine triphosphate (ATP) production

Category term	Effort perception	Relative intensity	Energy pathway predominating	Representative duration (min)	Other terminology
Endurance exercise activities					
Light exercise	Easy	<35% VO_{2max}	Aerobic	>30	Short-term submaximal
Moderate exercise	Modest difficulty	<70 >35% VO_{2max}	Aerobic	30–180+	Submaximal, prolonged
Heavy exercise	Difficult	<100 >70% VO_{2max}	Anaerobic-aerobic	≤120	Submaximal prolonged, high-intensity
Maximal exercise	Strenuous	100% VO_{2max}	Anaerobic-aerobic	<15	Maximal, max, high-intensity
Supramaximal exercise	Extremely strenuous	>100% VO_{2max}	Anaerobic	<1	All-out
Resistance exercise activities					
Light exercise	Easy	<35% 1-RM	Anaerobic-aerobic	>5	Submaximal
Submaximal exercise	Modest difficulty	<70 >35% 1-RM	Anaerobic-aerobic	<5	Submaximal
Maximal exercise	Strenuous	~100% 1-RM	Anaerobic	<0.2	All-out, max
Supramaximal exercise	Extremely strenuous	>100% 1-RM	Anaerobic	≤0.1	Negatives

Other terminology sometimes utilized to describe the activities is also listed. VO_{2max} maximal oxygen uptake, *1-RM* one repetition maximum

Researchers and coaches need to consider this point when working with female athletes. This topic is worthy of an extensive and lengthy monograph of its own, as such the reader is directed to the review article by Oosthuyse et al. to gain some insights into the complexity of this issue [41].

1. Physical exercise is a stress to the human body evoking physiological changes to accommodate the stressor, with the neuroendocrine system being the most reactive to stress
2. Critical changes in the endocrine system enable the body to adjust, to accommodate, and to adapt to the physiological needs of the exercise, thereby reducing the amount of hormone needed to bring about the physiological outcome/changes desired for the exercise
3. Exercise has an effect on the production of hormone-like substances ultimately influencing the processes that they regulate
4. Hormonal Exercise Response Model (HERM) focuses on the hormonal changes allowing for the individual to adjust to and mediate the homeostatic disruption of the exercise. It alludes to the interrelated and interdependence of the endocrine glands of the hormone system, and consists of three interactive phases; (1) immediate response (2) intermediate response, (3) prolonged response
5. Metabolically exercise places great demands on the ability of the body to utilize and in turn produce energy (ATP) within the skeletal muscle
6. Intensity and duration of an exercise session are responsible for which biochemical pathways are utilized (i.e., whether energy/use generation is dependent on ATP sources, or anaerobic- versus aerobic biochemical pathways).

References

1. Bunt JC. Hormonal alterations due to exercise. Sports Med. 1986;3(33):331–45.
2. Hackney AC. Exercise as a stressor to the neuroendocrine system. Medicina. 2006;42(10):788–97.
3. Galbo H. The hormonal response to exercise. Diabetes Metab Rev. 1986;1(4):385–408.
4. Urhausen A, Gabriel H, Kindermann W. Blood hormones as markers of training stress and overtraining. Sports Med. 1995;20:251–76.
5. Garber CE, Blissmer B, Deschenes MR, et al. American College of Sports Medicine (ACSM) position stand: quantity and quality of exercise for developing and maintaining cardiorespiratory, musculoskeletal, and neuromotor fitness in apparently healthy adults: guidance for prescribing exercise. Med Sci Sports Exerc. 2011;43(7):334–1359.
6. Joyner MJ. Physiologic limiting factors and distance running: influence of gender and age on record performances. Exerc Sport Sci Rev. 1993;21:103–33.
7. Joyner MJ, Coyle EF. Endurance exercise performance: the physiology of champions. J Physiol. 2008;586(1):35–44.

8. Rowell LB. Cardiovascular adjustments to thermal stress. In: Lindsey M, Hester R, editors. Handbook of physiology, the cardiovascular system, peripheral circulation and organ blood flow (supplement 8). New York: Wiley-Blackwell; 1983. p. 967–1023.
9. Saltin B, Astrand PO. Maximal oxygen uptake in athletes. J Appl Physiol. 1967;23:353–8.
10. Kraemer W, Fleck S. Optimizing strength training: designing nonlinear periodization workouts. Champaign: Human Kinetics Publishing; 2007.
11. Galbo H. Hormonal and metabolic adaptation to exercise. Stuttgart: Thieme Verlag; 1983.
12. Hackney AC. Stress and the neuroendocrine system: the role of exercise as a stressor and modifier of stress. Expert Rev Endocrinol Metab. 2006;1(6):783–92.
13. Viru A, Viru M. Biochemical monitoring of sport training. Champaign: Human Kinetics; 2001.
14. Viru A. Hormonal ensemble in exercise—hormones in muscular activity, vol. 1. Boca Raton: CRC Press; 1985.
15. Christensen NJ, Galbo H. Sympathetic nervous activity during exercise. Annu Rev Physiol. 1983;45:139–53. https://doi.org/10.1146/annurev.ph.45.030183.001035. PMID: 6342511.
16. Kjaer M, Secher NH, Galbo H. Physical stress and catecholamine release. Bailliere Clin Endocrinol Metab. 1987;1(2):279–98.
17. Viru A. Plasma hormones and physical exercise. Int J Sports Med. 1992;13(3):201–9.
18. Davies CT, Few JD. Effects of exercise on adrenocortical function. J Appl Physiol. 1973;35(6):887–91.
19. Steenberg A, Fischer CP, Keller C, Møller K, Pedersen BK. IL-6 enhances plasma IL-1ra, IL-10, and cortisol in humans. Am J Physiol Endocrinol Metab. 2003;285(2):E433–7.
20. Radomski MW, Cross M, Buguet A. Exercise-induced hyperthermia and hormonal responses to exercise. Can J Physiol Pharmacol. 1998;76:547–52.
21. Hackney AC, Viru A. Research methodology: issues with endocrinological measurements in exercise science and sport medicine. J Athl Train. 2008;43(6):631–9.
22. Hackney AC, Lane AR. Exercise and the regulation of endocrine hormones. Prog Mol Biol Transl Sci. 2015;135:293–311.
23. Goodman HM. Endocrinology concepts for medical students. Adv Physiol Educ. 2001;25:213–24.
24. Widmaier EP. Metabolic feedback in mammalian endocrine systems. Horm Metab Res. 1992;24:147–53.
25. Ball D. Metabolic and endocrine response to exercise: sympathoadrenal integration with skeletal muscle. J Endocrinol. 2015;224(2):R79–95.
26. Turner D, Hoppler H. Muscle energetic: aerobic strategies. In: Whipp BJ, Sargeant AJ, editors. Physiological determinants of exercise tolerance in humans. London: Portland Press; 1999. p. 29–49.
27. Hawley JA, Maughan RJ, Hargreaves M. Exercise metabolism: historical perspective. Cell Metab. 2015;22(1):12–7.
28. Greenhaff PL, Nevill ME, Soderland K, Bodin K, Boobis LH, Williams C, Hultman E. The metabolic responses of human type I and II muscle fibres during maximal treadmill sprinting. J Physiol. 1994;478:149–55.
29. Romijn JA, Coyle EF, Sidossis LS, Gastadelli A, Horowitz JA, Endert E, Wolfe RR. Regulation of endogenous fat and carbohydrate metabolism in relation to exercise intensity and duration. Am J Physiol Endocrinol Metab. 1993;265:E380–91.
30. Ahlborg G, Felig P, Hagenfeldt L, Hendler R, Wahren J. Substrate turnover during prolonged exercise in man: splanchnic and leg metabolism of glucose, free fatty acids and amino acids. J Clin Invest. 1974;53:1080–90.
31. Wahren J, Efendi S, Luft R, Hagenfeldt L, Bjorkman O, Felig P. Influence of somatostatin on splanchnic glucose metabolism in postabsorptive and 60-hour fasted humans. J Clin Invest. 1977;59:299–307.
32. Ahlborg G, Felig P. Lactate and glucose exchange across the forearm, legs, and splanchnic bed during and after prolonged leg exercise. J Clin Invest. 1982;69:45–54.

33. Van Loon LJ, Greenhaff PL, Constantin-Teodosiu D, Saris WH, Wagenmakers AJ. The effect of increasing exercise intensity on muscle fuel utilization in humans. J Physiol. 2001;536:295–304.
34. Stephens FB, Constantin-Teodosiu D, Greenhaff PL. New insights concerning the role of carnitine in the regulation of fuel metabolism in skeletal muscle. J Physiol. 2007;581:431–44.
35. Herzig S, Shaw R. AMPK: guardian of metabolism and mitochondrial homeostasis. Nat Rev Mol Cell Biol. 2018;19:121–35.
36. Egan B, Zierath JR. Exercise metabolism and the molecular regulation of skeletal muscle adaptation. Cell Metab. 2013;17:162–84.
37. Egan B, Sharples AP. Molecular response to acute exercise and their relevance for adaptation in skeletal muscle to exercise training. Physiol Rev. 2023;103(3):2057–170.
38. Hackney AC. Exercise, sport, and bioanalytical chemistry: principles and practice. New York: Elsevier Publishing; 2016.
39. Hackney AC, Kraemer WJ, Hooper DR. Hormonal changes associated with physical activity and exercise training. In: Vaamonde D, du Plessis SS, Agarwal A, editors. Exercise and human reproduction: induced fertility disorders and possible therapies. New York: Springer Publishing; 2016.
40. Elliott-Sale KJ, Minahan CL, de Jonge XAKJ, Ackerman KE, Sipilä S, Constantini NW, Lebrun CM, Hackney AC. Methodological considerations for studies in sport and exercise science with women as participants: a working guide for standards of practice for research on women. Sports Med. 2021;51(5):843–61. https://doi.org/10.1007/s40279-021-01435-8.
41. Oosthuyse T, Strauss JA, Hackney AC. Understanding the female athlete: molecular mechanisms underpinning menstrual phase differences in exercise metabolism. Eur J Appl Physiol. 2022;123:423–50. https://doi.org/10.1007/s00421-022-05090-3.
42. Hackney AC. Menstrual cycle hormonal changes and energy substrate metabolism in exercising women: a perspective. Int J Environ Res Public Health. 2021;18(19):10024. https://doi.org/10.3390/ijerph181910024.

Chapter 6
Microvascular Blood Flow in Skeletal Muscle in Health and Injury

Coral L. Murrant

Introduction

Blood flow to skeletal muscle can change from 0.24 L/min at rest to approximately 6 L/min under maximal exercise conditions in an effort to match blood flow with the heightened metabolic demand of contracting muscle [1]. This approximate 25-fold increase in blood flow is required to ensure adequate substrate and nutrient delivery and waste product removal to support muscle performance. Skeletal muscle blood flow is tightly matched to metabolic demand of the muscle whereby blood flow has been shown to be linearly related to many metrics of metabolism such as oxygen consumption [2], contraction strength [3, 4], muscular work [5] and running speed [6]. The change in blood flow in response to different metabolic rates are the result of the integration of multiple central and local processes that influence vascular function to allow blood flow to be delivered to active skeletal muscle fibres and the feedback required to maintain adequate blood flow delivery over time [7]. Much of the extensive integration between the central nervous system and local control of blood flow is to ensure that the local vasodilation needed to increase flow to satisfy the tissues needs does not interfere or outstrip the body's need to maintain blood pressure. Central processes increase flow to active muscles by increasing driving pressure, which occurs through an increase in cardiac output (achieved via increases in both heart rate and stroke volume), as well as a redistribution of blood flow from less active areas of the body to more active areas.

Although the central cardiovascular changes are crucial to the delivery of an adequate blood flow to the active muscles, the central nervous system influence over

C. L. Murrant (✉)
Department of Human Health and Nutritional Sciences, University of Guelph, Guelph, ON, Canada
e-mail: cmurrant@uoguelph.ca

N. C. Apostolopoulos et al. (eds.), *Fundamentals of Recovery, Regeneration, and Adaptation to Exercise Stress: An Integrated Approach*,
https://doi.org/10.1007/978-3-031-44270-4_6

the local muscle microvasculature, mostly through the sympathetic nervous system and related hormonal responses to exercise, are in place to limit blood flow to the tissues to preserve mean arterial pressure. However, there are processes local to the active muscles that help reduce the central impact (i.e., functional sympatholysis—for review, see [8]).

Processes that control the vasculature locally, i.e., to, and within, the active tissue, are in place to ensure that blood flow to the active tissue is adequate and distributed to the capillaries that feed the active skeletal muscle fibres. Interestingly, the linear relationship between blood flow and metabolic demand has been demonstrated in preparations where the nerve to the contracting muscle has been cut, severing the connection between the contracting skeletal muscle and the central nervous system influences [2]. This result shows that local mechanisms involved in the control of the microvasculature are able to coordinate blood flow to contracting muscle in a way that matches blood flow to metabolic demand in the absence of central nervous system influences and exposes the critical nature of localized vasodilatory processes in ensuring adequate blood flow to active skeletal muscle fibres.

Matching blood flow to the metabolic demand of a tissue such as skeletal muscle where metabolic rate can vary so widely is a complicated task. This task is made more difficult by the potential for the active skeletal muscle fibres to be heterogeneously spread throughout the muscle, amongst muscle fibres that are not contracting. Skeletal muscle cells are stimulated in motor units. A motor unit is comprised of an alpha motor neuron and all of the muscle fibres that the motor neuron innervates. The number of muscle fibres that each alpha motor neuron innervates can vary from 10 to greater than 1000 muscle fibres depending on the muscle [9] and the muscle fibres within each motor unit is dispersed throughout the muscle (for example, see [10–16]). Therefore, muscle fibres that reside beside each other within a muscle may not be in the same motor unit, which means that if one motor unit is active and another is not, a muscle fibre may be contracting beside another muscle fibre that is not active. Motor units are recruited individually depending on the amount of force required for the movement [9], and therefore, not all motor units are active during a given contraction. It has been estimated that the generation of 40–70% of maximal force will require the recruitment of all motor units, depending on the muscle [17]. This means that in activities that require forces below 40–70% of maximal, not all motor units will be active, and therefore, not all skeletal muscle fibres within the muscle will be contracting and those that are actively contracting will be "intermingled" with muscle fibres of motor units that are not active. In terms of metabolism, this means that while the metabolism of one muscle fibre is increasing, that of its neighbor will not be changing. The task of the microvasculature is to direct blood flow to capillaries feeding the active muscle fibres dispersed throughout the tissue, and not direct blood flow to the capillaries of all the muscle fibres within the tissue [18, 19].

Therefore, the vasculature network must be anatomically structured and physiologically controlled, in a way that when muscle fibres are activated, the vasculature is able to intercommunicate within the vascular network in order to be able to direct blood flow to the active skeletal muscle fibres that are heterogeneously distributed within the tissue.

> Skeletal muscle blood flow is tightly matched to metabolic demand, increasing from 240 ml/min at rest to over 6000 ml/min during heavy effort—a 25-fold increase! Matching blood flow to the metabolic demand of a tissue is a complicated task. The microvascular network must be anatomically structured and physiologically controlled to be able to direct blood flow principally to the heterogeneously distributed active skeletal muscle fibres.

Microvascular Network Structure

Microvascular networks have been described in many different muscles, from gluteus maximus [20], temporalis muscle [21], tibialis anterior [22, 23], extensor hallicus proprius [24], cremaster [18, 25–27], spinotrapezius [28], using vascular casting techniques or intravital microscopy to understand the microvascular architecture. And while the architecture of the network may vary depending on the architecture of the muscle, there are common elements found among these vascular beds. In general, an artery enters the muscle, paired with a vein. Within the muscle the artery and arteriolar architecture will remain roughly paired with the venous vasculature as the arterioles branch into smaller diameter daughter vessels (Fig. 6.1a). At some point in the network, the arteriolar microvasculature will branch in a way to separate itself from the branch order of the venous microvasculature. This unpaired arteriolar vasculature will continue to branch into smaller branches until a terminal arteriole is reached, the last branch of the arteriolar tree. The final ramification of the arteriolar tree gives rise to a group of 2–20 capillaries called a capillary unit [22, 23], capillary bundle [28], or capillary module [18]. Capillaries in a unit merge to converge into venules, and venules converge into larger veins. As the venular vessels converge they will begin to follow the architecture of the arterioles resulting in the observed paired architecture between larger veins and arteries.

> Each terminal arteriole gives rise to 2–20 capillaries. Each capillary has a fixed diameter of ≈5–6 μm, and consists of only a single endothelial cell layer, an optimal design for diffusion into and out of the muscle cells. Flow to capillaries is controlled by the terminal arteriole that gives rise to the capillaries in the unit, and is regulated by a balance between neural activity and local chemical mediators.

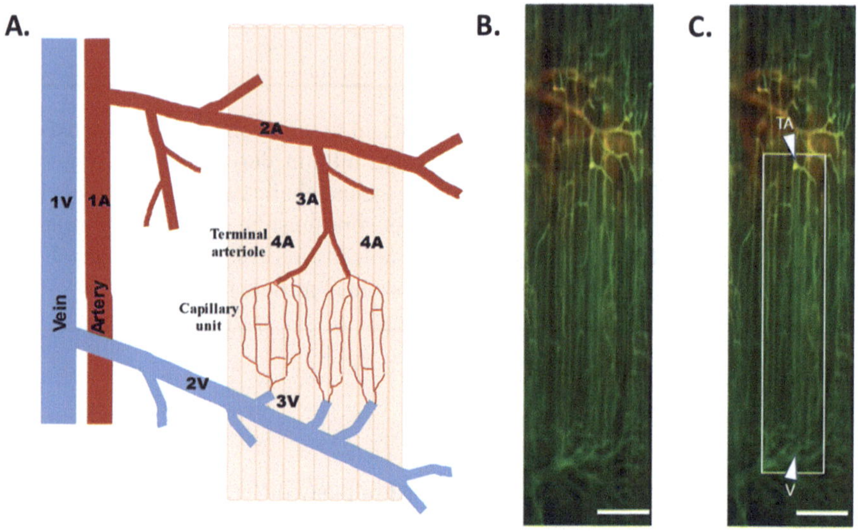

Fig. 6.1 The architecture of the microvascular network in skeletal muscle. (**a**) A representation of part of the microvasculature network in skeletal muscle showing branch orders of the arteriolar (red) and venous (blue) side of the vasculature and where the capillary units are situated in relation to the rest of the microvascular network and in relation to skeletal muscle cells. Only 2 capillary units and a few skeletal muscle fibres are depicted for clarity. This image highlights that capillaries are uniquely situated in close proximity to each skeletal muscle cell and uniquely situated to connect the arteriolar and venous vascular network. Schematics are not drawn to scale. (Schematic is a modified redrawing from Murrant and Fletcher [29]). (**b**) A fluorescent image of mouse diaphragm terminal arteriole, capillary units, and venules. This image is an overlay of two fluorescent images of the same microvasculature site where the microvasculature endothelial cells have been dual labeled: (1) labeled with wheat germ agglutinin (WGA) conjugated with Alexa Fluor 488 (Green) which will show all of the components of the microvasculature and (2) Isolectin GS-IB_4 conjugated to Alexa Fluor 568 (Red) which labels arteriolar endothelial cells and arteriolar capillaries preferentially when perfused. Where the two labels overlap display yellow. This allows for visualization and identification of the arteriolar end of the capillary unit (yellow) and the venular end of the capillary unit (green) and the capillary unit itself in between. Capillaries are running parallel to the skeletal muscle cells. This image highlights the reality of the density of the capillaries and the complexity of capillary modules that could not be conveyed in the schematic in **a**. (**c**) Shows the same capillary unit as in **b** but with a white box highlighting a capillary unit from a terminal arteriole (TA arrowhead) to the venule (V arrowhead) Scale bar = 100 μm. (This image was generated and generously donated by Mackenzie Charter and Barbara Hyde-Lay)

Arterioles and venules within the microvasculature consist of an endothelial cell layer circumferentially wrapped with smooth muscle cells, which control changes in diameter depending on the contractile state of the vascular smooth muscle cells. Capillaries, however, do not have a smooth muscle layer and therefore, cannot actively change diameter. Capillaries, as they consist of only a single endothelial cell layer attached to a basement membrane, represent a vessel optimized for diffusion and fluid flux between the blood compartment and the interstitial space surrounding skeletal muscle cells.

Because the diameter of the capillaries cannot be actively changed, the terminal arteriole holds a special place of control in terms of capillary perfusion. Given that each terminal arteriole gives rise to a capillary unit containing 2–20 individual capillaries, vasodilation of the terminal arteriole will result in the perfusion of all of the capillaries in the capillary unit it gives rise to. Individual capillaries within a unit cannot increase blood flow independently of other capillaries within the same capillary unit [30]. Precapillary sphincters that can control blood flow in individual capillaries have been observed in the gut but are not a common feature in the skeletal muscle microvasculature [31]. Therefore, the vasodilatory state of the terminal arteriole will dictate the perfusion of the entire capillary unit.

Capillaries are aligned parallel to the long axis of the skeletal muscle fibres, with small connecting segments. Since each fibre is much longer than a capillary, each muscle fibre will require several capillaries along its length to supply the fibre. In cross section, we see 3–6 capillaries supplying each fibre, depending on fibre type and training state.

Capillary units are considered the functional unit for skeletal muscle perfusion (Fig. 6.1b). Capillary units tend to align themselves parallel to the long axis of the skeletal muscle fibres with small branching, connecting capillaries crossing over, perpendicular to muscle fibres, connecting long capillary segments to each other. Capillaries vary in length depending on the definition of the capillary. If the capillary is considered to span the distance between the terminal arteriole and the venule, then capillaries are approximately 800 μm in length (range 500–1700 μm) [23, 24, 32–35]. There is a conserved architectural relationship between capillaries and skeletal muscle fibres. Each skeletal muscle cell is surrounded by 3–6 capillaries depending on the muscle fibre type and training state of the muscle (for example see [36–38]), and these capillaries will not necessarily be from the same capillary unit. Since skeletal muscle fibres are much longer than a capillary unit (for review, see [39]), each muscle fibre will require several capillary units along its length to supply the complete muscle fibre. For example, Emerson and Segal [40] estimated that more than 50 capillary units are required to feed the length of a muscle fibre in the hamster retractor muscle. Recently, it has been proposed that a series of capillary units are interconnected to form continuous columns that align with muscle fascicles, implying that capillary fascicles represent a potential higher order organizational structure for capillaries [41]. The functional implications of this arrangement are yet to be determined.

As capillaries are tethered to skeletal muscle cells [42], the longitudinal shape of capillaries can change dramatically depending on the contractile state of the muscle. During contraction, the shortening of the fibres also shortens the capillaries, but as they do not contract, the capillaries become more tortuous, with the degree of tortuosity being dependent on the degree of shortening (i.e., during muscle contraction) or stretch (i.e., during muscle lengthening) [43]. When skeletal muscle fibres contract, sarcomere length becomes very short (i.e., 1.6 μm), and the capillaries become extremely tortuous or serpentine along the length of the muscle. As muscle fibres lengthen, and sarcomere lengths get longer (i.e., 2.2 μm), the capillaries become less tortuous, i.e., straighter [43, 44]. When muscles are stretched even further, the capillaries can become straight and even stretched [43, 45–47], possibly reducing luminal diameter and affecting blood flow through the capillary [45, 47, 48].

Cells of the vasculature are interconnected to each other through gap junctions, which allow for longitudinal intercommunication between capillaries and arterioles and along the length of arterioles into different levels of the arteriolar vasculature. With this connectivity in place, the vasculature can respond rapidly to vasodilatory and vasoconstricting signals along the length of a vessel and into connected parent vessels.

Vascular Network Interconnectivity

Cells of the vasculature are interconnected to each other through gap junctions, which allow for longitudinal intercommunication between capillaries and arterioles and along the length of arterioles into different levels of the arteriolar vasculature (Fig. 6.2). Gap junctions between endothelial cells, between vascular smooth muscle cells, and between endothelial cells and vascular smooth muscle cells, connect each cell of the vasculature to each other, and provide the mechanism by which blood vessels can move signals longitudinally along the length of a vessel and between different levels of the vasculature, called conducted responses (for review, see [49, 50]). Gap junctions between cells are formed by a cluster of 6 connexin proteins on one cell (a hemichannel) that binds with a hemichannel on a neighboring cell to create a connection between the cytosols of each cell. The connexin proteins that make up the hemichannels and the gap junctions are comprised of different isoforms. Connexin isoform complements differ between tissues. The connexin isoforms common to the vasculature of skeletal muscle are Cx37, Cx40, Cx43 and Cx45 (for review, see [49, 50]). The gap junction connections between cells allows ions and small molecules to move from one cell to the other without moving through the extracellular environment. With this connectivity in place, the vasculature can be stimulated in one region and generate a signal that can move along the vessel wall to stimulate other connected parts of the vascular network. As the lowest resistance for transmission of a conducted response

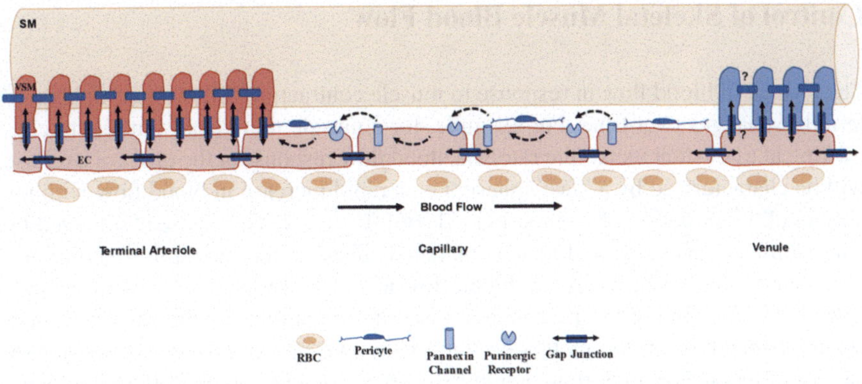

Fig. 6.2 Schematic of the relationship between the cells of the vasculature. and skeletal muscle cells and red blood cells. A magnified schematic of a single wall of the microvasculature depicting cells in the terminal arteriole, capillary and venule portion of the microvasculature and associated skeletal muscle fibres (SM) and red blood cells. The interrelation between endothelial cells (EC), vascular smooth muscle cells (VSM), skeletal muscle fibres and pericytes are depicted. This schematic highlights that capillary endothelial cells can communicate with upstream arteriolar network through gap junction connections and a pannexin and purinergic dependent paracrine signaling pathway. Whether gap junctions exist between VSM on venules and between EC's and VSM of venules is unknown. Schematics are not drawn to scale. (Schematic is a modified redrawing from Murrant and Fletcher [29])

is between endothelial cells, it is thought that the primary connections used for longitudinal movement of signals through the microvascular wall is via endothelial cell connections. Conducted electrical signaling between endothelial cells has been shown to rapidly move vasodilatory (hyperpolarization) and vasoconstricting (depolarization) signals along the length of a vessel and into connected parent vessels [51, 52]. Slower and shorter conducting of calcium waves have also been observed between endothelial cells [53, 54]. Conducted responses originating at the arteriolar level can move in both directions, moving into upstream and downstream arterioles [54–58]. More recently capillary endothelial cells have been shown to be able to move signals longitudinally to vasodilate upstream arterioles using a pannexin/purinergic dependent pathway [59]. This would represent a second pathway that capillaries would have in order to communicate with upstream arterioles that control their perfusion. Endothelial cells are also connected to vascular smooth muscle cells, called myoendothelial gap junctions, to facilitate the movement of vasoactive signals from endothelial cells to vascular smooth muscle cells directly [60].

> Blood flow in response to muscle contraction follows a tri-phasic pattern—a rapid increase in blood flow when contraction begins, which is followed by a transition to "ballpark" level, which is then fine-tuned to a new steady state level, where the blood flow is matched to metabolic demand.

Control of Skeletal Muscle Blood Flow

The change in blood flow in response to muscle contraction follows a complex pattern that changes over time. For example, for a muscle that goes from rest to contracting at a constant metabolic rate, the blood flow response at the onset of exercise will be characterized by a "fast" phase, i.e., a rapid increase in blood flow within a second of the initiation of contraction [61–66] (Fig. 6.3). This phase is followed by a transition or "seeking" and then a "matched" phase where blood flow settles at a new steady state level, where the blood flow is matched to metabolic demand [67] (Fig. 6.3). The timing of these phases and the magnitude of the responses will change depending on the metabolic rate of the muscle. To produce this complex pattern of blood flow over time, a number of processes, involving different components contribute to the response, i.e., there will be mechanisms to increase blood flow that has a large impact early, but not later, in any given contraction bout; similarly, there are processes that are important later in a contraction bout, but not as critical early. In addition, there may be processes that are important depending on the metabolic rate of the muscle, the amount of force being generated, and the number of motor units recruited. There are processes that will be primary drivers of changes in blood flow at a given time and processes that are secondary modifiers of blood flow. Unfortunately, at this time, it is not well understood which processes are critical and when they might be primary or secondary drivers of blood flow control.

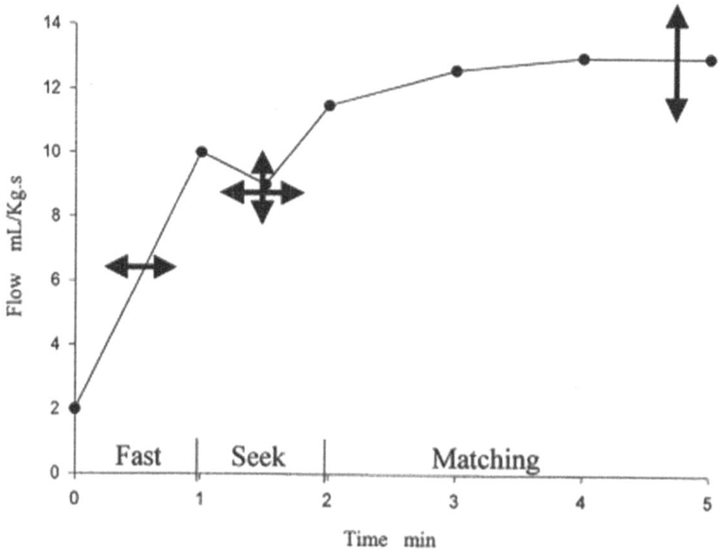

Fig. 6.3 The pattern of change in blood flow in response to muscle contraction. Muscle begins to contract at time 0 and the resulting blood flow response proceeds temporally through three phases (1) a fast phase, a rapid increase in flow at the onset of contraction, (2) a seeking phase as the blood flow begins to change to a new level that leads to phase 3, (3) matched phase, where blood flow tends to be matched to metabolic demand. (Graphic from Barclay et al. [67]). Arrows indicate the variability of each of the phases depending on the metabolic rate of the muscle

Local Processes That Can Change Blood Flow to Contracting Skeletal Muscle

Pressure-Induced Changes in Flow

The physical nature of shortening muscle, generates high intramuscular force and increases intramuscular pressure in proportion to contraction force [68, 69]. This increase in intramuscular pressure, coupled with a venous system that has one-way valves, can produce changes in flow. The intramuscular pressure during contraction will be translated to the vasculature within the tissue, resulting in an increased extravascular pressure and compression of vascular tissue within the skeletal muscle. Intramuscular pressures generated during maximal contractions can reach values over 550 mmHg pressure [69], which can cause the cessation of blood flow altogether for the duration of the contraction [64, 70, 71]. This physical compression of the venous vasculature, with their one-way valves, forces blood out of the venous system and, upon relaxation, will result in a reduced venous pressure that will help drive an increase in blood flow. This mechanism, known as the muscle pump, could help drive very early changes in flow [65, 72, 73].

External Compression

External pressure on the vascular wall will compress endothelial cells within the vessels causing endothelial cell deformation. The subsequent increase in wall shear stress of the endothelium has been shown to produce endothelial cell hyperpolarization [74] and vasodilation [75–77]. The extracellular matrix-dependent vasodilation [78] may be at the core of deformation-based vasodilation in response to the physicality of muscle contraction. The circumferential deformation of the skeletal muscle intra-muscle arteries produces nitric oxide dependent vasodilation that involves integrins as part of the mechanotransduction of the endothelial-dependent vasodilation [79]. Given that endothelial cell deformation would occur at the onset of contraction, vasodilation in response to endothelial cell deformation would be a process that could contribute to changes in blood flow early in the contraction bout.

Rapid changes in transmural pressure through external mechanical compression of the isolated arteries results in rapid vasodilation that was reduced by endothelial cell removal—but not abolished, indicating that while there is an endothelial cell dependent component of the response, there is also a vascular smooth muscle component of this response [80]. Therefore, deformation of the vascular smooth muscle directly contributes to the compression-induced vasodilation. Mechanically-induced compression resulted in vasodilation in humans *in vivo* as well [81], but when compared to contraction-induced compression, could not account completely for all the observed vasodilation. Therefore, vasodilation resulting from external compression of arterioles may account for some, but not all, of the early vasodilatory processes.

Products of Skeletal Muscle Excitation

Vasodilators, generated as a result of excitation of muscle cells, can be produced/released quickly, even before the increase of muscle metabolism begins, e.g., acetylcholine from the neuromuscular junction and potassium released from skeletal muscle due to the action potentials opening of voltage dependent potassium channels on the sarcolemma of the muscle. Nitric oxide also fits into this category. Nitric oxide is released from skeletal muscle at rest and increases during contraction, presumably resulting from the increase in intracellular calcium initiated by the processes of excitation-contraction coupling and the consequent stimulation of nitric oxide synthase [82, 83]. Adenosine can also fit into this category as it can be formed on the outside of the muscle fibres by ecto-5′-nucleotidase from the ATP, ADP and AMP released from the active skeletal muscle fibres [84–86]. The release of these products is separate from "metabolic vasodilators" which are products released from skeletal muscle as byproducts of metabolism. It has been demonstrated in models where vascular pressure and compression have been minimized, that these types of vasodilators are involved in the rapid vasodilation associated with muscle contraction. Acetylcholine has also been shown to be involved in the rapid vasodilation in response to muscle contraction [87, 88]. Potassium and adenosine were implicated as being involved in the earliest phase of vasodilation [89, 90]. These studies support the idea that muscle stimulation-derived products can contribute to the very early hyperaemic responses to exercise.

Rapid biphasic vasodilation has been shown to result from a single contraction, initiating vasodilation that started after 1 s and peaked at 4–10 s, plus a second vasodilation that peaked at 20–40 s, depending on the stimulus frequency that the muscle was stimulated with [87, 88]. The separate temporal peaks imply multiple vasodilator processes in response to a single contraction.

Movement-Induced Hyperaemia

Early hyperemia can occur in response to passive movement of a limb, in the absence of active muscle contraction. Movement-induced hyperemia as a result of passive limb movement has been shown to result in an increase in flow to the moving muscle within 4 s from the start of the movement and returned to control baseline levels 45 s later while the limb was still being moved [91]. An early and transient hyperemia is thought to result from multiple factors, including a contribution from the central cardiovascular changes (i.e., a heart rate-driven increase in cardiac output) [92], the result of mechanoreceptor feedback, peripheral effects due to mechanical compression [80, 81], and a nitric oxide-dependent component [93, 94] that may be part of the flow-mediated vasodilatory response [95]. Movement-induced hyperemia does not appear to be the result of a diffusible factor as a mechanism of communication between skeletal muscle cells and the vasculature. There have been reports that the extracellular matrix may be involved in signaling between skeletal muscle and the vasculature [78], but skeletal muscle fiber movement, in the absence of contraction, does not induce vasodilation or an increase in capillary perfusion to the moving muscle fiber [18]. Therefore, passive movement-induced hyperaemia may play a small, transient role in the early active hyperaemic process.

Flow-Mediated Vasodilation

Any phenomena that initially changes blood flow at the onset of contraction, whether it be compression-induced vasodilation, the muscle pump, or diffusible vasodilators, will change flow and result in an elevated wall shear stress secondary to the initial change in flow. The increased wall shear stress will then result in the induction of a flow-mediated vasodilation [96, 97]. Flow-mediated vasodilation in small arterioles is characterized by being slower to manifest (onset being approx. 8–20 s) and is mediated by endothelial cell derived nitric oxide and prostaglandins [98–100]. Conducted responses induced by muscle contraction have been shown to have a faster component that causes arteriolar vasodilation compared to the slower vasodilation that is dependent on the local endothelial cell production of nitric-oxide in response to luminal shear stress [52]. Thus, flow-mediated vasodilation in skeletal muscle microvasculature appears to be a secondary mechanism for vasodilation, secondary to primary contraction-induced vasodilation. Given that flow-mediated vasodilation is secondary to a primary change in flow, and is slower to manifest, it is not likely that flow-mediated vasodilation contributes to the early, fast phase of active hyperaemia.

Reduced Sympathetic Nervous System Vasoconstriction

Skeletal muscle contraction will result in a vasodilation of the tissues that have increased activity. If enough tissue becomes active at a given time, a large number of muscle vascular beds vasodilate. This vasodilation could act to reduce total peripheral resistance within the body, which could act to decrease mean arterial pressure. Muscle contraction is associate with an increased sympathetic nervous system vasoconstrictor output to the body [101, 102], which in turn, would act to increase total peripheral resistance and maintain mean arterial pressure. This vasoconstriction will also include the vascular bed of the active skeletal muscle fibres. The active tissue must now work to overcome this centrally-induced vasoconstriction by acting to reduce the impact of the sympathetic nervous system through local control mechanisms. Contraction-mediated inhibition of sympathetic vasoconstriction in the active muscle (termed functional sympatholysis) is thought to be mediated by nitric oxide release from skeletal muscle cells or endothelial cells or ATP from red blood cells, which acts pre-synaptically to inhibit norepinephrine release or post-synaptically to reduce the effectiveness of the adrenergic receptors [103–106]. Regardless of the mechanism(s), the overall effect is that release of local mediators reduces the effectiveness of the constriction, and thereby contributes to an overall vasodilation. The local inhibition of vasoconstriction will reduce the resistance to flow in the active tissue and help to increase blood flow to the contracting muscle. This local functional sympatholysis coupled with the sympathetic nervous system vasoconstriction to other tissues that are not active, will allow for blood volume in the body to be redistributed from inactive tissues of the body to active regions. Given that nitric oxide is produced by stimulated skeletal muscle, and that ATP can be present as soon as the oxygen gradient changes, the decrease in the sympathetic nervous system vasoconstriction resulting from these products, could happen very early in the hyperaemic response.

> Local factors involved in fine-tuning flow involve various metabolic mediators from muscle, such as acetylcholine, nitric oxide, adenosine, potassium, low oxygen levels, increased carbon dioxide, lactic acid, hydrogen, AMP, and mediators from other sources, such as nitric oxide, prostaglandins, and endothelial derived relaxing factor.

Vasodilators Derived from Skeletal Muscle Metabolism (Metabolic Vasodilators)

The onset of muscle contraction is associated with rapid, early changes in blood flow. If contraction persists after this initial fast phase, there will be an adjustment in blood flow, i.e., a seeking or interim phase, aimed at establishing a new blood flow level—one that is matched to the metabolic demand i.e., the matched phase. Given the linear relationship between blood flow and metabolic demand in skeletal muscle of this phase, much attention has been given to identify potential vasodilators resulting from skeletal muscle metabolism, that might establish this relationship. Vasodilatory products produced by skeletal muscle stimulation, products released as the result of muscle metabolic pathways, and/or products that exhibit a change in concentration with increases or decreases in the metabolic rate of the muscle, would support the linear relationship between muscle blood flow and metabolic rate in the matched phase.

Classic "metabolic vasodilators" are those resulting from increased metabolism and the byproducts from metabolic pathways required to generate the ATP needed for contraction, such as decreased oxygen, increased local carbon dioxide, lactate, lactic acid,, hydrogen, AMP, adenosine, and even generalized changes in osmolarity in the extracellular space as a result of skeletal muscle release of products, have all been investigated for their role in active hyperaemia. There are also products that skeletal muscle cells may release during contraction, such as prostaglandins and

ATP, that will result in vasodilation (for reviews, see [7, 107–109]). Each of these substances that are released relative to metabolic rate could result in blood flow changes that occur in proportion to skeletal muscle metabolic demand.

Given their relationship to metabolic rate, and the time involved in the generation and release of these substances (depending on the vasodilator), any, or all, would be ideal candidate(s) for generating, and maintaining, the matched phase of the active hyperaemic response.

Vasodilators Derived from Other Sources

Vasodilators from sources other than contracting skeletal muscle have been proposed to contribute to the active hyperaemic response. For example, substances released from the endothelium during muscle contraction. These processes can include nitric oxide, prostaglandins, and endothelial derived relaxing factor, possibly as a result of changes in wall shear stress. A further complication in assessing the role of nitric oxide is that nitric oxide is a component of the vasodilator response to adenosine, ATP, and prostaglandins, so whether nitric oxide is playing an independent role of its own, or is its role secondary to the action of one or more of the primary vasodilators, is not well understood [107].

The role of endothelial derived products in the active hyperaemic process is not clear. If products from endothelial cells play a primary role in the matched phase of the active the hyperaemic response—where blood flow is related to skeletal muscle metabolic demand, then endothelial cells would have to "know/sense" what the metabolic demand of the skeletal muscle cells is at any given time. For this to work, endothelial cell production of a vasodilator in relation to skeletal muscle metabolic rate would require a connection or communication between skeletal muscle cells and endothelial cells, and also a connection between skeletal muscle metabolism and endothelial cell release of vasodilatory products. At this point, whether endothelial cells are a primary source of vasodilators during exercise hyperemia or act as a modifier of the blood flow response is unclear.

Red blood cells may be a potential source of vasodilators in response to muscle contraction. Red blood cells release nitric oxide upon deoxygenation [110, 111] and release ATP under hypoxic conditions and when they are deformed (as would occur when moving through a capillary [112–116]). Oxygen consumption at the onset of muscle contraction is translated to a decrease in capillary PO_2 very quickly, within seconds [117]; therefore, the release of ATP and nitric oxide from the red blood cells could happen rapidly at the onset of muscle contraction, and may contribute to the early active hyperaemic response. Deoxygenation of the red blood cells would also be related to the oxygen consumption of the muscle over time, and therefore its metabolic rate over time. Therefore, red blood cell-derived products could contribute to the matched phase of the active hyperaemic response as well.

Paired Venular-Arteriolar Communication

Paired blood vessels may have unique vascular influences compared to unpaired arterioles. For example, vasoactive metabolites can diffuse from venules to neighbouring, paired arterioles to change their diameter [118–121]. These data suggest that, during an increase in metabolic rate of the skeletal muscle fibres, the release of metabolic byproducts can spill over into the capillary blood flow and then be carried into the post-capillary venules and veins. These vasoactive metabolites can then diffuse into paired arterioles to cause vasodilation and alter blood flow. In addition, Koller et al. demonstrated that an increase in shear stress stimulated the release of endothelium-derived prostaglandins [122] and venular prostaglandins have been shown to influence paired arteriolar diameter during active hyperaemia [123, 124]. This process would be secondary to the production of metabolic vasodilators and require the build-up of products that would result in significant spill-over into the blood stream. Therefore, this process would not likely contribute to the fast phase of active hyperaemia.

Redundancy

Multiple vasodilators and multiple processes have been identified and investigated for decades, but the identity of the primary vasodilator product or most important process critical to the development of active hyperaemia has not been identified. This inability to identify the critical vasodilatory product or process has led to the idea that there must be a more integrated, complex redundant control system involving several vasodilators and processes [125–128]. At the core of the concept of redundancy is the idea that increases in blood flow to match metabolic demand are so critical to our survival that we have evolved redundant mechanisms to ensure the blood flow is always matched to metabolic demand. Interactions between vasodilators have been observed, whereby it has been suggested that inhibition of one vasodilator experimentally will result in the upregulation of another vasodilator to ensure the matched vasodilation persists, for example, the relationship between prostaglandins and nitric oxide [129–132]. Further, a mechanism for redundancy has been identified, whereby the presence of one vasodilator inhibits the effects of a second. In this way, when the first vasodilator is removed, the effects of the second become expressed [133].

The Role of Pericytes in Bloodflow Control

Pericytes are present at the arteriolar and capillary level in skeletal muscle, but their function is not well understood (for review, see [134]). Pericytes reside abluminally within the blood vessel wall sharing a basement membrane with underlying endothelial cells. In the brain, three types of pericytes have been identified, ensheathing pericytes associated with arteriolar regions that wrap circumferentially around the vessel, mesh pericytes in the transition zone between arterioles and capillaries with processes that extend along the vessel wall, and thin-strand pericytes on capillaries with thin processes that extend along the longitudinal axis of the capillary wall [135]. Thin-strand pericytes and endothelial cells have been shown to interface through peg-and-socket connections [136].

In skeletal muscle, the pericyte population is less defined. Pericytes have been found on arterioles, collecting venules, and veins (for review, see [134]). These pericytes have finger-like projections that extend along the longitudinal axis of the vessels as well as a possible second population with secondary projections wrapping circumferentially around the vessel itself (similar in morphology to ensheathing pericytes). Capillaries also have a population of pericytes that appear similar to thin-strand pericytes that extend along the capillaries and make peg-and socket-like connections with the endothelial cells [137, 138].

The percent coverage of capillaries by pericytes has been estimated at 17–26% [139, 140], with a more recent report indicating what appears to be more extensive coverage [141]. It has been suggested that pericytes on capillaries may be contractile in nature [142]. Although changing the diameter of a small region of a capillary will not result in significant changes in capillary perfusion, these data raise the possibility that pericytes may be involved in blood flow control in skeletal muscle [141]. Much more work is required to determine the structural and functional relationship between capillaries and pericytes, and the relationship between skeletal muscle fibres and pericytes to determine whether pericytes play a role in skeletal muscle blood flow control.

Local Processes Needed to Direct Blood Flow to Active Skeletal Muscle Fibres and Match Blood Flow to Metabolic Demand

Muscle blood flow varies considerably, depending on how the muscle is contracting, and the metabolic state of the muscle. Based on the above-mentioned processes, vasodilation in response to muscle contraction is the result of the action of multiple products and several processes that will differ over time and change in terms of their contribution to increases in flow.

Although there are many potential vasodilatory processes, they are not the only processes required to ensure that blood flow is matched to metabolic demand. As mentioned previously, if only a small amount of force is being generated by a muscle, only a few motor units are being stimulated and the muscle fibres that are contracting will be dispersed throughout the whole muscle tissue, intermingled with muscle cells that are not contracting. Some of the processes that can increase flow cannot direct blood flow to specific muscle fibres (i.e., the muscle pump, wall shear stress, endothelial cell depression due to vascular compression, venous-arteriolar communication, SNS inhibition, movement-induced hyperaemia). These processes will promote an increase in flow throughout the whole muscle.

There are processes, however, that could help in directing blood flow to the active skeletal muscle fibres; these processes involve the action of products released from the active skeletal muscle itself (e.g., products of skeletal muscle excitation and/or metabolic vasodilators), or vasodilators from neighboring cells, and other sources (e.g., red blood cells). Since these products are derived in the proximity of the active muscle fibres, they could also be involved in communication with the local vasculature to direct blood to the active regions. This communication pathway would require that cells localized to the change in metabolism have the ability to communicate with the vasculature in that region, and then the vasculature signals the network using conducted responses to coordinate the vasodilation of specific arterioles in order to initiate an increase in perfusion of the capillary units that feed the active skeletal muscle fibres. Given the structural relationship between the microvasculature and skeletal muscle cells, each contracting muscle fibre would need to communicate its needs to many terminal arterioles, and in turn, the terminal arterioles would have to trigger an upstream dilation of the parent arterioles along their length in order to reduce resistance and increase flow to the many capillaries units that feed the contracting muscle fibre. This would have to happen for every active muscle fibre in that motor unit. Therefore, to increase the perfusion of the capillary units would require the vasodilation of many terminal arterioles and the initiation of conducted responses to coordinate the network response. In the example given earlier, where Emerson and Segal estimated that 50 capillary units were required to feed the length of a muscle fibre in the hamster retractor muscle [40], this would equate to needing to coordinate the vasodilation of 50 terminal arterioles. To vasodilate these terminal arterioles directly would require that the contracting skeletal muscle cells be in close proximity to each of these arterioles so they could communicate their metabolic need. Yet there is no conserved architectural relationship between arterioles and muscle fibres. There is, however, a conserved architectural relationship between capillaries and skeletal muscle fibres [143].

Historically, because of the need to vasodilate the resistance arterioles in the microvasculature in order to increase blood flow in response to muscle contraction, there has been a focus on communication between contracting skeletal muscle cells and arterioles directly. However, there is now a growing body of evidence to suggest that the vascular target for signaling may not be arterioles directly but rather it may be capillaries. Looking at muscle cross section, one sees that each skeletal muscle fibre is supplied by 3–5 capillaries that are approximately 2 μm from each fibre [144, 145]. This structural relationship would allow capillaries to be rapidly influenced by products released from contracting skeletal muscle. Skeletal muscle fiber contraction has been shown to stimulate associated capillaries [18, 146, 147] and result in an increased red blood cell flux in the stimulated capillaries [18]. Capillaries can also be stimulated to cause upstream vasodilation in response to pharmacologically applied vasoactive substances, i.e., the substances that have been investigated as communicating molecules in the active hyperemic response, e.g., reduced oxygen [148, 149], adenosine [59, 146, 147], and adenosine analog adenosine-50-N-ethyluronamide [150, 151], potassium [59, 147, 150, 152, 153], nitric oxide [59, 146, 147], ATP-dependent potassium channel opener pinacidil [59, 146, 154], ATP [155], and possibly hydrogen ions [150], although a response has not been demonstrated in all cases [156]. Once stimulated, capillaries can communicate to upstream terminal arterioles through gap junction connections [18, 59, 151, 157], as well as via a pannexin/purinergic signaling pathway [59] to promote vasodilation and increase their own perfusion.

Further, not only are capillary-induced vasodilatory signals transmitted upstream to their connected terminal arteriole, but the signal is also conducted to stimulate vasodilation of arterioles upstream of the terminal arteriole, i.e., to very specific arterioles—and not to arterioles associated with inactive muscle fibres [18, 154] (Fig. 6.4). This specific pathway for transmitting an upstream conducted response will result in perfusion of the stimulated capillaries and not perfusion of all capillaries within the region.

In summary, capillaries are ideally located to receive signals from the vicinity of contracting skeletal muscle fibres and are mechanistically equipped to communicate to upstream arterioles to control perfusion across the stimulated capillary unit. This makes capillaries the ideal level of the vasculature for active skeletal muscle cells to communicate with in order to differentially increase blood flow directly to active skeletal muscle fibres and not to inactive muscle fibres as well.

> Effective repair and regeneration of muscle following injury requires integration of various processes including inflammation/destruction, regeneration, and remodeling. The inflammation/ destruction phase occurs within 24–72 h post injury and involves cytokine and myokine release, and activation of satellite cells. The regeneration phase is characterized by phagocytosis of necrotized tissue, myocyte regeneration from stimulated satellite cells, and capillary ingrowth into the area. The final stage—remodeling, is marked by continued protein synthesis and maturation of the myocyte, and myocyte activation by an alpha motor neuron.

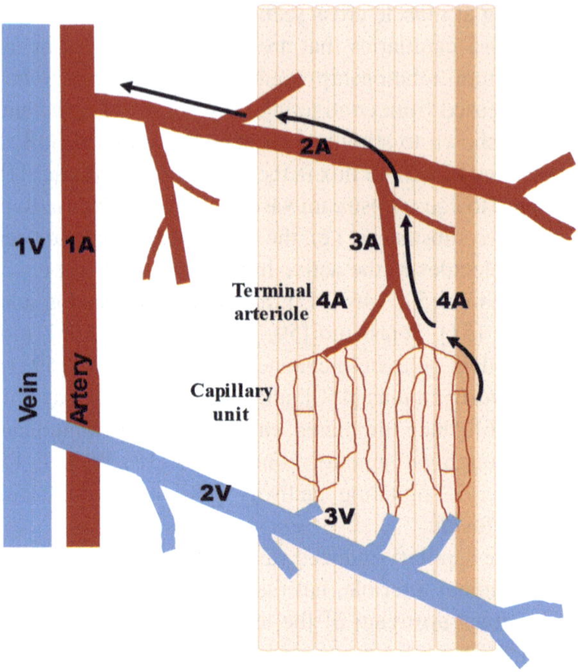

Fig. 6.4 The architecture of the microvascular network in skeletal muscle. A representation of part of the microvasculature network in skeletal muscle showing branch orders of the arteriolar (red) and venous (blue) side of the vasculature and where the capillary units are situated in relation to the rest of the microvascular network and in relation to skeletal muscle cells. Only 2 capillary units and a few skeletal muscle fibres are depicted for clarity. This image shows the vascular response to stimulating capillaries with contracting skeletal muscle. Of the many skeletal muscle cells depicted (orange) if the darker orange muscle fibre was the only muscle fibre contracting in this region, it would stimulate capillary endothelial cells closely associated with it. This stimulation would initiate a signal that would travel upstream to dilate the connected terminal arteriole. If the signal were stronger and traveled further upstream it would cause vasodilation of the upstream 3A arteriole and 2A arteriole in the direction of the 1A arteriole. This path of vasodilation would create a path of least resistance for blood flow to be directed to the active skeletal muscle fibres. Importantly, when the conducted vasodilation arrives in the 3A arteriole it does not spread to vasodilate all 4A arterioles branching from the 3A. This allows for blood flow to be directed very specifically to the stimulated capillary unit and the active skeletal muscle fibres. Schematics are not drawn to scale. (Schematic is a modified redrawing from Murrant and Fletcher [29])

Types of Skeletal Muscle Injury and Damage

Muscle Injury/Damage

Muscle injury is common during daily activities, through sport and as a result of accidents and trauma. Intensive, unaccustomed, and eccentric exercise can lead to structural damage and cellular disruption of myocytes in skeletal muscle. Strains for example, from sprinting and jumping can generate excessive tensile force that leads to a rupture of the interface of myocytes with tendinous connective tissue, the myotendinous junctions [158] and require repair and regeneration of muscle tissue. Sport and industrial work can result in blunt force, high-energy impact, and crush injuries to muscle. Muscle injuries can be caused by freezing, contusion, or laceration. Contusion results from a sudden, heavy compressive force, such as a direct blow to the muscle as in contact sports, vehicular and industrial workplace accidents [159]. Each of these types of injuries will have a range of severity of injury and a range of severity of impact on muscle cells themselves and, therefore, a range of responses for repair and regeneration.

Although the steps in the healing process will depend on the severity of the injury, there are common steps in the process of recovery from injury. Following injury, the common stages of healing include destruction/inflammation, regeneration, and remodeling. Injury to the muscle will involve a loss of calcium homeostasis and a disruption of force generation.

The destruction/inflammation stage occurs within 24–72 h post injury and involves cytokine and myokine release, and activation of satellite cells. There is monocyte rupture, necrosis of tissue, and edema at this stage. Delayed onset muscle soreness from contraction related injuries, a common effect of eccentric exercise, will begin in this early phase and may last for 5–7 days post injury.

The regeneration phase is characterized by the phagocytosis of necrotized tissue, myocyte regeneration from stimulated satellite cells, and capillary ingrowth into the area. The final stage—remodeling, is marked by continuous protein synthesis and maturation of the myocyte, and myocyte activation from the alpha motor neuron (for review, see [158, 160–162]).

Experimentally, muscle damage and recovery have been investigated through contraction-induced muscle damage models and models where skeletal muscle fibres are injured using chemical or physical methods. Across the range of damage models, the regenerative process appears similar, varying with the amount of damage accrued and the amount of muscle and tissue generation required. Contraction-induced damage causes more minor damage and a smaller numbers of fibres undergo necrosis. Intense exercise can produce damage to a larger number of fibres and result in structural damage. Crushing, freezing, and chemical damage (induced by cardiotoxin, notexin or barium chloride) can cause extensive damage, and significant areas of necrosis of nearly all muscle fibres at the site of the injury [162].

Contraction-Induced Muscle Injury Models

Given that it is well documented that eccentric (muscle lengthening during contraction) results in muscle damage, the majority of studies investigating muscle injury and repair have used eccentric exercise to induce muscle damage. Common markers for the amount of damage induced by the eccentric exercise bout in human subjects are the amount of delayed-onset muscle soreness, measurement of blood protein indicators of skeletal muscle membrane disruption (e.g., creatine kinase or lactate dehydrogenase), and the amount of decrease in contractile force following the eccentric exercise.

Following this damage, there are cellular and molecular processes required for myocyte regeneration after injury (for review, see [163, 164]), the involvement of different processes will depend on the type of exercise undertaken, intramuscular forces generated, and the severity of the damage incurred. For example, in eccentric exercise-induced muscle damage, maximum damage appears to occur 1–3 days following the exercise and is characterized by the presence of necrotic and regenerating muscle fibres, disruption of the sarcolemma and myofibrils, leakage of intracellular proteins into the blood stream, swollen mitochondria, stimulation and release of pro-inflammatory cytokines, leukocyte infiltration and edema, reduced force generation, and muscle soreness. The immune reaction is followed by the promotion of myofibril regeneration and the activation of satellite cells (for review, see [165]). Given the extensive changes occurring within damaged skeletal muscle, it is highly likely that other cells within the muscle will also be affected, such as cells of the vasculature.

Damage to the vasculature may occur concurrent with myocyte damage during the exercise or may be secondary to muscle injury and occur during the processes of myocyte repair or both. Vascular damage will impair delivery of O_2 and energy substrate to the muscle at rest and during subsequent exercise bouts. It is also possible that muscle regeneration is hampered by blood flow dysregulation. Along with myocyte repair, there will need to be angiogenic processes required to repair the vasculature in order to restore proper perfusion of the regenerated myocyte.

Destructive Models of Muscle Injury

Destructive injury usually results in a more severe injury to the muscle than contraction-induced injury and can be induced in multiple ways, usually either through physical injury (for example freeze or crush injury) or through the use of toxic chemical application. Each of these models will induce injury in a different way, and therefore, will result in different regeneration and repair mechanisms—although the general phases of recovery are similar to those outlined above. A freeze injury is usually invoked by freezing a small part of the muscle, thereby damaging a focal point with undamaged tissue surrounding the damaged region. Freezing can induce a greater strength loss and a slower recovery when compared to contraction-induced injury [166]. Crush injury can cause greater damage to the muscle than contraction-induced muscle damage. Muscle force can be reduced by 80% [167], compared to a 20–40% reduction in force induced by eccentric contraction-induced injury models [168–172]. Muscle damage can also be induced chemically by exposure of a number of chemicals, such as barium chloride, and muscle toxins, e.g., notexin and cardiotoxin. Cardiotoxin, a group of homologous proteins found in cobra snake venom, causes damage by inducing depolarization and contraction of muscle fibres, and disrupts cell membranes. Notexin, a toxin produced by the tiger snake, disrupts membranes, produces the loss of ion gradients, and produces contraction of muscle fibres [163]. The mechanism by which barium chloride induces muscle fibre necrosis is not well understood, but through its effects on potassium channels, it may elicit a prolonged depolarization and a contracture of the muscle fibres [173]. Toxins will produce more widespread muscle damage whereas barium chloride induces a milder injury [163].

Regardless of how skeletal muscle is damaged, the damage will also extend to other cells of the tissue, including the vasculature. While there has been longstanding interest in the processes of skeletal muscle myocyte regeneration, there has been much less focus on the vascular response to muscle damage and the restoration of the microvasculature in either contraction-induced or destructive muscle injury. Effective regeneration and repair of the skeletal muscle will require integration of the processes involved in both myogenesis and angiogenesis.

Changes in Skeletal Muscle Blood Flow Following Muscle Damage

Changes in Blood Flow Resulting from Exercise-Induced Muscle Damage

Most studies show no changes in resting blood flow to the injured muscle at early time points after exercise-induced muscle damage (i.e., 1 h post-injury [172]) or in the days following the injury (1–6 days [168, 171, 172, 174]). However, Sbriccoli et al., using Doppler ultrasound, have reported an increase in blood flow in the injured muscle immediately following eccentric exercise-induced injury that normalized by day 7 [170]. This technique estimated blood flow within the tissue rather than bulk flow to the tissue and may represent the dysfunctional intra-muscular flow that is occurring (discussed below). This study used eccentric exercise that produced more damage than the other studies discussed, and therefore, there may be a different flow response depending on the amount of damage done to the muscle.

Contraction-induced blood flow was decreased on day 2 post exercise induced muscle injury [171, 174], indicating an impaired perfusion of the eccentrically exercised muscle. On day 3 post-injury, Ahmadi et al., reported an increase in blood flow to the injured muscle during exercise, but the blood flow within the muscle was more homogenous, suggesting an altered distribution of blood flow within the muscle and suggestive of a dysfunctional microvasculature [168]. The change in contraction-induced blood flow was restored 6 days post injury [171].

The reduced blood flow to the contraction-induced injured muscle may be a product of a dysfunctional microvasculature. Intramuscular blood flow, monitored using ultrasound Doppler, was shown to be reduced by day 2 post-injury, indicating an impaired perfusion of the eccentrically exercised muscle [171]. Using intravital microscopy, Kano et al. observed that at rest and 1–3 days post injury, there was a significant reduction in the percentage of capillaries with sustained, continuous red blood cell flux, and capillary hematocrit was elevated on day 1 post-injury but reduced on day 3 [175]. These findings are consistent with microvascular dysfunction following eccentric exercise. The altered blood flow does not appear to affect the steady state microvascular PO_2 of the muscle [175] or peak oxygen extraction [176] during contractions 1–3 days post injury, but the kinetics of the fall in PO_2 at the onset of contraction was faster, suggesting that there may be an early compromise in the matching of oxygen delivery to oxygen uptake.

Contraction-induced vasodilation may be impacted by direct damage to the vasculature, or possibly due to the impact of the damage on the skeletal muscle itself. Eccentrically-induced muscle damage may cause perturbations in resting metabolic rate up to 48 h post-exercise [177]. Furthermore, long term effects of eccentric exercise on muscle metabolism, such as an elevation of the ratio between inorganic phosphate and phosphocreatine as measured by PNMR at rest have been observed for several days after eccentric exercise [178]. Further, Lieber and Friden observed that eccentric exercise induced more damage to fast-twitch fibres [179]. Subsequent bouts of contraction may require a different pattern of motor unit recruitment, or a greater recruitment of slow oxidative muscle fibres. The resulting switch in metabolism due to the change in fibre type recruitment may alter the metabolic products and vasoactive substances that are released during subsequent contractions. In turn, the changes in metabolic products may alter how skeletal muscle cells communicate with the vasculature which, ultimately, may affect the matching of blood flow with metabolic demand.

Changes in Blood Flow Resulting from Destruction-Induced Muscle Damage

There are very few studies that have investigated blood flow changes resulting from destructive muscle injury. Much of the work studying the vasculature and destructive injury has been at the microvascular level. These studies have shown destruction of the capillary network which could preclude blood flow in the damaged area. This is followed by a restoration of the vascular network on a timeline that differs depending on the amount of damage incurred. Jacobsen et al., using barium chloride to injure the muscle, observed fragmented capillaries and no perfusion within 24 h of the injury [180]. While perfusion of capillary networks was established by 5 days post injury, these early networks were disorganized and dilated, and resulted in disordered flow within the muscle. Restoration of normal looking capillary units was not established until day 21. Arpino et al. induced more severe damage by rendering the muscle ischemic [181]. They observed new capillaries emerging after day 3 post injury and capillary perfusion by day 10. The chaotic flow through the new vessels resulted in widespread hypoxia persisting until day 28 post injury and red blood cell flux remained low up to 120 days post injury.

Alterations in Vascular Structure and Function Control Following Skeletal Muscle Damage

Macrovascular Disruption Resulting from Muscle Damage

Increases in circulating levels of factors involved in inflammation and repair have been observed in response to eccentric exercise, including cytokines and myokines, such as IL-6 and TNF-alpha [165, 182, 183], reactive oxygen and nitrogen species [183], cortisol [184], and markers of muscle damage, such as creatine kinase and lactate dehydrogenase [185]. These changes in circulating substances have the ability to alter vascular function in areas distant and remote to the site of damage. For example, there have been reports of central vascular stiffness (carotid artery) following eccentric exercise muscle damage [169, 186]. Significant increases in brachial artery diameter were observed 1 and 48 h after exercise [172].

At the macrovascular level, eccentric exercise appears to cause a decrease in endothelial cell reactivity as demonstrated by a reduced flow-mediated dilation in large arteries associated with the exercising limb 1–96 h following the eccentric exercise [172, 176, 186]. Interestingly, some effects were observed on large arteries not associated with the contracting limb [187], although this has not been universally observed [176]. The reduced flow-mediated dilation is attributed to a reduced responsiveness of endothelial cells to changes in wall shear stress, which is a nitric oxide dependent pathway [95]. Stacy et al., observed a decreased nitroglycerin response of the brachial artery 1–96 h after eccentric exercise muscle damage, indicating a reduction in smooth muscle reactivity to nitric oxide [172]. Intriguingly, 4 days following crush injury, there was an increase in arterial blood flow that was nitric oxide dependent [188]. These data show that the effects on the microvasculature may be dependent on the injury incurred, and that the damage may include adversely affecting the nitric oxide system. Therefore, both large artery endothelial cells and smooth muscle cells may be affected by exercise-induced muscle damage, with the nitric oxide vasodilatory mechanisms being particularly sensitive to disruption. The idea that both endothelial and smooth muscle function are affected by the same products released from the injured muscle is noteworthy. Even more intriguing is the fact that the wide-ranging effects on the vasculature have been observed in areas distant from the site of muscle injury, which supports the idea that blood borne products could have vascular effects that reach far beyond the affected muscle.

> Since all components of the microvasculature lie within the damaged muscle, the arterioles, capillaries, and venules will be exposed to the damaging stimulus. The microvasculature will also be affected by the processes related to degeneration and regeneration of the injured skeletal fibres. Blood flow is decreased, and flow distribution to active fibres is altered—suggestive of a dysfunctional microvasculature. Therefore, an angiogenic process will be required to restore proper perfusion to support myocyte repair/regeneration.

Microvascular Disruption Resulting from Exercise-Induced Muscle Damage

Since all components of the microvasculature lie within the damaged muscle itself, the arterioles, capillaries, and venules will be exposed to the damaging stimulus as skeletal muscle cells degenerate and regenerate. Therefore, the microvasculature may be affected directly by the damaging stimulus, and also indirectly, by the processes related to degeneration and regeneration of the injured skeletal muscle fibres.

The microvasculature has also been shown to be disrupted during eccentric exercise-induced muscle damage. Large venules have been reported to have a reduced diameter [189] 48 h following eccentric exercise-induced muscle damage, but arterioles and smaller venular diameter was similar to controls and the capillary to fibre ratio was unchanged. Kano et al. reported that 1–3 days following exercise-induced muscle damage, collapsed or obstructed capillaries were not apparent, but capillary diameter was increased on day 3 following exercise-induced muscle damage. Capillary hematocrit was elevated on day 1, but was reduced on day 3, with a decrease in the percentage of capillaries with sustained or continuous blood flow [175]. Further, Kano et al. observed that microvascular PO_2 was not different during contractions subsequent to exercise-induced muscle damage, but PO_2 fell faster following exercise-induced muscle damage. Therefore, dynamic changes in capillary flow and altered microvascular PO_2 dynamics may indicate a compromised oxygen delivery/uptake system. Caldwell et al. observed no effects of exercise-induced muscle damage on muscle oxygen extraction during exercise bouts following exercise-induced muscle damage, suggesting that the extent to which changes to the microvasculature affected O_2 delivery were not readily observable [176]. Hotta et al. has reported a hyperpermeability of the microvasculature on days 1 and 3, which returned to control levels by day 7 [190]. While they found no differences in microvessel diameter, they reported a higher vessel tortuosity on day 3, but not on day 1 or 7, which they suggested may be related to the skeletal muscle structural changes given that capillary tortuosity is related to muscle length/morphology.

A post-occlusive reactive hyperemic response is commonly used to assess the responsiveness of the microvascular to changes in wall shear stress and is used as an indicator of endothelial cell function of the vasculature within the muscle. Rakobowchuk et al. reported no change in post-occlusive reactive hyperaemic response immediately following the eccentric exercise (5 and 40 min) [187] or 48 h post-exercise injury [176]. The reactive hyperaemic response has been shown to be dependent on inward-rectifier potassium channels with contributions from nitric oxide and prostaglandins [191, 192]; therefore, these data imply that neither of these processes appear to be affected by exercise-induced muscle damage or by the processes involved in muscle recovery. This is in stark contrast to the larger vessels where nitric oxide-dependent flow-mediated dilation was attenuated in response to exercise-induced muscle damage. The finding of no effect of exercise-induced muscle damage on endothelial cell wall shear stress responses in microvessels while observing an inhibition of the endothelial cell wall shear stress response in larger blood vessels, highlights potential differences in endothelial cell function between the macro- and the microvasculature.

Although the microvasculature showed little change in the peak magnitude of the active hyperaemic response in response to muscle contraction after the eccentric exercise induced muscle damage [193], the dynamics around reaching peak blood flow were longer and the blood flow recovery back to basal levels following contraction was prolonged [193]. These data suggest that, although steady-state blood flow appeared unaffected by exercise-induced muscle damage, the reactivity of the microvasculature may be impeded such that there may be a mismatch between blood flow and metabolic demand at the onset of contraction. Heap et al. showed that arterioles and venules dilate less in response to adenosine 48 h following exercise-induced muscle damage, indicating that the vasculature may not be as responsive to metabolic vasodilators following exercise-induced muscle damage [189]. This lack of vascular reactivity to adenosine following eccentric contraction could also lead to blood supply/demand imbalances during subsequent muscle contractions. Given that adenosine is a proposed metabolic vasodilator, subsequent muscle contractions that use adenosine to cause vasodilation during muscle contraction would result in a blunted vasodilatory response and reduced blood flow response during exercise. Yet, the active hyperemic response was not altered [193], meaning that either the reduced adenosine reactivity is not playing a large role in the blood flow response to the muscle stimulation parameters [194] or the redundant vasodilatory systems were able to compensate for the diminished vascular response to adenosine [133].

On the whole, after exercise-induced muscle damage, there is a 12–30% decrease in force production by an exercised muscle, and a significant disruption of the microvascular that appears limited to the response magnitude of the microvessels. Arteriolar vasculature responses may be reduced or slowed following exercise-induced muscle damage which may affect O_2 kinetics at the onset of exercise, and the changes in microvascular regulation may impede the ability of the vascular to rapidly adjust muscle blood flow both at the onset of exercise and during activities in which the metabolic needs change during the activity. While there is a reduction in sustained capillary flow through some capillaries, these changes do not seem to significantly alter the active hyperaemic response in matching blood flow to metabolic demand. Future work should assess the implications of the increased heterogeneity of red blood cell flux through the capillaries on oxygen delivery to contracting muscle following exercise-induced muscle damage to determine if there is a mismatch between oxygen delivery and muscle oxygen utilization.

> Direct examination of the capillary network 24–48 h post injury reveals fragmented capillaries. While perfusion of capillary networks is re-established 5–6 days post injury, restoration of "normal" looking capillary units was not established until day 21; however, widespread hypoxia persists until day 28 post injury, and red blood cell flux remains low up to 120 days post injury.

After 21 days, the capillary network begins to resemble uninjured muscle vasculature, the capillary unit structure is restored. By 28 days post-injury, the network resembles an uninjured, control network.

Microvascular Disruption Resulting from Destructive Muscle Injury

For the most part, a damaging injury to muscle results in an initial destruction of the capillary network within 18–24 h post-injury. Capillaries are fragmented and perfusion is abolished [180, 195]. Whether the destruction of the capillary network is directly due to the agent used to cause muscle injury or secondary to the unstable environment produced by damaged myocytes is unknown and may well vary depending on the method used to induce injury. It has been shown that barium chloride injury does not directly cause microvascular cell death, and therefore, damage to the vascular network is a result of the adverse environment created by the injury to the muscle fibres themselves [180]. Leukocyte recruitment may be part of the environment created by the damage that may contribute to damage of the vascular network. Within 12–24 h post injury, there is maximal leukocyte activation and recruitment to the damaged area and an increase in reactive oxygen species and pro-inflammatory cytokines [162]. Neutrophil depletion prior to barium chloride damage afforded partial protection from capillary damage as more capillaries remained perfused; however, neutrophil depletion did not prevent capillary damage from occurring as extensive capillary fragmentation still occurred [180]. Thus, leukocyte activity early in the injury process may contribute to capillary damage but is not the primary driver behind early capillary fragmentation. Crush injury showed a reduction in capillary density, an increased capillary diameter, and an increased capillary permeability that were delayed from the initial injury. These delayed effects have been used to suggest that capillary changes were not caused directly by the initial injury but occurred in response to skeletal muscle fibre damage [196]. Therefore, the extent of the initial vascular damage that occurs concurrent with the skeletal muscle injury is difficult to ascertain, there is strong evidence to support that the microvascular injury is occurring secondarily to the skeletal muscle damage.

Following this destructive phase, angiogenesis and budding of capillaries from existing capillaries begins 4–12 days post-injury, resulting in a vascular network that is partially restored and allowing for perfusion to be restored [181, 195, 197]. The new capillary network presents as disorganized, dense, and more tortuous than usual, presenting a chaotic network with very little hierarchy of arterioles, caps, and venules [163, 181, 197]. There is no capillary unit organization along myocytes [180]. Capillary density often exceeds normal capillary density [163, 180, 181], but not in all types of injury [163]. See Fig. 6.6 for the time course of recovery and the differences in capillary network as a result of different methods of muscle damage.

After 21 days, the capillary network begins to resemble uninjured muscle vasculature, the capillary unit structure is restored. By 28 days post-injury, the network resembles an uninjured, control network. The increased capillarization can persist for up to 6 months depending on the type of agent used to invoke the injury [163, 181].

Although the regenerated vascular network may appear structurally similar to an uninjured network, the functionality may not return to normal. Arpino et al., using an ischemic obliteration model of muscle damage, showed that the regenerated vascular network has a similar timeline of regeneration, with similar morphological changes during restoration [181]. Functionally, however, they observed a slower red blood cell transit through the new capillaries, with an increased heterogeneity in transit time, a lower red blood cell O_2 saturation, and the muscle remained hypoxic. These abnormalities remained for up to 120 days post-injury. They suggest that the new vascular networks are dysfunctional, resulting in ineffective oxygen delivery. Therefore, functionality of the new vascular networks resulting from destructive injuries may require up to 4 months to restore functionality.

Interestingly, while the capillary network is fragmented and perfusion abolished within 18 h of barium chloride-induced injury, the arterioles and venules were not damaged, and the vascular smooth muscle covering these vessels was not fragmented [180]. However, while morphologically they appeared normal, functionally they were not. At 5 days post injury, arterioles have an attenuated vasodilation in response to acetylcholine (endothelial dependent vasodilator) and a reduced vasoconstrictor response to phenylephrine (endothelial independent vasoconstrictor). Further, there is an impaired vasodilatory response to both twitch and tetanic contractions. Full responsiveness to acetylcholine and phenylephrine occurred within 21 and 35 days. Therefore, although the structure of the arteriolar network did not appear to be altered by muscle injury, the function of the blood vessels was adversely affected by the recovery environment, which took over a month to recover.

When the damage to the muscle is severe, as occurs with total muscle ischemia, the reformed arterioles are part of a chaotic network with little blood vessel hierarchy by day 14 post-injury [181]. By day 28, the network more closely resembles a control network, but the arterioles run more parallel to the muscle fibres, daughter branches are smaller in diameter, and trifurcations and quadrifications rather than just bifurcations are observed more often. Interestingly, not all arterioles terminate in capillaries, some lead directly to veins, and there are more arterio-venous connections. In addition, there is aberrant vascular smooth muscle wrapping of arterioles, where the middle of some arterioles is devoid of vascular smooth muscle. These differences can be observed as long as 120 days post-injury [181].

Crush injury results in a severe form of muscle damage, and as such, causes more extensive microvascular damage and follows a different timeline. The early microvascular response to a crush injury results in arteriolar and venular vasodilation, with a decreased flow velocity within 6 h and continues for at least 3 days post injury [198]. Microthrombi were observed in the microvasculature, and it was suggested that erythrocyte aggregation could decrease oxygen transport at rest and during subsequent exercise bouts [198]. There was a reduced capillary density and an increased capillary diameter 1.5–72 h post injury [167, 196] that persisted to day 4 post injury [188]. These crush-induced microvascular responses differ from those seen with chemical or toxin-induced damage. The observed differences highlight that the extent and type of damage to the vascular network and the time course of recovery from that damage will vary considerably depending on how the skeletal muscle was damaged initially.

Role of the Microcirculation in Recovery of Muscle from Injury

Given the importance of the microvasculature in maintaining proper blood flow to the muscle to maintain tissue health, it follows that the microvasculature will be critical in muscle cell recovery from injury. We do know that vascular injury appears to happen secondarily to muscle injury, and therefore, the recovery of the two may not happen in concert in the early days post injury. The expectation would be that the myogenic processes must happen in concert with angiogenic processes in order to restore proper delivery of oxygen and nutrient and removal of waste products as quickly and efficiently as possible, which will enhance the recovery of the muscle fibres themselves.

The research would suggest that the capillaries may be the vascular level most affected during muscle damage, whether it be through mild effects of altered endothelial cell function occurring with exercise-induced muscle damage, or the complete fragmentation that occurs with more destructive types of damage. Damaging capillary endothelial cell function will disrupt the coordination of the vascular network to respond to changes in skeletal muscle cell requirements, and result in a mismatch between supply and demand. If capillaries are the critical interface between skeletal muscle function and vascular coordination, then skeletal muscle will lose its ability to communicate with the vascular network when capillaries are not functioning correctly. This in turn, will result in a lack of coordination between the needs of the muscle and the supply from the vasculature. When capillaries are fragmented, all communications between skeletal muscle and the vascular network are lost.

With exercise-induced muscle damage, the vascular network, while affected, is not fragmented as with destructive-types of damage. Therefore, the capillary communication system remains interconnected and functioning to help deliver nutrients and remove waste. However, disruptions in PO_2 appear possible, therefore, subsequent contraction bouts may not experience proper matching of supply to metabolic demand in a timely fashion, which means that muscle fibres may operate under more hypoxic conditions. Critically, since blood flow remains intact, the microvasculature can help in the processes of recovery from muscle damage.

The fragmentation of the capillary network resulting from freezing and chemical injury (Figs. 6.5 and 6.6) would eliminate vascular network control that would have originated at the capillary level, which would disrupt oxygen and nutrient delivery.

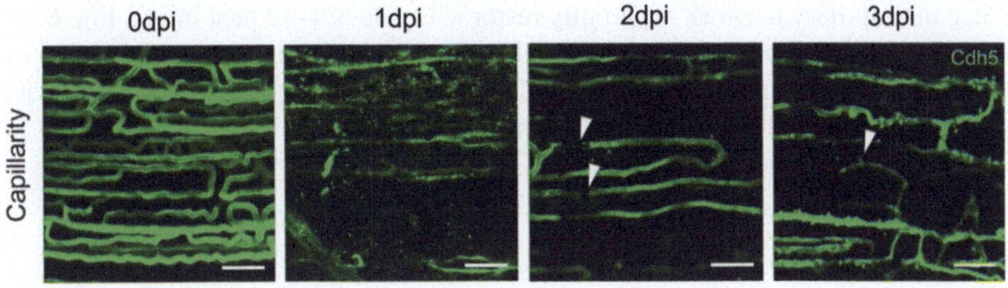

Fig. 6.5 Confocal microscopy images of capillaries in extensor digitorum longus of male mice under control conditions (0 days post injury (dpi)) and 1-, 2-, and 3-dpi injured using BaCl. Capillaries (in green; Green Fluorescent Protein labelled endothelial cells) in an uninjured muscle showing an orderly capillary network where the capillaries run in parallel to the skeletal muscle fibres. One day following BaCl injury resulted in fragmented and disrupted capillaries. By day 2 and 3 new capillary formation is observed. Arrowheads on 2-days post injury indicate fragmented and disrupted capillaries. Arrowheads on 3-days post injury indicate new vessel growth. Scale bar = 50 μm. (This image is modified from Morton et al. [173] with permissions through the Open Access Creative Commons license)

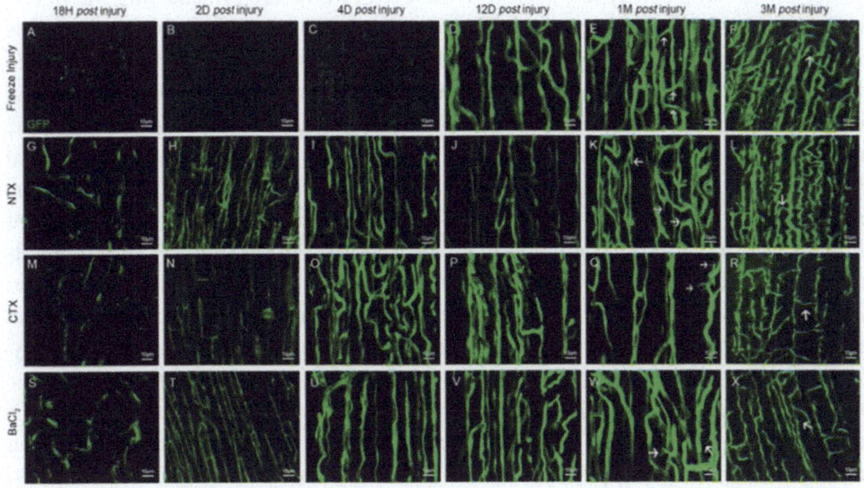

Fig. 6.6 Confocal microscopy images of capillaries in mouse tibialis anterior at different time-points and different models of muscle injury. The images depict capillaries (in green; Green Fluorescent Protein labelled endothelial cells) and the capillary damage and vessel organization resulting from freeze injury, notexin (NTX), cardiotoxin (CTX) and $BaCl_2$ injury at 18 h, 2 days, 4 days, 12 days, 1 and 3 months post injury. This figure shows that the damage to the capillaries and the timeline for recovery will depend on the method used to injury the muscle. Further, it emphasizes the fragmentation of the capillary network in early stages post injury and a longer lasting disorganization even after 3 months post injury especially when compared to the control capillary organization seen in Fig. 6.5. Scale bars represent 10 μm. (This image is adapted from Hardy et al. [163] with permissions through the Open Access Creative Commons license)

Once the capillary network is partially restored (by days 4–12 post injury, Fig. 6.6), regardless of the chaotic nature of the network, there is a potential for restoration of the longitudinal communication from the capillaries to the network, and at least partial restoration of the matching of supply to demand.

Conclusions

The matching of blood flow to the metabolic demand of skeletal muscle fibres is critical to ensure the health and function of skeletal muscle tissue. Central and local processes work to ensure that this matching occurs. To facilitate the local processes microvascular networks are structured and mechanistically equipped to sense metabolic activity of the skeletal muscle fibres and coordinate a network response in order to direct blood flow to active skeletal muscle fibres and capillaries are integral to this coordinated response. Muscle damage produces an interesting dilemma whereby damaged tissue needs flow to remove waste products, to deliver nutrients, to deliver specific cell types, such as leukocytes, to the area for healing the fibres, but often blood vessel damage occurs concurrently with muscle fibre damage. This vascular damage may be produced by the initial trauma that affected the muscle fibres, it may occur secondary to the initial muscle fibre damage, or both. Regardless of the etiology of the damage, the level of the microvasculature most affected are the capillaries. With capillaries being fundamental to network coordination and blood flow delivery to damaged areas, a functioning vascular network is integral to the repair and recovery of the damaged tissue. Contraction- and destruction-induced injury result in very different patterns of vascular disruption and very different patterns of recovery. The extent of the vascular disruption will depend on the extent and duration of the damaging injury, and therefore, recovery of the microvascular can take days to months and the network may never recover its original morphology. The long term effects that a permanently altered vascular morphology would have on oxygen delivery and blood flow control are not well understood. Critical to understanding the role that the microvasculature plays in the recovery process is an understanding of how capillary endothelial cell function is damaged and recovers, how the processes of vascular network restoration and recovery of myocyte function proceed temporally during recovery, and how the two processes interact.

Acknowledgments The author would like to thank Mackenzie Charter and Barbara Hyde Lay for contributing the capillary unit images of the diaphragm in Fig. 6.1.

References

1. Blomstrand E, Radegran G, Saltin B. Maximum rate of oxygen uptake by human skeletal muscle in relation to maximal activities of enzymes in the Krebs cycle. J Physiol. 1997;501(Pt 2):455–60.
2. Bockman EL. Blood flow and oxygen consumption in active soleus and gracilis muscles in cats. Am J Phys. 1983;244(4):H546–51.
3. Hamann JJ, Buckwalter JB, Clifford PS, Shoemaker JK. Is the blood flow response to a single contraction determined by work performed? J Appl Physiol (1985). 2004;96(6):2146–52.
4. VanTeeffelen JW, Segal SS. Effect of motor unit recruitment on functional vasodilatation in hamster retractor muscle. J Physiol. 2000;524(Pt 1):267–78.
5. Andersen P, Saltin B. Maximal perfusion of skeletal muscle in man. J Physiol. 1985;366:233–49.
6. Laughlin MH, Armstrong RB. Muscular blood flow distribution patterns as a function of running speed in rats. Am J Phys. 1982;243(2):H296–306.
7. Joyner MJ, Casey DP. Regulation of increased blood flow (hyperemia) to muscles during exercise: a hierarchy of competing physiological needs. Physiol Rev. 2015;95(2):549–601.
8. DeLorey DS, Clifford PS. Does sympathetic vasoconstriction contribute to metabolism: perfusion matching in exercising skeletal muscle? Front Physiol. 2022;13:980524.
9. Heckman CJ, Enoka RM. Motor unit. Compr Physiol. 2012;2(4):2629–82.
10. Burke RE, Tsairis P. Anatomy and innervation ratios in motor units of cat gastrocnemius. J Physiol. 1973;234(3):749–65.
11. Harris AJ, Duxson MJ, Butler JE, Hodges PW, Taylor JL, Gandevia SC. Muscle fiber and motor unit behavior in the longest human skeletal muscle. J Neurosci. 2005;25(37):8528–33.
12. Heroux ME, Brown HJ, Inglis JT, Siegmund GP, Blouin JS. Motor units in the human medial gastrocnemius muscle are not spatially localized or functionally grouped. J Physiol. 2015;593(16):3711–26.
13. Kanda K, Hashizume K. Factors causing difference in force output among motor units in the rat medial gastrocnemius muscle. J Physiol. 1992;448:677–95.
14. Kugelberg E. Histochemical composition, contraction speed and fatiguability of rat soleus motor units. J Neurol Sci. 1973;20(2):177–98.
15. Vieira TM, Loram ID, Muceli S, Merletti R, Farina D. Postural activation of the human medial gastrocnemius muscle: are the muscle units spatially localised? J Physiol. 2011;589(Pt 2):431–43.
16. Weijs WA, Juch PJ, Kwa SH, Korfage JA. Motor unit territories and fiber types in rabbit masseter muscle. J Dent Res. 1993;72(11):1491–8.
17. Kukulka CG, Clamann HP. Comparison of the recruitment and discharge properties of motor units in human brachial biceps and adductor pollicis during isometric contractions. Brain Res. 1981;219(1):45–55.
18. Berg BR, Cohen KD, Sarelius IH. Direct coupling between blood flow and metabolism at the capillary level in striated muscle. Am J Phys. 1997;272(6 Pt 2):H2693–700.
19. Mackie BG, Terjung RL. Blood flow to different skeletal muscle fiber types during contraction. Am J Phys. 1983;245(2):H265–75.
20. Bearden SE, Payne GW, Chisty A, Segal SS. Arteriolar network architecture and vasomotor function with ageing in mouse gluteus maximus muscle. J Physiol. 2004;561(Pt 2):535–45.
21. Cheung LK. The blood supply of the human temporalis muscle: a vascular corrosion cast study. J Anat. 1996;189(Pt 2):431–8.
22. Delashaw JB, Duling BR. A study of the functional elements regulating capillary perfusion in striated muscle. Microvasc Res. 1988;36(2):162–71.
23. Lund N, Damon DH, Damon DN, Duling BR. Capillary grouping in hamster tibials anterior muscles: flow patterns, and physiological significance. Int J Microcirc Clin Exp. 1987;5(4):359–72.

24. Myrhage R, Hudlicka O. The microvascular bed and capillary surface area in rat extensor hallucis proprius muscle (EHP). Microvasc Res. 1976;11(3):315–23.
25. Hutchins PM, Bond RF, Green HD. The response of skeletal muscle arterioles to common carotid occlusion. Microvasc Res. 1974;7(3):321–5.
26. Smaje L, Zweifach BW, Intaglietta M. Micropressures and capillary filtration coefficients in single vessels of the cremaster muscle of the rat. Microvasc Res. 1970;2(1):96–110.
27. Sweeney TE, Sarelius IH. Arteriolar control of capillary cell flow in striated muscle. Circ Res. 1989;64(1):112–20.
28. Skalak TC, Schmid-Schonbein GW. The microvasculature in skeletal muscle. IV. A model of the capillary network. Microvasc Res. 1986;32(3):333–47.
29. Murrant CL, Fletcher NM. Capillary communication: the role of capillaries in sensing the tissue environment, coordinating the microvascular, and controlling blood flow. Am J Physiol Heart Circ Physiol. 2022;323(5):H1019–36.
30. Segal SS. Microvascular recruitment in hamster striated muscle: role for conducted vasodilation. Am J Phys. 1991;261(1 Pt 2):H181–9.
31. Sakai T, Hosoyamada Y. Are the precapillary sphincters and metarterioles universal components of the microcirculation? An historical review. J Physiol Sci. 2013;63(5):319–31.
32. Bloch EH, Iberall AS. Toward a concept of the functional unit of mammalian skeletal muscle. Am J Phys. 1982;242(5):R411–20.
33. Eriksson E, Myrhage R. Microvascular dimensions and blood flow in skeletal muscle. Acta Physiol Scand. 1972;86(2):211–22.
34. Honig CR, Feldstein ML, Frierson JL. Capillary lengths, anastomoses, and estimated capillary transit times in skeletal muscle. Am J Phys. 1977;233(1):H122–9.
35. Mrazkova O, Grim M, Carlson BM. Enzymatic heterogeneity of the capillary bed of rat skeletal muscles. Am J Anat. 1986;177(2):141–8.
36. Plyley MJ, Groom AC. Geometrical distribution of capillaries in mammalian striated muscle. Am J Phys. 1975;228(5):1376–83.
37. Zoladz JA, Semik D, Zawadowska B, et al. Capillary density and capillary-to-fibre ratio in vastus lateralis muscle of untrained and trained men. Folia Histochem Cytobiol. 2005;43(1):11–7.
38. Ross M, Kargl CK, Ferguson R, Gavin TP, Hellsten Y. Exercise-induced skeletal muscle angiogenesis: impact of age, sex, angiocrines and cellular mediators. Eur J Appl Physiol. 2023;123:1415–32.
39. Sharafi B, Blemker SS. A mathematical model of force transmission from intrafascicularly terminating muscle fibers. J Biomech. 2011;44(11):2031–9.
40. Emerson GG, Segal SS. Alignment of microvascular units along skeletal muscle fibers of hamster retractor. J Appl Physiol (1985). 1997;82(1):42–8.
41. Mendelson AA, Milkovich S, Hunter T, et al. The capillary fascicle in skeletal muscle: structural and functional physiology of RBC distribution in capillary networks. J Physiol. 2021;599(8):2149–68.
42. Borg TK, Caulfield JB. Morphology of connective tissue in skeletal muscle. Tissue Cell. 1980;12(1):197–207.
43. Mathieu-Costello O. Capillary tortuosity and degree of contraction or extension of skeletal muscles. Microvasc Res. 1987;33(1):98–117.
44. Ledvina MA, Segal SS. Sarcomere length and capillary curvature of rat hindlimb muscles in vivo. J Appl Physiol (1985). 1995;78(6):2047–51.
45. Ellis CG, Mathieu-Costello O, Potter RF, MacDonald IC, Groom AC. Effect of sarcomere length on total capillary length in skeletal muscle: in vivo evidence for longitudinal stretching of capillaries. Microvasc Res. 1990;40(1):63–72.
46. Mermod L, Hoppeler H, Kayar SR, Straub R, Weibel ER. Variability of fiber size, capillary density and capillary length related to horse muscle fixation procedures. Acta Anat (Basel). 1988;133(2):89–95.

47. Poole DC, Musch TI, Kindig CA. In vivo microvascular structural and functional consequences of muscle length changes. Am J Phys. 1997;272(5 Pt 2):H2107–14.
48. Kindig CA, Poole DC. Sarcomere length-induced alterations of capillary hemodynamics in rat spinotrapezius muscle: vasoactive vs passive control. Microvasc Res. 2001;61(1):64–74.
49. Figueroa XF, Duling BR. Gap junctions in the control of vascular function. Antioxid Redox Signal. 2009;11(2):251–66.
50. Segal SS. Integration and modulation of intercellular signaling underlying blood flow control. J Vasc Res. 2015;52(2):136–57.
51. Sinkler SY, Fernando CA, Segal SS. Differential alpha-adrenergic modulation of rapid onset vasodilatation along resistance networks of skeletal muscle in old versus young mice. J Physiol. 2016;594(23):6987–7004.
52. Sinkler SY, Segal SS. Rapid versus slow ascending vasodilatation: intercellular conduction versus flow-mediated signalling with tetanic versus rhythmic muscle contractions. J Physiol. 2017;595(23):7149–65.
53. Domeier TL, Segal SS. Electromechanical and pharmacomechanical signalling pathways for conducted vasodilatation along endothelium of hamster feed arteries. J Physiol. 2007;579(Pt 1):175–86.
54. Tallini YN, Brekke JF, Shui B, et al. Propagated endothelial Ca2+ waves and arteriolar dilation in vivo: measurements in Cx40BAC GCaMP2 transgenic mice. Circ Res. 2007;101(12):1300–9.
55. Beach JM, McGahren ED, Duling BR. Capillaries and arterioles are electrically coupled in hamster cheek pouch. Am J Phys. 1998;275(4):H1489–96.
56. Duling BR, Berne RM. Propagated vasodilation in the microcirculation of the hamster cheek pouch. Circ Res. 1970;26(2):163–70.
57. Segal SS, Damon DN, Duling BR. Propagation of vasomotor responses coordinates arteriolar resistances. Am J Phys. 1989;256(3 Pt 2):H832–7.
58. Segal SS, Duling BR. Flow control among microvessels coordinated by intercellular conduction. Science. 1986;234(4778):868–70.
59. Lamb IR, Novielli-Kuntz NM, Murrant CL. Capillaries communicate with the arteriolar microvascular network by a pannexin/purinergic-dependent pathway in hamster skeletal muscle. Am J Physiol Heart Circ Physiol. 2021;320(4):H1699–711.
60. Garland CJ, Dora KA. Endothelium-dependent hyperpolarization: the evolution of myoendothelial microdomains. J Cardiovasc Pharmacol. 2021;78(Suppl 6):S3–S12.
61. Buckwalter JB, Ruble SB, Mueller PJ, Clifford PS. Skeletal muscle vasodilation at the onset of exercise. J Appl Physiol (1985). 1998;85(5):1649–54.
62. MacDonald MJ, Shoemaker JK, Tschakovsky ME, Hughson RL. Alveolar oxygen uptake and femoral artery blood flow dynamics in upright and supine leg exercise in humans. J Appl Physiol (1985). 1998;85(5):1622–8.
63. Naik JS, Valic Z, Buckwalter JB, Clifford PS. Rapid vasodilation in response to a brief tetanic muscle contraction. J Appl Physiol (1985). 1999;87(5):1741–6.
64. Radegran G, Saltin B. Muscle blood flow at onset of dynamic exercise in humans. Am J Phys. 1998;274(1):H314–22.
65. Sheriff DD, Rowell LB, Scher AM. Is rapid rise in vascular conductance at onset of dynamic exercise due to muscle pump? Am J Phys. 1993;265(4 Pt 2):H1227–34.
66. Tschakovsky ME, Shoemaker JK, Hughson RL. Vasodilation and muscle pump contribution to immediate exercise hyperemia. Am J Phys. 1996;271(4 Pt 2):H1697–701.
67. Barclay JK, Murrant CL, Woodley NE, Reading SA. Potential interactions among vascular and muscular functional compartments during active hyperemia. Can J Appl Physiol. 2003;28(5):737–53.
68. Sadamoto T, Bonde-Petersen F, Suzuki Y. Skeletal muscle tension, flow, pressure, and EMG during sustained isometric contractions in humans. Eur J Appl Physiol Occup Physiol. 1983;51(3):395–408.

69. Sejersted OM, Hargens AR, Kardel KR, Blom P, Jensen O, Hermansen L. Intramuscular fluid pressure during isometric contraction of human skeletal muscle. J Appl Physiol Respir Environ Exerc Physiol. 1984;56(2):287–95.
70. Barcroft H, Millen JL. The blood flow through muscle during sustained contraction. J Physiol. 1939;97(1):17–31.
71. Hirche H, Raff WK, Grun D. The resistance to blood flow in the gastrocnemius of the dog during sustained and rhythmical isometric and isotonic contractions. Pflugers Arch. 1970;314(2):97–112.
72. Laughlin MH. Skeletal muscle blood flow capacity: role of muscle pump in exercise hyperemia. Am J Phys. 1987;253(5 Pt 2):H993–1004.
73. Tschakovsky ME, Sheriff DD. Immediate exercise hyperemia: contributions of the muscle pump vs. rapid vasodilation. J Appl Physiol (1985). 2004;97(2):739–47.
74. Olesen SP, Clapham DE, Davies PF. Haemodynamic shear stress activates a K+ current in vascular endothelial cells. Nature. 1988;331(6152):168–70.
75. de Wit C, Schafer C, von Bismarck P, Bolz SS, Pohl U. Elevation of plasma viscosity induces sustained NO-mediated dilation in the hamster cremaster microcirculation in vivo. Pflugers Arch. 1997;434(4):354–61.
76. Huang A, Wu Y, Sun D, Koller A, Kaley G. Effect of estrogen on flow-induced dilation in NO deficiency: role of prostaglandins and EDHF. J Appl Physiol (1985). 2001;91(6):2561–6.
77. Watanabe S, Yashiro Y, Mizuno R, Ohhashi T. Involvement of NO and EDHF in flow-induced vasodilation in isolated hamster cremasteric arterioles. J Vasc Res. 2005;42(2):137–47.
78. Hocking DC, Titus PA, Sumagin R, Sarelius IH. Extracellular matrix fibronectin mechanically couples skeletal muscle contraction with local vasodilation. Circ Res. 2008;102(3):372–9.
79. Lu X, Kassab GS. Integrins mediate mechanical compression-induced endothelium-dependent vasodilation through endothelial nitric oxide pathway. J Gen Physiol. 2015;146(3):221–32.
80. Clifford PS, Kluess HA, Hamann JJ, Buckwalter JB, Jasperse JL. Mechanical compression elicits vasodilatation in rat skeletal muscle feed arteries. J Physiol. 2006;572(Pt 2):561–7.
81. Kirby BS, Carlson RE, Markwald RR, Voyles WF, Dinenno FA. Mechanical influences on skeletal muscle vascular tone in humans: insight into contraction-induced rapid vasodilatation. J Physiol. 2007;583(Pt 3):861–74.
82. Balon TW, Nadler JL. Nitric oxide release is present from incubated skeletal muscle preparations. J Appl Physiol (1985). 1994;77(6):2519–21.
83. Kobzik L, Reid MB, Bredt DS, Stamler JS. Nitric oxide in skeletal muscle. Nature. 1994;372(6506):546–8.
84. Cheng B, Essackjee HC, Ballard HJ. Evidence for control of adenosine metabolism in rat oxidative skeletal muscle by changes in pH. J Physiol. 2000;522(Pt 3):467–77.
85. Hellsten Y. The effect of muscle contraction on the regulation of adenosine formation in rat skeletal muscle cells. J Physiol. 1999;518(Pt 3):761–8.
86. Lynge J, Juel C, Hellsten Y. Extracellular formation and uptake of adenosine during skeletal muscle contraction in the rat: role of adenosine transporters. J Physiol. 2001;537(Pt 2):597–605.
87. VanTeeffelen JW, Segal SS. Rapid dilation of arterioles with single contraction of hamster skeletal muscle. Am J Physiol Heart Circ Physiol. 2006;290(1):H119–27.
88. Welsh DG, Segal SS. Coactivation of resistance vessels and muscle fibers with acetylcholine release from motor nerves. Am J Phys. 1997;273(1 Pt 2):H156–63.
89. Armstrong ML, Dua AK, Murrant CL. Potassium initiates vasodilatation induced by a single skeletal muscle contraction in hamster cremaster muscle. J Physiol. 2007;581(Pt 2):841–52.
90. Ross GA, Mihok ML, Murrant CL. Extracellular adenosine initiates rapid arteriolar vasodilation induced by a single skeletal muscle contraction in hamster cremaster muscle. Acta Physiol (Oxf). 2013;208(1):74–87.
91. McDaniel J, Fjeldstad AS, Ives S, Hayman M, Kithas P, Richardson RS. Central and peripheral contributors to skeletal muscle hyperemia: response to passive limb movement. J Appl Physiol (1985). 2010;108(1):76–84.

92. Hayman MA, Nativi JN, Stehlik J, et al. Understanding exercise-induced hyperemia: central and peripheral hemodynamic responses to passive limb movement in heart transplant recipients. Am J Physiol Heart Circ Physiol. 2010;299(5):H1653–9.
93. Trinity JD, Groot HJ, Layec G, et al. Passive leg movement and nitric oxide-mediated vascular function: the impact of age. Am J Physiol Heart Circ Physiol. 2015;308(6):H672–9.
94. Trinity JD, Groot HJ, Layec G, et al. Nitric oxide and passive limb movement: a new approach to assess vascular function. J Physiol. 2012;590(6):1413–25.
95. Kooijman M, Thijssen DH, de Groot PC, et al. Flow-mediated dilatation in the superficial femoral artery is nitric oxide mediated in humans. J Physiol. 2008;586(4):1137–45.
96. Koller A, Kaley G. Endothelium regulates skeletal muscle microcirculation by a blood flow velocity-sensing mechanism. Am J Phys. 1990;258(3 Pt 2):H916–20.
97. Koller A, Kaley G. Endothelial regulation of wall shear stress and blood flow in skeletal muscle microcirculation. Am J Phys. 1991;260(3 Pt 2):H862–8.
98. Koller A, Sun D, Huang A, Kaley G. Corelease of nitric oxide and prostaglandins mediates flow-dependent dilation of rat gracilis muscle arterioles. Am J Phys. 1994;267(1 Pt 2):H326–32.
99. Koller A, Sun D, Kaley G. Role of shear stress and endothelial prostaglandins in flow- and viscosity-induced dilation of arterioles in vitro. Circ Res. 1993;72(6):1276–84.
100. Sun D, Messina EJ, Koller A, Wolin MS, Kaley G. Endothelium-dependent dilation to L-arginine in isolated rat skeletal muscle arterioles. Am J Phys. 1992;262(4 Pt 2):H1211–6.
101. Boulton D, Taylor CE, Green S, Macefield VG. The metaboreflex does not contribute to the increase in muscle sympathetic nerve activity to contracting muscle during static exercise in humans. J Physiol. 2018;596(6):1091–102.
102. DiCarlo SE, Chen CY, Collins HL. Onset of exercise increases lumbar sympathetic nerve activity in rats. Med Sci Sports Exerc. 1996;28(6):677–84.
103. DeLorey DS. Sympathetic vasoconstriction in skeletal muscle: modulatory effects of aging, exercise training, and sex. Appl Physiol Nutr Metab. 2021;46(12):1437–47.
104. Just TP, Cooper IR, DeLorey DS. Sympathetic vasoconstriction in skeletal muscle: adaptations to exercise training. Exerc Sport Sci Rev. 2016;44(4):137–43.
105. Mueller PJ, Clifford PS, Crandall CG, Smith SA, Fadel PJ. Integration of central and peripheral regulation of the circulation during exercise: acute and chronic adaptations. Compr Physiol. 2017;8(1):103–51.
106. Thomas GD, Segal SS. Neural control of muscle blood flow during exercise. J Appl Physiol (1985). 2004;97(2):731–8.
107. Clifford PS, Hellsten Y. Vasodilatory mechanisms in contracting skeletal muscle. J Appl Physiol (1985). 2004;97(1):393–403.
108. Mortensen SP, Saltin B. Regulation of the skeletal muscle blood flow in humans. Exp Physiol. 2014;99(12):1552–8.
109. Sarelius I, Pohl U. Control of muscle blood flow during exercise: local factors and integrative mechanisms. Acta Physiol (Oxf). 2010;199(4):349–65.
110. Diesen DL, Hess DT, Stamler JS. Hypoxic vasodilation by red blood cells: evidence for an s-nitrosothiol-based signal. Circ Res. 2008;103(5):545–53.
111. Stamler JS, Jia L, Eu JP, et al. Blood flow regulation by S-nitrosohemoglobin in the physiological oxygen gradient. Science. 1997;276(5321):2034–7.
112. Bergfeld GR, Forrester T. Release of ATP from human erythrocytes in response to a brief period of hypoxia and hypercapnia. Cardiovasc Res. 1992;26(1):40–7.
113. Gonzalez-Alonso J. ATP as a mediator of erythrocyte-dependent regulation of skeletal muscle blood flow and oxygen delivery in humans. J Physiol. 2012;590(20):5001–13.
114. Jagger JE, Bateman RM, Ellsworth ML, Ellis CG. Role of erythrocyte in regulating local O2 delivery mediated by hemoglobin oxygenation. Am J Physiol Heart Circ Physiol. 2001;280(6):H2833–9.
115. Sprague RS, Ellsworth ML, Stephenson AH, Lonigro AJ. ATP: the red blood cell link to NO and local control of the pulmonary circulation. Am J Phys. 1996;271(6 Pt 2):H2717–22.

116. Sprague RS, Ellsworth ML, Stephenson AH, Lonigro AJ. Participation of cAMP in a signal-transduction pathway relating erythrocyte deformation to ATP release. Am J Physiol Cell Physiol. 2001;281(4):C1158–64.
117. Colburn TD, Hirai DM, Craig JC, et al. Transcapillary PO(2) gradients in contracting muscles across the fibre type and oxidative continuum. J Physiol. 2020;598(15):3187–202.
118. Falcone JC, Bohlen HG. EDRF from rat intestine and skeletal muscle venules causes dilation of arterioles. Am J Phys. 1990;258(5 Pt 2):H1515–23.
119. Hester RL. Venular-arteriolar diffusion of adenosine in hamster cremaster microcirculation. Am J Phys. 1990;258(6 Pt 2):H1918–24.
120. Hester RL. Uptake of metabolites by postcapillary venules: mechanism for the control of arteriolar diameter. Microvasc Res. 1993;46(2):254–61.
121. Tigno XT, Ley K, Pries AR, Gaehtgens P. Venulo-arteriolar communication and propagated response. A possible mechanism for local control of blood flow. Pflugers Arch. 1989;414(4):450–6.
122. Koller A, Dornyei G, Kaley G. Flow-induced responses in skeletal muscle venules: modulation by nitric oxide and prostaglandins. Am J Phys. 1998;275(3):H831–6.
123. McKay MK, Gardner AL, Boyd D, Hester RL. Influence of venular prostaglandin release on arteriolar diameter during functional hyperemia. Hypertension. 1998;31(1 Pt 2):213–7.
124. Xiang L, Naik JS, Hester RL. Functional vasodilation in the rat spinotrapezius muscle: role of nitric oxide, prostanoids and epoxyeicosatrienoic acids. Clin Exp Pharmacol Physiol. 2008;35(5–6):617–24.
125. Chilian WM, Koshida R. EDHF and NO: different pathways for production--similar actions. Circ Res. 2001;89(8):648–9.
126. Hellsten Y, Nyberg M, Jensen LG, Mortensen SP. Vasodilator interactions in skeletal muscle blood flow regulation. J Physiol. 2012;590(24):6297–305.
127. Joyner MJ, Wilkins BW. Exercise hyperaemia: is anything obligatory but the hyperaemia? J Physiol. 2007;583(Pt 3):855–60.
128. Rowell LB. Ideas about control of skeletal and cardiac muscle blood flow (1876–2003): cycles of revision and new vision. J Appl Physiol (1985). 2004;97(1):384–92.
129. Boushel R, Langberg H, Gemmer C, et al. Combined inhibition of nitric oxide and prostaglandins reduces human skeletal muscle blood flow during exercise. J Physiol. 2002;543(Pt 2):691–8.
130. Markwald RR, Kirby BS, Crecelius AR, Carlson RE, Voyles WF, Dinenno FA. Combined inhibition of nitric oxide and vasodilating prostaglandins abolishes forearm vasodilatation to systemic hypoxia in healthy humans. J Physiol. 2011;589(Pt 8):1979–90.
131. Mortensen SP, Gonzalez-Alonso J, Damsgaard R, Saltin B, Hellsten Y. Inhibition of nitric oxide and prostaglandins, but not endothelial-derived hyperpolarizing factors, reduces blood flow and aerobic energy turnover in the exercising human leg. J Physiol. 2007;581(Pt 2):853–61.
132. Schrage WG, Joyner MJ, Dinenno FA. Local inhibition of nitric oxide and prostaglandins independently reduces forearm exercise hyperaemia in humans. J Physiol. 2004;557(Pt 2):599–611.
133. Lamb IR, Murrant CL. Potassium inhibits nitric oxide and adenosine arteriolar vasodilatation via K(IR) and Na(+)/K(+) ATPase: implications for redundancy in active hyperaemia. J Physiol. 2015;593(23):5111–26.
134. Murray IR, Baily JE, Chen WCW, et al. Skeletal and cardiac muscle pericytes: functions and therapeutic potential. Pharmacol Ther. 2017;171:65–74.
135. Grant RI, Hartmann DA, Underly RG, Berthiaume AA, Bhat NR, Shih AY. Organizational hierarchy and structural diversity of microvascular pericytes in adult mouse cortex. J Cereb Blood Flow Metab. 2019;39(3):411–25.
136. Ornelas S, Berthiaume AA, Bonney SK, et al. Three-dimensional ultrastructure of the brain pericyte-endothelial interface. J Cereb Blood Flow Metab. 2021;41(9):2185–200.

137. Armulik A, Abramsson A, Betsholtz C. Endothelial/pericyte interactions. Circ Res. 2005;97(6):512–23.
138. Armulik A, Genove G, Betsholtz C. Pericytes: developmental, physiological, and pathological perspectives, problems, and promises. Dev Cell. 2011;21(2):193–215.
139. Sims DE. The pericyte--a review. Tissue Cell. 1986;18(2):153–74.
140. Tilton RG, Kilo C, Williamson JR. Pericyte-endothelial relationships in cardiac and skeletal muscle capillaries. Microvasc Res. 1979;18(3):325–35.
141. Attrill E, Ramsay C, Ross R, et al. Metabolic-vascular coupling in skeletal muscle: a potential role for capillary pericytes? Clin Exp Pharmacol Physiol. 2020;47(3):520–8.
142. Spiranec K, Chen W, Werner F, et al. Endothelial C-type natriuretic peptide acts on pericytes to regulate microcirculatory flow and blood pressure. Circulation. 2018;138(5):494–508.
143. Murrant CL, Lamb IR, Novielli NM. Capillary endothelial cells as coordinators of skeletal muscle blood flow during active hyperemia. Microcirculation. 2017;24(3):e12348.
144. Brodal P, Ingjer F, Hermansen L. Capillary supply of skeletal muscle fibers in untrained and endurance-trained men. Am J Phys. 1977;232(6):H705–12.
145. Gul N, Cebesoy S, Ozsoy N, Eskizengin H, Ozer C. The ultrastructure of skeletal muscle capillaries of streptozotocin diabetic rats and the therapeutic effect of benfluorex. Microsc Microanal. 2022:1–5.
146. Cohen KD, Sarelius IH. Muscle contraction under capillaries in hamster muscle induces arteriolar dilatation via K(ATP) channels and nitric oxide. J Physiol. 2002;539(Pt 2):547–55.
147. Lamb IR, Novielli NM, Murrant CL. Capillary response to skeletal muscle contraction: evidence that redundancy between vasodilators is physiologically relevant during active hyperaemia. J Physiol. 2018;596(8):1357–72.
148. Ghonaim NW, Fraser GM, Goldman D, Milkovich S, Yang J, Ellis CG. Evidence for role of capillaries in regulation of skeletal muscle oxygen supply. Microcirculation. 2021;28(6):e12699.
149. Sove RJ, Milkovich S, Nikolov HN, Holdsworth DW, Ellis CG, Fraser GM. Localized oxygen exchange platform for intravital video microscopy investigations of microvascular oxygen regulation. Front Physiol. 2021;12:654928.
150. Song H, Tyml K. Evidence for sensing and integration of biological signals by the capillary network. Am J Phys. 1993;265(4 Pt 2):H1235–42.
151. Yu J, Bihari A, Lidington D, Tyml K. Gap junction uncouplers attenuate arteriolar response to distal capillary stimuli. Microvasc Res. 2000;59(1):162–8.
152. McGahren ED, Beach JM, Duling BR. Capillaries demonstrate changes in membrane potential in response to pharmacological stimuli. Am J Phys. 1998;274(1):H60–5.
153. Tyml K, Song H, Munoz P, Ouellette Y. Evidence for K+ channels involvement in capillary sensing and for bidirectionality in capillary communication. Microvasc Res. 1997;53(3):245–53.
154. Twynstra J, Ruiz DA, Murrant CL. Functional coordination of the spread of vasodilations through skeletal muscle microvasculature: implications for blood flow control. Acta Physiol (Oxf). 2012;206(4):229–41.
155. Ellsworth ML, Forrester T, Ellis CG, Dietrich HH. The erythrocyte as a regulator of vascular tone. Am J Phys. 1995;269(6 Pt 2):H2155–61.
156. Charter ME, Lamb IR, Murrant CL. Arteriolar and capillary responses to CO(2) and H(+) in hamster skeletal muscle microvasculature: implications for active hyperemia. Microcirculation. 2018;25(7):e12494.
157. Cohen KD, Berg BR, Sarelius IH. Remote arteriolar dilations in response to muscle contraction under capillaries. Am J Physiol Heart Circ Physiol. 2000;278(6):H1916–23.
158. Jarvinen TA, Jarvinen TL, Kaariainen M, Kalimo H, Jarvinen M. Muscle injuries: biology and treatment. Am J Sports Med. 2005;33(5):745–64.
159. Singh DP, Barani Lonbani Z, Woodruff MA, Parker TJ, Steck R, Peake JM. Effects of topical icing on inflammation, angiogenesis, revascularization, and myofiber regeneration in skeletal muscle following contusion injury. Front Physiol. 2017;8:93.

160. Charge SB, Rudnicki MA. Cellular and molecular regulation of muscle regeneration. Physiol Rev. 2004;84(1):209–38.
161. Greising SM, Corona BT, Call JA. Musculoskeletal regeneration, rehabilitation, and plasticity following traumatic injury. Int J Sports Med. 2020;41(8):495–504.
162. Tidball JG. Regulation of muscle growth and regeneration by the immune system. Nat Rev Immunol. 2017;17(3):165–78.
163. Hardy D, Besnard A, Latil M, et al. Comparative study of injury models for studying muscle regeneration in mice. PLoS One. 2016;11(1):e0147198.
164. Karalaki M, Fili S, Philippou A, Koutsilieris M. Muscle regeneration: cellular and molecular events. In Vivo. 2009;23(5):779–96.
165. Paulsen G, Mikkelsen UR, Raastad T, Peake JM. Leucocytes, cytokines and satellite cells: what role do they play in muscle damage and regeneration following eccentric exercise? Exerc Immunol Rev. 2012;18:42–97.
166. Warren GL, Summan M, Gao X, Chapman R, Hulderman T, Simeonova PP. Mechanisms of skeletal muscle injury and repair revealed by gene expression studies in mouse models. J Physiol. 2007;582(Pt 2):825–41.
167. Rotter R, Menshykova M, Winkler T, et al. Erythropoietin improves functional and histological recovery of traumatized skeletal muscle tissue. J Orthop Res. 2008;26(12):1618–26.
168. Ahmadi S, Sinclair PJ, Foroughi N, Davis GM. Monitoring muscle oxygenation after eccentric exercise-induced muscle damage using near-infrared spectroscopy. Appl Physiol Nutr Metab. 2008;33(4):743–52.
169. Barnes JN, Trombold JR, Dhindsa M, Lin HF, Tanaka H. Arterial stiffening following eccentric exercise-induced muscle damage. J Appl Physiol (1985). 2010;109(4):1102–8.
170. Sbriccoli P, Felici F, Rosponi A, et al. Exercise induced muscle damage and recovery assessed by means of linear and non-linear sEMG analysis and ultrasonography. J Electromyogr Kinesiol. 2001;11(2):73–83.
171. Souza-Silva E, Christensen SW, Hirata RP, Larsen RG, Graven-Nielsen T. Blood flow after contraction and cuff occlusion is reduced in subjects with muscle soreness after eccentric exercise. Scand J Med Sci Sports. 2018;28(1):29–39.
172. Stacy MR, Bladon KJ, Lawrence JL, McGlinchy SA, Scheuermann BW. Serial assessment of local peripheral vascular function after eccentric exercise. Appl Physiol Nutr Metab. 2013;38(12):1181–6.
173. Morton AB, Norton CE, Jacobsen NL, Fernando CA, Cornelison DDW, Segal SS. Barium chloride injures myofibers through calcium-induced proteolysis with fragmentation of motor nerves and microvessels. Skelet Muscle. 2019;9(1):27.
174. Larsen RG, Thomsen JM, Hirata RP, et al. Impaired microvascular reactivity after eccentric muscle contractions is not restored by acute ingestion of antioxidants or dietary nitrate. Physiol Rep. 2019;7(13):e14162.
175. Kano Y, Padilla DJ, Behnke BJ, Hageman KS, Musch TI, Poole DC. Effects of eccentric exercise on microcirculation and microvascular oxygen pressures in rat spinotrapezius muscle. J Appl Physiol (1985). 2005;99(4):1516–22.
176. Caldwell JT, Wardlow GC, Branch PA, Ramos M, Black CD, Ade CJ. Effect of exercise-induced muscle damage on vascular function and skeletal muscle microvascular deoxygenation. Physiol Rep. 2016;4(22):e13032.
177. Dolezal BA, Potteiger JA, Jacobsen DJ, Benedict SH. Muscle damage and resting metabolic rate after acute resistance exercise with an eccentric overload. Med Sci Sports Exerc. 2000;32(7):1202–7.
178. Rodenburg JB, de Boer RW, Schiereck P, van Echteld CJ, Bar PR. Changes in phosphorus compounds and water content in skeletal muscle due to eccentric exercise. Eur J Appl Physiol Occup Physiol. 1994;68(3):205–13.
179. Lieber RL, Friden J. Selective damage of fast glycolytic muscle fibres with eccentric contraction of the rabbit tibialis anterior. Acta Physiol Scand. 1988;133(4):587–8.

180. Jacobsen NL, Norton CE, Shaw RL, Cornelison DDW, Segal SS. Myofibre injury induces capillary disruption and regeneration of disorganized microvascular networks. J Physiol. 2022;600(1):41–60.
181. Arpino JM, Nong Z, Li F, et al. Four-dimensional microvascular analysis reveals that regenerative angiogenesis in ischemic muscle produces a flawed microcirculation. Circ Res. 2017;120(9):1453–65.
182. Agarwal M, Singh S, Narayan J, Pandey S, Tiwari S, Sharma P. Cardiovascular response and serum interleukin-6 level in concentric vs. eccentric exercise. J Clin Diagn Res. 2017;11(4):CC04–8.
183. Zembron-Lacny A, Naczk M, Gajewski M, et al. Changes of muscle-derived cytokines in relation to thiol redox status and reactive oxygen and nitrogen species. Physiol Res. 2010;59(6):945–51.
184. Herrlinger KA, Chirouzes DM, Ceddia MA. Supplementation with a polyphenolic blend improves post-exercise strength recovery and muscle soreness. Food Nutr Res. 2015;59:30034.
185. Kerksick CM, Kreider RB, Willoughby DS. Intramuscular adaptations to eccentric exercise and antioxidant supplementation. Amino Acids. 2010;39(1):219–32.
186. Choi Y, Akazawa N, Zempo-Miyaki A, et al. Acute effect of high-intensity eccentric exercise on vascular endothelial function in young men. J Strength Cond Res. 2016;30(8):2279–85.
187. Rakobowchuk M, Ritter O, Wilhelm EN, et al. Divergent endothelial function but similar platelet microvesicle responses following eccentric and concentric cycling at a similar aerobic power output. J Appl Physiol (1985). 2017;122(4):1031–9.
188. Rubinstein I, Abassi Z, Coleman R, Milman F, Winaver J, Better OS. Involvement of nitric oxide system in experimental muscle crush injury. J Clin Invest. 1998;101(6):1325–33.
189. Heap SJ, Fulgenzi GL, Hudlicka O. Microcirculation in rat soleus muscle after eccentric exercise: the effect of nifedipine. Eur J Appl Physiol. 2006;97(6):687–94.
190. Hotta K, Behnke BJ, Masamoto K, et al. Microvascular permeability of skeletal muscle after eccentric contraction-induced muscle injury: in vivo imaging using two-photon laser scanning microscopy. J Appl Physiol (1985). 2018;125(2):369–80.
191. Crecelius AR, Richards JC, Luckasen GJ, Larson DG, Dinenno FA. Reactive hyperemia occurs via activation of inwardly rectifying potassium channels and Na+/K+-ATPase in humans. Circ Res. 2013;113(8):1023–32.
192. Engelke KA, Halliwill JR, Proctor DN, Dietz NM, Joyner MJ. Contribution of nitric oxide and prostaglandins to reactive hyperemia in human forearm. J Appl Physiol (1985). 1996;81(4):1807–14.
193. Larsen RG, Hirata RP, Madzak A, Frokjaer JB, Graven-Nielsen T. Eccentric exercise slows in vivo microvascular reactivity during brief contractions in human skeletal muscle. J Appl Physiol (1985). 2015;119(11):1272–81.
194. Dua AK, Dua N, Murrant CL. Skeletal muscle contraction-induced vasodilator complement production is dependent on stimulus and contraction frequency. Am J Physiol Heart Circ Physiol. 2009;297(1):H433–42.
195. Fernando CA, Pangan AM, Cornelison D, Segal SS. Recovery of blood flow regulation in microvascular resistance networks during regeneration of mouse gluteus maximus muscle. J Physiol. 2019;597(5):1401–17.
196. Schaser KD, Vollmar B, Menger MD, et al. In vivo analysis of microcirculation following closed soft-tissue injury. J Orthop Res. 1999;17(5):678–85.
197. Jacobsen NL, Morton AB, Segal SS. Angiogenesis precedes myogenesis during regeneration following biopsy injury of skeletal muscle. Skelet Muscle. 2023;13(1):3.
198. Zhang C, Wang X, Tang J. Correlation between microcirculation and contrast-enhanced ultrasonography after crush injury of limbs. J Med Ultrason (2001). 2018;45(2):307–13.

Chapter 7
Mechanotransduction in Skeletal Muscle

Eirini Chatzinikita, Maria Maridaki, and Anastassios Philippou

- Mechanotransduction describes the ability of a cell to sense, modify and translate a mechanical stimulus into biochemical signals.
- Mechanotransduction occurs at different components of the muscular system and includes the combined action of various elements of skeletal muscle, i.e., myofibers, myotendinous junction, tendons, muscle connective tissue and extracellular matrix.
- Isolated whole muscle ex vivo, and in vitro models have been developed to study the intracellular alterations that occur through mechanotransduction.
- Mechanotransduction, through specialized proteins, activates signaling pathways that contribute to the maintenance, regeneration, remodeling, and development of skeletal muscle.

Introduction

Mechanotransduction is a process that allows organisms to respond to their mechanical environments, it refers to the detection of tension or deformation and is essential to a wide range of physiological mechanisms [1]. Specifically, mechanotransduction describes the cellular processes that translate mechanical stimuli into biochemical signals, thus enabling cells to adapt to their physical surroundings [2]. Cells and subcellular structures experience mechanical stress during

E. Chatzinikita (✉) · A. Philippou
Department of Physiology, Medical School, National and Kapodistrian University of Athens, Athens, Greece
e-mail: echatzin@med.uoa.gr; tfilipou@med.uoa.gr

M. Maridaki
Faculty of Physical Education and Sport Science, National and Kapodistrian University of Athens, Athens, Greece
e-mail: mmarida@phed.uoa.gr

N. C. Apostolopoulos et al. (eds.), *Fundamentals of Recovery, Regeneration, and Adaptation to Exercise Stress: An Integrated Approach*,
https://doi.org/10.1007/978-3-031-44270-4_7

deformation from many sources ranging from whole-body to tissue-specific forces, while there are many different cellular responses to these forces and various conformations and mechanisms by which such forces are transduced linking mechanical stimulation with biochemical responses [3]. Numerous molecules and subcellular structures that mediate force sensation and mechano-electrochemical conversion have been identified [4], such as integrins, stretch-activated ion channels, G protein-coupled receptors and growth factor receptors, each of which can activate different mechanotransduction pathways [5], (Fig. 7.1).

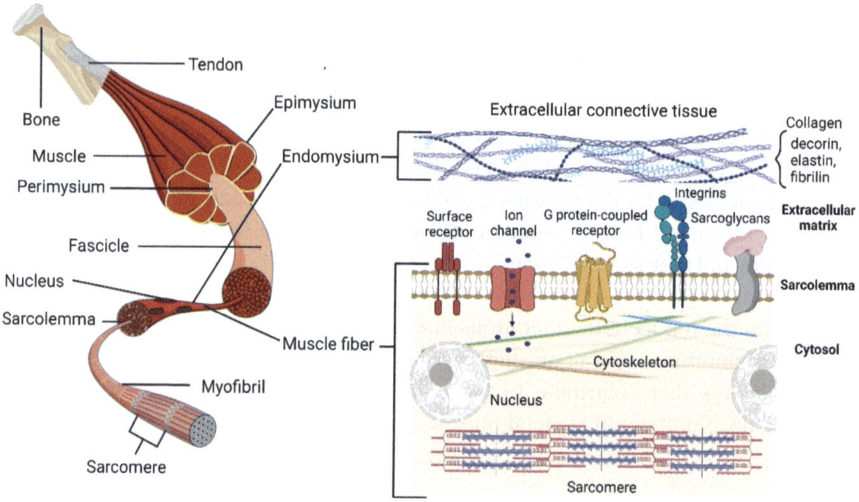

Fig. 7.1 Schematic representation illustrating the skeletal muscle structure from its macroscopic organization (left) to the muscle fiber extracellular microenvironment and intracellular structures (right). Each muscle consists of muscle fibers wrapped by (extracellular) connective tissue sheaths, typically depicted to exhibit a three-level structural organization. Specifically, the extracellular matrix (ECM) of skeletal muscle consists of glycoproteins, proteoglycans, collagen, and elastin, with collagen forming a network of intramuscular connective tissue, i.e., the fibrous components of the ECM. The intramuscular connective tissue is organized in three layers; The extracellular connective tissue endomysium surrounds each muscle fiber, the perimysium divides the muscle into bundles of fibers (fascicles), and the epimysium serves as the outer connective tissue layer of the muscle. Myofibrils are striated muscle organelles that are comprised of sarcomeres and are strictly aligned in series. The sarcomere is the structural and functional (contractile) unit of the myofibril and the whole muscle fiber. Myofibrils almost fill intracellular space (sarcoplasm) together with its liquid part (cytosol) and the cytoskeleton proteins within a muscle fiber. The muscle cell cytoskeleton is composed mainly of intermediate filaments surrounding the myofibrils and connecting the peripheral myofibrils with the sarcolemma and the nuclear membrane. When stressed, the cytoskeleton's ability to stretch is determined by its elastic tension and results in mechanical coupling between the cytoskeleton and the extracellular matrix. In addition, the interaction between the contractile proteins actin and myosin may generate intracellular constraints, which result in mechanical effects, like those from the extracellular matrix, on mechanosensitive complexes (mechanosensors) situated in the muscle cell membrane (sarcolemma). These mechanosensors, such as the stretch-activated ion channels, integrins, sarcoglycans, growth factor receptors and G protein-coupled receptors, may rely on both external and internal forces and serve to sense mechanical alterations and activate intracellular signaling mechanisms (mechanotransduction) in muscle fibers [6–8]

In physiological systems, the conversion of physical force into biological signals is fundamental and provides a simple means by which organisms and cells ensure structural stability and regulate morphogenetic movements to generate precise three-dimensional structures [9]. In particular, muscle tissue architecture and metabolism are particularly susceptible to mechanical stimulation [10]. Skeletal muscle plasticity is unique in that muscle can adapt to mechanical forces by modifying its mass and contractile phenotype [11]. Resistance or aerobic exercise training causes mechanical loading and metabolic stress, thus inducing adaptive mechanisms in the exercising muscle [12, 13]. Many studies, involving human subjects and animal models, have examined the mechanisms underlying muscle adaptation to exercise. However, defining the molecular responses of muscle cells to mechanical stimuli in vivo is technically very difficult, because skeletal muscle contains several cell types (e.g., muscle cells, endothelial cells, fibroblasts, and resident macrophages), which influence its responses to mechanical loading [14]. Interestingly, in the last decade ex vivo and in vitro models have been adopted to mimic in vivo muscle loading conditions, to reveal the cellular and molecular mechanisms of muscle adaption to exercise-induced mechanical loading [15]. A significant progress has been made in recent years regarding the underlying molecular responses of cellular mechanotransduction [7, 16, 17], and there is a growing body of evidence showing that mechanical cues modulate many fundamental aspects of muscle cell function, such as protein synthesis, proliferation, myogenic differentiation, and survival [18–22], (Fig. 7.3). The focus of this review has been directed mainly to the mechanisms through which mechanosensitive muscle cells sense external mechanical stimuli (mechanosensation) and convert them into electrochemical and biochemical signals (mechanotransduction), modulating transcriptional activities for a wide range of biological responses and crucial cellular functions.

Skeletal Muscle as the Central Driver of Mechanical Loading-Induced Responses and Adaptations

In response to physical exercise, both functional and structural adaptations occur in the body [23]. In particular, the architecture and metabolism of muscle tissue are extremely sensitive to exercise-induced mechanical and metabolic loading applied on it, and mechanotransduction is critical for the maintenance of muscle tissue homeostasis [11, 18, 24]. Skeletal muscle tissue accounts for over 35% of the adult human body and contains 50–75% of all body proteins [25]. Various factors, such as hormones, genetics, nutrition, growth factors and mainly mechanical stimuli regulate protein synthesis and skeletal muscle mass [26]. Indeed, mechanical stress (loading) is known to result in increased muscle mass (hypertrophy), while conversely, the absence of loading leads to decreased muscle mass (muscle wasting/atrophy) [27]. Moreover, alterations in mechanical loading have been shown to be important for the regulation of bone density and strength [28]. Skeletal muscle is mainly loaded by internal longitudinal and transverse forces of varying magnitude and velocity, generated either by active muscle contractions (shorting, isometric, or lengthening) or by passive stretching (Fig. 7.2). Evidence strongly indicates that

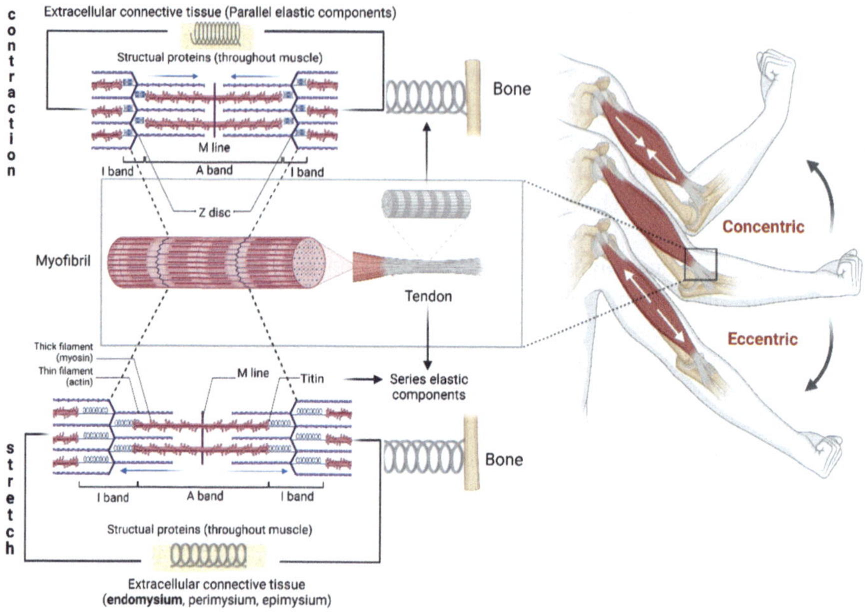

Fig. 7.2 A schematic diagram of skeletal muscle and its components under different mechanical conditions. In its details, the model illustrates the mechanical behavior of an individual sarcomere (located between two adjacent Z disks) and its contractile elements (actin and myosin) within a myofibril, as well as of the non-contractile components of the musculotendinous unit (including the myofibrillar/cytoskeletal protein titin and the extracellular connective tissue), during concentric (contraction) and eccentric (stretch) muscle action. Sarcomeres are repetitive structural and functional subunits of a myofibril extended along the length of the muscle fiber. A muscle fiber consists of multiple myofibrils, which contain the active proteins responsible for contraction within the fiber. The contractile proteins actin and myosin are the key myofilaments within the myofibril, arranged in unique structure and responsible for muscle contraction. This intricate process involves a complex physiological and mechanical interplay between the two proteins. In relaxed muscle fiber, myosin is partially superimposed by actin, while the degree of their overlapping varies during a concentric (dynamic sarcomere's shortening), or eccentric (active sarcomere's lengthening) muscle action. The A band, also known as the dark region within a sarcomere, is attributable to the presence of myosin (thick filament), while the luminous bands, known as I-bands, consist of actin thin (filaments). Myosin and actin, supported by intracellular structural proteins and aligned within the sarcomere, provide mechanical stability as the fiber contracts or elongates (stretch). Through the structural proteins of a muscle fiber and the extracellular connective tissue endomysium, myofiber eventually links with the tendon. This sophisticated connective network, established between intracellular structural proteins and extracellular connective tissue, distributes forces along the longitudinal and lateral/transverse axes of skeletal muscle. Specifically, the non-contractile elements (represented as extension coiled springs) are distinguished as either series or parallel elastic components of muscle. The tendon and the structural protein titin within the sarcomere are examples of series elastic components aligned with the contractile elements. Extracellular connective tissue, such as endomysium, and other structural proteins (e.g., perimysium and epimysium) located throughout the muscle represent parallel elastic constituents, aligned in parallel with the contractile elements. The differential mechanical behavior of the non-contractile elements during muscle contraction and stretching is exhibited [27, 59, 60]

skeletal muscle-derived mechanical loading is vital for bone mass maintenance [28, 29]. Specifically, skeletal muscles contribute to the mechanical loading applied on bones in several ways, including tensile, compression and bending forces produced by muscle contractions [29].Tensile forces produced by muscle contractions at the site of tendon-to-bone insertion (enthesis) is transmitted by the myotendinous junction (MTJ), a combination of muscle and tendon tissue that is located at the interface between bones and muscle and is a part of the musculotendinous unit, which is responsible for the production of skeletal movement. Despite being structurally distinct, muscle and tendons remain spatially and functionally connected. Tendons consist of a well-structured connective tissue capable of producing skeletal movement [30] through tension generated by muscle [31], (Fig. 7.2). In addition, the cytoskeleton and extracellular matrix (ECM) strength and stiffness modulate the muscle cells and tissue resistance in response to external forces. The mechanical connection between the ECM and the cytoskeleton across the cell surface enables the propagation of mechanical forces [5] (Fig. 7.1). Intracellularly, negligible forces, caused mainly by cytoskeleton's resistance and muscle contraction, result in strains on the ECM substrate [5, 32]. Tissue stresses are transmitted via deformation (stretch and/or shearing stress) from ECM to embedded cells via their matrix attachment [32]. Nevertheless, due to the ECM stiffness, resulting muscle cell strains or deformations are limited and the ECM safeguards enclosed cells against detrimental effects of excessive mechanical forces [33–35]. This small mechanical stimulation may elicit chemical signaling, mediated by mechanochemical coupling and leading to numerous cellular responses [36]. Moreover, skeletal muscle and its connective tissue respond effectively to external mechanical stresses by adapting to long-term loading conditions [5, 36]. In conditions where the ECM is weakened, such as following tissue injury, excessive tissue strain exerts forces on ECM fibroblasts, and may result in degradation, degeneration and apoptosis [37].

> A whole muscle is much more like a fiber–reiηforced composite than it is like a bundle of spaghetti [17].

Mechanical Loading-Induced Skeletal Muscle Deformation

Skeletal muscle deforms under tensile and compressive stress [38]. When muscle is subjected to one direction tension, it develops internal forces resulting in its contraction in the longitudinal axis. Conversely, in the transverse axis, compressive forces develop internal stresses that result in lengthening of muscle tissue. Muscle tissue deformation occurs due to either internal or external force application, while both the force application and the deformation may dramatically differ between myofibers and the whole muscle [17]. Interestingly, recent evidence identified a significant role of the ECM in establishing the mechanical response of skeletal muscle tissue and suggests that it contributes to the passive stress and balances the volume changes of muscle fibers during deformation, leading to more incompressible muscle tissue behavior [38].

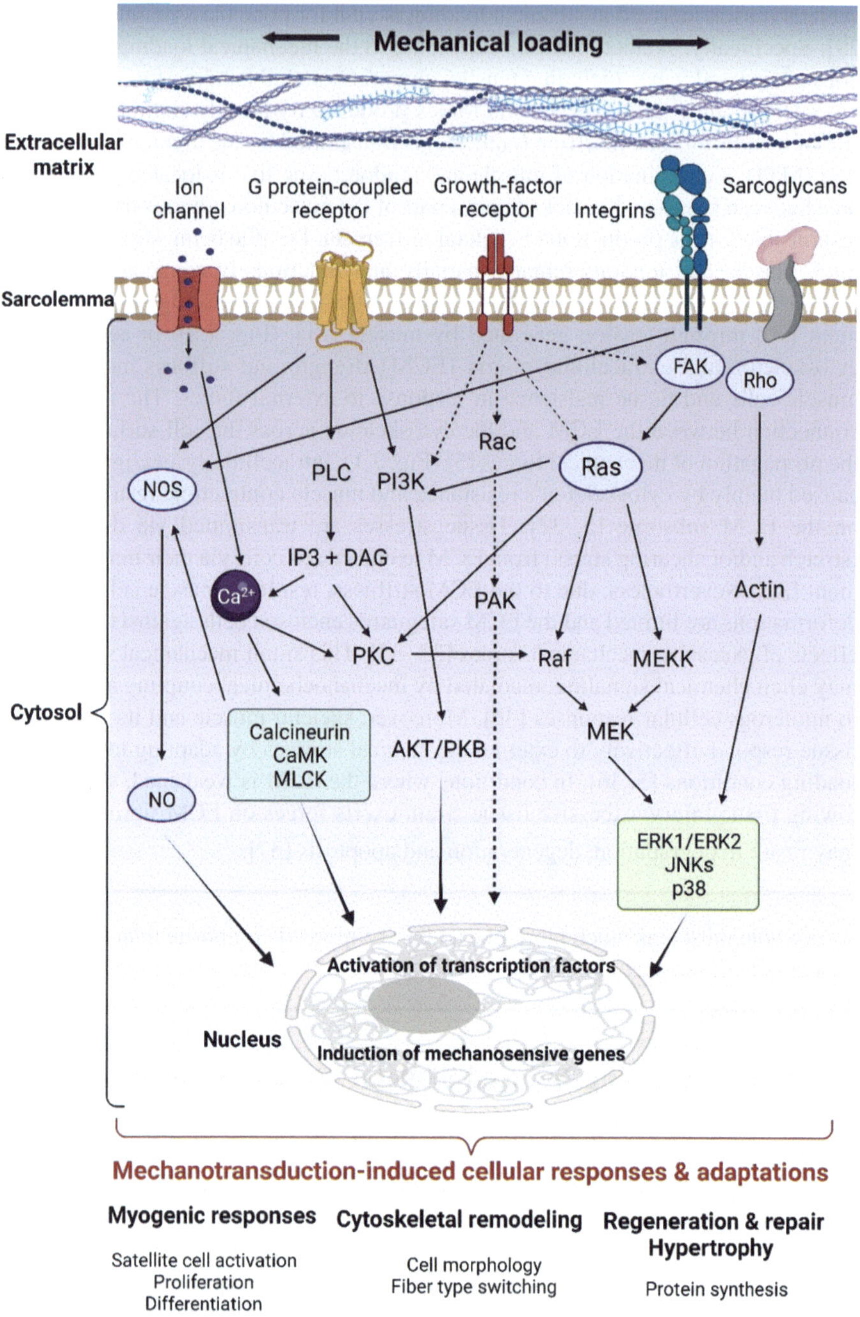

Fig. 7.3 Various intracellular signaling pathways involved in maintaining and regulating cell function are activated in muscle cells in response to mechanical loading. Mechanical force applied on muscle cells can be translated into biochemical signals via a complex group of mechanosensors

Models of Muscle Cell and Tissue Mechanical Deformation

Various models of force application and muscle cell/tissue deformation have been utilized to investigate their behavior under mechanical stimuli. Of exceptional value are the in vivo models of intact physiological systems. In addition, the ex vivo models of isolated skeletal muscle can closely mimic deformation of muscle tissue under various mechanical loading conditions. Moreover, single muscle fibers isolated either by microdissection or enzymatic dissociation, without neighboring cells or substrates' homogenizing influence, and embedded in a three-dimensional (3-D) culture can be used for force reconstruction and the measurement of 3-D matrix deformations and mechanical responses. Lastly, in vitro models of muscle cell lines or muscle precursor (satellite) cells that can be cultured on elastic membranes, differentiate into myotubes, and then be passively stretched, are also valuable for the study of cellular and molecular mechanisms of mechano-biochemical responses of muscle cells [14, 15, 19–22, 39, 40].

In the in vivo models, a muscle can be imposed to loads via its tendons (myotendinous junction), via distributed attachments to bones, via pressures imposed by bone, via fascial structures or other muscle's contacts and via connective tissue adhesions to adjacent tissues, resulting in myofascial force transmission [40]. In the in vivo models, the deformation characteristics vary among body muscles, and may be difficult to quantify depending on the conditions. For example, nearby muscles' activation and direct muscle deformation measurements under force generation are complex conditions [41]. Also, the loads and deformations of a muscle in an in vivo

Fig. 7.3 (continued) anchored on the cell membrane (sarcolemma) that includes stretch-gated ion channels, G protein-coupled receptors, growth factor receptors, integrins and sarcoglycans. These elements can transmit tension to intracellular proteins, activating a mechanical switch. Specifically, the mechanical stimulus is transformed into biochemical signals through an array of alternate intracellular cascades in the cytoplasm, triggered by the sequential phosphorylation and activation of multiple downstream molecules in the cytosol. These series of reactions include the activation, among others, of FAK, Rac and Ras molecules, NO and calcium signaling, as well as the PLC and PI3K pathways. These signaling cascades transmit information downstream, activating certain transcription factors that transit to the nucleus and result in the induction of mechanosensitive genes. Ultimately, the expression of these genes dictates not only the mechanotransduction-induced cellular responses and cell fate, but also the muscle cells' adaptations to mechanical stress. Indeed, mechanosensitivity and mechanotransduction regulate muscle cell homeostasis and growth balance. Muscle cells respond to the applied mechanical forces by reorganizing their cytoskeleton, synthesizing new proteins, and regulating various cellular functions and adaptations, such as satellite cell proliferation and differentiation, and muscle fiber type conversion, regeneration, and hypertrophy [72–79]. [Akt/PKB: Protein kinase b; Ca^{2+}: Calcium cation; CaMK: Ca2+/calmodulin-dependent protein kinase; DAG: Diacylglycerol; ERK: Extracellular-signal regulated kinase; FAK: Focal adhesion kinase; IP3: Inositol triphosphate; JNKs: c-Jun N-terminal kinases; MEK: Mitogen-activated protein kinase kinase; MEKK: Mammalian mitogen-activated protein kinase kinase kinase; MLCK: Myosin light-chain kinase; NO: Nitric oxide; NOS: Nitric-oxide synthase; p38: p38 mitogen-activated protein kinases; PAK: Family of serine/threonine p21-activated kinases; PKC: Protein kinase C; PLC: Phospholipase C; Rac: Subfamily of the Rho family of GTPases; Raf: Rapidly accelerated fibrosarcoma family kinases; Ras: Guanosine-nucleotide-binding protein; Rho: Rho-family GTPases]

model may be attenuated or amplified by shear forces associated with adjacent tissues. In contrast, in isolated muscle ex vivo models, bending and substantial transverse loads to which a muscle is subjected during in vivo conditions are not present.

Moreover, while single muscle fibers generally shorten under force generation, they may also rotate and twist [42]. Thus, the intramuscular deformations are difficult to be directly quantified [43], although muscle finite element models have been developed during force application [40–43]. These models have demonstrated that the deformation distribution within a muscle differs and shear strains, both perpendicular and parallel to the fiber axis, contribute to the overall deformation. Shear force transmission, commonly referred to as "lateral force", contributes to overall muscle force transmission, although there is notable morphological specialization at myotendinous junction suggesting its role in force transduction to ECM and tendon [42].

Deformation may also occur due to compressive loads, especially when skeletal muscle is relaxed. The structure of ECM is made up of a woven stiff collagen fibril network, and the mechanical behavior of a relaxed muscle is consistent with the matrix reorientation around myofibers of equal volume [44, 45]. Muscle fibers may experience deformation/elongation due to transverse compressive forces, rather than shear or tension on molecular bonds [45]. The volume fractions and the properties (e.g., stiffness) of the ECM and those of (active or passive fibers) affect the mechanical behavior of the entire muscle [44]. Interestingly, because of the generation of intramuscular pressure by active fibers [46], it is possible that passive fibers experience a transverse compression load as the primary force, rather than an axial tension load.

At a single fiber level, the force generated by contractile myofilaments actin and myosin is integrated into the cytoskeleton through tight junction at Z-disks parallel and perpendicular to the fiber axis [47], (Figs. 7.1 and 7.2). Moreover, the series and parallel elastic components of skeletal muscle, such as tendons and extracellular connective tissue (i.e., endomysium, perimysium, epimysium) in conjunction with its contractile elements of muscle fibers, contribute to passive force production [46, 48]. The distribution of passive tension between myofibers and ECM is nearly impossible to be estimated due to the need of determining the mechanical properties of the intramuscular connective tissue [42, 47] while, further, a differential mechanical behavior of the non-contractile elements occurs during concentric (contraction) and eccentric (stretch) muscle actions (Fig. 7.2).

In the in vitro models, the main distinction between cultured myotubes and whole muscle is the lack of integration of the former with a 3-D ECM and the lack of their extensive connections with nearby fibers. Such a mechanical interaction tends to homogenize distortions, meaning that the deformation gradients in cultured myotubes are smaller than they would be in an isolated fiber [46, 48]. The mechanotransduction in cell culture models variously differs from models of isolated muscle [49, 50]. Cell adhesions are considered dynamic structures, and their remodeling contributes to the rapid relaxation of cultured cells after static stretching [42, 45]. Although the contractile apparatus does not typically develop well in cell culture models [46], when cell adhesions are disturbed, the adherent cell rapidly contracts, suggesting that high tension is sustained on the culture surface, even when culture cells are not stressed [44, 48]. Cells cultured for long periods of time synthesize and assemble their own ECM [46], however the features of this ECM structure and its role in muscle cells stabilization under mechanical deformation, are not clear. In the in vitro models of cell mechanical loading (stretching), the 3-D deformation that cells undergo when their culture surface (elastic membrane) is stretched remains also unclear [51]. Nevertheless, the process by which force, and deformation are transmitted to adherent cells is well described, and various cell culture models have been effectively developed for studying mechanotransduction in vitro. Adhesion molecules like dystroglycans and integrins link the cytoskeletal system to membrane-bound ECM molecules (Figs. 7.1 and 7.3). As the elastic membrane is stretched, the ECM molecules move, causing shear forces on the adhesion molecules and cytoskeletal stresses that deform the cultured cells. This model is appealing because it largely replicates the force transmission mechanisms that are expected to occur during force generation in an intact muscle.

It should be mentioned, however, that there is a dramatic stiffness mismatch between the elastic cell culture surface and the cells themselves, in terms that the strain is determined by the culture surface and that culture-based strains, even uniaxial ones, limit both dimensions of the adherent interface. In culture, the media-facing side is not subject to shear stress but is deformed similarly to the elastic membrane and, thus, for the cells to maintain their isovolumetric state during stretching, they need to contract only through their free dimension. This contrasts sharply with the in vivo stretching, where the surrounding cells have a poor resistance to transverse constriction that causes deformation in one dimension. Overall, when placed under stretching in a two-dimensional culture system, cells are likely to experience unusual transverse strains [52].

Cultured myotubes exhibit different properties along their longitudinal axis compared to their diameter (lateral/transverse axis), and thus different cellular responses to strain would be expected along these different axes [53]. In biaxial systems, pressure applied to an elastic silicone membrane causes significant deformation in two dimensions. Resulting strain is complex and varies across the culture surface [54]. Interestingly, when a large flat indenter applies pressure, deformation is equal in radial and tangential directions, producing an equibiaxial pattern. Equiaxial deformation has the unique property that there is no orientation in the plane, resulting in myotubes attached in any direction experiencing the same deformation [53].

> It is rational to investigate exercise mimetics to balance the benefits and drawbacks of exercise [55].

Cellular Deformation

Cellular or molecular deformation is the critical event for mechanotransduction. Deformation of the adhesive complexes and contractile apparatuses are usually the most obvious mechanical changes associated with skeletal muscle loading. The contractile apparatus is a prominent structure within a muscle fiber and could serve as a useful marker for cell body deformation (Fig. 7.2). The M-line and Z-disks rigidly immobilize the interdigitating arrays of thin and thick filaments of a sarcomere, while titin tethers the thick filaments to the Z-disk. As the fiber elongates, the filaments slide relative to one another. The thick and thin filaments have a moderate level of extensibility. When cross-bridges are present, a reduction in the ease with which the A- and I-bands slide occurs. The sarcomere's complex structures M-line and Z-disks include various biologically active molecules, which can be modulated by mechanical signals [42, 43, 56]. When tension develops in the myofilaments, the M-line and Z-disks, as well as the molecules that anchor these structures to the ECM may be protected from deformation by other structures, such as desmin and titin [56]. Titin binds to both the thick filaments and the Z-disk, becoming loaded and elongated as the sarcomere extends [42, 43]. Moreover, titin and other structures may experience deformation due to kinase activation [56, 57]. Interestingly, myofibrils on the convex side experience slightly more tension than those on the concave side, creating a perpendicular shear force to the fiber axis. These loads can separate adjacent myofibrils, but the Z-disks of adjacent myofibrils consistently maintain a registry over a wide range of sarcomere lengths [42, 43]. The intermediate filament desmin, which connects adjacent Z-disks, is essential for preserving sarcomere alignment during force production [40]. This indicates that desmin undergoes loading and elongation during cellular deformation [57].

The sarcomere is attached to the ECM through adhesions that include both integrins and dystroglycans [58], (Figs. 7.1 and 7.3), while mechanical loading can stimulate the activation of signaling pathways associated with these complexes [43, 56, 58], (Figs. 7.3 and 7.4). The transmission of force from the ECM to the cytoskeleton and contractile apparatus has been an extensively reviewed topic [5, 37, 42, 44], and specialized structures such as costameres or focal adhesions, which are abundant in both adhesion and signaling molecules [37, 47], have been described as

Fig. 7.4 (continued) the Ca^{2+}/CaM/JNK axis. Inside the muscle cells, the sarcoplasmic reticulum (SR) holds the calcium stores, while its release is mainly controlled by InsP3R. Calcium's return into the SR is regulated by SERCA pumps, providing a steady concentration to the muscle cell, since it constitutes a vital component for the contraction of the sarcomeres. Actin interacts with the activity of calcium channels and its assembly, as part of the cytoskeletal scaffold, is calcium dependent. Actin metabolism is regulated by small G proteins through their control on cytoskeletal elements (dystrophin, filamin) reorganization. Muscle cell mechanosensation is attributed to extracellular laminin, which is connected to membrane-anchored sarcoglycans and activates intracellular FAK and calcium. The FAK pathway consists of Ras-activated Akt and MEK (through phosphorylation), and the latter in turn phosphorylates downstream ERK (also phosphorylated by G protein coupled-receptors), JNK/SAPK and p38. Intracellular FAK is also activated by integrins, which both link the extracellular matrix to the cytoskeleton [80–82]. [Akt: Protein kinase b; Ca^{2+}: Calcium cation; CaM: Calmodulin; ERK: Extracellular-signal regulated kinase; FAK: Focal adhesion kinase; InsP3R: Inositol trisphosphate receptor; JNK: Jun amino-terminal kinases; MEK: Mitogen-activated protein kinase; mTORC1: mTOR Complex 1; NFAT: Nuclear factor of activated T cells protein; nNOS: Neuronal NOS; NO: Nitric oxide; NOS: Nitric-oxide synthase; p38: p38 mitogen-activated protein kinases; PKG: Protein kinase G; Ras: Guanosine-nucleotide-binding protein; SAPK: Stress-activated protein kinases; SERCA: Sarcoendoplasmic reticulum calcium ATPase pump; α-DG: Alpha dystroglycan; β-DG: Beta dystroglycan]

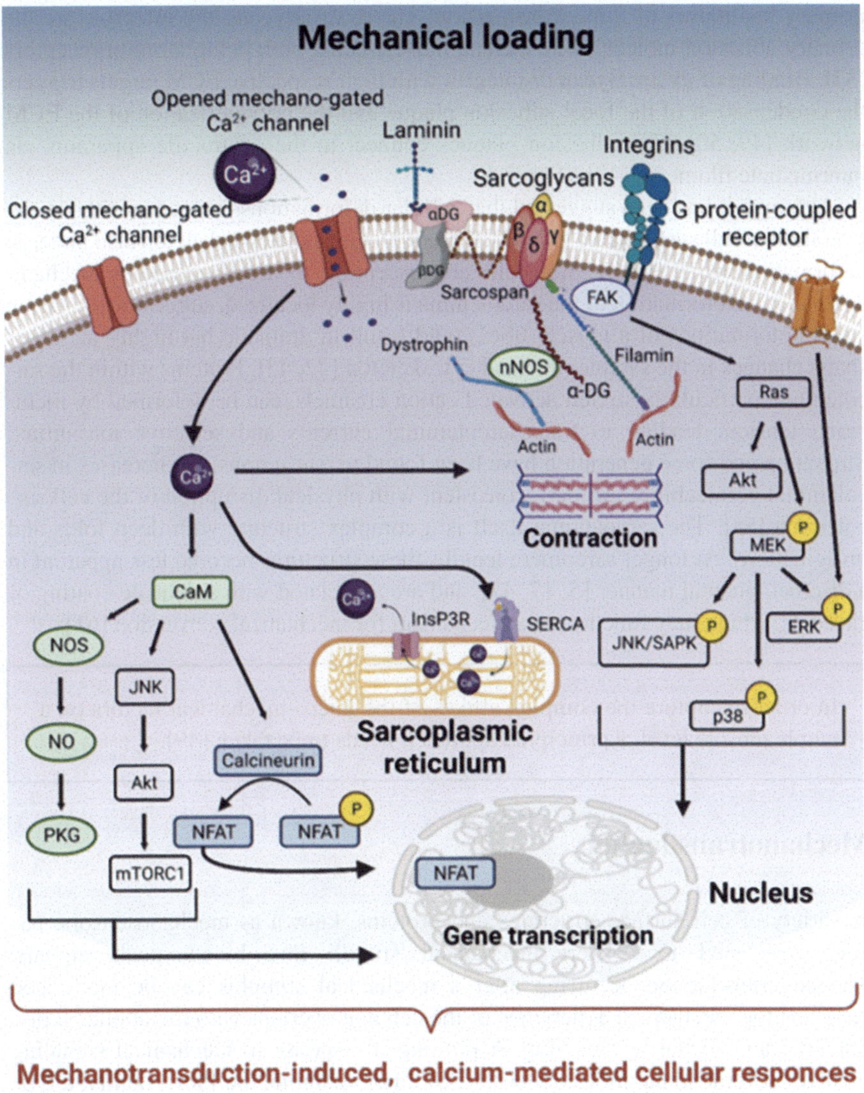

Fig. 7.4 Mechanotransduction-induced calcium-mediated cellular responses. Due to the application of mechanical forces, mechano-gated Ca^{2+} channels open, allowing the entrance of calcium ions inside the muscle cell and triggering the activation of downstream signaling pathways and intracellular processes. Specifically, after binding to calmodulin (CaM), calcium stimulates the calcineurin protein phosphatase activity, which in turn causes NFAT to dephosphorylate and translocate into the cell nucleus, initiating gene transcription. In addition, the production of NO through a calcium-dependent mechanism negatively regulates anabolic processes in skeletal muscle cells, utilizing the classic NO/PKG signaling pathway to impact protein and energy metabolism. Moreover, JNK is an essential mechanosensitive kinase that integrates with mTOR signaling responses to the applied forces on muscle cells. A possible signaling pathway that connects loading-induced ion channel activation and the upregulation of muscle protein synthesis involves

primary facilitators of force transmission. The dystroglycans and integrins are the primary adhesion molecules in skeletal muscle cells, both being laminin receptors [37]. Binding of dystroglycan or integrin with their respective ECM targets triggers the condensation of the focal adhesion plaque and the polymerization of the ECM network [49, 50]. The adhesion plaques connect to the contractile apparatus via intermediate filaments [40], (Fig. 7.1).

Although it has been suggested that cellular deformations can occur without any alteration in cell surface area [17, 43], the sarcolemma and cytoskeleton also undergo deformation, while specific molecules are susceptible to stretch caused by mechanical loading. Deformation of the sarcolemma is highly localized, suggesting that even small deformations of a muscle fiber could result in dramatic but highly localized shape changes in the sarcolemma and cytoskeleton [42, 43]. Proteins within the sarcolemma, particularly stretch activated cation channels, can be deformed by membrane tension, leading to trans-sarcolemmal currents and selective ion influx. Stretching and force generation have been found to cause transient increases in sarcolemmal permeability [56, 58], consistent with physical disruption of the cell sarcolemma [58]. The sarcolemma itself is a complex structure with deep folds and invaginations. At longer sarcomere lengths these structures become less apparent in a discreet, quantal manner [5, 17, 47], and are associated with a delicate coating of caveolae, which may function as a mechanism for mechanical conversion [61].

> In order to capture the complex effects of the micro-mechanical factors on a whole muscle level, a principled approach needs to be taken [49].

Mechanotransduction

A variety of cell surface structures and proteins, known as mechanosensors, has been suggested to convert mechanical stimuli into biochemicals signals. Mechanotransduction occurring after a mechanical stimulus can be modulated based on the mechanical experience of the cell and overlaps with metabolic, action potential and oxidative signaling. A prolonged response to mechanical signaling involves several signaling cascades [17, 57]. For example, the Piezo family, a type of mechanically activated ion channels, serve as critical mechanosensors in cells, detecting various forms of mechanical forces, such as membrane or fluid shear stretch and static pressure [62–64]. Through mechanosensors, several different cell types, including myocytes, demonstrate common features in their activation of sequential signaling pathways and the expression of downstream target genes [65]. Mechanotransduction can occur through direct physical connections between membranes, cytoskeletal structures and nuclei, which triggers gene expression and protein synthesis [66]. Extracellular mechanical forces are transmitted to the cell nucleus through the interplay between mechanosensors, their corresponding intracellular signaling pathways and a variety of second messengers which can modulate transcriptional activities for crucial cellular functions, such as growth and survival. In skeletal muscle, mechanical signaling has been linked to satellite cell activation, cell growth, protein synthesis, and cytoskeletal remodeling, regeneration and growth factor release [17, 67–70], (Figs. 7.3, 7.4 and 7.5). Indeed, mechanical loading-induced changes in muscle mass are closely linked to alterations in protein synthesis, which affect the net difference between the rates of protein breakdown and synthesis and ultimately result in muscle hypertrophy [27, 71], (Fig. 7.3).

Significant signaling cascades are engaged in the sustained cellular responses and adaptations to mechanical stimuli, including but not limited to Phosphoinositide 3-kinase (PI3-K), diverse Mitogen-activated protein kinases (MAPKs), calcium, and AMP-activated protein kinase (AMPK) [17, 56]. These signaling pathways may be regarded as downstream effectors of the mechanotransduction process (Figs. 7.3 and 7.4).

A critical role in the intracellular transmission of mechanical signals is played by focal adhesion kinase (FAK), an attachment protein related to integrin-based adhesion complexes [83], (Fig. 7.4). FAK, through contractile forces mediated by Ras homolog family member A (RhoA), regulates the transcriptional coactivators yes-associated protein (YAP) and transcriptional coactivator with PDZ-binding motif (TAZ), which are members of the Hippo signaling pathway and play a critical role in cell growth regulation, proliferation, and regeneration. In cells experiencing high levels of mechanical signaling, YAP and TAZ translocate from the cytoplasm to the nucleus [84] and this nuclear translocation is regulated by Hippo signaling pathway [85]. The inhibition of the Hippo pathway caused by multiple mechanical stimuli, promotes YAP/TAZ to enter the nucleus, thus activating genes involved in cell cycle and proliferation [86]. Suppression of Hippo signaling is important in promoting tissue regeneration and suggests a new mechanism of intervention for exercise-induced muscle regeneration [87].

Moreover, mechanosensitive Piezo and certain integrins sense mechanical loading that results in ligand transcription of signal sending cell and Notch signaling regulation in signal receiving cell, which in turn modulates proliferation and differentiation [88]. It has been postulated that impaired Notch signaling in regenerating aged skeletal muscle can be restored by exercise-induced mechanical stimuli [89]. In addition, the contribution of the PI3-K/Akt signaling pathway has been characterized as essential to exercise-induced regeneration and hypertrophy [90], and various growth factors, such as brain-derived neurotrophic factor (BDNF) [91] and Insulin-like Growth Factor 1 (IGF-1) [92] trigger these PI3K/Akt pathway-mediated cellular responses in skeletal muscle.

Specifically, IGF-1, one of the most studied exercise-induced factors, triggers the activation of PI3K/Akt signaling pathway, regulating myoblasts proliferation and differentiation [93–95]. A variety of downstream responses to PI3K/Akt pathway have been described. For example, mammalian target of rapamycin (mTOR), a serine/threonine kinase that exists in two distinct complexes (mTORC1 and mTORC2) constitutes a significant target of Akt [96]. More specifically, mTORC1 activation has been reported as an adaptive response to mechanical loading and the activation of the mTOR-axis is critical in the process of regeneration in different tissues, including skeletal muscle [97, 98]. Moreover, IGF-1-induced muscle hypertrophy requires FAK, acting through tuberous sclerosis complex 2 (TSC2)/mammalian target of rapamycin (mTOR)/ribosomal S6 kinase 1 (S6K1)–dependent signaling pathway [99].

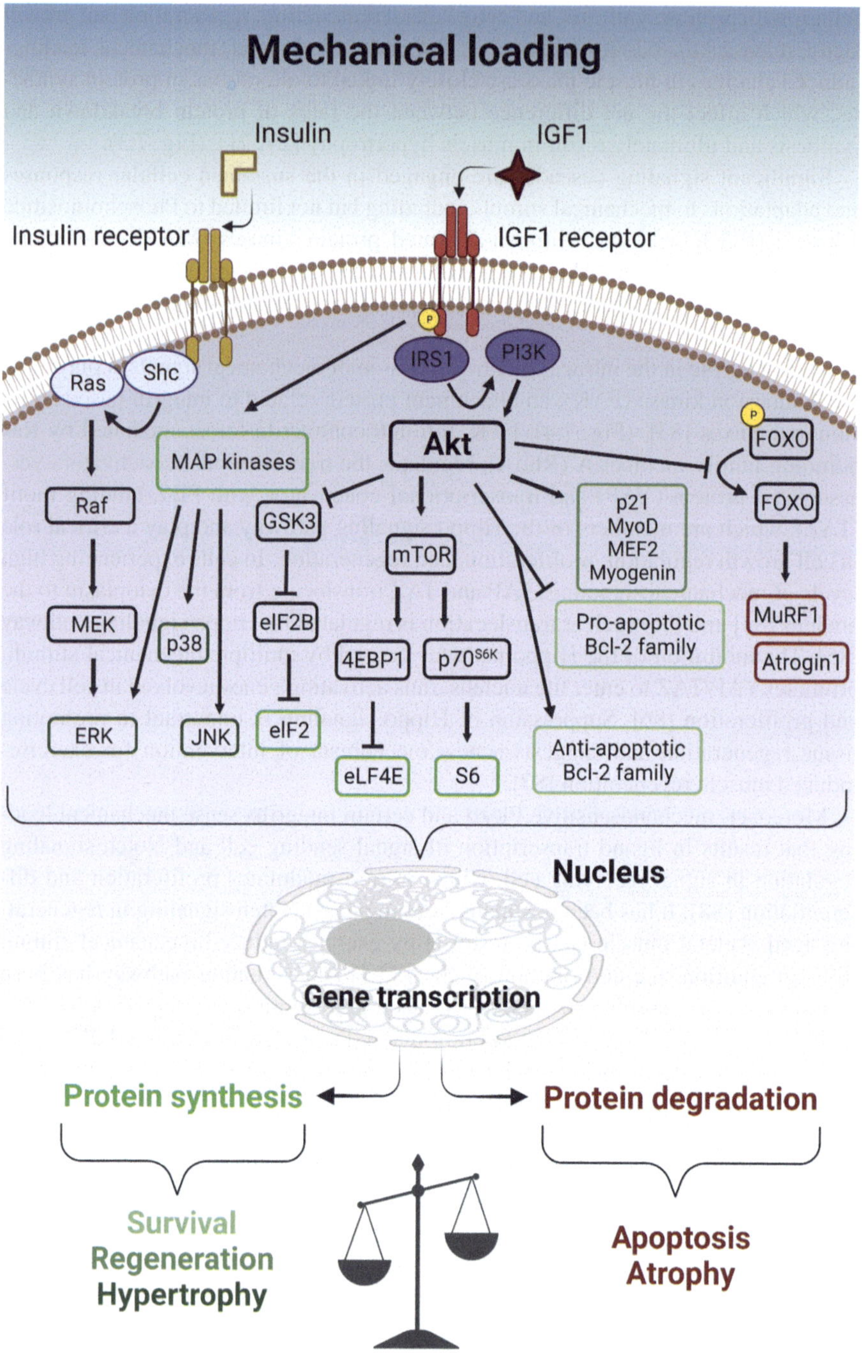

Fig. 7.5 Mechanotransduction-induced, growth factor-mediated cellular responses and adaptations. Skeletal muscle response to mechanical signals is a complex process regulated by multiple signaling pathways including those of growth factors. Mechanical loading-induced alterations in

In addition, Forkhead box class O family (FOXOs) present a potential as downstream mediators of the exercise-induced cell proliferation and growth (Fig. 7.5). Moreover, the Akt/FOXO3 pathway is activated in exercise-induced autophagy, which is effective in suppressing sarcopenia [100]. Additionally, it appears that Akt mediates FOXO inhibition in the regulation of stem cell proliferation; when the activity of FOXOs is inhibited, myostatin expression is decreased and satellite cell proliferation and fusion is increased, leading to muscle hypertrophy [100], (Fig. 7.5).

Also, several biological activities are mediated by the extracellular signal-regulated kinases 1 and 2 (ERK1/2), p38 mitogen-activated protein kinase (MAPK), and c-Jun N-terminal kinase (JNK), as they represent key members of the MAPK family [103]. In skeletal muscle, phosphorylation of JNK and p38 MAPK, similarly to PI3K/Akt signaling, leads to the overexpression of myogenic regulatory factors (MRFs), myogenin and MyoD [104]. In addition, skeletal muscle, as a high-energy demanding tissue, is considered a core regulator of metabolic pathways activated by exercise, such as the adenosine monophosphate-activated protein kinase (AMPK)-sirtuin 1 (SIRT1)—peroxisome proliferator-activated receptor-gamma (PPAR)-γ coactivator-1α (PGC-1a) pathway [55, 105]. PGC-1α promotes exercise-induced tissue regeneration and is involved in mitochondrial signaling [106]. Specifically, PGC-1α has been shown to act as a regulator of cell proliferation and differentiation

Fig. 7.5 (continued) protein synthesis seem to be regulated by a coordinated array of biochemical events. Using IGF-1 or insulin stimulation as examples, a schematic representation of the signaling events involved in the regulation of protein synthesis and degradation is shown. IGF-1 stimulates tissue growth by promoting protein synthesis, cell survival, proliferation, and differentiation, while its growth-promoting effects are primarily mediated via interaction with its type I receptor (IGF-1R). Binding of IGF-1 to its receptor induces its phosphorylation. Subsequently, acting via the MAP kinase pathway, it activates factors of cell cycle progression, as also insulin does after binding to its receptor and activating the Ras/Raf/MEK pathway. In addition, IGF-1 activates Akt via its action on PI3-K, which then modulates the expression of factors associated with muscle cell terminal differentiation, such as MyoD, Myogenin, MEF2, and P21. Moreover, pro-apoptotic Bcl-2 family proteins are inhibited due to the activation of PI3-K/Akt pathway, which also causes the induction of anti-apoptotic Bcl-2 family proteins (Bcl-X), implicating IGF-1 in cell survival. In addition, the activation of PI3-K/Akt/GSK3 and mTOR pathways lead to the induction of protein synthesis and muscle hypertrophy. Also, Akt inhibits dephosphorylation of FOXO and thus the transcription of the muscle atrophy genes atrogin-1 and MuRF1, as well as of E3 ubiquitin ligases responsible for protein degradation. The different frame colors of the factors and signaling proteins discussed correspond to the respective responses and biological effects, i.e., red color indicates protein degradation and muscle atrophy-inducing factors, while the green color represents signaling proteins and factors associated with protein synthesis and muscle hypertrophy [93, 94, 101, 102]. [4EBP1: Eukaryotic translation initiation factor 4E (eLF4E)-binding protein 1; Akt: Protein kinase b; Bcl-2: B-cell lymphoma 2; elF2: E74-like factor 2; elF2B: Eukaryotic initiation factor 2B; eLF4E: Eukaryotic translation initiation factor 4E; ERK: Extracellular-signal regulated kinase; FOXO: Forkhead box protein O; GSK3: Glycogen synthase kinase 3; IGF1: Insulin-like growth factor 1; IRS1: Insulin receptor substrate 1; JNK: Jun amino-terminal kinases; MAP kinase: Mitogen-activated protein kinase; MEF2: Myocyte enhancer factor-2 protein; MEK: Mitogen-activated protein kinase kinase; mTOR: Mechanistic Target of Rapamycin; MuRF1: Muscle RING-finger protein-1; MyoD: Myoblast determination protein; p38: p38 mitogen-activated protein kinases; $p70^{S6K}$: Ribosomal protein S6 kinase beta-1; PI3K: Phosphoinositide 3-kinase; Raf: Rapidly accelerated fibrosarcoma family kinases; Ras: Guanosine-nucleotide-binding protein; S6: Ribosomal protein S6]

and being involved in mitochondrial biosynthesis and cellular respiration [107]. Moreover, overexpression of PGC-1α is strongly induced by mechanical loading and may trigger satellite cell remodeling [108].

> Careful consideration of the mechanical environment is the first step to sort out the labyrinthine mechanotransduction signaling [17].

Conclusion

Exercise affects many vital processes and biological functions in the human body, acting as a multistressor for most of the organ systems and inducing their adaptive responses to exercise stress. Skeletal muscle is an organ on which exercise exerts multiple stress and which possesses the ability to adapt to mechanical stimuli by changing its mass and overall phenotype. When muscle tissue is subjected to mechanical deformation, mechanosensitive complexes (mechanosensors) situated in the muscle cell membrane sense mechanical alterations and activate intracellular signaling pathways (mechanotransduction). Mechanotransduction can occur through direct physical connections between membranes, cytoskeletal structures, and nuclei to trigger gene expression and protein synthesis. Mechanical signaling results in a plethora of acute molecular and cellular responses, such as growth factor release and satellite cell activation, proliferation, and differentiation, as well as in chronic adaptations like regeneration, cytoskeletal remodeling, and hypertrophy. A better understanding the complexity of mechanotransduction mechanisms that mediate this wide range of biological responses and crucial functions of skeletal muscle would result in the characterization of exercise stress-induced muscle adaptations and the development of new modalities in skeletal muscle performance and therapeutics.

- Skeletal muscle tissue exhibits a remarkable plasticity possessing the ability to adapt to exercise-induced mechanical and metabolic loading.
- Changes in cellular and molecular conformations, because of deformation imposed on muscle, link the mechanical forces with intracellular signaling pathways,
- The ability of mechanosensitive muscle cells to sense external mechanical stimuli and convert them into electrochemical and biochemical signals is critical in the regulation of essential cellular processes and adaptations, such as proliferation, myogenic differentiation, and hypertrophy.
- Defining the molecular mechanisms that mediate muscle adaptions to exercise-induced mechanical loading would contribute to the development of more effective modalities for optimizing muscle performance and treating various pathological conditions, such as sarcopenia, muscle dystrophy, and muscle damage due to exercise or myopathies.

References

1. French AS. Mechanotransduction. Annu Rev Physiol. 1992;54:135–52. https://doi.org/10.1146/annurev.ph.54.030192.001031.
2. Jaalouk DE, Lammerding J. Mechanotransduction gone awry. Nat Rev Mol Cell Biol. 2009;10(1):63–73. https://doi.org/10.1038/nrm2597.
3. Chen CS. Mechanotransduction – a field pulling together? J Cell Sci. 2008;121(Pt 20):3285–92. https://doi.org/10.1242/jcs.023507.
4. Ingber DE. Cellular mechanotransduction: putting all the pieces together again. FASEB J. 2006;20(7):811–27. https://doi.org/10.1096/fj.05-5424rev.
5. Martino F, Perestrelo AR, Vinarský V, Pagliari S, Forte G. Cellular mechanotransduction: from tension to function. Front Physiol. 2018;9:824. https://doi.org/10.3389/fphys.2018.00824.
6. De R, Zemel A, Safran SA. Theoretical concepts and models of cellular mechanosensing. Methods Cell Biol. 2010;98:143–75. https://doi.org/10.1016/S0091-679X(10)98007-2.
7. Wang N. Review of cellular mechanotransduction. J Phys D Appl Phys. 2017;50(23):233002. https://doi.org/10.1088/1361-6463/aa6e18.
8. Koenderink GH, Dogic Z, Nakamura F, et al. An active biopolymer network controlled by molecular motors. Proc Natl Acad Sci USA. 2009;106(36):15192–7. https://doi.org/10.1073/pnas.0903974106.
9. Orr AW, Helmke BP, Blackman BR, Schwartz MA. Mechanisms of mechanotransduction. Dev Cell. 2006;10(1):11–20. https://doi.org/10.1016/j.devcel.2005.12.006.
10. Lieber RL, Roberts TJ, Blemker SS, Lee SSM, Herzog W. Skeletal muscle mechanics, energetics and plasticity. J Neuroeng Rehabil. 2017;14(1):108. https://doi.org/10.1186/s12984-017-0318-y.
11. Tidball JG. Mechanical signal transduction in skeletal muscle growth and adaptation. J Appl Physiol (1985). 2005;98(5):1900–8. https://doi.org/10.1152/japplphysiol.01178.2004.
12. Steinbacher P, Eckl P. Impact of oxidative stress on exercising skeletal muscle. Biomol Ther. 2015;5(2):356–77. https://doi.org/10.3390/biom5020356.
13. Petriz BA, Gomes CPC, Almeida JA, et al. The effects of acute and chronic exercise on skeletal muscle proteome. J Cell Physiol. 2017;232(2):257–69. https://doi.org/10.1002/jcp.25477.
14. Boonen KJM, Langelaan MLP, Polak RB, van der Schaft DWJ, Baaijens FPT, Post MJ. Effects of a combined mechanical stimulation protocol: value for skeletal muscle tissue engineering. J Biomech. 2010;43(8):1514–21. https://doi.org/10.1016/j.jbiomech.2010.01.039.
15. Kasper AM, Turner DC, Martin NRW, Sharples AP. Mimicking exercise in three-dimensional bioengineered skeletal muscle to investigate cellular and molecular mechanisms of physiological adaptation. J Cell Physiol. 2018;233(3):1985–98. https://doi.org/10.1002/jcp.25840.
16. DuFort CC, Paszek MJ, Weaver VM. Balancing forces: architectural control of mechanotransduction. Nat Rev Mol Cell Biol. 2011;12(5):308–19. https://doi.org/10.1038/nrm3112.
17. Burkholder TJ. Mechanotransduction in skeletal muscle. Front Biosci. 2007;12:174–91. https://doi.org/10.2741/2057.
18. Wozniak MA, Chen CS. Mechanotransduction in development: a growing role for contractility. Nat Rev Mol Cell Biol. 2009;10(1):34–43. https://doi.org/10.1038/nrm2592.
19. Moustogiannis A, Philippou A, Zevolis E, Taso O, Chatzigeorgiou A, Koutsilieris M. Characterization of optimal strain, frequency and duration of mechanical loading on skeletal myotubes' biological responses. In Vivo. 2020;34(4):1779–88. https://doi.org/10.21873/invivo.11972.
20. Zevolis E, Philippou A, Moustogiannis A, Chatzigeorgiou A, Koutsilieris M. The effects of mechanical loading variations on the hypertrophic, anti-apoptotic, and anti-inflammatory responses of differentiated cardiomyocyte-like H9C2 cells. Cells. 2022;11(3):473. https://doi.org/10.3390/cells11030473.
21. Zevolis E, Philippou A, Moustogiannis A, Chatzigeorgiou A, Koutsilieris M. Optimizing mechanical stretching protocols for hypertrophic and anti-apoptotic responses in

cardiomyocyte-like H9C2 cells. Mol Biol Rep. 2021;48(1):645–55. https://doi.org/10.1007/s11033-020-06112-z.
22. Moustogiannis A, Philippou A, Zevolis E, et al. Effect of mechanical loading of senescent myoblasts on their myogenic lineage progression and survival. Cells. 2022;11(24):3979. https://doi.org/10.3390/cells11243979.
23. Pillon NJ, Gabriel BM, Dollet L, et al. Transcriptomic profiling of skeletal muscle adaptations to exercise and inactivity. Nat Commun. 2020;11(1):470. https://doi.org/10.1038/s41467-019-13869-w.
24. Laskou F, Dennison E. Interaction of nutrition and exercise on bone and muscle. Eur Endocrinol. 2019;15(1):11–2. https://doi.org/10.17925/EE.2019.15.1.11.
25. Janssen I, Heymsfield SB, Wang ZM, Ross R. Skeletal muscle mass and distribution in 468 men and women aged 18–88 yr. J Appl Physiol (1985). 2000;89(1):81–8. https://doi.org/10.1152/jappl.2000.89.1.81.
26. DiGirolamo DJ, Clemens TL, Kousteni S. The skeleton as an endocrine organ. Nat Rev Rheumatol. 2012;8(11):674–83. https://doi.org/10.1038/nrrheum.2012.157.
27. Goodman CA. The role of mTORC1 in regulating protein synthesis and skeletal muscle mass in response to various mechanical stimuli. Rev Physiol Biochem Pharmacol. 2014;166:43–95. https://doi.org/10.1007/112_2013_17.
28. Hart NH, Nimphius S, Rantalainen T, Ireland A, Siafarikas A, Newton RU. Mechanical basis of bone strength: influence of bone material, bone structure and muscle action. J Musculoskelet Neuronal Interact. 2017;17(3):114–39.
29. Lu TW, Taylor SJ, O'Connor JJ, Walker PS. Influence of muscle activity on the forces in the femur: an in vivo study. J Biomech. 1997;30(11–12):1101–6. https://doi.org/10.1016/s0021-9290(97)00090-0.
30. Mikolajewicz N, Sehayek S, Wiseman PW, Komarova SV. Transmission of mechanical information by purinergic signaling. Biophys J. 2019;116(10):2009–22. https://doi.org/10.1016/j.bpj.2019.04.012.
31. Felsenthal N, Zelzer E. Mechanical regulation of musculoskeletal system development. Development. 2017;144(23):4271–83. https://doi.org/10.1242/dev.151266.
32. Yan R, Zhang H, Ma Y, et al. Discovery of muscle-tendon progenitor subpopulation in human myotendinous junction at single-cell resolution. Research (Wash D C). 2022;2022:9760390. https://doi.org/10.34133/2022/9760390.
33. Jakobsen JR, Krogsgaard MR. The myotendinous junction – a vulnerable companion in sports. A narrative review. Front Physiol. 2021;12:635561. https://doi.org/10.3389/fphys.2021.635561.
34. Magnusson SP, Narici MV, Maganaris CN, Kjaer M. Human tendon behaviour and adaptation, in vivo. J Physiol. 2008;586(1):71–81. https://doi.org/10.1113/jphysiol.2007.139105.
35. Carpenter RD, Carter DR. The mechanobiological effects of periosteal surface loads. Biomech Model Mechanobiol. 2008;7(3):227–42. https://doi.org/10.1007/s10237-007-0087-9.
36. Di X, Gao X, Peng L, et al. Cellular mechanotransduction in health and diseases: from molecular mechanism to therapeutic targets. Signal Transduct Target Ther. 2023;8(1):282. https://doi.org/10.1038/s41392-023-01501-9.
37. Humphrey JD, Dufresne ER, Schwartz MA. Mechanotransduction and extracellular matrix homeostasis. Nat Rev Mol Cell Biol. 2014;15(12):802–12. https://doi.org/10.1038/nrm3896.
38. Böl M, Kohn S, Leichsenring K, Morales-Orcajo E, Ehret AE. On multiscale tension-compression asymmetry in skeletal muscle. Acta Biomater. 2022;144:210–20. https://doi.org/10.1016/j.actbio.2022.03.034.
39. Rausch M, Böhringer D, Steinmann M, et al. Measurement of skeletal muscle fiber contractility with high-speed traction microscopy. Biophys J. 2020;118(3):657–66. https://doi.org/10.1016/j.bpj.2019.12.014.
40. Ajimsha MS, Shenoy PD, Surendran PJ, Jacob P, Bilal MJ. Evidence of in-vivo myofascial force transfer in humans – a systematic scoping review. J Bodyw Mov Ther. 2022;32:183–95. https://doi.org/10.1016/j.jbmt.2022.05.006.

41. Zhang J, Zhao Y, Shone F, et al. Physics-informed deep learning for musculoskeletal modelling: predicting muscle forces and joint kinematics from surface EMG. IEEE Trans Neural Syst Rehabil Eng. 2022; https://doi.org/10.1109/TNSRE.2022.3226860.
42. Roberts TJ, Eng CM, Sleboda DA, et al. The multi-scale, three-dimensional nature of skeletal muscle contraction. Physiology (Bethesda). 2019;34(6):402–8. https://doi.org/10.1152/physiol.00023.2019.
43. Purslow PP. The structure and role of intramuscular connective tissue in muscle function. Front Physiol. 2020;11:495. https://doi.org/10.3389/fphys.2020.00495.
44. Lieber RL, Binder-Markey BI. Biochemical and structural basis of the passive mechanical properties of whole skeletal muscle. J Physiol. 2021;599(16):3809–23. https://doi.org/10.1113/JP280867.
45. He X, Taneja K, Chen JS, et al. Multiscale modeling of passive material influences on deformation and force output of skeletal muscles. Int J Numer Method Biomed Eng. 2022;38(4):e3571. https://doi.org/10.1002/cnm.3571.
46. Ando R, Suzuki Y. Positive relationship between passive muscle stiffness and rapid force production. Hum Mov Sci. 2019;66:285–91. https://doi.org/10.1016/j.humov.2019.05.002.
47. Wang Z, Raunser S. Structural biochemistry of muscle contraction. Annu Rev Biochem. 2023;92:411–33. https://doi.org/10.1146/annurev-biochem-052521-042909.
48. Wang AB, Perreault EJ, Royston TJ, Lee SSM. Changes in shear wave propagation within skeletal muscle during active and passive force generation. J Biomech. 2019;94:115–22. https://doi.org/10.1016/j.jbiomech.2019.07.019.
49. Konno RN, Nigam N, Wakeling JM. Modelling extracellular matrix and cellular contributions to whole muscle mechanics. PLoS One. 2021;16(4):e0249601. https://doi.org/10.1371/journal.pone.0249601.
50. Saraswathibhatla A, Indana D, Chaudhuri O. Cell-extracellular matrix mechanotransduction in 3D. Nat Rev Mol Cell Biol. 2023;24(7):495–516. https://doi.org/10.1038/s41580-023-00583-1.
51. Cantini M, Donnelly H, Dalby MJ, Salmeron-Sanchez M. The plot thickens: the emerging role of matrix viscosity in cell Mechanotransduction. Adv Healthc Mater. 2020;9(8):e1901259. https://doi.org/10.1002/adhm.201901259.
52. Apa L, Carraro S, Pisu S, Peruzzi B, Rizzuto E, Prete ZD. Development and validation of a device for in vitro uniaxial cell substrate deformation with real-time strain control. Meas Sci Technol. 2020;31(12):125702. https://doi.org/10.1088/1361-6501/aba011.
53. Furusawa K, Kawahana Y, Miyashita R. Construction of engineered muscle tissue consisting of myotube bundles in a collagen gel matrix. Gels. 2023;9(2):141. https://doi.org/10.3390/gels9020141.
54. Yamada KM, Sixt M. Mechanisms of 3D cell migration. Nat Rev Mol Cell Biol. 2019;20(12):738–52. https://doi.org/10.1038/s41580-019-0172-9.
55. Fan W, Evans RM. Exercise mimetics: impact on health and performance. Cell Metab. 2017;25(2):242–7. https://doi.org/10.1016/j.cmet.2016.10.022.
56. Mirzoev TM, Shenkman BS. Mechanosensory structures in the mechanotransduction system of muscle fibers. J Evol Biochem Phys. 2023;59(4):1341–59. https://doi.org/10.1134/S0022093023040269.
57. Iyer SR, Folker ES, Lovering RM. The nucleoskeleton: crossroad of mechanotransduction in skeletal muscle. Front Physiol. 2021;12:724010. https://doi.org/10.3389/fphys.2021.724010.
58. Wilson DGS, Tinker A, Iskratsch T. The role of the dystrophin glycoprotein complex in muscle cell mechanotransduction. Commun Biol. 2022;5(1):1022. https://doi.org/10.1038/s42003-022-03980-y.
59. Isaacson J, Brotto M. Physiology of mechanotransduction: how do muscle and bone "talk" to one another? Clin Rev Bone Miner Metab. 2014;12(2):77–85. https://doi.org/10.1007/s12018-013-9152-3.

60. Maurel DB, Jähn K, Lara-Castillo N. Muscle-bone crosstalk: emerging opportunities for novel therapeutic approaches to treat musculoskeletal pathologies. Biomedicines. 2017;5(4):62. https://doi.org/10.3390/biomedicines5040062.
61. Sotodosos-Alonso L, Pulgarín-Alfaro M, Del Pozo MA. Caveolae mechanotransduction at the interface between cytoskeleton and extracellular matrix. Cells. 2023;12(6):942. https://doi.org/10.3390/cells12060942.
62. Murthy SE, Dubin AE, Patapoutian A. Piezos thrive under pressure: mechanically activated ion channels in health and disease. Nat Rev Mol Cell Biol. 2017;18(12):771–83. https://doi.org/10.1038/nrm.2017.92.
63. Gudipaty SA, Lindblom J, Loftus PD, et al. Mechanical stretch triggers rapid epithelial cell division through Piezo1. Nature. 2017;543(7643):118–21. https://doi.org/10.1038/nature21407.
64. Kefauver JM, Ward AB, Patapoutian A. Discoveries in structure and physiology of mechanically activated ion channels. Nature. 2020;587(7835):567–76. https://doi.org/10.1038/s41586-020-2933-1.
65. Wackerhage H, Schoenfeld BJ, Hamilton DL, Lehti M, Hulmi JJ. Stimuli and sensors that initiate skeletal muscle hypertrophy following resistance exercise. J Appl Physiol (1985). 2019;126(1):30–43. https://doi.org/10.1152/japplphysiol.00685.2018.
66. Tsata V, Beis D. In full force. Mechanotransduction and morphogenesis during homeostasis and tissue regeneration. J Cardiovasc Dev Dis. 2020;7(4):40. https://doi.org/10.3390/jcdd7040040.
67. Herrmann M, Engelke K, Ebert R, et al. Interactions between muscle and bone-where physics meets biology. Biomol Ther. 2020;10(3):432. https://doi.org/10.3390/biom10030432.
68. Monemian Esfahani A, Rosenbohm J, Reddy K, et al. Tissue regeneration from mechanical stretching of cell-cell adhesion. Tissue Eng Part C Methods. 2019;25(11):631–40. https://doi.org/10.1089/ten.TEC.2019.0098.
69. Dolan CP, Imholt F, Yan M, et al. Digit specific denervation does not inhibit mouse digit tip regeneration. Dev Biol. 2022;486:71–80. https://doi.org/10.1016/j.ydbio.2022.03.007.
70. Schiaffino S, Reggiani C, Akimoto T, Blaauw B. Molecular mechanisms of skeletal muscle hypertrophy. J Neuromuscul Dis. 2021;8(2):169–83. https://doi.org/10.3233/JND-200568.
71. Adams GR, Bamman MM. Characterization and regulation of mechanical loading-induced compensatory muscle hypertrophy. Compr Physiol. 2012;2(4):2829–70. https://doi.org/10.1002/cphy.c110066.
72. Vogel V, Sheetz M. Local force and geometry sensing regulate cell functions. Nat Rev Mol Cell Biol. 2006;7(4):265–75. https://doi.org/10.1038/nrm1890.
73. Jabre S, Hleihel W, Coirault C. Nuclear mechanotransduction in skeletal muscle. Cells. 2021;10(2):318. https://doi.org/10.3390/cells10020318.
74. Linari M, Brunello E, Reconditi M, et al. Force generation by skeletal muscle is controlled by mechanosensing in myosin filaments. Nature. 2015;528(7581):276–9. https://doi.org/10.1038/nature15727.
75. Vanmunster M, Rojo Garcia AV, Pacolet A, et al. Mechanosensors control skeletal muscle mass, molecular clocks, and metabolism. Cell Mol Life Sci. 2022;79(6):321. https://doi.org/10.1007/s00018-022-04346-7.
76. Jani K, Schöck F. Molecular mechanisms of mechanosensing in muscle development. Dev Dyn. 2009;238(6):1526–34. https://doi.org/10.1002/dvdy.21972.
77. Kolahi KS, Mofrad MRK. Mechanotransduction: a major regulator of homeostasis and development. Wiley Interdiscip Rev Syst Biol Med. 2010;2(6):625–39. https://doi.org/10.1002/wsbm.79.
78. Herzog W, Powers K, Johnston K, Duvall M. A new paradigm for muscle contraction. Front Physiol. 2015;6:174. https://doi.org/10.3389/fphys.2015.00174.
79. Bershadsky AD, Balaban NQ, Geiger B. Adhesion-dependent cell mechanosensitivity. Annu Rev Cell Dev Biol. 2003;19:677–95. https://doi.org/10.1146/annurev.cellbio.19.111301.153011.

80. Benavides Damm T, Egli M. Calcium's role in mechanotransduction during muscle development. Cell Physiol Biochem. 2014;33(2):249–72. https://doi.org/10.1159/000356667.
81. LaGuardia JS, Shariati K, Bedar M, et al. Convergence of calcium channel regulation and mechanotransduction in skeletal regenerative biomaterial design. Adv Healthc Mater. 2023;12(27):e2301081. https://doi.org/10.1002/adhm.202301081.
82. Moorwood C, Philippou A, Spinazzola J, Keyser B, Macarak EJ, Barton ER. Absence of γ-sarcoglycan alters the response of p70S6 kinase to mechanical perturbation in murine skeletal muscle. Skelet Muscle. 2014;4:13. https://doi.org/10.1186/2044-5040-4-13.
83. Sato T, Verma S, Andrade CDC, et al. A FAK/HDAC5 signaling axis controls osteocyte mechanotransduction. Nat Commun. 2020;11(1):3282. https://doi.org/10.1038/s41467-020-17099-3.
84. Wang L, Luo JY, Li B, et al. Integrin-YAP/TAZ-JNK cascade mediates atheroprotective effect of unidirectional shear flow. Nature. 2016;540(7634):579–82. https://doi.org/10.1038/nature20602.
85. Zheng Y, Pan D. The hippo signaling pathway in development and disease. Dev Cell. 2019;50(3):264–82. https://doi.org/10.1016/j.devcel.2019.06.003.
86. Ma S, Meng Z, Chen R, Guan KL. The hippo pathway: biology and pathophysiology. Annu Rev Biochem. 2019;88:577–604. https://doi.org/10.1146/annurev-biochem-013118-111829.
87. Liu Q, Pan S, Liu S, et al. Suppressing hippo signaling in the stem cell niche promotes skeletal muscle regeneration. Stem Cells. 2021;39(6):737–49. https://doi.org/10.1002/stem.3343.
88. Stassen OMJA, Ristori T, Sahlgren CM. Notch in mechanotransduction – from molecular mechanosensitivity to tissue mechanostasis. J Cell Sci. 2020;133(24):jcs250738. https://doi.org/10.1242/jcs.250738.
89. Fujimaki S, Machida M, Wakabayashi T, Asashima M, Takemasa T, Kuwabara T. Functional overload enhances satellite cell properties in skeletal muscle. Stem Cells Int. 2016;2016:7619418. https://doi.org/10.1155/2016/7619418.
90. Feng L, Li B, Xi Y, Cai M, Tian Z. Aerobic exercise and resistance exercise alleviate skeletal muscle atrophy through IGF-1/IGF-1R-PI3K/Akt pathway in mice with myocardial infarction. Am J Physiol Cell Physiol. 2022;322(2):C164–76. https://doi.org/10.1152/ajpcell.00344.2021.
91. Yu T, Chang Y, Gao XL, Li H, Zhao P. Dynamic expression and the role of BDNF in exercise-induced skeletal muscle regeneration. Int J Sports Med. 2017;38(13):959–66. https://doi.org/10.1055/s-0043-118343.
92. Philippou A, Halapas A, Maridaki M, Koutsilieris M. Type I insulin-like growth factor receptor signaling in skeletal muscle regeneration and hypertrophy. J Musculoskelet Neuronal Interact. 2007;7(3):208–18.
93. Yoshida T, Delafontaine P. Mechanisms of IGF-1-mediated regulation of skeletal muscle hypertrophy and atrophy. Cells. 2020;9(9):1970. https://doi.org/10.3390/cells9091970.
94. Philippou A, Maridaki M, Halapas A, Koutsilieris M. The role of the insulin-like growth factor 1 (IGF-1) in skeletal muscle physiology. In Vivo. 2007;21(1):45–54.
95. Philippou A, Papageorgiou E, Bogdanis G, et al. Expression of IGF-1 isoforms after exercise-induced muscle damage in humans: characterization of the MGF E peptide actions in vitro. In Vivo. 2009;23(4):567–75.
96. Laplante M, Sabatini DM. mTOR signaling in growth control and disease. Cell. 2012;149(2):274–93. https://doi.org/10.1016/j.cell.2012.03.017.
97. Baraldo M, Geremia A, Pirazzini M, et al. Skeletal muscle mTORC1 regulates neuromuscular junction stability. J Cachexia Sarcopenia Muscle. 2020;11(1):208–25. https://doi.org/10.1002/jcsm.12496.
98. Chen Z, Li L, Wu W, et al. Exercise protects proliferative muscle satellite cells against exhaustion via the Igfbp7-Akt-mTOR axis. Theranostics. 2020;10(14):6448–66. https://doi.org/10.7150/thno.43577.

99. Crossland H, Kazi AA, Lang CH, et al. Focal adhesion kinase is required for IGF-I-mediated growth of skeletal muscle cells via a TSC2/mTOR/S6K1-associated pathway. Am J Physiol Endocrinol Metab. 2013;305(2):E183–93. https://doi.org/10.1152/ajpendo.00541.2012.
100. Zeng Z, Liang J, Wu L, Zhang H, Lv J, Chen N. Exercise-induced autophagy suppresses sarcopenia through Akt/mTOR and Akt/FoxO3a signal pathways and AMPK-mediated mitochondrial quality control. Front Physiol. 2020;11:583478. https://doi.org/10.3389/fphys.2020.583478.
101. Philippou A, Barton ER. Optimizing IGF-I for skeletal muscle therapeutics. Growth Hormon IGF Res. 2014;24(5):157–63. https://doi.org/10.1016/j.ghir.2014.06.003.
102. Yamada AK, Verlengia R, Bueno Junior CR. Mechanotransduction pathways in skeletal muscle hypertrophy. J Recept Signal Transduct Res. 2012;32(1):42–4. https://doi.org/10.3109/10799893.2011.641978.
103. Wen X, Jiao L, Tan H. MAPK/ERK pathway as a central regulator in vertebrate organ regeneration. Int J Mol Sci. 2022;23(3):1464. https://doi.org/10.3390/ijms23031464.
104. Valladares-Ide D, Peñailillo L, Collao N, Marambio H, Deldicque L, Zbinden-Foncea H. Activation of protein synthesis, regeneration, and MAPK signaling pathways following repeated bouts of eccentric cycling. Am J Physiol Endocrinol Metab. 2019;317(6):E1131–9. https://doi.org/10.1152/ajpendo.00216.2019.
105. Herzig S, Shaw RJ. AMPK: guardian of metabolism and mitochondrial homeostasis. Nat Rev Mol Cell Biol. 2018;19(2):121–35. https://doi.org/10.1038/nrm.2017.95.
106. Thirupathi A, de Souza CT. Multi-regulatory network of ROS: the interconnection of ROS, PGC-1 alpha, and AMPK-SIRT1 during exercise. J Physiol Biochem. 2017;73(4):487–94. https://doi.org/10.1007/s13105-017-0576-y.
107. Koulmann N, Richard-Bulteau H, Crassous B, et al. Physical exercise during muscle regeneration improves recovery of the slow/oxidative phenotype. Muscle Nerve. 2017;55(1):91–100. https://doi.org/10.1002/mus.25151.
108. Koves TR, Li P, An J, et al. Peroxisome proliferator-activated receptor-gamma co-activator 1alpha-mediated metabolic remodeling of skeletal myocytes mimics exercise training and reverses lipid-induced mitochondrial inefficiency. J Biol Chem. 2005;280(39):33588–98. https://doi.org/10.1074/jbc.M507621200.

Chapter 8
Muscle Damage Induced by Eccentric Exercise, Recovery and Adaptations

Kazunori Nosaka and Trevor C. Chen

Characteristics of Eccentric Muscle Contraction

Muscle Contraction Terminology

There are three types of muscle contractions that are performed in daily activities, exercises and sports. They are isometric (static) where the force produced by a muscle (force) is equal to the load to the muscle (load), concentric (shortening) in which force > load, and eccentric (lengthening) in which force < load. In 2003, Professor John Faulkner [1] stated; ***contraction is redefined as "to undergo activation and generate force" and the directionality of the movement is best described by the adjectives "shortening" and "lengthening," thus "concentric" and "eccentric" are inappropriate to describe the contractions of skeletal muscles***. This has to be respected, but "concentric" and "eccentric" are often used to describe types of muscle contractions in many articles still recently. Thus, this book chapter uses "concentric" for shortening, and "eccentric" for lengthening.

K. Nosaka (✉)
Exercise and Sports Science, School of Medical and Health Sciences, Edith Cowan University, Joondalup, WA, Australia
e-mail: k.nosaka@ecu.edu.au

T. C. Chen
Department of Physical Education and Sport Sciences, National Taiwan Normal University, Taipei, Taiwan

N. C. Apostolopoulos et al. (eds.), *Fundamentals of Recovery, Regeneration, and Adaptation to Exercise Stress: An Integrated Approach*,
https://doi.org/10.1007/978-3-031-44270-4_8

Force

Muscles can produce greater force in eccentric than isometric or concentric contractions (Fig. 8.1). "Eccentric exercises" are the ones that mainly consist of eccentric contractions, and descending stair walking, sitting on a chair slowly, and lowering a dumbbell are typical examples. In these movements, the force produced by muscles is smaller than the load (i.e., body mass, dumbbell), by which activated muscle fibers are lengthened.

The ratio between the forces generated in maximal voluntary concentric and eccentric contraction varies among studies and muscles, but overall approximately 40% greater force production is possible for eccentric than concentric contraction [2].

This gives advantages for eccentric exercise such that less effort is required to handle the same load in eccentric than concentric contraction, and greater mechanical stimulus could be imposed to muscles by eccentric contractions.

Fatiguability

Eccentric contractions are less fatigable than concentric contraction such that the number of contractions to failure (how many repetitions can be performed at 70%, 80%, 90% and 95% of respective one repetition maximum of eccentric-only and concentric-only contraction) is at least 60% greater for the eccentric-only than the concentric-only contractions of the elbow flexors using a dumbbell [3].

When eccentric-only and concentric-only maximal isokinetic contractions of the elbow flexors were compared for the decrease in torque over 30 contractions, the decrease in torque was smaller for the eccentric-only (−32%) than concentric-only (−50%) contractions [4].

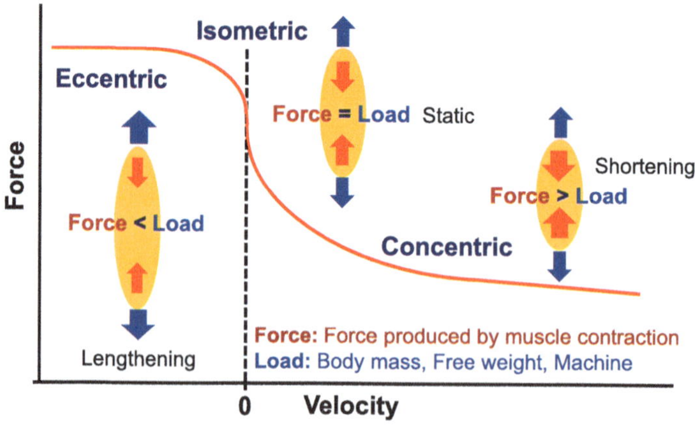

Fig. 8.1 Force-velocity relationship and muscle contraction types

Metabolic Demand

Eccentric exercise is less metabolically demanding than concentric exercise, as indicated by 50% less oxygen consumption and 20–30% lower heart rate during eccentric cycling (resisting against pedals rotating backward by a motor) than conventional cycling at the same workload [6].

> Abbot et al. [5] in a classic study published in 1952 stated that *the physiological cost of negative work is considerably less than that of positive work, and the difference increases with speed of movement.*

Cognitive Demand

It appears that cognitive demand is greater for eccentric than concentric exercise [7], such that a reaction time in Stroop test performed eccentric cycling was more than 15% longer than that in concentric cycling, and errors occur more for eccentric (20%) than concentric (10%) cycling. It is likely that greater attention and vigilance are required when walking down-stairs than walking up-stairs.

Cross-Education Effect

Eccentric resistance exercise training appears to induce greater cross-education effect than concentric or conventional resistance exercise training. Non-dominant arms of sedentary young men were immobilized for 3 weeks. During the immobilization period, 5 sets of 6 dumbbell curl eccentric-only or concentric-only contractions at 20–80% of maximal voluntary isometric contraction strength were performed by the non-immobilized arm, and a control group did not perform any exercise. The eccentric training increased elbow flexion muscle strength (17%) and cross-sectional area of biceps brachii (9%) greater than the concentric training (6%, 3%) for the trained arm. The control group showed decreases in muscle strength (−17%) and cross-sectional area (−12%) for the immobilized arm, but these changes were attenuated by the concentric training (−4%, −1%) but completely abolished by the eccentric training [8].

When comparing eccentric-only exercise and conventional resistance exercise in which both eccentric and concentric contractions were included, much greater cross-education effect was reported to be induced eccentric-only exercise, which even increased muscle strength of the immobilized arm [9].

Three types of muscle contractions are performed in daily activities, exercise and sports, isometric (static), concentric (shortening), and eccentric (lengthening), with muscles producing greater force in eccentric than both isometric and concentric contractions. Eccentric contractions less fatigable, and metabolically demanding, requiring greater cognitive demand, as well as inducing greater cross-education effect than concentric contraction or conventional resistance exercise training.

Muscle Damage

Muscle damage is induced by exercises that consist of eccentric contractions or isometric contractions at a long muscle length to a lesser extent. Muscle damage is minimal or not induced in exercises that consist of only concentric contractions [10]. The focus in this section is muscle damage induced by exercises that mainly consist of eccentric contractions (i.e., eccentric exercise).

Symptoms

Typical symptoms of muscle damage induced by eccentric exercises are delayed onset muscle soreness (DOMS) and a prolonged loss of muscle function, especially decreases in maximal voluntary contraction (MVC) strength [10]. It should be noted that the time course of DOMS is different from that of MVC strength changes. DOMS normally develops several hours after eccentric exercise, and peaks between 24-72 hours post-exercise [11]. In contrast, the largest decrease in MVC strength is generally evident at immediately after eccentric exercise, and gradually recover thereafter.

Additionally, increased muscle stiffness indicated by difficulty in extending a joint is often observed, especially after eccentric exercise of the elbow flexors [10, 12]. Muscle swelling is evident after both concentric and eccentric exercise, but when muscle damage is induced, swelling increases for several days after eccentric exercise, although swelling is subsided by next day after concentric exercise [13]. Due to the increased stiffness and swelling, muscles become harder after eccentric exercise [14].

DOMS

DOMS is considered to be mechanical hyperalgesia through nociceptor sensitization induced by nerve growth factor (NGF) and glial cell line-derived neurotrophic factor (GDNF) upregulated by cyclooxygenase-2 (COX-2) [15]. A recent study showed that an increase in NGF was associated with the development of DOMS in humans [16]. However, NGF is not the sole molecule that is involved in DOMS, and the cause-effect relationship between NGF and DOMS has not been clarified. Transient receptor potential vanilloid 1 (TRPV1) and transient receptor potential vanilloid 4 (TRPV4) have been shown to be involved in the upstream and/or downstream of NGF in animal studies [15, 16].

Signs of Inflammation

The typical signs of inflammation include pain, redness, immobility, swelling and heat. It is interesting to note that pain (DOMS), immobility (loss of muscle function) and swelling are evident in eccentric exercise-induced muscle damage (Fig. 8.2). An increased temperature is often felt over a damaged muscle. Thus, it is likely that the symptoms of muscle damage (DOMS, loss of muscle function, increased stiffness and swelling) are induced by inflammation rather than muscle damage itself. A loss of muscle function can be observed immediately after exercise, but it could be also due to neuromuscular fatigue than muscle damage [17]. Thus, it is difficult to distinguish muscle damage and neuromuscular fatigue right after exercise. Muscle damage becomes more evident at several hours to 24 hours after exercise.

> Muscle damage becomes more evident several hours to 24 hours post exercise with both eccentric and isometric contractions (at a long muscle length), compared to concentric contraction which is responsible for minimal muscle damage.

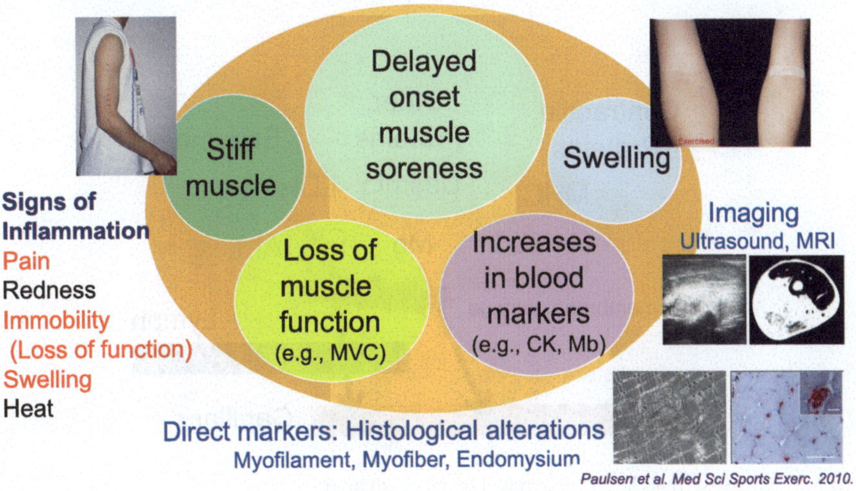

Fig. 8.2 Markers of eccentric exercise-induced muscle damage

Markers of Muscle Damage

Blood Markers

In addition to the symptoms described above, increases in muscle proteins in the blood such as creatine kinase (CK) activity and myoglobin concentration measured in plasma or serum samples are used as markers of muscle damage [18]. CK and other enzymes in muscle cells (fibers) are leaked when plasma membrane permeability is increased and/or plasma membrane is damaged (Fig. 8.3). Less typical to skeletal muscles, but other enzymes contained in muscle fibers such as lactate dehydrogenase, aspartate aminotransferase (AST) and alanine aminotransferase (ALT) activities are also increased in the blood [19].

These enzymes are large molecules (e.g., the molecular weight of CK is >80,000 kDa), thus cannot get to capillaries directly, but via lymph to circulation. On the other hand, myoglobin is a smaller molecule (<18,000 kDa), thus could get to the capillaries directly [20]. Because of this, increases in myoglobin after eccentric exercise are seen earlier than CK, and myoglobin concentration peaks 2–3 days, but CK activity peaks 4–5 days after eccentric exercise of the elbow flexors [18].

When muscle membrane is damaged and becomes leaky, calcium concentration (Ca^{2+}) in the cell increases, which activate protease activity to break down muscle contractile proteins such as myosin and troponin. They are also leaked out and increased in the circulation. In troponin I, there are two skeletal muscle isoforms, slow-twitch skeletal muscle troponin I (ssTnI) and fast-twitch skeletal muscle troponin I (fsTnI). It has been shown that only fsTnI increases after eccentric exercise in the blood, and the time course of changes in serum fsTnI concentration is similar to that of serum CK activity [21, 22]. This indicates that muscle fiber damage occurs predominantly to fast-twitch muscle fibers.

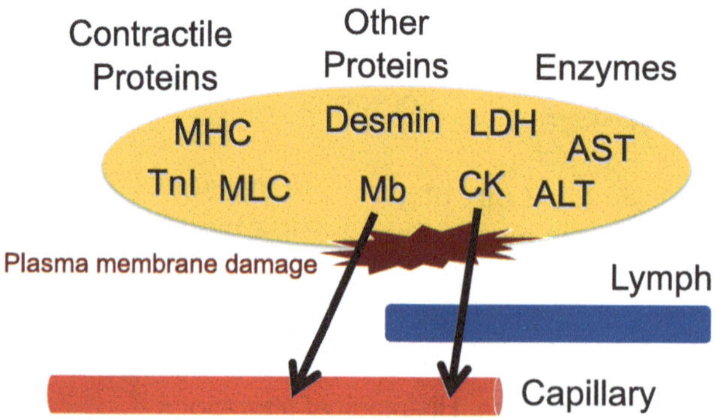

Fig. 8.3 Release of muscle specific protein to the circulation

Diagnostic Imaging

Muscle damage can be visualized by B-mode ultrasound and magnetic resonance imaging technique as an increase in echo-intensity and T2 relaxation time, respectively [23]. An increase in T2 relaxation time normally indicating an increase in intracellular and/or extracellular water, is coincided with swelling, thus greater changes in T2 relaxation time is observed when swelling peaks, which is 4–5 days post-exercise. This is also the case for the changes in echo-intensity after eccentric exercise. However, it is not necessarily clear why echo-intensity increases in the muscles that are damaged by eccentric exercises, since an increase in water normally decreases the echo-intensity. It may be that increased echo-intensity is caused by changes in extracellular matrix such as endomysium and perimysium surrounding muscle fibers and fascicles.

Histological Alterations

Muscle damage is defined as "physical harm that impairs normal function of skeletal muscle." According to a dictionary, "physical harm" is "physical injury", or "thus muscle damage could be described as "muscle injury" resulting in disruption of normal structure of muscle fibers and/or myofilaments. However, it is important to note that muscle injury may not be severe even if symptoms of muscle damage described above such as DOMS, strength loss, and swelling are severe [24, 25].

No muscle fiber disruption with 10% of intracellular disruption and destroyed Z-lines in the transmission electron microscope observation were observed at 5, 24, 96 and 192 hours after 240 maximal voluntary eccentric contractions of the knee extensors performed by untrained young men [24]. Interestingly, when electric stimulation was added to the voluntary contractions, myofilament disruption was greater (40%) and desmin-negative, dystrophin-negative and vimentin-positive muscle fibers were found in more than 15% of observed muscle fibers at 192 hours post-exercise. Furthermore, no significant difference in DOMS was found between the voluntary contraction and the voluntary + electrical stimulation conditions and decreases in force and power were greater for the voluntarily contraction than the voluntary + electrical stimulation condition. These indicate that histological alterations in muscle fibers are minimum after voluntary eccentric contractions, and the histological changes do not represent the magnitude of symptoms of muscle damage.

Little muscle fiber disruption after eccentric exercise was reported in another study [25], in which 300 maximal eccentric contractions of the knee extensors were performed by young men, and muscle biopsy samples were obtained from vastus lateralis at 0.5, 4, 8, 24, 96, and 168 hours after exercise. Infiltration of inflammatory cells ($CD16^+$ and/or $CD68^+$ cells) in muscle fibers was found only occasionally in 4 out of 11 participants, with 0.2%–1.35% of the muscle fibers, but larger number of $CD16^+$ cells were observed in the interstitial spaces at 96 and 168 hours after exercise. These confirmed that muscle fiber disruption is minimal after eccentric exercise of the knee extensors, and the histological changes do not necessarily relate to DOMS and loss of muscle function.

Mechanisms of Muscle Damage

The mechanisms underpinning the eccentric exercise-induced muscle damage have not been fully elucidated. As shown in Fig. 8.4, one of the main events has been considered as an increase in intracellular Ca^{2+}, which triggers phospholipase activity to induce plasma membrane damage, and protein kinase activity to breakdown muscle contractile proteins [26]. This eventually leads to necrosis of muscle fibers. However, as explained previously, a large number of necrotic muscle fibers are not found after voluntary eccentric exercises, at least in vastus lateralis muscle.

Instead, inflammatory cells are found in connective tissue surrounding muscle fibers and capillaries located in the extracellular matrix. This may suggest that damage and subsequent inflammation occur more in the extracellular matrix, especially endomysium and perimysium. It is also important to note that nociceptors such as C fibers and Aδ fibers are located in connective tissue such as endomysium, perimysium and epimysium [15]. Without muscle fiber damage, DOMS is still evident [25]. Thus, it seems reasonable to assume that DOMS is more associated connective tissue inflammation [27]. It is also possible that decreases in force generating capacity and range of motion (ROM) especially extending a joint caused by increased muscle stiffness and swelling of muscle are associated with damage and inflammation to extracellular matrix. In 1902, Hough [28] already stated in his paper about muscular soreness that ***it may even involve inflammation of the interstitial connective tissue.*** What a wonderful foresight he made!

However, when large increases in CK activity (>5000 IU/L) and/or myoglobin concentration (>1000 ng/ml) in the blood are observed several days after eccentric exercise, it is likely that muscle membrane damage is induced, resulting in muscle

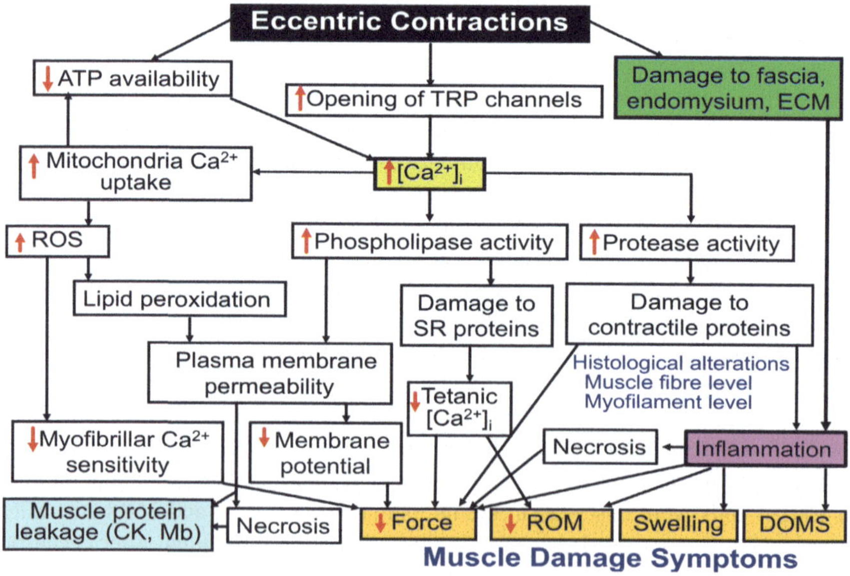

Fig. 8.4 Possible mechanisms of eccentric exercise-induced muscle damage

fiber necrosis eventually. It is important to note that the symptoms of muscle damage (DOMS, force loss, decreased ROM, and swelling) are not necessarily closely related each other [29, 30]. Then, what is the best indicator of eccentric exercise-induced muscle damage? It appears that muscle function measures such as MVC strength best represent muscle damage, and time course of it [31].

> Histological alterations in muscle fibers are minimum after voluntary eccentric contractions, and the histological changes do not represent the magnitude of symptoms of muscle damage. However, it is suggested the muscle damage, and the subsequent inflammation occur in the extracellular matrix, especially in the endomysium and perimysium.

Factors Influencing Muscle Damage

There are many factors influencing the magnitude of muscle damage induced by eccentric exercise (Fig. 8.5).

Eccentric Contraction Factors

First, intensity, the number of repetitions, velocity, and muscle length of eccentric contractions affect the magnitude of muscle damage. The magnitude of muscle damage is greater with higher intensity, larger repetitions, faster velocity, and long muscle length eccentric contractions than lower intensity, smaller repetitions, slower velocity and short muscle length eccentric contractions [32–38]. Thus, in order to avoid muscle damage, higher intensity, larger repetitions, faster velocity, and long muscle length eccentric contractions should be avoided at least for the first time.

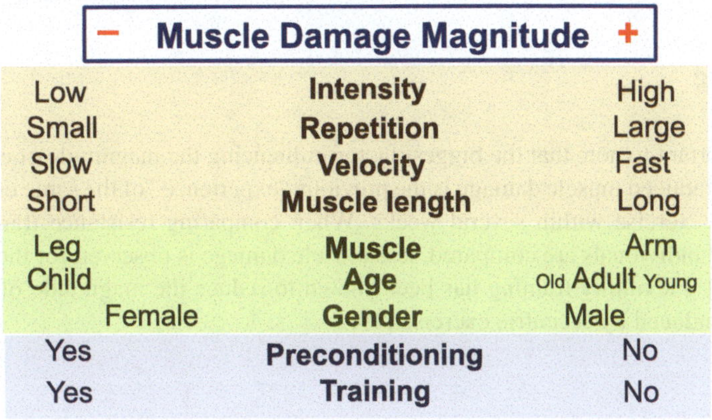

Fig. 8.5 Factors influencing the magnitude of eccentric exercise-induced muscle damage

Arm vs Leg Muscles

The magnitude of muscle damage induced by eccentric exercise is different among muscle groups. Generally speaking, arm muscles are more prone to muscle damage than leg muscles [39]. This is due to eccentric contractions performed in daily activities, which are normally greater for leg than arm muscles, thus leg muscles are more accustomed to eccentric contractions than arm muscles.

Age

Age also affects the magnitude of muscle damage. When comparing children (pre-adolescent, adolescent) and adults, muscle damage induced by maximal eccentric exercise of the elbow flexors is greater for adults than children, and adolescent than pre-adolescent children [40, 41]. It is speculated that this was due to the number of fast-twitch muscle fibers, which are more vulnerable than slow-twitch fibers, and are smaller for pre-adolescent children than adults. It is also possible that muscles and tendons are more compliant for children than adults. When comparing young and older adults, eccentric exercise-induced muscle damage is less for older than younger adults [42–44]. This may be associated to a decrease in fast-twitch muscle fibers with ageing.

Sex

Regarding the sex difference in eccentric exercise-induced muscle damage, animal studies showed greater muscle damage for male than female, which could be due to the muscle membrane protective effect by estrogen [45]. However, this is not necessarily confirmed in human studies [46, 47].

Training

It is important to note that the biggest factor influencing the magnitude of eccentric exercise-induced muscle damage is the previous "experience" of the same or similar eccentric exercise within several weeks. When comparing resistance trained and untrained individuals are compared, less muscle damage is observed for the trained ones [48]. Flexibility training has been shown to reduce the magnitude of muscle damage induced by eccentric exercise [49].

Intensity, number of repetitions, velocity, and muscle length of eccentric contractions affect the magnitude of muscle damage, with previous experience of same or similar eccentric exercise within several weeks biggest factor influencing magnitude of eccentric exercise-induced muscle damage.

The Repeated Bout Effect and Preconditioning Effect

Repeated Bout Effect

Less muscle damage is induced when the same eccentric exercise is repeated. This is referred to as the ipsilateral repeated bout effect (IL-RBE) [50]. For example, maximal eccentric exercise of the elbow flexors is repeated twice with an interval of 4, 8, and 12 weeks, the second bout exercise does not induce as much damage as the first bout indicated by faster recovery of muscle function, less DOMS, and smaller swelling and increases in CK activity and myoglobin concentration in the blood, but the magnitude of muscle damage is increased with increasing the interval between bouts [51]. When the interval is extended to 6, 9 and 12 months, muscle damage is still less at 6 months, and some reduction of muscle damage is still seen at 9 months, but not at 12 months [52].

It should be noted that substantial muscle damage is induced after the first bout of eccentric exercise in the repeated bout effect. However, muscle damage does not appear to be prerequisite for the muscle damage protective effects to be conferred [53–56].

"The ability of muscle to rapidly adapt following the damage from eccentric exercise raises the possibility of clinical applications of mild eccentric exercise, such as for protecting a muscle against more major injuries [57]".

Preconditioning Effect

As shown in Fig. 8.6, muscle damage protective effect is produced by submaximal eccentric exercise that does not induce any symptoms of muscle damage. For example, low-intensity eccentric contractions such as 10% and 20% of maximal eccentric exercise can reduce the symptoms of muscle damage after maximal eccentric exercise [58–60]. This is referred to as the preconditioning effect. The greater the intensity of the preconditioning exercise, the greater the protective effect [59]. It appears that the protective effects conferred by low-intensity eccentric contractions does not last as long as that by maximal eccentric contractions [61].

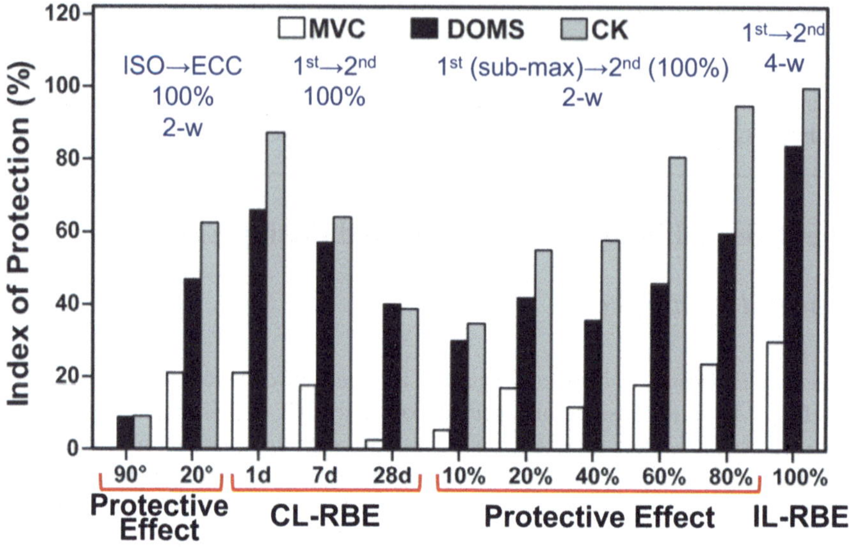

Fig. 8.6 Repeated bout effect and protective effects against muscle damage induced by maximal eccentric exercise

It is important to note that isometric contractions also provide the preconditioning effect against muscle damage induced by maximal eccentric exercise, but the effect is induced only by the isometric contractions at a long muscle length [62–64]. It has been reported that only two maximal isometric contractions at the elbow joint angle of 20° (elbow full extension = 0°), which do not induce any DOMS or strength loss can still provide potent protective effect against muscle damage induced by maximal eccentric exercise of the elbow flexors [62].

Contralateral Repeated Bout Effect

It has been reported that muscle damage induced by eccentric exercise is attenuated in the subsequent bouts of the same exercise performed by not only the same muscle but also by the homologous muscle of the contralateral limb, and this is termed as the contralateral repeated bout effect [64, 65]. In a study [64], untrained young men were allocated to either a control group that performed maximal eccentric exercise of the elbow flexors of the non-dominant arm twice with an interval of 2 weeks between bouts or one of seven contralateral repeated bout groups that performed exercise consisting of 30 maximal isokinetic eccentric contractions of the elbow flexors with either dominant or nondominant arm followed 0.5, 6, 12 hours, as well as 1, 7, 28, or 56 days by the same exercise using the opposite arm. Maximal voluntary concentric contraction torque, peak torque angle, range of motion, upper arm circumference, muscle soreness, ultrasound echo-intensity, and plasma creatine kinase activity and myoglobin concentration changed after the first exercise bout, without significant difference among the groups. Changes in all variables after the second exercise bout were smaller than those after the first bout for the control, 1 day, 7 days and 28 days groups, indicating the repeated bout effect. However, the changes were not significantly different between the first and second bouts for the 0.5, 6, 12 hours, and 56 days groups. The magnitude of the contralateral repeated bout was reduced with increasing the time between bouts from 1 day to 28 days [64]. These results suggest that the contralateral repeated bout effect lasts shorter than the ipsilateral repeated bout effect that was shown to last more than 8 weeks as mentioned in the previous section.

The contralateral repeated bout effect was also shown for the knee flexors, and the magnitude of the protective effect was similar between elbow flexors and knee flexors, but the ipsilateral repeated bout effect was slightly greater for the elbow flexors than knee flexors [66]. These demonstrate that the magnitude of the contralateral repeated bout effect relative to the ipsilateral repeated bout effect is similar between the elbow flexors and knee flexors (approximately 50%), regardless of the greater muscle damage for the elbow flexor than knee flexor eccentric exercise. It is important to note that such repeated bout effect is found for homologous muscles only, and no protective effect is evident from a leg muscle to an arm muscle [67], and probably flexors to extensors or vice versa of the same limb.

Mechanisms of the Repeated Bout Effect and Pre-conditioning Effect

Figure 8.7 depicts possible mechanisms underpinning the repeated bout effect and the preconditioning effect [68]. The mechanisms are complex, and probably a combination of neural (e.g., increased recruitment of slow-twitch motor units and activation of larger motor unit pool), mechanical (e.g., increased dynamic and passive muscle stiffness), cellular adaptations (e.g., longitudinal addition of sarcomeres, adaptation in inflammatory response, and strengthened extracellular matrix), and others (e.g., heat shock protein and strengthened weak sarcomeres) is involved in them. As for neural adaptations, it is highly possible that adaptations at the cortical and spinal levels are involved in the contralateral repeated bout effect. Additionally, it might be that muscle movement changes resulting in less muscle elongation in eccentric contractions are associated with the repeated bout effect [69, 70].

It is possible that the mechanisms underpinning the contralateral repeated bout effect is largely a function of acute neural adaptations similar to those seen in the cross-education effect. The potential sites of adaptations in the cross-education include changes in brains (supplementary motor area, primary motor cortex, middle temporal gyrus, interior temporal gyrus, occipital lobe, cerebellum), interhemispheric inhibition, spinal cord, motoneuron, and muscles [71]. Thus, changes within cortical motor and non-motor regions, including increased corticospinal excitability, reduced cortical inhibition, reduced interhemispheric inhibition, and changes in

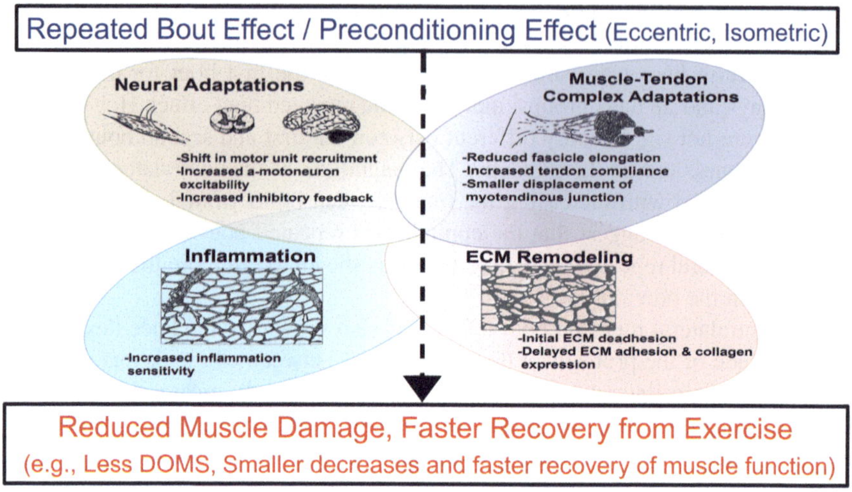

Fig. 8.7 Possible mechanisms of repeated bout effect and protective effect against eccentric exercise-induced muscle damage

voluntary activation are associated with the repeated bout effect, especially the contralateral one. Further research is necessary to understand the mechanisms.

> Mechanisms underpinning the repeated bout and precondition effect although complex in nature are probably a combination of neural, mechanical, cellular adaptions and others.

Strategies to Prevent and Minimize Muscle Damage

Muscle Damage Prevention

As mentioned above, the best strategy is to avoid the situation that could result in severe muscle damage induced by high-intensity, large number of repetitions, fast-velocity and long muscle length eccentric contractions for the first time to perform eccentric exercise. As shown in Fig. 8.8, it is better to start from eccentric contractions at short muscle lengths. Avoiding eccentric contractions at long muscle length reduce the magnitude of muscle damage. It has been shown in a study [72] that a bout of eccentric exercise consisting of short-muscle length eccentric contractions, which induces minimal damage, can attenuate muscle damage in maximal eccentric contractions at longer muscle lengths. Then, it is better to increase the intensity and the number of contractions gradually. When the intensity is increased, it may be better to reduce the number of contractions, as recommended by general resistance training protocol. Faster velocity eccentric contractions are better to be introduced last, once eccentric contractions can be performed in well controlled manner even at a fast velocity [73].

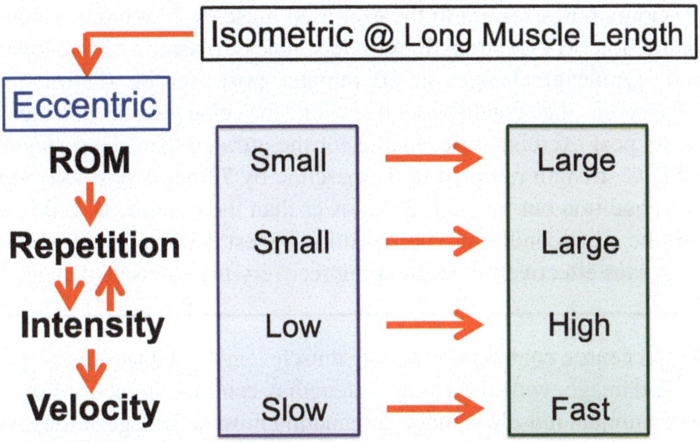

Fig. 8.8 Prescription of eccentric exercise to minimize muscle damage

This is not only for resistance exercises but also could be applied for hiking and mountain climbing. A short downhill walking could prevent muscle damage in long distance activities consisting of eccentric contractions, which could lead to more serious injuries [74, 75]. It should be also considered to use the contralateral repeated bout effect and cross-education effect to prevent muscle damage and induce additional adaptations [76, 77]. As one of the possible practical applications, the contralateral repeated bout effect should be considered for rehabilitation after immobilization and/or disuse. It is possible that the remobilization of an immobilized or disused limb induces muscle damage when eccentric contractions are included because disused muscles are susceptible to eccentric exercise-induced muscle damage [78]. To minimize this, it can be recommended that before retraining the injured limb muscles, eccentric exercise should be given to the contralateral "non-immobilized" limb muscles [8].

Prophylactic and Therapeutic Modalities

When muscle damage is induced, therapeutic modalities to reduce symptoms of muscle damage and enhance recovery are applied, and many modalities such as massage [79], vibration [80], light exercise [81], and nutritional supplements [82, 83] are just some examples that have been investigated. However, many of them do not appear to be effective, although some effects are found, and some interventions appear to be better than others. Thus, the best strategy is not to induce severe muscle damage.

Far-infrared radiation therapy, which is in the category of phototherapy is reported to enhance recover from maximal eccentric exercise of the elbow flexors and knee extensors [84]. In the study, participants received a 30-min far-infrared radiation lump therapy (wavelength: 8–14 μm) or sham treatment at 1, 25-, 49-,

73- and 97-hours post-exercise to the exercised muscles. Maximum voluntary isometric contraction (MVC) strength, muscle soreness, plasma creatine kinase activity showed significant changes at 30 minutes post-exercise (before treatment) similarly between the conditions; however, the changes in all measures at 24–120 hours post-exercise were smaller for the infrared than sham condition. For example, MVC strength returned to the baseline by 72 hours post-exercise for the far-infrared condition but was still 20% lower than the baseline at 120 hours post-exercise for the sham condition. These results suggest that the far-infrared radiation lamp therapy was effective for accelerating recovery from muscle damage.

Avoiding eccentric contractions at long muscle lengths reduces the magnitude of muscle damage, with short-muscle length eccentric contractions observed to induce minimal muscle damage, attenuating muscle damage at longer muscle lengths.

Muscle Damage in Sports

Even though athletes have had the repeated bout effect and protective effect from training and taking part in similar matches and competitions, they still experience muscle damage caused by eccentric contractions in training and matches and competitions. The magnitude of muscle damage varies among sports.

Team Sports

Figure 8.9 shows changes some muscle damage markers and performance measures after a 90-minute Loughborough Intermittent Shuttle Test (LIST) performed by 12 elite female soccer players. Muscle soreness of the knee extensors and flexors muscles develops, and MVC strength of the knee extensors and flexors was still more than 10% lower than the baseline even at 48 hours post-LIST, and many of the performance measures are not recovered to the baseline at 72 hours post-LIST [85]. It is assumed that an official match in which both teams are playing a close game could result in greater muscle damage than that after LIST. If this is the case, the recover from the match may take longer.

Muscle damage induced by rugby matches is related to the number of physical contacts and the total number of high-intensity accelerations and decelerations [86]. Impairment of neuromuscular function such as countermovement jump performance and muscle strength is reported to last for up to 4 days post-match [87, 88]. It appears that the magnitude of muscle damage induced in basketball is not as severe as that in soccer, and players can recover from a match within 2 days [89, 90].

A study comparing soccer, basketball, volleyball and handball at an elite competitive level for muscle damage and inflammatory induces [91], showed that soccer

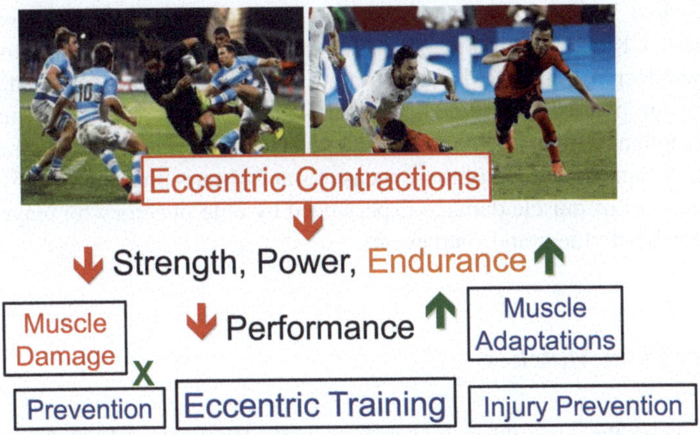

Fig. 8.9 Changes in muscle damage markers and performance parameters after a single bout of 90-minute Loughborough intermittent shuttle test (a simulated soccer match)

produced the greatest increase in inflammatory cytokines (tumor necrosis factor-alpha and interleukin-6), which were increased by 3–4-folds immediately after the game, as well as in C-reactive protein, which was increased by threefold in the next morning after the match. Metabolic stress (urea, ammonia and cortisol) and muscle damage indices (CK and lactate dehydrogenase activities) were also higher after soccer, with CK responses being almost 2–3 times higher than the other sports. Volleyball showed the smallest increase in inflammation and muscle damage markers compared with the other three sports.

Regarding field hockey, no significant changes in counter movement jump, MVC strength of the knee extensors and flexors were reported at 1, 24 and 48 hours after a match consisting of two 35-minutes period, although mild DOMS and small increases in serum CK activity were evident at 1 and 24 hours after the match [92].

Racket Sports

In racket sports, no significant changes in MVC strength of the knee extensors and badminton-specific running and movement velocity were found after a 45-min simulated badminton match, although some increases in plasma CK activity and myoglobin concentration were observed [93]. The authors concluded that the level of muscle damage was minor and did not affect muscle performance. A simulated single badminton match played by state-level male players decreased muscle strength, voluntary activation, and muscle soreness of knee extensors and flexors at immediately after and 1 hour after the match, but they returned to the baseline by 1 day after the match [94]. The authors conclude that muscle damage induced by singles badminton matches was minimal, but the more the lunges are performed, the greater the neuromuscular fatigue.

A simulated 3-hour tennis match on a hard court induced small increases in muscle soreness, CK activity and myoglobin concentration at 24–48 hours post-match, and marked decreases in 1-RM squat strength (35%), squat jump (7%), and counter movement jump (10%) at immediately post-match. At 24 hours post-match, the 1-RM strength and jump heights returned to the baseline values, but several players showed a sustained decrease of these measures at 24 hours post-match [95]. It is not known the level of muscle damage experienced by elite players who played longer singles matches in the grand slam events.

Recovery from Matches

Table 8.1 shows the interval between matches in the World Cup (2022–2023) and the Olympic Games (2020) for some sports, and the grand slam events for tennis. The interval between matches differs significantly among sports. For example, soccer matches are generally scheduled with 2–7 days of rest between matches in the World Cup, but 3–4 days during the Olympic Games. Rugby has a longer interval (5–15 days) between matches during the World Cup. The interval between matches is shorter for basketball, volleyball, badminton and tennis, when compared with soccer and rugby. Considering the time to recover from a previous match is shorter for basketball, volleyball, badminton and tennis than soccer and rugby, this appears to be reasonable.

It is important to know how many days are required for athletes to recover from a previous match to be ready for the next match. In high levels of competitions, it is interesting to examine the effects of different intervals between matches on readiness and performance of the players. For example, the interval between matches was 3 to 7 days for the group rounds, and that for the quarter finals, semifinals, third place and final was 3 to 6 days in the 2023 FIFA Women's World Cup. As mentioned above, it takes 4 days for players to recover from a simulated soccer match, thus if the interval between matches is shorter than 4 days, a team that has a shorter interval between matches may not perform well against a team that has a longer interval between matches. It is interesting that the interval between matches was not the same in 6 out of 8 matches in the quarter-final, semi-final, final and the third-place matches between teams, and 5 out of 6 matches (83%) were won by the teams

Table 8.1 Interval (days) between matches (range) in World Cups (2022–2023) and Olympic Games (2021) for some sports

	World Cup		Olympic	
Sport	Men	Women	Men	Women
Soccer	2–7	3–7	3–4	3–4
Rugby	5–15	6–8	–	–
Basketball	1–3	1–2	2–3	2–3
Volleyball	1–3	1–3	2	2
Field hockey	2–7	1–7	1–2	1–2
Badminton	1–2	1–2	1–4	1–4
Tennis	1–4[a]	1–4[a]	1–3	1–2

[a]Australia Open, French Open, Wimbledon, US Open

who had a longer interval from the previous match in the 2023 FIFA Women's World Cup. Thus, it is necessary to schedule matches to give the same interval between matches to the teams competing against. It is interesting that the interval between matches in rugby (7–9 days) is longer than that in soccer (3–7 days), even though the playing time is shorter for rugby (80 minutes) than soccer (90–120 minutes). This difference does not appear to be based on scientific studies.

Importance of Eccentric Training

Beneficial effects of eccentric training on athletic performance and health have been identified.

Athletic Performance

As shown in Fig. 8.10, eccentric contractions performed in exercises and sports impair muscle function such as strength and power, and affects endurance, resulting in decreased performance. This also induces DOMS and other symptoms of muscle damage, which could affect training and next match. Thus, it is important to perform eccentric training to make muscles more immune to muscle damage induced by eccentric contractions. Eccentric training also improves muscle strength and power, performance and reduces injury risks. Therefore, training focusing on eccentric contractions is important.

> "Using an accentuated eccentric [phase] might stimulate muscle building through another pathway that you wouldn't necessarily get during traditional training" according to Dr. Brad Schoenfeld, the hypertrophy specialist (https://www.lookgreatnaked.com/blog/).

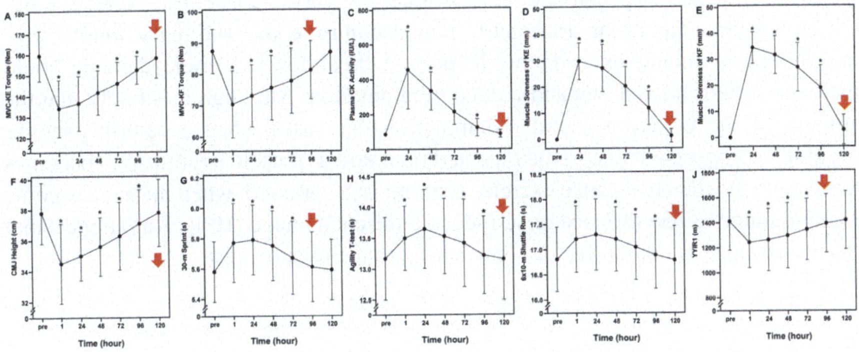

Fig. 8.10 Muscle damage induced in team sports and importance of eccentric exercise training

For example, one systematic review paper [96] summarized that eccentric training can improve muscle mechanical function to a greater extent than other modalities; and eccentric training may be especially beneficial in enhancing strength, power and speed performance. Another review article [97] states; (1) there is growing evidence from basic research indicating that eccentric strength training produce-specific muscular, neuromuscular, tendinous, and metabolic adaptations that could be beneficial for change of direction (CoD) speed performance, and (2) eccentric muscle strength is especially needed to decelerate and stabilize the body during CoD tasks.

Health and Fitness

Regarding more health aspects, a review article [98] summarized that eccentric exercise, characterized by the production of high mechanical forces at a low metabolic demand could be used with elderly adults, with or without clinical conditions, to improve quality of life. Particular attention needs to be taken when interventions avoid muscle damage, can similar adaptations be achieved with and without the presence of delayed onset muscle soreness and a decline in maximal strength. Another systematic review and meta-analysis paper [99] found that (1) eccentric exercises have more significant benefits for muscle strength and some markers of health than traditional exercise; and (2) eccentric exercise is an attractive alternative to traditional exercise in encouraging sedentary people to become more active.

No Pain, More Gain

To implement eccentric exercise training, it is better to avoid and minimize muscle damage, since muscle damage itself is not a factor necessary for muscle adaptations including hypertrophy and increased muscle function. ***"There are no gains without pains"*** is a quote of Benjamin Franklin, and Jane Fonda used ***"no pain no gain"*** in her 1982 video series of aerobic workouts. It is true that "effort" is necessary to improve health and fitness, but not "pain".

Figure 8.11 shows possible effects of eccentric versus concentric exercise training on muscle adaptations. In order to increase muscle size and mass, muscle protein synthesis should exceed muscle protein breakdown. Muscle damage could increase protein breakdown and induce inflammation. Although it is likely that the amount of the breakdown and inflammation decreases when eccentric exercise training is progressed, due to the repeated bout effect, muscle hypertrophy becomes more evident when resistance exercise training is progressed, when muscles become more immune to eccentric exercise-induced muscle damage [100]. This suggest that muscle damage is not the factor contributing to the hypertrophy.

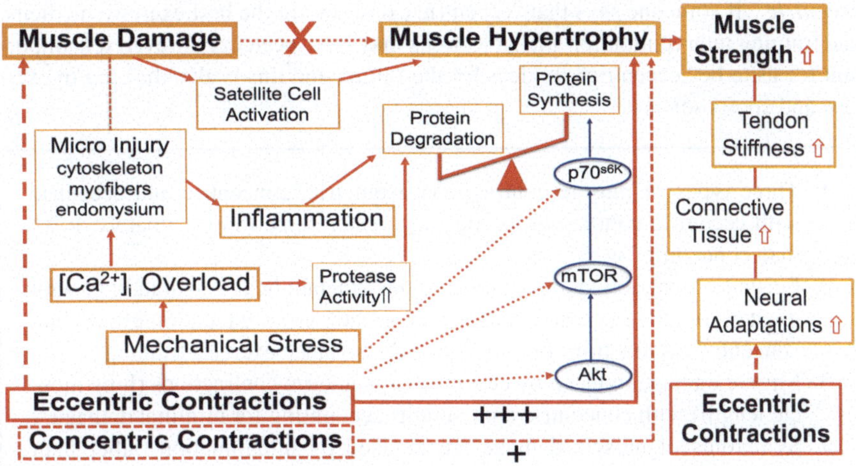

Fig. 8.11 Is muscle damage necessary for muscle hypertrophy and muscle strength increase?

Muscle protein synthesis is triggered by mechanical stimulus generated by muscle contractions, but it appears that eccentric contractions activate muscle protein synthesis signaling pathways (e.g., mTOR, $P70^{S6k}$) greater than concentric contractions. Thus, it is important to maximize the use of eccentric contraction while minimizing muscle damage. It is possible that mechanical stimulus generated in eccentric contractions affect the extracellular matrix surrounding muscle fibers, which could activate receptors relating to the signals.

Eccentric training also increases neural adaptations at the spinal and brain levels, and connective tissue and tendon strength that is important for force transmission from muscle fibers to the bone. This likely contributes to the greater improvement of muscle function after eccentric than concentric training.

Conclusion

Muscle damage is induced by eccentric exercises, and it is often inevitable, but it is necessary to minimize it. We often hear "no pain, no gain" but it does not necessarily mean that muscle damage is necessary for muscle adaptations. However, to improve health and fitness of general and clinical populations, and athletic performance, exercise interventions consisting of eccentric contractions should be considered, in which it is important to avoid muscle damage. Fortunately, there are strategies to prevent and/or minimize muscle damage and enhance recovery from it. We need to make the best use of eccentric exercises to make people including children to elderlies, and sedentary individuals to athletes, stronger, fitter, and healthier. Bertrand Russel, who was awarded the Novel Prize in Literature in 1950 quoted; "***Do not fear to be eccentric in opinion, for every opinion now accepted was once***

eccentric." If someone says that "eccentric exercises are the best exercise medicine and training intervention for all," objections may be expected. However, it is important for us to be "eccentric" in ideas for the future, and this is also the case in exercise and sports science.

1. Three types of muscle contractions, isometric, concentric and eccentric with eccentric contraction producing greater force (approximately 40%) than concentric.
2. Eccentric contractions less fatigable, less metabolically demanding, with greater cognitive demand, and induce greater cross-education effect than concentric contractions or conventional resistance exercise training.
3. Muscle damage induced by eccentric or isometric contractions (long muscle length) with concentric contractions accounting for minimal damage.
4. Symptoms of muscle damage are induced by inflammation rather than muscle damage itself.
5. DOMS is still evident with muscle fiber damage suggesting that DOMS is associated with connective tissue inflammation.
6. Muscle function measures such as MVC strength are best indicators of muscle damage and the time course of it.
7. Previous experience of same or similar eccentric exercise within several weeks is the biggest factor influencing magnitude of eccentric exercise-induced muscle damage.
8. To help muscle become more immune to muscle damage induced by eccentric contractions, it is important to include and perform eccentric training. This training also improves muscle strength and power, speed performance and reduces the risk of injury.
9. Eccentric training increases neural adaptations at the spinal and brain levels, and connective tissue and tendon strength, which is important for force transmission from muscle fibers to bone.

References

1. Faulkner J. Terminology for contractions of muscles during shortening, while isometric, and during lengthening. J Appl Physiol. 2003;95:455–9.
2. Nuzzo JL, Pinto MD, Nosaka K, Steele J. The eccentric:concentric strength ratio of human skeletal muscle in vivo: meta-analysis of the influences of sex, age, joint action, and velocity. Sports Med. 2023;53:1125–36.
3. Shibata K, Yamaguchi T, Takizawa K, Nosaka K. Comparison in repetitions to failure between concentric-only and eccentric-only dumbbell arm curl exercise at four different relative intensities. J Strength Cond Res. 2023;37:1754–60.
4. Yoshida R, Kasahara K, Murakami Y, Sato S, Nosaka K, Nakamura M. Less fatiguability in eccentric than concentric repetitive maximal muscle contractions. Eur J Appl Physiol. 2023;123:1553–65.

5. Abbott BC, Bigland B, Ritchie JM. The physiological cost of negative work. J Physiol. 1952;117:380–90.
6. Peñailillo L, Blazevich A, Numazawa H, Nosaka K. Metabolic and muscle damage profiles of concentric versus repeated eccentric cycling. Med Sci Sports Exerc. 2013;45:1773–81.
7. Kan B, Speelman C, Nosaka K. Cognitive demand of eccentric versus concentric cycling and its effects on post-exercise attention and vigilance. Eur J Appl Physiol. 2019;119:1599–610.
8. Chen TC, Wu S-H, Chen H-L, Tseng W-C, Tseng K-W, Kang H-Y, Nosaka K. Effects of unilateral eccentric versus concentric training of non-immobilized arm during immobilization. Med Sci Sports Exerc. 2023;55:1195–207.
9. Valdes O, Ramirez C, Perez F, Garcia-Vicencio S, Nosaka K, Penailillo L. Contralateral effects of eccentric resistance training on immobilized arm. Scand J Med Sci Sports. 2021;31:76–90.
10. Clarkson PM, Nosaka K, Braun B. Muscle function after exercise-induced muscle damage and rapid adaptation. Med Sci Sports Exerc. 1992;24:512–20.
11. Stožer A, Vodopivc P, Bombek LK. Pathphysiology of exercise-induced muscle damage and its structural, functional, metabolic, and clinical consequences. Physiol Res. 2020;69:565–98.
12. Hody S, Croisier J-L, Bury T, Rogister B, Leprince P. Eccentric muscle contractions: risks and benefits. Front Physiol. 2019;10:536.
13. Lavender AP, Nosaka K. Changes in steadiness of isometric force following eccentric and concentric exercise. Eur J Appl Physiol. 2006;96:235–40.
14. Murayama M, Nosaka K, Inami T, Shima N, Yoneda T. Biceps brachii muscle hardness assessed by a push-in meter in comparison to ultrasound strain elastography. Sci Rep. 2020;10:20308.
15. Mizumura K, Taguchi T. Delayed onset muscle soreness: involvement of neurotrophic factors. J Physiol Sci. 2016;66:43–52.
16. Koeda T, Mavropalias G, Mizumura K, Kitanosaka K, Nosaka K. Changes in nerve growth factor in vastus lateralis muscle after the first versus second bout of one-leg eccentric cycling. Scand J Med Sci Sports. 2023;34:e14497. https://doi.org/10.1111/sms.14497.
17. Nosaka K, Newton M, Chapman D, Sacco P. Is isometric strength loss immediately after eccentric exercise related to changes in indirect markers of muscle damage? Appl Physiol Nutr Metab. 2006;31:313–9.
18. Huang TH, Nosaka K, Chen TC. Changes in blood bone markers after the first and second bouts of whole-body eccentric exercises. Scand J Med Sci Sports. 2022;32:521–32.
19. Nosaka K, Clarkson PM, Apple F. Time course of serum protein changes after strenuous exercise of the forearm flexors. J Lab Clin Med. 1992;119:183–8.
20. Driessen-Kletter MF, Amelink GJ, Bar PR, van Gijn J. Myoglobin is a sensitive marker of increased muscle membrane vulnerability. J Neurol. 1990;237:234–8.
21. Chapman DW, Simpson JA, Iscoe S, Nosaka K. Changes in serum fast and slow skeletal troponin I concentration following maximal eccentric contractions. J Sci Med Sport. 2013;16:82–5.
22. Chen TC, Liu HW, Russel A, Barthel BL, Tseng KW, Huang MJ, Chou TY, Nosaka K. Large increases in plasma fast skeletal muscle troponin I after whole-body eccentric exercises. J Sci Med Sport. 2020;23:776–81.
23. Nosaka K, Clarkson PM. Changes in indicators of inflammation after eccentric exercise of the elbow flexors. Med Sci Sports Exerc. 1996;28:953–61.
24. Crameri RM, Aagaard P, Qvortrup K, Langberg H, Olesen J, Kjær M. Myofibre damage in human skeletal muscle: effects of electrical stimulation *versus* voluntary contraction. J Physiol. 2007;583:365–80.
25. Paulsen G, Crameri R, Benestad HB, Fjeld JG, Mørkrid L, Hallen J, Raastad T. Time course of leukocyte accumulation in human muscle after eccentric exercise. Med Sci Sports Exerc. 2010;42:75–85.
26. Armstrong RB, Warren GL, Warren JA. Mechanisms of exercise-induced muscle fibre injury. Sports Med. 1991;12:184–207.

27. Lau WI, Blazevich AJ, Newton MJ, Wu SSX, Nosaka K. Changes in electrical pain threshold of fascia and muscle after initial and secondary bouts of elbow flexors eccentric exercise. Eur J Appl Physiol. 2015;115:959–68.
28. Hough T. Ergographic studies in muscular soreness. Am J Physiol. 1902;7:76–92.
29. Damas F, Nosaka K, Libardi CA, Chen TC, Ugrinowitsch C. Susceptibility to exercise-induced muscle damage: a cluster analysis with a large sample. Int J Sports Med. 2016;37:633–40.
30. Nosaka K, Newton M, Sacco P. Delayed-onset muscle soreness does not reflect the magnitude of eccentric exercise-induced muscle damage. Scand J Med Sci Sports Exercise. 2002;12:337–46.
31. Chalchat E, Gaston A-F, Charlot K, Penailillo L, Valdes O, Tardo-Dino P-E, Nosaka K, Martin V, Garcia-Vicencio S, Siracusa J. Appropriateness of indirect markers of muscle damage following lower limbs eccentric-biased exercises: a systematic review with meta-analysis. PLoS One. 2022;17:e0271233.
32. Chen TC, Nosaka K, Chen H-L, Lin M-R, Wu C-J. Changes in running economy at different intensities following downhill running. J Sports Sci. 2009;27:1–8.
33. Nosaka K, Newton M. Differences in the magnitude of muscle damage between maximal and submaximal eccentric loading. J Strength Cond Res. 2002;16:202–8.
34. Chen TC, Huang G-L, Hsieh C-C, Tseng K-W, Tseng W-C, Chou T-Y, Nosaka K. Comparison among three different intensities of eccentric contractions of the elbow flexors resulting in the same strength loss at one day post-exercise for changes in indirect muscle damage markers. Eur J Appl Physiol. 2020;120:267–79.
35. Chapman D, Newton M, Sacco P, Nosaka K. Greater muscle damage induced by fast versus slow velocity eccentric exercise. Int J Sports Med. 2006;27:591–8.
36. Chapman D, Newton M, McGuigan M, Nosaka K. Effect of contraction velocity on muscle damage after maximal lengthening contractions. Med Sci Sports Exerc. 2008;40:926–33.
37. Nosaka K, Sakamoto K. Effect of elbow joint angle on the magnitude of muscle damage to the elbow flexors. Med Sci Sports Exerc. 2001;33:22–9.
38. Lau WY, Blazevich JA, Newton M, Wu SSX, Nosaka K. Effects of wrist position on eccentric exercise-induced muscle damage of the elbow flexors. Scand J Med Sci Sports. 2021;31:1290–300.
39. Chen TC, Lin KY, Chen H-L, Lin M-J, Nosaka K. Comparison in eccentric exercise-induced muscle damage among four limb muscles. Eur J Appl Physiol. 2011;111:211–23.
40. Chen TC, Chen H-L, Liu Y-C, Nosaka K. Eccentric exercise-induced muscle damage of pre-adolescent and adolescent boys in comparison to young men. Eur J Appl Physiol. 2014;114:1183–95.
41. Lin MJ, Nosaka K, Ho CC, Chen HL, Tseng KW, Ratel S, Chen TC. Influence of maturation status on eccentric exercise-induced muscle damage and the repeated bout effect in females. Front Physiol. 2018;8:1118.
42. Lavender AP, Nosaka K. Comparison of changes in markers of muscle damage between old and young men following voluntary eccentric exercise of the elbow flexors. Appl Physiol Nutr Metab. 2006;31:218–25.
43. Lavender A, Nosaka K. Responses of old men to repeated bouts of eccentric exercise of the elbow flexors in comparison to young men. Eur J Appl Physiol. 2006;97:619–26.
44. Chapman D, Newton M, McGuigan MR, Nosaka K. Comparison between old and young men for responses to fast velocity maximal lengthening contractions of the elbow flexors. Eur J Appl Physiol. 2008;104:531–9.
45. Tiidus PM. Can estrogens diminish exercise induced muscle damage? Can J Appl Physiol. 1995;20:26–38.
46. Rinard J, Clarkson PM, Smith LL, Grossman M. Response of males and females to high-force eccentric exercise. J Sports Sci. 2000;18:229–36.
47. Morawetz D, Blank C, Koller A, Arvandi M, Siebert U, Schobersberger W. Sex-related differences after a single bout of maximal eccentric exercise in response to acute effects: a systematic review and meta-analysis. J Strength Cond Res. 2020;34:2697–707.

48. Newton M, Morgan GT, Sacco P, Chapman D, Nosaka K. Comparison between resistance trained and untrained men for responses to a bout of strenuous eccentric exercise of the elbow flexors. J Strength Cond Res. 2008;22:597–607.
49. Chen C-H, Nosaka K, Chen H-L, Lin M-J, Tseng K-W, Chen TC. Effects of flexibility training on eccentric exercise-induced muscle damage. Med Sci Sports Exercise. 2011;43:491–500.
50. Nosaka K, Clarkson PM. Muscle damage following repeated bouts of high force eccentric exercise. Med Sci Sports Exerc. 1995;27:1263–9.
51. Nosaka K, Sakamoto K, Newton M, Sacco P. How long does the protective effect on eccentric exercise-induced muscle damage last? Med Sci Sports Exerc. 2001;33:1490–5.
52. Nosaka K, Newton MJ, Sacco P. Attenuation of protective effect against eccentric exercise-induced muscle damage. Can J Appl Physiol. 2005;30:529–42.
53. Nosaka K, Sakamoto K, Newton M, Sacco P. The repeated bout effect of reduced-load eccentric exercise on elbow flexor muscle damage. Eur J Appl Physiol. 2001;85:34–40.
54. Chen TC, Nosaka K, Sacco P. Intensity of initial eccentric exercise and the magnitude of repeated bout effect. J Appl Physiol. 2007;102:992–9.
55. Lavender A, Nosaka K. A light load eccentric exercise confers protection against a subsequent bout of more demanding eccentric exercise performed 2 days later. J Sci Med Sport. 2008;11:292–8.
56. Chen TC, Chen H-L, Peace A, Nosaka K. Attenuation of eccentric exercise-induced muscle damage by preconditioning exercises. Med Sci Sports Exerc. 2012;44:2090–8.
57. Proske U, Morgan DL. Muscle damage from eccentric exercise: mechanism, mechanical signs, adaptation and clinical applications. J Physiol. 2001;537:333–45.
58. Chen H-L, Nosaka K, Chen TC. Muscle damage protection by low-intensity eccentric contractions remains for 2 weeks but not 3 weeks. Eur J Appl Physiol. 2012;112:555–65.
59. Chen TC, Tseng W-C, Huang G-L, Chen H-L, Tseng K-W, Nosaka K. Low-intensity eccentric contractions attenuate muscle damage induced by subsequent maximal eccentric exercise of the knee extensors in the elderly. Eur J Appl Physiol. 2013;113:1005–15.
60. Lin M-J, Chen T, Chen H-L, Wu B-H, Nosaka K. Low-intensity eccentric contractions of the knee extensors and flexors protect against muscle damage. Appl Physiol Nutr Metab. 2015;40:1004–11.
61. Huang MJ, Nosaka K, Wang HS, Tseng KW, Chen HL, Chou TY, Chen TC. Damage protective effects conferred by low-intensity eccentric contractions on arm, leg and trunk muscles. Eur J Appl Physiol. 2019;119:1055–64.
62. Chen H-L, Nosaka K, Chen TC. Two maximal isometric contractions attenuate the magnitude of eccentric exercise-induced muscle damage. Appl Physiol Nutr Metab. 2012;37:680–9.
63. Chen TC, Chen H-L, Lin M-J, Peace AJ, Nosaka K. Effect of two maximal isometric contractions on eccentric exercise-induced muscle damage of the elbow flexors. Eur J Appl Physiol. 2013;113:1545–54.
64. Tseng K-W, Tseng W-C, Lin M-J, Chen H-L, Nosaka K, Chen TC. Protective effect by maximal isometric contractions against maximal eccentric exercise-induced muscle damage of the knee extensors. Res Sports Med. 2016;24:243–56.
65. Chen TC, Chen H-L, Lin M-L, Yu H-I, Nosaka K. Contralateral repeated bout effect of eccentric exercise of the elbow flexors. Med Sci Sports Exerc. 2016;48:2030–19.
66. Chen TC, Lin M-J, Chen H-L, Yu H-I, Nosaka K. Contralateral repeated bout effect of the knee flexors. Med Sci Sports Exerc. 2018;50:542–50.
67. Chen TC, Chen H-L, Tseng W-C, Chou T-Y, Tu J-H, Parcell AC, Nosaka K. Contralateral versus ipsilateral protective effect against muscle damage of the elbow flexors and knee extensors induced by maximal eccentric exercise. Scand J Med Sci Sports. 2023;33:2548–60.
68. Hyldahl RD, Chen TC, Nosaka K. Mechanisms and mediators of the skeletal muscle repeated bout effect. Exerc Sport Sci Rev. 2017;45:24–33.

69. Lau WI, Blazevich AJ, Newton MJ, Wu SSX, Nosaka K. Reduced muscle lengthening during eccentric contractions as a mechanism underpinning the repeated bout effect. Am J Physiol Regul Integr Comp Physiol. 2015;308:R879–86.
70. Ho CC, Nosaka K, Tseng KW, Tseng WC, Lau WY, Bogdanis GC, Chen TC. Effect of preconditioning exercise on biceps brachii myotendinous junction displacement during elbow flexor eccentric exercise. Scand J Med Sci Sports. 2021;31:813–25.
71. Frazer AK, Pearce AJ, Howatson G, Thomas K, Goodall S, Kidgell DJ. Determining the potential sites of neural adaptation to cross-education: implications for the cross-education of muscle strength. Eur J Appl Physiol. 2018;118:1751–72.
72. Nosaka K, Newton M, Sacco P, Chapman D, Lavender A. Partial protection against muscle damage by eccentric actions at short muscle lengths. Med Sci Sports Exerc. 2005;37:746–53.
73. Chapman D, Newton M, McGuigan M, Nosaka K. Effect of a slow velocity eccentric exercise on muscle damage induced by fast velocity eccentric exercise. J Strength Cond Res. 2011;25:211–9.
74. Maeo S, Yamamoti M, Kanehisa H, Nosaka K. Effect of a prior bout of preconditioning exercise on muscle damage from downhill walking. Appl Physiol Nutr Metab. 2015;40:274–9.
75. Maeo S, Yamamoto M, Kanehisa H, Nosaka K. Prevention of downhill walking-induced muscle damage by non-damaging downhill walking. PLoS One. 2017;12:e0173909.
76. Chen TC, Lin M-J, Chen H-L, Lai J-H, Yu H-I, Nosaka K. Muscle damage protective effect by two maximal isometric contractions on maximal eccentric exercise of the elbow flexors of the contralateral arm. Scand J Med Sci Sports. 2018;28:1354–60.
77. Tseng W-C, Nosaka K, Tseng K-W, Chen TC. Contralateral effects by unilateral eccentric versus concentric resistance training. Med Sci Sports Exerc. 2020;52:474–83.
78. Chen TC, Kang H-Y, Tseng W-C, Lin S-C, Chan C-W, Chen H-L, Chou T-Y, Wang H-H, Lau WY, Nosaka K. Muscle damage induced by maximal eccentric exercise of the elbow flexors after 3-week immobilization. Scand J Med Sci Sports. 2023;33:382–92.
79. Zainuddin Z, Newton M, Sacco P, Nosaka K. Massage reduces DOMS and swelling without effect on recovery of muscle function. J Athl Train. 2005;40:174–80.
80. Lau WY, Nosaka K. Effect of vibration treatment on symptoms associated with eccentric exercise-induced muscle damage. Am J Phys Med Rehabil. 2011;90:648–57.
81. Zainuddin Z, Sacco P, Newton M, Nosaka K. Light concentric exercise has a temporarily analgesic effect on DOMS but no effect on recovery from eccentric exercise. Appl Physiol Nutr Metab. 2006;31:126–34.
82. Nosaka K, Sacco P, Mawatari K. Effects of amino acid supplementation on muscle soreness and damage. Int J Sport Nutr Exerc Metab. 2006;16:620–35.
83. Tanabe Y, Maeda S, Akazawa N, Zempo-Miyaki A, Choi Y, Ra S-G, Imaizumi A, Okutsu Y, Nosaka K. Attenuation of indirect markers of eccentric exercise-induced muscle damage by curcumin. Eur J Appl Physiol. 2015;115:1949–57.
84. Chen TC, Huang Y-C, Chou T-Y, Hsu S-T, Chen M-Y, Nosaka K. Effects of far-infrared radiation lump therapy on recovery from muscle damage induced by eccentric exercise. Eur J Sport Sci. 2023;23:1638–46.
85. Chou T-Y, Nosaka K, Chen TC. Muscle damage and performance after single and multiple simulated matches in university elite female soccer players. Int J Environ Res Public Health. 2021;18:4134.
86. Jones MR, West DJ, Harrington BJ, Cook CJ, Bracken RM, Shearer DA, Kilduff LP. Match play performance characteristics that predict post-match creatine kinase responses in professional rugby union players. BMC Sports Sci Med Rehabil. 2014;6:38.
87. McLellan CP, Lovell DI, Gass GC. Markers of post-match fatigue in professional rugby league players. J Strength Cond Res. 2011;25:1030–9.
88. Tavares F, Smith TB, Driller M. Fatigue and recovery in rugby: a review. Sports Med. 2017;47:1515–30.
89. Moreira A, Nosaka K, Nunes JA, Viveiros L, Jamurtas AZ, Aoki MS. Changes in muscle damage markers in female basketball players. Biol Sport. 2014;31:3–7.

90. Doma K, Leicht A, Sinclair W, Schumann M, Damas F, Burt D, Woods C. Impact of exercise-induced muscle damage on performance test outcomes in elite female basketball players. J Strength Cond Res. 2018;32:1731–8.
91. Souglis A, Bogdanis GC, Giannopoulou I, Papadopoulos C, Apostolidis N. Comparison of inflammatory responses and muscle damage indices following a soccer, basketball, volleyball and handball game at an elite competitive level. Res Sports Med. 2015;23:59–72.
92. Burt D, Hayman O, Forsyth J, Doma K, Twist C. Monitoring indices of exercise-induced muscle damage and recovery in male field hockey: is it time to retire creatine kinase? Sci Sports. 2020;35:402–4.
93. Abián P, Coso JD, Salinero JJ, Gallo-Salazar C, Areces F, Ruiz-Vicente D, Lara B, Soriano L, Muñoz V, Lorenzo-Capella I, Abián-Vicén J. Muscle damage produced during a simulated badminton match in competitive male players. Res Sports Med. 2016;24:104–17.
94. Lin Z, Blazevich AJ, Abbiss CR, Chochrane-Wilkie J, Nosaka K. Neuromuscular fatigue and muscle damage following a simulated singles badminton match. Eur J Appl Physiol. 2023;123:1229–40.
95. Gomes RV, Santos RCO, Nosaka K, Moreira A, Miyabara EH, Aoki MS. Muscle damage after a tennis match in young players. Biol Sport. 2014;31:27–32.
96. Douglas J, Pearson S, Ross A, McGuigan M. Chronic adaptations to eccentric training: a systematic review. Sports Med. 2017;47:917–41.
97. Chaabene H, Prirske O, Negra Y, Granacher U. Changes of direction speed: toward a strength training approach with accentuated eccentric actions. Sports Med. 2018;48:1773–9.
98. Ansari M, Hardcastle S, Myers S, Williams AD. The health and functional benefits of eccentric versus concentric exercise training: a systematic review and meta-analysis. J Sports Sci Med. 2023;22:288–309.
99. Gault ML, Willems MET. Aging, functional capacity and eccentric exercise training. Aging Dis. 2013;4:351–63.
100. Dames F, Libardi CA, Ugrinowitsch C. The development of skeletal muscle hypertrophy through resistance training: the role of muscle damage and muscle protein synthesis. Eur J Appl Physiol. 2018;118:485–500.

Chapter 9
Inflammatory Response to Different Types of Exercise

Konstantinos Papanikolaou, Dimitrios Draganidis, Ioannis G. Fatouros, and Athanasios Z. Jamurtas

Exercise-Induced Inflammation: Mechanisms and Consequences

- Intense, unaccustomed exercise is associated with the development of exercise-induced muscle damage (EIMD)
- EIMD initiates a local and systemic inflammatory and immune response
- Inflammation is necessary for optimal muscle regeneration and recovery
- The time course of the inflammatory response is affected by controllable and uncontrollable factors

Exercise-Induced Muscle Damage

Strenuous, unaccustomed exercise and especially eccentric exercise (muscle lengthening contractions) has been associated with exercise-induced muscle damage (EIMD) of cellular structures, disruption of extracellular matrix organization, development of inflammation, dysfunction of intramuscular calcium (Ca^{2+}) release and excitation-contraction (E-C) coupling and impairment of muscle contractile activity [1–5]. Proske and Allen [6, 7] have proposed a mechanistic model in which some sarcomeres of a stretched muscle fiber may be more resilient than others due to differences in length-tension relationships, resulting in greater absorption of the stretch by the less resilient sarcomeres which, based on the magnitude of the imposed

K. Papanikolaou · D. Draganidis · I. G. Fatouros · A. Z. Jamurtas (✉)
Department of Physical Education & Sport Science, School of Physical Education, Sport Science & Dietetics, University of Thessaly, Trikala, Greece
e-mail: jamurtas@gmail.com

N. C. Apostolopoulos et al. (eds.), *Fundamentals of Recovery, Regeneration, and Adaptation to Exercise Stress: An Integrated Approach*,
https://doi.org/10.1007/978-3-031-44270-4_9

force, become weaker as the overlap between myosin and actin filaments is reduced at the end of the contraction [8]. After repeated cycles of intense contractions, more of the less resilient sarcomeres are affected and become overstretched. Since the myofilaments of these sarcomeres fail to return to their initial non-overlapping state during the uncoupling phase of the muscle contraction, a mechanical disruption develops, leading to subcellular damage of muscle fiber membrane, sarcoplasmic reticulum, T-tubules and terminal cisternae [2, 8]. This mechanical damage compromises the E-C coupling in the affected sarcomeres and inhibits muscle contractile activity via the uncontrollable release of Ca^{2+} as a consequence of damage in the T-tubule/triad system in the cytoplasm [6, 7]. Additionally, uncontrollable Ca^{2+} influx triggers sensitive degradative pathways that stimulate proteolytic enzymes promoting further degradation of muscle proteins [2, 8]. Specifically, the elevation of intracellular Ca^{2+} activates the non-lysosomal, calpain proteases, located in the I and Z regions of the sarcomere which are suggested to contribute to cytoskeletal protein cleavage and degradation of Z-disks (Z-line streaming) and desmin protein [9]. Moreover, the extracellular matrix (ECM) which is a complex of collagen proteins (i.e., integrins, dystroglycans and proteoglycans) associated with the fiber basement membrane is also affected by exercise-induced mechanical damage [8]. Histological observations suggested that EIMD results in a dissociation of perimysium and endomysium from the muscle fiber membrane [10] and these observations are supported by others indicating increased collagen breakdown and synthesis following EIMD [11, 12]. The mechanical damage observed in the muscle fiber membrane, cytoplasm and ECM, results in the release of muscle proteins in the blood circulation (i.e., creatine kinase, myoglobin) and several studies have indicated a positive correlation between these markers and the extent of EIMD [13]. Following the initial phase of mechanical stress events, EIMD leads to the development of an inflammatory response associated with the activation of immune cells, muscle swelling, delayed onset of muscle soreness (DOMS), increase in muscle temperature and impairment of muscle function [8, 14–16].

Exercise-Induced Inflammation and Immune Response

In the context of sports medicine, exercise-induced inflammation encompasses physiological, clinical, cellular and molecular signatures within the injured muscle [17]. It is now evident that under physiological conditions, exercise-induced inflammation is a tightly coordinated and dynamic process, which is necessary for adaptive remodeling and homeostasis [17]. EIMD initiates an aseptic inflammatory response with concomitant activation of the immune system and specific myeloid cell populations [18]. $Ly6C^+/F4/80^-$ neutrophils are the first immune cell type that is attracted to the injured muscle area within 2 hours of damage, peaking their concentration between 4 and 24 hours and subsiding thereafter [19, 20]. At rest, neutrophils are almost exclusively marginated in the blood vessels, but upon EIMD, conformational changes in the cell membrane receptors facilitate their infiltration to the damaged tissue [21]. Of note, neutrophil activity in the damaged muscle vicinity is biphasic. Initially, these cells are responsible for the removal of cellular debris (i.e., damaged proteins) by releasing reactive oxygen and nitrogen species (RONS) [22], such as superoxide and hydrogen peroxide [23], through the respiratory burst mechanism [24, 25]. This process is considered a prerequisite for optimal tissue healing as incomplete clearance of cellular debris from the damaged site can hamper the subsequent regeneration process of the muscle [18]. Furthermore, neutrophils as well as myocytes produce pro-inflammatory cytokines such as IL-1β, IL-6, IL-8 and TNF-α which attract pro-inflammatory M1 macrophages to the injured area [26]. M1 macrophages ($CD68^+/CD206^-$) are also responsible for the phagocytic removal of cellular debris but can also contribute to secondary damage and further muscle cell lysis by the production of cytotoxic levels of nitric oxide (NO) during the early phase of inflammation [18]. Proportionally with the magnitude of damage induced, M1 macrophages peak their concentration between 1 and 3 days following injury [27] and thereafter they are gradually replaced by a population of anti-inflammatory M2 macrophages ($CD68^+/CD206^+$) that attenuate the inflammatory response and initiate the regeneration of the injured tissue [18, 28]. The transition of tissue-resident macrophages from a pro-inflammatory to an anti-inflammatory phenotype is triggered by a shift in the production of anti-inflammatory cytokines such as IL-10, IL-12 and IL-1ra [29, 30]. However, it should be noted that the exercise-induced inflammatory response is not only evident within the injured muscle but also in the blood circulation, since the local mobilization of immune cells is also accompanied by systemic changes in the levels of chemotactic factors and pro-inflammatory cytokines such as IL-1β and TNF-α [19, 28].

The Fundamental Role of Inflammation in Muscle Regeneration and Performance Recovery

Skeletal muscle regeneration following injury depends on a heterogeneous population of tissue-specific resident stem cells, also known as satellite cells, representing ~30–35% of sublaminar nuclei in postnatal myofibers [31]. Animal studies have shown that conditional depletion of satellite cells impairs muscle regeneration and decreases myofiber size and accumulation of fibrotic tissue [32]. In resting muscle, satellite cells remain mitotically quiescent, but following EIMD, they are activated and enter the cell cycle as they undergo a series of proliferation, differentiation and fusion with injured myofibers to repair the damage [33]. During the proliferative phase, a number of activated satellite cells do not terminally differentiate to myoblasts but instead return to a quiescent state, in a process of self-renewal, in order to replenish the satellite cell pool for future damage [33]. In adult muscle, satellite cells express specific transcription factors such as paired box protein 7 (Pax7), myogenic factor 5 (Myf5), myogenic differentiation 1 (MyoD) and Myogenin [34]. Inflammation and muscle regeneration are two interconnected processes, as the inhibition of the former may hamper the latter and ultimately result in impaired muscle recovery [35]. During the first phase of the inflammatory response, infiltrating leukocytes release pro-inflammatory cytokines such as IL-6 and TNF-α, which at physiological concentrations stimulate satellite cell activation and proliferation [36, 37]. Moreover, the phagocytic potential of leukocytes and monocytes promotes the initiation of the myogenic process [33], while the effect on myogenesis largely depends on the type of injury [38]. This pro-inflammatory environment attracts the recruitment of classical ($Ly6C^+$) and nonclassical ($Ly6C^-$) monocyte populations which are predominant during the first few days following injury [19]. $Ly6C^+$ monocytes secrete high levels of IL-1β and TNF-α cytokines which promote satellite cell proliferation while $Ly6C^-$ monocytes produce anti-inflammatory cytokines such as IL-10 and TGF-β which downregulate myoblast fusion and upregulate differentiation and fusion activity [19]. Following the initial pro-inflammatory activity of monocytes and M1 macrophages, the polarization of M2 macrophages has been shown to tightly regulate the resolution of the inflammatory process [39]. M2 macrophages release growth factors and anti-inflammatory cytokines (i.e., IL-4, IL-10, and IGF-1) which influence the late stages of regeneration [33]. Interestingly, inhibition of M2 macrophages in vivo, results in reduced differentiation potential and blunted myofiber growth [33]. Importantly, M2 macrophages have been shown to produce large amounts of collagen proteins such as fibronectin and collagen VI, which are involved in satellite cell expansion, self-renewal and ECM remodeling following damage [40]. The uninterrupted inflammatory response is also important for optimal muscle performance recovery as it was shown that nutritional and/or pharmacological manipulation of the inflammatory response and inhibition of the proinflammatory phase may hamper muscle strength recovery at the late stage of regeneration [35]. Altogether, the inflammatory response plays a fundamental role in performance recovery through the induction of myogenesis following muscle injury.

Factors Affecting the Time Course of the Inflammatory Response During Recovery

The time course of the inflammatory responses after EIMD depends on several controllable and uncontrollable factors such as type of muscle contraction (i.e., concentric, eccentric, isometric), exercise overload characteristics (i.e., intensity, volume and unaccustomedness), participation in regular exercise, muscle group engaged, nutritional supplements consumption, gender and genetic profile [8, 16]. The most well-characterized factor that affects the time course of the inflammatory response and the muscle recovery following EIMD is previous exposure to muscle damage, known as the repeated bout effect. Following an initial muscle-damaging exercise bout, skeletal muscle initiates an adaptation process that acts prophylactically as the inflammatory symptoms are less pronounced and the return to homeostasis is more rapid after a subsequent muscle-damaging bout of exercise of the same overload characteristics [41–44]. Of note, the repeated bout effect can be also observed in the contralateral muscles such that the second eccentric exercise bout performed by the contralateral arm leads to attenuated EIMD and inflammation than the first bout performed by the opposite arm [17, 45]. It is evident that severe EIMD is the result of greater mechanical strain with high- vs. low-force muscle contractions and also a greater number of contractions that induce greater damage to contractile proteins and ECM components [46]. Moreover, some reports indicate that eccentric contractions at long muscle lengths, single-joint exercises and eccentric protocols performed in upper limbs, elicit greater amounts of muscle damage and inflammation [47–49] whereas there are reports indicating that short muscle length results in greater muscle damage and peak torque decline [50]. Low to moderate EIMD levels result in a controlled inflammatory response which is beneficial for promoting positive adaptations to the exercise stimulus, whereas high levels of EIMD trigger a more intense inflammatory response which, depending on the magnitude of EIMD, is accompanied by severe disruption of skeletal muscle fibers and reduced recovery capacity [51]. Furthermore, in extreme cases of EIMD or injury the propagation of the inflammatory response is augmented resulting in muscle fiber necrosis, impaired regeneration, scar tissue development and delayed recovery [52]. Other factors that may affect the predisposition to EIMD and the course of the inflammatory response are associated with biological differences between men and women. It has been observed that women experience lower levels of muscle damage and inflammation following EIMD. Additionally, data in the literature suggest that women demonstrate strength decrements, DOMS and creatine kinase elevations and myofiber disruption of lower magnitude compared to men, probably due to a protective effect of estrogens on muscle fiber membranes [53, 54]. Another factor that may attenuate the development of the inflammatory response following EIMD is the training status of the individual. Specifically, athletes are subjected to heavy training schedules which result in positive adaptations to skeletal muscles and controlled inflammatory responses following exercise programs of different characteristics. Of note, the repeated bout effect has been suggested to attenuate the EIMD and inflammatory responses when an athlete or a trained individual is subjected to the same exercise modality [55]. However, data in the literature also suggest the existence of high interindividual variability regarding the efficacy of the repeated bout effect and that the inflammatory gene expression levels associated with EIMD may not be attenuated following a second exercise bout of the same characteristics, pinpointing a role for specific genes in the adaptation process [55, 56]. The utilization of nutritional and/or pharmaceutical compounds such as sports supplements, antioxidants and anti-inflammatory agents may also interfere with the development of the inflammatory responses following EIMD [57, 58]. It is well documented that participation in elite sports can be quite demanding as athletes need to compete in a congested fixture with two or more competitions in the same week [59]. Thus, maximizing or accelerating the recovery time is of great significance with respect to elite sports participation. Nutritional strategies have been proposed in the literature to assist in this aim. For example, during a congested game fixture in soccer, protein supplementation was shown to efficiently restore football-specific performance and strength and provide antioxidant protection [59]. Moreover, it was shown that branched-chain amino acids (BCAA) administration before and following damaging resistance exercise reduces indices of muscle damage and accelerates recovery in resistance-trained males [60]. Additionally, creatine supplementation has been shown to effectively attenuate creatine kinase rise and EIMD symptoms following eccentric exercise [61]. The potential mechanisms of action for creatine kinase include a decrease in the circulating levels of inflammatory molecules (PGE_2 and TNF-a), increased satellite cell activation and proliferation, reduction of oxidative stress levels (i.e., lipid and protein oxidation) and improved calcium pump function [62].

- Strenuous, unaccustomed exercise is associated with exercise-induced muscle damage (EIMD)
- EIMD initiates an aseptic inflammatory response with concomitant activation of the immune system
- The inflammatory response plays a fundamental role in performance recovery through the induction of myogenesis following muscle injury

Inflammatory Response to Eccentric Exercise

- Eccentric exercise leads to the development of inflammation
- The inflammatory response is interconnected with the muscle regeneration process
- Eccentric exercise-induced muscle damage and inflammation are associated with favorable muscle and performance adaptations

Mechanisms and Time Course of Eccentric Exercise-Induced Inflammation

Human and animal research has shown a strong inflammatory response following eccentric exercise, which substantially contributes to EIMD [8]. The innate inflammatory response is driven by the action of Th1 cytokines, especially IFN-γ and TNF-α and can be evident immediately after eccentric exercise with leukocyte mobilization which peaks within the first 24 hours [63]. The systemic neutrophil accumulation is accompanied by an increase in plasma proteins which transfer to the damaged tissue and attract other inflammatory cell types such as natural killer (NK) cells and lymphocytes. Specifically, Ly6C/F4/80 neutrophils are attracted to the injured area by chemotactic and cleavage factors within the first two hours after an eccentric exercise bout [28]. Ly6C/F4/80 neutrophils have been shown to contribute to further damage in the tissue by promoting calpain-mediated protein breakdown but also assisting in the removal of cellular necrotic debris via phagocytosis along with tissue-resident macrophages [20, 28]. It has been shown that neutrophils levels can remain elevated in damaged tissue for approximately 5 days however, in most cases, tissue macrophages replace neutrophils and become the predominant immune cell type in the injured muscle for up to 2 weeks [17, 64]. Injured muscle cells produce pro-inflammatory cytokines (i.e., IL-1β, IL-6 and TNF-α) with their levels persisting for up to 5–7 days post-injury [65]. The increase in pro-inflammatory cytokines levels triggers the differentiation of macrophages to M1-like phenotype ($CD68^+/CD206^-$ cells) with their concentration peaking at ~2 days following damage [28]. M1 macrophages are pro-inflammatory macrophages and are essential for the phagocytosis of damaged proteins and cellular debris. Additionally, M1 macrophages express pro-inflammatory cytokines such as TNF-α that activate protein degradation pathways within skeletal muscle [66]. More specifically, TNF-α activates the ubiquitin-proteasome pathway through the induction of E3 ligases and key molecular targets such as muscle ring finger 1 (MuRF1) and muscle atrophy F-box (MAFbx or Atrogin1), thus facilitating tissue remodeling and adaptation following EIMD [66]. Moreover, it has been suggested that this cell type contributes to secondary muscle injury, by promoting the expression of proteolytic enzymes (i.e., calpains, caspases) as well as reactive oxygen and nitrogen species (RONS) during a process called respiratory burst [18]. The production of RONS is considered essential for the macrophage-mediated inflammatory processes in skeletal muscle following injury [67]. In general, low to moderate levels of RONS promote signal transduction, regulate gene expression and facilitate cell signaling resulting in RONS-mediated adaptation to exercise [68]. However, high RONS levels are toxic to most cell types and promote DNA degradation and apoptotic mechanisms [67]. In the time course of the inflammatory response, at approximately 2 days following EIMD, M1 macrophages are gradually replaced by anti-inflammatory M2 macrophages ($CD68^+/CD206^+$ cells). This shift in macrophage phenotype is partly regulated by the expression of Th2 anti-inflammatory cytokines (i.e., IL-4, IL-10 and IL-13), and data suggest that this switch plays a significant role in the regeneration process of the injured tissue through the activation of muscle stem cells (i.e., satellite cells) [18, 69]. M2 macrophages represent a complex cell type that reflects distinct molecular and functional properties. M2a macrophages are activated by IL-4 and IL-13 and promote tissue repair mechanisms. M2b macrophages are activated by immune complexes and Toll-like receptors in order to release anti-inflammatory cytokines. Lastly, M2c macrophages are activated by IL-10 and release anti-inflammatory cytokines that deactivate M1 macrophages and enhance the proliferation of satellite cells to initiate the regenerative response [18]. Apart from macrophages, skeletal muscle cells can play a pivotal role in inflammatory responses. During exercise, muscle cells produce cytokines, among which IL-6 with its concentration varying depending on the intensity and duration of the eccentric exercise [70]. This increase is evident locally in the injured tissue but also systemically. During exercise, circulating levels of both pro-inflammatory and anti-inflammatory cytokines are elevated, thus promoting a systemic response to eccentric EIMD [65]. To this point, it should be noted that circulating cytokine levels are the product of their kinetics from the tissue of origin to the bloodstream and can only partially explain the exact role of eccentric exercise-induced inflammation [8]. At the later stages of the inflammatory response, cytokine levels are subsiding followed by the resolution of inflammatory mediators and the regeneration processes.

Tissue Regeneration and Performance Recovery

Skeletal muscle regeneration is dependent on a heterogeneous population of muscle-specific stem cells called satellite cells [32, 33]. Satellite cells account for approximately ~35% of sublaminal myonuclei in adult fibers with oxidative fibers (type I) containing a higher proportion of satellite cells than glycolytic fibers (type II) [71]. Satellite cells are identified in mature skeletal muscle due to their anatomic location between the basement membrane and the sarcolemma of the fiber as well as by the expression of nuclear and surface transcriptional factors such as Pax7 and NCAM [33]. In resting muscle satellite cells reside in a quiescent state but upon muscle injury, they are activated by secretory factors such as cytokines, chemokines, growth factors and signaling molecules and begin to proliferate in order to assist in tissue repair [72]. Activated satellite cells ($Pax7^+/Myf5^+$ cells) enter the cell cycle and undergo cell division to provide daughter cells that translocate to the injured area of the muscle within the first hours following damage [73]. PI3K/Akt signaling pathway is known to promote satellite cell activation via ERK signaling [73]. During the proliferation state, satellite cells upregulate the expression of the myoblast determination factor MyoD which is a key regulator of the myogenic response [74]. Proliferative myoblasts ($Pax7^+/MyoD^+$ cells) increase their mitotic activity due to the higher metabolic demand during skeletal muscle regeneration while studies in animal models have shown that $MyoD^{-/-}$ myoblasts lead to incomplete regeneration of the tissue [75]. This proliferative process usually peaks at 2–3 days following muscle damage and can be sustained for up to 5 days depending on the magnitude of muscle injury. Following proliferation, proliferative myoblasts differentiate into myocytes by downregulating Pax7 and MyoD expression and upregulating Myogenin expression levels, which peak 8–10 days after skeletal muscle damage [76]. Recent evidence in the literature suggests that this process is under a redox-dependent regulation and control [77]. Differentiating myocytes undergo a series of early and late fusions to form mature myofibers within the damaged area [78]. In the stages of terminal differentiation, newly formed myotubes exhibit high proportions of genetic DNA material (myonuclei) and undergo a process of functional maturation which involves the expression of early myosin heavy chain (MHC) isoforms (i.e., embryonic and neonatal MHC) [79]. This late phase of the functional maturation process is tightly regulated by the Akt/mTOR signaling pathway which stimulates muscle protein synthesis through the phosphorylation of ribosomal protein S6 [80]. The last phase of regeneration includes angiogenesis, ECM remodeling and reinnervation processes [81]. Angiogenesis is driven by the transforming growth factor-β (TGF-β) which initiates a physiological fibrosis mechanism while ECM components gradually accumulate in the regenerative area to provide stability and building docks for neuromuscular junction formation [82]. Following neuromuscular junction formation, reinnervation is a crucial step for optimal tissue remodeling as it is tightly connected with the functional capacity and strength of the muscle [83]. During the time course of inflammation and the regeneration process, muscle performance is reduced, depending on the duration, intensity and total volume of the eccentric exercise stimulus [84]. Mild muscle performance decline leads to limited strength loss of approximately 20%, minimum inflammation and quick recovery (usually 12–48 hours post-exercise) [8]. Moderate muscle performance decrease (20%–50%) results in an increase of CK levels in serum (>1000 U/l), initiates the inflammatory response in a more intense way and is characterized by immune cell infiltration in the injured area. Under these circumstances, the recovery phase usually lasts from 5 to 7 days [8]. In severe muscle damage, strength performance deterioration exceeds >50%, while histological analyses have revealed extensive necrosis in myofibers which is followed by a pronounced accumulation of neutrophils, macrophages and cytotoxic T cells in the muscle tissue, soreness and swelling. In this scenario, circulating CK levels may exceed 10,000 U/l [8].

Muscle Adaptation Following Eccentric Exercise Training

Eccentric exercise-induced mechanical tension stimulates subcellular pathways involved in muscle protein synthesis and growth [85, 86]. During eccentric exercise, muscles are subjected to both stretch and overload, leading to subcellular damage to contractile proteins and structural components of the fibers [87]. This is accompanied by the stimulation of the mitogen-activated protein kinase pathway and its downstream targets which are associated with muscle hypertrophy [88]. Additionally, it has been suggested that eccentric exercise-associated mechanotransduction drives the increase in muscle cross-sectional area (CSA) and pennation angle [89]. In corroboration, previous research has indicated that chronic eccentric exercise can increase muscle fiber length and hypertrophy [90]. It was also observed that muscle performance recovery after immobilization is expedited following participation in chronic eccentric exercise compared to concentric or isometric exercise, probably due to the higher mechanical tension to the muscle tissue which leads to favourable functional adaptations [91]. Apart from the aforementioned adaptations, several studies have investigated the adaptive changes of the inflammatory response after repeated participation in eccentric exercise [17]. Eccentric training minimized the transcription of several genes involved in monocyte recruitment and activity, whereas it augmented the transcription of genes involved in the transition of macrophages from a pro-inflammatory to anti-inflammatory phenotype following, thus promoting an anti-inflammatory environment that is related to chronic eccentric exercise [92]. Additionally, evidence in the literature suggests that the expression of several cytokines is upregulated, while NF-kB signaling (pro-inflammatory pathway) is downregulated following chronic eccentric exercise [93]. Collectively, after muscle adaptation to eccentric training, the pro-inflammatory response is attenuated, while the greater responsiveness of anti-inflammatory pathways enhances the processes associated with muscle repair (i.e., satellite cell proliferation and differentiation) and performance recovery [94]. It is also important to pinpoint that the sequence of events regarding the muscle's inflammatory response is crucial for optimal muscle regeneration and recovery [17]. An initial inflammatory response is required for efficient muscle regeneration via satellite cell activation. However, the interplay between pro- and anti-inflammatory signals must be tightly regulated to avoid sustained inflammation, excessive tissue damage and attenuated performance recovery [18].

- Eccentric exercise results in a severe inflammatory response and cytokine production
- Tissue regeneration and performance recovery are dependent on the resolution of the inflammatory response
- The interplay between pro- and anti-inflammatory factors controls the time course of performance recovery

Inflammatory Response to Resistance Exercise

- Resistance exercise initiates the activation of the immune system
- Inflammatory cytokines are secreted both locally and systemically in response to resistance exercise
- Resistance exercise-induced muscle adaptations include the attenuation of the inflammatory response in subsequent bouts of exercise

Mechanisms and Time Course of Resistance Exercise-Induced Inflammation

The inflammatory immune responses to resistance exercise are typically defined as changes in the levels and activity of leukocytes in the circulation and muscle tissue. Neutrophils and basophils increase immediately post-exercise with sustained neutrophilia for up to 2 hours following resistance exercise [95]. Monocytes increase during resistance exercise with studies showing a sustained monocytosis for as long as 2 hours post-exercise [95]. Helper T cells ($CD4^+$) have been shown to moderately increase following resistance exercise with their values increasing for as long as 30 minutes post-exercise, while cytotoxic T cells ($CD8^+$) increase immediately post-exercise and return to baseline levels between 25–45 minutes post-exercise [96]. However, a report by Ramel et al. [95] indicated a decrease in $CD8^+$ cells during recovery from exercise. Mature $CD19^+$ B cells typically increase immediately post-exercise and return to baseline levels at approximately 2 hours following exercise [97]. Natural Killer (NK) cells also increase during and after acute resistance exercise with their levels returning to baseline by 30–45 minutes post-training [98]. Acute resistance exercise has been also shown to initiate a systemic and local inflammatory cytokine response. Resistance exercise of the elbow flexors resulted in an increase in IL-6 levels at 3 hours post-training and upregulation of soluble tumor necrosis factor-α receptor 1 (sTNF-αR1), which lasted 1 day following training [99]. Moreover, both low and high-intensity resistance exercise elevate IL-6 levels in the circulation in an intensity-dependent manner [100]. On the contrary, another study by Uchida et al. failed to report any changes in cytokine responses (IL-1β, IL-6 and IL-10) at 24 hours following resistance exercise of different intensity in untrained men probably due to the lack of early measurement timepoints [101]. Resistance exercise has been also shown to induce changes at the gene expression level. Specifically, resistance exercise upregulates the mRNA expression levels of IL-1β, IL-2, IL-5, IL-6, IL-8, IL-10 and TNF-α in muscle tissue [102]. Notably, these changes are not always followed by the elevation of these markers in the circulation, suggesting a conservative local cytokine response within the skeletal muscle [103]. The available literature indicates that the increase in IL-6 is the most consistent finding regarding the effect of resistance exercise on cytokines levels. However, methodological differences in these studies have provided different results regarding the timing of the peak in IL-6 levels following resistance exercise, which ranges from 1 to 24 hours post-exercise [104]. Collectively, it is plausible to suggest that immediate sampling time points are crucial in evaluating cytokine kinetics following resistance exercise in humans.

Muscle Adaptation Following Resistance Exercise Training

Chronic resistance exercise results in positive adaptations of the inflammatory responses. It has been suggested that the main factor leading to favourable adaptations is exercise intensity as greater intensity has a stronger effect on the cytokine response (i.e., IL-1ra, IL-10) [17]. Following resistance exercise training, pro-inflammatory cytokine levels are attenuated while the expression of anti-inflammatory cytokines is upregulated, thus facilitating exercise adaptations and recovery. In corroboration, it was shown that after 6 weeks of resistance exercise training, an increase was observed in the mRNA expression and receptor levels of the anti-inflammatory cytokines IL-4, IL-4rα, IL-13 and IL-13rα in healthy individuals [105]. IL-4 and IL-13 have been shown to promote muscle hypertrophic responses by modulating muscle regeneration and *de novo* myofiber formation through myoblast recruitment [106]. Moreover, the literature suggests a reduction in TNF-α mRNA expression and plasma circulating levels following resistance exercise training [107], while this effect was accompanied by an enhanced mobilization of NK cells, suggesting a key role for immune cells in regulating resistance exercise-induced adaptations.

- Resistance exercise provokes changes in the activity of immune cell populations and the inflammatory profile
- Acute resistance exercise has been shown to initiate a systemic and local inflammatory cytokine response
- Chronic resistance exercise results in positive adaptations of the inflammatory responses

Inflammatory Response to Endurance Exercise

- Endurance exercise increases pro- and anti-inflammatory cytokine levels
- Endurance exercise results in a transient immunosuppression
- Endurance exercise leads to favorable muscle adaptations and improves performance

Mechanisms and Time Course of Endurance Exercise-Induced Inflammation

Endurance exercise has also been shown to elicit profound changes in immune cell and inflammation responses [108]. A study by Millard et al. [109] investigated the responses of Natural Killer (NK) cells following a single bout of high-intensity running. In this study, running exercise increased the number of $CD3^-/CD56^+$ NK cells by sixfold, but did not alter baseline levels of IFNγ secretion or NK cell cytotoxicity. In vitro stimulation of pre-exercise isolated NK cells with IL-2 and Toll-like receptor 2 (TLR2) agonists enhanced NK cells' IFNγ production, degranulation, and cytotoxicity. On the contrary, minimal effects were observed after stimulation in NK cells isolated following exercise, suggesting an alteration of NK cell-mediated immunity by short-term high-intensity running. Moreover, an animal study by Shi et al. indicated that 5 weeks of moderate-intensity aerobic exercise training attenuated lipopolysaccharide-induced acute lung injury in mice via inhibition of neutrophil extracellular traps formation and pro-inflammatory macrophage phenotype polarization through downregulation of ERK1/2 and NF-κB signaling pathways [110]. In healthy males, continuous cycling at 80% of VO2max for 20 minutes was shown to increase total B cell count by 88% [111]. Specifically, immature ($CD27^-/CD10^+$), naïve ($CD27^-/CD10^-$), and memory ($CD27^+/CD38^-$) B cells significantly increased following acute cycling [111]. Profound changes in inflammation and innate immunity have also been reported after intense endurance exercise in humans. Nieman et al. showed that cycling for 1.75 hours at 60% watts combined with a 10-km time trial increased the plasma levels of IL-6, TNF-α, granulocyte-macrophage colony-stimulating factor, IFN-γ, IL-1b, IL-2, IL-8, IL-10 and IL-12p70 [112]. Finally, a study by Córdova Martínez et al. [113] reported an accumulation of pro-inflammatory cytokines such as IL6, TNF-α, interferon c (IFN-γ) and IL2 whereas IL1β remained unchanged following three consecutive-day cycling competitions in eight male well-trained semi-professional cyclists. Collectively, endurance exercise is associated with significant changes in the inflammatory response and immune cell recruitment both after acute and chronic exercise. The transient exercise-induced immunosuppression should be also taken into consideration when evaluating the effects of different endurance exercise modalities.

Muscle Adaptation Following Endurance Exercise Training

Endurance exercise training leads to adaptations in muscle tissue that facilitate an overall increase in exercise capacity and performance [114]. The local adaptations in skeletal muscle include increased mitochondrial content and function, improved capillary density and enhanced glucose transport. These effects are coupled with more efficient oxygen transport and usage for energy production, thus delaying muscle fatigue during prolonged endurance exercise [114]. 4 weeks of sprint interval training increased mitochondrial respiration by 25%, while this effect was accompanied by favorable changes in PGC-1α, p53, and PHF20 protein content [115]. Additionally, HIIT training alone compared to continuous single-leg cycling matched for total work, has been reported to increase markers of mitochondrial content and respiration in skeletal muscle following 2 weeks of training [116]. Endurance adaptations to skeletal muscle are also influenced by the intensity and volume of training. Granata et al. reported an increase in mitochondrial respiration and citrate synthase (CS) activity by ~50% during the high-volume training period. Apart from the increase in mitochondrial respiration, high-volume exercise training increased the levels of regulatory proteins, such as PGC-1α, p53, and PHF20 [115]. Following 2 weeks of decreased training volume, mitochondrial-specific respiration was sustained at high levels, with a slight decrease in CS activity being the only sign of detraining [114]. Another adaptation to endurance exercise training is the stiffness of the muscle-tendon ECM and the ability to store and utilize elastic energy efficiently. It has been shown that male and female distance athletes who display adaptations to the musculotendinous system (longer and stiffer apparatus) exhibit a lower oxygen uptake (VO_2) when performing exercise at submaximal running velocities [117]. Altogether, adaptations to endurance training increase overall muscle function leading to improved performance.

- Endurance exercise elicits profound changes in immune cell and inflammatory responses
- Endurance exercise leads to a transient suppression of the immune system
- Endurance exercise training leads to adaptations in muscle tissue that facilitate an overall increase in exercise capacity and performance

Inflammatory Response to Team Sports

- Team sport activity results in the activation of an inflammatory response
- Soccer elicits the greatest changes in the inflammatory profile
- Team sport-induced inflammation is intensity-dependent

Mechanisms and Time Course of Team Sport-Induced Inflammation

Participation in team sport activity results in exercise-induced inflammation and activation of an immune response. Ispirlidis et al. investigated the time course of inflammatory responses during a 6-day period following a soccer match in male athletes [118]. Leukocyte count and C-reactive protein levels increased in the circulation immediately post-match and for up to 24 hours following the match, while CK levels remained increased for up to 96 hours post-match. In addition, the soccer match resulted in a significant increase in pro-inflammatory cytokine IL-6 levels only immediately post-match, with IL-1b levels remaining unchanged. In a study by Poulios et al., 20 semi-professional soccer players participated in 2 soccer matches performed 2 days apart with and without protein supplementation before and after the matches [59]. In both trials, it was shown that white blood cells and granulocyte counts were increased immediately post-match and reached their baseline values 1-day post-match. This was accompanied by increased inflammation, as evident by the increase in CK activity after the matches and throughout the recovery period. In another study, Mohr et al. investigated the inflammatory responses to 3 soccer games in 1 week in male players [119]. C-reactive protein levels and white blood cells increased immediately post-match and for up to 1-day following recovery. Moreover, CK activity was elevated at post-matches and throughout the 1-week recovery period. Additionally, a transient increase was observed in IL-1β and IL-6 levels after all 3 soccer matches, which lasted for up to 24 hours. Another study compared the inflammatory responses following soccer, volleyball and handball games at elite competitive level [120]. Soccer elicited the greatest increase in inflammatory cytokine TNF-α and IL-6 levels (3–6-fold), which lasted for up to 13 hours post-match. In basketball, volleyball and handball only a transient post-match increase in TNF-α and IL-6 levels was observed. C-reactive protein increased only in soccer at post-match and across sports disciplines at 13 hours post-matches. Volleyball match-play elicited the smallest magnitude of increase in the inflammatory markers. Handball match-play has also been shown to induce an inflammatory response. Following a team handball game circulating leukocytes increased immediately post-match while C-reactive protein levels increased for up to 48 hours compared to the control group. Moreover, IL-6 was elevated for up to 24 hours post-match and IL-1β levels increased only immediately post-match [121]. These results indicate that handball matches can induce a short-lived and mild inflammatory response. A study by Chatzinikolaou et al. also investigated the microcycle of inflammation and performance changes after a basketball match [122]. In this study, the authors reported a significant rise in circulating leukocytes for up to 24 hours and an increase in C-reactive protein levels for up to 48 hours post-match. IL-1β and IL-6 levels increased immediately post-match and remained elevated for up to 24 hours post-match. It seems that, compared to handball, basketball matches can induce a more pronounced short-lived inflammatory response. Lastly, a recent study examined the time course of inflammation after a futsal match in female athletes [123]. In accordance with previous reports on team sports, a futsal match-play induced a mild inflammatory response as evidenced by the increase in TNF-α and IL-6 levels only immediately post-match. Interestingly, C-reactive protein increased immediately post-match and remained elevated for up to 72 hours post-match.

- Participation in team sport activity results in exercise-induced inflammation and activation of an immune response
- Among sport disciplines, soccer results in a more intense and prolonged inflammatory response which depends on the intensity of the activity

Concluding Remarks

In conclusion, different exercise modalities induce an inflammatory and immune response that is of paramount importance for optimal muscle remodeling, performance recovery, adaptation and growth. The time course of the inflammatory responses and the molecular events associated with innate immunity are strictly regulated by the overall characteristics of the exercise and/or training. It should be outlined that apart from a transient immune suppression response to acute high-intensity exercise, participation in regular exercise training of different characteristics (i.e., eccentric, resistance, endurance and team sport) has beneficial effects on the adaptive inflammatory profile by shifting the balance between pro- and anti-inflammatory cytokines in favour of the latter, thus promoting muscle health and optimal exercise performance.

References

1. Clarkson PM, Hubal MJ. Exercise-induced muscle damage in humans. Am J Phys Med Rehabil/Assoc Acad Physiatrists. 2002;81(11 Suppl):S52–69. https://doi.org/10.1097/00002060-200211001-00007.
2. Clarkson PM, Sayers SP. Etiology of exercise-induced muscle damage. Can J Appl Physiol = Revue canadienne de physiologie appliquee. 1999;24(3):234–48. https://doi.org/10.1139/h99-020.

3. Clarkson PM, Tremblay I. Exercise-induced muscle damage, repair, and adaptation in humans. J Appl Physiol (Bethesda, Md: 1985). 1988;65(1):1–6. https://doi.org/10.1152/jappl.1988.65.1.1.
4. Ebbeling CB, Clarkson PM. Exercise-induced muscle damage and adaptation. Sports Med (Auckland, NZ). 1989;7(4):207–34. https://doi.org/10.2165/00007256-198907040-00001.
5. Kuipers H. Exercise-induced muscle damage. Int J Sports Med. 1994;15(3):132–5. https://doi.org/10.1055/s-2007-1021034.
6. Proske U, Allen TJ. Damage to skeletal muscle from eccentric exercise. Exerc Sport Sci Rev. 2005;33(2):98–104. https://doi.org/10.1097/00003677-200504000-00007.
7. Proske U, Morgan DL. Muscle damage from eccentric exercise: mechanism, mechanical signs, adaptation and clinical applications. J Physiol. 2001;537(Pt 2):333–45. https://doi.org/10.1111/j.1469-7793.2001.00333.x.
8. Fatouros IG, Jamurtas AZ. Insights into the molecular etiology of exercise-induced inflammation: opportunities for optimizing performance. J Inflamm Res. 2016;9:175–86. https://doi.org/10.2147/jir.S114635.
9. Belcastro AN, Shewchuk LD, Raj DA. Exercise-induced muscle injury: a calpain hypothesis. Mol Cell Biochem. 1998;179(1–2):135–45. https://doi.org/10.1023/a:1006816123601.
10. Stauber WT, Clarkson PM, Fritz VK, Evans WJ. Extracellular matrix disruption and pain after eccentric muscle action. J Appl Physiol (Bethesda, Md: 1985). 1990;69(3):868–74. https://doi.org/10.1152/jappl.1990.69.3.868.
11. Heinemeier KM, Olesen JL, Haddad F, et al. Expression of collagen and related growth factors in rat tendon and skeletal muscle in response to specific contraction types. J Physiol. 2007;582(Pt 3):1303–16. https://doi.org/10.1113/jphysiol.2007.127639.
12. Tofas T, Jamurtas AZ, Fatouros I, et al. Plyometric exercise increases serum indices of muscle damage and collagen breakdown. J Strength Cond Res/Natl Strength Cond Assoc. 2008;22(2):490–6. https://doi.org/10.1519/JSC.0b013e31816605a0.
13. Baird MF, Graham SM, Baker JS, Bickerstaff GF. Creatine-kinase- and exercise-related muscle damage implications for muscle performance and recovery. J Nutr Metab. 2012;2012:960363. https://doi.org/10.1155/2012/960363.
14. Fehrenbach E, Schneider ME. Trauma-induced systemic inflammatory response versus exercise-induced immunomodulatory effects. Sports Med (Auckland, NZ). 2006;36(5):373–84. https://doi.org/10.2165/00007256-200636050-00001.
15. Jamurtas AZ, Theocharis V, Tofas T, et al. Comparison between leg and arm eccentric exercises of the same relative intensity on indices of muscle damage. Eur J Appl Physiol. 2005;95(2–3):179–85. https://doi.org/10.1007/s00421-005-1345-0.
16. Paschalis V, Koutedakis Y, Jamurtas AZ, Mougios V, Baltzopoulos V. Equal volumes of high and low intensity of eccentric exercise in relation to muscle damage and performance. J Strength Cond Res/Natl Strength Cond Assoc. 2005;19(1):184–8. https://doi.org/10.1519/r-14763.1.
17. Peake JM, Neubauer O, Della Gatta PA, Nosaka K. Muscle damage and inflammation during recovery from exercise. J Appl Physiol (Bethesda, Md: 1985). 2017;122(3):559–70. https://doi.org/10.1152/japplphysiol.00971.2016.
18. Tidball JG, Villalta SA. Regulatory interactions between muscle and the immune system during muscle regeneration. Am J Physiol Regul Integr Comp Physiol. 2010;298(5):R1173–87. https://doi.org/10.1152/ajpregu.00735.2009.
19. Arnold L, Henry A, Poron F, et al. Inflammatory monocytes recruited after skeletal muscle injury switch into antiinflammatory macrophages to support myogenesis. J Exp Med. 2007;204(5):1057–69. https://doi.org/10.1084/jem.20070075.
20. Tidball JG. Inflammatory cell response to acute muscle injury. Med Sci Sports Exerc. 1995;27(7):1022–32. https://doi.org/10.1249/00005768-199507000-00011.
21. Arnaout MA. Structure and function of the leukocyte adhesion molecules CD11/CD18. Blood. 1990;75(5):1037–50.

22. Jamurtas AZ, Fatouros IG, Deli CK, et al. The effects of acute low-volume HIIT and aerobic exercise on leukocyte count and redox status. J Sports Sci Med. 2018;17(3):501–8.
23. Brickson S, Hollander J, Corr DT, Ji LL, Best TM. Oxidant production and immune response after stretch injury in skeletal muscle. Med Sci Sports Exerc. 2001;33(12):2010–5. https://doi.org/10.1097/00005768-200112000-00006.
24. Raastad T, Risoy BA, Benestad HB, Fjeld JG, Hallen J. Temporal relation between leukocyte accumulation in muscles and halted recovery 10–20 h after strength exercise. J Appl Physiol (Bethesda, Md: 1985). 2003;95(6):2503–9. https://doi.org/10.1152/japplphysiol.01064.2002.
25. Hamada K, Vannier E, Sacheck JM, Witsell AL, Roubenoff R. Senescence of human skeletal muscle impairs the local inflammatory cytokine response to acute eccentric exercise. FASEB J. 2005;19(2):264–6. https://doi.org/10.1096/fj.03-1286fje.
26. Butterfield TA, Best TM, Merrick MA. The dual roles of neutrophils and macrophages in inflammation: a critical balance between tissue damage and repair. J Athl Train. 2006;41(4):457–65.
27. Brunelli S, Rovere-Querini P. The immune system and the repair of skeletal muscle. Pharmacol Res. 2008;58(2):117–21. https://doi.org/10.1016/j.phrs.2008.06.008.
28. Tidball JG. Inflammatory processes in muscle injury and repair. Am J Physiol Regul Integr Comp Physiol. 2005;288(2):R345–53. https://doi.org/10.1152/ajpregu.00454.2004.
29. Deng B, Wehling-Henricks M, Villalta SA, Wang Y, Tidball JG. IL-10 triggers changes in macrophage phenotype that promote muscle growth and regeneration. J Immunol. 2012;189(7):3669–80. https://doi.org/10.4049/jimmunol.1103180.
30. Saclier M, Cuvellier S, Magnan M, Mounier R, Chazaud B. Monocyte/macrophage interactions with myogenic precursor cells during skeletal muscle regeneration. FEBS J. 2013;280(17):4118–30. https://doi.org/10.1111/febs.12166.
31. Zammit P, Beauchamp J. The skeletal muscle satellite cell: stem cell or son of stem cell? Differ Res Biol Divers. 2001;68(4–5):193–204.
32. Lepper C, Partridge TA, Fan CM. An absolute requirement for Pax7-positive satellite cells in acute injury-induced skeletal muscle regeneration. Development. 2011;138(17):3639–46. https://doi.org/10.1242/dev.067595.
33. Dumont NA, Bentzinger CF, Sincennes MC, Rudnicki MA. Satellite cells and skeletal muscle regeneration. Compr Physiol. 2015;5(3):1027–59. https://doi.org/10.1002/cphy.c140068.
34. Fukada SI. The roles of muscle stem cells in muscle injury, atrophy, and hypertrophy. J Biochem. 2018;163:353–8. https://doi.org/10.1093/jb/mvy019.
35. Michailidis Y, Karagounis LG, Terzis G, et al. Thiol-based antioxidant supplementation alters human skeletal muscle signaling and attenuates its inflammatory response and recovery after intense eccentric exercise. Am J Clin Nutr. 2013;98(1):233–45. https://doi.org/10.3945/ajcn.112.049163.
36. Chen SE, Jin B, Li YP. TNF-alpha regulates myogenesis and muscle regeneration by activating p38 MAPK. Am J Physiol Cell Physiol. 2007;292(5):C1660–71. https://doi.org/10.1152/ajpcell.00486.2006.
37. Serrano AL, Baeza-Raja B, Perdiguero E, Jardí M, Muñoz-Cánoves P. Interleukin-6 is an essential regulator of satellite cell-mediated skeletal muscle hypertrophy. Cell Metab. 2008;7(1):33–44. https://doi.org/10.1016/j.cmet.2007.11.011.
38. Pizza FX, Peterson JM, Baas JH, Koh TJ. Neutrophils contribute to muscle injury and impair its resolution after lengthening contractions in mice. J Physiol. 2005;562(Pt 3):899–913. https://doi.org/10.1113/jphysiol.2004.073965.
39. Chazaud B, Chrétien F, Gherardi RK. Macrophages regulate skeletal muscle regeneration. Med Sci (Paris). 2007;23(10):794–5. Les macrophages régulent les différentes phases de la régénération musculaire. https://doi.org/10.1051/medsci/20072310794.
40. Bentzinger CF, Wang YX, von Maltzahn J, Soleimani VD, Yin H, Rudnicki MA. Fibronectin regulates Wnt7a signaling and satellite cell expansion. Cell Stem Cell. 2013;12(1):75–87. https://doi.org/10.1016/j.stem.2012.09.015.

41. Nosaka K, Sakamoto K, Newton M, Sacco P. How long does the protective effect on eccentric exercise-induced muscle damage last? Med Sci Sports Exerc. 2001;33(9):1490–5. https://doi.org/10.1097/00005768-200109000-00011.
42. Paulsen G, Lauritzen F, Bayer ML, et al. Subcellular movement and expression of HSP27, alphaB-crystallin, and HSP70 after two bouts of eccentric exercise in humans. J Appl Physiol (Bethesda, Md: 1985). 2009;107(2):570–82. https://doi.org/10.1152/japplphysiol.00209.2009.
43. Paschalis V, Nikolaidis MG, Giakas G, Jamurtas AZ, Owolabi EO, Koutedakis Y. Position sense and reaction angle after eccentric exercise: the repeated bout effect. Eur J Appl Physiol. 2008;103(1):9–18. https://doi.org/10.1007/s00421-007-0663-9.
44. Nikolaidis MG, Paschalis V, Giakas G, et al. Decreased blood oxidative stress after repeated muscle-damaging exercise. Med Sci Sports Exerc. 2007;39(7):1080–9. https://doi.org/10.1249/mss.0b013e31804ca10c.
45. Chen TC, Chen HL, Lin MJ, Yu HI, Nosaka K. Contralateral repeated bout effect of eccentric exercise of the elbow flexors. Med Sci Sports Exerc. 2016;48(10):2030–9. https://doi.org/10.1249/mss.0000000000000991.
46. Nosaka K, Newton M, Sacco P. Muscle damage and soreness after endurance exercise of the elbow flexors. Med Sci Sports Exerc. 2002;34(6):920–7. https://doi.org/10.1097/00005768-200206000-00003.
47. Nosaka K, Sakamoto K. Effect of elbow joint angle on the magnitude of muscle damage to the elbow flexors. Med Sci Sports Exerc. 2001;33(1):22–9. https://doi.org/10.1097/00005768-200101000-00005.
48. Chapman DW, Newton M, McGuigan M, Nosaka K. Effect of lengthening contraction velocity on muscle damage of the elbow flexors. Med Sci Sports Exerc. 2008;40(5):926–33. https://doi.org/10.1249/MSS.0b013e318168c82d.
49. Duchateau J, Enoka RM. Neural control of lengthening contractions. J Exp Biol. 2016;219(Pt 2):197–204. https://doi.org/10.1242/jeb.123158.
50. Paschalis V, Koutedakis Y, Baltzopoulos V, Mougios V, Jamurtas AZ, Giakas G. Short vs. long length of rectus femoris during eccentric exercise in relation to muscle damage in healthy males. Clinical biomechanics (Bristol, Avon). 2005;20(6):617–22. https://doi.org/10.1016/j.clinbiomech.2005.02.011.
51. Allen DG, Gervasio OL, Yeung EW, Whitehead NP. Calcium and the damage pathways in muscular dystrophy. Can J Physiol Pharmacol. 2010;88(2):83–91. https://doi.org/10.1139/y09-058.
52. Butterfield TA, Best TM. Stretch-activated ion channel blockade attenuates adaptations to eccentric exercise. Med Sci Sports Exerc. 2009;41(2):351–6. https://doi.org/10.1249/MSS.0b013e318187cffa.
53. Clarkson PM, Hubal MJ. Are women less susceptible to exercise-induced muscle damage? Curr Opin Clin Nutr Metab Care. 2001;4(6):527–31. https://doi.org/10.1097/00075197-200111000-00011.
54. Sewright KA, Hubal MJ, Kearns A, Holbrook MT, Clarkson PM. Sex differences in response to maximal eccentric exercise. Med Sci Sports Exerc. 2008;40(2):242–51. https://doi.org/10.1249/mss.0b013e31815aedda.
55. Hubal MJ, Chen TC, Thompson PD, Clarkson PM. Inflammatory gene changes associated with the repeated-bout effect. Am J Physiol Regul Integr Comp Physiol. 2008;294(5):R1628–37. https://doi.org/10.1152/ajpregu.00853.2007.
56. Chen TC. Variability in muscle damage after eccentric exercise and the repeated bout effect. Res Q Exerc Sport. 2006;77(3):362–71. https://doi.org/10.1080/02701367.2006.10599370.
57. McGinley C, Shafat A, Donnelly AE. Does antioxidant vitamin supplementation protect against muscle damage? Sports Med (Auckland, NZ). 2009;39(12):1011–32. https://doi.org/10.2165/11317890-000000000-00000.

58. Trombold JR, Reinfeld AS, Casler JR, Coyle EF. The effect of pomegranate juice supplementation on strength and soreness after eccentric exercise. J Strength Cond Res/Natl Strength Cond Assoc. 2011;25(7):1782–8. https://doi.org/10.1519/JSC.0b013e318220d992.
59. Poulios A, Fatouros IG, Mohr M, et al. Post-game high protein intake may improve recovery of football-specific performance during a congested game fixture: results from the PRO-FOOTBALL study. Nutrients. 2018;10(4) https://doi.org/10.3390/nu10040494.
60. Howatson G, Hoad M, Goodall S, Tallent J, Bell PG, French DN. Exercise-induced muscle damage is reduced in resistance-trained males by branched chain amino acids: a randomized, double-blind, placebo controlled study. J Int Soc Sports Nutr. 2012;9:20. https://doi.org/10.1186/1550-2783-9-20.
61. Jiaming Y, Rahimi MH. Creatine supplementation effect on recovery following exercise-induced muscle damage: a systematic review and meta-analysis of randomized controlled trials. J Food Biochem. 2021;45(10):e13916. https://doi.org/10.1111/jfbc.13916.
62. Kim J, Lee J, Kim S, Yoon D, Kim J, Sung DJ. Role of creatine supplementation in exercise-induced muscle damage: a mini review. J Exercise Rehabil. 2015;11(5):244–50. https://doi.org/10.12965/jer.150237.
63. Close GL, Ashton T, Cable T, Doran D, MacLaren DP. Eccentric exercise, isokinetic muscle torque and delayed onset muscle soreness: the role of reactive oxygen species. Eur J Appl Physiol. 2004;91(5–6):615–21. https://doi.org/10.1007/s00421-003-1012-2.
64. Peake JM, Suzuki K, Wilson G, et al. Exercise-induced muscle damage, plasma cytokines, and markers of neutrophil activation. Med Sci Sports Exerc. 2005;37(5):737–45.
65. Peake JM, Suzuki K, Hordern M, Wilson G, Nosaka K, Coombes JS. Plasma cytokine changes in relation to exercise intensity and muscle damage. Eur J Appl Physiol. 2005;95(5–6):514–21. https://doi.org/10.1007/s00421-005-0035-2.
66. Murton AJ, Constantin D, Greenhaff PL. The involvement of the ubiquitin proteasome system in human skeletal muscle remodelling and atrophy. Biochim Biophys Acta. 2008;1782(12):730–43. https://doi.org/10.1016/j.bbadis.2008.10.011.
67. Powers SK, Duarte J, Kavazis AN, Talbert EE. Reactive oxygen species are signalling molecules for skeletal muscle adaptation. Exp Physiol. 2010;95(1):1–9. https://doi.org/10.1113/expphysiol.2009.050526.
68. Crane JD, Abadi A, Hettinga BP, et al. Elevated mitochondrial oxidative stress impairs metabolic adaptations to exercise in skeletal muscle. PLoS One. 2013;8(12):e81879. https://doi.org/10.1371/journal.pone.0081879.
69. Tidball JG. Mechanisms of muscle injury, repair, and regeneration. Compr Physiol. 2011;1(4):2029–62. https://doi.org/10.1002/cphy.c100092.
70. Hennigar SR, McClung JP, Pasiakos SM. Nutritional interventions and the IL-6 response to exercise. FASEB J. 2017;31(9):3719–28. https://doi.org/10.1096/fj.201700080R.
71. Gibson MC, Schultz E. Age-related differences in absolute numbers of skeletal muscle satellite cells. Muscle Nerve. 1983;6(8):574–80. https://doi.org/10.1002/mus.880060807.
72. Ceafalan LC, Popescu BO, Hinescu ME. Cellular players in skeletal muscle regeneration. Biomed Res Int. 2014;2014:957014. https://doi.org/10.1155/2014/957014.
73. Perez-Ruiz A, Gnocchi VF, Zammit PS. Control of Myf5 activation in adult skeletal myonuclei requires ERK signalling. Cell Signal. 2007;19(8):1671–80. https://doi.org/10.1016/j.cellsig.2007.03.003.
74. Hollenberg SM, Cheng PF, Weintraub H. Use of a conditional MyoD transcription factor in studies of MyoD trans-activation and muscle determination. Proc Natl Acad Sci USA. 1993;90(17):8028–32. https://doi.org/10.1073/pnas.90.17.8028.
75. Megeney LA, Kablar B, Garrett K, Anderson JE, Rudnicki MA. MyoD is required for myogenic stem cell function in adult skeletal muscle. Genes Dev. 1996;10(10):1173–83.
76. Forcina L, Miano C, Pelosi L, Musaro A. An overview about the biology of skeletal muscle satellite cells. Curr Genomics. 2019;20(1):24–37. https://doi.org/10.2174/1389202920666190116094736.

77. Papanikolaou K, Veskoukis AS, Draganidis D, et al. Redox-dependent regulation of satellite cells following aseptic muscle trauma: implications for sports performance and nutrition. Free Radic Biol Med. 2020;161:125–38. https://doi.org/10.1016/j.freeradbiomed.2020.10.001.
78. Schwander M, Leu M, Stumm M, et al. Beta1 integrins regulate myoblast fusion and sarcomere assembly. Dev Cell. 2003;4(5):673–85.
79. Almeida CF, Fernandes SA, Ribeiro Junior AF, Keith Okamoto O, Vainzof M. Muscle satellite cells: exploring the basic biology to rule them. Stem Cells Int. 2016;2016:1078686. https://doi.org/10.1155/2016/1078686.
80. Rommel C, Bodine SC, Clarke BA, et al. Mediation of IGF-1-induced skeletal myotube hypertrophy by PI(3)K/Akt/mTOR and PI(3)K/Akt/GSK3 pathways. Nat Cell Biol. 2001;3(11):1009–13. https://doi.org/10.1038/ncb1101-1009.
81. Musarò A. The basis of muscle regeneration. Adv Biol. 2014;2014:1–16. https://doi.org/10.1155/2014/612471.
82. Li Y, Foster W, Deasy BM, et al. Transforming growth factor-beta1 induces the differentiation of myogenic cells into fibrotic cells in injured skeletal muscle: a key event in muscle fibrogenesis. Am J Pathol. 2004;164(3):1007–19. https://doi.org/10.1016/s0002-9440(10)63188-4.
83. Mozdziak PE, Pulvermacher PM, Schultz E. Muscle regeneration during hindlimb unloading results in a reduction in muscle size after reloading. J Appl Physiol (Bethesda, Md: 1985). 2001;91(1):183–90. https://doi.org/10.1152/jappl.2001.91.1.183.
84. Paulsen G, Mikkelsen UR, Raastad T, Peake JM. Leucocytes, cytokines and satellite cells: what role do they play in muscle damage and regeneration following eccentric exercise? Exerc Immunol Rev. 2012;18:42–97.
85. Aronson D, Violan MA, Dufresne SD, Zangen D, Fielding RA, Goodyear LJ. Exercise stimulates the mitogen-activated protein kinase pathway in human skeletal muscle. J Clin Invest. 1997;99(6):1251–7. https://doi.org/10.1172/jci119282.
86. Douglas J, Pearson S, Ross A, McGuigan M. Chronic adaptations to eccentric training: a systematic review. Sports Med (Auckland, NZ). 2017;47(5):917–41. https://doi.org/10.1007/s40279-016-0628-4.
87. Hornberger TA, Chien S. Mechanical stimuli and nutrients regulate rapamycin-sensitive signaling through distinct mechanisms in skeletal muscle. J Cell Biochem. 2006;97(6):1207–16. https://doi.org/10.1002/jcb.20671.
88. Coffey VG, Hawley JA. The molecular bases of training adaptation. Sports Med (Auckland, NZ). 2007;37(9):737–63. https://doi.org/10.2165/00007256-200737090-00001.
89. Narici MV, Roi GS, Landoni L, Minetti AE, Cerretelli P. Changes in force, cross-sectional area and neural activation during strength training and detraining of the human quadriceps. Eur J Appl Physiol Occup Physiol. 1989;59(4):310–9. https://doi.org/10.1007/bf02388334.
90. Aagaard P, Andersen JL, Dyhre-Poulsen P, et al. A mechanism for increased contractile strength of human pennate muscle in response to strength training: changes in muscle architecture. J Physiol. 2001;534(Pt. 2):613–23. https://doi.org/10.1111/j.1469-7793.2001.t01-1-00613.x.
91. Hortobágyi T, Dempsey L, Fraser D, et al. Changes in muscle strength, muscle fibre size and myofibrillar gene expression after immobilization and retraining in humans. J Physiol. 2000;524(Pt 1):293–304. https://doi.org/10.1111/j.1469-7793.2000.00293.x.
92. Gordon PM, Liu D, Sartor MA, et al. Resistance exercise training influences skeletal muscle immune activation: a microarray analysis. J Appl Physiol (Bethesda, Md: 1985). 2012;112(3):443–53. https://doi.org/10.1152/japplphysiol.00860.2011.
93. Xin L, Hyldahl RD, Chipkin SR, Clarkson PM. A contralateral repeated bout effect attenuates induction of NF-κB DNA binding following eccentric exercise. J Appl Physiol (Bethesda, Md: 1985). 2014;116(11):1473–80. https://doi.org/10.1152/japplphysiol.00133.2013.
94. Yahiaoui L, Gvozdic D, Danialou G, Mack M, Petrof BJ. CC family chemokines directly regulate myoblast responses to skeletal muscle injury. J Physiol. 2008;586(16):3991–4004. https://doi.org/10.1113/jphysiol.2008.152090.

95. Ramel A, Wagner KH, Elmadfa I. Acute impact of submaximal resistance exercise on immunological and hormonal parameters in young men. J Sports Sci. 2003;21(12):1001–8. https://doi.org/10.1080/0264041031000164l395.
96. Stock C, Schaller K, Baum M, Liesen H, Weiss M. Catecholamines, lymphocyte subsets, and cyclic adenosine monophosphate production in mononuclear cells and CD4+ cells in response to submaximal resistance exercise. Eur J Appl Physiol Occup Physiol. 1995;71(2–3):166–72. https://doi.org/10.1007/bf00854975.
97. Freidenreich DJ, Volek JS. Immune responses to resistance exercise. Exerc Immunol Rev. 2012;18:8–41.
98. Simonson SR, Jackson CG. Leukocytosis occurs in response to resistance exercise in men. J Strength Cond Res/Natl Strength Cond Assoc. 2004;18(2):266–71. https://doi.org/10.1519/r-12572.1.
99. Peake JM, Nosaka K, Muthalib M, Suzuki K. Systemic inflammatory responses to maximal versus submaximal lengthening contractions of the elbow flexors. Exerc Immunol Rev. 2006;12:72–85.
100. Phillips MD, Mitchell JB, Currie-Elolf LM, Yellott RC, Hubing KA. Influence of commonly employed resistance exercise protocols on circulating IL-6 and indices of insulin sensitivity. J Strength Cond Res/Natl Strength Cond Assoc. 2010;24(4):1091–101. https://doi.org/10.1519/JSC.0b013e3181cc2212.
101. Uchida MC, Nosaka K, Ugrinowitsch C, et al. Effect of bench press exercise intensity on muscle soreness and inflammatory mediators. J Sports Sci. 2009;27(5):499–507. https://doi.org/10.1080/02640410802632144.
102. Buford TW, Cooke MB, Willoughby DS. Resistance exercise-induced changes of inflammatory gene expression within human skeletal muscle. Eur J Appl Physiol. 2009;107(4):463–71. https://doi.org/10.1007/s00421-009-1145-z.
103. Nielsen AR, Mounier R, Plomgaard P, et al. Expression of interleukin-15 in human skeletal muscle effect of exercise and muscle fibre type composition. J Physiol. 2007;584(Pt 1):305–12. https://doi.org/10.1113/jphysiol.2007.139618.
104. Smith LL, Anwar A, Fragen M, Rananto C, Johnson R, Holbert D. Cytokines and cell adhesion molecules associated with high-intensity eccentric exercise. Eur J Appl Physiol. 2000;82(1–2):61–7. https://doi.org/10.1007/s004210050652.
105. Prokopchuk O, Liu Y, Wang L, Wirth K, Schmidtbleicher D, Steinacker JM. Skeletal muscle IL-4, IL-4Ralpha, IL-13 and IL-13Ralpha1 expression and response to strength training. Exerc Immunol Rev. 2007;13:67–75.
106. Horsley V, Jansen KM, Mills ST, Pavlath GK. IL-4 acts as a myoblast recruitment factor during mammalian muscle growth. Cell. 2003;113(4):483–94. https://doi.org/10.1016/s0092-8674(03)00319-2.
107. Kirwan JP, del Aguila LF. Insulin signalling, exercise and cellular integrity. Biochem Soc Trans. 2003;31(Pt 6):1281–5. https://doi.org/10.1042/bst0311281.
108. Nieman DC. The immune response to prolonged cardiorespiratory exercise. Am J Sports Med. 1996;24(6 Suppl):S98–103.
109. Millard AL, Valli PV, Stussi G, Mueller NJ, Yung GP, Seebach JD. Brief exercise increases peripheral blood NK cell counts without immediate functional changes, but impairs their responses to ex vivo stimulation. Front Immunol. 2013;4:125. https://doi.org/10.3389/fimmu.2013.00125.
110. Shi Y, Liu T, Nieman DC, et al. Aerobic exercise attenuates acute lung injury through NET inhibition. Front Immunol. 2020;11:409. https://doi.org/10.3389/fimmu.2020.00409.
111. Turner JE, Spielmann G, Wadley AJ, Aldred S, Simpson RJ, Campbell JP. Exercise-induced B cell mobilisation: preliminary evidence for an influx of immature cells into the bloodstream. Physiol Behav. 2016;164(Pt A):376–82. https://doi.org/10.1016/j.physbeh.2016.06.023.
112. Nieman DC, Konrad M, Henson DA, Kennerly K, Shanely RA, Wallner-Liebmann SJ. Variance in the acute inflammatory response to prolonged cycling is linked to exercise intensity. J Interferon Cytokine Res. 2012;32(1):12–7. https://doi.org/10.1089/jir.2011.0038.

113. Córdova Martínez A, Martorell Pons M, Sureda Gomila A, Tur Marí JA, Pons BA. Changes in circulating cytokines and markers of muscle damage in elite cyclists during a multi-stage competition. Clin Physiol Funct Imaging. 2015;35(5):351–8. https://doi.org/10.1111/cpf.12170.
114. Hughes DC, Ellefsen S, Baar K. Adaptations to endurance and strength training. Cold Spring Harb Perspect Med. 2018;8(6) https://doi.org/10.1101/cshperspect.a029769.
115. Granata C, Oliveira RS, Little JP, Renner K, Bishop DJ. Training intensity modulates changes in PGC-1α and p53 protein content and mitochondrial respiration, but not markers of mitochondrial content in human skeletal muscle. FASEB J. 2016;30(2):959–70. https://doi.org/10.1096/fj.15-276907.
116. MacInnis MJ, Zacharewicz E, Martin BJ, et al. Superior mitochondrial adaptations in human skeletal muscle after interval compared to continuous single-leg cycling matched for total work. J Physiol. 2017;595(9):2955–68. https://doi.org/10.1113/jp272570.
117. Barnes KR, McGuigan MR, Kilding AE. Lower-body determinants of running economy in male and female distance runners. J Strength Cond Res/Natl Strength Cond Assoc. 2014;28(5):1289–97. https://doi.org/10.1519/jsc.0000000000000267.
118. Ispirlidis I, Fatouros IG, Jamurtas AZ, et al. Time-course of changes in inflammatory and performance responses following a soccer game. Clin J Sport Med. 2008;18(5):423–31. https://doi.org/10.1097/JSM.0b013e3181818e0b.
119. Mohr M, Draganidis D, Chatzinikolaou A, et al. Muscle damage, inflammatory, immune and performance responses to three football games in 1 week in competitive male players. Eur J Appl Physiol. 2016;116(1):179–93. https://doi.org/10.1007/s00421-015-3245-2.
120. Souglis A, Bogdanis GC, Giannopoulou I, Papadopoulos C, Apostolidis N. Comparison of inflammatory responses and muscle damage indices following a soccer, basketball, volleyball and handball game at an elite competitive level. Res Sports Med (Print). 2015;23(1):59–72. https://doi.org/10.1080/15438627.2014.975814.
121. Chatzinikolaou A, Christoforidis C, Avloniti A, et al. A microcycle of inflammation following a team handball game. J Strength Cond Res/Natl Strength Cond Assoc. 2014;28(7):1981–94. https://doi.org/10.1519/jsc.0000000000000330.
122. Chatzinikolaou A, Draganidis D, Avloniti A, et al. The microcycle of inflammation and performance changes after a basketball match. J Sports Sci. 2014;32(9):870–82. https://doi.org/10.1080/02640414.2013.865251.
123. Souglis A, Bourdas DI, Gioldasis A, et al. Time course of performance indexes, oxidative stress, inflammation, and muscle damage markers after a female futsal match. Sports (Basel, Switzerland). 2023;11(7) https://doi.org/10.3390/sports11070127.

Chapter 10
Addition and Loss of Serial Sarcomeres—Implications for Muscle Mechanical Function

Avery Hinks and Geoffrey A. Power

Overview of Serial Sarcomere Number and the Influence on Mechanical Function

> A muscle's serial sarcomere number adapts to conditions imposed upon it to maintain optimal actin-myosin overlap and regulate passive tension development. Serial sarcomere number is important because it directly influences a muscle's active force-length relationship, maximum shortening velocity, and ability to produce work and power through a joint's range of motion. Because of these associations with biomechanical performance, the use of interventions promoting an increase in serial sarcomere number are important in conditions in which serial sarcomere number is reduced. Studies on humans are unfortunately limited to assumptions of serial sarcomere number by measuring fascicle length with ultrasound, therefore, we currently rely mostly on studies using animal models to understand how serial sarcomere number adaptations influence performance.

Muscle architecture is a hierarchy, with sarcomeres, the smallest unit of muscle contraction, forming the basic building blocks. Many sarcomeres aligned in series comprise a myofibril, many parallel myofibrils comprise a muscle fibre, many muscle fibres comprise a muscle fascicle, and many fascicles comprise a whole muscle (Fig. 10.1) [1].

A. Hinks · G. A. Power (✉)
Human Health and Nutritional Sciences, University of Guelph, Guelph, ON, Canada
e-mail: ahinks@uoguelph.ca; gapower@uoguelph.ca

N. C. Apostolopoulos et al. (eds.), *Fundamentals of Recovery, Regeneration, and Adaptation to Exercise Stress: An Integrated Approach*,
https://doi.org/10.1007/978-3-031-44270-4_10

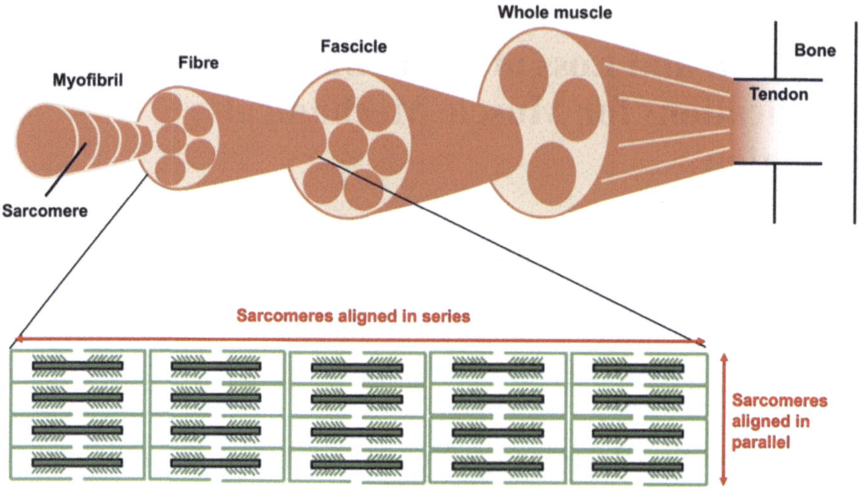

Fig. 10.1 The hierarchy of muscle architecture, demonstrating how many sarcomeres aligned in series comprise a myofibril, many parallel myofibrils comprise a whole muscle fiber, many muscle fibers comprise a muscle fascicle, and many muscle fascicles comprise a whole muscle. The enlarged image at the bottom is a two-dimensional sideview of sarcomeres aligned in parallel and in series

Skeletal muscle has the remarkable ability to undergo architectural remodeling in response to changes in the external conditions imposed upon it, and remodeling can occur at each level of the hierarchy. Remodeling occurs in both radial (outward from the centre of the fibre) and longitudinal (along the length of the fibre) directions, with the restructuring being driven by different events [1]. Radial growth is mainly driven by the addition of myofibrils in parallel to one another, which increases fibre cross-sectional area. Conversely, longitudinal growth is driven by the addition of sarcomeres in series along a myofibril, which increases myofibril length. Owing to the hierarchical nature of muscle architecture, parallel sarcomere addition manifests as an increase in whole muscle cross-sectional area, thickness, and fascicle pennation angle [1], while serial sarcomere addition results in an increase in fibre length and consequently fascicle length [1, 2]. The latter process, termed "serial sarcomerogenesis," will be the focus of this chapter, and has been explored in numerous studies on animals and humans.

The most accepted hypothesis for how serial sarcomerogenesis is initiated depends upon the muscle force-length relationship: active force is suboptimal in overstretched and overshortened positions based on actin-myosin overlap, and passive force increases with stretch as titin and the extracellular matrix bear increasing tension [3–5] (Fig. 10.2).

Serial sarcomere number is optimized in fibres of healthy muscle to assure that a muscle's average sarcomere operating range favours force production for required tasks, and to ensure passive force is not chronically elevated [6–8]. Therefore, when undergoing repeated or chronic stretch, muscle may adapt by adding serial

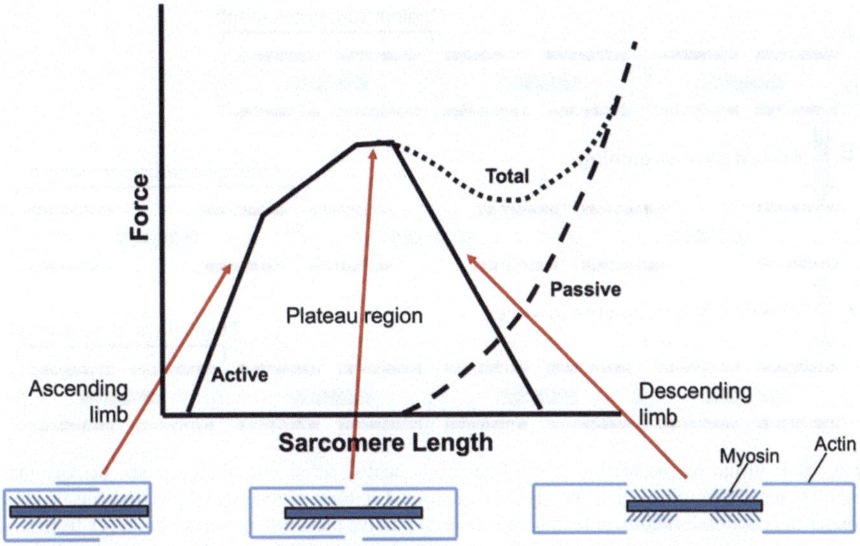

Fig. 10.2 The sarcomere force-length relationship, showing how active force (solid black line) is suboptimal in stretched and shortened positions due to limited actin-myosin interaction, and how passive force (dashed black line) increases with increasing sarcomere stretch beyond optimal length

sarcomeres to maintain optimal actin-myosin overlap [2, 5] or alleviate passive tension [9, 10] at the stretched length (Fig. 10.3). The addition of new sarcomeres is signaled by structures intrinsic to the muscle fibre that sense the change in mechanical tension (i.e., suboptimal active force production or increased passive tension), and begin the process of transducing that mechanical signal into biochemical events that regulate sarcomeric gene transcription in the nucleus and subsequent protein synthesis [11–14]. A muscle's serial sarcomere number is tightly regulated, likely due to its profound impact on a muscle's biomechanical function—most notably the active force-length relationship, the force-velocity relationship, and work and power output [3].

Theoretical Influence of Sarcomerogenesis on the Active Force-Length Relationship

The active force-length relationship includes an ascending limb (force increases with increasing length), a plateau region (where the length for optimal force production is located), and a descending limb (force decreases with increasing length) (Fig. 10.2). These regions of the force-length relationship are determined by the amount of overlap of thick and thin filaments in the sarcomere, with the most overlap yielding the maximum number of actin-myosin cross-bridge attachments, thereby producing the greatest possible active force [15]. As the muscles crossing a

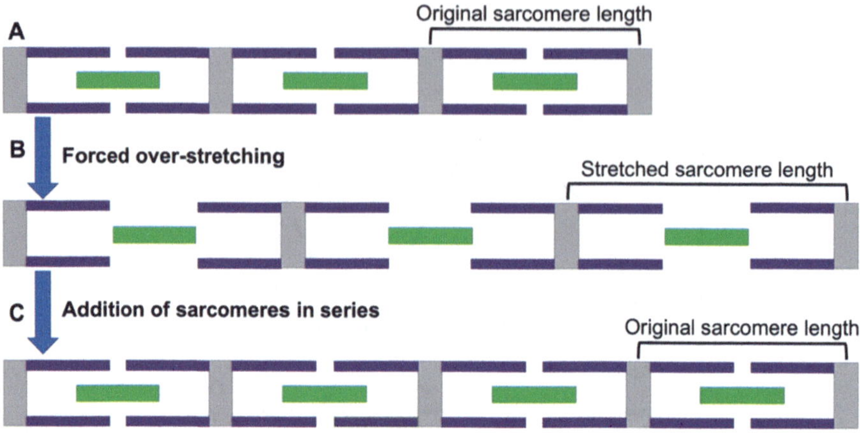

Fig. 10.3 Visual representation of the hypothesis behind serial sarcomerogenesis. (**a**) Original (non-stretched) myofibril. (**b**) The myofibril is forced into an overstretched position (i.e., with a longer than optimal sarcomere length and elevated passive tension) via intervention. (**c**) To regain optimal actin-myosin overlap and/or alleviate passive tension in the stretched position, serial sarcomere number increases in the myofibril

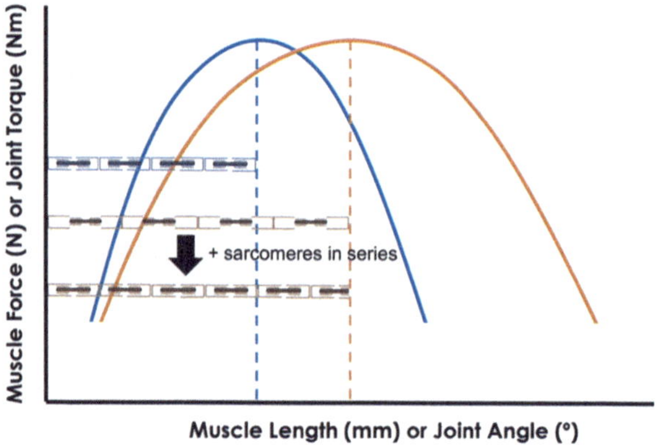

Fig. 10.4 Hypothesis illustrating how longitudinal skeletal muscle growth increases the optimal muscle length or joint angle for active force or torque production. A hypothetical intervention causes overstretching of sarcomeres, causing sarcomeres to be added in series to maintain the optimal sarcomere length, which in turn leads to an increase in the range of the active force-length relationship, and shifts the optimal muscle length to the right. Dashed lines denote the location of optimal muscle length or joint angle. Blue curve represents the original non-stretched muscle, and orange curve represents the overstretched muscle before and after the addition of serial sarcomeres

joint will lengthen and shorten with increases and decreases in joint angle, respectively, the force-length relationship's shape also applies to assessments of joint torque and angle (i.e., the torque-angle relationship) (Fig. 10.4). As mentioned above, maintaining the integrity of this relationship is often hypothesized to drive longitudinal muscle fascicle growth [2, 16]. Consequently, a shift in the plateau region to a muscle length or joint angle previously on the descending limb (i.e., an increase in optimal length), and widening of the whole force-length relationship (i.e., an increase in the range of muscle lengths for active force production) are expected with serial sarcomerogenesis (Fig. 10.4).

Theoretical Influence of Sarcomerogenesis on the Force-Velocity Relationship

The force-velocity relationship dictates that force decreases hyperbolically with increasing shortening velocity [17, 18]. This hyperbolic shape arises because with increasing speed of shortening, myofilaments slide past each other faster, reducing the probability of actin-myosin binding and the force produced per crossbridge, thereby limiting force production at higher velocities [18, 19]. With sarcomeres arranged in series, their velocities and length changes are additive while their forces are shared; hence, a muscle with a greater serial sarcomere number is capable of a greater total excursion for the same sarcomere shortening. Velocity equals excursion (i.e., the change in length) divided by time, therefore, when shortening under zero load, a muscle can theoretically achieve maximum shortening velocity (V_{max}) [19–22]. In other words, a greater serial sarcomere number can allow individual sarcomeres to shorten less for a given muscle excursion, increasing its possible V_{max} (Fig. 10.5). It is important to note, however, that a muscle's fibre type composition (i.e., fast twitch versus slow twitch) also largely influences its shortening velocity. Therefore, the relationship between serial sarcomere number and V_{max} assumes a similar fibre type composition among whichever muscles are being compared.

Theoretical Influence of Sarcomerogenesis on Work and Power Output

Muscle power production is defined as the product of contractile force and velocity. Therefore, per the force-velocity relationship, peak power occurs when force and velocity are both submaximal [23, 24]. Power is also quantified as the rate of mechanical work, with work being the product of the muscular contraction's linear or angular displacement and the force or torque causing that displacement [25]. These definitions can be applied to work and power obtained from isokinetic (constant velocity) and isotonic (constant load) shortening contractions [26, 27], sinusoidal contractions (i.e., work loops) that simulate real-world locomotion *in-vitro* [28, 29], and Wingate or countermovement jump tests in humans [30, 31]. Increasing serial sarcomere number may allow individual sarcomeres to shorten slower and remain closer to the region of optimal force production during shortening (Fig. 10.6).

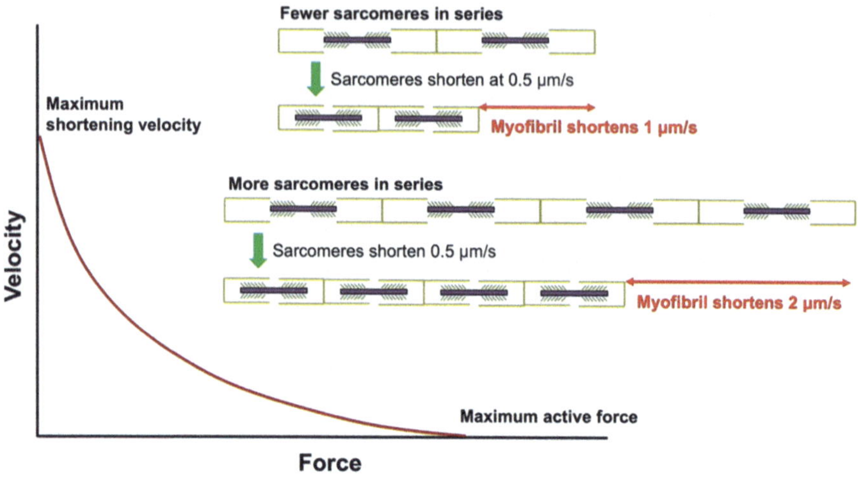

Fig. 10.5 The muscle force-velocity relationship, with a demonstration of how a myofibril (or muscle fiber, fascicle, or whole muscle) with more sarcomeres in series can shorten faster for a given sarcomere shortening velocity. These myofibril lengths are theoretical, as typical myofibrils would have thousands of sarcomeres in series

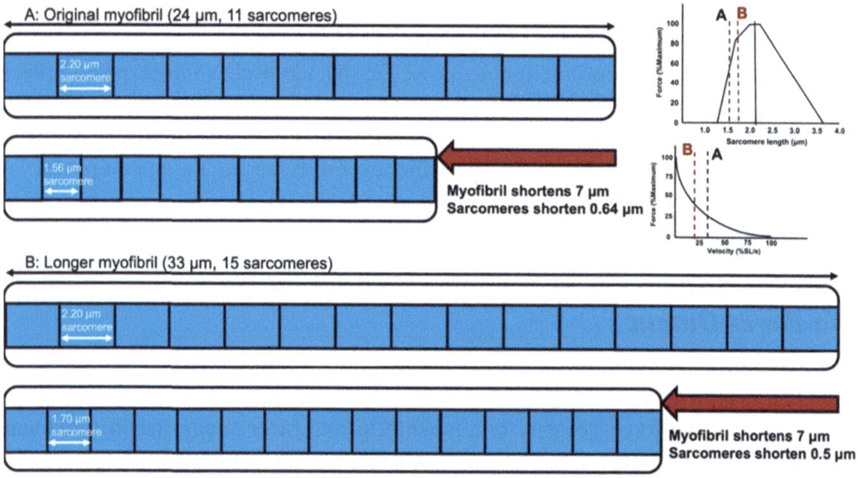

Fig. 10.6 Visual representation of how sarcomerogenesis may allow for greater force production for the same absolute myofibril excursion. Note that the longer myofibril with more sarcomeres in series (B) falls closer to the optimal position for force production on both the sarcomere force-length and force-velocity relationships

This optimization of force on the sarcomere force-length and force-velocity relationships would consequently increase mechanical work (and thus power) during shortening for a given whole-muscle excursion and velocity [32–34] (Fig. 10.6).

> Understanding the functional structure of the plantar flexors is important to link an isolated tendon injury with the complex muscle-tendon pathology that limits patient function. This also has direct clinical relevance because the heel-raise test is a primary clinical assessment of function and reflects the muscle's ability to do positive work. A shorter muscle with fewer serial sarcomeres will be able to do less mechanical work for a given cross section.
>
> – Dr. Josh Baxter, Assistant Professor of Orthopedic Surgery, University of Pennsylvania, Perelman School of Medicine.

Methodological Considerations of Serial Sarcomere Number Measurements

Correct determination of serial sarcomere number requires measurement of sarcomere length for use in the equation: $SSN = FL \div SL$, where SSN is serial sarcomere number, FL is fascicle length, and SL is sarcomere length.

In animals, sarcomere length is usually measured using a system called laser diffraction, in which a laser is shone through a dissected fascicle, producing a diffraction pattern from the sarcomere's I-bands. Laser diffraction is only possible in humans intraoperatively [35], and other methods, such as microendoscopy, are limited by availability and prohibitive cost [36]; therefore, most studies on serial sarcomere number adaptation in humans are limited to measuring fascicle length via ultrasound. However, with ultrasound, increases or decreases in fascicle length may also reflect longer or shorter sarcomere lengths, respectively, at the joint angle in which fascicle length was measured. For example, Pincheira et al. [37] observed an increase in biceps femoris fascicle length following 3 weeks of eccentric training as measured with the leg in full extension; however, microendoscopy revealed this increase in fascicle length was due to resting sarcomere lengths being longer at that joint angle following training, not training-induced serial sarcomerogenesis. A recent study from our lab used rodents to assess the accuracy of ultrasonographic measurements of fascicle length in a neutral position for estimating actual serial sarcomere number adaptation [38]. Following 2 weeks of casting the soleus in a stretched position, we observed an 11% increase in fascicle length as measured by ultrasound. However, the actual increase in serial sarcomere number was only 6%, indicating ultrasonographic measurements of fascicle length overestimated serial sarcomere number adaptations by 5%. What, then, is currently recognized as the ideal joint angle for measuring fascicle length in humans? A study by Reeves et al. [39] may demonstrate the best practice. By measuring vastus lateralis fascicle length at the optimal knee extension angle (determined by constructing an active torque-angle curve), they obtained a fascicle length at which sarcomere length may be assumed as optimal [40]. This method of obtaining an assumed optimal fascicle length allows for estimation of serial sarcomere number using the known optimal sarcomere length in humans of 2.64 μm [41].

In summary, as much of the literature concerning serial sarcomerogenesis in humans has used increases in fascicle length measured by ultrasound as a proxy of

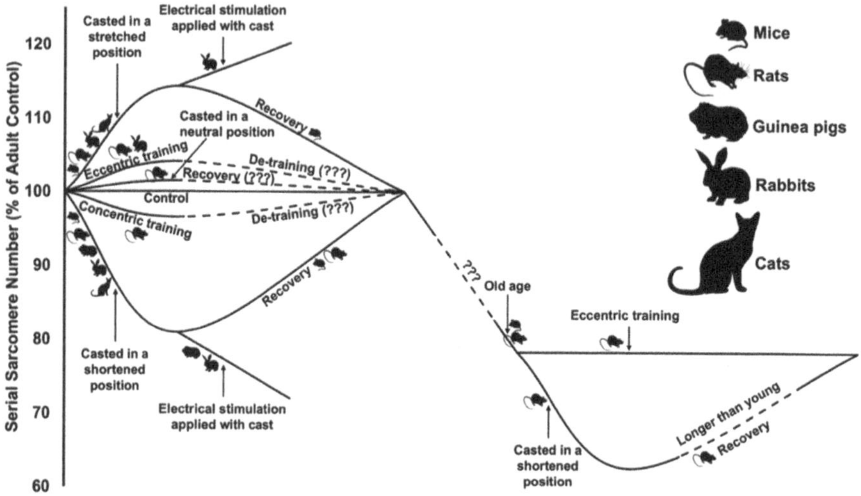

Fig. 10.7 Data on serial sarcomere number adaptations following immobilization and training interventions and with age. Data on percent changes in serial sarcomere number compared to control were averaged across studies on mice, rats, guinea pigs, rabbits, and cats. Dashed lines represent areas of uncertainty, as they have not been investigated by any studies

serial sarcomerogenesis [42–53], and given the limitations of that technique, the following material will focus primarily on the evidence of serial sarcomere number adaptations in animals (compiled in Fig. 10.7), but will include a stronger focus on translation to humans (i.e., aging, clinical settings) toward the end.

The idea that muscle adds and subtracts sarcomeres in a way that optimizes muscle function is generally accepted, but animal models of orthopaedic procedures such as tendon transfer and tenotomy as well as sarcomere data collected intraoperatively in children with cerebral palsy suggest that muscle's structural response can be more complex in response to injury, disease, and clinical interventions. The primary bottleneck limiting progress in the field for understanding this type of adaptation in human muscle in clinical populations is that sarcomere length is required to calculate serial sarcomere number. Because sarcomeres are at the micron length scale, sarcomere length measurements in living human subjects have traditionally been limited to biopsy or intraoperative studies. Integrating multiscale imaging techniques in people with chronic, hemiparetic stroke, including the Zebrascope – a two-photon microendoscopic system – we observed significantly fewer sarcomeres in series in the paretic biceps compared to the contralateral muscle. These data provide support for the common interpretation that the shorter muscle fascicles that are commonly observed via ultrasound in paretic muscles in the stroke population are consistent with a loss of sarcomeres in series. We hope that these techniques become more accessible and enable direct testing of the links between muscle structure, function, and adaptation in clinical populations, as needed to effectively translate our understanding of muscle adaptation from animal studies to improved clinical outcomes.

– Dr. Wendy Murry, Professor of Biomedical Engineering, Director, Applied Research in Musculoskeletal Simulation Lab, Northwestern University.

- Serial sarcomere number adapts to conditions imposed upon it to maintain optimal sarcomere length
- Serial sarcomere number influences the shape of a muscle's active force-length relationship, maximum shortening velocity, and work and power production during muscle shortening
- Most studies on humans are currently limited to an assumption of serial sarcomere number via ultrasonographic measurements of fascicle length
- Ultrasound measurement may overestimate actual serial sarcomere number adaptations

Serial Sarcomere Number Adaptations in Response to Casting Interventions and the Impact on Mechanical Function

When immobilized in a shortened position, muscles lose sarcomeres in series, and when immobilized in a lengthened position, muscles add sarcomeres in series. Immobilization in a neutral position results in little change in serial sarcomere number as that is already close to the position for optimal sarcomere length. In these conditions of immobilization, serial sarcomere number is urgently regulated, emphasizing its importance for biomechanical function. A decrease in serial sarcomere number corresponds to a leftward shift and narrowing of the active force-length relationship, while an increase in serial sarcomere number corresponds to a rightward shift and widening of the active force-length relationship. Changes in dynamic performance with immobilization seem to depend more on changes in fibre type composition than serial sarcomere number adaptations.

Serial Sarcomere Number Adaptations to Immobilization at Short and Long Muscle Lengths

Investigation of the regulation of serial sarcomere number in muscle was first pioneered by Dr. Geoffrey Goldspink in the 1970s in cats and mice. Goldspink saw that when muscles were casted in a shortened position, they lost sarcomeres in series, and when muscles were casted in a lengthened position, sarcomeres were added in series. Specifically, serial sarcomere number losses of 10–40% and serial sarcomere number increases of 16–26% were observed after 2–4 weeks of casting. Furthermore, serial sarcomere number returned to control values within 4 weeks following cast removal regardless of whether the muscle was casted in a shortened or stretched position, emphasizing that serial sarcomere number adapts to best suit a muscle's present force-producing conditions. Since Goldspink's original work, numerous other studies have replicated these findings, all of which are displayed in Table 10.1.

Table 10.1 Adaptations in serial sarcomere number and mechanical function in response to casting interventions

Study	Species	Intervention	Muscles	Change in Serial Sarcomere Number	Functional outcomes
Tabary et al. (1972)	Cats	4 weeks immobilization in a shortened or stretched position	Soleus	40% ↓SSN for immobilization in a shortened position 19% ↑SSN for immobilization in a stretched position.	
Williams and Goldspink (1973)	Mice	1–3.5 weeks of immobilization in a shortened or stretched position, followed by 4 weeks of recovery	Soleus	~10% ↓SSN after 3.5 weeks immobilization in a shortened position. Full return to control values after 4 weeks recovery. ~20% ↑SSN after 2 weeks immobilization in a stretched position. Full return to control values after 4 weeks recovery.	
Goldspink et al. (1974)	Cats	4 weeks denervation while immobilized in a stretched or shortened position	Soleus	25–36% ↓SSN following immobilization in a shortened position. 25–26% ↑SSN following immobilization in a stretched position.	No change in joint angle-passive tension curve compared to control
Williams and Goldspink (1978)	Mice	2–3 weeks immobilization in a stretched or shortened position	Soleus	16% ↑SSN following immobilization in a stretched position. 20% ↓SSN following immobilization in a shortened position.	Rightward shift in active force-length relationship following immobilization in a stretched position. Leftward shift in active force-length relationship following immobilization in a stretched position.

(continued)

Table 10.1 (continued)

Study	Species	Intervention	Muscles	Change in Serial Sarcomere Number	Functional outcomes
De la Tour et al. (1979)	Guinea pigs	2–3 days of immobilization in a shortened or stretched position with injection of tetanus toxin (inducing constant muscle activation)	Soleus	32% ↓SSN following immobilization in a shortened position with tetanus toxin. No change in SSN following immobilization in a stretched position with tetanus toxin.	
De la Tour et al. (1979)	Guinea pigs	6 days of immobilization in a shortened position	Soleus	~6%↓SSN following immobilization in a shortened position.	
Tabary et al. (1981)	Guinea pigs	12 hours of immobilization in a shortened or lengthened position with and without electrical stimulation, then recovery for 36–48 hours (following only immobilization in a shortened position)	Soleus	No SSN change from 12 hours of immobilization alone. 25% ↓SSN following immobilization in a shortened position with stimulation. No change in SSN with immobilization in a lengthened position with stimulation.	
Witzmann et al. (1982)	Rats	42 days of immobilization in a shortened position (soleus and extensor digitorum longus) or neutral position (vastus lateralis)	Soleus, extensor digitorum longus, vastus lateralis	27% ↓SSN for soleus following immobilization in a shortened position. No changes in SSN for extensor digitorum longus or vastus lateralis.	Leftward shift in active force-length relationship for all three muscles. ↑V_{max} for all three muscles when normalized to SSN.

(continued)

Table 10.1 (continued)

Study	Species	Intervention	Muscles	Change in Serial Sarcomere Number	Functional outcomes
Spector et al. (1982)	Rats	4 weeks of immobilization in a shortened, neutral, or lengthened position	Soleus, medial gastrocnemius, tibialis anterior	Soleus: 18% ↑SSN when immobilized in stretched position, no difference in SSN when immobilized in a neutral position, 16% ↓SSN when immobilized in shortened position. Medial gastrocnemius: 20% ↑SSN when immobilized in stretched position, no change in SSN when immobilized in a neutral position, 13% ↓SSN when immobilized in shortened position. Tibialis anterior: 4% ↑SSN when immobilized in stretched position, 9% ↑SSN when immobilized in a neutral position, 7% ↓SSN when immobilized in shortened position.	
Williams et al. (1986)	Rabbits	4 days of immobilization in a stretched position with and without electrical stimulation	Tibialis anterior	8% ↑SSN following immobilization in a stretched position. 14% ↑SSN following immobilization in a stretched position combined with electrical stimulation.	
Williams et al. (1988)	Rabbits	7 days of immobilization in a shortened position with or without electrical stimulation, and immobilization in a stretched position	Soleus	38% ↓SSN when immobilized in a shortened position. 47% ↓SSN when immobilized in a shortened position with electrical stimulation. 13% ↑SSN when immobilized in a stretched position.	

(continued)

Table 10.1 (continued)

Study	Species	Intervention	Muscles	Change in Serial Sarcomere Number	Functional outcomes
Heslinga and Huijing (1993)	Rats	4 and 6 weeks of immobilization in a shortened position	Medial gastrocnemius and soleus	Medial gastrocnemius: 11% ↓SSN after 4 weeks, no difference after 6 weeks of immobilization in a shortened position. Soleus: 19–30% ↓SSN after 4 weeks, 26–29% ↓SSN after 6 weeks of immobilization in a shortened position.	Leftward shift in force-length relationship at 4 weeks and further at 6 weeks.
Koh and Tidball (1999)	Rats	4 weeks of immobilization in a shortened position followed by 3 weeks of remobilization	Soleus	42% ↑SSN after 3 weeks of remobilization.	
Gomes et al. (2004)	Rats	3 weeks of immobilization in a shortened position with or without 40 min stretching once/week	Soleus	23–24% ↓SSN following immobilization in a shortened position regardless of stretching.	
Shah et al. (2004)	Mice	4 weeks immobilization in a shortened (soleus) or stretched (tibialis anterior and extensor digitorum longus) position	Soleus, tibialis anterior, EDL	Soleus: 26% ↓SSN following immobilization in a shortened position Tibialis anterior: 9% ↑SSN following immobilization in a stretched position. Extensor digitorum longus: 17% ↑SSN following immobilization in a stretched position.	
Soares et al. (2007)	Rats	1–4 days of immobilization in a stretched position	Soleus, tibialis anterior	Soleus: 19% ↑SSN at 2 days, 29% ↑SSN at 4 days of immobilization in a stretched position. Tibialis anterior: No changes in SSN after 4 days of immobilization in a stretched position.	

(continued)

Table 10.1 (continued)

Study	Species	Intervention	Muscles	Change in Serial Sarcomere Number	Functional outcomes
Aoki et al. (2009)	Rats	1–4 days of immobilization in a stretched position	Soleus	15% ↑SSN at 2 days, 27% ↑SSN at 4 days of immobilization in a stretched position.	
Kinney et al. (2017)	Mice	2 weeks of immobilization in a stretched (soleus) and shortened (tibialis anterior) position	Soleus, tibialis anterior	Soleus: 19% ↑SSN following immobilization in a stretched position. Tibialis anterior: 22% ↓SSN following immobilization in a shortened position.	
Dayanidhi et al. (2020)	Mice	30 days of immobilization in a shortened (soleus) or stretched (tibialis anterior and extensor digitorum longus) position followed by 30 days of remobilization	Soleus, tibialis anterior, extensor digitorum longus	Soleus: 20% ↓SSN following immobilization in a shortened position, SSN returned to control after remobilization. Extensor digitorum longus: No difference in casted and un-casted SSN following remobilization. Tibialis anterior: No difference in casted and un-casted SSN following remobilization.	
Hinks et al. (2023)	Rats	2 weeks immobilization in a stretched (soleus) position or neutral (medial gastrocnemius and lateral gastrocnemius) position	Soleus, medial gastrocnemius, lateral gastrocnemius	Soleus: 6% ↑SSN following immobilization in a stretched position. Medial gastrocnemius and lateral gastrocnemius: No changes in SSN following immobilization in a neutral position.	

SSN serial sarcomere number, V_{max} maximum shortening velocity

What Is the Stimulus for Serial Sarcomere Number Adaptation During Immobilization: Active Tension, Passive Tension, or Both?

Of course, the question arose regarding whether these serial sarcomere number adaptations were more associated with the passive tension generated in the muscle at shortened (low passive tension) and stretched (high passive tension) positions, or from the animal attempting to generate active force in the immobilized muscle when in those positions. Goldspink and others went forth to address this question in mice, guinea pigs, and rabbits. First, Goldspink repeated his previous experiment immobilizing the mouse soleus in a shortened or stretched position for 4 weeks, but this time also cut off the nerve supply to the muscle, meaning it could no longer generate active force [9]. Goldspink still observed a 25–36% decrease in serial sarcomere number following immobilization in a shortened position and a 25–26% increase when immobilized in a stretched position in these denervated muscles, suggesting the passive tension developed at a particular muscle length was more important than generating active tension at that muscle length for serial sarcomere number adaptation. Subsequently, de la Tour [54] injected the guinea pig soleus with tetanus toxin to induce constant contracture while immobilized in a shortened or stretched position for 2 to 3 days. Muscles immobilized in a shortened position with tetanus toxin lost 32% of their serial sarcomere number, while those immobilized in a stretched position experienced no serial sarcomere number change, illustrating the protective ability of muscle stretch on the loss of serial sarcomere number with contracture. Two studies in the 1980s by Williams [55, 56] in rabbits further refined the above findings from mice and guinea pigs. Williams found that when electrically stimulating the rabbit tibialis anterior while immobilized in a shortened position for 7 days, it loses 11% more sarcomeres in series than when immobilized in a shortened position without stimulation [56]. Conversely, when immobilized in a stretched position with electrical stimulation for 4 days, the rabbit soleus added 6% more sarcomeres in series than when immobilized in a stretched position without stimulation [55]. Therefore, the development of passive and active tension at a particular muscle length each have independent influences on the direction and magnitude of serial sarcomere number adaptation.

> Sarcomerogenesis, the addition of serial sarcomeres in a muscle, is ill-understood, and experiments using similar approaches have resulted in opposite outcomes. Eccentric exercise, for example, has been shown to promote sarcomerogenesis in some studies (e.g., Butterfield et al. 2005), but not in others (e.g., Morais et al. 2020; Pincheira et al. 2022). The negative findings might be caused by eccentric training intervention periods that were too short, or stimuli that were not sufficient (Herzog and Fontana, 2022). When testing if sarcomerogenesis is dependent on muscle force or excursion, Koh and Herzog (1998) developed a model of tibialis anterior retinacular release that caused a decrease in active and passive forces in rabbits during everyday movements, and an increase in excursion. The retinacular release resulted in sarcomerogenesis, suggesting that excursion is a strong positive regulator of sarcomerogenesis, while a decrease in force did not produce the antic-

ipated loss of serial sarcomeres. This positive relationship between excursion and sarcomerogenesis seems to only hold in growing but not in adult animals (Burkholder and Lieber, 1998). The loss of normal sarcomerogenesis in some diseases like spastic cerebral palsy, has dramatic functional implications (Howard and Herzog, 2021), therefore it seems essential to study, and hopefully to come to fully understanding, the complex mechanisms that underly sarcomerogenesis in healthy and diseased muscle.

– Dr. Walter Herzog, Professor, Canada Research Chair in Cellular and Molecular Biomechanics, Human Performance Lab, University of Calgary.

Serial Sarcomere Number Adaptations to Immobilization in a Neutral Position

Immobilization of a muscle in a neutral position (typically a 90° joint angle) results in less drastic shifts in serial sarcomere number. Spector et al. [57] immobilized the rat soleus, medial gastrocnemius, and tibialis anterior in a neutral position for 4 weeks and only observed a serial sarcomere number change in the tibialis anterior, which increased by 9%. For the soleus and medial gastrocnemius, the lack of change in serial sarcomere number when immobilizing in a neutral position was in contrast to serial sarcomere number increases of 18–20% when immobilizing in a stretched position and serial sarcomere number decreases of 13–16% when immobilizing in a shortened position for the same duration [57]. Given that a neutral position is less likely to deviate far from the actin-myosin overlap normally assumed by a muscle, it is understandable that there would be less of a stimulus to alter serial sarcomere number.

The Rapid Regulation of Serial Sarcomere Number Compared to Parallel Sarcomere Number in Conditions of Immobilization

A perhaps overlooked feature of serial sarcomere number adaptation is the urgency in which it is regulated (hours to days) in immobilized muscle. As noted above, a 32% decrease in serial sarcomere number could occur in 2–3 days in the guinea pig soleus with injection of tetanus toxin while immobilized in a shortened position [54]. With electrical stimulation while immobilized in a shortened position, a 25% decrease in guinea pig soleus serial sarcomere number could also occur in as little as 12 hours [58]. The 8–14% increases in rabbit tibialis anterior serial sarcomere number with immobilization in a stretched position were also observed after just 4 days [55]. In the early 2000s, another group observed that a 19–25% increase in rat soleus serial sarcomere number could occur after only 2 days of immobilization in a maximally stretched position, which increased further to 27–29% after 4 days [59, 60]. These large-magnitude changes in serial sarcomere number seem to be able to occur faster than comparable magnitude changes in fibre cross-sectional area (i.e., parallel sarcomere number). For example, when immobilizing the rat rectus femoris in a shortened position, the earliest decrease in fibre cross-sectional area was detected at 3 days—a decrease of ~2%, and after 1 week, a decrease of ~6% [61]. In another study immobilizing the soleus in a stretched position for 4 days, cross-sectional area of type I and II fibres only trended toward decreasing, with no statistically significant difference [62]. It seems that only with long time periods are comparably large decreases in fibre cross-sectional area observed, such as a 53–67% decrease in the rat soleus following 4 weeks of immobilization in a shortened position [63]. These rapid changes in serial sarcomere number allude to an urgency for muscle to preserve optimal actin-myosin overlap and resting passive tension, however, in cases of immobilization in a shortened position or contracture, this urgency can be considered maladaptive, as overall biomechanical performance is negatively impacted.

Changes in Mechanical Function Alongside Immobilization-Induced Serial Sarcomere Number Adaptations

The addition and subtraction of serial sarcomeres with immobilization has been shown to have some of the expected impacts on biomechanical function. Most notably, one of Goldspink's early studies in the mouse soleus observed a more leftward, narrower active force-length relationship with a reduction in serial sarcomere number, but a more rightward, wider active force-length relationship with an increase in serial sarcomere number [4] (Fig. 10.8), meaning there were smaller and larger 'windows' (i.e., muscle excursion, or joint range of motion), respectively, in which force could be generated. Others since have also observed leftward-shifted active force-length relationships alongside an immobilization-induced decrease in serial sarcomere number in rats [64, 65].

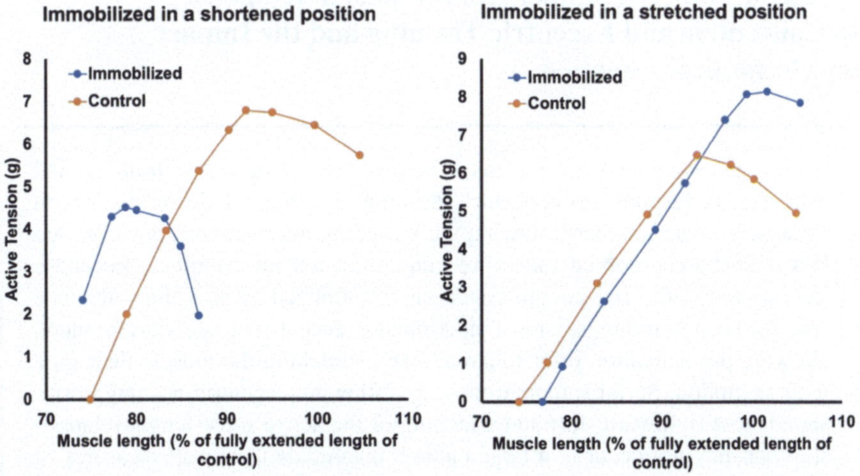

Fig. 10.8 Reconstructed data based on results from Williams and Goldspink (1978) showing a leftward shift and narrowing of the active force-length relationship following immobilization in a shortened position, and a rightward shift and widening of the force-length relationship following immobilization in a stretched position

Whether an increase in serial sarcomere number following immobilization in a stretched position corresponds to increases in V_{max}, work, and power output has been less investigated. However, due to other changes that occur within the muscle during immobilization, such as changes in fibre type composition and the loss of fibre cross-sectional area, we would not necessarily expect an increase in serial sarcomere number following immobilization in a stretched position to positively improve these variables. In fact, a study by Witzman [64] observed an increase in V_{max} alongside a *decrease* in serial sarcomere number induced by immobilization in a shortened position, attributed to a shift toward more fast-twitch fibres. As immobilization involves muscle disuse, it does not provide the growth stimuli that accompany physical exercise interventions (discussed in the next section). Therefore, immobilization in a stretched position provides more insight on a muscle's general sarcomere-adding ability, but has limited scope in prompting improvements in performance. With that said, due to their ability to alter force-length properties, these immobilization interventions have been adapted in orthopaedics with bone lengthening procedures to address limb length inequalities [66, 67], so their impact from a clinical standpoint is not unfounded.

- Serial sarcomere number adaptations during immobilization are muscle length dependent, and are more exaggerated by adding chronic electrical stimulation
- Serial sarcomere number adapts rapidly to immobilization
- The whole muscle active force-length relationship changes in accordance to increases or decreases in serial sarcomere number following immobilization

Serial Sarcomere Number Adaptations in Response to Concentric and Eccentric Training and the Impact on Mechanical Function

Serial sarcomere number increases in response to eccentric training and decreases in response to concentric training. The magnitude of changes in serial sarcomere number following these training interventions, however, are less than those observed following immobilization interventions. Eccentric training-induced serial sarcomerogenesis is optimized by providing adequate rest between training sessions, maximizing eccentric force development, allowing pre-activation prior to stretch, and stretching the muscle through a long excursion. Serial sarcomerogenesis following eccentric training corresponds to a rightward shift and widening of the active force-length relationship but may not occur at a large enough magnitude to impact measures of dynamic muscle function.

Serial Sarcomere Number Adaptations in Response to Concentric and Eccentric Training

During physical exercise, muscles undergo three types of contractions: (1) isometric, in which the muscle is actively generating force without a change in joint rotation; (2) concentric, in which the muscle is actively generating force while shortening; and (3) eccentric, in which the muscle is actively generating force while lengthening (Fig. 10.9).

While exercise is of course an opposite stimulus to immobilization, the same serial sarcomere number related principles as described in the casting studies also apply to training interventions. Training biased to concentric contractions tends to induce the subtraction of sarcomeres, while training biased to eccentric contractions tends to induce addition of serial sarcomeres (Table 10.2). Importantly, the decrease in serial sarcomere number with concentric training is different from that observed with immobilization in a shortened position because concentric training does not also result in an overall loss of muscle mass, and instead induces an increase in parallel sarcomere number [44, 68]. These contraction-type dependent serial sarcomere number adaptations were first noted in 1994 by Lynn and Morgan [69] in the rat vastus intermedius, with a 4% decrease in serial sarcomere number following incline running training (which emphasizes concentric contractions) and an 8% increase in serial sarcomere number following decline running training (which emphasizes eccentric contractions). The rationale behind these adaptations is that a

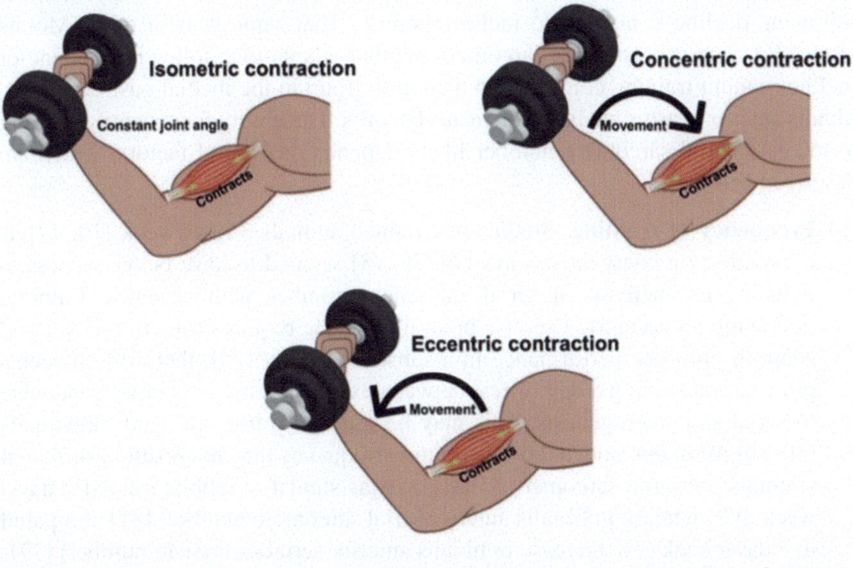

Fig. 10.9 The three types of muscle contractions: isometric (muscle generates active force without joint rotation), concentric (muscle generates active force while shortening), and eccentric (muscle generates active force while lengthening)

typical concentric contraction would end with a muscle near fully shortened, and a typical eccentric contraction would end with a muscle near fully lengthened. Subtraction and addition of serial sarcomeres, respectively, would then better optimize actin-myosin overlap for concentric and eccentric contractions [3, 4, 16]. For eccentric exercise, muscle damage must also be considered as a stimulus for serial sarcomerogenesis. Repeated, high-intensity eccentric contractions can stretch sarcomeres to the point that Z-lines are damaged, and adapting by adding sarcomeres in series allows a muscle to stretch further without that sarcomere damage occurring on subsequent bouts [70–73].

Factors Affecting Training-Induced Serial Sarcomerogenesis

The expected serial sarcomere number adaptations with eccentric training do not occur as consistently or to as great a magnitude as they do in casting interventions (Table 10.2). For example, while Butterfield et al. [74] managed to replicate Lynn and Morgan's findings in the rat vastus intermedius (observing an 8% increase in serial sarcomere number following 10 days of decline walking and an 11% decrease in serial sarcomere number following 10 days of incline walking), the vastus lateralis only saw a 4% decrease in serial sarcomere number with incline walking and no change in serial sarcomere number with decline walking. Subsequently, Chen et al. [75] observed no differences in rat soleus or extensor digitorum longus serial sarcomere number following 4 weeks of incline or decline running training compared to a control group, only noting a significantly greater soleus serial sarcomere number following decline compared to incline running. That same year in mice, Morais et al. [76] observed no serial sarcomere number adaptations following decline or incline running training compared to a control group in the medial gastrocnemius, tibialis anterior, vastus medialis, or vastus lateralis. The ability for eccentric training to increase serial sarcomere number likely depends on several factors, which are described below.

(a) **Frequency of training.** Studies that trained animals 3 days/week [10, 77] as opposed to on consecutive days [74–76, 78] seemed to have better success in inducing an increase in serial sarcomere number with eccentric training. Following an eccentric exercise bout, the muscle requires time to recover and adapt to optimize performance in a subsequent bout [79]; therefore, it seems prudent that leaving a day of rest between exercise sessions is more conducive to serial sarcomerogenesis. This may be especially true for aged individuals [50, 80]. With this said, a too-low training frequency may not result in sufficient stimulus for serial sarcomerogenesis, as was shown in rabbits trained 2 days/week (3% increase in tibialis anterior serial sarcomere number) [81] compared to 3 days/week (7% increase in tibialis anterior serial sarcomere number) [77].
(b) **Force developed during eccentric contractions.** Butterfield and Herzog [77] subjected rabbit dorsiflexors to three different eccentric training protocols in

Table 10.2 Adaptations in serial sarcomere number and mechanical function in response to concentric and eccentric training

Study	Species	Intervention	Muscles	Change in Serial Sarcomere Number	Functional outcomes
Lynn and Morgan (1994)	Rats	7 days of incline or decline running training compared to control	Vastus intermedius	4% ↓SSN following incline training compared to sedentary. 8% ↑SSN following decline training compared to sedentary.	
Lynn et al. (1998)	Rats	5 days of decline running training vs incline running training	Vastus intermedius	9% greater SSN in decline-trained compared to incline-trained.	~25% rightward shift in torque-angle relationship in decline-trained compared to incline-trained.
Koh and Herzog (1998)	Rabbits	12 weeks of eccentric training 2 days/week	Tibialis anterior	3% ↑SSN	
Butterfield et al. (2005)	Rats	5 or 10 days of incline or decline walking compared to control	Vastus intermedius and vastus lateralis	Only changes in SSN at 10 days: Vastus intermedius: 8% ↑SSN following incline training, 11% ↓SSN following decline training. Vastus lateralis: No change in SSN following decline training, 4% ↓SSN following incline training.	
Butterfield and Herzog (2006)	Rabbits	6 weeks eccentric training with stimulation at onset of stretch to a short muscle length, stimulation preceding stretch to a short muscle length, and stimulation preceding stretch to a long muscle length	Tibialis anterior	3–7% ↑SSN, greatest with activation preceding stretch.	Rightward shift in torque-angle relationship with all three interventions, greatest with stimulation preceding stretch and at long muscle length.

(continued)

Table 10.2 (continued)

Study	Species	Intervention	Muscles	Change in Serial Sarcomere Number	Functional outcomes
Chen et al. (2020)	Rats	4 weeks (5 days/ week) of decline running training vs incline running training vs control	Soleus and extensor digitorum longus	Soleus: 6% ↑SSN in decline-trained compared to incline-trained. No differences in SSN compared to control for incline or decline-trained. EDL: No differences in SSN among groups.	No differences in optimal muscle length or work of shortening among groups.
Morais et al. (2020)	Mice	8 weeks of moderate training (3 days/week) or overtraining (5 days/week) decline or incline running compared to control	Medial gastrocnemius, tibialis anterior, vastus medialis, vastus lateralis	Medial gastrocnemius: No differences in SSN among groups. Tibialis anterior: Moderate decline training 11% >incline overtraining, no other differences among groups. Vastus medialis: No differences in SSN among groups. Vastus lateralis: Incline overtraining 7% < moderate incline training, 8% < moderate decline training, and 9% < decline overtraining, no other differences between groups.	
Hinks et al. (2022)	Rats	4 weeks of weighted downhill running training 3 days/week compared to control	Soleus	8% ↑SSN	Trend toward rightward shift in active force-length relationship. No relationships between SSN adaptations and improvements in work loop performance.
Hinks et al. (2023)	Rats	4 weeks of isokinetic eccentric training	Soleus, medial gastrocnemius, lateral gastrocnemius	4–6% ↑SSN in the soleus and medial gastrocnemius. No change in SSN for lateral gastrocnemius.	Widening of the torque-angle relationship plateau region. No change in V_{max} or peak power.

SSN serial sarcomere number, V_{max} maximum shortening velocity

which varying levels of force were developed during the eccentric contractions. Among the three protocols, they found that the eccentric force developed was directly proportional to the observed increase in serial sarcomere number. A study by Hinks et al. [10] sought to provide further insight by adding weighted vests to rats during decline running, which would increase the forces developed in the muscles during eccentric contractions. The researchers found that 4 weeks of weighted downhill running training induced an 8% increase in rat soleus serial sarcomere number, up from an average increase of 3% in rat soleus serial sarcomere number following 4 weeks bodyweight downhill running training [75].

(c) **Timing of muscle activation.** In the rabbit tibialis anterior, Butterfield and Herzog [77] compared two eccentric training protocols: one in which muscle activation occurred at the onset of muscle stretch; and one in which activation preceded muscle stretch. Activating the muscle prior to stretch better assures that the muscle will be at maximum force development when the stretch occurs, and therefore maximizes the eccentric stimulus. Accordingly, the researchers saw a 7% increase in rabbit tibialis anterior serial sarcomere number when muscle activation preceded stretch, but only a 3% increase when muscle activation occurred at stretch onset.

(d) **Muscle excursion.** Butterfield and Herzog [77] also compared eccentric training when the muscle was stretched to a short length versus a longer length. While it may seem intuitive that stretching to a longer muscle length would provide a stronger sarcomerogenic stimulus, these interventions did not differ in their serial sarcomere number increase. It should be noted, however, that the researchers matched the degrees of ankle rotation between these protocols (i.e., changed the ankle angle by ~45° during the eccentric contraction whether stretching to a short or long muscle length). Perhaps, then, the magnitude of muscle stretch, rather than the length the muscle is stretched to, is more important for serial sarcomerogenesis. Koh and Herzog [82] supported that perspective by showing that, during developmental growth, more serial sarcomeres were added in the rabbit tibialis anterior when undergoing a greater excursion. Additionally, by comparing the findings of Lynn and Morgan [69] and Butterfield et al. [74] to Chen et al. [75], we can gain some insight. The first two studies [69, 74] observed an 8% increase in rat vastus intermedius serial sarcomere number following decline running training, while the last study [75] observed an average increase of 3% in rat soleus serial sarcomere number. During locomotion, rat vastus muscles undergo an eccentric excursion of ~25% of optimal length [83], compared to ~10% in the soleus [84, 85]. Therefore, the greater magnitude of serial sarcomerogenesis in the rat vastus intermedius following decline running training may be due to its longer excursion.

Changes in Mechanical Function Alongside Training-Induced Serial Sarcomere Number Adaptations

Training-induced serial sarcomere number adaptation results in expected alterations to the active force-length relationship. Rats that completed 5 days of decline running had a greater serial sarcomere number [78] and a rightward shifted torque-angle relationship compared to rats that completed incline running, (Fig. 10.10). Similarly, following isokinetic eccentric training of the rabbit dorsiflexors, there was both a rightward shift and a widening of the active torque-angle relationship, along with an increase in serial sarcomere number in the tibialis anterior [77]. Recent data from our lab has shown that 4 weeks of isokinetic eccentric training broadened the torque-angle relationship's plateau region in the rat plantar flexors alongside increases in soleus and medial gastrocnemius serial sarcomere number [86].

Compared to the isometric force-length relationship, dynamic muscle performance seems to be less affected by training-induced serial sarcomere number adaptations. Alterations to work output during cyclic muscle length changes following 4 weeks of weighted downhill running training were more related to adaptations in muscle physiological cross-sectional area than serial sarcomere number [10]. As well, recent data from our lab showed 4 weeks of isokinetic eccentric training of the rat plantar flexors did not alter maximum shortening velocity nor peak isotonic power despite the occurrence of serial sarcomerogenesis (Fig. 10.11) [86]. The reason for these discrepancies between how we expect training-induced serial sarcomerogenesis to alter muscle mechanical performance compared to what we actually observed is unknown. Eccentric training only seems to increase serial sarcomere number by ~6–8%, which is a smaller magnitude than what we know is possible from immobilization in a stretched position (up to a 29% increase). Training-induced adaptations in other aspects of the muscle-tendon unit such as tendon compliance and radial muscle growth (i.e., addition of sarcomeres in parallel, and changes in fascicle pennation angle) also impact dynamic muscle performance [3]. Perhaps the concurrent changes in these other variables negatively influences the magnitude of serial sarcomerogenesis provided by eccentric training, which is not enough to cause the expected impact on V_{max}, work, and power.

- Serial sarcomere number increases following eccentric training and decreases following concentric training
- Eccentric training-induced serial sarcomerogenesis induces a rightward shift and widening of the active force-length relationship (improved isometric force production throughout the range of motion), however, does not seem to improve maximum shortening velocity or peak power
- Adequate rest, maximal eccentric force development, and a long excursion are important for maximizing eccentric training-induced serial sarcomerogenesis

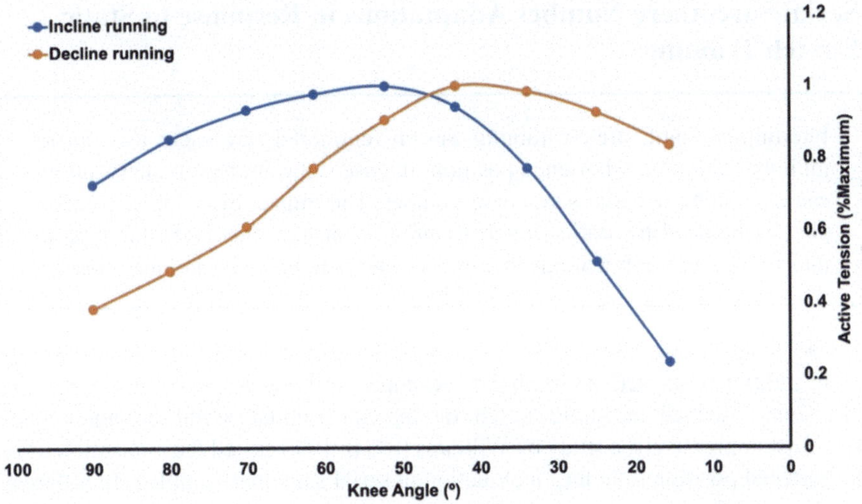

Fig. 10.10 Representative knee extension torque-angle curves from an incline-trained rat and a decline-trained rat based on reconstructed data from Lynn et al. (1998). Note that the decline-trained rat achieves optimal torque production at a smaller knee angle, which corresponds to a longer vastus intermedius muscle length

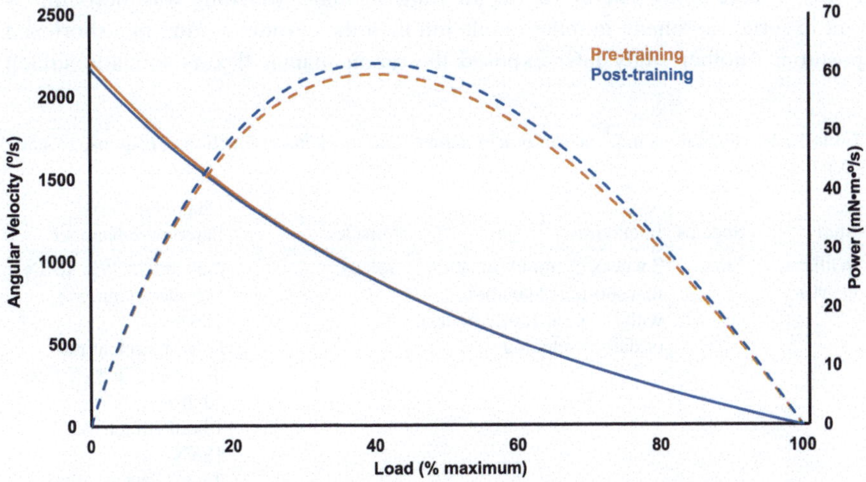

Fig. 10.11 Relative torque-velocity (solid lines) and torque-power (dashed lines) relationships before and after 4 weeks of isokinetic eccentric training in rats. There were no significant differences in maximum shortening velocity or peak power from pre- to post-training

Serial Sarcomere Number Adaptations in Response to Static Stretch Training

Intermittent static stretch training can prevent serial sarcomere loss during immobilization in a shortened position. Intense static stretching interventions can also increase serial sarcomere number. The impact of serial sarcomerogenesis induced by static stretch training is unclear, however, this type of intervention has implications in clinical settings to prevent sarcomere loss.

In addition to the active stretch (i.e., eccentric) training described above, there is also some evidence that static stretch training can stimulate serial sarcomerogenesis. Most impressively, a study by Williams [87] in 1990 casted the mouse soleus in a shortened position to induce a reduction in serial sarcomere number. In different cohorts of mice, however, the researchers removed the casts and subjected the soleus to 15 minutes, 30 minutes, 1 hour, or 2 hours of stretching per day. As expected, mice exposed to only casting lost 19% of their serial sarcomere number. Just 15 minutes/day of stretching reduced this loss to 5%, and 1 and 2 hours of stretching actually increased serial sarcomere number beyond control values by 8–10% (Table 10.3). Therefore, the intermittent static stretching was more important in serial sarcomere number regulation than the chronic casting in a shortened position. Another study later exposed the rabbit plantar flexors to static stretch

Table 10.3 Adaptations in serial sarcomere number and mechanical function in response to static stretch training

Study	Species	Intervention	Muscles	Change in Serial Sarcomere Number
Williams (1990)	Mice	2 weeks of immobilization in a shortened position, with 0, ¼, ½, 1, or 2 hours of daily stretching	Soleus	No stretch: 19% ↓SSN ¼ hour stretch: 5% ↓SSN ½ hour stretch: no difference in SSN from control 1 hour stretch: 8% ↑SSN 2 hour stretch: 10% ↑SSN
De Jaeger et al. (2015)	Rabbits	4 weeks static stretch training 3 days/week	Soleus, plantaris, medial gastrocnemius	Soleus: 9% ↑SSN Plantaris and medial gastrocnemius: no change in SSN compared to control.
Peviana et al. (2018)	Rats	10 or 15 days of static stretch training	Soleus	No change in SSN at 10 days, 11% ↑SSN at 15 days.

SSN serial sarcomere number

training a total of 12.5 minutes 3 days/week for 4 weeks [88]. The soleus underwent a 9% increase in serial sarcomere number while the plantaris and medial gastrocnemius underwent no change, demonstrating the ability for some muscles to add sarcomeres in series in response to intense static stretching. It seems longer periods of static stretch training are required to induce serial sarcomerogenesis, as 15 but not 10 days of static stretching for 10 minutes/day induced an increase in serial sarcomere number in the rat soleus [89]. The ability for static stretch training-induced serial sarcomerogenesis to alter mechanical function is understudied, but will be explored briefly in the section Clinical Applications of Serial Sarcomere Number.

- Intermittent static stretch training prevents immobilization-induced serial sarcomere loss
- Intent static stretch training can increase serial sarcomere number
- The impact of static stretch training-induced serial sarcomerogenesis on muscle mechanical function is currently unclear

Changes in Serial Sarcomere Number with Age and the Impact on Mechanical Function

Aging is associated with the loss of sarcomeres in series, which contributes to the loss of muscle fascicle length and cross-sectional area, and the decline in mechanical performance. The ability for eccentric training-induced serial sarcomerogenesis is limited in old age due to alterations to mechanotransduction pathways, a greater susceptibility to muscle damage, and extracellular matrix remodelling that resembles fibrosis.

Aging causes impairments in skeletal muscle mechanical performance including reductions in strength, power output, and shortening velocity [20, 90–93] – all of which are closely associated with serial sarcomere number. These functional declines with age are caused by alterations in both neuromuscular function and muscle-tendon unit morphology [90, 93]. The loss of muscle cross-sectional area comprises a large portion of the morphological changes, and occurs at a rate of at least 0.5–0.8% per year starting at age 65 according to cross-sectionally designed studies [92, 94, 95], but likely at an even faster rate of 1% per year or more according to longitudinal studies [96, 97]. Geometrically, age-related declines in fascicle pennation angle and fascicle length can both contribute to the loss of muscle cross-sectional area [98, 99]. Studies comparing specifically fascicle length between old and young healthy humans have shown that fascicle length decreases by 7–35% in the vastus lateralis [100–103], 7–19% in the medial gastrocnemius [24, 99, 101, 104–108], and up to 15% in the soleus [109] from young to old age (Table 10.4).

Table 10.4 Differences in muscle fascicle length or serial sarcomere number between healthy young and older individuals

Study	Participants	Activity level	Muscle	Measurement technique used	% Difference[a] in Fascicle Length or Serial Sarcomere Number
Hooper (1981)	Young (26 wk) and old (70 wk) male mice	Sedentary	Sternomastoid, biceps brachii, tibialis anterior	Eyepiece micrometer	–25% (sternomastoid), −37% (biceps brachii), and −18% (tibialis anterior) fibre length driven by −26%, −35%, and −16% SSN.
Narici et al. (2003)	Young (27–42) and old (70–81 y) men	Recreationally active	Medial gastrocnemius	Ultrasound, 38 mm probe, linear extrapolation	−10% FL
Kubo et al. (2003)	Young (20–39 y) and old (60–85 y) men and women	Sedentary or mildly active	Vastus lateralis, medial gastrocnemius, long head of triceps brachii	Ultrasound, probe width not provided, trigonometric estimation	Vastus lateralis: −7% FL for men; no difference for women Medial gastrocnemius: −10% FL for both men and women Triceps brachii: No difference for men or women
Karamanidis and Arampatzis (2005)	Young (21–32) and old (60–69) men	Physically inactive group and runner group	Vastus lateralis and medial gastrocnemius	Ultrasound, probe width not provided, length of fascicular path	No differences in FL between groups
Morse et al. (2005)	Young (25 ± 5 y) and old (74 ± 4 y) men	Physically active but not participating in structured training	Medial gastrocnemius, lateral gastrocnemius, soleus	Ultrasound, probe width not provided, linear extrapolation	Medial gastrocnemius: −16% FL No differences for other muscles (but trends toward shorter in old)

(continued)

Table 10.4 (continued)

Study	Participants	Activity level	Muscle	Measurement technique used	% Difference[a] in Fascicle Length or Serial Sarcomere Number
Thom et al. (2007)	Young (25 ± 5 y) and old (75 ± 4 y) men	Recreationally active	Medial gastrocnemius	Ultrasound, probe width not provided, linear extrapolation	−19% FL
Baudry et al. (2012)	Young (23–36 y) and old (68–83 y) men and women	Not given	Medial gastrocnemius	Ultrasound, 60 mm probe, trigonometric estimation	−24% FL
Stenroth et al. (2012)	Young (24 ± 2 y) and old (75 ± 3 y) men and women	Equal number of sedentary and active subjects	Medial gastrocnemius and soleus	Ultrasound, 60 mm probe, length of fascicle from aponeurosis to aponeurosis	Medial gastrocnemius: −7% FL No difference in soleus FL
Baroni et al. (2013)	Young (21–40 y) and old (61–80 y) men	No enrollment in systematic training program	Vastus lateralis	Ultrasound, 38 mm probe, linear extrapolation	−35% FL
Panizzolo et al. (2013)	Young (26 ± 4 y) and old (66 ± 2 y) men and women	Not given	Soleus	Ultrasound, 60 mm probe, length of fascicle from aponeurosis to aponeurosis	−14% to −15% FL throughout gait cycle
Power et al. (2013)	Young (26 ± 3 y) and old (77 ± 6 y) men	Recreationally active	Vastus lateralis	Ultrasound, 49 mm probe, trigonometric estimation	At rest: −6% (short MTU length) to −11% (long MTU length) FL During isometric contraction: −3% (short MTU length) to −13% (long MTU length) FL

(continued)

Table 10.4 (continued)

Study	Participants	Activity level	Muscle	Measurement technique used	% Difference[a] in Fascicle Length or Serial Sarcomere Number
Barber et al. (2013)	Young (27 ± 3 y) and old (70 ± 3 y) men and women	Not given	Medial gastrocnemius	Ultrasound, 60 mm probe, semi-automated fascicle tracking	No differences in FL across range of motion
Wu et al. (2016)	Young (24 ± 4 y) and old (67 ± 4 y) men and women	No engagement in strength training or sports in last 5 years	Vastus lateralis	Ultrasound, 40 mm probe, length of fascicle from aponeurosis to aponeurosis	−14% FL
Erskine et al. (2017)	Young (24 ± 8 y) and old (66 ± 8 y) men and women	Less than 2 hours/week of physical activity	Medial gastrocnemius	Ultrasound, 40 mm probe, length of fascicle from aponeurosis to aponeurosis	No difference in FL
Gerstner et al. (2017)	Young (20 ± 2 y) and old (80 ± 8 y) males	Recreationally active	Medial gastrocnemius	Panoramic ultrasound, 38 mm probe, length of fascicle from aponeurosis to aponeurosis	No difference in FL
Franchi et al. (2019)	Young (25 ± 4 y) and old (70 ± 3 y) men	Recreationally active	Vastus lateralis	Ultrasound, 100 mm probe, linear extrapolation	No differences in FL between old and young
Power et al. (2020)	Young (9 mo) and old (32 mo) rats	Sedentary	Medial gastrocnemius	Sarcomere lengths: laser diffraction Fascicle lengths: Digital camera and imaging software	−14% FL driven by −10% SSN

(continued)

Table 10.4 (continued)

Study	Participants	Activity level	Muscle	Measurement technique used	% Difference[a] in Fascicle Length or Serial Sarcomere Number
Conway and Franz (2020)	Young (25 ± 5 y) and old (75 ± 3 y) men and women	Not given	Medial gastrocnemius during walking	Ultrasound, 60 mm probe, semi-automated fascicle tracking	−7% FL on average during walking
Quinlan et al. (2021)	Young (18–35 y) and old (65–73 y) men	Recreationally active	Vastus lateralis	Ultrasound, 100 mm probe, linear extrapolation	No difference in FL
Kelp et al. (2021)	Young (21 ± 2 y) and old (70 ± 3 y) men	At least moderately active	Medial gastrocnemius and lateral gastrocnemius	Ultrasound, 56 mm probe, length of fascicle from aponeurosis to aponeurosis	Medial gastrocnemius: −8% FL No difference in lateral gastrocnemius FL
Pinel et al. (2021)	Young (25 ± 4 y) and old (70 ± 2 y) men and women	Physically active	Medial gastrocnemius, lateral gastrocnemius, soleus	Ultrasound, probe width not provided, length of fascicle from aponeurosis to aponeurosis	No difference in FL of any muscles
Hinks et al. (2023)	Young (9 months) and old (33 months) rats	Sedentary	Medial gastrocnemius, lateral gastrocnemius, soleus	Sarcomere lengths: laser diffraction Fascicle lengths: Digital camera and imaging software	Medial gastrocnemius: −14% SSN Lateral gastrocnemius: −7% Soleus: −9%

[a]calculated as [(old − young)/young] × 100%
FL fascicle length, *MTU* muscle-tendon unit, *SSN* serial sarcomere number

As mentioned earlier, correct determination of serial sarcomere number requires measurement of sarcomere length for use in the equation: $SSN = FL \div SL$, where SSN is serial sarcomere number, FL is fascicle length, and SL is sarcomere length. By measuring sarcomere lengths, studies in rodents have supported the assumption that shorter fascicle lengths with old age are indeed driven by a loss of serial sarcomeres. In mice, Hooper [110] showed 18–25% shorter sternomastoid, biceps brachii, and tibialis anterior fibre lengths in old compared to young mice, driven by a 16–35% smaller serial sarcomere number. Power et al. [111] observed similar results in old rats, showing that a 14% shorter medial gastrocnemius fascicle length compared to young rats, was driven by having 10% fewer sarcomeres in series.

> With aging, there is a loss of serial sarcomere number. Whether that is owing to a reduction in active force production or changes to muscle excursion during ambulation is unknown. Understanding adaptability of the aged neuromuscular system will offer insight into how we can improve muscle function and ultimately improve performance in older adults.
>
> – Dr. Geoffrey Power, Associate Professor, Director of the Neuromechanical Performance Research Lab, University of Guelph.

Can Training-Induced Serial Sarcomerogenesis Mitigate the Age-Related Decline in Mechanical Performance?

While studies on humans suggest training-induced serial sarcomerogenesis appears to be possible in old age, it may not occur as easily as in young muscle. Reeves et al. [39] observed an 11% increase in vastus lateralis fascicle length following conventional (i.e., the combination of concentric and eccentric contractions) resistance training in older men (~74 years). Importantly, by obtaining a measurement of optimal fascicle length, they verified that the increase in fascicle length could have been driven by an increase in serial sarcomere number, assuming a constant optimal sarcomere length. In a follow-up study, Reeves et al. [112] observed a greater magnitude fascicle length increase (+20%) following eccentric compared to conventional resistance training (+8%) in older men (~70 years). Quinlan et al. [50] later refined these observations. They showed a 4% increase in vastus lateralis fascicle length 2 weeks into moderate-load eccentric training in young men (~25 years), but an increase in fascicle length of comparable magnitude did not occur until 4 weeks in older men (~68 years). Raj et al. [113] also observed no changes in vastus lateralis or medial gastrocnemius fascicle length in older men (~68 years) following 16 weeks of eccentric resistance training. The only investigation of actual serial sarcomere number adaptations following eccentric training in old compared to young animals is from recent data from our lab in rats [86]. After 4 weeks of maximal isokinetic eccentric training of the plantar flexors, young rats had a 6% increase in soleus and medial gastrocnemius serial sarcomere number while those muscles in old rats did not change serial sarcomere number. Furthermore, whereas young rats experienced beneficial adaptations in mechanical function, old rats saw reductions in strength, maximum shortening velocity, and peak power (Fig. 10.12).

Older individuals may experience blunted serial sarcomerogenesis due to their muscles being more susceptible to eccentric exercise-induced damage, thus requiring longer periods to recover between exercise sessions. A hallmark of eccentric exercise-induced damage is a long-lasting deficit in maximum strength. Studies in rats [80], mice [114–117], and humans [118–120] have consistently observed that: (1) older individuals remain weaker for longer durations than young following

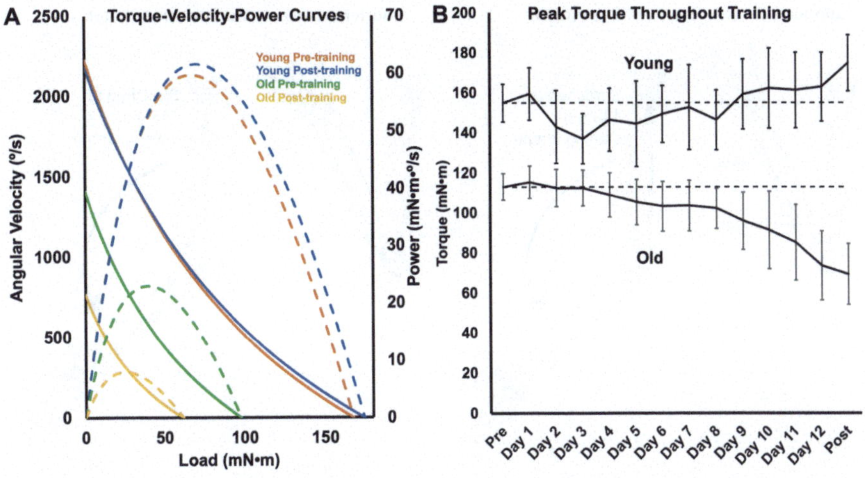

Fig. 10.12 (**a**) Changes in the torque-velocity (solid lines) and torque-power (dashed lines) relationships from pre- to post-4 weeks of isokinetic eccentric training 3 days/week in young and old rats. (**b**) Changes in peak isometric torque throughout the training period in young and old rats. Dashed lines represent the pre-training isometric torque values. Note that young rats either improved (for isometric strength) or did not change (the torque-velocity-power curves) while old rats were negatively impacted following training

high-intensity eccentric exercise; and (2) following a subsequent eccentric exercise bout performed several weeks later, older individuals exhibited a similar drop in strength to the first bout, while younger individuals exhibited no deficit in strength. Therefore, there is indeed an age-related impaired ability to both recover and adapt following eccentric exercise, limiting the ability for serial sarcomerogenesis. This may occur due to older individuals having a lower serial sarcomere number at baseline. Assuming similar series elastic compliance in the muscle-tendon unit, with fewer serial sarcomeres, older individual's sarcomeres would undergo a greater stretch amplitude for the same joint rotation, thereby incurring more damage [80, 121–123] (Fig. 10.13). Additionally, a recent review identified age-related changes in several mechanotransduction pathways that may limit serial sarcomerogenesis [124]. These include impaired mechanistic target of rapamycin and insulin-like growth factor 1 signaling, increased myostatin activity, reduced activation of serum response factor, increased protein degradation by muscle ring finger protein, and reduced satellite cell content and function. Furthermore, following 50 maximal eccentric contractions once a week for 5 weeks, both young and old mice exhibit an upregulation of genes associated with sarcomere organization and assembly, but only old have an upregulation of genes associated with extracellular matrix remodeling, likely indicative of fibrosis [125]. This remodeling of the local connective tissue environment in old individuals likely blunts the mechanical stimulus on the actual contractile tissue, limiting the activation of mechanotransduction pathways and subsequent serial sarcomerogenesis [124]. To minimize muscle damage and the development of fibrosis, moderate intensity or lower-frequency eccentric training may be required to induce serial sarcomerogenesis in older adults, albeit over a

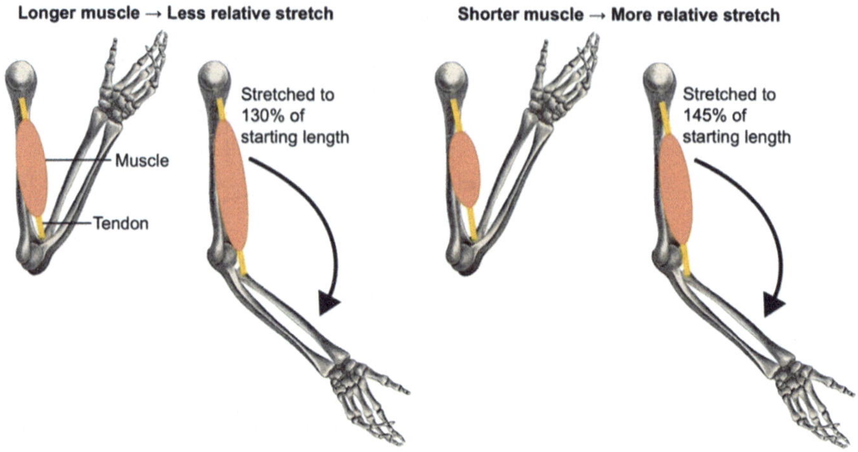

Fig. 10.13 Schematic showing how shorter muscles (right) of older individuals may undergo a greater relative stretch for the same magnitude of joint rotation, which would put them at a greater risk of incurring damage during eccentric exercise

longer time period, as suggested by Quinlan et al. [50] above. However, this area of research requires further investigation.

- Serial sarcomere number decreases in old age, which contributes to functional deficits in older adults
- Muscles of older adults have a heightened susceptibility to damage, which seems to limit the ability to add sarcomeres in series with eccentric exercise at the same duration/intensity as young

Clinical Applications of Serial Sarcomere Number

A reduction in serial sarcomere number causes functional deficits in people with cerebral palsy, people recovering from a hemiparetic stroke, and people submitted to bed rest. A lower serial sarcomere number also increases the risk of hamstring strain injury in athletes. Interventions such as eccentric training and static stretch training may mitigate the impacts of these conditions on muscle mechanical function by increasing serial sarcomere number.

Serial sarcomere number has important clinical applications in scenarios where a reduced serial sarcomere number is a limiting factor for everyday function and injury risk. There is a particular focus on the role of serial sarcomere number in cerebral palsy, stroke, bed rest, and hamstring strain injury.

> The study of sarcomere longitudinal (length and number) adaptations is essential to develop and improve muscle-involved treatments for important clinical disorders like spinal deformity.
>
> – Dr. Stephen Brown, Associate Professor, Director of the Spine and Muscle Biomechanics lab, University of Guelph.

Cerebral Palsy

Cerebral palsy places muscles in a state of contracture while in a shortened position, not unlike the conditions of immobilization in a shortened position with electrical stimulation described earlier, in which rapid sarcomere loss occurred in rabbits and guinea pigs. In children with cerebral palsy, 43% shorter medial gastrocnemius fascicles have been observed throughout the range of motion, along with a shorter muscle excursion and a 92% longer relative fascicle excursion compared to controls [126]. Since fascicles in children with cerebral palsy are shorter but go through the same range of motion as typically developing children, their sarcomeres are forced to act over a wider range. Correspondingly, longer sarcomere lengths have also been observed in patients with cerebral palsy [127–129]. A shorter fascicle length and a longer sarcomere length amount mathematically to having fewer serial sarcomeres, which contributes to the functional limitations associated with cerebral palsy [130]. Static stretching interventions have been mostly ineffective for improving muscle function in children with cerebral palsy [131]. Instead, static stretching combined with strength training [132] and eccentric strength training alone [133] have led to positive functional adaptations in children and adolescents with cerebral palsy. Eccentric strength training in particular widened the torque-angle relationship such that their strength throughout the range of motion resembled that of typically developing children [133]. This widening of the torque-angle relationship could have been due to the occurrence of serial sarcomerogenesis, however, further research is required to confirm this notion.

Stroke Patients

The paretic biceps brachii of patients recovering from hemiparetic stroke experiences chronic disuse combined with placement in a shortened position, which, as described earlier, results in the loss of serial sarcomeres in animals. Accordingly, Adkins et al. [134] found that these patients have 21% shorter biceps brachii optimal fascicle length of the paretic compared to the non-paretic side, which was caused by having 21% fewer sarcomeres in series. This reduction in serial sarcomere number is believed to contribute to a reduced muscle length range for active force generation in stroke patients [134]. Interventions aimed at targeting this considerable loss of serial sarcomeres are likely to mitigate inequalities between paretic and non-paretic muscles and improve functional capacity.

Bed Rest

Humans are sometimes submitted to prolonged periods of bed rest due to illness, injury, or surgery. When confined to bed rest, individuals are in a state of complete inactivity, which has profound impacts on muscle architecture, including fascicle length. Five weeks of horizontal bed rest results in 5% and 6% reductions in fascicle length of the medial gastrocnemius and vastus lateralis, respectively [135], and 10 days of horizontal bed rest results in a 3% reduction in biceps femoris fascicle length [136]. Bed rest seems to have a stronger impact on fascicle length of these weight-bearing muscles, as the tibialis anterior and biceps brachii (non-weight-bearing muscles) do not show changes in fascicle length following 5 weeks of horizontal bed rest [135]. It is important to note among all these measurements, however, that fascicle length was measured via ultrasound without any measurement of sarcomere length, thus the impact on actual serial sarcomere number is uncertain. Regardless, these studies provide insight on which muscles should be targeted during the recovery period following bed rest to treat the functional impairments associated with serial sarcomere number loss. Considering the findings of Williams [87], in which 1 hour/day of stretching inhibited the immobilization-induced loss of serial sarcomeres in mice, future research should investigate whether intermittent stretching during bed rest can help preserve serial sarcomere number in these weight-bearing muscles.

Hamstring Strain Injury

- Children with cerebral palsy have reduced serial sarcomere numbers compared to typically developing children
- Stroke patients and people submitted to bed rest lose serial sarcomeres
- Interventions aimed at increasing serial sarcomere number can potentially mitigate the negative impacts on muscle function associated with the above conditions
- Eccentric hamstring training can decrease the risk of hamstring strain injury by increasing serial sarcomere number, but may require longer than 3 weeks of training

During the late swing phase of running, hip flexion and knee extension together stretch the hamstring to a long length, and as a result, hamstring strain injuries are among the most prevalent across a variety of sports [137]. The risk of re-injuring the hamstrings is even higher, therefore, minimizing risk for this injury in athletes is a high priority. Having a greater serial sarcomere number, theoretically, will keep myofilaments more on the plateau region than the descending limb of the force-length relationship during eccentric contractions, which would allow for better stability and protection from damage and strain at longer muscle lengths [137]. It follows that many have pointed to training-induced sarcomerogenesis as a way to minimize risk of hamstring strain injuries [137, 138]. One study in humans used microendoscopy to investigate serial sarcomere number adaptations following 3 weeks of eccentric resistance training of the hamstrings in healthy young adults [37]. Surprisingly, the researchers observed no change in biceps femoris long head serial sarcomere number. However, fascicle length appeared longer in ultrasound images taken with the knee and hips in full extension due to longer sarcomere lengths at that position. They concluded this may have represented a transition period in which sarcomeres are beginning to stretch, preceding the addition of new sarcomeres along the muscle fascicle. Therefore, eccentric training may be beneficial for hamstring strain injury prevention by inducing serial sarcomerogenesis, but appears to require training beyond 3 weeks.

Conclusion

A muscle's serial sarcomere number is tied to its mechanical performance, particularly its optimal length, maximum shortening velocity, and ability to generate active force and power throughout the range of motion. Serial sarcomere number is highly adaptable to the conditions imposed upon it, with the addition and subtraction of sarcomeres occurring to restore optimal actin-myosin overlap regions and reduce resting passive tension. Muscles immobilized in a shortened position subtract sarcomeres in series while muscles immobilized in a lengthened position add sarcomeres in series, with the latter being adapted in orthopedics to correct limb length inequalities. Serial sarcomere number also adapts in a contraction-type dependent manner during exercise training, with training emphasizing concentric contractions tending to reduce serial sarcomere number and training emphasizing eccentric contractions tending to increase serial sarcomere number. Researchers have begun to apply interventions that promote serial sarcomerogenesis to clinical populations including older adults, children with cerebral palsy, patients recovering from hemiparetic stroke, people submitted to bed rest, and athletes at risk of hamstring strain injury. Further research is required to refine the use of serial sarcomere number as a marker of musculoskeletal health and a target for improvement of muscle function in these populations.

References

1. Jorgenson KW, Phillips SM, Hornberger TA. Identifying the structural adaptations that drive the mechanical load-induced growth of skeletal muscle: a scoping review. Cells. 2020;9(7):1658. https://doi.org/10.3390/cells9071658.
2. Koh TJ. Do adaptations in serial sarcomere number occur with strength training? Hum Mov Sci. 1995;14(1):61–77. https://doi.org/10.1016/0167-9457(94)00047-I.
3. Hinks A, Franchi MV, Power GA. The influence of longitudinal muscle fascicle growth on mechanical function. J Appl Physiol. 2022;133(1):87–103. https://doi.org/10.1152/japplphysiol.00114.2022.
4. Williams PE, Goldspink G. Changes in sarcomere length and physiological properties in immobilized muscle. J Anat. 1978;127(3):459–68.
5. Herring SW, Grimm AF, Grimm BR. Regulation of sarcomere number in skeletal muscle: a comparison of hypotheses. Muscle Nerve. 1984;7(2):161–73. https://doi.org/10.1002/mus.880070213.
6. Lieber RL, Ward SR. Skeletal muscle design to meet functional demands. Philos Trans R Soc B Biol Sci. 2011;366(1570):1466–76. https://doi.org/10.1098/rstb.2010.0316.
7. Lutz GJ, Rome LC. Built for jumping: the design of the frog muscular system. Sci New Ser. 1994;263(5145):370–2.
8. Rome LC. The mechanical design of the fish muscular system. In: Maddock L, Bone Q, Rayner JMV, editors. The mechanics and physiology of animal swimming. Cambridge University Press; 1994. p. 75–98. Accessed June 16, 2021. https://www.cambridge.org/core/product/identifier/CBO9780511983641A014/type/book_part.
9. Goldspink G, Tabary C, Tabary JC, Tardieu C, Tardieu G. Effect of denervation on the adaptation of sarcomere number and muscle extensibility to the functional length of the muscle. J Physiol. 1974;236(3):733–42. https://doi.org/10.1113/jphysiol.1974.sp010463.
10. Hinks A, Jacob K, Mashouri P, et al. Influence of weighted downhill running training on serial sarcomere number and work loop performance in the rat soleus. Biol Open. 2022;11(7):bio059491. https://doi.org/10.1242/bio.059491.
11. Franchi MV, Atherton PJ, Maganaris CN, Narici MV. Fascicle length does increase in response to longitudinal resistance training and in a contraction-mode specific manner. Springerplus. 2016;5(1):94. https://doi.org/10.1186/s40064-015-1548-8.
12. Herzog W, de Fontana HB. Does eccentric exercise stimulate sarcomerogenesis? J Sport Health Sci. 2021;11(1):40–2. https://doi.org/10.1016/j.jshs.2021.10.001.
13. Rindom E, Vissing K. Mechanosensitive molecular networks involved in transducing resistance exercise-signals into muscle protein accretion. Front Physiol. 2016;7:547. https://doi.org/10.3389/fphys.2016.00547.
14. Wackerhage H, Schoenfeld BJ, Hamilton DL, Lehti M, Hulmi JJ. Stimuli and sensors that initiate skeletal muscle hypertrophy following resistance exercise. J Appl Physiol. 2019;126(1):30–43. https://doi.org/10.1152/japplphysiol.00685.2018.
15. Gordon AM, Huxley AF, Julian FJ. The variation in isometric tension with sarcomere length in vertebrate muscle fibres. J Physiol. 1966;184(1):170–92. https://doi.org/10.1113/jphysiol.1966.sp007909.
16. Davis JF, Khir AW, Barber L, et al. The mechanisms of adaptation for muscle fascicle length changes with exercise: implications for spastic muscle. Med Hypotheses. 2020;144:110199. https://doi.org/10.1016/j.mehy.2020.110199.
17. Hill AV, Sec RS. The heat of shortening and the dynamic constants of muscle. Proc R Soc Lond B Biol Sci. 1938;126(843):136–95.
18. Alcazar J, Csapo R, Ara I, Alegre LM. On the shape of the force-velocity relationship in skeletal muscles: the linear, the hyperbolic, and the double-hyperbolic. Front Physiol. 2019;10:769. https://doi.org/10.3389/fphys.2019.00769.
19. Wickiewicz TL, Roy RR, Powell PL, Perrine JJ, Edgerton VR. Muscle architecture and force-velocity relationships in humans. J Appl Physiol. 1984;57:435–43.

20. Narici M, Franchi M, Maganaris C. Muscle structural assembly and functional consequences. J Exp Biol. 2016;219(2):276–84. https://doi.org/10.1242/jeb.128017.
21. Spector SA, Gardiner PF, Zernicke RF, Roy RR, Edgerton VR. Muscle architecture and force-velocity characteristics of cat soleus and medial gastrocnemius: implications for motor control. J Neurophysiol. 1980;44(5):951–60. https://doi.org/10.1152/jn.1980.44.5.951.
22. Bodine SC, Roy RR, Meadows DA, et al. Architectural, histochemical, and contractile characteristics of a unique biarticular muscle: the cat semitendinosus. J Neurophysiol. 1982;48(1):192–201. https://doi.org/10.1152/jn.1982.48.1.192.
23. Hinks A, Hess A, Debenham MIB, et al. Power loss is attenuated following a second bout of high intensity eccentric contractions due to the repeated bout effect's protection of rate of torque and velocity development. Appl Physiol Nutr Metab. Published online October 30, 2020. 2020; https://doi.org/10.1139/apnm-2020-0641.
24. Thom JM, Morse CI, Birch KM, Narici MV. Influence of muscle architecture on the torque and power–velocity characteristics of young and elderly men. Eur J Appl Physiol. 2007;100(5):613–9. https://doi.org/10.1007/s00421-007-0481-0.
25. Kosterina N, Westerblad H, Eriksson A. Mechanical work as predictor of force enhancement and force depression. J Biomech. 2009;42(11):1628–34. https://doi.org/10.1016/j.jbiomech.2009.04.028.
26. Hinks A, Davidson B, Akagi R, Power GA. Influence of isometric training at short and long muscle-tendon unit lengths on the history dependence of force. Scand J Med Sci Sports. 2021;31(2):325–38. https://doi.org/10.1111/sms.13842.
27. Davidson B, Hinks A, Dalton BH, Akagi R, Power GA. Power attenuation from restricting range of motion is minimized in subjects with fast RTD and following isometric training. J Appl Physiol. 2022;132:497–510. https://doi.org/10.1152/japplphysiol.00688.2021.
28. Cox VM, Williams PE, Wright H, et al. Growth induced by incremental static stretch in adult rabbit latissimus dorsi muscle. Exp Physiol. 2000;85(2):193–202.
29. Ahn AN. How muscles function- the work loop technique. J Exp Biol. 2012;215(7):1051–2. https://doi.org/10.1242/jeb.062752.
30. van der Zwaard S, Laarse WJ, Weide G, et al. Critical determinants of combined sprint and endurance performance: an integrative analysis from muscle fiber to the human body. FASEB J. 2018;32(4):2110–23. https://doi.org/10.1096/fj.201700827R.
31. Methenitis SK, Zaras ND, Spengos KM, et al. Role of muscle morphology in jumping, sprinting, and throwing performance in participants with different power training duration experience. J Strength Cond Res. 2016;30(3):807–17. https://doi.org/10.1519/JSC.0000000000001147.
32. Lieber RL, Fridén J. Functional and clinical significance of skeletal muscle architecture. Muscle Nerve. 2000;23(11):1647–66. https://doi.org/10.1002/1097-4598(200011)23:11<1647::AID-MUS1>3.0.CO;2-M.
33. Baxter JR, Hullfish TJ, Chao W. Functional deficits may be explained by plantarflexor remodeling following Achilles tendon rupture repair: preliminary findings. J Biomech. 2018;79:238–42. https://doi.org/10.1016/j.jbiomech.2018.08.016.
34. Drazan JF, Hullfish TJ, Baxter JR. Muscle structure governs joint function: linking natural variation in medial gastrocnemius structure with isokinetic plantar flexor function. Biol Open. 2019;8(12):bio.048520. https://doi.org/10.1242/bio.048520.
35. Lieber RL, Ljung BO, Fridén J. Intraoperative sarcomere length measurements reveal differential design of human wrist extensor muscles. J Exp Biol. 1997;200(Pt 1):19–25. https://doi.org/10.1242/jeb.200.1.19.
36. Lichtwark GA, Farris DJ, Chen X, Hodges PW, Delp SL. Microendoscopy reveals positive correlation in multiscale length changes and variable sarcomere lengths across different regions of human muscle. J Appl Physiol. 2018;125(6):1812–20. https://doi.org/10.1152/japplphysiol.00480.2018.
37. Pincheira PA, Boswell MA, Franchi MV, Delp SL, Lichtwark GA. Biceps femoris long head sarcomere and fascicle length adaptations after three weeks of eccentric exercise training. J Sport Health Sci. 2021;11(1):43–9. https://doi.org/10.1016/j.jshs.2021.09.002.

38. Hinks A, Franchi MV, Power GA. Ultrasonographic measurements of fascicle length overestimate adaptations in serial sarcomere number. Exp Physiol. Published online August 23, 2023. 2023; https://doi.org/10.1113/EP091334.
39. Reeves ND, Narici MV, Maganaris CN. In vivo human muscle structure and function: adaptations to resistance training in old age: muscle adaptations to training in old age. Exp Physiol. 2004;89(6):675–89. https://doi.org/10.1113/expphysiol.2004.027797.
40. de Brito FH, Won HS, Sawatsky A, Herzog W. The mechanics of agonistic muscles. J Biomech. 2018;79:15–20. https://doi.org/10.1016/j.jbiomech.2018.07.007.
41. Walker SM, Schrodt GR. I segment lengths and thin filament periods in skeletal muscle fibers of the rhesus monkey and the human. Anat Rec. 1974;178(1):63–81. https://doi.org/10.1002/ar.1091780107.
42. Coratella G, Longo S, Cè E, et al. Sex-related responses to eccentric-only resistance training in knee-extensors muscle strength and architecture. Res Q Exerc Sport. 2018;89(3):347–53. https://doi.org/10.1080/02701367.2018.1472734.
43. Duclay J, Martin A, Duclay A, Cometti G, Pousson M. Behavior of fascicles and the myotendinous junction of human medial gastrocnemius following eccentric strength training. Muscle Nerve. 2009;39(6):819–27. https://doi.org/10.1002/mus.21297.
44. Franchi MV, Atherton PJ, Reeves ND, et al. Architectural, functional and molecular responses to concentric and eccentric loading in human skeletal muscle. Acta Physiol. 2014;210(3):642–54. https://doi.org/10.1111/apha.12225.
45. Franchi MV, Wilkinson DJ, Quinlan JI, et al. Early structural remodeling and deuterium oxide-derived protein metabolic responses to eccentric and concentric loading in human skeletal muscle. Physiol Rep. 2015;3(11):e12593. https://doi.org/10.14814/phy2.12593.
46. Franchi MV, Ruoss S, Valdivieso P, et al. Regional regulation of focal adhesion kinase after concentric and eccentric loading is related to remodelling of human skeletal muscle. Acta Physiol. 2018;223(3):e13056. https://doi.org/10.1111/apha.13056.
47. Guex K, Degache F, Morisod C, Sailly M, Millet GP. Hamstring architectural and functional adaptations following long vs. short muscle length eccentric training. Front Physiol. 2016:7. https://doi.org/10.3389/fphys.2016.00340.
48. Marušič J, Vatovec R, Marković G, Šarabon N. Effects of eccentric training at long-muscle length on architectural and functional characteristics of the hamstrings. Scand J Med Sci Sports. 2020;30(11):2130–42. https://doi.org/10.1111/sms.13770.
49. Potier TG, Alexander CM, Seynnes OR. Effects of eccentric strength training on biceps femoris muscle architecture and knee joint range of movement. Eur J Appl Physiol. 2009;105(6):939–44. https://doi.org/10.1007/s00421-008-0980-7.
50. Quinlan JI, Franchi MV, Gharahdaghi N, et al. Muscle and tendon adaptations to moderate load eccentric vs. concentric resistance exercise in young and older males. GeroScience. 2021;43(4):1567–84. https://doi.org/10.1007/s11357-021-00396-0.
51. Stasinaki AN, Zaras N, Methenitis S, Bogdanis G, Terzis G. Rate of force development and muscle architecture after fast and slow velocity eccentric training. Sports. 2019;7(2):41. https://doi.org/10.3390/sports7020041.
52. Vianna KB, Rodrigues LG, Oliveira NT, Ribeiro-Alvares JB, Baroni BM. A preseason training program with the Nordic hamstring exercise increases eccentric knee flexor strength and fascicle length in professional female soccer players. Int J Sports Phys Ther. 2021;16(2):459–67. https://doi.org/10.26603/001c.19452.
53. Zacharia E, Spiliopoulou P, Methenitis S, et al. Changes in muscle power and muscle morphology with different volumes of fast eccentric half-squats. Sports. 2019;7(7):164. https://doi.org/10.3390/sports7070164.
54. de la Tour EH, Tabary JC, Tabary C, Tardieu C. The respective roles of muscle length and muscle tension in sarcomere number adaptation of Guinea-pig soleus muscle. J Physiol Paris. 1979;75(5):589–92.

55. Williams P, Watt P, Bicik V, Goldspink G. Effect of stretch combined with electrical stimulation on the type of sarcomeres produced at the ends of muscle fibers. Exp Neurol. 1986;93:500–9.
56. Williams PE, Catanese T, Lucey EG, Goldspink G. The importance of stretch and contractile activity in the prevention of connective tissue accumulation in muscle. J Anat. 1988;158:109–14.
57. Spector SA, Simard CP, Fournier M, Sternlicht E, Edgerton' VR. Architectural alterations of Rat Hind-Limb skeletal muscles immobilized at different lengths. Exp Neurol. 1982;76:94–110.
58. Tabary JC, Tardieu C, Tardieu G, Tabary C. Experimental rapid sarcomere loss with concomitant hypoextensibility. Muscle Nerve. 1981;4(3):198–203. https://doi.org/10.1002/mus.880040305.
59. Soares AG, Aoki MS, Miyabara EH, et al. Ubiquitin-ligase and deubiquitinating gene expression in stretched rat skeletal muscle. Muscle Nerve. 2007;36(5):685–93. https://doi.org/10.1002/mus.20866.
60. Aoki MS, Soares AG, Miyabara EH, Baptista IL, Moriscot AS. Expression of genes related to myostatin signaling during rat skeletal muscle longitudinal growth: Myostatin and longitudinal growth. Muscle Nerve. 2009;40(6):992–9. https://doi.org/10.1002/mus.21426.
61. Zhou CX, Wang F, Zhou Y, Fang QZ, Zhang QB. Formation process of extension knee joint contracture following external immobilization in rats. World J Orthop. 2023;14(9):669–81. https://doi.org/10.5312/wjo.v14.i9.669.
62. Aoki MS, Miyabara EH, Soares AG, Saito ET, Moriscot AS. mTOR pathway inhibition attenuates skeletal muscle growth induced by stretching. Cell Tissue Res. 2006;324(1):149–56. https://doi.org/10.1007/s00441-005-0081-4.
63. Ansved T. Effects of immobilization on the rat soleus muscle in relation to age. Acta Physiol Scand. 1995;154(3):291–302. https://doi.org/10.1111/j.1748-1716.1995.tb09913.x.
64. Witzmann FA, Kim DH, Fitts RH. Hindlimb immobilization: length-tension and contractile properties of skeletal muscle. J Appl Physiol. 1982;53(2):335–45. https://doi.org/10.1152/jappl.1982.53.2.335.
65. Heslinga JW, Huijing PA. Muscle length-force characteristics in relation to muscle architecture: a bilateral study of gastrocnemius medialis muscles of unilaterally immobilized rats. Eur J Appl Physiol. 1993;66(4):289–98. https://doi.org/10.1007/BF00237771.
66. Boakes JL, Foran J, Ward SR, Lieber RL. Muscle adaptation by serial sarcomere addition 1 year after femoral lengthening. Clin Orthop. 2007;456:250–3. https://doi.org/10.1097/01.blo.0000246563.58091.af.
67. Caiozzo VJ, Utkan A, Chou R, et al. Effects of distraction on muscle length: mechanisms involved in sarcomerogenesis. Clin Orthop. 2002;403(Suppl):S133–145. https://doi.org/10.1097/00003086-200210001-00016.
68. Franchi MV, Reeves ND, Narici MV. Skeletal muscle remodeling in response to eccentric vs. concentric loading: morphological, molecular, and metabolic adaptations. Front Physiol. 2017;8:447. https://doi.org/10.3389/fphys.2017.00447.
69. Lynn R, Morgan DL. Decline running produces more sarcomeres in rat vastus intermedius muscle fibers than does incline running. J Appl Physiol. 1994;77(3):1439–44. https://doi.org/10.1152/jappl.1994.77.3.1439.
70. Yu JG, Thornell LE. Desmin and actin alterations in human muscles affected by delayed onset muscle soreness: a high resolution immunocytochemical study. Histochem Cell Biol. 2002;118(2):171–9. https://doi.org/10.1007/s00418-002-0427-x.
71. Yu JG, Fürst DO, Thornell LE. The mode of myofibril remodelling in human skeletal muscle affected by DOMS induced by eccentric contractions. Histochem Cell Biol. 2003;119(5):383–93. https://doi.org/10.1007/s00418-003-0522-7.
72. Yu JG, Carlsson L, Thornell LE. Evidence for myofibril remodeling as opposed to myofibril damage in human muscles with DOMS: an ultrastructural and immunoelectron

microscopic study. Histochem Cell Biol. 2004;121(3):219–27. https://doi.org/10.1007/s00418-004-0625-9.
73. Faulkner JA, Brooks SV, Zerba E. Muscle atrophy and weakness with aging: contraction-induced injury as an underlying mechanism. J Gerontol A Biol Sci Med Sci. 1995;50 Spec No:124–9. https://doi.org/10.1093/gerona/50a.special_issue.124.
74. Butterfield TA, Leonard TR, Herzog W. Differential serial sarcomere number adaptations in knee extensor muscles of rats is contraction type dependent. J Appl Physiol. 2005;99:7.
75. Chen J, Mashouri P, Fontyn S, et al. The influence of training-induced sarcomerogenesis on the history dependence of force. J Exp Biol. 2020;223(Pt 15):jeb218776. https://doi.org/10.1242/jeb.218776.
76. Morais GP, da Rocha AL, Neave LM, et al. Chronic uphill and downhill exercise protocols do not lead to sarcomerogenesis in mouse skeletal muscle. J Biomech. 2020;98:109469. https://doi.org/10.1016/j.jbiomech.2019.109469.
77. Butterfield TA, Herzog W. The magnitude of muscle strain does not influence serial sarcomere number adaptations following eccentric exercise. Pflüg Arch - Eur J Physiol. 2006;451(5):688–700. https://doi.org/10.1007/s00424-005-1503-6.
78. Lynn R, Talbot JA, Morgan DL. Differences in rat skeletal muscles after incline and decline running. J Appl Physiol. 1998;85(1):98–104. https://doi.org/10.1152/jappl.1998.85.1.98.
79. Hyldahl RD, Chen TC, Nosaka K. Mechanisms and mediators of the skeletal muscle repeated Bout Effect. Exerc Sport Sci Rev. 2017;45(1):24–33. https://doi.org/10.1249/JES.0000000000000095.
80. McBride TA, Gorin FA, Carlsen RC. Prolonged recovery and reduced adaptation in aged rat muscle following eccentric exercise. Mech Ageing Dev. 1995;83(3):185–200. https://doi.org/10.1016/0047-6374(95)01629-e.
81. Koh TJ, Herzog W. Eccentric training does not increase sarcomere number in rabbit dorsiflexor muscles. J Biomech. 1998;31(5):499–501. https://doi.org/10.1016/s0021-9290(98)00032-3.
82. Koh TJ, Herzog W. Excursion is important in regulating sarcomere number in the growing rabbit tibialis anterior. J Physiol. 1998;508(Pt 1):267–80. https://doi.org/10.1111/j.1469-7793.1998.267br.x.
83. Gillis GB, Biewener AA. Effects of surface grade on proximal hindlimb muscle strain and activation during rat locomotion. J Appl Physiol. 2002;93(5):1731–43. https://doi.org/10.1152/japplphysiol.00489.2002.
84. Hodson-Tole EF, Wakeling JM. The influence of strain and activation on the locomotor function of rat ankle extensor muscles. J Exp Biol. 2010;213(2):318–30. https://doi.org/10.1242/jeb.031872.
85. Bernabei M, van Dieën JH, Maas H. Longitudinal and transversal displacements between triceps surae muscles during locomotion of the rat. J Exp Biol. 2017;220(4):537–50. https://doi.org/10.1242/jeb.143545.
86. Hinks A, Patterson MA, Njai BS, Power GA. Age-related blunting of serial sarcomerogenesis and mechanical adaptations following 4 weeks of maximal eccentric resistance training. bioRxiv. Published online November 10, 2023. https://doi.org/10.1101/2023.11.07.566004.
87. Williams PE. Use of intermittent stretch in the prevention of serial sarcomere loss in immobilised muscle. Ann Rheum Dis. 1990;49(5):316–7. https://doi.org/10.1136/ard.49.5.316.
88. De Jaeger D, Joumaa V, Herzog W. Intermittent stretch training of rabbit plantarflexor muscles increases soleus mass and serial sarcomere number. J Appl Physiol. 2015;118(12):1467–73. https://doi.org/10.1152/japplphysiol.00515.2014.
89. Peviani SM, Guzzoni V, Pinheiro-Dardis CM, et al. Regulation of extracellular matrix elements and sarcomerogenesis in response to different periods of passive stretching in the soleus muscle of rats. Sci Rep. 2018;8:9010. https://doi.org/10.1038/s41598-018-27239-x.
90. Macaluso A, De Vito G. Muscle strength, power and adaptations to resistance training in older people. Eur J Appl Physiol. 2004;91(4):450–72. https://doi.org/10.1007/s00421-003-0991-3.
91. Narici MV, Maganaris CN. Adaptability of elderly human muscles and tendons to increased loading. J Anat. 2006;208(4):433–43. https://doi.org/10.1111/j.1469-7580.2006.00548.x.

92. Power GA, Dalton BH, Rice CL. Human neuromuscular structure and function in old age: a brief review. J Sport Health Sci. 2013;2(4):215–26. https://doi.org/10.1016/j.jshs.2013.07.001.
93. Wu R, Ditroilo M, Delahunt E, De Vito G. Age related changes in motor function (II). Decline in motor performance outcomes. Int J Sports Med. 2021;42(03):215–26. https://doi.org/10.1055/a-1265-7073.
94. Lexell J, Henriksson-Larsén K, Winblad B, Sjöström M. Distribution of different fiber types in human skeletal muscles: effects of aging studied in whole muscle cross sections. Muscle Nerve. 1983;6(8):588–95. https://doi.org/10.1002/mus.880060809.
95. Vandervoort AA. Aging of the human neuromuscular system. Muscle Nerve. 2002;25(1):17–25. https://doi.org/10.1002/mus.1215.
96. Frontera WR, Hughes VA, Fielding RA, Fiatarone MA, Evans WJ, Roubenoff R. Aging of skeletal muscle: a 12-yr longitudinal study. J Appl Physiol. 2000;88(4):1321–6. https://doi.org/10.1152/jappl.2000.88.4.1321.
97. Delmonico MJ, Harris TB, Visser M, et al. Longitudinal study of muscle strength, quality, and adipose tissue infiltration. Am J Clin Nutr. 2009;90(6):1579–85. https://doi.org/10.3945/ajcn.2009.28047.
98. Narici MV, Maffulli N. Sarcopenia: characteristics, mechanisms and functional significance. Br Med Bull. 2010;95(1):139–59. https://doi.org/10.1093/bmb/ldq008.
99. Narici MV, Maganaris CN, Reeves ND, Capodaglio P. Effect of aging on human muscle architecture. J Appl Physiol. 2003;95(6):2229–34. https://doi.org/10.1152/japplphysiol.00433.2003.
100. Baroni BM, Geremia JM, Rodrigues R, et al. Functional and morphological adaptations to aging in knee extensor muscles of physically active men. J Appl Biomech. 2013;29(5):535–42. https://doi.org/10.1123/jab.29.5.535.
101. Kubo K, Kanehisa H, Azuma K, et al. Muscle architectural characteristics in young and elderly men and women. Int J Sports Med. 2003;24(2):125–30. https://doi.org/10.1055/s-2003-38204.
102. Power GA, Makrakos DP, Rice CL, Vandervoort AA. Enhanced force production in old age is not a far stretch: an investigation of residual force enhancement and muscle architecture. Physiol Rep. 2013;1(1):e00004. https://doi.org/10.1002/phy2.4.
103. Wu R, Delahunt E, Ditroilo M, Lowery M, De Vito G. Effects of age and sex on neuromuscular-mechanical determinants of muscle strength. Age Dordr Neth. 2016;38(3):57. https://doi.org/10.1007/s11357-016-9921-2.
104. Baudry S, Lecoeuvre G, Duchateau J. Age-related changes in the behavior of the muscle-tendon unit of the gastrocnemius medialis during upright stance. J Appl Physiol Bethesda Md 1985. 2012;112(2):296–304. https://doi.org/10.1152/japplphysiol.00913.2011.
105. Conway KA, Franz JR. Shorter gastrocnemius fascicle lengths in older adults associate with worse capacity to enhance push-off intensity in walking. Gait Posture. 2020;77:89–94. https://doi.org/10.1016/j.gaitpost.2020.01.018.
106. Kelp NY, Gore A, Clemente CJ, Tucker K, Hug F, Dick TJM. Muscle architecture and shape changes in the gastrocnemii of active younger and older adults. J Biomech. 2021;129:110823. https://doi.org/10.1016/j.jbiomech.2021.110823.
107. Morse CI, Thom JM, Birch KM, Narici MV. Changes in triceps surae muscle architecture with sarcopenia. Acta Physiol Scand. 2005;183(3):291–8. https://doi.org/10.1111/j.1365-201X.2004.01404.x.
108. Stenroth L, Peltonen J, Cronin NJ, Sipilä S, Finni T. Age-related differences in Achilles tendon properties and triceps surae muscle architecture in vivo. J Appl Physiol Bethesda Md 1985. 2012;113(10):1537–44. https://doi.org/10.1152/japplphysiol.00782.2012.
109. Panizzolo FA, Green DJ, Lloyd DG, Maiorana AJ, Rubenson J. Soleus fascicle length changes are conserved between young and old adults at their preferred walking speed. Gait Posture. 2013;38(4):764–9. https://doi.org/10.1016/j.gaitpost.2013.03.021.
110. Hooper AC. Length, diameter and number of ageing skeletal muscle fibres. Gerontology. 1981;27(3):121–6. https://doi.org/10.1159/000212459.

111. Power GA, Crooks S, Fletcher JR, Macintosh BR, Herzog W. Age-related reductions in the number of serial sarcomeres contribute to shorter fascicle lengths but not elevated passive tension. J Exp Biol. 2021;224(10):jeb242172. https://doi.org/10.1101/2020.12.21.423814.
112. Reeves ND, Maganaris CN, Longo S, Narici MV. Differential adaptations to eccentric *versus* conventional resistance training in older humans: eccentric resistance training in old age. Exp Physiol. 2009;94(7):825–33. https://doi.org/10.1113/expphysiol.2009.046599.
113. Raj IS, Bird SR, Westfold BA, Shield AJ. Effects of eccentrically biased versus conventional weight training in older adults. Med Sci Sports Exerc. 2012;44(6):1167–76. https://doi.org/10.1249/MSS.0b013e3182442ecd.
114. Brooks SV, Faulkner JA. Contraction-induced injury: recovery of skeletal muscles in young and old mice. Am J Phys. 1990;258(3 Pt 1):C436–42. https://doi.org/10.1152/ajpcell.1990.258.3.C436.
115. Zerba E, Komorowski TE, Faulkner JA. Free radical injury to skeletal muscles of young, adult, and old mice. Am J Physiol-Cell Physiol. 1990;258(3):C429–35. https://doi.org/10.1152/ajpcell.1990.258.3.C429.
116. Rader EP, Faulkner JA. Effect of aging on the recovery following contraction-induced injury in muscles of female mice. J Appl Physiol Bethesda Md 1985. 2006;101(3):887–92. https://doi.org/10.1152/japplphysiol.00380.2006.
117. Rader EP, Faulkner JA. Recovery from contraction-induced injury is impaired in weight-bearing muscles of old male mice. J Appl Physiol Bethesda Md 1985. 2006;100(2):656–61. https://doi.org/10.1152/japplphysiol.00663.2005.
118. Dedrick ME, Clarkson PM. The effects of eccentric exercise on motor performance in young and older women. Eur J Appl Physiol. 1990;60(3):183–6. https://doi.org/10.1007/BF00839156.
119. Lavender AP, Nosaka K. Responses of old men to repeated bouts of eccentric exercise of the elbow flexors in comparison with young men. Eur J Appl Physiol. 2006;97(5):619–26. https://doi.org/10.1007/s00421-006-0224-7.
120. Chapman DW, Newton M, McGuigan MR, Nosaka K. Comparison between old and young men for responses to fast velocity maximal lengthening contractions of the elbow flexors. Eur J Appl Physiol. 2008;104(3):531–9. https://doi.org/10.1007/s00421-008-0806-7.
121. Proske U, Morgan DL. Muscle damage from eccentric exercise: mechanism, mechanical signs, adaptation and clinical applications. J Physiol. 2001;537(Pt 2):333–45. https://doi.org/10.1111/j.1469-7793.2001.00333.x.
122. Clarkson PM, Hubal MJ. Exercise-induced muscle damage in humans. Am J Phys Med Rehabil. 2002;81(Supplement):S52–69. https://doi.org/10.1097/00002060-200211001-00007.
123. Butterfield TA, Herzog W. Effect of altering starting length and activation timing of muscle on fiber strain and muscle damage. J Appl Physiol. 2006;100(5):1489–98. https://doi.org/10.1152/japplphysiol.00524.2005.
124. Hinks A, Hawke TJ, Franchi MV, Power GA. The importance of serial sarcomere addition for muscle function and the impact of aging. J Appl Physiol. Published online July 6, 2023, 2023; https://doi.org/10.1152/japplphysiol.00205.2023.
125. Baumann CW, Deane CS, Etheridge T, Szewczyk NJ, Willis CRG, Lowe DA. Adaptability to eccentric exercise training is diminished with age in female mice. J Appl Physiol. Published online October 12, 2023. 2023; https://doi.org/10.1152/japplphysiol.00428.2023.
126. Matthiasdottir S, Hahn M, Yaraskavitch M, Herzog W. Muscle and fascicle excursion in children with cerebral palsy. Clin Biomech. 2014;29(4):458–62. https://doi.org/10.1016/j.clinbiomech.2014.01.002.
127. Lieber RL, Fridén J. Spasticity causes a fundamental rearrangement of muscle–joint interaction. Muscle Nerve. 2002;25(2):265–70. https://doi.org/10.1002/mus.10036.
128. Pontén E, Gantelius S, Lieber RL. Intraoperative muscle measurements reveal a relationship between contracture formation and muscle remodeling. Muscle Nerve. 2007;36(1):47–54. https://doi.org/10.1002/mus.20780.

129. Mathewson MA, Ward SR, Chambers HG, Lieber RL. High resolution muscle measurements provide insights into equinus contractures in patients with cerebral palsy. J Orthop Res. 2015;33(1):33–9. https://doi.org/10.1002/jor.22728.
130. Lieber RL, Steinman S, Barash IA, Chambers H. Structural and functional changes in spastic skeletal muscle. Muscle Nerve. 2004;29(5):615–27. https://doi.org/10.1002/mus.20059.
131. Kalkman BM, Bar-On L, O'Brien TD, Maganaris CN. Stretching interventions in children with cerebral palsy: why are they ineffective in improving muscle function and how can we better their outcome? Front Physiol. 2020;11:131. https://doi.org/10.3389/fphys.2020.00131.
132. Kalkman BM, Holmes G, Bar-On L, et al. Resistance training combined with stretching increases tendon stiffness and is more effective than stretching alone in children with cerebral palsy: a randomized controlled trial. Front Pediatr. 2019;7:333. https://doi.org/10.3389/fped.2019.00333.
133. Reid S, Hamer P, Alderson J, Lloyd D. Neuromuscular adaptations to eccentric strength training in children and adolescents with cerebral palsy. Dev Med Child Neurol. 2010;52(4):358–63. https://doi.org/10.1111/j.1469-8749.2009.03409.x.
134. Adkins AN, Dewald JPA, Garmirian LP, Nelson CM, Murray WM. Serial sarcomere number is substantially decreased within the paretic biceps brachii in individuals with chronic hemiparetic stroke. Proc Natl Acad Sci. 2021;118(26):e2008597118. https://doi.org/10.1073/pnas.2008597118.
135. de Boer MD, Seynnes OR, di Prampero PE, et al. Effect of 5 weeks horizontal bed rest on human muscle thickness and architecture of weight bearing and non-weight bearing muscles. Eur J Appl Physiol. 2008;104(2):401–7. https://doi.org/10.1007/s00421-008-0703-0.
136. Franchi MV, Sarto F, Simunič B, Pišot R, Narici MV. Early changes of hamstrings morphology and contractile properties during 10 d of complete inactivity. Med Sci Sports Exerc. 2022;54(8):1346. https://doi.org/10.1249/MSS.0000000000002922.
137. Brughelli M, Cronin J. Altering the length-tension relationship with eccentric exercise: implications for performance and injury. Sports Med. 2007;37(9):807–26. https://doi.org/10.2165/00007256-200737090-00004.
138. Timmins RG, Shield AJ, Williams MD, Lorenzen C, Opar DA. Architectural adaptations of muscle to training and injury: a narrative review outlining the contributions by fascicle length, pennation angle and muscle thickness. Br J Sports Med. 2016;50(23):1467–72. https://doi.org/10.1136/bjsports-2015-094881.

Part II
Applied Sciences

Chapter 11
Training Adaptation and Maladaptation in Athletes

Jonathon Weakley, Shona L. Halson, and Iñigo Mujika

Physiological Bases of Training-Induced Adaptation

Background

Fatigue and adaptations from training are underpinned by the acute training stimulus that is applied to the athlete. Furthermore, possibly one of the most important principles of training is that of overload. Understanding and suitably applying the principle of overload is essential for appropriate training prescription and recovery, as it can support athletes by facilitating adaptation to occur following acute and

J. Weakley
School of Behavioural and Health Sciences, Australian Catholic University, Brisbane, QLD, Australia

Sports Performance, Recovery, Injury and New Technologies (SPRINT) Research Centre, Australian Catholic University, Brisbane, QLD, Australia

Carnegie Applied Rugby Research (CARR) Centre, Carnegie School of Sport, Leeds Beckett University, Leeds, UK

S. L. Halson
School of Behavioural and Health Sciences, Australian Catholic University, Brisbane, QLD, Australia

Sports Performance, Recovery, Injury and New Technologies (SPRINT) Research Centre, Australian Catholic University, Brisbane, QLD, Australia

I. Mujika (✉)
Department of Physiology, Faculty of Medicine and Nursing, University of the Basque Country, Leioa, Basque Country, Spain

Exercise Science Laboratory, School of Kinesiology, Faculty of Medicine, Universidad Finis Terrae, Santiago, Chile
e-mail: inigo.mujika@inigomujika.com

N. C. Apostolopoulos et al. (eds.), *Fundamentals of Recovery, Regeneration, and Adaptation to Exercise Stress: An Integrated Approach*,
https://doi.org/10.1007/978-3-031-44270-4_11

chronic periods of training, and in achieving optimal performance during periods of competition. While it is not crucial for coaches and athletes to have an encyclopaedic knowledge of the biological underpinnings of overload and adaptation, understanding the broad concepts can help guide the planning and implementation of training programmes.

Training prescription and its subsequent outcomes can be explained and justified by the physiological stress that is imposed on the athlete. During exercise, homeostasis is disturbed, and homeostatic adjustments occur to maintain the body's internal milieu. For example, increased body temperature, rate of respiration, and heart rate are all adjustments that occur so that exercise can continue. However, following exercise, each physiological response has a different time course of recovery. For example, body temperature and respiration rate may take minutes to return to pre-exercise levels, but neuromuscular recovery make take hours or days to return to baseline. Furthermore, if substantial morphological damage is induced (e.g., to muscle fibres), it may take weeks or even months to fully recover and regenerate completely. These responses are all dependent upon acute programming variables, with the exercise modality, intensity, and duration influencing the physiological stress that is imposed and their subsequent recovery [1]. Therefore, the careful programming of these variables is of importance.

Processes

Type of Exercise

Long term training adaptations occur as a result of cumulative exposure to acute exercise bouts. When individuals are repeatedly exposed to physiological stress, they will respond to that stress through adapting. The specific adaptations and the magnitude of change are associated with the type of exercise stimulus and the individual's genetic makeup [2, 3]. The adaptations enable greater tolerance to exercise and help to reduce the physiological response to the next exposure. This means that a later bout of exercise disturbs the homeostatic milieu less than the previous bout and it is why progressive overload is required within training programmes. It should also be noted that the response to training is exercise specific, with different forms of exercise (e.g., resistance training *vs* endurance training) inducing different responses and subsequent adaptations.

The mechanisms of adaptation from different forms of exercise are physiologically distinct, with each form of training either activating or repressing specific molecular signalling pathways [3, 4]. Helping to induce these changes is the mechanical loading upon the skeletal muscle, with this loading influencing the calcium flux and metabolic response [4]. All modalities of exercise apply some type of mechanical loading to skeletal muscle, with different forms of exercise and contractions inducing different responses [5]. Furthermore, several protein complexes have been identified as important mechanosensors that act as molecular transducers during myofiber contraction [3], with chronic changes in these pathways leading to alterations in protein synthesis and muscle protein breakdown [6, 7].

Metabolic stress during exercise is associated with the rate of adenosine triphosphate (ATP) utilisation [4]. Furthermore, the rate of utilisation is associated with the substrate availability within the working muscle cell. Consequently, the training stimulus for a muscle cell that has low levels of glycogen is different from that which has adequate levels of glycogen [4]. Repeated exposure to training that causes high rates of ATP turnover can induce greater mitochondrial content and oxidative enzyme activity [8], which in turn can lead to adaptations which allow reduced glycogen use and increased fat oxidation during submaximal exercise [4, 8]. This allows enhanced exercise capacity through improved fatigue resistance and economy.

Calcium flux refers to the stimulus associated with calcium concentration in the sarcoplasm [9]. Calcium plays a crucial role in the translation of forces to biochemical signals that control various biological processes fundamental in muscle development [10]. During muscle contraction, there are transient increases in calcium within the sarcoplasm and this calcium binds to troponin C which allows the contraction to occur. If there are sustained, high rates of muscle contractions, the sarcoplasmic calcium concentration is maintained at a high level, and this is associated with increased mitochondrial biogenesis and an increase in the glucose transporter type 4 (GLUT4) [11].

As a result of different forms of exercise, different adaptive processes occur. These processes are often an intricate interplay between external stimulus and internal outcomes. Resistance and endurance-based exercise are both commonly used as potent tools for developing a diverse range of favourable adaptations. However, underpinning the external stimulus that help drive internal responses are acute programming variables, such as modality, duration, and intensity, with these central to the training prescription of athletes [1]. Consequently, having a broad appreciation of the physiological basis of different exercise methods can lead to greater understanding of subsequent fatigue and adaptive outcomes.

> To induce different training adaptations, the external stimulus (e.g., forces applied to the individual) help drive the internal response. To augment or attenuate the internal response, the modality, duration, and intensity of the prescribed exercise can be modified. These changes in internal response may also lead to different levels of fatigue and amount of recovery time required by the athlete.

Adaptive Outcomes

Type of Exercise

Generally, exercise can be separated into endurance and resistance training types. From these forms of training, there are distinct morphological, metabolic, and neuromuscular adaptations, with each of these types of adaptation tending to support the physical performance of the type of training that was completed. For the endurance athlete, the objective is often to increase the ability to produce and sustain the highest average power output or speed of movement for a predetermined time or distance [8]. Alternatively, for the strength athlete, the goal is often to be able to produce the maximum amount of force and power during brief (e.g., 30 seconds) periods of exercise [12, 13]. Consequently, considering the different objectives of the exercise, the adaptations that occur from that form of training often cause vastly different physical outcomes.

Endurance capacity can be improved through a range of training methods, and it is well established that the tolerance to prolonged exercise can be markedly increased with consistent aerobic training. For many years it was believed that improvements in endurance performance were strictly associated with improvements in delivery of oxygen to the working muscles and a concomitant increase in maximal oxygen consumption (VO_{2max}). However, regular endurance exercise has been shown to induce substantial cardiovascular, endocrine, neuromuscular, and skeletal muscle adaptations, with these changes undoubtedly playing a key role in the ability to continuously perform exercise [14]. For example, a host of skeletal muscle changes can occur that can lead to improved endurance capacity (refer to Fig. 11.1) [8]. It should be noted that endurance performance is strongly influenced by cardiorespiratory changes though, with improvements in VO_{2max} often being considered the factor that explains the large difference in performance between elite level athletes and active or moderately trained individuals [15].

Systematic strength training, on the other hand, can lead to substantial improvements in strength, power, and hypertrophy of the muscle groups trained [13, 16]. These improvements are irrespective of sex and age when the intensity (as a percentage of one repetition maximum (1RM)), volume (total tonnage), and proximity to concentric failure during the training programme are appropriately applied [17–19]. Depending on the initial training base of the individual, strength improvements can be >30% in the first weeks of beginning training, but following these rapid improvements strength development usually occurs at a diminished rate [16, 20]. Underlying these favourable changes are not only peripheral changes (e.g., increased cross sectional area and changes in fibre type), but also beneficial neural adaptations [20]. These adaptations include improvements in agonist motor unit recruitment, firing rate, synchronisation of motor unit discharge [20]. Furthermore, there are reductions in Golgi tendon reflex and antagonist muscle activity [20]. Collectively these changes improve neural drive and allow the working muscle to express greater force in shorter periods of time (i.e., rate of force development).

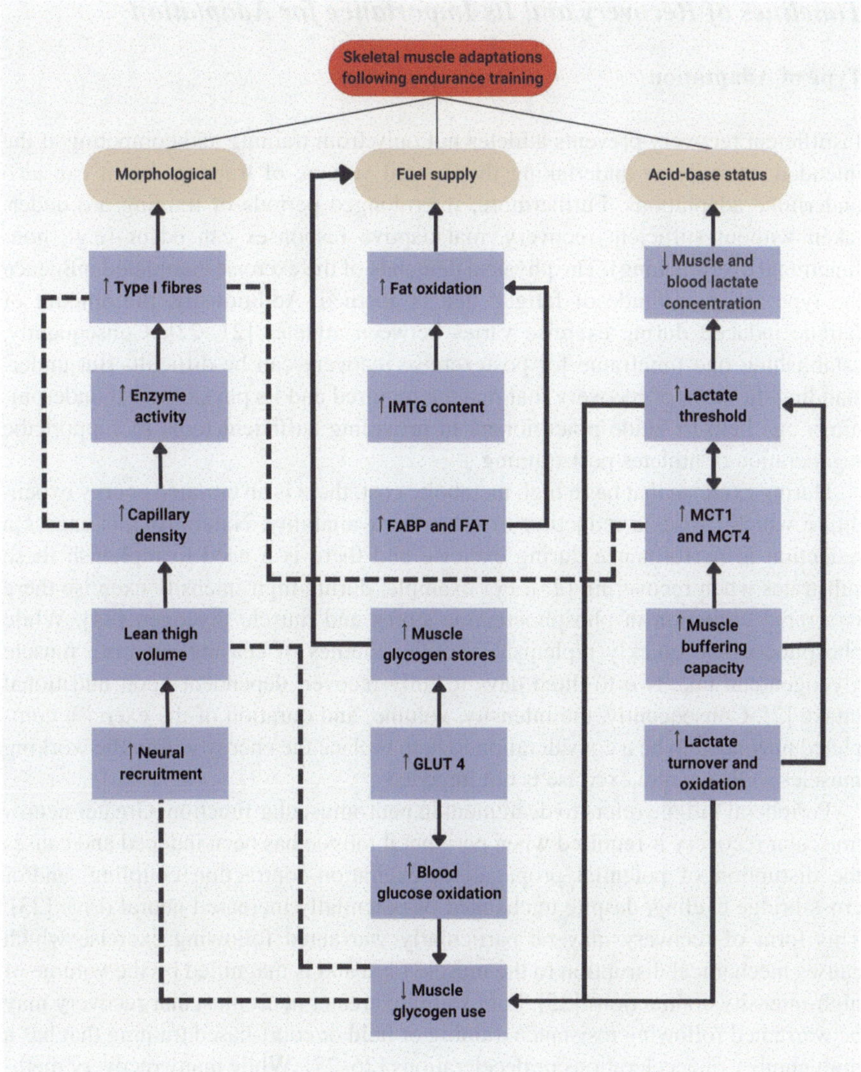

Fig. 11.1 Outline of some of the adaptations that occur in skeletal muscle in response to high-volume endurance training. FABP fatty acid binding protein, FAT fatty acid translocase, GLUT4 glucose transporter protein, IMTG intramuscular triglyceride, MCT monocarboxylate transporter. (Adapted from Hawley and Stepto [8])

Timelines of Recovery and Its Importance for Adaptation

Type of Adaptation

Insufficient recovery prevents athletes not only from training and competing at the intended intensity or undertaking the desired volume of training, but it can also undermine adaptations. Furthermore, if prolonged periods of training are undertaken without sufficient recovery, maladaptive responses can occur (e.g., non-functional overreaching). The physical demands of the exercise completed influence the type and magnitude of fatigue that is accrued. Additionally, the amount of fatigue induced during exercise varies between athletes [21, 22]. Consequently, establishing one timeframe for postexercise recovery can be difficult. But understanding the form of recovery that may be required and its physiological underpinnings can help to guide practitioners in providing sufficient tools to support the regeneration of athletes post-training.

During exercise that has a high metabolic cost, there is an elevated energy expenditure which causes a reduction in substrate availability. Naturally, this causes a reduction in performance during exercise and there is a need to replenish these substrates when recovering [23]. For example, during high-intensity exercise there is a rapid reduction in phosphocreatine stores and muscle glycogen [24]. While phosphocreatine quickly replenishes within minutes of ceasing exercise, muscle glycogen can take two to three days to fully recover, dependent upon nutritional intake [2]. Consequently, the intensity, volume, and duration of the exercise completed may need to be a consideration to help replace the energy within the working muscles so that future exercise is not impeded.

Peripheral fatigue refers to decrement in neuromuscular function. Greater neuromuscular recovery is required when peripheral fatigue has been induced and causes the disruption of potential propagation, excitation-contraction coupling, and/or cross-bridge cycling, despite unchanged or potentially increased neural drive [23]. This form of recovery may be particularly warranted following exercise which causes mechanical disruption to the muscle fibre and is magnified by the volume of high-intensity contractions [25]. For example, greater neuromuscular recovery may be warranted following resistance training or field or court-based training that has a high number of accelerations or decelerations [26, 27]. While many recovery methods focus on returning function to homeostasis and mitigating inflammatory responses within the muscle, there is often a disassociated time course of recovery between markers of inflammation and the return of muscle performance. For example, the ability to maximally produce force has been shown to return to baseline anywhere between five and 96 hours following exercise despite indices of muscle damage (e.g., C-reactive protein, interleukin-6) remaining elevated for at least 96 hours [28]. Thus, when deciding upon the recovery time to allow the full recuperation of force generating capacity, it may be prudent to carefully consider the mechanical disruption of the previous session and whether the kinetics of the markers of neuromuscular damage that are being used to guide assessment truly reflect the ability to perform.

During and following exercise, reductions at all levels of the nervous system (including the brain, spinal cord, motor output, sensory input, and autonomic function) occur [23]. These changes are often termed 'central fatigue' and despite the exact mechanisms not being well understood, it has been proposed that these changes occur in response to a change in the metabolism and synthesis of brain monoamines [29]. These neurotransmitters play an important role in the modulation of functions such as motivation, attention, and motor control [30]. When there is insufficient output from the motor cortex, this deficit is termed supraspinal or cortical fatigue and it is known that this deficit can reduce skeletal muscle recruitment and consequently a reduction in function occurs [31]. It has been shown through nerve stimulation that following soccer match play, there is reduced voluntary activation which suggest that not only is neuromuscular recovery required, but also recovery of the central nervous system [32]. However, due to the complexity of monitoring and quantifying central fatigue, further research is still required to fully understand the role of the central nervous system during recovery and how much time is required to optimise training outcomes.

It should be remembered that, irrespective of the physiological underpinning of fatigue, the ability to perform is of primary importance to athletes and coaches. Indeed, the decrement in sport specific performance should be the 'gold standard measure' for the assessment of short-term fatigue in athletes. While the mechanisms involved in performance fatigue can be multifaceted, the recovery timeline of performance may not match that of the physiological underpinning. For example, Skorski et al. [33], demonstrated that despite a six-day intensive training camp being undertaken that caused significant increases in several measures of fatigue, performance could be maintained in some athletes. Therefore, while changes in physiological and psychological measures of fatigue may be detected, these shifts may not impede actual performance. Thus, when planning training, practitioners should carefully consider whether additional recovery is truly necessary, and that multivariate assessment of fatigue may be more useful than monitoring changes in a single measure.

Training Responses and Maladaptation

> There are many ways to quantify fatigue (e.g., neuromuscular, performance) and it is important to understand how a prior training session may impact upon recovery. However, perhaps of greatest consequence for athletes is to consider whether overall performance is influenced by the fatigue that has been accrued.

The development of physical qualities in human athletes is a primary objective of training. To induce these adaptations, training needs to be prescribed that provides an adequate stimulus that allows fatigue and, subsequently, recovery. However, there is a fine line between prescribing exercise that adequately supports training adaptations and prescribing exercise that causes unnecessary fatigue. Within the scientific literature there has been considerable documentation and discussion of the need to balance fitness and fatigue, with 'functional overreaching', 'non-functional overreaching', and 'overtraining' often being used to describe changes in fatigue and subsequent training adaptations [34–37]. However, these terms are frequently misused, misunderstood, and, in the case of overtraining syndrome (OTS), rarely (if ever) objectively observed. This section details the differences in these terms and specifically focuses in the OTS, why it is so difficult to establish, and recommendations to guide its proper diagnosis.

Functional and Non-functional Overreaching

Functional overreaching can be defined as "an accumulation of training and/or non-training stress resulting in short-term decrement in performance capacity with or without related physical and psychological signs and symptoms of maladaptation in which restoration of performance capacity may take from several days to several weeks" [37]. Furthermore, non-functional overreaching has been described as a state of "extreme overreaching that can result from the continuation of extended periods of overload training which leads to a stagnation or decrease in performance which will not resume for several weeks or months" [34, 38]. Consequently, the differences in these forms of overreaching are based on the duration of time required for performance restoration, rather than the magnitude of the exercise impairment. However, it is clear that these definitions do not clearly differentiate these two conditions, given that the time required for performance restoration may take "weeks" in either state. Consequently, throughout the literature [34–36], studies with varying degrees of performance suppression have reported different forms of overreaching.

It is well established that athletes require a certain amount of training overload to provide a sufficient stimulus and to induce physiological adaptations and enhance physical performance. However, whether there is a need to induce a state of functional or non-functional overreaching to maximise training outcomes has been debated [38]. Previous research has demonstrated that increases in training load above normal volumes can cause a state of functional overreaching followed by supercompensation [39]. However, further research by Aubry et al. [40], has demonstrated that improvements in performance and VO_{2max} occur only in athletes who did not demonstrate signs of overreaching following an intensified training period. Furthermore, given that functional overreaching has been associated with a range of negative consequences [41–43], such as increased muscle soreness, disturbed sleep, reduced immunological function, and impaired training intensity, it is likely that practitioners do not need to purposely induce a state of overreaching and it may be suboptimal for long term physical adaptation and performance.

Within the scientific literature, there is a lack of differentiation between functional and non-functional overreaching. Consequently, due to the ambiguity of the definitions, it is likely that researchers and practitioners have frequently misdiagnosed their athletes as (non-)functionally overreached. Additionally, practitioners should carefully consider whether inducing a state of overreaching is required to optimise training adaptations.

Overtraining Syndrome

The overtraining syndrome is a "sports-specific decrease in performance together with disturbances in mood state. Underperformance persists despite a period of recovery lasting weeks or months" [34]. OTS was first reported in the scientific literature in the 1930s and has been proposed to have an extremely wide range of ailments and affect up to 60% of athletes throughout their careers [44–47]. Compared to functional and non-functional overreaching, OTS is the most severe condition and reflects a chronic accumulation of training. While often considered a more extreme version of overreaching, consensus definitions imply that the main difference between OTS and overreaching is the amount of time needed for performance restoration [34, 35, 37]. However, due to the vague terminology and definitions that are often used in conjunction with OTS, there is difficulty in identifying whether athletes are nonfunctionally overreached or experiencing OTS.

While OTS is undoubtedly a severe and complex syndrome, understanding, diagnosing, and treating it can be difficult [48]. To date, very little scientific evidence in athletes that details the effects of OTS is available, with this due to several reasons associated with the definition, ethics of inducing OTS, and the practicalities of monitoring the syndrome. For example, the vague definition and timelines associated with OTS has made it difficult to ascertain when the onset of it occurs. Currently, the definition states that 'several weeks' is required but it is known that fatigue is transient and common tapering strategies last two to four weeks [48]. Furthermore, intensive training strategies commonly used by athletes (e.g., altitude training) [49], have been shown to suppress performance prior to supercompensation beyond the timeframe associated with the syndrome. Additionally, it is hard to comprehend that severe training and limited recovery that causes such substantial physiological responses would not cause changes in mood and psychological state. However, these are not always included in definitions and required for the diagnosis of OTS.

Additional to the lack of a clear definition that limits our understanding of OTS, are the difficulties associated with capturing the necessary data to establish its presence. Unexplained decrement in performance is one of the criteria for the diagnosis of OTS, however measuring 'performance' in sport can often be difficult. For example, reductions in sports performance where outcomes are dictated by an opponent or subjectivity (e.g., wrestling), inherently invites bias and does not allow the objective assessment of changes in physical capacity. Furthermore, the requirement to complete standardised, consistent maximal testing, is likely untenable for an athlete who is indeed truly suffering from OTS. Additionally, normal variance in performance naturally occurs and basing changes in performance off a previous best will nearly always suggest an athlete is not performing at their peak. Consequently, from a practical standpoint, the criteria just to establish the physical component of OTS is likely a limiting factor and may not allow practitioners or researchers to establish whether OTS is occurring.

Within the scientific literature, 15 studies have stated that objective methods were used to objectively identify suppression in performance that lasted more than two weeks along with psychological symptoms [50–64]. However, none of these studies provided baseline data and it was not possible to elucidate the actual change in performance. Additionally, it is important to acknowledge that 80% of these papers came from the same cohort and research group [50, 52–62]. This suggests that only four different cohorts of athletes are available that have provided objective assessment of performance and psychological state, with two of these being individual case studies [63, 64]. Thus, while it is commendable that these authors have attempted to objectively assess and report upon OTS, without the provision of data prior to their diagnosis, it is still not possible to understand how OTS presents over time.

Recommendations for the Monitoring and Assessment of the Overtraining Syndrome

To objectively establish the presence of the OTS, three steps should be taken. These are: (1) testing athletes when they are healthy; (2) testing again at the initial point of suspected overtraining; and (3) testing at least four weeks post-initial suspicion of onset. Furthermore, tests that are selected should be relevant to the athlete, and changes in performance must be outside those that are seen with normal performance variation. Finally, valid objective assessment of the athlete's psychological state should be established and should be assessed at the second and third testing occasions to verify symptoms of psychological maladaptation. Figure 11.2 provides recommendations on the objective assessment of OTS.

To support the assessment of the OTS, several recommendations can be provided to both researchers and practitioners. First, there is a need to establish baseline performance measures that are both valid and reliable assessments in performance for a given athlete. This helps to verify when a real change in performance has occurred. Second, standardised regular testing must occur, as this helps elucidate whether performance has declined. However, it should be noted that if the OTS is present, completing maximal physical testing may be undesirable, ethically questionable, and may exacerbate fatigue and psychological burden on the athlete. Third, due to the transient nature of fatigue, it should be clearly established that any decreases in performance are not acute responses to training. Timelines of performance

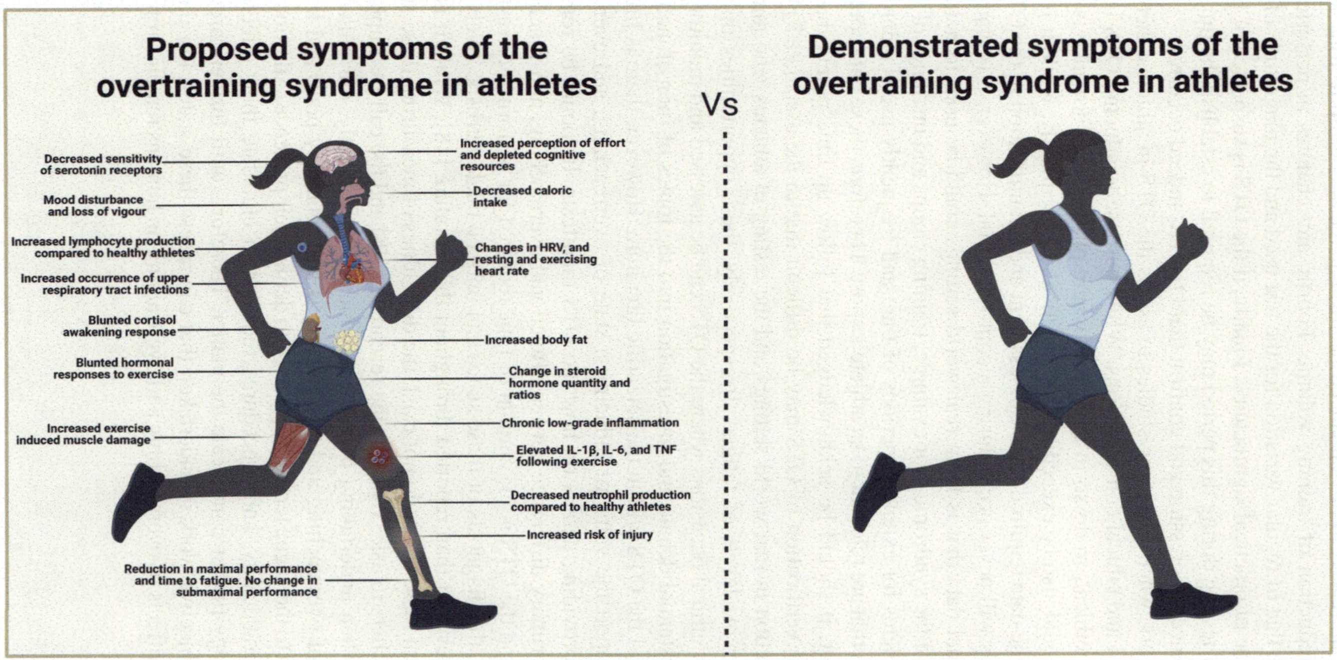

Fig. 11.2 Proposed and objectively demonstrated symptoms of the overtraining syndrome. (Adapted from Weakley et al. [48])

supercompensation can differ between athletes, despite undertaking identical training loads, and the realisation of enhanced performance can take up to four weeks following the reduction of training volume. Fourth, any changes in performance must be outside that of regular variation for the test used and the magnitude of decrement should be of practical significance. Finally, if the OTS is suspected of being present, regular testing during this period may be required to establish the duration of the performance suppression and confirm that OTS has indeed occurred.

For researchers, when designing studies to assess the OTS in athletes, there are several practical and ethical considerations. While purposely inducing OTS in any individual is unethical, to do so in an athlete where physical performance is their livelihood, it would also be exceedingly impractical. Consequently, while it is feasible to gather the data required for the assessment and quantification of the OTS, it is likely that this will never occur during a controlled trial. However, it is plausible that observational data that is from real world settings could be used to help our understanding of the syndrome. For example, training camps and institutes of sport often house athletes for extended periods of time and frequently use standardised testing batteries that are relevant to an athlete's sport. Therefore, these data could be of use. However, it should be acknowledged that follow up time points that are required for the verification of OTS may be difficult due to the associated issues with data collection in real world settings and the testing of athletes who may be undergoing extreme levels of fatigue. Irrespectively, these opportunities may be some of the few ethical occasions where the OTS can be assessed and monitored.

There is substantial debate and investigation into the topics of (non-)functional overreaching and the OTS within the scientific literature. However, there is limited evidence to support the value of inducing a state of overreaching, and even less objective data providing details on how the OTS manifests. It should be remembered that the primary difference between functional overreaching, non-functional overreaching, and the OTS is the amount of time required for performance restoration. Therefore, the diagnosis of these states can only be made retrospectively and is likely less prevalent than reported throughout the literature [48]. While it is not being argued that these states do not exist, the difficulties of measuring these states have led researchers to ask 'Where is the research?' [48] and the call for improved consideration when undertaking their measurement. Consequently, our collective understanding of overreaching and overtraining is still relatively poor, and this is likely attributed to the vague terminology used to define their presence. Additionally, the volatility, practicality, and procedures required to elucidate their effects on human physiology further complicate the matter. Therefore, until further carefully standardised, objective work is undertaken that can assess these states, our understanding of the effects of overreaching and overtraining on humans remains limited.

Frame 4

To assess whether an athlete has the OTS, practitioners must:

1. testing athletes when they are healthy
2. testing again at the initial point of suspected overtraining
3. testing at least four weeks post-initial suspicion of onset.

Additionally, tests should be/demonstrate/have:

1. relevant to the athlete,
2. changes in performance outside those that are seen with normal performance variation.

Valid objective assessment of the athlete's psychological state at the second and third testing occasions to verify symptoms of psychological maladaptation.

Proactive Recovery Methods

Periodisation of Recovery

The role of recovery in long-term training models and adaptation is a contentious issue. Fatigue and muscle damage from training and competition can influence the ability to train and perform, and the addition of recovery is thought to allow overall adaptation as a consequence of supercompensation. However, it has been suggested that recovery methods that dampen inflammatory responses following exercise may mitigate subsequent adaptations [65, 66]. Consequently, implementing recovery appropriately within the annual training plan so that training stress and fatigue are appropriately distributed is necessary to maximise training responses. As previously detailed by Mujika et al. [67], the manipulation of recovery within the training programme can incorporate the below themes:

- Withholding recovery at certain times to maximise adaptation.
- Using recovery during specific phases of preparation to prepare for certain training sessions.
- Implementing increased recovery to decrease fatigue during competition.
- Incorporating recovery during travel, during injury, and in the management of psychological stress.

While some of the scientific literature indicates that proactive recovery strategies (e.g., compression garments, hydrotherapy) can acutely improve recovery following exercise, there is still a lack of understanding of the long-term effects of these methods on athletic performance and adaptation. Thus, considering the influence of acute and chronic recovery methods may aid in the periodisation of training.

Acute Effects of Recovery Methods

Much of the scientific literature has demonstrated that specific recovery interventions have equivocal evidence to support their use [68, 69]. For example, a review by Weakley and colleagues [68] that investigated the effects of compression garments on recovery largely demonstrated negligible changes in performance and physiological outcomes. This may be due in part to differences in their implementation and the compression applied across the body. Furthermore, differences in individual study protocols may have further shrouded the true effect of their use, as some studies were substantially underpowered, or the actual exercise protocol did not induce sufficient fatigue to begin with [68]. It should be acknowledged that the authors did note that the most consistent positive changes that occurred were improvements in perception of muscle soreness (>50% of studies) [68]. Consequently, irrespective of whether there were changes in measures of muscle damage, most individuals reported improvements in how they felt.

Cold water immersion (CWI), or ice baths, have also received significant attention and a recent meta-analysis demonstrates performance benefits when CWI is performed acutely [70]. However, longer term use of CWI has recently been questioned and will be discussed below.

Thus, if the 'recovery' intervention causes no harm and the athlete feels that they are indeed more recovered, then this may provide a benefit to wellbeing and subsequent training outcomes. This may be an important consideration when implementing proactive recovery methods in athletes.

Chronic Effects of Recovery Methods

Complicating the discussion of when and where enhanced recovery methods should be used is that there is often still a lack of evidence supporting their chronic implementation. While recovery is often used to enhance training performance, some evidence indicates that greater recovery may alter subsequent adaptation. For example, work by Yamane et al. [66], indicated that chronic exposure to CWI following exercise may negate changes in maximal muscle strength and endurance, while additional work by the same group indicated that muscle hypertrophy responses can also be retarded [65]. Despite these important findings, several methodological limitations were present including the practical relevance of the imposed protocol (e.g., prolonged exposure to the cold water that would not likely occur in athletic populations). Contrasting these outcomes, Halson et al. [71], investigated the effects of CWI four times per week in well trained cyclists across a 21-day intensified training period and showed that regular exposure to cold water following training may not hinder adaptations but actually enhance performance. It should also be noted that this study is one of the few that investigates ongoing exposure to a recovery method in well-trained athletes, using a protocol that is practically feasible in the real world. Therefore, it is recommended that prior to implementing recovery strategies that may or may not alter adaptation, practitioners should consider the real-world application of the protocol, effectiveness in these conditions, the need to enhance acute recovery, and whether these benefits can outweigh any potential reduction in adaptation.

Periodising Recovery to Enhance Adaptations

Like training, periods of greater rest and regeneration may be warranted. Consequently, periodising and strategically planning phases which have greater recovery may allow the optimal distribution of training stress and fatigue [67]. While it is acknowledged that it is not possible to provide strict guidelines for all athletes due to their heterogenous environments and needs, several recommendations of how recovery can be periodised can be made based on previous work by Mujika et al. [67]. First, in the general preparatory phase where competition frequency is low and there is a need to maximise adaptation, withholding certain recovery modalities may be beneficial. Alternatively, there may be phases of training where higher levels of fatigue are acceptable, and therefore, greater recovery may not be essential or may be harmful. Second, acutely, there may be benefits of providing greater recovery prior to sessions that have a skill emphasis or require prolonged high intensities of exercise. Furthermore, improved recovery may be beneficial after particularly fatiguing sessions or those that induce substantial muscle damage (e.g., a session that has considerable eccentric forces applied to the musculature). Finally, allowing greater recovery following injury, travel, and psychologically stressful periods may be beneficial in helping to reduce the risks of athletes being exposed to unnecessary burden. While there is limited scientific evidence to support this recommendation, it is well established that athletes travel extensively [72], which can interfere with training schedules, and athletes are often exposed to low training loads following injury. Furthermore, if an athlete is experiencing substantial psychological stress, it can be beneficial to allow greater recovery and regeneration. Strategically, these recommendations may help avoid an athlete being overwhelmed by their daily activities and reduce the risk of inducing overreaching or the OTS.

Conclusion

Training is fundamental for the physical development of an athlete. It is a process that requires repeated exposure to stimuli, with the goal of inducing favourable adaptations. Differing adaptations (e.g., muscle hypertrophy, maximal aerobic capacity) often occur as a response to the stimulus that the athlete experiences and the practitioner should carefully consider how the underpinning physiological response occurs. However, it should be noted that if the training stimulus is not appropriately applied or sufficient recovery is not provided, maladaptive responses such as non-functional overreaching and overtraining may occur. These maladaptive responses are still poorly understood though and further research is still required to fully elucidate their manifestation. An individualised and periodised approach to proactive recovery strategies may facilitate an athlete's acute recovery and subsequent adaptations, with a positive impact on sports performance.

1. An effective training programme should carefully account for the effects of fatigue and allow for adequate periods of recovery.
2. Recovery periods are important for the optimal adaptation and well-being of the athlete and can help stave off maladaptive outcomes such as overtraining.
3. Principle of overload is essential for appropriate training prescription and recovery. It can support athletes by facilitating adaptation to occur following acute and chronic periods of training.
4. Specific adaptations and magnitude of change are associated with the type of exercise stimulus and the individual's genetic makeup.
5. Broad appreciation of the physiological basis of different exercise methods can lead to a greater understanding of subsequent fatigue and adaptive outcomes.
6. Insufficient recovery prevents athletes not only from training and competing at the intended intensity or undertaking the desired volume of training, but it can also undermine adaptations.
7. Understanding the form of recovery required and its physiological underpinnings can help to guide practitioners in providing sufficient tools to support the regeneration of athletes post-training.

References

1. Bird SP, Tarpenning KM, Marino FE. Designing resistance training programmes to enhance muscular fitness. Sports Med. 2005;35(10):841–51. https://doi.org/10.2165/00007256-200535100-00002.
2. Ivy JL. Regulation of muscle glycogen repletion, muscle protein synthesis and repair following exercise. J Sport Sci Med. 2004;3(3):131.
3. Lim C, Nunes EA, Currier BS, Mcleod JC, Thomas AC, Phillips SM. An evidence-based narrative review of mechanisms of resistance exercise-induced human skeletal muscle hypertrophy. Med Sci Sports Ex. 2022; https://doi.org/10.1249/MSS.0000000000002929.
4. Baar K. The signaling underlying FITness. App Physiol Nut Metab. 2009;34(3):411–9. https://doi.org/10.1139/H09-035.

5. Farthing JP, Chilibeck PD. The effects of eccentric and concentric training at different velocities on muscle hypertrophy. Eur J App Physiol. 2003;89(6):578–86. https://doi.org/10.1007/s00421-003-0842-2.
6. Lim C, Kim HJ, Morton RW, et al. Resistance exercise–induced changes in muscle phenotype are load dependent. Med Sci Sport Ex. 2019;51(12):2578–85. https://doi.org/10.1249/mss.0000000000002088.
7. Morton RW, Oikawa SY, Wavell CG, et al. Neither load nor systemic hormones determine resistance training-mediated hypertrophy or strength gains in resistance-trained young men. J Appl Physiol. 2016;121(1):129–38. https://doi.org/10.1152/japplphysiol.00154.2016.
8. Hawley JA, Stepto NK. Adaptations to training in endurance cyclists. Sports Med. 2001;31(7):511–20. https://doi.org/10.2165/00007256-200131070-00006.
9. Lambert MI. General adaptations to exercise: acute versus chronic and strength versus endurance training. In: Exercise and human reproduction. Springer; 2016. p. 93–100.
10. Damm TB, Egli M. Calcium's role in mechanotransduction during muscle development. Cell Physiol Biochem. 2014;33(2):249–72. https://doi.org/10.1159/000356667.
11. Ojuka EO. Role of calcium and AMP kinase in the regulation of mitochondrial biogenesis and GLUT4 levels in muscle. Proc Nut Soc. 2004;63(2):275–8. https://doi.org/10.1079/PNS2004339.
12. Suchomel TJ, Nimphius S, Bellon CR, Stone MH. The importance of muscular strength: training considerations. Sports Med. 2018;48(4):765–85. https://doi.org/10.1007/s40279-018-0862-z.
13. Suchomel TJ, Nimphius S, Stone MH. The importance of muscular strength in athletic performance. Sports Med. 2016;46(10):1419–49. https://doi.org/10.1007/s40279-016-0486-0.
14. Hughes DC, Ellefsen S, Baar K. Adaptations to endurance and strength training. Cold Spring Harb Perspect Med. 2018;8(6):a029769. https://doi.org/10.1101/cshperspect.a029769.
15. Bassett DR, Howley ET. Limiting factors for maximum oxygen uptake and determinants of endurance performance. Med Sci Sports Ex. 2000;32(1):70–84. https://doi.org/10.1097/00005768-200001000-00012.
16. Till K, Darrall-Jones J, Weakley JJ, Roe GA, Jones BL. The influence of training age on the annual development of physical qualities within academy rugby league players. J Strength Cond Res. 2017;31(8):2110–8. https://doi.org/10.1519/JSC.0000000000001546.
17. Jones MD, Wewege MA, Hackett DA, Keogh JW, Hagstrom AD. Sex differences in adaptations in muscle strength and size following resistance training in older adults: a systematic review and meta-analysis. Sports Med. 2021;51(3):503–17. https://doi.org/10.1007/s40279-020-01388-4.
18. Martel GF, Roth SM, Ivey FM, et al. Age and sex affect human muscle fibre adaptations to heavy-resistance strength training. Exp Physiol. 2006;91(2):457–64. https://doi.org/10.1113/expphysiol.2005.032771.
19. Kraemer WJ, Ratamess NA. Fundamentals of resistance training: progression and exercise prescription. Med Sci Sports Ex. 2004;36(4):674–88. https://doi.org/10.1249/01.mss.0000121945.36635.61.
20. Aagaard P. Training-induced changes in neural function. Ex Sport Sci Rev. 2003;31(2):61–7. https://doi.org/10.1097/00003677-200304000-00002.
21. Weakley J, McLaren S, Ramirez-Lopez C, et al. Application of velocity loss thresholds during free-weight resistance training: responses and reproducibility of perceptual, metabolic, and neuromuscular outcomes. J Sports Sci. 2020;38(5):477–85. https://doi.org/10.1080/02640414.2019.1706831.
22. Weakley JJ, Till K, Read DB, et al. The effects of traditional, superset, and tri-set resistance training structures on perceived intensity and physiological responses. Eur J App Physiol. 2017;117(9):1877–89. https://doi.org/10.1007/s00421-017-3680-3.
23. Skorski S, Mujika I, Bosquet L, Meeusen R, Coutts AJ, Meyer T. The temporal relationship between exercise, recovery processes, and changes in performance. Int J Sports Physiol Perf. 2019;14(8):1015–21. https://doi.org/10.1123/ijspp.2018-0668.

24. Morton RW, Sonne MW, Farias Zuniga A, et al. Muscle fibre activation is unaffected by load and repetition duration when resistance exercise is performed to task failure. J Physiol. 2019;597(17):4601–13. https://doi.org/10.1113/JP278056.
25. Duffield R, Murphy A, Snape A, Minett GM, Skein M. Post-match changes in neuromuscular function and the relationship to match demands in amateur rugby league matches. J Sci Med Sport. 2012;15(3):238–43. https://doi.org/10.1016/j.jsams.2011.10.003.
26. Moore DA, Jones B, Weakley J, Whitehead S, Till K. The field and resistance training loads of academy rugby league players during a pre-season: comparisons across playing positions. PLoS One. 2022;17(8):e0272817. https://doi.org/10.1371/journal.pone.0272817.
27. Morrison M, Martin DT, Talpey S, et al. A systematic review on fitness testing in adult male basketball players: tests adopted, characteristics reported and recommendations for practice. Sports Med. 2022;52(7):1491–532. https://doi.org/10.1007/s40279-021-01626-3.
28. Nedelec M, McCall A, Carling C, Legall F, Berthoin S, Dupont G. The influence of soccer playing actions on the recovery kinetics after a soccer match. J Strength Cond Res. 2014;28(6):1517–23. https://doi.org/10.1519/JSC.0000000000000293.
29. Roelands B, Meeusen R. Alterations in central fatigue by pharmacological manipulations of neurotransmitters in normal and high ambient temperature. Sports Med. 2010;40(3):229–46. https://doi.org/10.2165/11533670-000000000-00000.
30. Meeusen R, Watson P, Hasegawa H, Roelands B, Piacentini MF. Central fatigue: the serotonin hypothesis and beyond. Sports Med. 2006;36(10):881–909. https://doi.org/10.2165/00007256-200636100-00006.
31. Klass M, Roelands B, Levenez M, et al. Effects of noradrenaline and dopamine on supraspinal fatigue in well-trained men. Med Sci Sports Ex. 2012;44(12):2299–308. https://doi.org/10.1249/MSS.0b013e318265f356.
32. Rampinini E, Bosio A, Ferraresi I, Petruolo A, Morelli A, Sassi A. Match-related fatigue in soccer players. Med Sci Sports Exerc. 2011;43(11):2161–70. https://doi.org/10.1249/MSS.0b013e31821e9c5c.
33. Skorski S, Hammes D, Schwindling S, et al. Effects of training-induced fatigue on pacing patterns in 40-km cycling time trials. Med Sci Sports Ex. 2015;47(3):593–600. https://doi.org/10.1249/mss.0000000000000439.
34. Meeusen R, Duclos M, Foster C, et al. Prevention, diagnosis and treatment of the overtraining syndrome: joint consensus statement of the European College of Sport Science (ECSS) and the American College of Sports Medicine (ACSM). Review. Eur J Sport Sci. 2013;13(1):1–24. https://doi.org/10.1080/17461391.2012.730061.
35. Meeusen R, Duclos M, Gleeson M, Rietjens G, Steinacker J, Urhausen A. Prevention, diagnosis and treatment of the overtraining syndrome. ECSS position statement 'task force'. Article. Eur J Sport Sci. 2006;6(1):1–14. https://doi.org/10.1080/17461390600617717.
36. Urhausen A, Kindermann W. Current markers for the diagnosis of overtraining syndrome in the practice of training. Review. Dtsch Z Sportmed. 2000;51(7–8):226–33.
37. Urhausen A, Kindermann W. Diagnosis of overtraining. Sports Med. 2002;32(2):95–102. https://doi.org/10.2165/00007256-200232020-00002.
38. Bellinger P. Functional overreaching in endurance athletes: a necessity or cause for concern? Sports Med. 2020;50(6):1059–73. https://doi.org/10.1007/s40279-020-01269-w.
39. Thomas L, Busso T. A theoretical study of taper characteristics to optimize performance. Med Sci Sports Ex. 2005;37(9):1615. https://doi.org/10.1249/01.mss.0000177461.94156.4b.
40. Aubry A, Hausswirth C, Louis J, Coutts AJ, Le Meur Y. Functional overreaching: the key to peak performance during the taper. Med Sci Sports Exerc. 2014;46(9):1769–77.
41. Le Meur Y, Louis J, Aubry A, et al. Maximal exercise limitation in functionally overreached triathletes: role of cardiac adrenergic stimulation. J Appl Physiol. 2014;117(3):214–22. https://doi.org/10.1152/japplphysiol.00191.2014.
42. Woods AL, Rice AJ, Garvican-Lewis LA, et al. The effects of intensified training on resting metabolic rate (RMR), body composition and performance in trained cyclists. PLoS One. 2018;13(2):e0191644. https://doi.org/10.1371/journal.pone.0191644.

43. Woods AL, Garvican-Lewis LA, Lundy B, Rice AJ, Thompson KG. New approaches to determine fatigue in elite athletes during intensified training: resting metabolic rate and pacing profile. PLoS One. 2017;12(3):e0173807. https://doi.org/10.1371/journal.pone.0173807.
44. Morgan W, Brown D, Raglin J, O'connor P, Ellickson K. Psychological monitoring of overtraining and staleness. Br J Sport Med. 1987;21(3):107–14. https://doi.org/10.1136/bjsm.21.3.107.
45. Morgan WP, O'Connor PJ, Sparling PB, Pate RR. Psychological characterization of the elite female distance runner. Int J Sports Med. 1987;8(Suppl 2):124–31. https://doi.org/10.1055/s-2008-1025717.
46. O'Connor PJ, Morgan WP, Raglin JS, Barksdale CM, Kalin NH. Mood state and salivary cortisol levels following overtraining in female swimmers. Psychoneuroendocrinology. 1989;14(4):303–10. https://doi.org/10.1016/0306-4530(89)90032-2.
47. Hooper S, Mackinnon L, Hanrahan S. Mood states as an indication of staleness and recovery. Int J Sport Psychol. 1997;28:1–12.
48. Weakley J, Halson SL, Mujika I. Overtraining syndrome symptoms and diagnosis in athletes: where is the research? A systematic review. Int J Sports Physiol Perf. 2022;17(5):675–81. https://doi.org/10.1123/ijspp.2021-0448.
49. Mujika I, Sharma AP, Stellingwerff T. Contemporary periodization of altitude training for elite endurance athletes: a narrative review. Sports Med. 2019;49(11):1651–69. https://doi.org/10.1007/s40279-019-01165-y.
50. Anderson T, Wideman L, Cadegiani FA, Kater CE. Effects of overtraining status on the cortisol awakening response—endocrine and metabolic responses on overtraining syndrome (EROS-CAR). Int J Sports Physiol Perf. 2021;16(7):965–73. https://doi.org/10.1123/ijspp.2020-0205.
51. Brown JF, Wilson MA, Sharp NC. "Down but Not out": an exploration of the psychological factors that impact the unexplained underperformance syndrome (UPS). Int J Sport Sci Coach. 2006;1(2):163–76. https://doi.org/10.1260/174795406777641212.
52. Cadegiani FA, da Silva PHL, Abrao TCP, Kater CE. Diagnosis of overtraining syndrome: results of the endocrine and metabolic responses on overtraining syndrome study: eros-diagnosis. J Sport Med. 2020;2020 https://doi.org/10.1155/2020/3937819.
53. Cadegiani FA, Kater CE. Hypothalamic-pituitary-adrenal (HPA) axis functioning in overtraining syndrome: findings from endocrine and metabolic responses on overtraining syndrome (EROS)—EROS-HPA Axis. Sport Med Open. 2017;3(1):1–11. https://doi.org/10.1186/s40798-017-0113-0.
54. Cadegiani FA, Kater CE. Body composition, metabolism, sleep, psychological and eating patterns of overtraining syndrome: results of the EROS study (EROS-PROFILE). J Sports Sci. 2018;36(16):1902–10. https://doi.org/10.1080/02640414.2018.1424498.
55. Cadegiani FA, Kater CE. Hormonal response to a non-exercise stress test in athletes with overtraining syndrome: results from the endocrine and metabolic responses on overtraining syndrome (EROS)—EROS-STRESS. J Sci Med Sport. 2018;21(7):648–53. https://doi.org/10.1016/j.jsams.2017.10.033.
56. Cadegiani FA, Kater CE. Basal hormones and biochemical markers as predictors of overtraining syndrome in male athletes: the EROS-BASAL study. J Ath Train. 2019;54(8):906–14. https://doi.org/10.4085/1062-6050-148-18.
57. Cadegiani FA, Kater CE. Inter-correlations among clinical, metabolic, and biochemical parameters and their predictive value in healthy and overtrained male athletes: the EROS-CORRELATIONS study. Front Endocrinol. 2019;10:858. https://doi.org/10.3389/fendo.2019.00858.
58. Cadegiani FA, Kater CE. Novel causes and consequences of overtraining syndrome: the EROS-DISRUPTORS study. BMC Sport Sci Med Rehab. 2019;11(1):1–13. https://doi.org/10.1186/s13102-019-0132-x.
59. Cadegiani FA, Kater CE. Novel insights of overtraining syndrome discovered from the EROS study. BMJ Open Sport Ex Med. 2019;5(1):e000542. https://doi.org/10.1136/bmjsem-2019-000542.

60. Cadegiani FA, Kater CE. Eating, sleep, and social patterns as independent predictors of clinical, metabolic, and biochemical behaviors among elite male athletes: the EROS-PREDICTORS study. Front Endocrinol. 2020;11:414. https://doi.org/10.3389/fendo.2020.00414.
61. Cadegiani FA, Kater CE, Gazola M. Clinical and biochemical characteristics of high-intensity functional training (HIFT) and overtraining syndrome: findings from the EROS study (the EROS-HIFT). J Sports Sci. 2019;37(11):1296–307. https://doi.org/10.1080/02640414.2018.1555912.
62. Cadegiani FA, Silva PHL, Abrao TC, Kater CE. Novel markers of recovery from overtraining syndrome: the EROS-LONGITUDINAL study. Int J Sports Physiol Perf. 2021;1(aop):1–10. https://doi.org/10.1123/ijspp.2020-0248.
63. Hedelin R, Wiklund U, Bjerle P, Henriksson-Larsén K. Cardiac autonomic imbalance in an overtrained athlete. Med Sci Sports Ex. 2000;32(9):1531–3. https://doi.org/10.1097/00005768-200009000-00001.
64. Solli GS, Tønnessen E, Sandbakk Ø. The multidisciplinary process leading to return from underperformance and sustainable success in the world's best cross-country Skier. Int J Sports Physiol Perf. 2020;15(5):663–70. https://doi.org/10.1123/ijspp.2019-0608.
65. Yamane M, Ohnishi N, Matsumoto T. Does regular post-exercise cold application attenuate trained muscle adaptation? Int J Sports Med. 2015;36(8):647–53. https://doi.org/10.1055/s-0034-1398652.
66. Yamane M, Teruya H, Nakano M, Ogai R, Ohnishi N, Kosaka M. Post-exercise leg and forearm flexor muscle cooling in humans attenuates endurance and resistance training effects on muscle performance and on circulatory adaptation. Eur J Appl Physiol. 2006;96(5):572–80. https://doi.org/10.1007/s00421-005-0095-3.
67. Mujika I, Halson S, Burke LM, Balagué G, Farrow D. An integrated, multifactorial approach to periodization for optimal performance in individual and team sports. Int J Sports Physiol Perf. 2018;13(5):538–61. https://doi.org/10.1123/ijspp.2018-0093.
68. Weakley J, Broatch J, O'Riordan S, Morrison M, Maniar N, Halson SL. Putting the squeeze on compression garments: current evidence and recommendations for future research: a systematic scoping review. Sports Med. 2021;52:1141–60. https://doi.org/10.1007/s40279-021-01604-9.
69. Poppendieck W, Wegmann M, Ferrauti A, Kellmann M, Pfeiffer M, Meyer T. Massage and performance recovery: a meta-analytical review. Sports Med. 2016;46(2):183–204. https://doi.org/10.1007/s40279-015-0420-x.
70. Moore E, Fuller JT, Buckley JD, et al. Impact of cold-water immersion compared with passive recovery following a single bout of strenuous exercise on athletic performance in physically active participants: a systematic review with meta-analysis and meta-regression. Sports Med. 2022;52(7):1667–88. https://doi.org/10.1007/s40279-022-01644-9.
71. Halson SL, Bartram J, West N, et al. Does hydrotherapy help or Hinder adaptation to training in competitive cyclists? Med Sci Sports Ex. 2014;46(8):1631–9. https://doi.org/10.1249/mss.0000000000000268.
72. Read DB, Williams S, Fullagar HHK, Weakley JJS. The effects of travel on performance: a 13-year analysis of the National Rugby League (NRL) competition. Sci Med Footb. 2021;6(1):60–5. https://doi.org/10.1080/24733938.2021.1876243.

Chapter 12
Heart Rate Variability as a Means of Assessing Workload Effects and Recovery During Exercise Training

Fabio Y. Nakamura, Petrus Gantois, Christoforos Giannaki, Angelos Vlahoyiannis, and Gregory C. Bogdanis

What Is Heart Rate Variability?

Mechanism of Heart Rate Variability

In a healthy cardiovascular system, the heart does not beat evenly like a metronome; instead, it fluctuates constantly in response to multiple heart-brain interactions to maintain a "flexible" steady state and rapidly adapt to environmental or behavioral changes [1]. Heart rate variability (HRV) is a non-invasive measurement of the normal variation in the intervals between consecutive heartbeats (R-R intervals) of the electrocardiogram (ECG) over a given period (e.g., 1 minute to 24 hours) [2]. The variability in R-R intervals reflects the neural control of the heart and the circulatory system via sympathetic and parasympathetic modulations. The changes in the sympathovagal balance of the autonomic nervous system (ANS) underpinning the regulation of heart rate (HR) are necessary to meet the body's physical demands (i.e., metabolic and thermoregulatory adjustments) in response to a challenging task [3].

F. Y. Nakamura (✉)
Research Center in Sports Sciences, Health Sciences and Human Development (CIDESD), University of Maia, Maia, Portugal
e-mail: fabioy_nakamura@yahoo.com.br

P. Gantois
Associate Graduate Program in Physical Education, Federal University of Paraíba, João Pessoa, Brazil

C. Giannaki · A. Vlahoyiannis
School of Life and Health Sciences, University of Nicosia, Nicosia, Cyprus

G. C. Bogdanis
Physical Education and Sport Science, National and Kapodistrian University of Athens, Athens, Greece

N. C. Apostolopoulos et al. (eds.), *Fundamentals of Recovery, Regeneration, and Adaptation to Exercise Stress: An Integrated Approach*,
https://doi.org/10.1007/978-3-031-44270-4_12

An increase in HR following the start of exercise is caused by the augmented cardiac sympathetic activity along with reduced parasympathetic drive, whereas the decline in HR immediately after exercise is characterized by parasympathetic activation and sympathetic withdrawal [4]. In this regard, the changes in HRV provide indirect information about the autonomic nervous system's (ANS) ability to adapt to multiple psychophysiological and environmental stimuli. Since the ANS is related to multiple physiological systems (i.e., neuroendocrine, cardiovascular), its changes, as measured objectively by HRV, could be useful for understanding the body's response to acute or chronic exercise. Indeed, training-induced adaptations that improve athletic performance are generally associated with increased resting vagal cardiac activity [5]. Having a high HRV is generally a good sign of adaptability, with vagally-mediated HRV indices being closely linked to global markers of health, wellness, and athletic "readiness to perform". The exception is the parasympathetic hyperactivity caused by overreaching in endurance athletes [6].

The accumulated knowledge regarding the importance of heart rate variability has made its monitoring a popular strategy among elite athletes, coaches, and supporting staff.

Heart Rate Variability Measurement

Heart rate variability has many clinical applications, and it is calculated in clinical settings by obtaining HR data from an ECG or other monitoring devices capable of measuring R-R intervals with high accuracy. Many accurate and low-cost portable devices and mobile applications are now available making HRV measurement a relatively simple and accessible physiological marker of training adaptations. Irrespective of the recording method used to acquire HRV data, there is a need for researchers and practitioners to standardize some internal and external conditions (i.e., time of day, body position, and nutritional intake).

For monitoring purposes, HRV has been assessed at rest (i.e., upon waking, nocturnal, upon arrival at training facilities, or after exercise), during exercise, and in various body positions (i.e., seated, supine, standing). In addition, HRV indices can be obtained from long-term (24 hours), short-term (2, 5, and 15 minutes), or ultra-short-term (1 minute) recordings [2]. All of these methodological features seem to have an impact on HRV measures and may affect the compliance of recreational and elite athletes due to multiple constraints (e.g., finding a quiet room or even a chair) [7]. The lack of standardization of these procedures during HRV recordings will result in inconsistent data and interpretation of the athletes' "readiness to perform" status. Therefore, in this chapter, we will discuss key features of the HRV assessment and interpretation in order to improve decisions made by coaches and sports scientists.

Given the various conditions, durations, and technologies available for collecting HRV data, it is recommended that practitioners or clinicians consult the relevant literature to make informed decisions and maintain consistency in settings across repeated measurements with the same athlete. This will ensure correct interpretation of potential changes due to training and/or other stressors.

Analysis and Interpretation of Heart Rate Variability

Analysis of HRV can be performed using linear (i.e., time and frequency domains) and non-linear methods. Under field conditions, time-domain HRV measures are considered a more suitable and attractive approach for monitoring stress-recovery processes [2]. Although many HRV metrics exist, the natural logarithm of the root mean square of successive differences between normal heartbeats (LnRMSSD)—a vagally-related HRV index obtained through time-domain analysis—is widely used by sports scientists and practitioners as a fatigue-sensitive marker for monitoring daily changes in ANS status [2, 8]. Since the RMSSD values are highly variable between athletes (40–200 ms), the natural log transformation is recommended to reduce data spread. Researchers found that LnRMSSD is more sensitive to training adaptations, highly reliable, less influenced by breathing patterns, requires shorter measurement time ($\leq$5 minutes), and is relatively easier to calculate and interpret than time-frequency analyses. In addition, LnRMSSD could discriminate competitive levels and correlate with performance among athletes with different backgrounds [9, 10]. Notably, the LnRMSSD (or at least RMSSD) is available in all devices that calculate HRV measures, including mobile applications. For that reason, when appropriately interpreted, this vagal-related HRV index is suitable for monitoring positive and negative responses to variations in training loads or other stressors on a daily/weekly basis.

Increased LnRMSSD at rest and/or after exercise is a generally good sign of adaptation. However, indices of HRV display a naturally high day-to-day fluctuation [7, 8]. Consequently, single-day HRV measures might not accurately reflect "true" training-induced changes in ANS status. Le Meur et al. [6] found that weekly means of HRV values better reflected chronic changes in ANS status associated with a prolonged overload in functionally overreached triathletes. In this regard, we recommend recording LnRMSSD data over 3–5 days per week (i.e., weekly average) to assess the day-to-day variation in cardiac autonomic function by calculating its coefficient of variation ($LnRMSSD_{CV}$) [11]. This approach is more sensitive for monitoring changes in HRV indices related to training-induced fatigue and adaptations than single-day measures [6, 8], with higher CV values associated with lower fitness and greater fatigue.

Empirical data suggest that low $LnRMSSD_{CV}$ reflects greater resilience to training loads and may be interpreted as an "acclimation" of athletes to training stimuli. During the first 3-weeks of a 5-week conditioning program, changes in $LnRMSSD_{CV}$

were inversely correlated ($r = -0.74$) with improvements in intermittent running ability in female soccer players [12]. On the other hand, high LnRMSSD_{CV} ($\geq 10\%$) represents perturbations to cardiac-autonomic homeostasis and may be interpreted as a potential sign of cardiac autonomic disturbances. Therefore, LnRMSSD_{CV} can be used by coaches as an earlier sign of inadequate stress-recovery status and might point to an alteration in training loads (associated with recovery strategies), especially when athletes are also feeling stressed and fatigued [13, 14].

It remains to be determined whether other HRV indices can match the practicality, user-friendliness, and informativeness of lnRMSSD in practical contexts. Additionally, both valid and reliable objective measures (e.g., hormones) and subjective measures (e.g., wellness) need to be tested alongside HRV to provide a more comprehensive understanding of training stress and adaptation across different sports.

HRV as a Means of Monitoring Sleep and Recovery

There is a myriad of physiological and behavioral measures that can be taken to monitor sleep and recovery in individuals who exercise regularly, including athletes. Heart rate variability is an option, and jointly with other variables (e.g., sleepiness scales, neuromuscular performance) might inform clinicians and exercise professionals about the need to extend recovery periods, methods to optimize recovery itself (e.g., cold water immersion), extended overnight sleep duration or even including naps during the day, aiming to improve performance acutely. Indices reflecting the cardiac autonomic modulation (described below) can be used at rest, after sub-maximal exercise, or during sleep to make inferences about fatigue, recovery, and readiness to perform. Concerning the latter, consistent evidence shows that a "flexible" training program based on HRV results in better adaptation than a more "rigid" predefined program [15–17]. The flexible program consists of performing low-intensity training every time vagal-related HRV indices at rest are below the average obtained in the last 10 days. High-intensity training is only performed when HRV is increased or unchanged compared to the 10-day reference. In a recent study, the training group that was guided by the HRV response improved maximal oxygen uptake, while the predefined training group did not [15]. This suggests that HRV may be a physiological measure that needs to be constantly monitored before each training session to help in adjusting the pre-planned load. Later in the chapter, the methods used for HRV monitoring after controlled exercise and during sleep will be discussed, together with the interpretation of the results and the use of HRV as a tool to gauge training adaptations.

Monitoring heart rate variability in athletes can be useful for early detection and intervention, before significant declines in performance and negative health outcomes are observed.

Assessment of Heart Rate Variability During Sleep

Measuring HRV during sleep provides a standardized assessment condition, free of disruptive events typically encountered in most hours of the day when the individual is awake. Since variations in cardiac autonomic regulation are observed during sleep because of the alternation of sleep stages, some authors [18, 19] argue that HRV should be quantified during "slow wave sleep" (SWS), which is a quiet sleep stage with high vagal dominance. The SWS "episode" method considers (1) the first 10 minutes of the first low and regular HR episode lasting at least 15 minutes (i.e., visual inspection of the whole nighttime R-R tachogram); (2) the lowest SDNN (the standard deviation of "normal" beats) throughout the period of interest; (3) a round Poincaré plot (i.e., a round cluster of points in R-R intervals plotted as a function of the previous one); and (4) a low interbeat autocorrelation between successive R-R intervals (i.e., the correlation coefficient of the Poincaré plot <.5). The vagal-related indices calculated during this time window (e.g., HF and RMSSD) are positively influenced in moderately trained subjects compared to sedentary peers [18]. However, highly trained athletes might display lower vagal activity during the SWS compared to moderately trained participants (bell-shaped relationship), suggesting that intensive training might not be beneficial regarding nighttime HRV [18]. In fact, it was shown that a single session of high-intensity interval training could disturb vagally-mediated indices during the following night's sleep [20] and that high training load undertaken at night, such as a competitive soccer match, might have the same effect. This is why highly trained athletes must balance training stress and recovery to maintain health and performance.

To gain insight into the cardiac autonomic regulation throughout the sleeping hours, disregarding stages of sleep, one can continuously assess HRV using the "hour-by-hour" approach [21, 22]. Briefly, sleep initiation can be recorded using an accelerometer or estimated using a sleep diary. After starting sleep, HRV can be calculated in consecutive 1-hour periods. This approach is limited because some individuals might have sleep disruptions and insomnia, meaning that some measures might not represent sleep time. Using this approach, Myllymäki et al. [21] could not detect differences in HRV between a control day without exercise and a day with vigorous late-night exercise, although higher HR was noticed during the first 3 hours of sleep. Similar results were reported for female soccer players while comparing a rest day and a day with a late-night training session [22]. Nonetheless, it is interesting to note that vagally-mediated indices in the "hour-by-hour" approach tend to increase towards the 5th–6th hour of sleep in the training days, suggesting a trend of parasympathetic reactivation. The circadian pattern cannot explain this, as the same trend was not observed on the rest day. Therefore, caution should be taken while using the "hour-by-hour" approach to detect acute exercise-stress-related changes in HRV, although at least one study showed differences between a low-intensity training day and a day with a maximal 3000-m test in runners [23]. In that study, authors used either an entire night or a 4-hour analysis window. This field of knowledge deserves more research to establish standardized, but at the same time, practical and accurate methods to assess HRV during sleep, and detect post-exercise and competition recovery. Fortunately, such measures are reliable when comparing consecutive nights after low-intensity training [23].

> Practitioners are encouraged to seek technologies that allow for measuring HRV during sleep without compromising its quality or quantity. Numerous options are available on the market, and it is essential to test and explore these options to identify the most effective solutions for each specific case.

Assessment of Heart Rate Variability Before Exercise

Heart rate variability measured before exercise is a valuable and reliable marker of fatigue/readiness in practical exercise settings. Studies demonstrated that suppression in vagally-mediated HRV indices at rest may reflect inadequate recovery, heightened psychophysiological stress, or a sign of non-functional overreaching (if sustained for weeks). After high-intensity exercises, cardiac parasympathetic reactivation is not fully restored to baseline for at least 48 hours [24], with fitter athletes being more resilient to training stress and requiring less time to recover [24, 25]. Therefore, one of the basic principles of measuring HRV before exercise is understanding the athlete's "readiness to perform". This information can be used to finely adjust the training load (i.e., train at high intensity only when vagal-related HRV indices have returned to normal levels) [2].

In sport-field settings, resting HRV before training can be assessed upon waking or upon arrival at training facilities, with large fluctuations in intra-day HRV metrics between conditions (CV ~23%) [26]. For practical reasons, coaches might prefer the assessment of HRV at training facilities due to the unsupervised nature of the upon-waking measures (i.e., home-based setting) [27]. However, daily HRV immediately after waking and bladder emptying is an optimal opportunity to monitor individual changes in ANS status since the conditions are highly standardized. Consequently, HRV measurement after waking minimizes many confounding factors (i.e., food, caffeine, water intake, and travel) [28]. In addition, HRV measures upon waking give coaches more time to design or modify the individual daily training content according to the results. Indeed, it is important to recall that an HRV-guided training approach is more effective for eliciting optimal physiological adaptations than "traditional" predefined training prescriptions [29]. Therefore, HRV measure upon waking is preferred as a best practice to monitor athletes' stress-recovery status when it is possible to incorporate it into the monitoring routine. The adequacy and usefulness of HRV upon waking derived from self-recordings on smartphone applications (e.g., ithlete™) have already been demonstrated in individual [30] and team sport athletes [14]. When HRV measurement upon waking is not feasible, measuring HRV at the standard time of day prior to training is also a useful option. The only restriction is that the athlete must abstain from physical exercise and beverages containing stimulants, such as coffee, before the HRV assessment.

> The HRV recordings can be easily performed in real-world settings, such as at home, ensuring high ecological validity that supports decisions made by coaches and support staff.

Assessment of HRV After Exercise: Relationship with Training Load and Fatigue

Post-exercise parasympathetic reactivation is dependent on the exercise intensity and, to a lesser extent, on exercise duration. Vagally-related HRV indices recover very quickly (i.e., less than 10 minutes) after exercise performed below the lactate threshold, while exercise above the lactate threshold will substantially delay the return to the baseline to approximately 30 minutes [31]. All-out exercises such as repeated sprints or supramaximal exercises (i.e., above the intensity corresponding to the maximal oxygen uptake) will further decrease post-exercise HRV [32, 33], and complete restoration of vagal modulation might take several hours to occur [20]. It is important to highlight that highly trained subjects will display faster parasympathetic reactivation, independent of exercise intensity [31, 34].

For practical reasons, assessment of HRV after a standardized exercise aiming to assess training load effects and fatigue is preferably completed using mild exercise intensity, which can be incorporated into warm-up routines. For instance, Buchheit et al. [35] proposed the following protocol for endurance runners: a 5-minutes submaximal run at 60% of maximal aerobic speed, followed by 5-minutes standing rest. LnRMSSD was calculated during the last 3 minutes of the 5-minutes recovery period. This test also allows the recording of the exercise HR during the last 30-seconds of the run, as well as the calculation of the time constant of the HR decay during recovery by fitting a first-order exponential decay curve. After an 8-week training program, post-exercise HRV was improved from 2.74 ± 0.64 to 3.05 ± 0.60 ms, and the change in this index was significantly correlated with change in the maximal aerobic speed ($r = 0.52$) and in the 10-km time ($r = -0.73$) [35].

In team sports, it is possible to improve this protocol's practicality by setting a common running speed of 9 km/h for all the players. When this protocol was applied to futsal players, the post-exercise LnRMSSD adaptation in response to training was again significantly correlated with changes in endurance. This means that post-exercise HRV can track positive performance changes, indicating whether the training load is appropriate for the athlete. Conversely, reductions in post-exercise LnRMSSD are associated with fatigue, meaning that the training load should be reduced, along with giving more opportunities for the athletes to recover. In order to allow the application of this test to court sports, we have used a 20 × 20 m square with the pace guidance (9 km/h) provided by audio bleeps. The test can be adapted to the specific conditions and requirements of different sports or even in non-sport contexts. The most important aspect of testing is keeping the protocol constant across the assessments during the year or the season. This will ensure reliable comparisons and interpretation of the positive or negative response to training stimuli over time. Due to the test's characteristics, assessments can be performed weekly, allowing for finely workload manipulation/adjustment in the following week.

> It is highly recommended that practitioners develop monitoring and visualization systems to aid decision-making based on physiological, biochemical, neuromuscular, and psychological responses to training and competition. While such systems can include HRV, relying solely on HRV recordings will offer only a partial view of adaptation responses.

Methods for Measuring and Evaluating Heart Rate Variability

Traditionally, HRV data are recorded in laboratory settings (i.e., ECG) using time-consuming methods (i.e., 10 minutes) [36]. Standard HRV procedures are complex for coaches due to the lengthy HRV recording time, issues related to data export and analysis, and the necessity of expensive equipment, limiting their use in applied settings. These time and logistical constraints may often prevent timely feedback to the coaching staff, hindering adjustments in training loads in the subsequent training sessions [37]. In the last years, HRV measurement tools have been simplified due to the advances related to new technologies (i.e., wearable devices and smartphone applications) [14], using time-efficient ultra-short recordings (i.e., 1 minute).

The availability of valid portable devices and mobile applications connected to either a chest strap or photoplethysmography (PPG) to measure HRV (e.g., ithlete™, Elite HRV) has made measuring HRV metrics more attractive and accessible than ever. The use of heart rate monitors such as wearable devices provides accurate R-R interval recordings [38] which are comparable to an ECG, even though they require manual export of the raw data to specialized software for further analysis and interpretation. Recently, smartphone HRV applications using PPG technology are growing in popularity due to their low cost and enhanced practicality. The PPG technique can accurately detect vagal-related HRV indices without needing a chest strap or specialized software to analyze HRV data. It is particularly relevant in applied field settings for daily ultra-short LnRMSSD self-recordings, which could increase compliance of coaches and athletes who wish to obtain frequent HRV data. In addition, PPG has been shown to provide accurate measures of HRV (e.g., HRV4Training) [39].

Several studies have shown that vagally-mediated HRV indices (i.e., LnRMSSD) can be accurately assessed in shorter durations (i.e., 1 minute) compared to traditional and more time-consuming measures (≥5 minutes) [40–42]. Using the ultra-short method, Nakamura et al. [42] demonstrated that LnRMSSD was equally sensitive to training-induced adaptations in elite futsal players as the standard HRV criterion (i.e., 10 minutes). In addition, it has been already demonstrated that ultra-short LnRMSSD from smartphone-derived HRV (i.e., self-recording) is sensitive to variations in training loads in individual [30] and team sports [14] athletes. Therefore, in applied field settings, the ultra-short HRV method derived from a smartphone application could be implemented in the work routine of athletes before training sessions on a weekly basis (3–5 days), being a user-friendly, affordable, and time-efficient approach.

> Non-invasive and time-efficient methods/equipment such as wearable beat-to-beat heart rate monitors and smartphone applications can provide detailed information about positive and negative adaptions over short and long periods throughout the competitive calendar.

Methodological Issues When Measuring Heart Rate Variability

The ideal external conditions to measure HRV include stable and comfortable room temperature, quiet room, lying position and relatively long recording time of R-R intervals. However, in field settings where exercise, training, and competition occur, it is often too difficult or impossible to control the environmental conditions and get prolonged HRV measurements. Nevertheless, HRV measurements in sub-optimal conditions are useful and may provide valuable information for controlling training load. In fact, our group has already recorded HRV in female basketball players on the bench inside the court, with the noise of other players training and shouting [43]. Although not ideal, this is often the only possible condition to measure HRV in several sports settings.

However, one condition that can significantly affect HRV values and should be standardized is body position. The supine position generates the highest values of vagal HRV indicators, followed by the seated and standing positions. We opt for the seated position to avoid the confounding effect of saturation of vagal-related parameters [44], especially in highly trained subjects [45]. However, others also use the supine [46, 47] and standing [47] positions to measure the R-R intervals for HRV calculation. In any case, the most important thing to consider is keeping the same body position across all assessments.

Another confounding factor when measuring HRV is breathing. In the field, it is often not practical to control breathing rate. Although this control is recommended when calculating HRV variables in the frequency domain, in practice, if we ask the subject to breathe spontaneously without accelerating or slowing down the normal rhythm, we get reproducible HF measurements [48].

In summary, it is not easy in sports settings to control all the external conditions to obtain a reliable HRV measure. The ones which are possible and relevant to control are body position and breathing rate. One other condition that deserves attention is that the subjects should not have performed vigorous physical activity before recording the R-R intervals.

Future research, along with technological advancements, may provide opportunities for reliably quantifying HRV in athletes during standardized resting activities such as reading, watching TV, or using smartphones. This could enhance adherence to daily HRV monitoring and potentially enable multiple measurements per day.

Perspectives/Future of HRV on Assessing Fatigue and Recovery

Heart rate variability can be a reliable and sensitive method to monitor changes in performance and fatigue. However, HRV alone is not sufficient to provide all the necessary information to track fatigue and accelerate recovery. The cardiac autonomic system is among a myriad of systems disturbed by a single training session or competition [49–51]. Each system responds and recovers according to the type, duration and intensity or the exercise stress. For example, after a plyometric training session, HR recovers quickly, while muscle stiffness and explosiveness take much longer time to return to baseline. After a soccer match, muscle soreness can peak later than the reduction in HRV. For this reason, training centers and clubs need to choose relevant monitoring variables that can genuinely bring insights into the training and recovery status of athletes. Besides this, subjective measures which help to identify overreaching should be more thoroughly developed in the future, as the current instruments lack validity and do not allow reliable examination of psychometric properties [52]. In addition, it is essential to highlight that HRV can identify "extreme" states, such as favorable vagal modulation after training in healthy children [53] and adults [54] and, on the other side, maladaptation caused by overtraining [55]. More evidence is necessary concerning daily variations in fitness/fatigue, so that we can establish the extent to which HRV may provide helpful information for daily decision-making. This is one of the future avenues we must follow to integrate HRV into more complex and comprehensive monitoring systems.

Conclusions

Recent advances have significantly enhanced our understanding of using Heart Rate Variability (HRV) to assess pre-exercise readiness, recovery, and chronic adaptation to training. Despite these advancements, there remain several gaps that need to be addressed to increase the adoption of HRV monitoring among practitioners and athletes. Nevertheless, HRV is currently recognized as a valuable physiological tool for monitoring athletes across various sports under different conditions.

References

1. Shaffer F, McCraty R, Zerr CL. A healthy heart is not a metronome: an integrative review of the heart's anatomy and heart rate variability. Front Psychol. 2014;5:1040. https://doi.org/10.3389/fpsyg.2014.01040.
2. Buchheit M. Monitoring training status with HR measures: do all roads lead to Rome? Front Physiol. 2014;5:73. https://doi.org/10.3389/fphys.2014.00073.
3. Acharya UR, Joseph KP, Kannathal N, Lim CM, Suri JS. Heart rate variability: a review. Med Biol Eng Comput. 2006;44(12):1031–51. https://doi.org/10.1007/s11517-006-0119-0.
4. Daanen HAM, Lamberts RP, Kallen VL, Jin A, Van Meeteren NLU. A systematic review on heart-rate recovery to monitor changes in training status in athletes. Int J Sports Physiol Perform. 2012;7(3):251–60. https://doi.org/10.1123/ijspp.7.3.251.
5. Bellenger CR, Fuller JT, Thomson RL, Davison K, Robertson EY, Buckley JD. Monitoring athletic training status through autonomic heart rate regulation: a systematic review and meta-analysis. Sports Med. 2016;46(10):1461–86. https://doi.org/10.1007/s40279-016-0484-2.
6. Le Meur Y, Pichon A, Schaal K, et al. Evidence of parasympathetic hyperactivity in functionally overreached athletes. Med Sci Sports Exerc. 2013;45(11):2061–71. https://doi.org/10.1249/MSS.0b013e3182980125.
7. Aubert E, Seps B, Beckers F, Aubert AE, Seps B, Beckers F. Heart rate variability in athletes. Sports Med. 2003;33(12):889–919. https://doi.org/10.1089/152702903769192331.
8. Plews DJ, Laursen PB, Stanley J, Kilding AE, Buchheit M. Training adaptation and heart rate variability in elite endurance athletes: opening the door to effective monitoring. Sports Med. 2013;43(9):773–81. https://doi.org/10.1007/s40279-013-0071-8.
9. Proietti R, di Fronso S, Pereira LA, et al. Heart rate variability discriminates competitive levels in professional soccer players. J Strength Cond Res. 2017;31(6):1719–25. https://doi.org/10.1519/JSC.0000000000001795.
10. Nakamura FY, Torres VBC, Da Silva LS, et al. Monitoring heart rate variability and perceived well-being in Brazilian Elite Beach volleyball players: a single-tournament pilot study. J Strength Cond Res. 2022;36(6):1708–14. https://doi.org/10.1519/JSC.0000000000003654.
11. Plews DJ, Laursen PB, Le Meur Y, Hausswirth C, Kilding AE, Buchheit M. Monitoring training with heart-rate variability: how much compliance is needed for valid assessment? Int J Sports Physiol Perform. 2014;9(5):783–90. https://doi.org/10.1123/IJSPP.2013-0455.
12. Flatt AA, Esco MR. Evaluating individual training adaptation with smartphone-derived heart rate variability in a collegiate female soccer team. J Strength Cond Res. 2016;30(2):378–85. https://doi.org/10.1519/JSC.0000000000001095.
13. Flatt AA, Esco MR, Nakamura FY, Plews DJ. Interpreting daily heart rate variability changes in collegiate female soccer players. J Sports Med Phys Fitness. 2017;57(6):907–15. https://doi.org/10.23736/S0022-4707.16.06322-2.
14. Flatt AA, Esco MR. Smartphone-derived heart-rate variability and training load in a women's soccer team. Int J Sports Physiol Perform. 2015;10(8):994–1000. http://www.embase.com/search/results?subaction=viewrecord&from=export&id=L611519606%0Ahttp://dx.doi.org/10.1123/ijspp.2014-0556.
15. Kiviniemi AM, Hautala AJ, Kinnunen H, Tulppo MP. Endurance training guided individually by daily heart rate variability measurements. Eur J Appl Physiol. 2007;101(6):743–51. https://doi.org/10.1007/s00421-007-0552-2.

16. Hautala AJ, Kiviniemi AM, Mäkikallio TH, et al. Individual differences in the responses to endurance and resistance training. Eur J Appl Physiol. 2006;96(5):535–42. https://doi.org/10.1007/s00421-005-0116-2.
17. Kiviniemi AM, Hautala AJ, Kinnunen H, et al. Daily exercise prescription on the basis of HR variability among men and women. Med Sci Sports Exerc. 2010;42(7):1355–63. https://doi.org/10.1249/MSS.0b013e3181cd5f39.
18. Buchheit M, Simon C, Piquard F, Ehrhart J, Brandenberger G. Effects of increased training load on vagal-related indexes of heart rate variability: a novel sleep approach. Am J Physiol Heart Circ Physiol. 2004;287(6):2813–8. https://doi.org/10.1152/ajpheart.00490.2004.
19. Brandenberger G, Buchheit M, Ehrhart J, Simon C, Piquard F. Is slow wave sleep an appropriate recording condition for heart rate variability analysis? Auton Neurosci Basic Clin. 2005;121(1–2):81–6. https://doi.org/10.1016/j.autneu.2005.06.002.
20. Al Haddad H, Laursen PB, Ahmaldl S, Buchhelt M. Nocturnal heart rate variability following supramaximal intermittent exercise. Int J Sports Physiol Perform. 2009;4(4):435–47. https://doi.org/10.1123/ijspp.4.4.435.
21. Myllymäki T, Kyröläinen H, Savolainen K, et al. Effects of vigorous late-night exercise on sleep quality and cardiac autonomic activity. J Sleep Res. 2011;20(1 PART II):146–53. https://doi.org/10.1111/j.1365-2869.2010.00874.x.
22. Costa JA, Brito J, Nakamura FY, Oliveira EM, Rebelo AN. Effects of late-night training on "slow-wave sleep episode" and hour-by-hour-derived nocturnal cardiac autonomic activity in female soccer players. Int J Sports Physiol Perform. 2018;13(5):638–44. https://doi.org/10.1123/ijspp.2017-0681.
23. Nuuttila OP, Seipäjärvi S, Kyröläinen H, Nummela A. Reliability and sensitivity of nocturnal heart rate and heart-rate variability in monitoring individual responses to training load. Int J Sports Physiol Perform. 2022;17(8):1296–303. https://doi.org/10.1123/ijspp.2022-0145.
24. Stanley J, Peake JM, Buchheit M. Cardiac parasympathetic reactivation following exercise: implications for training prescription. Sports Med. 2013;43(12):1259–77. https://doi.org/10.1007/s40279-013-0083-4.
25. Buchheit M, Mendez-Villanueva A, Quod MJ, Poulos N, Bourdon P. Determinants of the variability of heart rate measures during a competitive period in young soccer players. Eur J Appl Physiol. 2010;109(5):869–78. https://doi.org/10.1007/s00421-010-1422-x.
26. Coyne JOC, Coutts AJ, Fomin R, French DN, Newton RU, Haff GG. Heart rate variability and direct current measurement characteristics in professional mixed martial arts athletes. Sports. 2020;8(8):109. https://doi.org/10.3390/sports8080109.
27. Sherman SR, Holmes CJ, Hornikel B, MacDonald HV, Fedewa MV, Esco MR. Heart-rate variability recording time and performance in collegiate female rowers. Int J Sports Physiol Perform. 2021;16(4):550–6. https://doi.org/10.1123/IJSPP.2019-0587.
28. Schneider C, Hanakam F, Wiewelhove T, et al. Heart rate monitoring in team sports—a conceptual framework for contextualizing heart rate measures for training and recovery prescription. Front Physiol. 2018;9:639. https://doi.org/10.3389/fphys.2018.00639.
29. Düking P, Zinner C, Reed JL, Holmberg HC, Sperlich B. Predefined vs data-guided training prescription based on autonomic nervous system variation: a systematic review. Scand J Med Sci Sports. 2020;30(12):2291–304. https://doi.org/10.1111/sms.13802.
30. Flatt AA, Hornikel B, Esco MR. Heart rate variability and psychometric responses to overload and tapering in collegiate sprint-swimmers. J Sci Med Sport. 2017;20(6):606–10. https://doi.org/10.1016/j.jsams.2016.10.017.
31. Seiler S, Haugen O, Kuffel E. Autonomic recovery after exercise in trained athletes: intensity and duration effects. Med Sci Sports Exerc. 2007;39(8):1366–73. https://doi.org/10.1249/mss.0b013e318060f17d.
32. Buchheit M, Laursen PB, Ahmaidi S. Parasympathetic reactivation after repeated sprint exercise. AJP Heart Circ Physiol. 2007;293(1):H133–41. https://doi.org/10.1152/ajpheart.00062.2007.
33. Nakamura FY, Soares-Caldeira LF, Laursen PB, Polito MD, Leme LC, Buchheit M. Cardiac autonomic responses to repeated shuttle sprints. Int J Sports Med. 2009;30(11):808–13. https://doi.org/10.1055/s-0029-1234055.

34. Buchheit M, Millet GP, Parisy A, Pourchez S, Laursen PB, Ahmaidi S. Supramaximal training and postexercise parasympathetic reactivation in adolescents. Med Sci Sports Exerc. 2008;40(2):362–71. https://doi.org/10.1249/mss.0b013e31815aa2ee.
35. Buchheit M, Chivot A, Parouty J, et al. Monitoring endurance running performance using cardiac parasympathetic function. Eur J Appl Physiol. 2010;108(6):1153–67. https://doi.org/10.1007/s00421-009-1317-x.
36. Task Force of the European Society of Cardiology. The North American Society of Pacing Electrophysiology. Heart rate variability. Standards of measurement, physiological interpretation, and clinical use. Eur Heart J. 1996;17(3):354–81.
37. Perrotta AS, Jeklin AT, Hives BA, Meanwell LE, Warburton DER. Validity of the elite HRV smartphone application for examining heart rate variability in a field-based setting. J Strength Cond Res. 2017;31(8):2296–302. https://doi.org/10.1519/JSC.0000000000001841.
38. Dobbs WC, Fedewa MV, MacDonald HV, et al. The accuracy of acquiring heart rate variability from portable devices: a systematic review and meta-analysis. Sports Med. 2019;49(3):417–35. https://doi.org/10.1007/s40279-019-01061-5.
39. Stone JD, Ulman HK, Tran K, et al. Assessing the accuracy of popular commercial technologies that measure resting heart rate and heart rate variability. Front Sports Active Living. 2021;3(March 2021):585870. https://doi.org/10.3389/fspor.2021.585870.
40. Nakamura FY, Pereira LA, Esco MR, et al. Intraday and interday reliability of ultra-short-term heart rate variability in rugby union players. J Strength Cond Res. 2017;31(2):548–51. https://doi.org/10.1519/JSC.0000000000001514.
41. Esco MR, Flatt AA. Ultra-short-term heart rate variability indexes at rest and post-exercise in athletes: evaluating the agreement with accepted recommendations. J Sports Sci Med. 2014;13(3):535–41.
42. Nakamura FY, Flatt AA, Pereira LA, Ramirez-Campillo R, Loturco I, Esco MR. Ultra-short-term heart rate variability is sensitive to training effects in team sports players. J Sports Sci Med. 2015;14(3):602–5. https://doi.org/10.1128/AAC.03728-14.
43. Nakamura FY, Pereira LA, Cal Abad CC, et al. Adequacy of the ultra-short-term HRV to assess adaptive processes in youth female basketball players. J Hum Kinet. 2017;56(1):73–80. https://doi.org/10.1515/hukin-2017-0024.
44. Kiviniemi AM, Hautala AJ, Mäkikallio TH, Seppänen T, Huikuri HV, Tulppo MP. Cardiac vagal outflow after aerobic training by analysis of high-frequency oscillation of the R-R interval. Eur J Appl Physiol. 2006;96(6):686–92. https://doi.org/10.1007/s00421-005-0130-4.
45. Costa JA, Brito J, Nakamura FY, Dores H, Rebelo A. Associations between 24-h heart rate variability and aerobic fitness in high-level female soccer players. Scand J Med Sci Sports. 2022;32(S1):140–9. https://doi.org/10.1111/sms.14116.
46. Tanner V, Faiss R, Saugy J, Bourdillon N, Schmitt L, Millet GP. Similar supine heart rate variability changes during 24-h exposure to normobaric vs. hypobaric hypoxia. Front Neurosci. 2021;15(December):1–9. https://doi.org/10.3389/fnins.2021.777800.
47. Schmitt L, Regnard J, Coulmy N, Millet GP. Influence of training load and altitude on heart rate variability fatigue patterns in elite nordic skiers. Int J Sports Med. 2018;39(10):773–81. https://doi.org/10.1055/a-0577-4429.
48. Bloomfield DM, Magnano A, Bigger JTJ, Rivadeneira H, Parides M, Steinman RC. Comparison of spontaneous vs. metronome-guided breathing on assessment of vagal modulation using RR variability. Am J Physiol Heart Circ Physiol. 2001;280(3):H1145–50. http://ajpheart.physiology.org/content/280/3/H1145.
49. González-Badillo J, Rodríguez-Rosell D, Sánchez-Medina L, et al. Short-term recovery following resistance exercise leading or not to failure. Int J Sports Med. 2015;37(04):295–304. https://doi.org/10.1055/s-0035-1564254.
50. Chen J-L, Yeh D-P, Lee J-P, et al. Parasympathetic nervous activity mirrors recovery status in weightlifting performance after training. J Strength Cond Res. 2011;25(6):1546–52. https://doi.org/10.1519/JSC.0b013e3181da7858.

51. Hautala A, Tulppo MP, Mäkikallio TH, Laukkanen R, Nissilä S, Huikuri HV. Changes in cardiac autonomic regulation after prolonged maximal exercise. Clin Physiol. 2001;21(2):238–45. https://doi.org/10.1046/j.1365-2281.2001.00309.x.
52. Jeffries AC, Wallace L, Coutts AJ, McLaren SJ, McCall A, Impellizzeri FM. Athlete-reported outcome measures for monitoring training responses: a systematic review of risk of bias and measurement property quality according to the COSMIN guidelines. Int J Sports Physiol Perform. 2020;15(9):1203–15. https://doi.org/10.1123/IJSPP.2020-0386.
53. Buchheit M, Platat C, Oujaa M, Simon C. Habitual physical activity, physical fitness and heart rate variability in preadolescents. Int J Sports Med. 2007;28(3):204–10. https://doi.org/10.1055/s-2006-924296.
54. Iwasaki KI, Zhang R, Zuckerman JH, Levine BD. Dose-response relationship of the cardiovascular adaptation to endurance training in healthy adults: how much training for what benefit? J Appl Physiol. 2003;95(4):1575–83. https://doi.org/10.1152/japplphysiol.00482.2003.
55. Laurent M, Malika B, Stephane P, et al. Decrease in heart rate variability with overtraining: assessment by the Poincaré plot analysis. Clin Physiol Funct Imaging. 2004;24(1):10–8. http://www.embase.com/search/results?subaction=viewrecord&from=export&id=L38195530.

Chapter 13
Key Considerations in the Preparation of the Elite Athlete: Planning, Performance, and Recovery

Andrew N. Guard

Introduction

> The management of athlete's training and recovery has never been more important. With increased physical demands and congestion of sporting events facing athletes, there is a need for coaches to acutely understand planning, periodization, and recovery-regeneration in relation to the individual athlete's under their care.

The aim of this chapter is to detail how the use of various recovery and injury prevention strategies can aid training program prescription and subsequent athletic performance. The existing theory and literature pertaining to the optimal preparation of athletes will be discussed, as well as the implications for ensuing exercise program planning and the wider contextual factors that may be influential in training program design for an individual or team. 'Regeneration' has been defined by Kellmann and Beckmann to entail the physiological recovery from the stress of sports performance [1]. There is indeed a myriad of factors that impact an athlete's training regimens throughout the training calendar which may also influence the recovery time-course.

It is hoped that this chapter will provide the reader with a comprehensive, and applied, understanding of the necessary factors to consider when constructing training and recovery cycles for athletes.

A. N. Guard (✉)
Atlanta United Football Club, Atlanta, GA, USA
e-mail: andy_guard@yahoo.co.uk

N. C. Apostolopoulos et al. (eds.), *Fundamentals of Recovery, Regeneration, and Adaptation to Exercise Stress: An Integrated Approach*,
https://doi.org/10.1007/978-3-031-44270-4_13

With the continuous evolution of global sporting prowess and achievements, calendars are also subject to an exponential increase in regular domestic and international competitions. Accompanying this is an inevitable increase in media coverage, marketing interest, and overall economic investment. With these elements in mind, and the fine margins required to succeed, there is a necessity to prepare the contemporary athlete with a sound knowledge of the physical and bio-motor demands of the sport. This entails the physiological adaptations and recovery time-courses of the different biological systems as well as the type of training employed, and appropriate periodization models to attain peak athletic and cognitive performance.

> As scientific knowledge grows, so must the coaches' systematic and methodological approach towards program application.
>
> –Dan Pfaff

Superimposed on the above are several other areas having implications for the planning and prescription of exercise training. The competitive calendar and the density of events present numerous opportunities for training or recovery, influencing the training volumes and intensities of specific training phases. Therefore, the training-competition calendar is subject to detailed planning with regards to periodization and the training loads to which an athlete is subjected. The physical components needed to succeed in competition are influenced by the balance of load and recovery thereby reducing the risk of injury, allowing for peak performance.

In addition, today's athletes are exposed to numerous additional factors such as, frequent air travel to competitions and training camps [2], academic obligations for collegiate athletes [3], congested fixture periods [4], sub-optimal nutritional intake [5] and sleep hygiene [6]. All of these have been shown to be influential on performance and responsible for guiding the planning of exercise programs. Despite the aforementioned factors, with each athlete being different, the notion of 'individualization' should be honored, accounting for one's age and stage of maturation (if appropriate), gender, training status, injury history, and their experience as an athlete.

Sporting Demands

> It is imperative to consider the inherent variability in physical demands asked of an athlete. This will depend on the sport or task itself, format and duration as well as positional role and specific tactical constraints.

When constructing training programs, it is essential to have a good understanding of the demands of the sport. This could mean the extent to which biological systems of the body are taxed in terms of cardiovascular endurance, musculo-skeletal strength and power and neuromuscular proficiency. What type of movements are required in a field-, court- or track and field-based sport? What does the competition structure look like in terms of overall intensity, duration, work-rest periods, and spatial constraints? With these in mind, one needs to consider the typical demands associated with competition to better inform training prescription. For instance, a 90-min soccer match elicits different physical outputs to those of a American Football comprising four 15-min quarters. In addition, team sports involve position-specific requirements dictating their training prescription. The demands required of a wide midfielder differing from those of a central defender [7] as do those of an offensive lineman and a wide receiver in American Football [8, 9]. Individual events in track and field also adhere to the principle of specificity where the energy system requirements for sprinters, middle- and long-distance runners contrast [10]. The predominant energy system(s) and their contributions in producing energy via Adenosine Triphosphate (ATP) need to be considered. Therefore, knowing when to tax each energy system and to what extent, whilst allowing for sufficient recovery between exercise bouts and training sessions is important. In very short, high-intensity bouts of exercise, the Adenosine Triphosphate-Phosphcreatine, (ATP-PC) energy system is the main source of energy for the first 5–6 s (subsequently supported by the anaerobic alactic component), with high-energy phosphates made available in small amounts within skeletal muscle and as such are utilized to fuel brief anaerobic actions. Complete resynthesis of ATP stores has been reported to be attained after 3–5 min of recovery, with phosphocreatine (PC) resynthesis complete after 8 min post exercise [11–13]. The glycolytic (0–2 min) and aerobic-oxidative (2 min to 3 h) energy systems are relied upon for longer duration events. In these instances, glycogen in the muscle is the preferred energy substrate and is usually fully restored within 24 h post-exercise [11–13] (Table 13.1).

Lastly, the movement profile of each athlete should be considered within the sport. Comfort and Matthews emphasized how the design and prescription of the athlete's training must account for unique motion profiles, limb-dominance, and specificity with regards to muscle actions utilized in different planes of motion [14]. Field-based team sports often command the ability to perform repeated-effort jump, acceleration-deceleration, and sprint efforts at different intensities and orientations in key actions influencing goal-scoring scenarios. Swimming involves full-body muscular strength, power and endurance, and depending on the intensity and duration of the event(s), tax different energy systems, akin to track athletes [15]. Combat sports, such as boxing, warrant both aerobic and anaerobic energy provision, strength, and power training. These help to produce high forces repeatedly without fatigue, while withstanding impacts themselves [16].

By having a holistic understanding of what is needed for successful performance in a sport or event, and the individual athlete's characteristics, the training program is better served to produce the desired adaptations and goals [17] (Fig. 13.1).

Table 13.1 Contribution of energy systems to sports performance

Activities	ATP-PC	Glycolytic	Aerobic
Soccer	10%	70%	20%
American football (lineman)	90%	10%	–
100 m Track sprint	50%	50%	–
Middle distance running	10%	20%	70%
Marathon	–	2%	98%
Boxing	6%	9%	85%

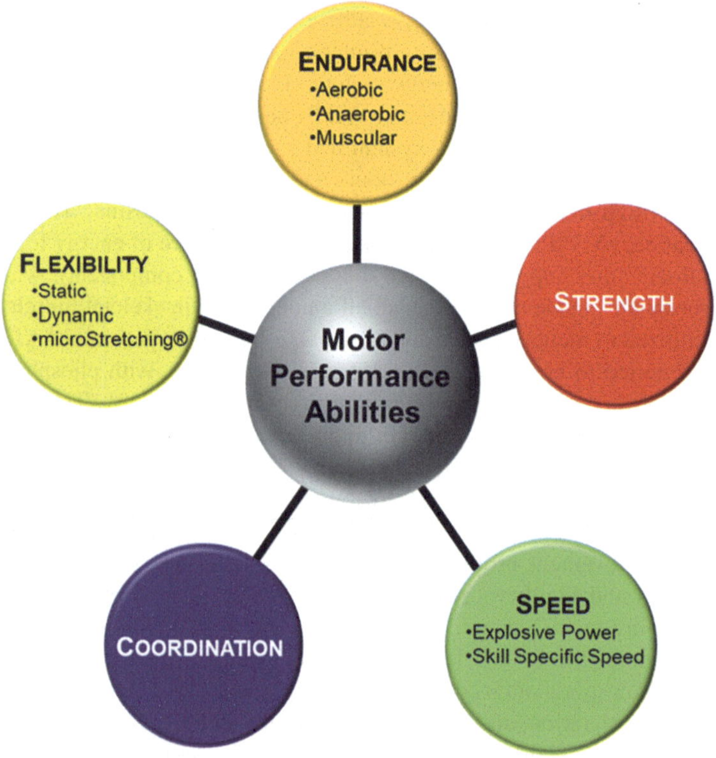

Fig. 13.1 Sporting bio-motor abilities. (Adapted from Lammle [18])

Principles of Training

> If you can't explain it simply, you don't understand it well enough.
>
> –Albert Einstein

When planning exercise and training programs, it is important to have a sound knowledge of the physical and cognitive demands of the sport. A clear understanding of what the athlete is being prepared to withstand in competition, and the extent to which certain motor performance abilities are taxed. Coaches must ensure that the program design is specific to the task in question. The training programs need to elicit adaptations relevant to the sporting performance. This is referred to as the SAID (specific adaptation to imposed demands) principle. For instance, an athlete who requires strength, power or speed in their discipline will focus their training to enhance these qualities. The training program must incorporate an element of progressive overload. SAID ensures that an athlete will be subjected to increased demands in their sessions through manipulation of working sets and repetitions, work-to-rest ratios, and increased intensity or training load. Securing a safe and gradual overload, based on the athlete's current capabilities, ensures the training stimulus will induce the desired positive physiological adaptations. Finally, the athlete must see progression in their training program. The program should involve safe progressions in intensity, the overload of certain stimuli when appropriate in the training cycle, as well as offering a variety of presentation to avoid training monotony, staleness, and lack of adaptation.

A thorough appreciation for the requirements of successful performance in relation to the sport or task is critical to be able to support the training process. Such an understanding will underpin the rationale and provision of a specific, periodized preparatory program.

Periodization

To be able to prescribe a safe and effective training program, coaches should have a working knowledge of the different models of training periodization and the consequences of the manipulation of training load. Paired with the annual competition calendar, it is then possible to construct specific training phases for the athlete(s) in order optimize their physical and psychological preparation for competition, whilst also being intentional about periods of planned recovery and regeneration.

For the design of training programs, it is important to incorporate an appropriate periodization strategy. Successful sports performance requires a well-thought-out training plan consisting of physical and technical-tactical elements of the sport. The challenge is to allot dedicated time to these various facets in a time-efficient way within the annual training cycle.

The concept of periodization offers a framework that enables the division of the training and competitive season distinct periods; *Preparatory*, *Competition* and *Transition*. Within these periods are defined cycles, mesocycles, that may last months and are made up of several microcycles, which are 1 week in length [19,

Table 13.2 Composition of training cycles [20]

Phase	Description
Training session (unit)	Single isolated session with specific training content
Microcycle	Consisting of multiple training units lasting 1–7 days
Mesocycle	Made up of multiple microcycles, totaling several weeks
Macrocycle	Seasonal or annual training cycle

20]. Specific mesocycles that comprise the aforementioned periods are termed phases: *General Preparatory, Special Preparatory, Pre-Competition, Competition,* and *Transition phases* [13, 19]. Within these phases, coaches can plan the training to systematically develop selected abilities to improve performance. This is achieved through the manipulation of training intensity, volume, density, and recovery in preparation for and during the competition phase [13]. Early research in periodization started with the Russian scientist Leo Matveyev in the 1960s. He proposed the traditional method of periodization, which was cyclical in nature, the 'purposeful sequencing of different training units'. This consisted of alternating load and recovery within microcycles in order to obtain desired training adaptations, whilst minimizing monotony and fatigue (Table 13.2).

In traditional, linear periodization, the training volume is high. This is gradually replaced by an increase in training intensity as competition approaches. With non-linear or undulating periodization models, both load and taper exposure (in team sports, for example) are varied [22]. Block periodization, conceived by Bondarchuk, referred to three mesocycle blocks; developmental (gradual increase to maximum training load), competitive (training is controlled to allow for focus on competing), and a restoration block (targets recovery of the body before the cycle repeats). Issurin [23] modified this further by referring to the three blocks as follows: accumulation (development of basic bio-motor abilities); transformation (exercising event-specific physical and technical abilities); the realization block, which entailed competition prep and maximal event rehearsal coupled with adequate recovery before competing.

The limitations of trying to train mixed abilities for a sport concurrently results in conflicting physical responses (energy pathways for example) reducing the effect of the training stimulus. Linear periodization involves progressive increases in intensity over time. In contrast, the non-linear or undulating approach [24] sees fluctuations in training intensity and targeted motor performance abilities between days over the course of a microcycle [24, 25]. A variation in low, moderate, and high intensity exposure to training stress benefits sports involving multiple peaks of competition such as a soccer season (macrocycle). The idea is to utilize load-recovery cycles where acute fatigue is induced by training load. This is followed by sufficient recovery to improve training capacities and obtain adaptations through means of supercompensation [23]. This has been demonstrated by Gilliam [26] who observed

Table 13.3 Contextual factors influencing the periodization process

Contextual	Environmental	Social
Sport physical requirements (and team style of play)	Temperature extremes: cold, heat & humidity	Age
Playing position	Altitude	Stage of maturation
Training status (fitness fatigue)	Playing surface (grass, turf)	Gender
Stage of season		Academic stressors
Proximity to previous/next competition (schedule)		Other team or sport commitments
Injury history		
Travel demands & time zone change		
Standard of opposition		
Strength training programming		

greater strength gains with high training frequency when mixing hard and easy days. Such an approach avoided overtraining, neural fatigue, and maintained the rate of force development.

Each periodization model; linear, undulating and block [22, 23, 27] consist of varying approaches to physical preparation, Coaches need to decide which method will lend itself best to the competition calendar and the individual athlete(s) under their care (Table 13.3).

Training Load

It is commonplace in today's elite sporting environment to ensure meticulously planned training programs are tolerated and are within the capabilities of the athlete, ensuring training adaptation and performance improvement. To evaluate the success of such programs, it is necessary to elucidate the training load of the team or individual athlete. Training load encompasses an athlete's response to a given workload to which they are exposed. This may be described as a dose-response relationship, with optimal training load stimuli eliciting positive training adaptations for enhanced performance. Inappropriately high, or low training loads lead to deleterious effects whereby injury or illness can occur. External load is the term given to the work prescribed for an athlete and the work they completed during a resistance or sport training session, such as the resistance lifted, or the distance covered [28]. Internal load refers to the athlete's pathophysiological response to the given external load [28, 29].

The model proposed by Hans Seyle in 1946, describes the General Adaptation Syndrome (GAS) [30] in which an external stimulus (load) elicits and internal load response. Following the cessation of exercise, the application of such a stimulus induced a degree of fatigue in the biological system targeted. The model describes the 'alarm' phase, when unaccustomed exercise disrupts the body's systems, resulting in feelings of soreness and tiredness. This is followed by the 'resistance' phase.

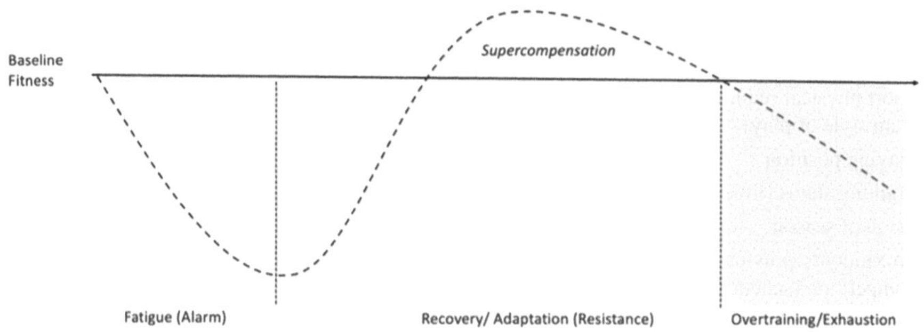

Fig. 13.2 Stimulus-Recovery cycle pertaining to physiological adaptation

When coupled with an appropriate period of recovery the body undergoes positive adaptions, in order to cope with the new stressor experienced through supercompensation. The final phase, 'exhaustion', occurs when the training load experienced becomes too great for the body's current capacities or too frequent, suggesting insufficient time for proper recovery. This results in symptoms of the Overtraining Syndrome (Fig. 13.2).

It is key to ensure that the programming an athlete is subjected to, entails an optimal balance of load and recovery during the training cycle(s) ensuring a positive adaptation, while avoiding excessive fatigue or injury. Homeostasis of the biological system(s) stressed must be achieved periodically, with sufficient recovery, so as to not induce negative consequences and maladaptation [31]. The process of the increased presence of stress from stimuli is termed the overtraining continuum [32]. With continued exposure to fatiguing environments, (functional) overreaching may occur. The athlete experiences short-term reductions in performance capabilities as a result of an intensified period of training for positive adaptations, obtained with supercompensation [33]. Non-functional overreaching is characterized by a persistence of intensified training stimuli, insufficient recovery, possibly accompanied by negative psychophysiological symptoms [33]. Overtraining occurs when the training stimuli induces long-term reductions in health, mood, and performance. This accumulation effect could result in consequences lasting several weeks to months [34].

Training load is manipulated to provide a stimulus for an acute response at the appropriate time and intensity, contributing to chronic physiological adaptation. Moreover, as detailed by the acute-chronic workload ratio (ACWR) [35], the training load applied should be within the tolerance capacity of the athlete without causing adverse consequences.

As a result, there are several scenarios to be aware of when planning and monitoring athlete workloads. With effectively periodized training it is important to ensure that the week-to-week change in training load is accounted for. The acute training load (current microcycle) should not significantly exceed what the athlete has been previously exposed to, on average, across the preceding 3–4 weeks (i.e. the chronic training load). Recent research has shown increases in acute weekly training load of >15% significantly increased the odds of soft tissue injury by 3–4 times in baseball, cricket, volleyball, soccer, and Australian rules football [35–38]. A latent period following such 'spikes' in training load, of 1–2 weeks may exist

> The data informs, it doesn't decide.
>
> –Sir David Brailsford

before injury occurs from these programming errors [39–41].

The occurrence of high training loads over long periods has been identified as a contributor to increased risk of both injury and/or illness. Athletes experiencing sustained high or accumulated weekly training loads across consecutive microcycles do not have sufficient recovery or 'de-load' periods. This significantly exposes them to a greater risk of injury [42]. The periodization of progressive load and recovery exposures are therefore pivotal in athlete load management.

With the principles of training in mind, load tolerance can be progressively increased (with appropriate interspersing recovery), resulting in positive training adaptations. This creates a more resilient athlete with greater training and physical qualities. This approach has a protective effect against injury and illness in elite sporting populations [35, 43] (Fig. 13.3).

> The most effective training programs will be underpinned by appropriate periodization. Optimizing the balance of training load exposure and planned recovery for adaptation will be key to ensuring high performance execution in competition(s). To achieve this, the multidisciplinary team should work in an interdisciplinary fashion to plan, program and review all elements of the training session, phase and cycle.

Biological Systems and Physiological Responses

> The periodized training plan of an athlete or team may not always bring about the desired physiological response or adaptation. To minimize unexpected consequences, practitioners should be aware of the numerous biological and physiological processes that impact the success of a training session or phase. The responses to the prescribed stimuli may be highly individual, and as such induce fatigue and varied time-course of recovery of biological systems.

When constructing exercise programs understanding the physiological consequences associated with stressing different biological systems is important for and sequencing their development, with sufficient recovery. An understanding of their contribution to sports performance, acute and chronic training responses, and time-course of recovery, are vital in the periodization of the athlete, ensuring a state of fitness and freshness.

Internal Load: Psycho-physiological response to training stimulus

OBJECTIVE:

- Heart Rate (% maximum heart rate, % heart rate recovery, heart rate variability (Rmssd))
- Blood and Muscle Lactate
- Maximal Oxygen Consumption (VO_{2max})

SUBJECTIVE:

- Session Rating of Perceived Exertion (sRPE)
- Differential RPE (Cardiorespiratory, Muscular, Technical)
- Wellness Responses (Sleep Quality, Fatigue, Energy, Muscle Soreness)
- Recovery Status (Recovery-Stress Questionnaire, Profile of Mood States (POMS)

External Load: Musculoskeletal-Biomechanical work completed

- Distance Covered (Global Positioning System-Derived Walking, Running, Sprinting)
- Accelerations and Impacts (Accelerometer-Derived)
- Pitches Thrown
- Plays Run
- Kicking Strikes
- Resistance Lifted

Fig. 13.3 Types of internal and external training load monitoring

Fatigue

Athletes who undertake regimented training programs during the training year experience periods of fatigue, a temporary disruption of the homeostasis of the body's systems [44]. With acute fatigue, one has an inability to sustain the required work-rate of force production required for the task at hand. Such fatigue occurs because of metabolic disturbances [45] after hard training sessions, sustained high training loads over a period of time, or a short period of purposefully intensified training or competition. Fatigue may be apparent in numerous biological systems within the body as a result of the aforementioned scenarios.

Origins of muscle fatigue are believed to be from both central and peripheral pathways [46], with the latter involving a reduction in force production occurring away from the neuromuscular junction [47]. Central fatigue involves an alteration to neural pathways of the central nervous system, namely a reduced drive of motor neurons and their ability to initiate maximal voluntary contractions [48, 49].

Peripheral fatigue entails a decline of muscular work capacity at the cellular level. For example, several metabolic factors are reported to contribute to peripheral fatigue including; energy substrate depletion of ATP and Phosphocreatine, necessary for cross-bridge cycling in muscle contraction [50]. Athletes experience limited performance due to the inability to sustain the required exercise intensity and energetic demands, though it has been reported that ATP levels may not decline below 60–70% of resting values in humans [51]. High-intensity and longer duration events may further deplete muscle glycogen levels, influencing calcium release in the sarcoplasmic reticulum and ATP re-synthesis necessary for excitation-contraction coupling and cross-bridge formation for muscle contraction [51, 52].

Other consequences of high-intensity exercise influencing peripheral fatigue are the accumulation of metabolites in the blood, such as hydrogen ions, reducing pH levels to give a more acidic environment and inhibiting the ability to produce force within the muscle [53]. Similarly, there may be an alteration to the sodium-potassium pump, in the absence of the required gradient necessary for maintaining action potentials at the membrane to stimulate required muscle force production [54].

Exercise involving high-intensity work and gross muscular contractions have been widely reported to elicit exercise-induced muscle damage (EIMD) and trauma at the cellular level, elevating the presence of muscle protein creatine kinase (CK) [55]. The high forces associated with eccentric contractions are known to cause greater damage to the muscle fibers, especially if an athlete is unaccustomed to such actions [56]. Increased permeability of the muscle cell membrane sees the leaking of CK into the extracellular space, lowering cell integrity, thus reducing force transmission and resulting delayed onset muscle soreness (DOMS), peaking in the 48–72 h post exercise [56, 57].

Energy Substrates (Glycogen)

Sports of longer durations (>60 min) consisting of high energetic requirements, rely on adequate supplies of glycogen, found within muscle and liver, to sustain the required intensity and work-rate and delay the deleterious effects of fatigue. Studies of high-level soccer players observed that post-match muscle glycogen levels are depleted compared to a control group in 24 h (−27%) and 48 h (−9%) compared to pre-match levels [58–60]. Other studies purported that glycogen replenishment could take over 2 days after a soccer match and may be fiber-type dependent, with type II fibers taking upwards to 72 h [61]. Saltin previously highlighted the importance of carbohydrate fueling in the days prior to competition to boost glycogen levels in the body, coupled with a concomitant taper in training load [58].

The type, intensity and duration of exercise or sport is highly influential in determining the preferred energy systems and bioenergetic requirements supporting optimal performance [62]. The magnitude of these requirements determines the recovery time-course of substrate resynthesis before recommencement of training and competition [63].

Endocrine Markers

Endocrine responses to exercise and the subsequent time-course of recovery of specific hormones must be understood in relation to exercise program planning. These hormones play a key role in facilitating responses to training and exercise by creating either an anabolic (adaptation) or catabolic (stress) environment [64, 65].

Testosterone

Despite being a key marker of anabolic activity, the research around Testosterone remains unclear [64]. A significant increase in salivary testosterone of 33% has been observed post-match in rugby union, with values returning to pre-match levels at 24 h [66], with others observing a continued increase until 38 h, well into the recovery period [69]. In competitive soccer, testosterone changes post-game have not been consistent either, with Ispiridis et al. [57], reporting a small decrease in serum testosterone, with Rowell et al. [67] reporting a large increase before returning to baseline at 66 h in a high game-exposure group in serum sample testosterone. In high-level endurance running (ultramarathon) athletes, serum testosterone levels were observed to decrease and remain below pre-race levels for the next 48 h [68].

Cortisol

This stress hormone and marker of a catabolic environment, has been observed to increase immediately after a soccer match before returning to baseline within 24 h [57]. Similar results were found in studies of rugby union athletes, where serum cortisol increased significantly immediately after matches until 14 h, before declining to 33% less than at pre-match at 38 h [69]. Serum Cortisol has also been reported to increase significantly after ultra-marathon events 24 h post, with the eventual return to pre-race levels at 48 h [68].

Testosterone-Cortisol (T:C) Ratio

The ratio of these two hormones can help indicate the success of training programs and adaptation. Based on opposite metabolic status, this ratio reflects the protein status of muscle, as measured by both anabolic and catabolic markers of recovery [31]. Studies in rugby union observed a decline in T:C ratio immediately post-game, before increasing significantly to pre-match levels (24–48 h), possibly reflecting a rebound anabolic stimulus following intense, collision sport matches [66]. The catabolic nature of high-intensity, intermittent exercise is reflected in the immediate decrease of T:C ratio as has also been observed in elite soccer, with an ensuing increase seen 24–42 h post-match [67].

It has been proposed that long-term changes in T:C ratio may be indicative of an overtraining state in athletes [70], with a decline of over 30% suggested to represent such a change in training status [71]. However, uncertainty exists as to the reliability of this marker with long-term training decrement, with a study observing no change in T:C ratio in athletes displaying overtraining symptoms [64], and decreased in those showing no measurable changes to performance [72].

Creatine Kinase and Exercise-Induced Muscle Damage (EIMD)

Creatine kinase (CK) increases significantly following high-intensity exercise involving forceful contractions of the musculature, such as the eccentric nature of deceleration actions. CK increases peak at 24–48 h post-match in rugby union, with values higher than those reported in collegiate-level American Football (800 $U.L^{-1}$ vs 350 $U.L^{-1}$) [73]. Increases of 91% from pre-to post soccer and rugby competition have also been reported, with peak values occurring at 24–120 h [57, 69]. However, it should be noted that blood biomarker monitoring is influenced by numerous factors affecting their reliability in sporting applications [74]. For example, biochemical monitoring is subject to large inter-individual variations due to the individual nature of responses to a given exercise stimulus, and as a result, may differ from normative reference values [75]. Additionally, collision sport athletes have persistently high levels of muscle damage compared to other sports due to the high-intensity collisions and associated trauma [69]. This is associated with high levels of inflammation and muscle damage [69]. Athletes in these sports may become accustomed to such actions and therefore display more resistance due to a 'repeated bout effect' [76]. The muscle is pre-conditioned to eccentric actions with this desensitization reflected in blood biomarker analysis [76].

Inflammatory Markers

It has been suggested that high intensity, intermittent sports with multi directional movement profiles (i.e., soccer, field hockey, basketball etc.), induce heightened inflammatory responses with a subsequent pro-antioxidant environment [63, 77, 78]. Several studies reported elevated white blood cell counts following team sport competition with higher counts of neutrophils and lymphocytes indicative of the inflammatory response, in the 30 min to 24 h post exercise, before returning to baseline [57, 79]. A delay effect has been suggested to occur in the time-course of white blood cell increases until 38 h [69] (Fig. 13.4).

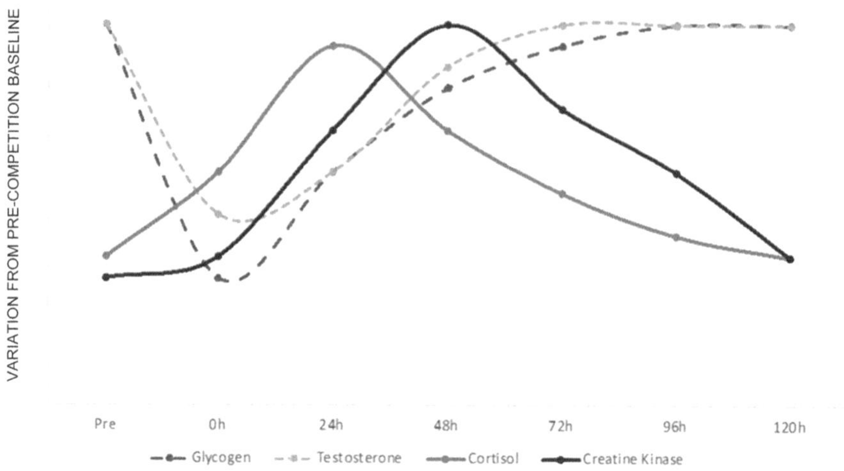

Fig. 13.4 Time-course of recovery of physiological processes. (Adapted from Skorski et al. [44])

Performance Measures

Performance assessments are increasingly popular in evaluating the time-course of recovery of specific bio-motor abilities related to sports performance. Obtaining a better understanding of an individual's ability to recover after intense exercise is beneficial for subsequent training prescriptions and the necessary recovery emphases. This information is important given the increased scheduling of competitions and congested periods, where athletes are asked to perform multiple times, with minimal interspersing recovery.

Measures of lower limb strength and power in vertical jump performance have shown decrements in jump height at 24 and 42 h before returning to baseline levels at 72 h. Similar findings were observed by Rowell who reported an effect of magnitude of match exposure and physical load in elite soccer players regarding impaired jump height and Flight-time:Contraction-time (FT:CT), suggesting altered jump strategy as a result of neuromuscular fatigue in high-load exposures in 42 h after a soccer game [67]. Numerous studies have shown concentric knee extensor and flexor maximal voluntary strength to be reduced by 5–10% for several days post soccer match [79]. The extent of decrements in voluntary muscle strength were exacerbated in the flexors of the knee, particularly in team-sports where this muscle group is susceptible to fatigue and injury [80]. Magalhaes et al., observed that maximal voluntary contraction of the knee extensors is compromised up to 72 h following soccer match play [81]. In addition, 20 m sprint performance was reduced by 10% at 72 h, normalizing by 120 h [81]. Decrements in repeated-sprint performance attributed to neuromuscular and peripheral fatigue has been observed in cycling. It has been suggested that this is due to a reduction in glycogen levels and muscle damage. This results in a reduction in knee extensor performance in the initial 24 h recovery period between repeated-sprint sessions, with maximum voluntary contraction returning to baseline by 48 h [82].

Sports, such as cycling, have seen significant decrements in maximum power output, oxygen consumption, with increases in the subjective perception of intensity following 2 weeks of intense training [83]. Time trial performance was reduced by 10% compared to baseline [83]. Another study on repeated sprint cycling performance described peripheral fatigue impacting maximal voluntary contraction of knee extensors in the 24–48 h recovery period, with sprint performance remaining unaffected [82].

These collective findings, summarized by Minett and Duffield [84], highlight the contrast in recovery time-course of markers of inflammation and muscle damage, with measures of strength and physical performance returning to pre-competition levels more rapidly. The subsequent programming of specific training stimuli should account for these variations between individual athletes recovery. To facilitate this, coaches should utilize multiple modalities monitoring the subjective and objective recovery time-courses of individual athletes [85] (Fig. 13.5).

> Individualized approaches to training prescription and monitoring will enhance the training process to obtain the desired training adaptation. Differences in physiological cost to a given training stimulus should be respected, with flexibility in subsequent programming important. For example, the addition of further strategies to aid the recovery of a specific biological system or motor quality after intensified training may be warranted.

Strength Training Recovery

> The inclusion of strength training sessions should not be overlooked in support of both the physical development and recovery of an athlete. The structural and endocrine responses may both play a role in sport-specific performance when part of an appropriately periodized program. The make-up of specific micro- and meso-cycles, however, should also recognize nuances of the athlete and the sport that may impact the training process.

Structured resistance training has been shown by Kraemer and Ratamess [86] to induce large acute hormonal changes. This influences anabolic adaptations related to tissue growth and remodeling [86]. In a study by Moran-Navarro [87] acute responses of growth hormone, cortisol, and CK were higher immediately post resistance training with 3 × 10 repetitions of bench press and back squat exercises compared to 3 × 5 and 6 × 5 set and repetition schemes, all at 75% 1RM. Testosterone values were not different between the different loading schemes, with all protocols

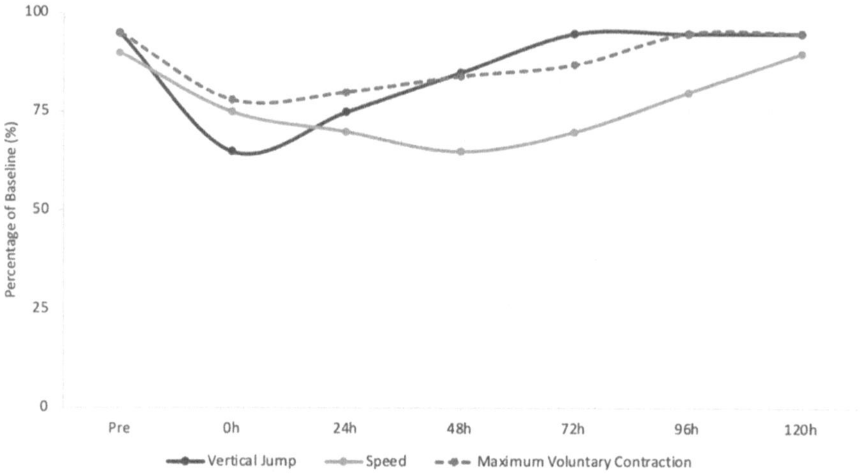

Fig. 13.5 Time-course of recovery of bio-motor abilities

observing that testosterone remained elevated between 24 and 72 h, except for the 3 × 5 repetitions. Strength and power performance were reduced with 3 × 10 repetitions in assessments of countermovement jump (CMJ) and movement velocity against 75% 1RM load within 72 h. Muscular performance recovery appeared to be slower following loading schemes to failure (3 × 10) due to a higher degree of neuromuscular fatigue. Resistance training influences muscle failure due to the magnitude of motor unit recruitment [87, 88], affecting mechanical and neuromuscular recovery, influencing the subsequent programming of the next training stimulus.

Individualization

When planning an athlete's exercise program(s) it is important that the program(s) are specific, catering to the athlete's individual characteristics. Factors such as age (biological and chronological), training age and history, gender, athletic and bio-motor competencies, stage of the season, and injury history need to be considered. Individualization is an important process in optimizing the training process to attain desired adaptation and performance results.

As mentioned previously, the coaching professional must have a sound knowledge of the physical demands (maximal and sub-maximal), the physiological responses associated with competition, and the 'worst case scenarios' that may be faced. These include absolute and relative distances covered at various speed thresholds and intensities, work duration, the scheduling of rest intervals and the density of sport-specific actions. Furthermore, a working knowledge of the metabolic demands of the predominant energy systems utilized during competition and the recovery rates of each system regarding substrate resynthesis is important. Mechanical demands on the musculoskeletal system need to considered relative to competition, for this may impact the time-course of recovery after high force eccentric loads. Acceleration and deceleration efforts place joints in different planes, with changes of direction and jumping actions increasing exercise induced muscle damage and delayed onset muscle soreness, as discussed above. This is especially pertinent in sports involving dense fixture periods (basketball, ice-hockey, baseball) as well as those collision sports consisting of high-force trauma (American football, rugby codes, Gaelic football, Australian rules football).

Finally, an appreciation of common injuries that athletes are susceptible to within their chosen sport, and their specific injury history, is important. This information impacts the design of a training program and the inclusion of supplementary work surrounding injury prevention (joint mobility and range of motion, proprioception, aerobic and anaerobic fitness, reactive strength, rotational strength, explosive power). The aim is to develop a resilient athlete who has a holistic training and strength and conditioning program to withstand the physical demands of their sport. This is coupled with an intentional recovery prescription before recommencing training or the next competitive event.

Contributing Stressors

In addition to planning specific exercise programs, contextual factors should be considered within the training year and its component cycles. The physical and cognitive demands on contemporary elite athletes have grown exponentially in recent years with the growth of organized competitions and events. Associated with this are the ramifications of calendar congestion, the evolution of physical demands [89, 90] and external pressures such as commercial interest, financial investment, rewards, and sociological themes.

The quality of big recovery pillars such as sleep and nutrition should be considered in light of these pressures. The greater frequency of national and international events means a greater reliance on regular air travel in the National Basketball Association [91], National Football League, and National Hockey League [92]. Interruption of the sleep-wake cycle, is itself influenced by the duration and direction of travel, with Fowler et al. [2], showing a greater effect of eastward travel on an athlete's performance. When traveling eastward, sleep onset, time in bed and physical performance in sprint and team-sport specific endurance tests all decreased within the initial 72 h period upon arrival, compared to westward travel. Travel strategies detailing optimal times for dark-light exposure need to be considered in the days preceding and the days after arrival at the competition location. For optimal performance, the advancing or delaying of the biological clock influences the reintroduction to training at the appropriate times based on the athlete's circadian rhythm [91]. In addition, the scheduling of start times for sporting events, as a result of television schedules, needs to be considered. With events timetabled any time between early morning or 11 pm, this affects the timing and quality of sleep. Recent research has reported a significant hindrance in sleep following late night sporting events [93]. This study observed that night matches in elite German soccer players had negative effects on subsequent sleep (later going to bed, greater sleep latency, less time in bed and overall hindered recovery) [93].

Travelling long distances also impacts the ability for proper nutritional intake, exacerbated by sub-optimal choices available at airports. Additionally, the pressurized cabin on the plane may increase dehydration [91]. Analogous to the planning and periodization of exercise programs, one can apply similar concepts of micro- and meso-cycles to nutritional intake, depending on the training and competition schedules of the athlete. Furthermore, the energetic, intensity and duration requirements of an exercise session may differ between sports training sessions and cycles. Similar to the variation of physical training loads programmed, periodized nutrition strategies will support athlete's through varied energy demands, primarily ensuring adequate carbohydrate availability and a positive protein balance. There are now various ways to approach optimal fueling for training and competition in elite sports, reviewed by Bartlett and colleagues [94].

It is also pertinent to consider stressors external to the sporting environment. For example, the student-athlete may have to balance heavy academic workloads with devoting time to training and competition. These additional academic and social elements should be considered when planning the weekly load prescription for training.

Transition between domestic and national team competitive environments are a common occurrence with team-sport athletes. Participating in international friendly games and major tournaments adds further congestion to an already busy domestic (club) calendar. The physical loads, global travel and psychological stress associated with these events are further heightened with a change of the training environment. In this respect, players are exposed to differing training physical preparation and sports science methodologies, which to some, may be unaccustomed, thereby increasing the risk of injury [95]. In this light, the staff of each sport environment (domestic club and national team) need to be transparent sharing pertinent information in support of the player ensuring a seamless transition between training settings [96, 97]. Such information may include training loads, routines, strength and conditioning and pre-habilitation programs and nutritional regimens. A reciprocal exchange of information facilitates the appropriate recovery and safe reintegration of the athlete to their domestic training environment when the national team event is complete.

> There are a myriad of factors to account for in the preparation and recovery of an athlete. The harmonious interplay of physical, psychological, social and financial elements is important to consider when managing the health and wellbeing of an athlete.

Recovery Principles

> The programming of planned recovery sessions and de-load or neutral weeks are essential in allowing desired physiological adaptations to occur ahead of the subsequent training phase. Training sessions will entail varying energy systems targeted, movement patterns and muscle actions all of varying intensity and duration. As a result, the recommended recovery modalities will differ according to predominance of these factors.

When overseeing the periodization, programming, and delivery of exercise programs one needs an awareness of the biological systems being stressed, in order to select the most appropriate modes of recovery. The stresses experienced may be mechanical, neuromuscular, metabolic, or cognitive in nature with each one having a different time-course of recovery [21]. The origin and magnitude of fatigue induced is largely task-(sport) dependent and vary with the intensity, duration, and mode of exercise [44, 47]. For example, strength and endurance training stress both the neuromuscular and aerobic energy systems, though to differing extents, influencing the post-exercise recovery necessary to return to pre-exercise states [21, 99]. Insufficient recovery for proper biological homeostasis could lead to reduced performance capabilities and overtraining symptoms [100] (Fig. 13.6).

This understanding enables practitioners to construct periodized recovery strategies, by addressing the type of stimulus the athlete is recovering from before the next training session or competition. In this light, it is important to remember that

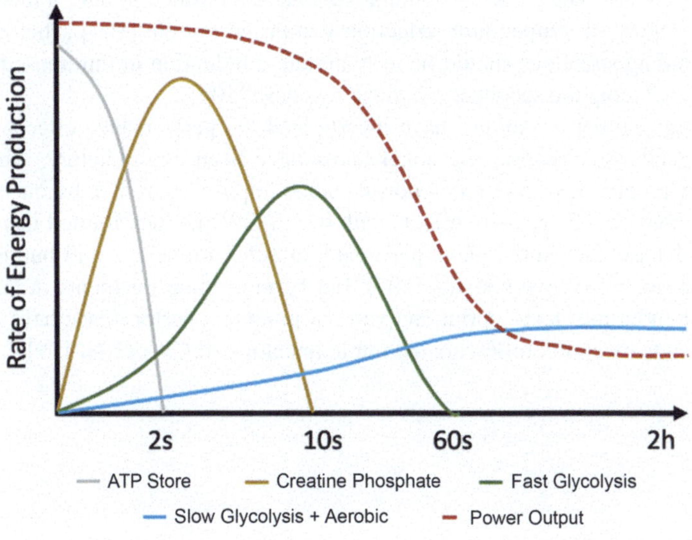

Fig. 13.6 Rate of energy production and subsequent resynthesis of biological energy systems

the internal (physiological) response(s) to a given workload may vary widely between individuals, determining a specific recovery profile or time-course [85], which should be included as part of the exercise plan [44]. This may require the use of various modalities to address the different types of stressors generated, such as; mechanical (EIMD, muscle protein accumulation, inflammatory markers), metabolic (ATP, Glycogen, CP substrate reductions), and biochemical (disruption calcium, sodium-potassium muscle pump affecting muscular contraction) [60, 98, 101, 102].

> The required time for recovery from training-induced fatigue and stress may differ within and between the different organismic systems of the human body.
>
> –Michael Kellman

Cooling Strategies

Thorpe [98] recently proposed that recovery strategies may be influenced with the use of thermoregulatory processes to negate the negative effects of fatigue at different time points post-exercise. Other authors have corroborated this by showing that reducing muscle temperature via cold water immersion (CWI) or cryotherapy at 10–15 °C can alleviate mechanical muscle damage by reducing local blood flow and metabolism [103, 104]. Furthermore, by decreasing the core and skin temperatures with the use of CWI, this reduces both the thermal strain and the inflammatory response post-exercise [105, 106]. The duration of cold application should be between 2 and 30 min in order to sustain reduced intramuscular temperature for optimal recovery. The duration of immersion has been found to have a relationship with the degree of temperature reduction within the muscle [106]. Therefore, the aim of cooling strategies should be to 'maintain a reduction in intramuscular temperature' reducing the secondary damage response [104].

Although numerous studies have investigated the performance effects of CWI following intense exercise, the conclusions have been contradictory, with some observing no effect, with others reporting some improvement in subsequent physical performance. The positive effects with use of CWI include limited increases in creatine kinase and subsequent perceived muscle soreness, contributing to an improved overall recovery status [107]. Furthermore, improvements in YoYo and time trial endurance tests, sprint and vertical jump test performance have all been reported in team sports following the implementation of CWI [108, 109].

Heating Strategies

Conversely, to combating metabolic symptoms of fatigue, the use of heat-application for recovery has become more prominent. The use of hot water immersion (36–44 °C) [110, 111], is believed to promote an anabolic response via the release of calcium, hydrostatic pressure from immersion aiding blood flow, stimulation of tissue metabolism, reduce inflammation, and promotion of glycogen resynthesis following endurance exercise [110, 112, 113]. However, further studies need to be conducted as currently limited evidence exists supporting any consistent, beneficial performance effects of post-exercise hot water immersion.

Several studies investigated the efficacy of contrast water therapy (CWT), alternating hot water (HWI) and cold water (CWI) immersion. Protocols have suggested immersion durations between 5 and 15 min, with temperatures typically between 10–15 °C and >38 °C for CWI and HWI respectively [110]. Although some studies of CWT have not observed any significant effect on performance, others report enhanced activity of the skeletal-muscle pump, promoting improvements in recovery of performance-related parameters in the subsequent 4–5 days [110]. These include improved sprint and repeated-sprint performance in rugby union compared to a control group [114, 115], increased average power output and total work done in field hockey players [116], and accelerated recovery of squat jump and isometric squat in strength trained males [117]. Alternating hot and cold water immersion has been reported to improve subjective perception of muscle soreness [118], as well as a reduction in blood biomarkers, lactate and creatine kinase, following repeated sprint training in netball players and a resistance training regimen in healthy males [119, 120]. Despite an absence of any dose-response relationship with CWT, one study suggested it is more beneficial than HWI and thermoneutral water immersion (TWI) for accelerating recovery from fatiguing exercise [110].

Strength Training

In recent times, strength training has been utilized to take advantages of acute hormonal changes in accelerating neuromuscular recovery from intense exercise. Changes in endocrine secretion has been reported to support physiological adaption in the days after sporting competition(s). The intention is to increase testosterone and growth hormone (GH) levels to contribute to the repair of muscle damage via the promotion of an anabolic environment [86]. The significant hormonal response induced has been shown to occur within 15–30 min following a strength training protocol [86].

Testosterone increases acute GH and insulin-growth factor-1 (IGF-1), enhancing anabolic recovery, as well as contributing to the recovery of neural pathways and force production [121]. Strength training increases acute testosterone levels which are influenced by greater early muscle mass involvement in exercise selection, as well as the volume and intensity of programming [86, 122]. Numerous studies have shown significantly increased testosterone levels with protocols of exercises consisting of 2–4 sets of 4 exercises to failure at 60–100% 1RM, suggesting that a threshold must be met to see a greater hormone response [123–125].

Similarly, GH increases appear to be in response to multi-set protocols of a high intensity [126, 127], with concentric muscle actions [128] and shorter rest periods [122, 129] highlighted to augment the acute response to strength training. Comparing protocols of strength (4 × 5 repetitions at 88% 1RM with 3-min recovery), hypertrophy (4 × 10 repetitions at 75% 1RM with 2-min recovery) and endurance (4 × 15 repetitions at 60% 1RM with 1 min recovery), studies by Zafeiridis [130] and Smilios [131] found that the most significant GH response is derived from endurance protocols when matching total work performed.

Psychological

As part of the planning of periods of physical recovery, the psychological recovery of the athlete should also be considered. Deliberate allotment of time away from sporting facilities for self-care is a key component of holistic recovery before the recommencing of training and/or competition. Whether during the Competition or the Off-season/Transition period, it is important to have a degree of separation from the stressors of structured sports training to which the athlete will hopefully return re-energized and motivated [132]. Spending time with family and friends, practicing mindfulness and relaxation techniques and alternative modes of exercise are some examples of how to achieve this. Psychological recovery is not only important for athletes but their coaches and support staff as well. With sporting calendars being long, demanding, and monotonous at times, family time is diminished, with certain life events missed due to the relentless nature of contemporary elite sport. As a result, there should be a shared belief in psychological recovery for all involved in the high-performance sport process to ensure optimal mood state and continued drive to perform (Fig. 13.7).

> Athlete recovery can be directed according to individual symptoms. The makeup of the previous training session or training phase will dictate the extent of exercise-induced muscle damage, tissue inflammation, depletion of energy substrates and psychological toll, for example. The nature of these responses should be addressed in an individual manner to ensure sufficient recovery before the same training stimulus is repeated in the program.

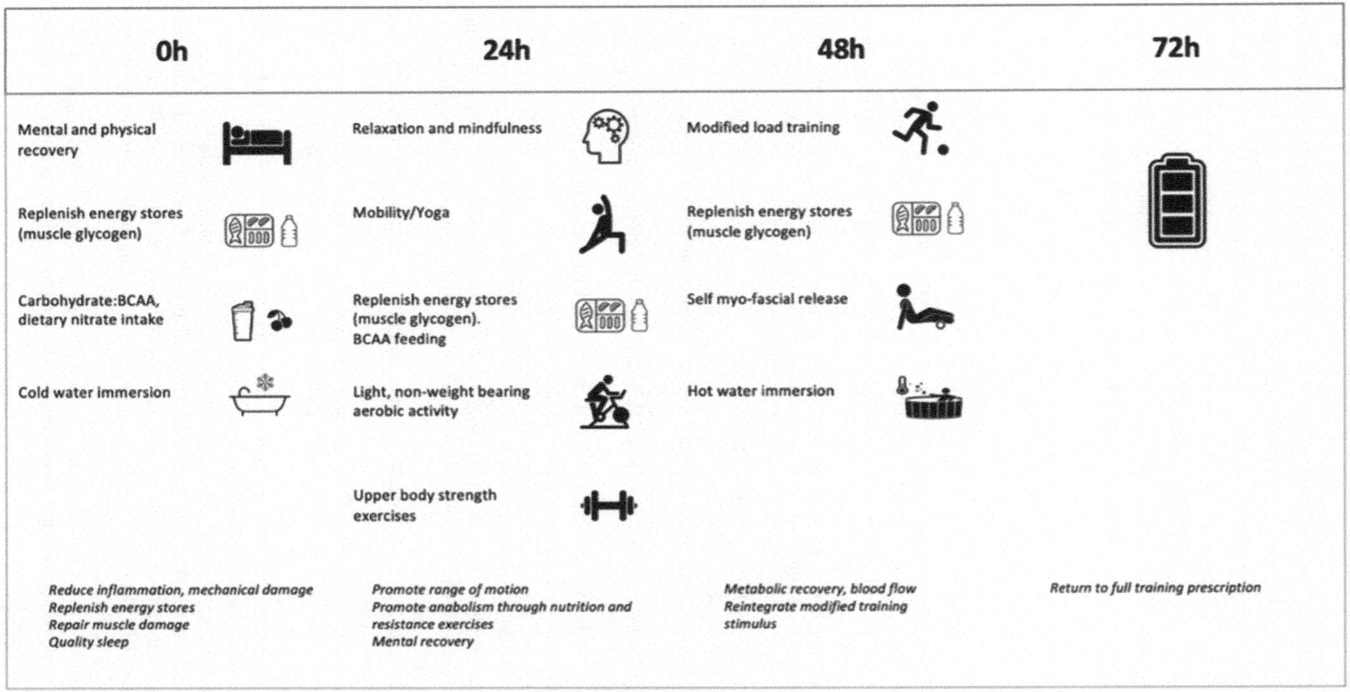

Fig. 13.7 Schematic representation of physiological recovery timeline, modalities advocated and supporting rationale

Conclusion

A sound knowledge of the training process, as well as the types and mechanisms of associated fatigue and their recovery characteristics in each athlete is essential in constructing safe and appropriate training programs. Moreover, coaches should be aware of acute physiological responses to both training and competition when planning training or recovery sessions. Depending on the intensity, frequency, and duration of exercise, the time-course of recovery of specific endocrine, metabolic and performance markers are pertinent to programming in the ensuing 96 h.

References

1. Kellmann M, Beckmann J, editors. Sport, recovery and performance. London: Routledge; 2018.
2. Fowler PM, Knez W, Crowcroft S, Mendham AE, Miller J, Sargent C, Halson S, Duffield R. Greater effect of east versus west travel on jet lag, sleep, and team sport performance. Med Sci Sports Exerc. 2017;49(12):2548–61.
3. Mann JB, Bryant KR, Johnstone B, Ivey PA, Sayers SP. Effect of physical and academic stress on illness and injury in division 1 college football players. J Strength Cond Res. 2016;30(1):20–5.
4. Ekstrand J, Walden M, Hagglund M. Risk for injury when playing in a national football team. Scand J Med Sci Sports. 2004;14(1):34–8.
5. Halson SL. Sleep in elite athletes and nutritional interventions to enhance sleep. Sports Med. 2014;44(1):S13–23.
6. Halson SL. Monitoring training load to understand fatigue in athletes. Sports Med. 2014;44(2):139–47.
7. Bradley PS, Sheldon W, Wooster B, Olsen P, Boanas P, Krustrup P. High-intensity running in English FA premier league soccer matches. J Sports Sci. 2009:1–10.
8. Hoffman JR. The applied physiology of American football. Int J Sports Physiol Perform. 2008;3(3):387–92.
9. Ward PA, Ramsden S, Coutts AJ, Hulton AT, Drust B. Positional differences in running and nonrunning activities during elite American football training. J Strength Cond Res. 2018;32(7):2072–84.
10. Baechle TR, Earle RW. Essentials of strength training and conditioning. 3rd ed. Champaign: Human Kinetics; 2008.
11. Harris RC, Essén B, Hultman E. Glycogen phosphorylase activity in biopsy samples and single muscle fibres of musculus quadriceps femoris of man at rest. Scand J Clin Lab Invest. 1976;36(6):521–6.
12. Hultman E, Sjoholm H. Biochemical causes of fatigue. In: Human muscle power. Champaign: Human Kinetics; 1986. p. 215–38.
13. Bompa TO. Periodization training for sports. Champaign: Human Kinetics; 1999.
14. Comfort P, Matthews M. Assessment and needs analysis. In: Comfort P, Abrahamson, editors. Sports rehabilitation and injury prevention. Chichester: Wiley-Blackwell; 2010.
15. Pyne D, Sharp RL. Physical and energy requirements of competitive swimming events. Int J Sports Physiol Perform. 2014;24:351–9.
16. Ruddock AD, Wilson DC, Thompson SW, Hembrough D, Winter EM. Strength and conditioning for professional boxing: recommendations for physical preparation. Strength Cond J. 2016;38(3):81–90.

17. Rosenblatt B. Planning a performance programme. In: Joyce D, Lewindon D, editors. High-performance training for sports. Human Kinetics: Champaign; 2014.
18. Lämmle L, Tittlbach S, Oberger J, Worth A, Bös K. A two-level model of motor performance ability. J Exerc Sci Fit. 2010;8(1):41–9.
19. Matveyev LP. Fundamental of sport training. Moscow: Progress Publishers; 1981.
20. Issurin VB. New horizons for the methodology and physiology of training periodization. Sports Med. 2010;40(3):189–206.
21. Lambert MI, Mujika I. Physiology of exercise training. In: Hausswirth C, Mujika I, editors. Recovery for performance in sport. Champaign: Human Kinetics; 2013.
22. Issurin VB. Benefits and limitations of block periodized training approaches to athletes' preparation: a review. Sports Med. 2016;46:329–38.
23. Issurin V. Block periodization versus traditional training theory: a review. J Sports Med Phys Fitness. 2008;48:65–75.
24. Fleck SJ, Kraemer WJ. Designing resistance training programs. 2nd ed. Champaign: Human Kinetics; 1997.
25. Bradley-Popovich GE. Nonlinear versus linear periodization models. Strength Cond J. 2009;23(1):42–4.
26. Gilliam GM. Effects of frequency of weight training on muscle strength training. J Sports Med. 1981;21:432–6.
27. Bondarchuk AP. Constructing a training system. Track Tech. 1988;102:254–69.
28. Impellizzeri FM, Rampinini E, Marcora SM. Physiological assessment of aerobic training in soccer. J Sports Sci. 2005;23(6):583–92.
29. Borreson J, Lambert MI. The quantification, of training load, the training response and the effect on performance. Sports Med. 2009;39(9):779–95.
30. Selye H. The general adaptation syndrome and the diseases of adaptation. J Clin Endocrinol Metab. 1946;6:117–230.
31. Halson SL, Jeukendrup AE. Does overtraining exist? An analysis of overreaching and overtraining research. Sports Med. 2004;34(14):967–81.
32. Fry RW, Morton AR, Keast., D. Overtraining in athletes. An update. Sports Med. 1991;12(1):32–65.
33. Meeusen R, Duclos M, Foster C, Fry A, Gleeson M, Nieman D, Raglin J, Rietjens G, Steinacker J, Urhausen A. Prevention, diagnosis and treatment of the overtraining syndrome: ECSS Position Statement 'Task Force'. Eur J Sport Sci. 2006;6(1):1–14.
34. Kreider R, Fry AC, O'Toole M. Overtraining in sport: terms, definitions, and prevalence. In: Kreider R, Fry AC, O'Toole M, editors. Overtraining in sport. Champaign: Human Kinetics; 1998.
35. Gabbett TJ. The training—injury prevention paradox: should athletes be training smarter and harder? Br J Sports Med. 2016:1–9.
36. Lyman S, Fleisig GS, Andrews JR, Osinski ED. Effect of pitch type, pitch count, and pitching mechanics on risk of elbow and shoulder pain in youth baseball pitchers. Am J Sports Med. 2002;30(4):463–8.
37. Visnes H, Bahr R. Training volume and body composition as risk factors for developing jumper's knee among young elite volleyball players. Scand J Med Sci Sports. 2012;23(5):607–13.
38. Hulin BT, Gabbett TJ, Blanch P, Chapman P, Bailey D, Orchard JW. Spikes in acute workload are associated with increased injury risk in elite cricket fast bowlers. Br J Sports Med. 2014;48:708–12.
39. Orchard JW, James T, Portus M, Kountouris A, Dennis R. Fast bowlers in cricket demonstrate up to 3- to 4-week delay between high workloads and increased risk of injury. Am J Sports Med. 2009;37(6):1186–92.
40. Hulin BT, Gabbett TJ, Lawson DW, Caputi P, Sampson JA. The acute: chronic workload predicts injury: high chronic workload may decrease injury risk in elite rugby league players. Br J Sports Med. 2015:1–7.
41. Drew MK, Finch CF. The relationship between training load and injury, illness and soreness: a systematic and literature review. Sports Med. 2016;46(6):861–83.

42. Bowen L, Gross AS, Gimpel M, Li F-X. Accumulated workloads and the acute: chronic workload ratio relate to injury risk in elite youth football players. Br J Sports Med. 2016;51:452–9.
43. Malone S, Owen A, Newton M, Mendes B, Collins KD, Gabbett TJ. The acute: chronic workload ratio in relation to injury risk in professional soccer. J Sci Med Sport. 2017;20(6):561–5.
44. Skorski S, Mujika I, Bosquet L, Meeusen R, Coutts AJ, Meyer T. The temporal relationship between exercise, recovery processes and changes in performance. Int J Sports Physiol Perform. 2019;14:1015–21.
45. Wu PP, Sterkenburg N, Everett K, Chapman DW, White N, Mengersen K. Predicting fatigue using countermovement jump force-time signatures: PCA can distinguish neuromuscular versus metabolic fatigue. PLoS One. 2019;14(7):1–16.
46. Mujika I, Halson S. Models of fatigue. In: Schwellness MP, editor. Olympic textbook of medicine in sport: volume XIV of the encyclopaedia of sports medicine. An IOC Medical Commission Publication. West Sussex: Wiley-Blackwell; 2008.
47. Carroll TJ, Taylor JL, Gandevia SC. Recovery of central and peripheral neuromuscular fatigue after exercise. J Appl Physiol. 2017;122:1068–76.
48. Green JH. The autonomic nervous system and exercise. London: Chapman and Hall; 1990.
49. Lattier G, Millet GY, Martin A, Martin V. Fatigue and recovery after high-intensity exercise part I: neuromuscular fatigue. Int J Sports Med. 2004;25(6):450–6.
50. Allen DG, Lamb GD, Westerblad H. Skeletal muscle fatigue: cellular mechanisms. Physiol Rev. 2008;88(1):287–332.
51. Bangsbo J, Madsen K, Kiens B, Richter EA. Effect of muscle acidity on muscle metabolism and fatigue during intense exercise in man. J Physiol. 1996;495(2):587–96.
52. Ortenblad N, Westerblad H, Nielsen J. Muscle glycogen stores and fatigue. J Physiol. 2013;591(18):4405–13.
53. Fitts RH. Cellular mechanisms of muscle fatigue. Physiol Rev. 1994;74(1):49–94.
54. McKenna MJ, Bangsbo J, Renaud JM. Muscle K+, Na+, and Cl disturbances and Na+-K+ pump inactivation: implications for fatigue. J Appl Physiol. 2008;104(1):288–95.
55. Urhausen A, Kindermann W. Diagnosis of overtraining: what tools do we have? Sports Med. 2002;32(2):95–102.
56. Brancaccio P, Lippi G, Maffulli N. Biochemical markers of muscular damage. Clin Chem Lab Med. 2010;48(6):757–67.
57. Ispirlidis I, Fatouros IG, Jamurtas AZ, Nikolaidis MG, Michailidis I, Douroudos I, Margonis K, Chatzinikolaou A, Kalistratos E, Katrabasas I, Alexiou V, Taxildaris K. Time-course of changes in inflammatory and performance responses following a soccer game. Clin J Sports Med. 2008;18(5):423–31.
58. Saltin B. Metabolic fundamentals in exercise. Med Sci Sports Exerc. 1973;5:137–46.
59. Bangsbo J, Mohr M, Krustrup P. Physical and metabolic demands of training and match-play in the elite soccer player. J Sports Sci. 2006;24(7):665–74.
60. Rustrup P, Ortenblad N, Nielsen J, Nybo L, Gunnarsson TP, Iaia FM, Madsen K, Stephens F, Greenhaff P, Bangsbo J. Maximal voluntary contraction force, SR function and glycogen resynthesis during the first 72 h after a high-level competitive soccer game. Eur J Appl Physiol. 2011;111(12):2987–95.
61. Gunnarsson TP, Bendiksen M, Bischoff R, Christensen PM, Lesivig B, Madsen K, Stephens F, Greenhaff P, Krustrup P, Bangsbo J. Effect of whey protein- and carbohydrate-enriched diet on glycogen resynthesis during the first 48 h after a soccer game. Scand J Med Sci Sports Exerc. 2013;23(4):508–15.
62. Bangsbo J. The physiology of soccer – with special reference to intense intermittent exercise. Acta Physiol Scand. 1994;151(supplement 619):1–155.
63. Nedelec M, McCall A, Carling C, Legall F, Berthoin S, Dupont G. Recovery in soccer: part I-post-match fatigue and time course of recovery. Sports Med. 2012;42(12):997–1015.
64. Urhausen A, Gabriel H, Kindermann W. Blood hormones as markers of training stress and overtraining. Sports Med. 1995;20(4):251–76.

65. Gleeson M. Biochemical and immunological markers of overtraining. J Sports Sci Med. 2002;1:31–41.
66. McLellan CP, Lovell DI, Gass GC. Creatine kinase and endocrine responses of elite players pre, during and post rugby league match play. J Strength Cond Res. 2010;24(11):2908–19.
67. Rowell AE, Aughey RJ, Hopkins WG, Stewart AM, McCormick SJ. Identification of sensitive measures of recovery after external load from football match play. Int J Sports Physiol Perform. 2017;12:969–76.
68. Kupchak BR, Kraemer WJ, Hoffman MD, Phinney SD, Volek JS. The impact of an ultramarathon on hormonal and biochemical parameters in men. Wilderness Environ Med. 2014;25(3):278–88.
69. Cunniffe B, Hore AJ, Whitcombe DM, Jones KP, Baker JS, Davies B. Time course of changes in immunoendocrine markers following an international rugby game. Eur J Appl Physiol. 2010;108:113–22.
70. MacKinnon LT, Hooper SL, Jones S, Gordon RD, Bachmann AW. Hormonal, immunological, and hematological responses to intensified training in elite swimmers. Med Sci Sports Exerc. 1996;29(12):1637–45.
71. Adlercreutz H, Härkönen M, Kuoppasalmi K, Näveri H, Huhtaniemi I, Tikkanen H, Remes K, Dessypris A, Karvonen J. Effect of training on plasma anabolic and catabolic steroid hormones and their response during physical exercise. Int J Sports Med. 1986;7(Suppl 1):27–8.
72. Vervoorn C, Quist AM, Vermulst LJ, Erich WB, de Vries WR, Thijssen JH. The behaviour of the plasma free testosterone/cortisol ratio during a season of elite rowing training. Int J Sports Med. 1991;12(3):257–63.
73. Kraemer WJ, Spiering BA, Volek JS, Martin GJ, Howard RL, Ratamess NA, Hatfield DL, Vingren JL, Ho JY, Fragala MS, Thomas GA, French DN, Anderson JM, Häkkinen K, Maresh CM. Recovery from a national collegiate athletic association division I football game: muscle damage and hormonal status. J Strength Cond Res. 2009;23(1):2–10.
74. Pedlar C, Newell J, Lewis NA. Blood biomarker profiling and monitoring for high-performance physiology and nutrition: current perspectives, limitations and recommendations. Sports Med. 2019;49(Suppl 2):S185–98.
75. Heisterberg MF, Fahrenkrug J, Krustrup P, Storskov A, Kjaer M, Andersen JL. Extensive monitoring through multiple blood samples in professional soccer players. J Strength Cond Res. 2013;27(5):1260–71.
76. McHugh M. Recent advances in the understanding of the repeated bout effect: the protective effect against muscle damage from a single bouts of eccentric exercise. Scand J Med Sci Sports. 2003;13:88–97.
77. Chatzinikolaou A, Draganidis D, Avloniti A, Karipidis A, Jamurtas AZ, Skevaki CL, Tsoukas D, Sovatzidis A, Theodorou A, Kambas A, Papassotiriou I, Taxildaris K, Fatouros I. The microcycle of inflammation and performance changes after a basketball match. J Sports Sci. 2014;32(9):870–82.
78. Hovsepian A, Esfarjani F, Bambaeichi E, Zolaktaf V. The effect of high intensity functional training on the oxidative status, muscle damage and performance of basketball players. J Sports Med Phys Fitness. 2021;61(2):188–98.
79. Ascencao A, Rebelo A, Oliveria E, Marques F, Pereira L, Magalhaes J. Biochemical impact of a soccer match – analysis of oxidative stress and muscle damage markers throughout recovery. Clin Biochem. 2008;41:841–51.
80. Small K, McNaughton LR, Greig M, Lohkamp M, Lovell R. Soccer fatigue, sprinting and hamstring injury risk. Int J Sports Med. 2010;30:573–8.
81. Magalhaes J, Rebelo A, Oliveira E, Silva JR, Marques F, Ascensao A. Impact of Loughborough Intermittent Shuttle Test versus soccer match on physiological, biochemical and neuromuscular parameters. Eur J Appl Physiol. 2010;108(1):39–48.
82. Milioni F, Azevedo RA, Zagatto AM, Millet GY. Time course of recovery after cycling repeated sprints. Med Sci Sports Exerc. 2020;53(2):413–20.

83. Halson S, Bridge MW, Meeusen R, Busschaert B, Gleeson M, Jones DA, Jeukendrup A. Time course of performance changes and fatigue markers during intensified training in trained cyclists. J Appl Physiol. 2002;93:947–56.
84. Minett GM, Duffield R. Is recovery driven by central or peripheral factors? A role for the brain in recovery following intermittent-sprint exercise. Front Physiol. 2014;5(24):2–9.
85. Brownstein CG. Etiology and recovery of neuromuscular fatigue following competitive soccer match-play. Front Physiol. 2017;8:1–14.
86. Kraemer WJ, Ratamess NA. Hormonal responses and adaptation to resistance exercise and training. Sports Med. 2005;35(4):339–61.
87. Moran-Navarro R, Perez CE, Mora-Rodriguez R, de la Cruz-Sanchez E, Gonzalez-Badillo JJ, Sanchez-Medina LS, Pallares JG. Time course of recovery following resistance training leading or not to failure. Eur J Appl Physiol. 2017;117(12):2387–99.
88. Ahtiainen JP, Pakarinen A, Alen M, Kraemer WJ, Häkkinen K. Muscle hypertrophy, hormonal adaptations and strength development during strength training in strength-trained and untrained men. Eur J Appl Physiol. 2003;89(6):555–63.
89. Barnes C, Archer DT, Hogg B, Bush M, Bradley PS. The evolution of physical and technical performance parameters in the English Premier League. Int J Sports Med. 2014;35(13):1095–100.
90. Bush M, Barnes C, Archer DT, Hogg B, Bradley PS. Evolution of match performance parameters for various playing positions in the English Premier League. Hum Mov Sci. 2015;39:1–11.
91. Huyghe T, Calleja-Gonzalez J. Long-distance traveling in basketball: practical applications based on scientific evidence. In: Laver L, Kocaoglu B, Cole B, Arundale AJH, Bytomski J, Amendola A, editors. Basketball sports medicine and science. Berlin, Heidelberg: Springer; 2020.
92. Roy J, Forest G. Greater circadian disadvantage during evening games for the National Basketball Association (NBA), National Hockey League (NHL) and National Football League (NFL) teams travelling westward. J Sleep Res. 2018;27(1):86–9.
93. Fullagar HHK, Skorski S, Duffield R, Julian R, Bartless J, Meyer T. Impaired sleep and recovery after nigh matches in elite football players. J Sports Sci. 2016;34(14):1333–9.
94. Bartlett JD, Hawley JA, Morton JP. Carbohydrate availability and exercise training adaptation: too much of a good thing? Eur J Sport Sci. 2015;15(1):3–12.
95. Ekstrand J, Lundqvist D, Lagerbäck L, Vouillamoz M, Papadimitiou N, Karlsson J. Is there a correlation between coaches' leadership styles and injuries in elite football teams? A study of 36 elite teams in 17 countries. Br J Sports Med. 2018;52(8):527–31.
96. Buchheit M, Dupont G. Elite clubs and national teams: sharing the same party? Sci Med Footb. 2018;2(2):83–5.
97. Noor D, McCall A, Jones M, Duncan C, Ehrmann F, Meyer T, Duffield R. Transitioning from club to national teams: training and match load profiles of international footballers. J Sci Med Sport. 2019;22(8):948–54.
98. Thorpe RT. Post-exercise recovery: cooling and heating, a periodized approach. Front Sports Active Living. 2022;3:1–6.
99. Buchheit M, Laursen PB. High-intensity interval training, solutions to the programming puzzle: part I: cardiopulmonary emphasis. Sports Med. 2013;43(5):313–38.
100. Barnett A. Using recovery modalities between training sessions in elite athletes: does it help? Sports Med. 2006;36(9):781–96.
101. Kellmann M, Bertollo M, Bosquet L, Brink M, Coutts AJ, Duffield R, Erlacher D, Halson SL, Hecksteden A, Heidari J, Kallus KW, Meeusen R, Mujika I, Robazza C, Skorski S, Venter R, Beckmann J. Recovery and performance in sport: consensus statement. Int J Sports Physiol Perform. 2018;13(2):240–5.
102. Mohr M, Vigh-Larsen JF, Krustrup P. Muscle glycogen in elite soccer – a perspective on the implication for performance, fatigue, and recovery. Front Sports Active Living. 2022;4:1–4.

103. Gregson W, Black MA, Jones H, Milson J, Morton J, Dawson B, Atkinson G, Green DJ. Influence of cold water immersion on limb and cutaneous blood flow at rest. Am J Sports Med. 2011;39(6):1316–23.
104. Kwiecien SY, McHugh MP. The cold truth: the role of cryotherapy in the treatment of injury and recovery from exercise. Eur J Appl Physiol. 2021;121(8):2125–42.
105. Vaile J, Halson S, Gill N, Dawson B. Effect of cold water immersion on repeat cycling performance and thermoregulation in the heat. J Sports Sci. 2008;26(5):431–40.
106. Peiffer JJ, Abbiss CR, Nosaka K, Peake JM, Laursen PB. Effect of cold water immersion after exercise in the heat on muscle function, body temperatures, and vessel diameter. J Sci Med Sport. 2009;12(1):91–6.
107. Ascensão A, Leite M, Rebelo AN, Magalhäes S, Magalhäes J. Effects of cold water immersion on the recovery of physical performance and muscle damage following a one-off soccer match. J Sports Sci. 2011;29(3):217–25.
108. Montgomery PG, Pyne DB, Hopkins WG, Dorman JC, Cook K, Minahan CL. The effect of recovery strategies on physical performance and cumulative fatigue in competitive basketball. J Sports Sci. 2008;26(11):1135–45.
109. Brophy-Williams N, Landers G, Wallman K. Effect of immediate and delayed cold water immersion after a high intensity exercise session on subsequent run performance. J Sports Sci Med. 2011;10(4):665–70.
110. Versey NG, Halson SL, Dawson BT. Water immersion recovery for athletes: effect on exercise performance and practical recommendations. Sports Med. 2013;43(11):1101–30.
111. Skurvydas A, Kamandulis S, Stanislovaitis A, Streckis V, Mamkus G, Drazdauskas A. Leg immersion in warm water, stretch-shortening exercise, and exercise-induced muscle damage. J Athl Train. 2008;43(6):592–9.
112. Cheng AJ, Willis SJ, Zinner C, Chaillou T, Ivarsson N, Ørtenblad N, Lanner JT, Holmberg HC, Westerblad H. Post-exercise recovery of contractile function and endurance in humans and mice is accelerated by heating and slowed by cooling skeletal muscle. J Physiol. 2017;595(24):7413–26.
113. McGorm H, Roberts L, Coombes J, Peak J. Turning up the heat: an evaluation of the evidence for heating to promote exercise recovery, muscle rehabilitation and adaption. Sports Med. 2018;48(6):1311–28.
114. King M, Duffield R. The effects of recovery interventions on consecutive days of intermittent sprint exercise. J Strength Cond Res. 2009;23(6):1795–802.
115. Higgins TR, Heazlewood IT, Climstein M. A random control trial of contrast baths and ice baths for recovery during competition in U/20 rugby union. J Strength Cond Res. 2011;25(4):1046–51.
116. Sayers M, Calder A, Sanders J. Effect of whole-body contrast- water therapy on recovery from intense exercise of short duration. Eur J Sport Sci. 2011;11(4):293–302.
117. Vaile J, Halson S, Gill N, Dawson B. Effect of hydrotherapy on signs and symptoms of delayed onset muscle soreness. Eur J Appl Physiol. 2008;102(4):447–55.
118. Ingram J, Dawson B, Goodman C, Wallman K, Beilby J. Effect of water immersion methods on post-exercise recovery from simulated team sport exercise. J Sci Med Sport. 2009;12(3):417–21.
119. Hamlin MJ. The effect of contrast temperature water therapy on repeated sprint performance. J Sci Med Sport. 2007;10(6):398–402.
120. French DN, Thompson KG, Garland SW, Barnes CA, Portas MD, Hood PE, Wilkes G. The effects of contrast bathing and compression therapy on muscular performance. Med Sci Sports Exerc. 2008;40(7):1297–306.
121. Giustina A, Veldhuis JD. Pathophysiology of the neuroregulation of growth hormone secretion in experimental animals and the human. Endocr Rev. 1998;19:717–97.
122. Kraemer WJ, Marchitelli L, Gordon SE, Harman E, Dziados JE, Mello R, Frykman P, McCurry D, Fleck SJ. Hormonal and growth factor responses to heavy resistance exercise protocols. J Appl Physiol. 1990;69(4):1442–50.

123. Weiss LW, Cureton KJ, Thompson FN. Comparison of serum testosterone and androstenedione responses to weight lifting in men and women. Eur J Appl Physiol. 1983;50:413–9.
124. Schwab R, Johnson GO, Housh TJ, Kinder JE, Weir JP. Acute effects of different intensities of weight lifting on serum testosterone. Med Sci Sports Exerc. 1993;25(12):1381–5.
125. Raastad T, Bjoro T, Hallen J. Hormonal responses to high- and moderate-intensity strength exercise. Eur J Appl Physiol. 2000;82:121–8.
126. Hansen S, Kvorning T, Kjaer M, Sjøgaard G. The effect of short-term strength training on human skeletal muscle: the importance of physiologically elevated hormone levels. Scand J Med Sci Sports. 2001;11(6):347–54.
127. Ahtiainen JP, Pakarinen A, Kraemer WJ, Häkkinen K. Acute hormonal and neuromuscular responses and recovery to forced vs maximum repetitions multiple resistance exercises. Int J Sports Med. 2003;24(6):410–8.
128. Durand RJ, Castracane VD, Hollander DB, Tryniecki JL, Bamman MM, O'Neal S, Hebert EP, Kraemer RR. Hormonal responses from concentric and eccentric muscle contractions. Med Sci Sports Exerc. 2003;35(6):937–43.
129. Kraemer WJ, Gordon SE, Fleck SJ, Marchitelli LJ, Mello R, Dziados JE, Friedl K, Harman E, Maresh C, Fry AC. Endogenous anabolic hormonal and growth factor responses to heavy resistance exercise in males and females. Int J Sports Med. 1991;12(2):228–35.
130. Zafeiridis A, Smilios I, Considine RV, Tokmakidis SP. Serum leptin responses after acute resistance exercise protocols. J Appl Physiol. 2003;94(2):591–7.
131. Smilios I, Pilianidis T, Karamouzis M, Tokmakidis SP. Hormonal responses after various resistance exercise protocols. Med Sci Sports Exerc. 2003;35(4):644–54.
132. Mujika I, Halson S, Burke LM, Balagué G, Farrow D. An integrated, multifactorial approach to periodization for optimal performance in individual and team sports. Int J Sports Physiol Perform. 2018;13(5):538–61.

Chapter 14
Interactions Between Sleep and Exercise Performance

Angelos Vlahoyiannis, Gregory C. Bogdanis, Fabio Y. Nakamura, Petrus Gantois, and Christoforos D. Giannaki

Sleep Physiology

Definition of Sleep

Sleep is an actively modulated and metabolically discrete state, and may be characterized as the cornerstone of health, recovery, and regeneration. Notably, humans spend about one-third of their lifespans either sleeping or trying to fall asleep. For many years, sleep was under extended investigation regarding its effects on brain function [1] and its consequences on cardiovascular health (especially in the case of sleep disorders). In the last decades, research revealed evidence that sleep directly impacts muscle recovery [2, 3] and can also affect a wide range of physiological parameters that could affect muscle recovery indirectly, such as hormone secretion, immune system, and exercise performance. It is incorrect to identify sleep as the absence of wakefulness and consciousness or the suspension of sensorial processes, which could also occur during pathological states. Instead, sleep is more accurately

A. Vlahoyiannis · C. D. Giannaki (✉)
School of Life and Health Sciences, University of Nicosia, Nicosia, Cyprus
e-mail: giannaki.c@unic.ac.cy

G. C. Bogdanis
Physical Education and Sport Science, National and Kapodistrian University of Athens, Athens, Greece

F. Y. Nakamura
Research Center in Sports Sciences, Health Sciences and Human Development (CIDESD), University of Maia, Maia, Portugal

P. Gantois
Associate Graduate Program in Physical Education, Federal University of Paraíba, João Pessoa, Brazil

N. C. Apostolopoulos et al. (eds.), *Fundamentals of Recovery, Regeneration, and Adaptation to Exercise Stress: An Integrated Approach*,
https://doi.org/10.1007/978-3-031-44270-4_14

described as the combination of the activation of specific neurons in selective brain areas and the passive withdrawal of afferent stimuli to the brain. Hence, in contrast to several pathological states characterized by decreased consciousness, sleep is a constantly reversible and actively regulated state that enforces rest and fasting, supporting the optimization of metabolic processes at the appropriate phase of the 24-h cycle.

Sleep Initiation, Maintenance and Architecture

The circadian effects are entrained to the 24 h cycle and several homeostatic influences -inversely related to time, spent wake-regulate sleep initiation and maintenance. This is widely described as a "two-process" model consisting of the circadian and homeostatic processes. The circadian process is mainly referred to as a 24-hour physiological response to environmental factors and endogenous neural oscillators. On the other hand, the homeostatic process consists of mechanisms that accumulate during wakefulness and dissipate during sleep. This two-process model simulates, to a great extent, sleep initiation and maintenance in diverse experimental protocols.

Sleep initiation and maintenance require suppression of the ascending reticular activating system, a subcortical neural network critical for consciousness. Similarly, a well-defined subcortical network of brain structures regulates the transitions between sleep and wake states. Even though the molecular triggers to sleep onset have not been fully defined, extracellular adenosine could be a potential candidate that triggers sleep initiation. Adenosine accumulates during wakefulness and decreases as sleep progress [4]. For example, caffeine and theophylline are commonly known for their alarming effects due to their acting on adenosine receptors as adenosine antagonists. Nevertheless, it is almost certain that other molecules play important signaling roles in sleep initiation and maintenance.

Following sleep onset, sleep maintenance is mediated by several sleep stages, composing the "sleep architecture". However, sleep by itself is not a homogenous construct. During sleep maintenance, the human body switches among four sleep stages and sleep usually progresses as follows: N1 and N2 sleep stages (commonly referred to as "light sleep"), N3 sleep stage (commonly referred to as "deep sleep" and REM sleep stage (from the words: rapid eye movement) [5]. This typical sleep architecture is characterized by a rhythmic alternation between non-REM (NREM) and REM stages that are controlled by the switching between activity and inhibition of specific neurons within the brainstem [6, 7].

Sleep Patterns in Athletes

Even though regular exercise is essential for good sleep and is considered an effective non-pharmacological approach for the management of sleep disorders [8, 9], athletic populations often experience poor sleep and sleep disturbances [10].

Sleep Quantity

The prevalence of inadequate sleep quantity in athletes is well documented [11]. In most cases, athletes do not achieve the suggested 8 hours of sleep per night [10]. This inability to maintain sleep for an adequate duration is met across all athletes' lifespan spectrum, disturbing their recovery. A great body of evidence suggests that athletes' sleep duration is limited to approximately 7.2 hours per night. Adolescent athletes also seem to experience sleep deprivation as they sleep approximately 6.3 hours per night, falling below the suggested American Academy of Sleep Medicine guidelines for pediatric populations of 8 hours per night [5]. During early adulthood, sleep increases at 7.1 hours per night, but again this falls into the lowest value of the suggested limit of 7 hours per night [10]. Nevertheless, these suggested cut-offs for the general population may underestimate athletes' sleep needs to recover, since regeneration demand is increased in athletes, and sleep extension can accelerate recovery and athletes' performance. Indeed, athletes have reported requiring more than 8 hours of sleep per night to feel fully recovered [12].

Elite athletes commonly experience insufficient sleep and poor sleep quality.

Sleep Quality

Sleep efficiency, defined as the percentage of the ratio of sleep duration to time in bed, is a key measure of athletes' sleep quality [10]. Similar to sleep duration, sleep quality is also lower in athletes, independent of athletic expertise or age group. Most importantly, decreased sleep duration is not attributed to the difficulty of initiating sleep, but rather to maintaining sleep. This, in turn, suggests that athletes have longer or more frequent wake bouts during night sleep. The combination of inadequate sleep duration and decreased sleep quality could be detrimental to both athletes' health and performance, and it could be a substantial limiting factor for regeneration and growth.

Sleep Architecture

As already mentioned, sleep is not a homogenous construct. Sleep progresses from NREM to REM sleep stages and vice versa, with each sleep stage promoting recovery and regeneration of specific tissues. The transition from light sleep to deep sleep usually takes 30–60 minutes, and NREM alternates with the REM sleep stages cyclically, four to six times per night. Commonly, the first two sleep cycles consist mainly of deep sleep. This pattern is altered in the following cycles, in which deep sleep decreases and REM sleep increases. As in the general population, athletes' NREM sleep stages constitute approximately 75–80% of adult sleep duration. Specifically, athletes sleep about 23.1% in REM, 9.7% N1, 48.7% N2, and 18.5% N3 sleep stage. However, sleep architecture can be internally modified acutely in response to environmental factors or strenuous exercise [13]. For example, ultra-endurance competitions increase N3 sleep stage, potentially as a compensatory mechanism to facilitate recovery [14].

Exercise Timing and Intensity on Sleep Physiology

Exercise is a potent, safe and accessible non-pharmacological approach to improve sleep duration and quality. There is a reciprocal relationship between exercise and sleep [15]. Exercise is usually associated with good sleep quality in the general population [16]. Furthermore, exercise may decrease the severity of symptoms of patients with sleep disorders [8]. Nevertheless, the effect of exercise on athletes' sleep and its subsequent impact on recovery may vary according to exercise type, intensity, volume, frequency, and timing.

Exercise Type and Intensity

In this part, exercise type is discussed in relation to energy metabolism (i.e., aerobic, anaerobic and resistance training). Data regarding resistance training are fewer than aerobic exercise, although it seems that this form of exercise can positively impact symptoms of anxiety and depression, and can also improve sleep quality [17]. Sleep seems to be unaffected by increasing strength training frequency up to twice per day [18].

Similarly, aerobic exercise does not seem to negatively affect sleep duration in acute [19] and long-term [20] interventions. There is a plethora of evidence showing that aerobic exercise intensities up to 75% of maximum oxygen uptake (VO_2max) do not affect sleep-related parameters [21, 22]. In controlled experiments, medium-intensity continuous training does not affect sleep initiation and continuity [19]. These results are consistent after a 4-week intervention [20]. However, increasing exercise intensity seems to improve sleep quality [20]. This is often attributed to the increase in body temperature during exercise which causes vasodilation and increases blood flow to the body's peripheral tissues. Similar vasodilation occurs during sleep, helping to decrease body temperature and attain homeostasis [23]. Another reason could be that sleep efficiency, sleep initiation and maintenance are consistently shown to be highly related to post-exercise fatigue, underlying once again the great importance of exercise on body sleep [24]. High-intensity interval training (HIIT) improves sleep quality and initiation, making it a promising tool for improving sleep and potentially its restoration effects [25]. Also, there is evidence that high-intensity anaerobic exercise, such as Sprint Interval Training (SIT), has been shown to acutely optimize sleep initiation and continuity in controlled environments [19].

However, further increases in training frequency within the microcycle, such as training for 6 days per week, may have a negative effect on sleep efficiency in athletes [18]. Similar results have also been reported for cyclists with increased training frequency [26]. In a real-life case of prolonged intensified exercise and overreaching, sleep efficiency and duration decrease gradually [26, 27]. It appears that overreached athletes have difficulty to recover because the sympathetic tone is increased during deep sleep [28]. Hence, this could be a reason why overreached and overstrained athletes report decreased self-perceived sleep quality and impaired exercise performance [29]. Considering all the above evidence, sleep disturbances along with decreased exercise performance in athletes are proposed as an indicator of overreaching during periods of intensified training [30].

In the case of an extreme duration aerobic event (e.g. ultra-marathons), it is consistently shown that sleep continuity is disturbed, and sleep architecture is altered, favoring the N3 sleep stage, rather than REM sleep [14, 31], indicating a key role of N3 on recovery after extreme endurance events. In contrast, in the evening of a "day-off", N3 sleep stage decreases and REM stage increases compared to a training day containing high-intensity exercise [32]. This reduction in N3 sleep may occur because of reduced recovery needs [32, 33], indicating a connection between training load and N3 sleep stage [34].

Recent evidence suggests that high-intensity exercise forms do not have a detrimental effect on sleep, even when performed in the late evening.

Timing of Exercise

Exercise timing plays a key role in sleep modulation. Data shows that morning exercise has little or no direct effect on sleep initiation and duration compared to evening or late-evening workouts. Generally, in clinical trials, training in the morning does not affect sleep quality or architecture [34, 35]. Nevertheless, in a population with sleep disorders, 2 weeks of morning exercise may improve sleep duration [36]. Scheduling exercise to the early evening hours, does not seem to significantly affect nocturnal sleep quantity [22, 37]. However, sleep architecture, and specifically REM sleep, may decrease [34]. Scheduling exercise up to 3 h prior to the usual bedtime has no impact or can even improve sleep efficiency [19]. Evening workouts at that time seem to increase deep sleep in the first sleep cycle and reduce daytime sleepiness [35]. This increase in N3 sleep after late-night exercise is consistent in both males and females without altering other sleep parameters [38]. However, the effect of exercise timing on sleep may be mediated by other factors, such as the individuals' chronotype. Timing-depended response to exercise, is shown to be different in individuals with morning or evening chronotypes [39].

Effects of Sleep on Recovery of Muscle Injuries

Sleep plays a vital role in maintaining and restoring skeletal muscle physiology by affecting various molecules, such as hormones and cytokines. After exercise-induced muscle damage or muscle injury, adequate sleep may accelerate tissue restoration [40]. It is continuously shown that recovery and adaptation to exercise can be disrupted if sleep is inadequate following a stimulus that results in skeletal muscle damage [40]. This could be attributed to the fact that sleep debt is coupled by decreased secretion of anabolic hormones and increased secretion of catabolic hormones [40, 41], disrupting muscle restoration. For example, testosterone is widely known for its effect on muscle hypertrophy [41] and post-exercise muscle regeneration [42, 43]. Sleep restriction can decrease basal testosterone levels by up to 15%. This decrease is quite large, considering that testosterone reduction in normal ageing is approximately 1% per year [44, 45]. Although acute increases in anabolic hormones, such as testosterone, have no effect on muscle protein synthesis [46], chronic low testosterone levels, along with increases in blood cortisol levels, can largely contribute to insufficient muscle restoration and may promote muscle degradation [47–49]. It should be highlighted that this cascade of physiological alterations also occurs in older people and in specific pathological states and conditions, such as cachexia [40]. The negative effects of sleep debt on muscle restoration are magnified with increasing duration and intensity of exercise [2, 3]. Partial sleep restriction to 4 hours per night over 5 days reduces skeletal muscle myofibrillar protein synthesis rates [50]. Remarkably, these rates are sustained at normal levels when high-intensity interval exercise is performed consistently during the sleep restriction period [50]. There is also an association between sleep loss and exercise-induced injuries, with decreases in the immune system role [51] and increases in increasing local IGF-I [52], playing an important role.

Effects of Exercise and Sleep Interaction on Inflammatory Responses

During nocturnal sleep, both sleep and circadian rhythm affect inflammatory activity [53]. However, the interpretation of relevant data is somehow confusing. There are studies showing that neither total nor partial sleep deprivation results consistently in increased circulating markers of inflammation [54]. In contrast, other studies show that even partial sleep deprivation may activate inflammatory signaling pathways, although the translation to increased inflammatory peptides may not always occur acutely [53, 55–57]. Inflammatory responses due to sleep deprivation can be enhanced when inadequate sleep is combined with exercise-induced muscle damage. For example, inadequate sleep or complete sleep restriction cause an increase in blood interleukin-6 (IL-6) levels, especially when combined with an exercise-induced muscle damage exercise protocol [58]. Moreover, sleep disorders and increased sleep duration variability are shown to be associated with higher levels of C-reactive protein (CRP) and IL-6 [59]. This is important as IL-6 plays a key role in hypertrophic signaling and muscle restoration [60]. There is also a link between peripheral tissues and the central nervous system, since IL-6 induces central fatigue and insomnia [61]. Also, long-term elevated levels of IL-6 may contribute to skeletal muscle atrophy [62] and as such, targeted inhibition of IL-6 signaling is a potentially vital option for clinical populations, such as cachectic patients, to increase muscle mass [62]. Overall, chronic inflammation is of great importance since it is well-known to be related to several major diseases [63].

Effects of Sleep on Sports and Exercise Performance and Recovery

Research has indicated that sleep can have a significant impact, either beneficial or detrimental, on sports and exercise performance. While sleep extension can enhance various aspects of athletic performance [64], sleep restriction, partial or total, has been negatively associated with exercise that requires sustained and stable performance, including cardiorespiratory efforts [65–69]. In particular, partial sleep deprivation has been shown to decrease athletic performance due to the increased physiological demand, increased lactate accumulation and decreased time to exhaustion [67]. Consistent with this, partial sleep restriction can reduce maximum work rate during cycling [70], indicating that both aerobic and anaerobic endurance performances are impaired after partial sleep restriction [70–73]. However, this decrease in performance may be attributed either to impairment of the aerobic metabolism [66] or may be due to neuromuscular fatigue or due to altered perception of effort [74]. Energy substrate utilization may partially explain the decreased athletic performance, since muscle glycogen repletion is disturbed after sleep restriction due to impaired glucose uptake [75]. Considering that muscle glycogen stores are important for athletic performance, reduced glycogen repletion may lead to decreased muscle function and stamina [76].

In addition, anaerobic power [77], isometric force [78] and accuracy [79] seem to be adversely affected by sleep restriction. However, submaximal athletic performance may be more affected by sleep deprivation than short-effort bouts like weightlifting [80], because of the prolonged duration and the subsequent effort perception, which is exponentially related to exercise duration after sleep restriction [81]. Sports-specific skills, such as tennis serving accuracy, seem to be affected by sleep restriction [79], and even caffeine supplementation fails to revert the detrimental effects of sleep restriction on serving accuracy [79].

Long-term sleep deprivation can alter the nervous system's activity by increasing sympathetic and decreasing parasympathetic activity and by altering spontaneous baroreflex sensitivity, which is an important component of cardiovascular regulation of homeostasis [82]. Interestingly, similar alterations in sympathetic and parasympathetic nervous system activities occur with overtraining [83], and as such, overtraining and sleep deprivation share common mechanisms affecting the nervous system [84]. Despite that there are established mechanisms explaining the negative effects of sleep deprivation on athletic performance, there are several studies showing minimal or no impact of partial sleep deprivation on exercise performance, and this may be dependent on the characteristics of the exercise bout [85]. Although more studies are needed to examine the exact mechanisms of reduced exercise performance after sleep deprivation, its negative effects on perceptual, cognitive and psychomotor performance are well -established [85].

Numerous studies have demonstrated that sleep can affect sports and exercise performance, either positively or negatively.

Effects of Nutrition on Sleep

Specific hormones and peptides can entrain the body's circadian rhythm and affect sleep initiation and maintenance, promoting recovery [86]. Since nutrients' distribution and periodization affects various hormones, nutrition can both, directly and indirectly facilitate recovery via tissue restoration and sleep optimization, respectively [87]. Several nutrient-sleep interactions are mediated by molecules that are actively affected by nutrition (Table 14.1), such as L-tryptophan, 5-hydroxytryptamine, cholecystokinin and gamma-aminobutyric acid [88]. Protein and carbohydrate consumption fire a cascade of physiological responses that lead to an alteration of L-tryptophan availability in the brain, which is a precursor of 5-hydroxytryptamine [89]. The later can be converted to melatonin, a hormone with a key role in sleep initiation and maintenance [90]. Hence, both dietary melatonin intake from specific foods, such as tart cherries [91], as well as macronutrient modification through diet have sleep-promoting effects [92, 93]. Increased L-tryptophan availability, is also related to increased REM sleep stage [94], promoting synthetic processes of brain reorganization and repair [95]. For example, a recent study

Table 14.1 Nutritional factors and their influence on sleep quantity, continuation and stages

	Effect on sleep quantity and continuation				Effect on sleep stages	
Dietary Factor	TST	% SE	SOL	WASO	N3	REM
HGI pre-bed meal	↑	↑	↓			
HCHO Diet						↑
LCHO Diet					↑	
Low Sodium Diet	↓			↑		
Tart Montmorency Cherry Juice	↑	↑	↓	↓		
Caffeine	↓	↓	↑			↓
GABA			↓			
Alcohol	↓	↓	↓	↑	↑	↓
Hydration	↓					

TST Total Sleep Time, *SE* Sleep Efficiency, *SOL* Sleep Onset Latency, *WASO* Wake After Sleep Onset, *N3* Sleep Stage, referred as "Deep Sleep", *REM* Rapid Eye Movement Sleep Stage, *HGI* High Glycemic Index, *HCHO* High Carbohydrates, *LCHO* Low Carbohydrates, *GABA* Gamma-aminobutyric acid

explored whether eating a high-glycemic index meal before bedtime could influence sleep [96]. The findings indicated a shorter time to fall asleep and a longer overall sleep period. These benefits were associated with better visual reaction time the following day, suggesting that sufficient rest can support this element of performance.

Macronutrient availability and distribution can potentially modify not only sleep initiation but also sleep architecture. For example, dietary fat and protein intake can stimulate cholecystokinin release [97, 98], increasing the subjective feeling of sleepiness [98] and N3 sleep stage [99, 100]. Interestingly, approximately 70% of growth hormone pulses coincide with the N3 sleep stage, and growth hormone secretion is proportionally related to the N3 stage duration [101]. This is highlighting the effect of nutrition on sleep optimization and body restoration. In addition, there are several dietary supplements and herbs, such as valerian and glycine, that may increase subjective sleep quality but have no significant effect on quantitative sleep-related measures [102, 103].

Napping Effects on Performance and Recovery

Naps may be used to increase overall sleep duration, especially in athletes with inadequate sleep traits. Daytime naps are usually taken during the afternoon slight "dip" of the circadian rhythms, taking advantage of this to initiate sleep [104]. These naps are very common in athletes with strict and heavy schedules [105], and could potentially aid recovery after sleep deprivation [106]. However, while naps may positively impact athletic performance, particularly during periods of unavoidable sleep restriction [107], the evidence regarding optimal nap duration remains ambiguous [108]. Longer naps, lasting between 35 to 90 minutes, might offer enhanced performance benefits [108]. Additionally, athletes should consider allocating 30 minutes to alleviate sleep inertia before engaging in training or competition for optimal performance [109]. Nevertheless, in order not only to maximize their effectiveness but to avoid impairing nocturnal sleep, both the duration and timing of naps must be carefully managed [107]. This is especially important since late-afternoon naps may have a negative impact on nighttime sleep [104]. Overall, when sleep restriction cannot be avoided, available evidence suggests that naps could be a valuable strategy to augment total sleep time and facilitate recovery, provided that their duration and timing is carefully considered.

Sleep Hygiene Routines

Acute sleep hygiene strategies have been shown to improve athletes' sleep [110, 111]. For example, a structured post-game sleep routine in elite athletes includes a specific bed time in a cool environment, with reduced blue light exposure, with no access to technology for at least 15–30 minutes prior to bedtime [110] and a warm pre-bed shower [111]. Notably, evidence shows that if these hygiene strategies are followed for longer periods, they may consistently improve sleep quality and exercise performance, while shifting athletes- chronobiological phenotype to an earlier chronotype (i.e. morning type) [112]. Other strategies, such as post-exercise cryostimulation and non-red light irradiation, have been shown in some studies to preserve athletes' sleep during intensified training [113] and to improve sleep quality during a regular training period [114].

Conclusion

In recent years, the scientific community has shown significant interest in sleep among athletes. Sleep is one of the main components for the recovery of athletes and can positively or negatively affect certain aspects of sports and exercise performance. A significant percentage of athletes do not sleep well, which is why there should be monitoring of athletes' sleep so that sleep improvement techniques can be applied, thus avoiding the negative consequences of poor sleep on athletes' health and performance

> If sleep does not serve an absolutely vital function, then it is the biggest mistake the evolutionary process has ever made.
>
> –Allan Rechtschaffen 1927–2021

> Sleep deprivation is the most common brain impairment.
>
> –William C. Dement 1928–2020

Take home message: Many athletes have problems with their sleep and do not sleep well. Good sleep is very important for the recovery, performance, and health of athletes. Research in recent years has shown ways in which we can help our athletes sleep well to avoid the negative consequences of poor sleep and to try to create the appropriate environment for good rest and performance for athletes.

References

1. Hobson JA. Sleep is of the brain, by the brain and for the brain. Nature. 2005;437(7063):1254–6. https://doi.org/10.1038/nature04283.
2. Chase JD, Roberson PA, Saunders MJ, Hargens TA, Womack CJ, Luden ND. One night of sleep restriction following heavy exercise impairs 3-km cycling time-trial performance in the morning. Appl Physiol Nutr Metab. 2017;42(9):909–15. https://doi.org/10.1139/apnm-2016-0698.
3. Rae DE, Chin T, Dikgomo K, et al. One night of partial sleep deprivation impairs recovery from a single exercise training session. Eur J Appl Physiol. 2017;117(4):699–712. https://doi.org/10.1007/s00421-017-3565-5.
4. Strecker RE, Morairty S, Thakkar MM, et al. Adenosinergic modulation of basal forebrain and preoptic/anterior hypothalamic neuronal activity in the control of behavioral state. Behav Brain Res. 2000;115(2):183–204. https://doi.org/10.1016/s0166-4328(00)00258-8.
5. Berry RB, Gamaldo CE, Harding SM, et al. AASM scoring manual version 2.2 updates: new chapters for scoring infant sleep staging and home sleep Apnea testing. J Clin Sleep Med. 2015;11(11):1253–4. https://doi.org/10.5664/jcsm.5176.
6. Dunmyre JR, Mashour GA, Booth V. Coupled flip-flop model for REM sleep regulation in the rat. PLoS One. 2014;9(4):e94481. https://doi.org/10.1371/journal.pone.0094481.
7. Hobson JA, McCarley RW, Wyzinski PW. Sleep cycle oscillation: reciprocal discharge by two brainstem neuronal groups. Science. 1975;189(4196):55–8. https://doi.org/10.1126/science.1094539.
8. Sakkas GK, Giannaki CD, Karatzaferi C, et al. Current trends in the management of uremic restless legs syndrome: a systematic review on aspects related to quality of life, cardiovascular mortality and survival. Sleep Med Rev. 2015;21:39–49. https://doi.org/10.1016/j.smrv.2014.07.006.
9. Xie Y, Liu S, Chen XJ, Yu HH, Yang Y, Wang W. Effects of exercise on sleep quality and insomnia in adults: a systematic review and meta-analysis of randomized controlled trials. Front Psych. 2021;12:664499. https://doi.org/10.3389/fpsyt.2021.664499.
10. Vlahoyiannis A, Aphamis G, Bogdanis GC, Sakkas GK, Andreou E, Giannaki CD. Deconstructing athletes' sleep: a systematic review of the influence of age, sex, athletic expertise, sport type, and season on sleep characteristics. J Sport Health Sci. 2021;10(4):387–402. https://doi.org/10.1016/j.jshs.2020.03.006.
11. Gupta L, Morgan K, Gilchrist S. Does elite sport degrade sleep quality? A systematic review. Sports Med. 2017;47(7):1317–33. https://doi.org/10.1007/s40279-016-0650-6.

12. Sargent C, Lastella M, Halson SL, Roach GD. How much sleep does an elite athlete need? Int J Sports Physiol Perform. 2021;16(12):1746–57. https://doi.org/10.1123/ijspp.2020-0896.
13. Park I, Diaz J, Matsumoto S, et al. Exercise improves the quality of slow-wave sleep by increasing slow-wave stability. Sci Rep. 2021;11(1):4410. https://doi.org/10.1038/s41598-021-83817-6.
14. Shapiro CM, Bortz R, Mitchell D, Bartel P, Jooste P. Slow-wave sleep: a recovery period after exercise. Science. 1981;214(4526):1253–4. https://doi.org/10.1126/science.7302594.
15. Dolezal BA, Neufeld EV, Boland DM, Martin JL, Cooper CB. Interrelationship between sleep and exercise: a systematic review. Adv Prev Med. 2017;2017:1364387. https://doi.org/10.1155/2017/1364387.
16. Demirel H. Sleep quality differs between athletes and non-athletes. Clin Invest Med. 2016;39(6):27525.
17. Kovacevic A, Mavros Y, Heisz JJ, Fiatarone Singh MA. The effect of resistance exercise on sleep: a systematic review of randomized controlled trials. Sleep Med Rev. 2018;39:52–68. https://doi.org/10.1016/j.smrv.2017.07.002.
18. Kolling S, Wiewelhove T, Raeder C, et al. Sleep monitoring of a six-day microcycle in strength and high-intensity training. Eur J Sport Sci. 2016;16(5):507–15. https://doi.org/10.1080/17461391.2015.1041062.
19. Vlahoyiannis A, Aphamis G, Eddin DA, Giannaki CD. The effect of evening cycling at different intensities on sleep in healthy young adults with intermediate chronobiological phenotype: a randomized, cross-over trial. J Sports Sci. 2021;39(2):192–9. https://doi.org/10.1080/02640414.2020.1812194.
20. Ezati M, Keshavarz M, Barandouzi ZA, Montazeri A. The effect of regular aerobic exercise on sleep quality and fatigue among female student dormitory residents. BMC Sports Sci Med Rehabil. 2020;12:44. https://doi.org/10.1186/s13102-020-00190-z.
21. Youngstedt SD, Kripke DF, Elliott JA. Is sleep disturbed by vigorous late-night exercise? Med Sci Sports Exerc. 1999;31(6):864–9. https://doi.org/10.1097/00005768-199906000-00015.
22. Miura A, Myouken S, Yamada M, et al. Effects of aerobic exercise in early evening on the following nocturnal sleep and its haemodynamic response. Res Sports Med. 2016;24(1):16–29. https://doi.org/10.1080/15438627.2015.1076415.
23. Harding EC, Franks NP, Wisden W. The temperature dependence of sleep. Front Neurosci. 2019;13:336. https://doi.org/10.3389/fnins.2019.00336.
24. Brand S, Kalak N, Gerber M, Kirov R, Puhse U, Holsboer-Trachsler E. High self-perceived exercise exertion before bedtime is associated with greater objectively assessed sleep efficiency. Sleep Med. 2014;15(9):1031–6. https://doi.org/10.1016/j.sleep.2014.05.016.
25. Min L, Wang D, You Y, Fu Y, Ma X. Effects of high-intensity interval training on sleep: a systematic review and meta-analysis. Int J Environ Res Public Health. 2021;18(20) https://doi.org/10.3390/ijerph182010973.
26. Killer SC, Svendsen IS, Jeukendrup AE, Gleeson M. Evidence of disturbed sleep and mood state in well-trained athletes during short-term intensified training with and without a high carbohydrate nutritional intervention. J Sports Sci. 2017;35(14):1402–10. https://doi.org/10.1080/02640414.2015.1085589.
27. Hausswirth C, Louis J, Aubry A, Bonnet G, Duffield R, Y LEM. Evidence of disturbed sleep and increased illness in overreached endurance athletes. Med Sci Sports Exerc. 2014;46(5):1036–45. https://doi.org/10.1249/MSS.0000000000000177.
28. Dupuy O, Bherer L, Audiffren M, Bosquet L. Night and postexercise cardiac autonomic control in functional overreaching. Appl Physiol Nutr Metab. 2013;38(2):200–8. https://doi.org/10.1139/apnm-2012-0203.
29. Cadegiani FA, Kater CE. Hormonal response to a non-exercise stress test in athletes with overtraining syndrome: results from the Endocrine and metabolic Responses on Overtraining Syndrome (EROS) - EROS-STRESS. J Sci Med Sport. 2018;21(7):648–53. https://doi.org/10.1016/j.jsams.2017.10.033.

30. Lastella M, Vincent GE, Duffield R, et al. Can sleep be used as an indicator of overreaching and overtraining in athletes? Front Physiol. 2018;9:436. https://doi.org/10.3389/fphys.2018.00436.
31. Driver HS, Rogers GG, Mitchell D, et al. Prolonged endurance exercise and sleep disruption. Med Sci Sports Exerc. 1994;26(7):903–7.
32. Hague JF, Gilbert SS, Burgess HJ, Ferguson SA, Dawson D. A sedentary day: effects on subsequent sleep and body temperatures in trained athletes. Physiol Behav. 2003;78(2):261–7. https://doi.org/10.1016/s0031-9384(02)00975-7.
33. Taylor SR, Rogers GG, Driver HS. Effects of training volume on sleep, psychological, and selected physiological profiles of elite female swimmers. Med Sci Sports Exerc. 1997;29(5):688–93. https://doi.org/10.1097/00005768-199705000-00016.
34. Yamanaka Y, Hashimoto S, Takasu NN, et al. Morning and evening physical exercise differentially regulate the autonomic nervous system during nocturnal sleep in humans. Am J Physiol Regul Integr Comp Physiol. 2015;309(9):R1112–21. https://doi.org/10.1152/ajpregu.00127.2015.
35. Yoshida H, Ishikawa T, Shiraishi F, Kobayashi T. Effects of the timing of exercise on the night sleep. Psychiatry Clin Neurosci. 1998;52(2):139–40. https://doi.org/10.1111/j.1440-1819.1998.tb00994.x.
36. Morita Y, Sasai-Sakuma T, Inoue Y. Effects of acute morning and evening exercise on subjective and objective sleep quality in older individuals with insomnia. Sleep Med. 2017;34:200–8. https://doi.org/10.1016/j.sleep.2017.03.014.
37. Robey E, Dawson B, Halson S, Gregson W, Goodman C, Eastwood P. Sleep quantity and quality in elite youth soccer players: a pilot study. Eur J Sport Sci. 2014;14(5):410–7. https://doi.org/10.1080/17461391.2013.843024.
38. Myllymaki T, Kyrolainen H, Savolainen K, et al. Effects of vigorous late-night exercise on sleep quality and cardiac autonomic activity. J Sleep Res. 2011;20(1 Pt 2):146–53. https://doi.org/10.1111/j.1365-2869.2010.00874.x.
39. Vitale JA, Bonato M, Galasso L, et al. Sleep quality and high intensity interval training at two different times of day: a crossover study on the influence of the chronotype in male collegiate soccer players. Chronobiol Int. 2017;34(2):260–8. https://doi.org/10.1080/07420528.2016.1256301.
40. Dattilo M, Antunes HK, Medeiros A, et al. Sleep and muscle recovery: endocrinological and molecular basis for a new and promising hypothesis. Med Hypotheses. 2011;77(2):220–2. https://doi.org/10.1016/j.mehy.2011.04.017.
41. Kraemer WJ, Ratamess NA. Hormonal responses and adaptations to resistance exercise and training. Sports Med. 2005;35(4):339–61. https://doi.org/10.2165/00007256-200535040-00004.
42. Otzel DM, Lee J, Ye F, Borst SE, Yarrow JF. Activity-based physical rehabilitation with adjuvant testosterone to promote neuromuscular recovery after spinal cord injury. Int J Mol Sci. 2018;19(6) https://doi.org/10.3390/ijms19061701.
43. White JP, Baltgalvis KA, Sato S, Wilson LB, Carson JA. Effect of nandrolone decanoate administration on recovery from bupivacaine-induced muscle injury. J Appl Physiol (1985). 2009;107(5):1420–30. https://doi.org/10.1152/japplphysiol.00668.2009.
44. Leproult R, Van Cauter E. Effect of 1 week of sleep restriction on testosterone levels in young healthy men. JAMA. 2011;305(21):2173–4. https://doi.org/10.1001/jama.2011.710.
45. Jauch-Chara K, Schmid SM, Hallschmid M, Oltmanns KM, Schultes B. Pituitary-gonadal and pituitary-thyroid axis hormone concentrations before and during a hypoglycemic clamp after sleep deprivation in healthy men. PLoS One. 2013;8(1):e54209. https://doi.org/10.1371/journal.pone.0054209.
46. West DW, Kujbida GW, Moore DR, et al. Resistance exercise-induced increases in putative anabolic hormones do not enhance muscle protein synthesis or intracellular signalling in young men. J Physiol. 2009;587(Pt 21):5239–47. https://doi.org/10.1113/jphysiol.2009.177220.

47. Guyon A, Balbo M, Morselli LL, et al. Adverse effects of two nights of sleep restriction on the hypothalamic-pituitary-adrenal axis in healthy men. J Clin Endocrinol Metab. 2014;99(8):2861–8. https://doi.org/10.1210/jc.2013-4254.
48. Leproult R, Copinschi G, Buxton O, Van Cauter E. Sleep loss results in an elevation of cortisol levels the next evening. Sleep. 1997;20(10):865–70.
49. Spiegel K, Leproult R, Van Cauter E. Impact of sleep debt on metabolic and endocrine function. Lancet. 1999;354(9188):1435–9. https://doi.org/10.1016/S0140-6736(99)01376-8.
50. Saner NJ, Lee MJ, Pitchford NW, et al. The effect of sleep restriction, with or without high-intensity interval exercise, on behavioural alertness and mood state in young healthy males. J Sleep Res. 2023:e13987. https://doi.org/10.1111/jsr.13987.
51. Luke A, Lazaro RM, Bergeron MF, et al. Sports-related injuries in youth athletes: is overscheduling a risk factor? Clin J Sport Med. 2011;21(4):307–14. https://doi.org/10.1097/JSM.0b013e3182218f71.
52. Chennaoui M, Vanneau T, Trignol A, et al. How does sleep help recovery from exercise-induced muscle injuries? J Sci Med Sport. 2021;24(10):982–7. https://doi.org/10.1016/j.jsams.2021.05.007.
53. Irwin MR, Opp MR. Sleep health: reciprocal regulation of sleep and innate immunity. Neuropsychopharmacology. 2017;42(1):129–55. https://doi.org/10.1038/npp.2016.148.
54. Irwin MR, Olmstead R, Carroll JE. Sleep disturbance, sleep duration, and inflammation: a systematic review and meta-analysis of cohort studies and experimental sleep deprivation. Biol Psychiatry. 2016;80(1):40–52. https://doi.org/10.1016/j.biopsych.2015.05.014.
55. Irwin MR, Wang M, Ribeiro D, et al. Sleep loss activates cellular inflammatory signaling. Biol Psychiatry. 2008;64(6):538–40. https://doi.org/10.1016/j.biopsych.2008.05.004.
56. Irwin MR, Wang M, Campomayor CO, Collado-Hidalgo A, Cole S. Sleep deprivation and activation of morning levels of cellular and genomic markers of inflammation. Arch Intern Med. 2006;166(16):1756–62. https://doi.org/10.1001/archinte.166.16.1756.
57. Irwin MR, Witarama T, Caudill M, Olmstead R, Breen EC. Sleep loss activates cellular inflammation and signal transducer and activator of transcription (STAT) family proteins in humans. Brain Behav Immun. 2015;47:86–92. https://doi.org/10.1016/j.bbi.2014.09.017.
58. Dattilo M, Antunes HKM, Galbes NMN, et al. Effects of sleep deprivation on acute skeletal muscle recovery after exercise. Med Sci Sports Exerc. 2020;52(2):507–14. https://doi.org/10.1249/MSS.0000000000002137.
59. Park H, Tsai KM, Dahl RE, et al. Sleep and inflammation during adolescence. Psychosom Med. 2016;78(6):677–85. https://doi.org/10.1097/PSY.0000000000000340.
60. Serrano AL, Baeza-Raja B, Perdiguero E, Jardi M, Munoz-Canoves P. Interleukin-6 is an essential regulator of satellite cell-mediated skeletal muscle hypertrophy. Cell Metab. 2008;7(1):33–44. https://doi.org/10.1016/j.cmet.2007.11.011.
61. Vgontzas AN, Papanicolaou DA, Bixler EO, et al. Circadian interleukin-6 secretion and quantity and depth of sleep. J Clin Endocrinol Metab. 1999;84(8):2603–7. https://doi.org/10.1210/jcem.84.8.5894.
62. Narsale AA, Carson JA. Role of interleukin-6 in cachexia: therapeutic implications. Curr Opin Support Palliat Care. 2014;8(4):321–7. https://doi.org/10.1097/SPC.0000000000000091.
63. Couzin-Frankel J. Inflammation bares a dark side. Science. 2010;330(6011):1621. https://doi.org/10.1126/science.330.6011.1621.
64. Mah CD, Mah KE, Kezirian EJ, Dement WC. The effects of sleep extension on the athletic performance of collegiate basketball players. Sleep. 2011;34(7):943–50. https://doi.org/10.5665/SLEEP.1132.
65. Edwards BJ, Waterhouse J. Effects of one night of partial sleep deprivation upon diurnal rhythms of accuracy and consistency in throwing darts. Chronobiol Int. 2009;26(4):756–68. https://doi.org/10.1080/07420520902929037.
66. Mougin F, Davenne D, Simon-Rigaud ML, Renaud A, Garnier A, Magnin P. Disturbance of sports performance after partial sleep deprivation. C R Seances Soc Biol Fil.

1989;183(5):461–6. Perturbations de la performance sportive apres privation partielle de sommeil
67. Mougin F, Simon-Rigaud ML, Davenne D, et al. Effects of sleep disturbances on subsequent physical performance. Eur J Appl Physiol Occup Physiol. 1991;63(2):77–82. https://doi.org/10.1007/BF00235173.
68. Martin BJ. Effect of sleep deprivation on tolerance of prolonged exercise. Eur J Appl Physiol Occup Physiol. 1981;47(4):345–54. https://doi.org/10.1007/BF02332962.
69. Horne JA, Ostberg O. A self-assessment questionnaire to determine morningness-eveningness in human circadian rhythms. Int J Chronobiol. 1976;4(2):97–110.
70. Mougin F, Bourdin H, Simon-Rigaud ML, Nguyen NU, Kantelip JP, Davenne D. Hormonal responses to exercise after partial sleep deprivation and after a hypnotic drug-induced sleep. J Sports Sci. 2001;19(2):89–97. https://doi.org/10.1080/026404101300036253.
71. Abedelmalek S, Souissi N, Chtourou H, et al. Effects of partial sleep deprivation on proinflammatory cytokines, growth hormone, and steroid hormone concentrations during repeated brief sprint interval exercise. Chronobiol Int. 2013;30(4):502–9. https://doi.org/10.3109/07420528.2012.742102.
72. Souissi N, Chtourou H, Aloui A, et al. Effects of time-of-day and partial sleep deprivation on short-term maximal performances of judo competitors. J Strength Cond Res. 2013;27(9):2473–80. https://doi.org/10.1519/JSC.0b013e31827f4792.
73. Souissi N, Souissi M, Souissi H, et al. Effect of time of day and partial sleep deprivation on short-term, high-power output. Chronobiol Int. 2008;25(6):1062–76. https://doi.org/10.1080/07420520802551568.
74. Abbiss CR, Laursen PB. Models to explain fatigue during prolonged endurance cycling. Sports Med. 2005;35(10):865–98. https://doi.org/10.2165/00007256-200535100-00004.
75. Skein M, Duffield R, Edge J, Short MJ, Mundel T. Intermittent-sprint performance and muscle glycogen after 30 h of sleep deprivation. Med Sci Sports Exerc. 2011;43(7):1301–11. https://doi.org/10.1249/MSS.0b013e31820abc5a.
76. Costill DL, Flynn MG, Kirwan JP, et al. Effects of repeated days of intensified training on muscle glycogen and swimming performance. Med Sci Sports Exerc. 1988;20(3):249–54. https://doi.org/10.1249/00005768-198806000-00006.
77. Taheri M, Arabameri E. The effect of sleep deprivation on choice reaction time and anaerobic power of college student athletes. Asian J Sports Med. 2012;3(1):15–20. https://doi.org/10.5812/asjsm.34719.
78. Ben Cheikh R, Latiri I, Dogui M, Ben SH. Effects of one-night sleep deprivation on selective attention and isometric force in adolescent karate athletes. J Sports Med Phys Fitness. 2017;57(6):752–9. https://doi.org/10.23736/S0022-4707.16.06323-4.
79. Reyner LA, Horne JA. Sleep restriction and serving accuracy in performance tennis players, and effects of caffeine. Physiol Behav. 2013;120:93–6. https://doi.org/10.1016/j.physbeh.2013.07.002.
80. Charest J, Grandner MA. Sleep and athletic performance: impacts on physical performance, mental performance, injury risk and recovery, and mental health: an update. Sleep Med Clin. 2022;17(2):263–82. https://doi.org/10.1016/j.jsmc.2022.03.006.
81. Oliver SJ, Costa RJ, Laing SJ, Bilzon JL, Walsh NP. One night of sleep deprivation decreases treadmill endurance performance. Eur J Appl Physiol. 2009;107(2):155–61. https://doi.org/10.1007/s00421-009-1103-9.
82. Zhong X, Hilton HJ, Gates GJ, et al. Increased sympathetic and decreased parasympathetic cardiovascular modulation in normal humans with acute sleep deprivation. J Appl Physiol (1985). 2005;98(6):2024–32. https://doi.org/10.1152/japplphysiol.00620.2004.
83. Achten J, Jeukendrup AE. Heart rate monitoring: applications and limitations. Sports Med. 2003;33(7):517–38. https://doi.org/10.2165/00007256-200333070-00004.
84. Hynynen E, Uusitalo A, Konttinen N, Rusko H. Heart rate variability during night sleep and after awakening in overtrained athletes. Med Sci Sports Exerc. 2006;38(2):313–7. https://doi.org/10.1249/01.mss.0000184631.27641.b5.

85. Watson AM. Sleep and athletic performance. Curr Sports Med Rep. 2017;16(6):413–8. https://doi.org/10.1249/JSR.0000000000000418.
86. Saper CB, Scammell TE, Lu J. Hypothalamic regulation of sleep and circadian rhythms. Nature. 2005;437(7063):1257–63. https://doi.org/10.1038/nature04284.
87. Grandner MA, Kripke DF, Naidoo N, Langer RD. Relationships among dietary nutrients and subjective sleep, objective sleep, and napping in women. Sleep Med. 2010;11(2):180–4. https://doi.org/10.1016/j.sleep.2009.07.014.
88. Halson SL. Sleep in elite athletes and nutritional interventions to enhance sleep. Sports Med. 2014;44(Suppl 1):S13–23. https://doi.org/10.1007/s40279-014-0147-0.
89. Stamatakis KA, Punjabi NM. Effects of sleep fragmentation on glucose metabolism in normal subjects. Chest. 2010;137(1):95–101. https://doi.org/10.1378/chest.09-0791.
90. Herrera CP, Smith K, Atkinson F, et al. High-glycaemic index and -glycaemic load meals increase the availability of tryptophan in healthy volunteers. Br J Nutr. 2011;105(11):1601–6. https://doi.org/10.1017/S0007114510005192.
91. Pigeon WR, Carr M, Gorman C, Perlis ML. Effects of a tart cherry juice beverage on the sleep of older adults with insomnia: a pilot study. J Med Food. 2010;13(3):579–83. https://doi.org/10.1089/jmf.2009.0096.
92. Brzezinski A. Melatonin in humans. N Engl J Med. 1997;336(3):186–95. https://doi.org/10.1056/NEJM199701163360306.
93. Noorwali E, Hardie L, Cade J. Bridging the reciprocal gap between sleep and fruit and vegetable consumption: a review of the evidence, potential mechanisms, implications, and directions for future work. Nutrients. 2019;11(6) https://doi.org/10.3390/nu11061382.
94. Wurtman RJ, Wurtman JJ, Regan MM, McDermott JM, Tsay RH, Breu JJ. Effects of normal meals rich in carbohydrates or proteins on plasma tryptophan and tyrosine ratios. Am J Clin Nutr. 2003;77(1):128–32. https://doi.org/10.1093/ajcn/77.1.128.
95. Boyle PJ, Scott JC, Krentz AJ, Nagy RJ, Comstock E, Hoffman C. Diminished brain glucose metabolism is a significant determinant for falling rates of systemic glucose utilization during sleep in normal humans. J Clin Invest. 1994;93(2):529–35. https://doi.org/10.1172/JCI117003.
96. Vlahoyiannis A, Aphamis G, Andreou E, Samoutis G, Sakkas GK, Giannaki CD. Effects of high vs. low Glycemic index of post-exercise meals on sleep and exercise performance: a randomized, double-blind, counterbalanced polysomnographic study. Nutrients. 2018;10(11) https://doi.org/10.3390/nu10111795.
97. Chaudhri O, Small C, Bloom S. Gastrointestinal hormones regulating appetite. Philos Trans R Soc Lond Ser B Biol Sci. 2006;361(1471):1187–209. https://doi.org/10.1098/rstb.2006.1856.
98. Wells AS, Read NW, Uvnas-Moberg K, Alster P. Influences of fat and carbohydrate on postprandial sleepiness, mood, and hormones. Physiol Behav. 1997;61(5):679–86. https://doi.org/10.1016/s0031-9384(96)00519-7.
99. Kapas L, Obal F Jr, Alfoldi P, Rubicsek G, Penke B, Obal F. Effects of nocturnal intraperitoneal administration of cholecystokinin in rats: simultaneous increase in sleep, increase in EEG slow-wave activity, reduction of motor activity, suppression of eating, and decrease in brain temperature. Brain Res. 1988;438(1–2):155–64. https://doi.org/10.1016/0006-8993(88)91334-0.
100. Kapas L, Obal F Jr, Opp MR, Johannsen L, Krueger JM. Intraperitoneal injection of cholecystokinin elicits sleep in rabbits. Physiol Behav. 1991;50(6):1241–4. https://doi.org/10.1016/0031-9384(91)90589-g.
101. Steiger A, Herth T, Holsboer F. Sleep-electroencephalography and the secretion of cortisol and growth hormone in normal controls. Acta Endocrinol. 1987;116(1):36–42. https://doi.org/10.1530/acta.0.1160036.
102. Fernandez-San-Martin MI, Masa-Font R, Palacios-Soler L, Sancho-Gomez P, Calbo-Caldentey C, Flores-Mateo G. Effectiveness of Valerian on insomnia: a meta-analysis of randomized placebo-controlled trials. Sleep Med. 2010;11(6):505–11. https://doi.org/10.1016/j.sleep.2009.12.009.

103. Bannai M, Kawai N, Ono K, Nakahara K, Murakami N. The effects of glycine on subjective daytime performance in partially sleep-restricted healthy volunteers. Front Neurol. 2012;3:61. https://doi.org/10.3389/fneur.2012.00061.
104. Borbely AA. A two process model of sleep regulation. Hum Neurobiol. 1982;1(3):195–204.
105. Sargent C, Halson S, Roach GD. Sleep or swim? Early-morning training severely restricts the amount of sleep obtained by elite swimmers. Eur J Sport Sci. 2014;14(Suppl 1):S310–5. https://doi.org/10.1080/17461391.2012.696711.
106. Sargent C, Lastella M, Halson SL, Roach GD. The impact of training schedules on the sleep and fatigue of elite athletes. Chronobiol Int. 2014;31(10):1160–8. https://doi.org/10.3109/07420528.2014.957306.
107. Brooks A, Lack L. A brief afternoon nap following nocturnal sleep restriction: which nap duration is most recuperative? Sleep. 2006;29(6):831–40. https://doi.org/10.1093/sleep/29.6.831.
108. Botonis PG, Koutouvakis N, Toubekis AG. The impact of daytime napping on athletic performance - a narrative review. Scand J Med Sci Sports. 2021;31(12):2164–77. https://doi.org/10.1111/sms.14060.
109. Lastella M, Halson SL, Vitale JA, Memon AR, Vincent GE. To nap or not to nap? A systematic review evaluating napping behavior in athletes and the impact on various measures of athletic performance. Nat Sci Sleep. 2021;13:841–62. https://doi.org/10.2147/NSS.S315556.
110. Fullagar H, Skorski S, Duffield R, Meyer T. The effect of an acute sleep hygiene strategy following a late-night soccer match on recovery of players. Chronobiol Int. 2016;33(5):490–505. https://doi.org/10.3109/07420528.2016.1149190.
111. Whitworth-Turner C, Di Michele R, Muir I, Gregson W, Drust B. A shower before bedtime may improve the sleep onset latency of youth soccer players. Eur J Sport Sci. 2017;17(9):1119–28. https://doi.org/10.1080/17461391.2017.1346147.
112. Harada T, Wada K, Tsuji F, et al. Intervention study using a leaflet entitled 'three benefits of "go to bed early! get up early! and intake nutritionally rich breakfast!" a message for athletes' to improve the soccer performance of university soccer team. Sleep Biol Rhythms. 2016;14(1):65–74.
113. Schaal K, Lem Y, Louis J, et al. Whole-body Cryostimulation limits overreaching in elite synchronized swimmers. Med Sci Sports Exerc. 2015;47(7):1416–25. https://doi.org/10.1249/MSS.0000000000000546.
114. Zhao J, Tian Y, Nie J, Xu J, Liu D. Red light and the sleep quality and endurance performance of Chinese female basketball players. J Athl Train. 2012;47(6):673–8. https://doi.org/10.4085/1062-6050-47.6.08.

Chapter 15
Neurophysiology of Mindfulness in Sports—From Performance to Rehabilitation

Ann Friedman, Robin Jump, and Alejandro Chaoul

The Introduction of Mindfulness to Sports

Over the last five decades, the practice of mindfulness has been introduced in schools, companies, medical institutions, the US Military, British Parliament, the internet, and professional and amateur sports. Mindfulness—paying attention in a particular way: on purpose, in the present moment, and non-judgmentally—was introduced by Jon Kabat-Zinn, founder of the University of Massachusetts Medical Center's Stress Reduction Clinic [1]. His 8-week Mindfulness Based Stress Reduction (MBSR) program for those dealing with disease, illness or stress is now offered in hundreds of clinical settings. Based upon the secular mind-body practices of ancient Buddhism, this training has been found to have a scientific basis and is rapidly being incorporated into sports programs worldwide.

Prior to Kabat-Zinn's popularization of the term, mindful approaches to sports had been described in *Zen and the Art of Archery* [2] and *The Inner Game of Tennis*. [3] Kabat-Zinn introduced mindfulness to U.S. Olympic rowers in 1984 and U.S. professional sports encountered mindfulness when Chicago Bulls' Coach Phil Jackson engaged mindfulness coach George Mumford to train the team [4]. (Fig. 15.1) Mumford also worked with individual professional basketball players, privately coaching Kobe Bryant, Shaquille O'Neil, and Michael Jordan. Other mindfulness teachers continue in his footsteps.

A. Friedman (✉)
Mind Body Spirit Institute, The Jung Center of Houston, Houston, TX, USA

Mental Health America of Greater Houston, Houston, TX, USA
e-mail: Afriedman@mhahouston.org

R. Jump · A. Chaoul
Mind Body Spirit Institute, The Jung Center of Houston, Houston, TX, USA

N. C. Apostolopoulos et al. (eds.), *Fundamentals of Recovery, Regeneration, and Adaptation to Exercise Stress: An Integrated Approach*,
https://doi.org/10.1007/978-3-031-44270-4_15

Fig. 15.1 George Mumford, a renowned mindfulness coach known as the "Mindfulness Guru" and more recently as the "Performance Whisperer" is pictured here with acclaimed basketball player Kobe Bryant. Bryant attributed much of his success on the basketball court to the mental focus and ability to remain calm under pressure he gained from Mumford's teachings in the 1990s

Many individual athletes have talked about their own mindfulness practices. Golfer Phil Michelson attributed his PGA win at age 50 in part to meditation [5]. Gymnast Simone Biles mentioned using mindfulness when she experienced the twisties in the Olympics [6], and it was reported as instrumental in helping Kenley Jansen of the American baseball team the Atlanta Braves come out of his slump [7]. Players in the Seattle Seahawks, an American football team in the National Football League (NFL), also practice mindfulness [8]. Coach Steve Kerr of the National Basketball Association (NBA) team, the Golden State Warriors, promotes four key values—joy, mindfulness, competition, and compassion [9]—three of which are directly related to mindfulness. Dozens of other athletes are on board with the practices too.

Specific curriculum has been used, developed, and studied, specifically for the needs of athletes, as described in Table 15.1. Other curricula have been developed for use with athletes and are awaiting further study through intervention trials, including *A Still Quiet Place for Athletes* [10], the *Mindful-Acceptance-Insight-Commitment Program* [11], and *Mindfulness for Performance* [12].

Table 15.1 Curricula developed and tested for mindfulness training among athletes

Curriculum	Year of origin	Pain & injury	Performance	Flow	Emotional health	Psychological measures	Suggested timeframe
Mindfulness Based Stress Reduction (MBSR)	1974	X [13]	X [14]		X [15]	X [14]	8 weekly, 2.5 h sessions and one all day intensive retreat
Mindful Acceptance Commitment in Sport (MAC)	2004	X [16, 17]	X [16]		X [16, 18–21]		7 weekly, 50-min sessions
Mindful Sport Performance Enhancement (MSPE)	2005		X [22–25]	X [25, 26]	X [27]	X [26]	4 weekly, 2.5–3.0 h sessions
Mindfulness Mediation Training for Sport (MMT for Sport)	2014		X [28, 29]		X [30]	X [28, 29]	12 30 min sessions twice per week over 6 weeks
Berlin Mindfulness Based Training Athletes (BALT)	2017					X [31, 32]	8 weekly 90-min sessions
Acceptance Commitment Training in Sport (ACT in Sport)	2019	X [33, 34]	X [35–37]	X [35]	X [33–38]		4 weekly 60-min sessions

The Research Arena on Performance Benefits

Today, there are thousands of scientific articles on mindfulness. While growing, the research on mindfulness is suggestive, not conclusive, particularly around sports where research is limited. Unfortunately, interventions are not uniform in the length of training and content of curricula, making it difficult to make clear recommendations for athletes. In addition, most studies are not randomized controlled trials, the gold standard for research.

In sports, performance matters. Research is mixed for mindfulness-based interventions on performance, just as it is for other psychological skills trainings, such as thought suppression [39], self-talk [40], and goal setting [41]. A systemic review of the literature on sports psychology showed that interventions designed to enhance performance, including various psychological constructs (cohesion, confidence, and mindfulness), showed a moderately beneficial effect, with mindfulness having the largest impact [42]. Though limited in number, the studies and meta-analyses on sporting performance and mindfulness deserve closer scrutiny.

Research on Mindfulness and Sports Performance

Performance in sports is key. A review of 66 studies of 3908 athletes of all levels produced 17 studies with randomized clinical trials of mindfulness and acceptance-based interventions. Seven showed improvements in athletes' reported performance, with effect sizes varying from small to very large [43]. Improvements in performance compared to a control group have been seen in golf putting [44], with another golf study showing the mindfulness group outperforming typical psychological skills training on activation regulation skills and enhancement of national rankings [22]. Shooting [28], rowing with an ergometer test [14], free throws in basketball [45], and exhaustion times in a graded exercise test [46] have likewise improved in studies of mindfulness. Even novice dart throwers' scores were boosted after mindfulness training when compared to a control group without training [47].

A randomized clinical trial of 34 junior elite ice hockey players allocated into either an Acceptance and Commitment Training (ACT) group, incorporating a mindfulness framework, or a wait list control group, showed significant intervention effects on both objective performance outcomes (goals, assists, and taken shots) and blinded coach ratings of players' performance and focus. Effects continued at 3-month follow-up for the coach ratings, but not for the objective performance measures [36]. In another randomized clinical trial, expert observers rated athletic performance as significantly better for forty-five male amateur soccer players who received mindfulness training [16].

Ideally, mindfulness improves flow which then improves performance. When humans are in flow states, as defined by Csikszentmihalyi [48], they are intensely focused so that every possible distraction falls away. Action and awareness merge and the sense of self and of time vanish, often causing performance to improve dramatically. While the concepts of mindfulness and flow state differ, research is now suggesting that increases in mindfulness correlate with increases in flow [25, 26, 44, 49].

This chapter seeks to provide the most current neuropsychological and neurobiological research on mindfulness in sports and sports rehabilitation, ending in recommendations for future research to boost evidence and better understand the mechanisms by which mindfulness may benefit athletes.

- While growing, the research on mindfulness, particularly around sports, is limited.
- Ideally, mindfulness improves flow which then improves performance.
- Out of various psychological constructs studied in sports designed to enhance performance, mindfulness had the largest impact.

Mindfulness: The Mental Gym for Athletes

> Mindfulness encompasses a set of practices that can be integrated into athletic training. Accustomed to the demands of daily practice, athletes can hone their skills in focus on the present moment, mental rehearsal, and coping with setbacks through self-compassion.

Everything we do is either done mindfully or mindlessly, functioning on autopilot. Mindfulness is a natural quality, which varies among individuals as a trait and state, but it can be developed through systematic practices. [50] As defined earlier, mindfulness focuses on two key areas: (1) training to focus on the present moment; and (2) paying attention with open, curious, kind (non-judgmental) attention [51]. Mindfulness here is used as an umbrella term which includes diverse practices that can be employed to train for attention and acceptance, including meditations or focused attention or open monitoring, body scans, walking and moving practices, compassion, equanimity, and kindness practices, cognitive practices, daily activity practices, and gratitude practices, among others.

Paying attention involves awareness of thoughts, emotions, body sensations, needs, and the environment. Training to pay attention requires learning to observe and release distracting thoughts both with focused attention and with open monitoring while going about daily activities. Rather than replacing or resisting thoughts, practitioners notice them and refocus on the present.

> - Mindfulness practices help practitioners pay attention; to be more aware of thoughts, emotions, body sensations, needs, and the environment.
> - Over time, those who practice mindfulness become less reactive to unpleasant sensations, whether from physical injury or difficult emotions.
> - Mindfulness improves cohesion in sports teams.
> - Mindfulness can be practiced through formal meditation as well as informal practices during daily activities.

While psychological skills training often tries to alter physiological and psychological states, mindfulness alters *our relationship* with these states. Letting go of upsetting thoughts can be particularly challenging with resentments, disappointments, hurts, and grievances, which is why learning practices of equanimity, compassion, and kindness practices are essential. Imagery, also called visualization or mental rehearsal, is often used in sports successfully [52]. Similar practices are utilized in mindfulness to soothe intense feelings, allowing for a calm, more direct response, as opposed to being reactive, to challenging situations and people. Practicing self-compassion is necessary as setbacks, mistakes, and failures in life are inevitable. When these experiences are met with a harsh inner critic, it can cause decreased motivation and self-destructive behavior [53].

Mindfulness involves body awareness of strong and subtle sensations. For instance, practitioners practice sitting with unpleasant body sensations from injury or emotion, observing their thoughts without judging or trying to change them, developing resilience. Mindfulness has also been shown to be effective in management of physical pain. [54–56] Over time, practitioners become less reactive to unpleasant sensations.

Relational mindfulness can be beneficial in deepening personal relationships and building team cohesion in sports [57, 58].

Mindfulness is meant to be a lifestyle, practiced daily through formal meditation and informal awareness practices, requiring a strong commitment to practice, with which athletes are familiar. Since research into which mindfulness practices work best is growing, some investigators are calling for specific mindfulness practices for enhancing various athletic abilities. [59] More research will guide better integration of these practices.

> With more applied research of mindfulness training in sports, the benefits of incorporating mindfulness into the daily lives of athletes will be better understood.

Mindfulness Appears to Positively Affect Cortisol, Immune System Functioning, Respiration Rates, Blood Pressure, and Heart Rate Variability (HRV)

> The increasing volume of research on mindfulness in sports spans the essential topics of attention training, stress reduction for mental health, injury prevention, pain management and more.

A meta-analysis of mindfulness' effectiveness in sports performance examined nine studies of 290 athletes in twelve different sports. Athletes participated in mindfulness trainings ranging from four weeks to two years with practices being conducted at a frequency ranging from twice daily to once a week. Six of the nine studies employed randomized clinical trials. The studies examined physiological indicators, such as salivary cortisol level, immune response, resting heart rate and maximal oxygen uptake, as well as psychological indicators, such as flow, goal-directed energy, anxiety, and apprehension. Very large effects were noted on physiological performance and moderate effects on psychological indicators. [60]

Mindfulness May Reduce Cortisol

When perceiving challenges and threats, the brain initiates events involving the hypothalamic-pituitary-adrenal axis (HPA), which activates the endocrine stress response, releasing glucocorticoids. [61] One of those, cortisol, a reliable biological marker of stress, increases energy and body readiness to deal with challenges. Prolonged stress can result in impaired communication between the immune system and the HPA axis, causing serious health issues. [61, 62] Mindfulness calms the stress response, lowering cortisol levels.

A meta-analysis of five randomized controlled trials on the effects of mindfulness interventions on cortisol showed some beneficial improvements in healthy adults suggesting more research is needed as findings were not consistent. [63] In another review of six studies of various designs, no difference in cortisol levels was observed between the control and the mindfulness intervention groups in the randomized clinical trials, but significant changes in cortisol measures were observed in two studies comparing baseline values with repeat measures after a mindfulness intervention. [64]

Cortisol has been measured to be higher in pre-competition in various studies. [65] In a mindfulness randomized clinical trial with 96 male elite shooters, mean cortisol values significantly decreased from baseline with mindfulness, as did pre-competition stress compared with the control group. [28]

With competition being stressful, a study researching wheelchair basketball players during competition observed an attenuation in the increase of cortisol with use of mindfulness training compared with the control group. [66] Likewise, a randomized clinical trial of elite Wushu athletes showed decreased daily cortisol levels in competition [67], with a study consisting of 17 soccer players observing lower salivary cortisol concentration levels in the mindfulness group versus the control group. This was associated with shorter reaction time on the Stroop test, a test of cognitive interference, suggesting that higher cortisol worsens cognition. [68] Finally, a study with novice endurance runners showed a significant reduction in salivary cortisol concentration in the mindfulness group from pretest to post test and again on follow-up. [69] These studies on cortisol provide evidence for recommendations integrating mindfulness into athlete training programs.

Could Athletes Have Boosts to Immune Functioning?

Athletes are under ongoing stress to perform. Chronic stress is associated with immunosuppression, with the body losing its ability to regulate the inflammatory response. Stress raises catecholamine and suppressor T cells levels suppressing the immune system and raising the risk of viral infection. [62] Catecholamines, neurohormones such as dopamine, adrenaline, and noradrenaline, are responsible for the body's "fight-or-flight" response. Unusually high or low levels of catecholamines can cause medical issues; and imbalanced across multiple catecholamines may indicate serious health conditions. Chronic stress may also result in glucocorticoid receptor resistance (GCR) that, in turn, fails to decrease inflammatory response. GCR appears to predict production of proinflammatory proteins, called *cytokines*. [61] Cytokines are regulators of responses to infection, immune responses, inflammation, and trauma. Proinflammatory cytokines, specifically (IL-1β, TNF-α, and IL-6), act to make disease worse. [61] Meanwhile, modalities like mindfulness that stimulate the vagus nerve and thus target the parasympathetic nervous system inhibit the production of such cytokines. [70]

In a comprehensive review of randomized clinical trials examining the effects of mindfulness meditation on immune system, including pro-inflammatory cytokines, immune cell count, and antibody response, researchers found reductions in circulating levels of C-reactive protein, a protein produced during bacterial infections and inflammation. They also found improvement in cell-mediated immunity, important to destroying virus-infected cells, intracellular bacteria, and cancers. Researchers also saw increases in telomerase activity, which elongate cell life [71]. Other mindfulness studies of immune system function have shown increases in hemagglutination-inhibition influenza antibody titers in a four-week period from a Mindfulness Based Stress Reduction intervention [71] and decreased interleukin-6 (IL-6) [72], demonstrating biological boosts to the immune system.

Another study comparing mindfulness with a health enhancement program in reducing experimentally induced inflammation found both interventions had comparable post-training stress-evoked cortisol responses, but that the mindfulness had a smaller post-stress inflammatory response than the health enhancement program, suggesting that mindfulness may be more efficacious for inflammation symptom relief [73].

In sports, Solberg and colleagues studied the effects of mindfulness on runners' immune response by looking at counts of CD8 killer cells, indicative of stress in the body. By measuring the increase in CD8 killer cells after a maximum oxygen uptake test, they found that runners who meditated did not experience the expected increase in these immune cells, as did the runners who did not meditate [74]. They repeated the study five years later with ten longtime meditators vs a control group, finding that the runners who practiced meditation had lower CD8 killer cells count at rest before the race, but there was no difference after exercise and two hours of recovery. This suggests that longtime meditation may influence CD8 cell counts at rest. [74] More research is needed on these factors with athletes before any determinations are made.

Mindfulness Has Been Shown to Lower Blood Pressure

Despite their healthy lifestyles, many athletes struggle with high blood pressure or hypertension, consistent with the prevalence of hypertension in 45% of adults in the United States [75]. While reviews of studies of mindfulness on blood pressure have shown slightly positive, but mixed, results [76]; more recently, mindfulness meditation has been positively linked with reductions in blood pressure in programs developed specifically for this purpose [77]. Patients often struggle to adhere to medical recommendations, but these mindfulness programs showed increased compliance with protocols designed to reduce blood pressure, including nutrition and exercise guidelines. These programs also demonstrated improvements in participant attention control, self-awareness, and emotional regulation. Additional research on participants' brains showed changes in white matter neural tracts associated with hypertension [78]. Mindfulness has also been shown to be superior to progressive muscle relaxation exercises for lowering blood pressure [79].

- Cortisol is a reliable biological marker of stress that has helped assess the effect of mindfulness on athlete well-being.
- Meditation may influence CD8T (immune killer) cell counts, correlating to decrease oxidative stress and inflammation in the body.
- While many athletes struggle with high blood pressure, the research on mindfulness has contributed an increasing number of studies showing that mindfulness interventions to reduce blood pressure are effective.
- Mindfulness provides similar results as biofeedback has for increasing focused attention and modifying breathing for maximum benefit, without equipment.

Studies Show Improvement in Heart Rate Variability (HRV)

Heart rate variability (HRV) in athletes is being monitored closely with many turning to HRV biofeedback systems that provide auditory or visual feedback, fully engaging the attention of the subject. Evidence suggests that people are more physically and mentally resilient when HRV oscillation amplitudes—measures of sympathetic-parasympathetic influences on the heart- are higher and more complex [80]. While biofeedback can be helpful in learning to focus attention and modify breathing for maximum benefit, mindfulness provides similar effects without equipment. A systematic review and meta-analysis of slow breathing studies demonstrated comparable results with or without biofeedback [81].

The first training focus in mindfulness is usually the breath, as breathing serves as an anchor to the "here and now." Mindful breathing focuses on observing the breath and allowing the breath to become deeper and more abdominal. Many psycho-physiological changes in brain-body interactions from mindfulness meditation rely on the slowing down of breath frequency, which happens naturally with meditation (less than 10 breaths/min, as opposed to the usual rate of 12–14). At the beginning of meditations and in other meditation practices, mindfulness includes paced breathing or diaphragmatic breathing, slowing rates to about 5–7 breaths/min. [82] A review of 15 studies on slow breathing (less than 6 breaths/min) showed that these techniques increase HRV [83]. The inclusion of a 5-min rest period following 6 min of slow, diaphragmatic breathing has been shown to increase HRV still further [84]. It now appears that these changes not only happen during formal practice but are maintained for months and years in experienced meditators. Respiratory practices such as breathing during meditation are shown to influence the parasympathetic nervous system by stimulating the vagus nerve [85]. Stimulation of the vagus nerve is a known antagonist of sympathetic nervous system activation, suggesting vagal tone function is a mediator between mindfulness modalities and improved health [86].

Exercise-induced diaphragmatic breathing can cause fatigue when it competes with the locomotor muscles for available cardiac output, yet diaphragmatic breathing at rest following exercise appears to increase antioxidant defense [87]. In a study of 16 male cyclists, eight engaged in 40 min of diaphragmatic breathing and eight relaxed by reading magazines before and after a 900-calorie meal. Investigators found lower blood glucose and higher insulin and antioxidants in the breathing group. The diaphragmatic breathing intervention attenuated oxidative stress and down-regulated the sympathetic nervous system after an 8-hour period of strenuous exercise. The intervention group showed lower levels of reactive oxygen metabolites and cortisol and higher levels of nocturnal melatonin levels, leading to better sleep, compared to the control group. [87]

A recent randomized clinical trial on mindfulness meditation through an app vs. relaxation with music found increases with HRV while meditating that persisted 48 hours after the 10-day training. The mindfulness group showed a significant 13 millisecond increase in HRV over baseline while meditating, while the music group had a mean 9-millisecond increase. The mindfulness group slowed respiration by 4 seconds per minute, while the music group did not [88].

> Attention is an outcome of mindfulness training with relevance to sports as it is key for both performance and injury prevention. An understanding of neurological processes around attention facilitates our understanding of the importance of attention training as part of an athlete's practice routine.

> Athletes may benefit from mindfulness through the pathways of improved stress response, cardiovascular health, increased immune function, improved antioxidant defense, and other psycho-physiological changes.

Mindfulness Improves Attention

Attention and focus are critical for sports performance and the prevention of injury. In an early study of Olympic athletes, focus of attention proved to be a significant predictor of ranking. Attentional focus during an event for females and immediately before competition for males was the only predictor of performance in one study [89].

Humans find attentional focus challenging with research suggesting that the human mind wanders at least 30% of the time, with consequences for human well-being [90]. In a worldwide study, researchers found that about half the time our thoughts are drifting and tend to go to the negative [91].

While mind wandering can be beneficial for creativity and problem-solving, negative thoughts cause mental and physical stress. Attention researchers have identified three other experiences that can affect attention negatively: stress, poor mood, and threats. Through mindfulness training, participants learn to observe thoughts throughout the day and to disengage from negative thinking, returning to focus on the moment. According to researcher Amishi Jha, who has done extensive attention studies, including with the US Military, *mindfulness is the only brain training tool that consistently works to strengthen attention* [92]. It appears that even a brief mindfulness training can improve attention in novice meditators [93].

Attention has been viewed broadly as three separate processes—alerting, orienting, and executive attention/control—from three separate brain regions. *Alerting* reflects the readiness for incoming stimuli (phasic alertness) and the maintenance of an aroused state (tonic alertness). *Orienting* mediates selective prioritization of targets; and *executive attention* involves conflict monitoring when different stimuli compete for attention. A review of 57 studies suggests that mindfulness training shows promise for improving attention, including alerting, orienting, and guiding on-task behavior [94]. A recent meta-analysis of 87 studies suggests that generalized attention and the alerting and executive control networks are improved by mindfulness [95].

Sport excellence requires an alerting network that will quickly attend to novel stimulus and be prepared for the occurrence of new and different stimuli. Studies support the idea that mindfulness assists with perceptual detection and visual discrimination. In one study of athletes, 67% of those who were introduced to mindfulness remained injury free as compared to 40% of the control group [17].

- Humans find attentional focus challenging with research suggesting that the human mind wonders at least 30% of the time, with consequences on well-being.
- Attention and focus are critical for sports performance and the prevention of injury.
- Focus of attention has been shown to be a predictor of performance in athletes, even predicting rankings in the Olympics.
- Sports excellence requires an alerting network that will quickly attend to novel stimuli.
- Sports proficiency requires an alerting network that can sustain attention, such as in the high demand period of pre-season training.
- While athletes must tune-out, they must also tune-in(wards).
- While most athletes have learned to tune out external distractions, such as fans yelling, media or sound overhead, executive control of attention could be particularly important in sports for internal monitoring of emotional distress.

Sports proficiency requires an alerting network that can sustain attention necessary for performing more complicated functions, such as working memory and executive control. It is also critical for high-demand preseason training over a prolonged period. A study of mindfulness with 100 college football players on sustained attention showed that mindfulness practices protected the athletes from the usual decline in attention expected during preseason training [96]. A recent study using the Berlin Mindfulness-Based Training (BATL) showed improvements in both sustained and selective attention in young athletes, with gains positively correlated to both practice frequency and continued engagement over time [32].

Executive control is critical for athletes who have a lot of conflicting stimuli to which they must attend. As demonstrated in the Sumantry & Stewart meta-analysis cited earlier, mindfulness appears to improve executive control [95], with meditators being more skilled at prioritizing focus and inhibiting distracting stimuli than nonmeditators. While most athletes have learned to tune out external distractions, such as fans yelling, media or sounds overhead, executive control of attention could be particularly important in monitoring emotional distress and internal distractions.

Athletes are faced with the complex challenge of gaining attention control over the tasks at hand and internal signs of emotional distress, etc., while blocking external stimuli (i.e., fans yelling); a challenge mindfulness training can meet.

Mindfulness May Improve Psychological Distress and Mental Health, Including Injury-Related Stress, Anxiety, Depression, and Burnout, As Well As Physical Health

With athletes faced with high stress environments, a review of mental health issues and whether they are amenable to intervention with mindfulness training, is warranted.

The International Olympic Committee has stated that mental health cannot be separated from physical health [97]. Psychological distress and mental health disorders, anxiety, depression, substance use, eating and sleep disorders all appear to be as common in athletes as the general population [98] or possibly higher. [99]

Many researchers have reviewed factors of psychological stress that are challenged by the framework of competitive sports. Stress can arise from perfectionism [100], early maladaptive coping strategies and avoidance behaviors [101], overtraining [102], competition anxiety [103], and burnout, among other issues [104]. Injuries and concussions also affect mental health. [105]

Mindfulness research has shown benefits to mental health and psychological distress in the general population. In a meta-analysis of studies, mindfulness-based therapy was especially effective for reducing anxiety, depression, and stress [106], with another observing that mindfulness showed the most consistent evidence on specific mental disorders, such as depression, anxiety, pain, smoking, and addiction [107]. It is clear that mindfulness-based protocols promote mental health in nonclinical settings, but other prevention interventions may be equally effective. [108] Many athletes would prefer not to take medications and find natural ways to deal with anxiety. A randomized clinical trial of 276 adults with anxiety disorders comparing an anxiety medication with mindfulness showed mindfulness to be as effective as the drug [109].

Mindfulness can help athletes orient themselves toward emotional experience, even if that means being with difficult emotions. There is no need for escape through alcohol or drugs, or to lessen or control emotions. Mindfulness influences the trajectory of the emotional experience [110]. For example, it can buffer anger-related responses and inhibit aggression [111]. While the studies utilized untreated control groups, rather than placebo control groups, a review of mindfulness and acceptance-based interventions in sports concluded that they were effective at reducing competition anxiety [112]. In fact, recent studies using randomized clinical trials showed improvements in stress, anxiety, burnout, and other psychological issues, as described in the Table 15.2.

Non-randomized studies using mindfulness-based curricula also show improvements in mental health in athletes. DePetrillo and colleagues found decreases in sports related worries and perfectionism in 25 long-distance runners using the Mindful Sports Performance Enhancement (MSPE) curriculum [24]. A mindfulness

Table 15.2 Randomized controlled trials assessing the effect of mindfulness training on athletes[a]

Study	Participant description	Modality tested	Primary outcome
Aherne et al. (2011) [49]	Athletes competing at international or national level in various sports	A mindfulness-based intervention with exercise domains	Mindfulness, flow
Ajilchi et al. (2019) [113]	Amateur basketball players	Mindfulness Acceptance Commitment (MAC)	Mindfulness
Ajilchi et al. (2021) [114]	Female university athletes	Mindfulness Acceptance Commitment (MAC)	Mindfulness, mental toughness, psychological well-being
Dehghani et al. (2018) [18]	Female university basketball players	Mindfulness Acceptance Commitment (MAC)	Anxiety
Glass et al. (2019) [25]	National Collegiate Athletic Division III athletes in various sports	Mindfulness Sport Performance Enhancement (MSPE)	Depression
Gross et al. (2018) [20]	National Collegiate Athletic Division III female student athletes	Mindfulness Acceptance Commitment (MAC)	Psychological distress, generalized anxiety
Jekauc et al. (2017) [31]	Sport science students	Berlin Mindfulness Based Training for Athletes (BALT)	Mindfulness
John et al. (2011) [28]	Elite shooters	Mindfulness meditation	Salivary cortisol (stress response)
MacDonald et al. (2017) [66]	Wheelchair basketball players	A mindfulness-based intervention	Salivary cortisol (stress response)
Mehrsafar et al. (2019) [67]	Elite Wushu athletes	A mindfulness-based intervention	Mindfulness
Mohebi et al. (2021) [19]	Female athletes competing at national level in various sports	Mindfulness Acceptance Commitment (MAC)	Grit, self-compassion
Rothlin et al. (2016) [115]	Athletes from tennis, curling, football, and badminton federations	A mindfulness-based intervention	Mindfulness, emotion control, attention control
Samadi et al. (2020) [69]	Beginner endurance runners	A mindfulness-based intervention	Salivary cortisol, emotion regulation
Scott Hamilton (2016) [116]	Competing cyclists	A mindfulness-based intervention incorporating mindful cycling	Mindfulness, flow, anxiety
Sparks (2022) [117]	Collegiate rowers	Rowing specific mindfulness intervention	Mindfulness, flow
Wolch (2020) [118]	Basketball	A mindfulness-based intervention	Anxiety
Zadeh et al. (2019) [16]	Amateur soccer players	A mindfulness-based intervention	Mindfulness

[a]This comprehensive list of randomized clinical trials incorporated mixed methods with significant results across mental health outcomes between intervention and control groups, and also within groups pre- and post-intervention, suggesting mindfulness training, in various forms tested, was effective for psychological well-being of athletes. Some of them also assessed other metrics such as adherence, performance, and injury rate

training showed a progressive decrease in stress with an increase in athletic coping skills over the course of the intervention with a collegiate women's basketball team [119]. Mindfulness Acceptance and Commitment (MAC) reduced sports anxiety, as well as eating concerns (i.e. binge eating, emotional eating, and other disordered eating patterns), and psychological distress post-intervention with 18 National Collegiate Athletic Association Division II athletes [20]. Dispositional mindfulness, or a naturally mindful state of being, was shown to have direct effects on rumination and emotional regulation in athletes [120].

Casali et al. studied 263 athletes who completed measures of self-compassion, negative rumination, and competition anxiety. Those who had greater dispositional mindfulness, or trait mindfulness, were likely to score higher in self-compassion, ruminate less, and feel less competition anxiety. Regression models showed the effects of self-compassion on competition anxiety and worry (the cognitive component of anxiety), as well as the effects of worry on somatic anxiety, or anxiety resulting in physical symptoms. While competition anxiety decreased with greater levels of self-compassion, it increased with worry, while worry also positively correlated with somatic anxiety [121].

Mindfulness Can Reduce Stress to Decrease Sports Injuries

Stress occurs when the demands of a situation exceed an individual's available resources to respond and cope effectively. When athletes experience high levels of stress, either on or off the field, they are at a greater risk of injury [122]. Stress causes attentional changes, including the narrowing of attention, general distraction, and increased self-consciousness. It also increases muscle tension and coordination difficulties contributing to injury [123].

Many factors contribute to athlete injury as noted in the previous section; but the only factor which has been consistently demonstrated to relate to athlete injury is stress [124]. Stress and anxiety can increase the risk of injury in many sports. Experts recommend cognitive techniques, along with body-based methods like slow breathing, progressive muscle relaxation, and biofeedback, to help manage them [122].

The expanding evidence suggests that mindfulness practices should be added to the list of recommendations for athletes to reduce stress and risk of injury.

Ongoing stress is serious, as it is mediated by the neurohormones known as catecholamines, which are responsible for the body's fight-flight-freeze response necessary for human survival. Unusually high or low levels of catecholamines can contribute to medical issues when stress becomes long-term or chronic, causing muscle tension, exacerbating breathing problems, and contributing to migraine headaches, hypertension, heart attack, stroke, inflammatory bowel disease and irritable bowel syndrome, among other health issues [125].

Mixed, scientific studies on mindfulness suggest that mindfulness can reduce stress and burnout, [86] create an overall sense of well-being [125], and build resiliency [126]. A randomized clinical trial of 168 elite soccer players ages 16–19 using mindfulness acceptance and commitment (MAC) training explored the effects of attention, stress and anxiety, and mindfulness on sporting injuries. The control group attended seven sessions on sports-injury psychology, or the tendency of injury to reveal or trigger mental health issues such as anxiety. The mindfulness group had lower trait and state group anxiety following training and their frequency of injuries were significantly lower. Additionally, the number of days lost to injury were lower in the mindfulness group [127]. In another randomized clinical trial of amateur soccer players with MAC training, mindfulness was again found to help prevent injury, with the effects persisting up to three months [16].

Mindfulness to Reduce Burnout

Burnout in sports has been conceptualized as a three-dimensional construct which includes emotional and physical exhaustion, feelings of inefficacy and a reduced sense of accomplishment, and negative feelings about the benefits of sports involvement [128].

Perfectionism has also been identified as a cause of athlete burnout. It can include self-perfectionism, and parental and coach pressure. A study of 199 junior sports participants showed that perceptions of parental and coach pressure can be especially important, leaving the athlete without a sense of support to fulfill psychological needs [129]. In a study of male rugby players, the frequency of the perfectionistic cognitions of the athlete was positively associated with athlete burnout [130].

An exploration of a mindfulness attention training technique on junior elite athletes' level of burnout showed a significant decrease in burnout among the intervention group, but not the control group [131]. A meta-analysis of 10 mindfulness studies and athlete burnout showed that mindfulness-based interventions had positive effects in burnout prevention, indicating an inverse relationship between mindfulness and burnout [132].

Mindfulness Helps with Pain Management

Pain is modulated by complex interactions of body sensations, emotions, and thoughts. Mindfulness has been shown to reduce pain in experimental and clinical settings and the resulting mood disturbances and psychological symptomatology [54]. It has also been shown to create lower pain sensitivity and reduce respiration rate [133].

Regarding chronic pain, a recent systematic review meta-analysis of 30 randomized clinical trials found small decreases in pain compared to all types of pain controls (treatment as usual for pain management and support/educational groups). Another systematic review found a trend suggesting that mindfulness may be beneficial for managing affective and sensory pain [134]. Both studies found statistically significant improvement in depressive symptoms from pain, as well as quality of life. A randomized clinical trial of Mindfulness Based Stress Reduction (MBSR), Cognitive Behavioral Therapy (CBT) and standard pain management with 342 people of all ages suffering from chronic low back pain found that both mindfulness and CBT improved patients' ability to function independently and resulted in less back pain compared to usual treatments. The findings from this study persisted for a year [135]. Similarly, a randomized clinical trial with 143 rheumatoid arthritis patients found that mindfulness training led to greater reductions in daily pain catastrophizing, morning disability, fatigue, and anxiety compared to both Cognitive Behavioral Therapy and arthritis education [55].

Placebo effects may also contribute to the *relaxation response* in pain management [136]. Zeidan and colleagues demonstrated that mindfulness meditation, but not placebo, resulted in pain relief via greater activation in the brain's sensory processing regions. These brain regions are responsible for the cognitive control and appraisal of pain. When receiving painful leg heat, participants were asked to acknowledge their thoughts and feelings without judgment and to return attention back to the breath. They were told that sensory and affective events were momentary and fleeting, not requiring interpretation or evaluation. The sham mindfulness training group showed brain activity in many of the brain's relaxation response areas; but the mindfulness-related relief is mechanistically distinct from the placebo analgesia. Compared to both placebo mindfulness training and passive book listening, the mindfulness meditation group showed significantly greater reductions in pain intensity (−27%) and pain unpleasantness (−44%), whereas the placebo training resulted in only an 8% reduction in pain intensity and a 27% reduction in pain unpleasantness [137].

In one randomized clinical trial of 20 injured athletes, Mindfulness Based Stress Reduction (MBSR) training created an increase in pain tolerance for injured athletes as compared to the control group, while both normal physiotherapy (control group) and mindfulness training provided boosts in positive mood. [13] Another study of 142 recreational cyclists found mindfulness was positively associated with mental toughness and negatively associated with pain catastrophizing [14]. In addition to improving perception of pain, a study of female university runners who completed mindfulness training also improved endurance performances. They demonstrated higher thresholds for exhaustion and less negative feelings of fatigue [138]. A randomized clinical trial with recreational runners with knee pain found mindfulness training, in conjunction with traditional physical therapies, reduced pain while running, improved coping strategies, and decreased pain catastrophizing [139].

Mindfulness curricula for sports are being used for rehabilitation [140], with at least one curriculum in the development phases with a Division III public university [141]. There is little research on the effects of rehabilitation at this point, although a handful of studies suggests that mindfulness, in addition to assisting with pain management, may help with recovery and rehabilitation from injury [142].

Despite limited research, the positive experiences of those who practice mindfulness have inspired curricula in sports for athletes from aspiring university students to former NFL players focused on recovery from injury and on overall well-being. In 2015, Meditation Specialist Chad McGehee collaborated with former NFL linebacker and Neuroscientist Chris Borland to launch a meditation program for former NFL players. Chad McGehee then became the University of Wisconsin's Director of Meditation Training, a title with historical significance for both sports and the field of mindfulness (See Figs. 15.2 and 15.3).

> The existing literature on mindfulness training programs in various populations of athletes supports further investigation of how to maximize the benefits, both physical and mental, found thus far among athletes.

Conclusion

It is important for sports educators, physical therapists, trainers, athletes, and coaches to be aware of the positive impact that mindfulness training can have in their individual and team performance. Studies reveal that mindfulness may positively influence focus and attention, emotion regulation, performance, and pain management. Components of mindfulness curricula, such as proper breathing techniques, engage the parasympathetic nervous system, disengage the sympathetic

Fig. 15.2 As the University of Wisonsin's Director of Meditation Training, Chad McGhee trains across athletic teams, including the Wisconsin Badgers in Football and the Women's Volleyball team

Fig. 15.3 Pictured here in 2022, Chad McGee attributed the first NCAA Wisconsin Volleyball championship in program history to both the physical and mental skill of the team's players

nervous system, support the optimal functioning of the vagus nerve, and reduce cardiovascular risks and markers of stress, such as cortisol and inflammatory cytokines. [70]

This chapter does not address other issues relevant to athletes where there is growing evidence that mindfulness may be beneficial, including improved sleep, eating disorders, alcohol and substance abuse, and anger management.

It should be noted that not all individuals are suitable for mindfulness and a small percentage of individuals have reported negative results [143]. Like most trainings, it must be individualized and monitored by an experienced coach. As in sports, continued practice is key. While study findings are promising and show both before and after and between group (control vs. mindfulness intervention) differences, indicative of the large effect mindfulness can have on outcomes, adherence to the training programs and interventions is a key factor not discussed in detail in this chapter. In addition to the varying degrees of feasibility of integrating mindfulness across different sports, there is significant variability between individuals in terms of how they adapt to mindfulness training, the attention devoted to it, and specific time management challenges.

Studies continue to support positive outcomes, and implementing mindfulness into athlete training is relatively low cost. While the more developed programs depicted in Table 15.1 have standardized schedules ranging from four to eight weeks, many of the intervention trials and randomized clinical trials reveal significant improvements in desired outcomes after even as short an intervention as a single 15-min mindfulness training. More randomized clinical trials in each of the areas discussed are necessary to fill in the remaining, but narrowing, gaps.

References

1. Kabat-Zinn J. Wherever you; go, there you are: Mindfulness meditation for everyday life; 1994.
2. Herrigel E. Zen and the art of archery; 1953.
3. Gallwey W. The inner game of tennis; 1974.
4. Mumford G. The mindful athlete: Secrets to pure performance; 2016.
5. Dabbs R. Meditation helps mickelson win PGA championship. Golf Monthly. https://golf-monthly.com/news/tour-news/meditation-helps-mickelson-win-pga-championship-234191. 2023;24.
6. Nardino M. Simone biles' most honest quotes about mental health and wellness. US Weekly. https://www.usmagazine.com/celebrity-news/pictures/simone-biles-most-honest-quotes-about-mental-health-wellness/. 2023;29.
7. McCullough A. How therapy helped kenley jansen reclaim his place among the game's elite. The Athletic. https://theathletic.com/3300104/2022/05/09/atlanta-braves-kenley-jansen-therapy/. Accessed Jan 2023;8.
8. Delehanty H. The game changer. Mindful. https://www.mindful.org/the-game-changer/. 2023;25.
9. Freiberg K, Freiberg J. 4 things NBA champion steve kerr can teach CEOS about team chemistry. Forbes. https://www.forbes.com/sites/kevinandjackiefreiberg/2018/06/21/4-things-nba-champion-steve-kerr-can-teach-ceos-about-team-chemistry/. 2023;24.
10. Saltzman A. Still quiet place for athletes: mindfulness skills for achieving peak performance and finding flow. In: Sports and life; 2018.
11. Su N, Si G, Zhang C. Mindfulness and acceptance-based training for chinese athletes: The mindfulness-acceptance-insight-commitment (MAIC) program. Hong Kong Baptist University. https://scholars.hkbu.edu.hk/en/publications/mindfulness-and-acceptance-based-training-for-chinese-athletes-th. 2023;2.
12. Tebourski K, Bernier M, Salha B, Souissi N, Fournier JF. Effects of mindfulness for performance programme on actual performance in ecological sport context: two studies in basketball and table tennis. Int J Environ Res Public Health. 2022;19(19):12950.
13. Mohammed W. Effect of mindfulness-based stress reduction (MBSR) in increasing pain tolerance and improving the mental health of injured athletes. Front Psychol. 2018;9
14. Jones BJ, Kaur S, Miller M, Spencer R. Mindfulness-based stress reduction benefits psychological well-being, sleep quality, and athletic performance in female collegiate rowers. Front Psychol. 2020;11
15. di Fronso S, Robazza C, Bondar Z, Bertollo M. The effects of mindfulness-based strategies on perceived stress and psychobiosocial states in athletes and recreationally active people. Int J Environ Res Public Health. 2022;19
16. Zadeh MM, Ajilchi B, Salman Z, Kisely S. Effect of a mindfulness programme training to prevent the sport injury and improve the performance of semi-professional soccer players. Aust Psychiatry. 2019;27(6):589–95. https://doi.org/10.1177/1039856219859288.

17. Ivarsson A, Johnson U, Andersen M, Fallby J, Altemyr M. It pays to pay attention: a mindfulness-based program for injury prevention with soccer players. J Appl Sport Psychol. 2015;27:319–34.
18. Dehghani M, Saf A, Tebbenouri G, Zarnagh H. Effectiveness of the mindfulness-acceptance-commitment-based approach on athletic performance and sports competition anxiety: a randomized clinical trial. Electron Physician. 2018;10:6749–55.
19. Mohebi M, Sadeghi-Bahmani D, Zarei S, Gharayagh H, Andi Z, Brand S. Examining the effects of mindfulness-acceptance-commitment training on self-compassion and grit among elite female athletes. Int J Environ Res Public Health. 2021;19(1):134.
20. Gross M, Moore Z, Gardner F, Wolanin A, Pess R, Marks D. An empirical examination comparing the mindfulness-acceptance-commitment approach and psychological skills training for the mental health and sport performance of female student athletes. Int J Sport Exerc Psychol. 2016;16:431–51.
21. Oguntuase S, Sun Y. Effects of mindfulness training on resilience, self-confidence and emotion regulation of elite football players: the mediating role of locus of control. Asian J Sport Exerc Psychol. 2022;2:198–205.
22. Thompson R, Kaufman K, De Petrillo L, Glass C, Arnkoff D. One year follow-up of mindful sport performance enhancement (MSPE) with archers, golfers, and runners. J Clin Sport Psychol. 2011;5:99–116.
23. Hut M, Glass C, Ka D, Minkler T. The effects of mindfulness training on mindfulness, anxiety, emotion dysregulation, and performance satisfaction among female student-athletes: the moderating role of age. Asian J Sport Exerc Psychol. 2021;1:75–82.
24. DePetrillo L, Kaufman K, Glass C, Arnkoff D. Mindfulness for long-distance runners: an open trial using mindful sport performance enhancement (MSPE). J Clin Sport Psychol. 2009;4(4):357–76.
25. Glass C, Spears C, Perskaudas R, Kaufman K. Mindful sport performance enhancement: randomized controlled trial of a mental training program with collegiate athletes. J Clin Sport Psychol. 2017;
26. Chen MA, Meggs J. The effects of mindful sport performance enhancement (MSPE) training on mindfulness, and flow in national competitive swimmers. J Hum Sport Exercise. 2021;16
27. Widhi Harita A, Suryanto S, Ardi R. Effect of mindfulness sport performance enhancement (MSPE) to reduce competitive state anxiety on karate athletes, vol. 8. Jurnal Penelitian Pembelajaran; 2022. p. 169–88.
28. John S, Verma S, Khanna G. The effect of mindfulness meditation on the HPA-axis in pre-competition stress in sports performance with elite shooters. Nat J Intl Res Med. 2011;2(3)
29. Kumar P. An analysis on the effect of meditation on performance of shooting players. Int J Phys Educ Sport Sci. 2018;10
30. Baltzell A, Summers J. The power of mindfulness: Mindfulness meditation training in sport (MMTS). Springer; 2017. https://doi.org/10.1007/978-3-319-70410-4_3.
31. Jekauc D, Kittler C, Schlagheck M. Effectiveness of a mindfulness-based intervention for athletes. Psychology. 2017;08:1–13.
32. Kittler C, Arnold M, Jekauc D. Effects of a mindfulness-based intervention on sustained and selective attention in young top-level athletes in a school training setting: a randomized control trial study. Sport Psychol. 2021;36:77–88.
33. Mahoney J, Hanrahan S. A brief educational intervention using acceptance and commitment therapy: four injured athletes' experiences. J Clin Sport Psychol. 2011;5:252–73.
34. Perret KA. Can acceptance and commitment therapy increase rehabilitation adherence for the treatment of sport injury? ProQuest information & learning. US; 2014. p. 76.
35. Bernier M, Thienot E, Codron R, Fournier J. Mindfulness and acceptance approaches in sport performance. J Clin Sport Psychol. 2009;3:320–33.
36. Lungdren T, Reinebo G, Frojmark M, et al. Acceptance and commitment training for ice hockey players: a randomized controlled trial. Front Psychol. 2021;12
37. Kettunen A, Valimaki V. Acceptance and value-based psychological coaching intervention for elite female floorball players. Psychology. 2014;

38. Doron J, Rouault Q, Jubeau M, Bernier M. Integrated mindfulness-based intervention: effects on mindfulness skills, cognitive interference and performance satisfaction of young elite badminton players. Psychol Sport Exercise. 2020;47:101638.
39. Wegner D. Setting free the bears: escape from thought suppression. Am Psychol. 2011; https://doi.org/10.1037/a0024985.
40. Hatzigeorgiadis A, Zourbanos N, Galanis E, Theodorakis Y. Self-talk and sports performance. Perspect Psychol Sci. 2011;6(4):348–56. https://doi.org/10.1177/1745691611413136.
41. Jeong YH, Healy L, Ewan D. The application of goal setting theory to goal setting interventions in sport: a systematic review. The application of goal setting theory to goal setting interventions in sport: a systematic review. Int Rev Sport Exerc Psychol. 2021;16:474–99.
42. Lochbaum M, Stoner E, Hefner T, Cooper S, Lane A, Terry P. Sport psychology and performance meta-analyses: a systematic review of the literature. PLoS One. 2022;17(2):e0263408.
43. Noetal M, Ciarrochi J, Van Zanden B, Lonsdale C. Mindfulness and acceptance approaches to sporting performance enhancement: a systematic review. Int Rev Sport Exerc Psychol. 2017;3:1–37.
44. Perry J, Ross M, Weinstock J, Weaver T. Efficacy of a brief mindfulness intervention to prevent athletic task performance deterioration: a randomized controlled trial. Sport Psychol. 2016;31:410–21.
45. Shaabani F, Naderi A, Borella E, Calmeiro L. Does a brief mindfulness intervention counteract the detrimental effects of ego depletion in basketball free throw under pressure? semantic scholar. Sport, Exercise and Performance Psychology. https://www.semanticscholar.org/paper/Does-a-brief-mindfulness-intervention-counteract-of-Shaabani-Naderi/8d25901215893da3bfb45ed4324cfb36c8fb0242. 2023;1.
46. Nien J, Wu C, Yang K. Mindfulness training enhances endurance performance and executive functions in athletes: an event-related potential study. Hindawi. 2020;2020:1–12.
47. Zhang CQ, Si G, Duan Y, Lyu Y, Keathley DA, Chan K. The effects of mindfulness training on beginners' skill acquisition in dart throwing: a randomized controlled trial. Psychol Sport Exerc. 2015;22:279–85.
48. Csikszentmihalyi M. Flow: the psychology of optimal experience; 1991.
49. Aherne C, Moran A, Lonsdale C. The effect of mindfulness training on athletes flow: an initial investigation. Sport Psychol. 2011;25:177–89. https://doi.org/10.1123/tsp.25.2.177.
50. Creswell J. Mindfulness interventions. Annu Rev Psychol. 2017;68:491–516.
51. Center UMAR. UCLA health system. https://www.uclahealth.org/programs/marc. Accessed January. 2023;8.
52. Simonsmeier BA, Andronie M, Buecker S, Frank C. The effects of imagery interventions in sports: a meta-analysis. Int Rev Sport Exerc Psychol. 2020;
53. Parker H. Facing your inner critic. Psychol Today. 2023:29.
54. Kabat-Zinn J. An outpatient program in behavioral medicine for chronic pain patients based on the practice of mindfulness meditation: theoretical considerations and preliminary results. Gen Hosp Psychiatry. 1982;4(1):33–47.
55. Davis M, Zautra AJ, Wolf L, Tennen H, Yeung EW. Mindfulness and cognitive-behavioral interventions for chronic pain: differential effects on daily pain reactivity and stress reactivity. J Consult Clin Psychol. 2015;83(1):24–35.
56. Hilton L, Hempel S, Ewing B, et al. Mindfulness meditation for chronic pain: systematic review and meta-analysis. Ann Behav Med. 2017;51(2):199–213.
57. Aristegui R. Relational mindfulness: fundamentals and applications. Google Books. https://books.google.com/books/about/Relational_Mindfulness.html?id=aUOqzQEACAAJ. 2023;26.
58. Piasecki P, Loughead T, Paradis K, Munroe-Chandler K. Using a personal-disclosure mutual-sharing approach to deliver a team-based mindfulness meditation program to enhance cohesion. Hum Kinetics J. 2019;35(1):22–9.
59. Colzato L, Kibele A. How different types of meditation can enhance athletic performance depending on the specific sport skills. J Cogn Enhanc. 2017;1:122–6.

60. Buhlmayer L, Birrer D, Rothlin P, Faude O, Donath L. Effects of mindfulness practice on performance-relevant parameters and performance outcomes in sports: a meta-analytical review. 2017;47:11–1.
61. Cohen S, Janicki-Deverts D, Doyle W. Chronic stress, glucocorticoid receptor resistance, inflammation, and disease risk. Proc Natl Acad Sci. 2012;109:5995–9.
62. Salleh M. Life event, stress and illness. Malays J Med Sci. 2008;15(4):9–18.
63. Sanada K, Montero-Marin J, Alda D. Effects of mindfulness-based interventions on salivary cortisol in healthy adults: a meta-analytical review. Front Psychol. 2016;7:471.
64. O'Leary K, O'Neil S, Dockray S. A systematic review of the effects of mindfulness interventions on cortisol. J Health Psychol. 2016;21(9):2108–21. https://doi.org/10.1177/1359105315569095.
65. Filaire E, Sagnol M, Ferrand C, Maso F, Lac G. Psychophysiological stress in judo athletes during competitions. J Sport Med Phys Fit. 2001;41:263–8.
66. MacDonald L, Minahan C. Mindfulness training attenuates the increase in salivary cortisol concentration associated with competition in highly trained wheelchair-basketball players. J Sports Sci. 2018;36(4):378–83.
67. Mehrsafar A, Strahler J, Gazerani P. The effects of mindfulness training on competition-induced anxiety and salivary stress markers in elite wushu athletes: a pilot study. Physiol Behav. 2019:210.
68. Zhu Y, Sun F, Li C. Acute effects of mindfulness-based intervention on athlete cognitive function: an fNIRS investigation. J Exerc Sci Fit. 2022;20(2):90–9.
69. Samadi H, Maleki B, Sohbatiha M. The effectiveness of mindfulness-based cognitive-behavioral strategy training on cognitive emotion regulation strategies and salivary cortisol levels in endurance runners: a three-month follow-up. Middle East J Rehabil Health Stud. 2020;7(4) https://doi.org/10.5812/mejrh.100953.
70. Breit S, Kupferberg A, Rogler A, Hasler G. Vagus nerve as modulator of the brain-gut axis in psychiatric and inflammatory disorders. Front Psych. 2018;9:44.
71. Black D, Slavich G. Mindfulness meditation and the immune system: a systematic review of randomized controlled trials. Ann N Y Acad Sci. 2016;1373:13–24.
72. Creswell J, Taren A, Lindsay E. Alterations in resting-state functional connectivity link mindfulness meditation with reduced interleukin-6: a randomized controlled trial. Biol Psychiatry. 2016;80:53–61.
73. Rosenkranz M, Davidson R, MacCoon D, Sheridan J, Kalin N, Lutz A. A comparison of mindfulness-based stress reduction and an active control in modulation of neurogenic inflammation. Brain Behav Immun. 2013;27(1):174–84.
74. Solberg E, Halvorsen R, Holen A. Effect of meditation on immune cells. Stress Med. 2000;16(3):185–90. https://doi.org/10.1002/(sici).
75. Berge H, Isern C, Berge E. Blood pressure and hypertension in athletes: a systematic review. Br J Sports Med. 2014;49:716–23.
76. Abbott R, Whear R, Rodgers LR. Effectiveness of mindfulness-based stress reduction and mindfulness based cognitive therapy in vascular disease: a systematic review and meta-analysis of randomised controlled trials. J Psychosom Res. 2014;76:341–51.
77. Loucks E, Nardi W, Gutman R. Mindfulness-based blood pressure reduction (MB-BP): stage 1 single-arm clinical trial. PLoS One. 2019;14:e0223095.
78. Polcari J, Cali R, Nephew B. Effects of the mindfulness-based blood pressure reduction (MB-BP) program on depression and neural structural connectivity. medRxiv. 2023:1. https://doi.org/10.1101/2021.08.18.21262240.
79. Hughes J, Fresco D, Myerscough R, Dulmen M, Carlson L, Josephson R. Randomized controlled trial of mindfulness-based stress reduction for prehypertension. Accessed Jan 8; 2023.
80. An E, Nolty A, Amano S, Rizzo A, Buckwalter J, Rensberger J. Heart rate variability as an index of resilience. OUP Academic. https://academic.oup.com/milmed/article/185/3-4/363/5586497; 2023. p. 23.
81. Laborde S, Allen MS, Borges U, et al. Psychophysiological effects of slow-paced breathing at six cycles per minute with or without heart rate variability biofeedback. Psychophysiology. 2022;59

82. Brenner J, LeBlang S, Lizotte-Waniewski M, et al. Mindfulness with paced breathing reduces blood pressure. Med Hypotheses. 2020;142:109780.
83. Zaccaro A, Piarulli A, Laurino M, et al. How breath-control can change your life: a systematic review on psycho-physiological correlates of slow breathing. Front Hum Neurosci. 2018;12:353.
84. Russell M, Scott A, Boggero I, Carlson C. Inclusion of a rest period in diaphragmatic breathing increases high frequency heart rate variability: implications for behavioral therapy. Psychophysiology. 2017;54(3):358–65.
85. Wielgosz J, Schuyler BS, Lutz A, Davidson RJ. Long-term mindfulness training is associated with reliable differences in resting respiration rate. Sci Rep. 2016;6:27533.
86. Fendel J, Burkle J, Goritz A. Mindfulness-based interventions to reduce burnout and stress in physicians: a systematic review and meta-analysis. Acad Med. 2021;96(5):751–64.
87. Martarelli D, Cocchioni M, Scuri S, Pompei P. Diaphragmatic breathing reduces exercise-induced oxidative stress. Evid Based Complement Alternat Med. 2011;2011:932430.
88. Kirk U, Axelsen J. Heart rate variability is enhanced during mindfulness practice: a randomized controlled trial involving a 10-day online-based mindfulness intervention. PLoS One. 2020;15(12):e0243488.
89. Orlick T, Partington J. Mental links to excellence. Sport Psychol. 1988;2(2):105–30. The Sports Psychologist. 1982;2(2):105–130
90. Smallwood J, Schooler JW. The science of mind wandering: empirically navigating the stream of consciousness. Annu Rev Psychol. 2015;66:487–518.
91. Killingsworth M, Gilbert D. A wandering mind is an unhappy mind. Science. 2010;330(6006):932.
92. Jha A. Peak mind. Piatkus Books; 2021.
93. Norris CJ, Creem D, Hendler R, Kober H. Brief mindfulness meditation improves attention in novices: evidence from ERPs and moderation by neuroticism. Front Hum Neurosci. 2018;12:315.
94. Prakash R, Fountain-Zaragoza S, Kramer A, Samimy S, Wegman J. Mindfulness and attention: current state-of-affairs and future considerations. J Cogn Enhanc. 2020;4(3):340–67.
95. Sumantry D, Stewart K. Meditation, mindfulness, and attention: a meta-analysis. Mindfulness. 2021;12(6):1332.
96. Rooks J, Morrison A, Goolsarran M, Rogers S, Jha A. We are talking about practice: the influence of mindfulness vs. relaxation training on athletes' attention and well-being over high-demand intervals. J Cogn Enhanc. 2017;1(2):141.
97. Reardon C, Hainline B, Aron C. Mental health in elite athletes: international olympic committee consensus statement (2019). Br J Sports Med. 2019;53(11):667–99.
98. Gulliver A, Griffiths KM, Mackinnon A, Batterham PJ, Stanimirovic R. The mental health of australian elite athletes. J Sci Med Sport. 2015;18(3):255–61.
99. Gouttebarge V, Castaldelli-Maia JGP, et al. Occurrence of mental health symptoms and disorders in current and former elite athletes: a systematic review and meta-analysis. Br J Sports Med. 2019;53(11):700–6.
100. Mallinson S, Hill A. The relationship between multidimensional perfectionism and psychological need thwarting in junior sports participants. Psychol Sport Exerc. 2011;12(6):676.
101. Gallagher BV GF. An examination of the relationship between early maladaptive schemas, coping, and emotional response to athletic injury. J Clin Sport Psychol 2007;1(1):47.
102. Winsley R, Matos N. Overtraining and elite young athletes. Med Sport Sci. 2011;56:97–105.
103. Ford J, Ildefonso K, Jones M, Arvinen-Barrow M. Sport-related anxiety: current insights. Open Access J Sports Med. 2017;8:205–12.
104. De Francisco C, Arce C, Vilchez M, Vales A. Antecedents and consequences of burnout in athletes: perceived stress and depression. Int J Clin Health Psychol. 2016;16(3):239–46.
105. Wiese-Bjornstal DM. Psychology and socioculture affect injury risk, response, and recovery in high-intensity athletes: a consensus statement. Scand J Med Sci Sports. 2010;20(Suppl 2):103–11.
106. Khoury B, Lecomte T, Fortin G, et al. Mindfulness-based therapy: a comprehensive meta-analysis. Clin Psychol Rev. 2013;33(6):763–71.

107. Goldberg S, Tucker R, Greene P. Mindfulness-based interventions for psychiatric disorders: a systematic review and meta-analysis. Clin Psychol Rev. 2018;59:52.
108. Galante J, Friedrich C. Mindfulness-based programmes for mental health promotion in adults in nonclinical settings: a systematic review and meta-analysis of randomised controlled trials. 2021;18:1.
109. Hoge E, Bui E, Mete M. Mindfulness-based stress reduction vs escitalopram for the treatment of adults with anxiety disorders: a randomized clinical trial. JAMA Psychiatry. 2023;80(1):13–21.
110. Beames J, O'Dean S, Grisham J, Moulds M, Denson T. Anger regulation in interpersonal contexts: anger experience, aggressive behavior, and cardiovascular reactivity. J Soc Pers Relat. 2019;36(5):1441.
111. DeSteno D, Lim D, Duong F, Condon P. Meditation inhibits aggressive responses to provocations. Mindfulness. 2017;9(4):1117–22.
112. Gardner FL, Moore ZE. Mindfulness in sport contexts. In: Tenenbaum G, Eklund RC, Boiangin N, editors. Handbook of sport psychology: social perspectives, cognition, and applications. Wiley; 2020. p. 738. https://doi.org/10.1002/9781119568124.ch35.
113. Ajilchi B, Amini HR, Ardakani ZP, Zadeh MM, Kisely S. Applying mindfulness training to enhance the mental toughness and emotional intelligence of amateur basketball players. Australas Psychiatry. 2019;27(3):291–6.
114. Ajilchi B, Mohebi M, Zarei S, Kisely S. Effect of a mindfulness programme training on mental toughness and psychological well-being of female athletes. Australas Psychiatry. 2022;30(3):352–6.
115. Rothlin P, Horvath S, Trosch S, Holtforth MG, Birrer D. Differential and shared effects of psychological skills training and mindfulness training on performance-relevant psychological factors in sport: a randomized controlled trial. BMC Psychol. 2020;8(1):80–020.
116. Scott-Hamilton J, Schutte NS, Brown RF. Effects of a mindfulness intervention on sports-anxiety, pessimism, and flow in competitive cyclists. Appl Psychol Health Well Being. 2016;8(1):85–103.
117. Sparks KV, Ring C. A rowing-specific mindfulness intervention: effects on mindfulness, flow, reinvestment, and performance. Front Psychol. 2022;13:871804.
118. Wolch N, Arthur-Cameselle J, Keeler L, Suprak D. The effects of a brief mindfulness intervention on basketball free-throw shooting performance under pressure. J Appl Sport Psychol. 2020;33(5):510–26.
119. Vidic Z, St. Martin M, Oxhandler R. Mindfulness intervention with a U.S. women's NCAA division I basketball team: impact on stress, athletic coping skills and perceptions of intervention. Sport Psychol. 2017;31(2):147.
120. Josefsson T, Ivarsson A, Gustafsson H, Stenling A, Lindwall M, Tornberg R. Effects of mindfulness-acceptance-commitment (MAC) on sport-specific dispositional mindfulness, emotion regulation, and self-rated athletic performance in a multiple-sport population: an RCT study. Mindfulness. 2019;10(8):1518.
121. Casali N, Ghisi M, Jansen P, Feraco T, Meneghetti C. What can affect competition anxiety in athletes? The role of self-compassion and repetitive negative thinking. Psychol Rep. 2021;125(4):2009.
122. American College of Sports Medicine, American Academy of Family Physicians, American Academy of Orthopaedic Surgeons, American Medical Society for Sports Medicine, American Orthopaedic Society for Sports Medicine, American Osteopathic Academy of Sports Medicine. Psychological issues related to injury in athletes and the team physician: a consensus statement. Med Sci Sport Exercise. 2006;38(11):2030.
123. Herring S, Boyajian-O'Neill LA, Coppel D, Daniels J, Gould D. Psychological issues related to injury in athletes and the team physician. Med Sci Sport Exercise. 2016;38(11):2030.
124. Pal S, Kalra S, Awasthi S. Influence of stress and anxiety on sports injuries in athletes. J Clin Diagn Res. 2021;15(4)
125. Lomas T, Medina J, Ivtzan I, Rupprect S, Hart R, Elroa-Orsa F. The impact of mindfulness on wellbeing and performance in the workplace: an inclusive systematic review of the empirical literature. Eur J Work Organ Psy. 2017;26(4):492–513.

126. Greater Good Science Center. Mindfulness: greater good; 2023.
127. Naderi A, Gharayagh Z, Calmerio L, Brewer BL. The effects of a mindfulness-based program on the incidence of injuries in young male soccer players. J Sport Exerc Psychol. 2020:1–11.
128. Lee RT, Ashforth BE. On the meaning of maslach's three dimensions of burnout. J Appl Psychol. 1990;75(6):743–7.
129. Hill A, Hall H, Duda J, Appleton P. The cognitive, affective and behavioural responses of self-oriented perfectionists following successive failure on a muscular endurance task. Int J Sport Exerc Psychol. 2011;9(2):189.
130. Hill A, Appleton P. The predictive ability of the frequency of perfectionistic cognitions, self-oriented perfectionism, and socially prescribed perfectionism in relation to symptoms of burnout in youth rugby players. J Sports Sci. 2011;29(7):695.
131. Moen F. Can the attention training technique reduce burnout in junior elite athletes? Int J Coach Sci. 2016;10(1):53.
132. Li C, Yuxin Z, Mengge G, Henrik C, Chen T. Mindfulness and athlete burnout: a systematic review and meta-analysis. Int J Environ Res Public Health. 2019;16(3):449. https://doi.org/10.3390/ijerph16030449.
133. Grant JARP. Pain sensitivity and analgesic effects of mindful states in zen meditators: a cross-sectional study. Psychosom Med. 2009;71(1):106–14.
134. Ball E, Muhamme N, Franklin G, Rogozinska E. Does mindfulness meditation improve chronic pain? A systematic review. Obstet Gynecol. 2017;29(6):359.
135. Cherkin D, Sherman K, Balderson B. Effect of mindfulness-based stress reduction vs cognitive behavioral therapy or usual care on back pain and functional limitations in adults with chronic low back pain. JAMA J Am Med Assoc. 2016;315(12):1240.
136. Benson H, Friedman R. Harnessing the power of the placebo effect and renaming it "remembered wellness". Annu Rev Med. 1996;47:193–9.
137. Zeidan F, Emerson N, Farris M, et al. Mindfulness meditation-based pain relief employs different neural mechanisms than placebo and sham mindfulness meditation-induced analgesia. J Neurosci. 2015;35(46):15307–25.
138. Wang Y, Yang Q. On mindfulness training for promoting mental toughness of female college students in endurance exercise. Evid Based Complement Alternat Med. 2021;2021:5596111.
139. Bagheri S, Naderi A, Mirali S, Calmeiro L, Brewer B. Adding mindfulness practice to exercise therapy for female recreational runners with patellofemoral pain: a randomized controlled trial. J Athl Train. 2021;56(8):902–11.
140. Cooper B. The mindfulness meditation training for sport and injury rehabilitation with high school athletes: a pilot study; 2017.
141. Shortway K, Wolanin A, Block-Lerner J, Marks D. Acceptance and commitment therapy for injured athletes: development and preliminary feasibility of the return to action protocol. J Clin Sport Psychol. 2018;12(1):4.
142. Anderson S, Haraldsdottir K, Watson D. Mindfulness in athletes. Curr Sports Med Rep. 2021;20(12):655.
143. Britton W, Lindahl J, Cooper D, Canby N, Palitsky R. Defining and measuring meditation-related adverse effects in mindfulness-based programs. Clin Psychol Sci. 2021;9(6):1185–204.

Chapter 16
Application of Sport Nutrition and Sports Psychology to Recovery

Ian Rollo, Clyde Williams, and Leslie Podlog

Introduction

Professional sport requires athletes to dedicate significant amount of time to their training and subsequent competition. As competitive schedules intensify, athletes and coaches recognize the importance of recovery strategies to optimise the restoration of physiological and psychological processes, allowing the athlete to train or compete again without a loss in performance. However, successful recovery from training and competition is a complex process. The various factors which influence recovery are not independent but are intertwined at different levels (Table 16.1). In professional sport, the allocation of financial resources will determine the facility and availability of relevant equipment as well as access to experienced sports medical/science professionals to lead the recovery programme.

For professional and recreational athletes, optimal recovery provides the benefit of allowing repeated high-level training and competition. Conversely, inadequate recovery may result in sub-optimal performance and increased risk of illness and

I. Rollo (✉)
Gatorade Sports Science Institute, PepsiCo Life Sciences, Leicestershire, UK

School of Sports Exercise and Health Sciences, Loughborough University, Loughborough, UK
e-mail: ian.rollo@pepsico.com

C. Williams
School of Sports Exercise and Health Sciences, Loughborough University, Loughborough, UK

L. Podlog
School of Kinesiology and Physical Activity Sciences, Faculty of Medicine, Université de Montréal, Montreal, QC, Canada

CHU Sainte-Justine Research Center, Montreal, QC, Canada

N. C. Apostolopoulos et al. (eds.), *Fundamentals of Recovery, Regeneration, and Adaptation to Exercise Stress: An Integrated Approach*,
https://doi.org/10.1007/978-3-031-44270-4_16

Table 16.1 Factors influencing recovery strategies and practical considerations

Recovery factor	Considerations			
Sport	Competitive schedule	Domestic and international competition	Team or individual pursuit	Contact/ non-contact
Training/ competition	Volume, intensity, duration, type of training/ sport, degree of fatigue, recovery from previous training/competition	Sport physical and technical demands	Time available	Culture of sport/ experience of coach
Nutrition	Carbohydrate, protein and other nutrient intake, fluid, and electrolyte balance	Food preparation and hygiene. Budget available	Education of athletes and staff	Avoid indiscriminate use of dietary supplement
Psychological stress	Stress and anxiety from competition	Use of appropriate coping strategies	Athletes' history and personality predispositions	Self-regulation strategies
Lifestyle	Quality and amount of sleep, schedule, housing situation, leisure/social activities, relationship with team members, coach, friends and family, job, or schooling situation	Athlete and coach cultural background Beliefs and superstitions	Religious considerations	Exposure to conflicting advice from influencers
Health	Illnesses, infection, current injury, muscle soreness and damage	Injury history	Underlying medical condition	Poor hygiene
Environment	Temperature, humidity, altitude	Country, annual season	Pollutants	Allergies

injury [1, 2]. Thus, it follows that adhering to frequent, quality recovery strategies during the post exercise period are extremely important for athlete health and performance.

Any recovery strategy employed is not exclusive to the physiological or psychological domain. Practitioners should not only be aware of physiological and psychological recovery approaches but how the two may complement each other. Biopsychosocial factors are also important in the recovery process [3]. Exploring complementary approaches may enhance athletes' perceptions of well-being and recovery experience. However, specific guidance on how to integrate sports psychology and sports nutrition in ways that benefit the recovery of athletes is lacking. Thus, we believe efforts to ensure that athletes are physically and psychologically ready to exercise again, when required to do so, are best guided by taking an interdisciplinary approach. An interdisciplinary approach is understood to achieve outcomes (including new approaches) that could not be achieved within the framework of a single discipline [4]. An interdisciplinary approach is one in which the experience of scholars and practitioners from different disciplines are focused on the research or clinical problem in question, in this case, recovery.

To this end, the purpose of this chapter is to discuss how sport psychology and sport nutrition can be integrated to support an athlete's successful recovery from training and competition. Of course, practical experience of working with athletes on recovery strategies is clearly of benefit [5]. Therefore, personal commentary will be added to the available scientific research in both sports psychology and sports nutrition.

> Any recovery strategy employed is not exclusive to the physiological or psychological domain.

An Interdisciplinary Approach to Recovery

Based on our collective experience, we recommend having an appropriate leader who will establish a clear direction for an appropriate recovery strategy, after hearing the experiences and views of athletes [6]. The recovery strategy should be agreed well in advance of training and especially before competition, so that processes can be in place for implementation immediately after exercise is completed.

Key attributes of this leader include a mastery of communication strategies and an ability to foster collaborative decision-making [7]. The leader will be informed by the cultural background of the team, league and/or sport. It is advised the leader should be an experienced professionally trained individual with a daily presence in the athlete/team environment [3, 8].

In addition to identifying an appropriate leader, the composition of the decision-making team and individual roles and responsibilities should be made before holding discussions about the recovery strategy [3]. The decision-making process should be athlete centered. An athlete-centered approach is characterized by empowering athletes to be involved in the planning of recovery programs [9]. Understanding the "why" of recovery is as important as the "how". Guided by an athlete-centered approach, the ambition is to foster athlete autonomy (giving athletes choices and options) and ensure that all voices, perspectives, and experiences remain at the forefront of decision-making processes [10, 11]. Key factors to facilitate transparent communication processes, data collection and sharing, and collaborative, athlete-centered decision-making are captured in Fig. 16.1. Finally, decision-making should occur within a culture of respect, consensus, and one supporting training and understanding across multiple disciplines [9, 10]. Establishing a mutually supportive environment will facilitate a successful interdisciplinary approach to planning and implementation of a recovery strategy [12].

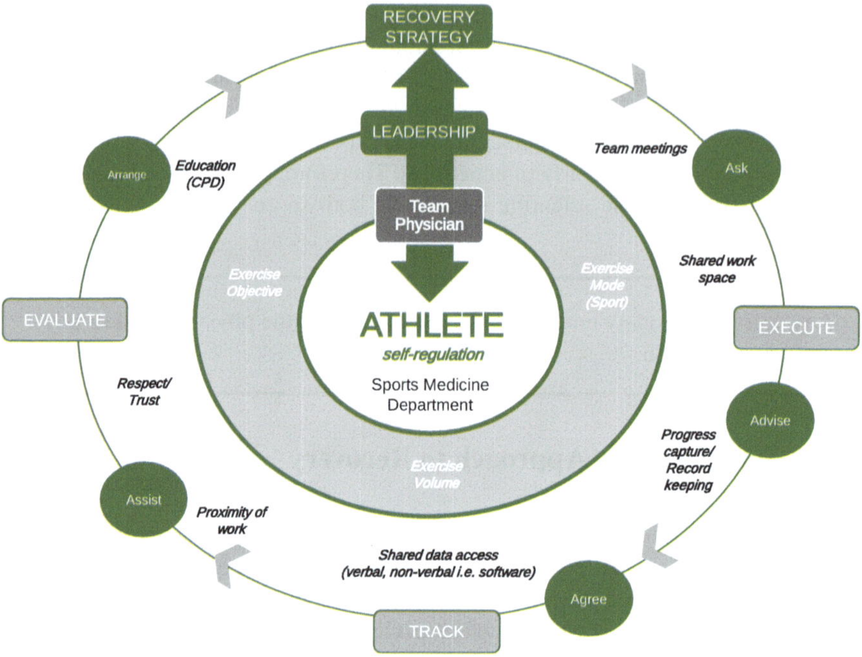

Fig. 16.1 An interdisciplinary model of an athlete-centered post-competition recovery program. Sports Medicine departments should include, but not limited to, physiotherapist, coach, sports scientist, sports nutritionist, ports psychologist and strength and conditioning experts. Inner circle includes key factors to consider for recovery. The outer circle includes "facilitating factors" to promote an interdisciplinary approach. The 5 As, represent the communication approach of: Assess, Advising about, Agreeing upon, Assisting and Arranging

Recovery programs should take an athlete-centred approach. Athletes should be empowered to understand and lead in the planning of their own recovery schedules.

Recovery

Recovery strategies are a fundamental component of athlete health and performance in professional sport. The recovery process—presents frequent opportunities for both the sport nutritionist and sports psychologist to engage with the athlete, as well as the sports medicine department. Other recovery interventions and their effects on fatigue, muscle injury, recovery and performance are covered in previous chapters. Thus, the following discussion of sports psychology and sports nutrition recommendations should be taken in context of other effective recovery approaches such as foam rolling, ice baths, sleep, compression garments and massage [13–16].

Sports Nutrition

Athletes' bodies are in a constant state of change as they adapt to the challenges of training, competition and, not infrequently, recovery from injury [17]. For example, after exercise the dynamic balance between protein synthesis and protein breakdown in skeletal muscles is upregulated to improve its capacity to better cope with subsequent physical challenges. It is well established that manipulation of dietary intake to include specific types and quantities of food, can positively influence the recovery and regeneration processes.

The importance of dietary intervention post-exercise is, in part, dictated by the time between exercise sessions. In individual pursuits, such as track and field or combat sport, individuals can periodize their training to "peak" for a specific event. Thus, recovery strategies are primarily focused on maximising gains from training [18]. In team sports, the competition schedule is more frequent than in many other sports. For example, elite soccer players may be required to compete on 2–3 occasions in a week. Thus, recovery strategies are primarily designed to help athletes sustain their level of performance over the duration of a competitive season.

The fundamental elements of a sports nutrition recovery program can be categorized into the "3 Rs"– Rehydrate, Rebuild, Refuel that are summarized below:

Rehydrate The objective of rehydration is to completely replace fluid and electrolyte losses prior to the start of the next exercise occasion. Fluid and electrolyte losses are specific to the individual, the intensity of exercise and the environmental conditions [19]. Methods used to determine the fluid and electrolyte losses of individuals are described by Barker et al. [20]. Athletes intuitively begin to rehydrate immediately after exercise by sipping on water, sports beverages, or protein shakes. In most situations, normal eating and drinking behaviour will replace lost water and sodium. However, if dehydration is severe (>5% of pre-exercise body mass) or if rapid rehydration is needed (<24 hours before next training session or match), it is recommended to drink 1.5 L of fluid for each 1 kg of body mass lost.

Rebuild The aim is to provide substrate (essential amino acids) for repairing and remodeling skeletal muscle and energy generating mitochondria as well as for protein synthesis [21]. Consuming approximately 20–30 grams of protein after exercise, as well as every couple of hours at regular intervals is recommended [22]. In addition, protein ingestion prior to sleep, is an effective recovery strategy to support muscle remodeling [23].

Refuel The aim of refueling is to 'refill' depleted glycogen stores in skeletal muscles and liver [24]. It can take up to ~24 hours to fully replenish glycogen stores after intensive or prolonged exercise. Therefore, athletes usually begin the refueling process by the immediate ingestion of a carbohydrate-electrolyte beverage, protein shake or carbohydrate-based foods. Meals ingested 2–3 hours after exercise, help ensure the athlete ingests enough carbohydrates (CHO) to maximize glycogen resynthesis. Furthermore, it is important to stress that athletes can meet the guidelines for recover using many combinations of appropriate foods and fluids [25–27].

> After exercise there is a dynamic interaction between breakdown and resynthesis that leads to regeneration of several tissues, the most prominent being skeletal muscles. Adjusting dietary intake to include specific types and quantities of food, can positively influence the recovery and regeneration processes.

Sports Psychology

Adhering to the dietary recommendations requires self-regulation skills on the part of the athlete. Self-regulation "refers to the exercise of control over oneself, especially with regard to bringing the self into line with preferred (thus, regular) standards" [28]. Self-regulation is essential for many health behaviors—including engaging in specific dietary recommendations and/or restrictions—that may require self-restraint and/or sustained effort before benefits can be derived. Three self-regulatory skills are relevant for post-exercise recovery: 1. self-monitoring, 2. regulation of thoughts and emotions, and 3. self-control [29, 30].

Self-monitoring pertains to the ability to identify potential discrepancies between one's current physiological or psychological state with a desired level or state [29]. As athletes' regularly experience disruptions to their desired state (e.g., fatigue, muscle strains, stiffness), developing an awareness of imbalances between current and desired states/levels may be critical in facilitating recovery. Therefore, self-monitoring requires ongoing vigilance on the part of the athlete. Such monitoring however, can prove challenging even among elite athletes who may struggle to contend with chronic mental and physical depletion following acute bouts of intense exercise or over prolonged training periods (e.g., a lengthy football season) [30]. As athletes may fail to recognize the need for post-exercise recovery procedures (e.g., a massage, a light stretch, a dietary supplement, sport psychologists and nutritionists can support athletes' self-monitoring efforts by listening to and observing the latter's statements and behaviors following exercise/training.

A second component—the regulation of post-exercise cognitions and emotions—also has important implications for the recovery process. The ability to mentally detach from or avoid perseverating about a mistake in training or competition and to distance oneself from negative sport or exercise-related emotions may be vital to achieving recovery [31]. The ability to let go of frustrations and negative emotions has been referred to as emotional detachment and appears to be important for replenishing athlete resources and achieving psychological recovery. In particular, the capacity to emotionally detach from the implications of a recent poor performance or exercise session may facilitate psychophysiological recovery by reducing worries about subsequent performance, cortisol levels, and vagal activity and facilitating enhanced sleep and motivation to recover or engage in sport [32].

A third component of self-regulation—self-control—is also pivotal for athlete recovery. Self-control refers to "the capacity to resist a temptation that is in conflict with a desired, long-term goal, in order to protect this valued goal" [33]. Thus, self-control requires three elements: a desire, a long-term goal, and a conflict between the two (i.e. desire–goal conflict; [33]). When a desire conflicts with a long-term goal, a cognitive disruption associated with self-control occurs [33]. Self-control is critical to countless exercise situations [29], since the ability to resist immediate urges in specific situations (e.g., to relieve perceptions of discomfort by diminishing effort or ceasing an exertive exercise task altogether) allows athletes to achieve sought-after long-term goals rather than giving into short-term desires [34]. Take for instance, a football player who becomes dehydrated and has significant sweat loss over the course of a match, that is about to go into extra time. Despite an intense thirst, the player is advised to slowly imbibe an appropriate volume of fluid to reduce the risk of gastrointestinal disturbance. If unable to resist and the hypohydrated footballer guzzles an excessive volume of fluid, it may result in bloating, vomiting, and subsequent loss of fluid, carbohydrate and electrolytes needed to support performance. As this example illustrates, individuals in a state of duress may exhibit lower levels of self-control strength. This argument is supported by Englert and Rummel [34], who found that daily stress was associated with lower levels of self-control strength. In particular, the researchers found that daily stress was associated with diminished levels of state self-control, which in turn, predicted lower exercise completion.

The above findings highlight the need for athletes to cultivate their recovery self-regulation skills. Fortunately, just as athletes can improve their physical skills with practice, so to regular engagement in self-regulation exercises can enhance athletes' self-regulatory capacities [35]. When compared with a control group, participants who practice self-regulation, such as monitoring and improving posture, regulating mood or monitoring and recording eating, improve performance on other self-regulation tasks [35]. Moreover, self-regulation could improve athletes' ability to dynamically self-organize their behavior when experiencing recovery-stress imbalances and homeodynamically find the new steady-state as a response to the training process [36]. Sport psychologists and neuroscientists have shown that training of psychological skills such as self-talk, goal-setting, and imagery may improve self-regulation behaviors [37]. For instance, in a longitudinal study, Elbe et al. [38] found that engaging in competitive sports alone, without specific psychological skills training, increased volitional skills in adolescents compared to a non-athletic control group. Further, Siekanska et al. [39], found that integrating technology into psychological skills training could facilitate athlete self-regulation and subsequent performance. Although further empirical investigation is warranted, these studies indicate that interventions aimed at improving self-regulation may also facilitate athlete engagement in post-exercise recovery regimens.

Once established with a qualified sports nutritionist/dietician, nutrition offerings can be delivered by various members of the sports medical department (Physiotherapist, Fitness coach, Masseuse, Sports Psychologist). While sport psychologists should not attempt to provide detailed nutritional information, nutrition professionals can advise sport psychologists to reinforce messages to athletes to follow recommended nutritional guidelines discussed above. Additionally, sport psychologists can help educate nutrition experts and athletes on strategies to increase the likelihood of behavioral compliance with nutritional programs. For instance, optimizing goal-setting processes and applying psychological principles may be vital for optimizing adoption of nutrition information. Along these lines, sport psychologists can work with nutrition experts to optimize their communication with athletes. One such approach, with empirical backing is the '5A Communication' intervention that highlights the importance of practitioners "Asking about" (or Assessing), "Advising", "Agreeing upon", "Assisting", and "Arranging" follow-up regarding athlete's behavioral changes such as nutritional intake [40, 41]. Additionally, athlete centered communication can be facilitated when nutritionists ask athletes for background information, for example, about their values, expectations, and beliefs relevant to the nutrition behaviors; provide rationales for nutrition provision; and support athlete choice in adoption of nutritional supplementation. The 5A approach may increase the likelihood of behavioral adoption by promoting athlete participation, satisfaction, trust, and adherence to nutrition plans [42–44].

> Self-regulation is important in the adherence to sports nutrition advice for recovery. Three self-regulatory skills relevant for post-exercise recovery, include: 1. self-monitoring, 2. regulation of thoughts and emotions, and 3. self-control.

An Interdisciplinary Approach

One interdisciplinary approach is to offer a suit of recovery options which meet nutrition, psychology and other recovery objectives but cater for individual preference and encourage self-regulation. Figure 16.2 provides an example of how various recovery modalities can be completed and presented to athletes to achieve a total recovery "score". Athletes are therefore encouraged to take personal responsibility for their recovery which is orientated towards a task-goal structure. In the authors' experience providing athletes with appropriate nutrition and psychological skills to manage the recovery process can also influence their engagement in other aspects of performance, for instance, preparation and planning, and game-day execution.

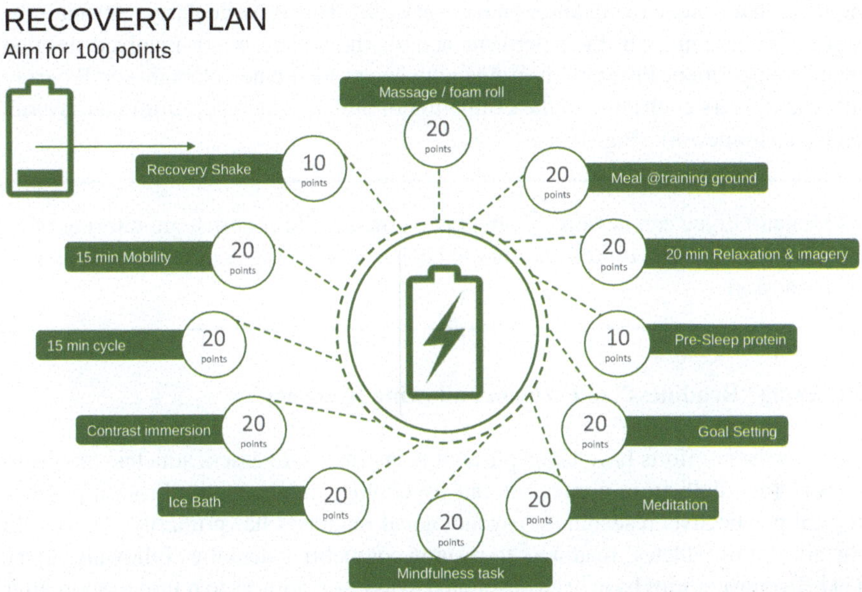

Fig. 16.2 An athlete-centered approach to recovery, providing options and ownership of the strategies. Options and points weighting can be modified based on priorities directed by sports medical team. Points weighting can also be modified based on the individual athlete

Mental Fatigue

Mental fatigue is defined as a psychobiological state that arises during prolonged demanding cognitive activity and results in an acute feeling of tiredness and/or a decreased cognitive ability [45, 46]. Mental fatigue is acknowledged as a key consideration in sports, due to the associated negative effect on physical, technical, decision-making, and tactical performance [47, 48]. As mental fatigue has been shown to fluctuate throughout a competitive team sport season, it provides reason to discuss corresponding recovery strategies alluded to previously [49].

From a nutrition perspective, carbohydrate ingestion appears to be an effective dietary counter measure to mental fatigue. Immediately after exercise, mouth rinsing and expectorating a carbohydrate (CHO) beverage has been reported to be a countermeasure to mental fatigue [50]. The recognition of CHO in the mouth, when administered immediately after a mentally fatiguing task was linked to increased excitability of corticomotor pathways [51]. However, this observation is an additional benefit to CHO ingestion in the recovery occasion, where substate delivery for the replenishment of glycogen stores is the priority [52–55]. Thus, athletes are advised to increase daily CHO intake, during intensified periods of training or competition which may help maintain the general mood of the athlete [56, 57].

An athletes' mood is directly relevant when considering the risk of recovery interventions becoming monotonous, when being completed routinely over a season. A key issue during a competitive season is an athlete's ability to summon motivation to adhere to guidance on recovery [58]. This is particularly relevant if the team is struggling with their performance, or the athlete is not involved in "first team" competition. Progress meetings with sports medicine colleagues will ensure all practitioners contribute to the evaluation of athlete adherence within the overall recovery framework (Fig. 16.1).

> Mental fatigue can negatively affect performance. Appropriate monitoring of mental fatigue is required and should be discussed within the overall recovery framework.

Assessing "Readiness" to Perform Following Recovery

Once recovery efforts have taken place, it is instructive to assess athletes' readiness to perform—both from a physical and psychological standpoint. From a psychological perspective, research on psychological readiness has primarily occurred in the context of athletes' readiness to resume competitive activities following injury [11]. Two inventories have been developed to assess readiness to return to sport after injury—the Injury Psychological Readiness to Return to Sport (IPRRS; [59]) Scale and the ACL-Return to Sport after Injury (ACL-RSI; [60]). While some items from these scales may be relevant for post-exercise recovery (e.g., "My overall confidence to play"? "My overall confidence to play without pain"; "My confidence to give 100% effort") other injury-specific items (e.g., "My confidence in the injured body part to handle the demands of the situation"; "My confidence to not concentrate on the injury") may not be appropriate. From a clinical standpoint, recovery relevant questions such as the examples mentioned above may be used to complement traditional sport medicine assessments focusing exclusively on physical/functional parameters (e.g., GPS parameters, muscle strength, endurance). As has been recently argued [61–63], holistic assessment of athletes' readiness to perform after recovery should include psychological test batteries, working knowledge of and/or observations of athlete behaviours by coaches and members of the sports medicine team, and shared decision-making involving key stakeholders. High levels of psychological readiness to perform may serve as an indicator of the effectiveness of the recovery strategies implemented. Future research is needed, however, to further validate this contention.

> High levels of psychological readiness to perform may serve as an indicator of the effectiveness of the sports nutrition regimen as well as all the recovery strategies implemented.

Group Functions and Recovery

A number of social psychological and motivational theories have long recognized the importance of feelings of connection and affiliation with relevant others [64]. Team functions and meetings may provide athletes with opportunities to bond and connect with relevant others and to provide social support—both of which can facilitate psychological and physical recovery. For example, team meals provide social connections whilst also delivering effective nutritional habits in service of recovery. Having sports psychologists and appropriately trained sports nutritionists present during meals may help reinforce effective dietary choices and/or promote effective recovery strategies. The team restaurant can also serve as an arena for distraction away from the focus of the sport.

Of interest is research that has shown that supportive interactions can influence chemical changes in the body known to positively influence physical healing and diminish stress. Social support helps release oxytocin, the same chemical mothers release during breast feeding to help them bond with their newborns. Oxytocin has been shown to have anxiolytic effects, that is stress-protective effects [65]. In a study of couples who had skin blister wounds, individuals who had more positive interactions with their marital partner during a social support task had higher plasma oxytocin levels. Higher circulating oxytocin levels were in turn associated with faster healing of the standard blister wounds [65]. Whether these observations transfer to a sports environment warrants further investigation.

> Mealtime occasions provide the opportunity to integrate both Sports Nutrition and Sports Psychology recovery strategies.

Alcohol

During the recovery period it is advisable to reduce, if not eliminate, alcohol consumption [66]. Drinking alcohol in sport is a common cultural activity to celebrate success. Of course, wins should be celebrated by athletes and alcohol can be enjoyed sensibly and in moderation. However, it is important to note that alcohol ingestion has been shown to reduce muscle protein synthesis during recovery from exercise, even if co-ingested with protein, potentially impairing recovery [67]. Furthermore, alcohol intake increases the risk of excessive energy intake, poor dietary choices and psychological coping strategies failing.

> Excessive alcohol intake increases the risk of poor dietary choices and psychological coping strategies failing.

Conclusion

The completion of any recovery strategy impacts both the physiology and psychology of the athlete. The current chapter highlights opportunities and benefits of incorporating sports psychology and sports nutrition into the recovery process. It is the authors opinion that an interdisciplinary approach will be advantageous in accelerating processes to return the athlete to a normal or new adapted/improved health and performance state (psychological, metabolic, neural) after fatigue caused by exercise. However, further empirical evidence to support our suggestions is warranted.

Conflict of Interest IR is an employee of the Gatorade Sports Science Institute, a division of PepsiCo, Inc. The views expressed in this article are those of the authors and do not necessarily reflect the position or policy of PepsiCo Inc.

References

1. Tipton KD. Nutrition for acute exercise-induced injuries. Ann Nutr Metab. 2010;57(Suppl 2):43–53.
2. Wall BT, Morton JP, van Loon LJ. Strategies to maintain skeletal muscle mass in the injured athlete: nutritional considerations and exercise mimetics. Eur J Sport Sci. 2015;15(1):53–62.
3. Ardern CL, Glasgow P, Schneiders A, Witvrouw E, Clarsen B, Cools A, Gojanovic B, Griffin S, Khan KM, Moksnes H, Mutch SA, Phillips N, Reurink G, Sadler R, Silbernagel KG, Thorborg K, Wangensteen A, Wilk KE, Bizzini M. 2016 Consensus statement on return to sport from the First World Congress in Sports Physical Therapy, Bern. Br J Sports Med. 2016;50(14):853–64.
4. Piggott B, Muller S, Chivers P, Papaluca C, Hoyne G. Is sports science answering the call for interdisciplinary research? A systematic review. Eur J Sport Sci. 2019;19(3):267–86.
5. McCall A, Lewin C, O'Driscoll G, Witvrouw E, Ardern C. Return to play: the challenge of balancing research and practice. Br J Sports Med. 2017;51(9):702–3.
6. Kellmann M, Bertollo M, Bosquet L, Brink M, Coutts AJ, Duffield R, Erlacher D, Halson SL, Hecksteden A, Heidari J, Kallus KW, Meeusen R, Mujika I, Robazza C, Skorski S, Venter R, Beckmann J. Recovery and performance in sport: consensus statement. Int J Sports Physiol Perform. 2018;13(2):240–5.
7. Rollo I, Carter JM, Close GL, Leyes JY, Gomez A, Leal DM, Duda JL, Holohan D, Erith SJ, Podlog L. Role of sports psychology and sports nutrition in return to play from musculoskeletal injuries in professional soccer: an interdisciplinary approach. Eur J Sport Sci. 2020;21:1–19.
8. Herring SA, Bergfeld J, Boyd AJ, Duffey T, Fields DOK, Grana BW, Indelicato AP, Kibler WB, Pallay R, Putukian M, Sallis RE. The team physician and return-to-play issues: a consensus statement. Med Sci Sports Exerc. 2002;34(7):1212–4.
9. King J, Roberts C, Hard S, Ardern CL. Want to improve return to sport outcomes following injury? Empower, engage, provide feedback and be transparent: 4 habits! Br J Sports Med. 2019;53(9):526–7.
10. Ahmed OH, Defoe J, West LR, Blake T. Creating the dream team: introducing participatory sports and exercise medicine via 'Patient Voices'. Br J Sports Med. 2018;52(24):1547–8.
11. Podlog L, Eklund RC. Psychosocial considerations of return to sport following Injury. In: Johnson U, Ivarsson A, editors. Psychological bases of sports injury. 4th ed. Fitness Information Technology; 2020. p. 101–14.

12. Van Raalte JL, Vincent A, Dickens YL. Dialogical consciousness and descriptive experience sampling: implications for the study of intrapersonal communication in sport. Front Psychol. 2019;10:653.
13. Nedelec M, McCall A, Carling C, Legall F, Berthoin S, Dupont G. Recovery in soccer: part II-recovery strategies. Sports Med (Auckland, N.Z). 2013;43(1):9–22.
14. Nedelec M, McCall A, Carling C, Legall F, Berthoin S, Dupont G. Recovery in soccer: part I – post-match fatigue and time course of recovery. Sports Med (Auckland, N.Z). 2012;42(12):997–1015.
15. Halson SL. Recovery techniques for athletes. Sports Sci Exch. 2013;26(120):1–6.
16. Versey NG, Halson SL, Dawson BT. Water immersion recovery for athletes: effect on exercise performance and practical recommendations. Sports Med (Auckland, N.Z). 2013;43(11):1101–30.
17. Rollo I, VanLoon LJC. Recovery nutrition for football: building a team. 2018. https://barcainnovationhub.com/recovery-nutrition-for-footballl/.
18. Stellingwerff T, Boit MK, Res PT, International Association of Athletics Federations. Nutritional strategies to optimize training and racing in middle-distance athletes. J Sports Sci. 2007;25(Suppl 1):S17–28.
19. Rollo I, Randell RK, Baker L, Leyes JY, Medina Leal D, Lizarraga A, Mesalles J, Jeukendrup AE, James LJ, Carter JM. Fluid balance, sweat Na(+) losses, and carbohydrate intake of elite male soccer players in response to low and high training intensities in cool and hot environments. Nutrients. 2021;13(2):401.
20. Baker BL. Sweat testing methodology in the field: challenges and best practices. Sports Sci Exch. 2016;28(161):1–6.
21. Churchward-Venne TA, Pinckaers PJM, Smeets JSJ, Peeters WM, Zorenc AH, Schierbeek H, Rollo I, Verdijk LB, van Loon LJC. Myofibrillar and mitochondrial protein synthesis rates do not differ in young men following the ingestion of carbohydrate with milk protein, whey, or micellar casein after concurrent resistance- and endurance-type exercise. J Nutr. 2019;149(2):198–209.
22. Phillips SM, Van Loon LJ. Dietary protein for athletes: from requirements to optimum adaptation. J Sports Sci. 2011;29(Suppl 1):S29–38.
23. Trommelen J, Kouw IWK, Holwerda AM, Snijders T, Halson SL, Rollo I, Verdijk LB, van Loon LJC. Presleep dietary protein-derived amino acids are incorporated in myofibrillar protein during postexercise overnight recovery. Am J Physiol. 2018;314(5):E457–67.
24. Collins J, Maughan RJ, Gleeson M, Bilsborough J, Jeukendrup A, Morton JP, Phillips SM, Armstrong L, Burke LM, Close GL, Duffield R, Larson-Meyer E, Louis J, Medina D, Meyer F, Rollo I, Sundgot-Borgen J, Wall BT, Boullosa B, Dupont G, Lizarraga A, Res P, Bizzini M, Castagna C, Cowie CM, D'Hooghe M, Geyer H, Meyer T, Papadimitriou N, Vouillamoz M, McCall A. UEFA expert group statement on nutrition in elite football. Current evidence to inform practical recommendations and guide future research. Br J Sports Med. 2021;55(8):416.
25. van Vliet S, Burd NA, van Loon LJ. The skeletal muscle anabolic response to plant- versus animal-based protein consumption. J Nutr. 2015;145(9):1981–91.
26. Churchward-Venne TA, Pinckaers PJM, Smeets JSJ, Peeters WM, Zorenc AH, Schierbeek H, Rollo I, Verdijk LB, van Loon LJC. Myofibrillar and mitochondrial protein synthesis rates do not differ in young men following the ingestion of carbohydrate with whey, soy, or leucine-enriched soy protein after concurrent resistance- and endurance-type exercise. J Nutr. 2019;149(2):210–20.
27. van Vliet S, Shy EL, Abou Sawan S, Beals JW, West DW, Skinner SK, Ulanov AV, Li Z, Paluska SA, Parsons CM, Moore DR, Burd NA. Consumption of whole eggs promotes greater stimulation of postexercise muscle protein synthesis than consumption of isonitrogenous amounts of egg whites in young men. Am J Clin Nutr. 2017;106(6):1401–12.
28. Baumeister RF, Vohs KD. Revisiting our reappraisal of the (surprisingly few) benefits of high self-esteem. Perspect Psychol Sci. 2018;13(2):137–40.

29. Balk YA, Englert C. Recovery self-regulation in sport: theory, research, and practice. Int J Sports Sci Coach. 2020;15:273–81.
30. Podlog L, Di Fronso S, Conti C, Bertollo M. Underreocvery in elite athletes: antecedents, implications and prevention strategies. In: Kellman M, Beckmann J, editors. Recovery: the underestimated factor in physical and mental health. Advances in recovery and stress research. Routledge (2023).
31. Balk YA, Tamminen KA, Eccles DW. Too tired to switch off? How post-training physical fatigue impairs mental recovery through increased worry. Sport Exerc Perform Psychol. 2021;10:489–503.
32. Eccles DW, Kazmier AW. The psychology of rest in athletes: an empirical study and initial model. Psychol Sport Exerc. 2019;44:90–8.
33. Fishbach A, Woolley K. Combatting temptation to promote health and well-being. In: de Ridder D, Adriaanse M, Fujita K, editors. International handbook of self-control in health and well-being. Routledge; 2018. p. 167–79.
34. Englert C, Rummel J. I want to keep on exercising but I don't: the negative impact of momentary lacks of self-control on exercise adherence. Psychol Sport Exerc. 2016;26:24–31.
35. Baumeister RF, Gailliot M, DeWall CN, Oaten M. Self-regulation and personality: how interventions increase regulatory success, and how depletion moderates the effects of traits on behavior. J Pers. 2006;74(6):1773–801.
36. Lloyd D, Aon MA, Cortassa S. Why homeodynamics, not homeostasis? Sci World J. 2001;1:133–45.
37. Hardy J. Speaking clearly: a critical review of the self-talk literature. Psychol Sport Exerc. 2006;7:81–97.
38. Elbe AM, Szymanski B, Beckmann J. The development of volition in young elite athletes. Psychol Sport Exerc. 2005;6:559–69.
39. Siekańska M, Bondar RZ, di Fronso S, Blecharz J, Bertollo M. Integrating technology in psychological skills training for performance optimization in elite athletes: a systematic review. Psychol Sport Exerc. 2021;57:102008.
40. Carroll JK, Fiscella K, Epstein RM, Sanders MR, Williams GC. A 5A's communication intervention to promote physical activity in underserved populations. BMC Health Serv Res. 2012;12:374.
41. Whitlock EP, Orleans CT, Pender N, Allan J. Evaluating primary care behavioral counseling interventions: an evidence-based approach. Am J Prev Med. 2002;22(4):267–84.
42. Fiscella K, Meldrum S, Franks P, Shields CG, Duberstein P, McDaniel SH, Epstein RM. Patient trust: is it related to patient-centered behavior of primary care physicians? Med Care. 2004;42(11):1049–55.
43. Williams GC, Grow VM, Freedman ZR, Ryan RM, Deci EL. Motivational predictors of weight loss and weight-loss maintenance. J Pers Soc Psychol. 1996;70(1):115–26.
44. Williams DA, Cary MA, Groner KH, Chaplin W, Glazer LJ, Rodriguez AM, Clauw DJ. Improving physical functional status in patients with fibromyalgia: a brief cognitive behavioral intervention. J Rheumatol. 2002;29(6):1280–6.
45. Habay J, Van Cutsem J, Verschueren J, De Bock S, Proost M, De Wachter J, Tassignon B, Meeusen R, Roelands B. Mental fatigue and sport-specific psychomotor performance: a systematic review. Sports Med (Auckland, N.Z). 2021;51(7):1527–48.
46. Roelands B, Kelly V, Russell S, Habay J. The physiological nature of mental fatigue: current knowledge and future avenues for sport science. Int J Sports Physiol Perform. 2022;17(2):149–50.
47. Smith MR, Thompson C, Marcora SM, Skorski S, Meyer T, Coutts AJ. Mental fatigue and soccer: current knowledge and future directions. Sports Med (Auckland, N.Z). 2018;48(7):1525–32.
48. Marcora SM, Staiano W, Manning V. Mental fatigue impairs physical performance in humans. J Appl Physiol (1985). 2009;106(3):857–64.

49. Russell S, Jenkins DG, Halson SL, Juliff LE, Kelly VG. How do elite female team sport athletes experience mental fatigue? Comparison between international competition, training and preparation camps. Eur J Sport Sci. 2022;22(6):877–87.
50. Proost M, Habay J, De Wachter J, De Pauw K, Rattray B, Meeusen R, Roelands B, Van Cutsem J. How to tackle mental fatigue: a systematic review of potential countermeasures and their underlying mechanisms. Sports Med (Auckland, N.Z). 2022;52(9):2129–58.
51. Bailey SP, Harris GK, Lewis K, Llewellyn TA, Watkins R, Weaver MA, Roelands B, Van Cutsem J, Folger SF. Impact of a carbohydrate mouth rinse on corticomotor excitability after mental fatigue in healthy college-aged subjects. Brain Sci. 2021;11(8):972.
52. Erith S, Williams C, Stevenson E, Chamberlain S, Crews P, Rushbury I. The effect of high carbohydrate meals with different glycemic indices on recovery of performance during prolonged intermittent high-intensity shuttle running. Int J Sport Nutr Exerc Metab. 2006;16(4):393–404.
53. Williams C, Rollo I. Carbohydrate nutrition and team sport performance. Sports Med (Auckland, N.Z). 2015;45(Suppl 1):S13–22.
54. Rollo I, Williams C, Nevill M. Influence of ingesting versus mouth rinsing a carbohydrate solution during a 1-h run. Med Sci Sports Exerc. 2011;43(3):468–75.
55. Rollo I, Gonzalez JT, Fuchs CJ, van Loon LJC, Williams C. Primary, secondary, and tertiary effects of carbohydrate ingestion during exercise. Sports Med (Auckland, N.Z). 2020;50(11):1863–71.
56. Achten J, Halson SL, Moseley L, Rayson MP, Casey A, Jeukendrup AE. Higher dietary carbohydrate content during intensified running training results in better maintenance of performance and mood state. J Appl Physiol. 2004;96(4):1331–40.
57. Killer SC, Svendsen IS, Jeukendrup AE, Gleeson M. Evidence of disturbed sleep and mood state in well-trained athletes during short-term intensified training with and without a high carbohydrate nutritional intervention. J Sports Sci. 2017;35(14):1402–10.
58. Brewer BW, Redmond C. Psychology of sport injury. Champaign: Human Kinetics; 2017.
59. Glazer DD. Development and preliminary validation of the Injury-Psychological Readiness to Return to Sport (I-PRRS) scale. J Athl Train. 2009;44(2):185–9.
60. Webster KE, Feller JA, Lambros C. Development and preliminary validation of a scale to measure the psychological impact of returning to sport following anterior cruciate ligament reconstruction surgery. Phys Ther Sport. 2008;9(1):9–15.
61. Forsdyke D, Smith A, Jones M, Gledhill A. Psychosocial factors associated with outcomes of sports injury rehabilitation in competitive athletes: a mixed studies systematic review. Br J Sports Med. 2016;50(9):537–44.
62. Forsdyke D, Gledhill A, Ardern C. Psychological readiness to return to sport: three key elements to help the practitioner decide whether the athlete is REALLY ready? Br J Sports Med. 2017;51(7):555–6.
63. Ardern CL, Taylor NF, Feller JA, Webster KE. A systematic review of the psychological factors associated with returning to sport following injury. Br J Sports Med. 2013;47(17):1120–6.
64. Ryan RM, Deci EL. Self-determination theory and the facilitation of intrinsic motivation, social development, and well-being. Am Psychol. 2000;55(1):68–78.
65. Gouin JP, Carter CS, Pournajafi-Nazarloo H, Glaser R, Malarkey WB, Loving TJ, Stowell J, Kiecolt-Glaser JK. Marital behavior, oxytocin, vasopressin, and wound healing. Psychoneuroendocrinology. 2010;35(7):1082–90.
66. Tipton KD. Nutritional support for exercise-induced injuries. Sports Med (Auckland, N.Z). 2015;45(Suppl 1):S93–104.
67. Parr EB, Camera DM, Areta JL, Burke LM, Phillips SM, Hawley JA, Coffey VG. Alcohol ingestion impairs maximal post-exercise rates of myofibrillar protein synthesis following a single bout of concurrent training. PLoS One. 2014;9(2):e88384.

Chapter 17
Vitamins, Minerals, and Antioxidants

Anatoli Petridou and Vassilis Mougios

Introduction

Engaging in physical activity, particularly at a competitive sport level, places considerable stress on the body, particularly its metabolic processes, owing to increased energy demands. To optimize both health and athletic performance, it is crucial to prioritize efficient recovery and regeneration of the body's resources, ensuring sustained adaptations for enhanced performance. Nutrition plays a pivotal role in enhancing post-exercise recovery and facilitating adaptation to training. An adequate and well-balanced diet not only replenishes energy stores but also supports the body in repairing and rebuilding tissues that were stressed during exercise. Thus, adequate intake of essential nutrients is of paramount importance, as it directly influences recovery rates and the ability to withstand upcoming physical challenges. This ensures that the physiological adaptations triggered by exercise contribute to enduring improvements in performance, rather than leading to fatigue and potential setbacks.

In this chapter, we delve into the impactful role of vitamins, minerals, and antioxidants in recovery, regeneration, and adaptation to exercise stress. Our exploration commences with an introduction to these nutrients, followed by an examination of their roles in bodily functions and the various food sources through which they can be obtained. Subsequently, our focus shifts to a thorough examination of their ergogenic effects, as they actively contribute to recovery, regeneration, and adaptation to the stress imposed by exercise.

Vitamins and minerals are the two micronutrient classes that, alongside macronutrients (carbohydrates, fats, and proteins), constitute the repertoire of our

A. Petridou (✉) · V. Mougios
School of Physical Education and Sport Science at Thessaloniki, Aristotle University of Thessaloniki, Thessaloniki, Greece
e-mail: apet@phed.auth.gr

N. C. Apostolopoulos et al. (eds.), *Fundamentals of Recovery, Regeneration, and Adaptation to Exercise Stress: An Integrated Approach*,
https://doi.org/10.1007/978-3-031-44270-4_17

nutrients. They are characterised as micronutrients, given that the daily consumption of each typically amounts to just a few grams, down to micrograms, as opposed to the tens or hundreds of each macronutrient. Antioxidants are substances that prevent the damaging effects of (free) radicals in the body. Radicals are highly reactive chemical entities that can cause oxidative stress, potentially leading to cellular damage and contributing to various health issues. Antioxidants counteract this process by donating electrons to stabilize radicals and eliminate their reactivity.

A term often used in conjunction with vitamins, minerals, and antioxidants is 'nutraceutical,' a hybrid of nutrient and pharmaceutical. Although there is no consensus as to its definition, people usually think of a nutraceutical as a food, food additive, or dietary supplement that has a medical or health benefit. In this sense, many things could fit under the term. Indeed, it has been argued that this is a portmanteau term that serves commercial, rather than scientific, purposes and should therefore be abandoned in favor of more accurate terms [1].

> 'Nutraceuticals' and 'functional foods' are vague, nondiscriminatory, unhelpful terms; the evidence suggests that they should be abandoned in favour of more precise terms. The term 'dietary supplement' is widely used to designate formulations that are also called 'nutraceuticals' but it would be better restricted to individual compounds used to treat or prevent deficiencies.
>
> Jeffrey K. Aronson

Navigating Through Vitamins, Minerals, and Antioxidants

Vitamins constitute a group of organic compounds found in minute quantities in food and actively participating in metabolic regulation. While not directly supplying energy, they play essential roles as regulators in various bodily functions, including energy provision during exercise. The term "vitamin" is derived from the Latin word for life (vita), appended with "amine", which reflects the presence of amino groups in many, though not all, vitamins [2]. Most vitamins are not synthesized in the human body. Upon closer examination, it becomes evident that a vitamin may not be a single compound but, rather, a collection of compounds sharing similar structures and functions. These compounds are linked by metabolic reactions (that is, one is the product of another), with one typically exhibiting the highest biological activity.

Currently, there are 13 recognized vitamins, categorized into two groups: 9 water soluble and 4 fat soluble. Water-soluble vitamins are abundant in most foods, given their substantial water content. In contrast, fat-soluble vitamins do not readily dissolve in water but are instead found in dietary fats. Water-soluble vitamins undergo conversion into coenzymes, engage in the active sites of enzymes, or otherwise participate in metabolic processes involving carbohydrates, lipids, proteins, and nucleic acids. Fat-soluble vitamins, on the other hand, play essential roles in specialized functions like vision, antioxidant protection, mineral

Table 17.1 Biological roles and common food sources of vitamins

Vitamin	Biological role	Good food sources
Water soluble		
Vitamin B_1 (thiamin)	Energy metabolism, nervous system function	Whole grains, pulses, pork, nuts, seafood
Vitamin B_2 (riboflavin)	Energy metabolism, cell growth	Dairy products, eggs, leafy greens, seafood
Vitamin B_3 (niacin)	Energy metabolism	Meat, poultry, seafood, nuts, mushrooms
Vitamin B_5 (pantothenate)	Energy metabolism	All foods
Vitamin B_6 (pyridoxin)	Protein and carbohydrate metabolism, neurotransmitter and heme synthesis	Chicken, seafood, nuts, potatoes, bananas
Vitamin B_7 (biotin)	Metabolism of fats, carbohydrates, and proteins	Meat, poultry, seafood, eggs, nuts, sweet potatoes
Vitamin B_9 (folate)	DNA synthesis, cell division	Leafy greens, pulses, liver, seafood, nuts
Vitamin B_{12} (cobalamin)	Erythrocyte formation, nervous system function	Meat, liver, poultry, seafood, dairy products
Vitamin C (ascorbate)	Antioxidant, collagen synthesis, immune support	Citrus fruits, berries, kiwi, peppers, leafy greens
Fat soluble		
Vitamin A (retinol, carotenoids)	Vision, immune function, skin health, antioxidant	Liver, seaffod, carrots, leafy greens
Vitamin D (cholecalciferol)	Calcium absorption, bone health, immune support	Fatty fish, meat, egg, mushrooms
Vitamin E (α-tocopherol)	Antioxidant, cell membrane protection	Nuts, vegetable oils, seafood, egg, vegetables
Vitamin K (phylloquinone)	Blood clotting, bone health	Leafy greens, avocado, vegetable oils

absorption, and blood clotting [2]. Table 17.1 presents the biological roles and food sources of each vitamin.

Vitamin deficiency can lead to health issues of varying severity, impacting the nervous system, muscles, heart, connective tissue, eyes, bones, blood, and energy production. Like other micronutrients, vitamins have recommended dietary intakes for optimal health. One may come across the terms population reference intake (PRI) in the European Union and recommended dietary allowance (RDA) in the USA. Both signify the scientifically documented amount that is sufficient to meet the nutritional requirements of nearly all (specifically, 97.5%) of the healthy individuals within a specific population group [3]. Groups are defined by shared characteristics such as age, gender, or states like pregnancy. Tables 17.2 and 17.3 provide indicative recommended vitamin intakes (for adults aged 18–50), according to the European Food Safety Authority [4] and the US Food and Nutrition Board of the National Institutes of Health [5], respectively.

Table 17.2 Recommended vitamin intakes in adults aged 18–50, according to European Food Safety Authority (for other populations refer to [4])

Vitamin	Females	Males
Water soluble		
Vitamin B_1, or thiamin (mg/MJ)	0.1	0.1
Vitamin B_2, or riboflavin (mg/d)	1.6	1.6
Vitamin B_3, or niacin (mg NE/MJ)[a]	1.6	1.6
Vitamin B_5, or pantothenate (mg/d)*	5	5
Vitamin B_6, or pyridoxin (mg/d)	1.6	1.7
Vitamin B_7, or biotin (μg/d)*	40	40
Vitamin B_9, or folate (μg DFE/d)[b]	330	330
Vitamin B_{12}, or cobalamin (μg/d)*	4	4
Vitamin C, or ascorbate (mg/d)	95	110
Fat soluble		
Vitamin A, or retinol (μg RE/d)[c]	650	750
Vitamin D, or cholecalciferol (μg)*[d]	15	15
Vitamin E, or α-tocopherol (mg)*	11	13
Vitamin K, or phylloquinone (μg)*	70	70

Values are population reference intakes (PRI) except for *adequate intakes (AI), that is, a level of intake presumed to guarantee nutritional adequacy (used when evidence is insufficient to develop a PRI)

[a]As niacin equivalents (NE). 1 NE = 1 mg niacin = 60 mg tryptophan

[b]As dietary folate equivalents (DFE). 1 DFE = 1 μg food folate = 1.7 μg supplemental folic acid

[c]As retinol equivalent (RE). 1 μg RE = 1 μg retinol = 6 μg β-carotene = 12 μg other carotenoids with provitamin A activity

[d]Assuming minimal vitamin D synthesis in the skin. In the presence of endogenous vitamin D synthesis, the requirement for dietary vitamin D is lower or may even be zero

Chemical elements, present in the human diet and playing important roles in the human body, are categorized as minerals. Minerals, like vitamins, are involved in metabolic regulation and some serve as structural components in the body. Nutrition science recognizes a total of 15 minerals, further divided into metals (10) and nonmetals (5) on the basis of their chemical character. Metals play diverse roles, including maintaining body water content, transmitting nerve signals, forming bones and teeth, acting as enzyme cofactors, and regulating gene expression. Nonmetals play roles in fluid balance, digestion, bone and tooth formation, antioxidant defense, as well as growth and development [2]. Table 17.4 presents the biological roles and food sources of each mineral.

Mineral deficiency can lead to neurological, muscular, and cardiac issues, along with conditions like weak bones, anemia, and hindered growth, development, and energy production. Most minerals are needed in quantities below 100 mg per day and are referred to as trace elements; they include chromium, copper, fluoride, iodine, iron, manganese, molybdenum, selenium, and zinc. Table 17.5 provides indicative recommended mineral intakes (for adults aged 18–50), according to the European Food Safety Authority [4] and the US Food and Nutrition Board of the National Institutes of Health [6].

Table 17.3 Recommended daily vitamin intakes in adults aged 18–50, according to US National Academies of Health (for other populations refer to [5])

Vitamin	Females	Males
Water soluble		
Vitamin B_1, or thiamin (mg)	1.1	1.2
Vitamin B_2, or riboflavin (mg)	1.1	1.3
Vitamin B_3, or niacin (mg)[a]	14	16
Vitamin B_5, or pantothenate (mg)*	5	5
Vitamin B_6, or pyridoxin (mg)	1.3	1.3
Vitamin B_7, or biotin (μg)*	30	30
Vitamin B_9, or folate (μg)[b]	400	400
Vitamin B_{12}, or cobalamin (μg)	2.4	2.4
Vitamin C, or ascorbate (mg)	75	90
Fat soluble		
Vitamin A, or retinol (μg)[c]	700	900
Vitamin D, or cholecalciferol (μg)	15	15
Vitamin E, or α-tocopherol (mg)	15	15
Vitamin K, or phylloquinone (μg)*	90	120

Values are recommended dietary allowances (RDA) except for *adequate intakes (AI), that is, a level of intake presumed to guarantee nutritional adequacy (used when evidence is insufficient to develop an RDA)

[a]As niacin equivalents (NE). 1 mg niacin = 60 of tryptophan

[b]As dietary folate equivalents (DFE). 1 DFE = 1 μg food folate = 0.6 μg folic acid from fortified food or as supplement consumed with food = 0.5 μg of supplement taken on empty stomach

[c]As retinol activity equivalents (RAEs). 1 RAE = 1 μg retinol =12 μg β-carotene =24 μg α-carotene = 24 μg β-cryptoxanthin

Antioxidants include substances that are either produced in the body (such as glutathione) or are obtained from the diet. Dietary antioxidants include some vitamins (vitamin C, β-carotene, and vitamin E) and phytochemicals (polyphenols, flavonoids, and lycopene) [7]. Table 17.6 presents the biological roles and food sources of these phytochemicals (vitamins have been presented in Table 17.1).

Vitamins, Minerals, and Antioxidants in Recovery, Regeneration, and Adaptation to Exercise Stress

In contemporary health recommendations, exercise training is advocated as a key strategy to uphold well-being and counteract inflammatory processes. However, excessive training (common among athletes) raises energy demands and micronutrient loss, potentially intensifying inflammation compromising immune surveillance, and increasing susceptibility to disease. Therefore, it is crucial for an athlete's nutrition to meet the increased energy requirements and cope with increased physical and metabolic stress. While the influence of micronutrient status on exercise performance and training adaptations is admittedly small compared to that of

Table 17.4 Biological roles and common food sources of minerals

Mineral	Biological role	Good food sources
Metals		
Sodium (Na)	Fluid balance, nerve transmission, muscle contraction	Table salt, processed foods, meat, poultry, seafood, cheese
Potassium (K)	Fluid balance, nerve transmission, muscle contraction	Nuts, seafood, vegetables, mushrooms, pulses, fruits
Calcium (Ca)	Bone and teeth formation, nerve function, muscle contraction	Dairy products, leafy greens, seafood, nuts, pulses, egg
Magnesium (Mg)	Muscle and nerve function, bone health	Nuts, grains, leafy greens, pulses, seafood
Iron (Fe)	Oxygen transport, energy metabolism	Liver, meat, poultry, seafood, grains, nuts, leafy greens, egg
Copper (Cu)	Energy metabolism, cellular health	Liver, seafood, nuts, grains, mushrooms, pulses, vegetables, fruits
Zinc (Zn)	Immune function, gene expression	Liver, meat, seafood, pulses, grains, cheese, egg
Chromium (Cr)	Insulin sensitivity, glucose metabolism	Broccoli, whole grains, nuts, fruits
Manganese (Mn)	Bone formation, antioxidant defense	Seafood, nuts, whole grains, pulses, fruits, tea
Molybdenum (Mo)	Energy metabolism	Soy milk, nuts, grains, leafy greens, fruits
Nonmetals		
Chlorine (Cl)	Fluid balance, ion balance	Table salt, seafood, cheese
Phosphorus (P)	Bone and teeth formation, energy metabolism	Dairy products, meat, poultry, seafood, nuts, pulses
Selenium (Se)	Antioxidant, thyroid function, enzyme function	Brazil nuts, liver, meat, poultry, seafood, grains, egg, dairy products
Fluorine (F)	Dental health, bone strength	Fluoridated water, tea, seafood, grape juice
Iodine (I)	Thyroid hormone production	Iodized salt, grains, egg, dairy products, seafood

macronutrients, micronutrient deficiencies can still affect overall health, potentially disrupting training volume and impacting adaptation and performance. Thus, athletes should follow nutrient-dense diets that, as a minimum, meet the recommended daily intakes for all micronutrients, avoiding deficiencies.

Many individuals dedicated to regular exercise (or to everyday training if athletes) often include vitamin, mineral, and antioxidant supplementation in their training routines to ensure the adequate replenishment of these nutrients, given the increased demand during exercise, and to feel confident about their nutritional needs being met [8]. It has also been reported that elite athletes use dietary supplements more than non-elite counterparts, whereas supplement use is similar between male and female athletes, except for higher iron use by females and higher vitamin E use by males [9]. Nevertheless, the effectiveness of dietary supplementation is often not scientifically documented, as most of the literature indicates that a well-balanced diet, adapted to their increased energy demands, can meet the augmented nutritional

Table 17.5 Recommended daily mineral intakes in adults aged 18–50, according to European and US Agencies (for other populations refer to [3, 4])

Mineral	European		US	
	Females	Males	Females	Males
Metals				
Sodium (mg)	2000	2000	1500*	1500*
Potassium (mg)	3500*	3500*	2600*	3400*
Calcium (mg)	1000	1000	1000	1000
Magnesium (mg)	300*	350*	310–320[a]	400–420[a]
Iron (mg)	16	11	18	8
Copper (mg)	1.3*	1.6*	0.9	0.9
Zinc (mg)	7.5–12.7[b]	9.4–16.3[b]	8	11
Chromium (μg)	–	–	25*	35*
Manganese (mg)	3*	3*	1.8*	2.3*
Molybdenum (μg)	65*	65*	45	45
Nonmetals				
Chlorine (mg)	3100*	3100*	2300*	2300*
Phosphorus (mg)	550*	550*	700	700
Selenium (μg)	70*	70*	55	55
Fluorine (mg)	2.9*	3.4*	3*	4*
Iodine (μg)	150*	150*	150	150

Values are population reference intakes (PRI) for Europe and recommended dietary allowances (RDA) for US except for *adequate intakes (AI), that is, a level of intake presumed to guarantee nutritional adequacy (used when evidence is insufficient to develop a RPI or RDA)
[a]The first value refers to ages 18–30 and the second to ages 31–50
[b]Depending on the level of phytate (a plant component that inhibits zinc absorption) intake

Table 17.6 Biological roles and common sources of phytochemicals

Antioxidant	Biological role	Good food sources
Lutein and zeaxanthin	Eye health, antioxidant	Leafy greens (spinach, kale), corn, eggs
Flavonoids	Anti-inflammatory, cardiovascular health	Berries, citrus fruits, onions, dark chocolate, tea
Anthocyanins	Antioxidant	Berries
Polyphenols	Cardiovascular health, anti-inflammatory	Grapes, red wine, berries, dark chocolate, green tea
Catechins	Antioxidant	Fruits, tea, cacao
Coenzyme Q10	Energy production in cells, antioxidant	Fatty fish, meat, whole grains, nuts, spinach

requirements of athletes [10]. In contrast, the consumption of supplements frequently elicits adverse effects on training adaptations and recovery [11, 12].

Although a well-rounded diet incorporating lean proteins, whole grains, fruits, and vegetables typically suffices to meet nutritional needs, athletes facing malabsorption or nutrient deficiencies after limiting energy intake, adopting extreme weight loss methods, excluding entire food groups, engaging in long and exhausting training sessions, or suffering from injuries or infections may find value in supplementation with some micronutrients [11]. There is evidence for the need to pay particular attention to the intake of some vitamins, minerals, and antioxidants to aid recovery and facilitate adaptations to training [12]. Thus, there are situations in which a "food only" approach could be detrimental in terms of an athlete's overall health and sporting performance. In line with this is the statement, "food first, but not always food only" [13, 14], which implies complementing the diet with targeted and evidence-based supplementation.

This brings us to the matter of "evidence." Prior to incorporating vitamin, mineral, or antioxidant supplements into their dietary regimens, athletes should ascertain the true extent of the benefits these supplements may offer to their overall health and exercise performance. The beneficial effects of a supplement or nutritional intervention are usually assessed through a four-level scale reflecting the quality of scientific evidence. Of the several available systems that decode the hierarchy of evidence by using letters, numbers, symbols, or words [15–17], we have adopted the following as being most transparent:

- High certainty of evidence is attributed to situations where there is considerable confidence that the true effect closely aligns with the estimated effect.
- Moderate certainty of evidence is assigned when there is a reasonable degree of confidence, though the true effect might differ substantially.
- Low certainty of evidence is associated with limited confidence, suggesting the true effect could deviate significantly from the estimate.
- Very low certainty of evidence is indicative of minimal confidence, with the true effect likely to differ substantially from the estimated effect.

In this classification, high and moderate certainty of evidence is based on randomized controlled trials and meta-analyses, whereas low and very low certainty of evidence is based on observational studies [18]. Prior to considering the use of vitamin, mineral, or antioxidant supplementation, an athlete should identify potential deficiencies through hematologic, biochemical, and nutritional assessment with the aid of qualified professionals (for details, refer to [19]).

> Overall, the assessment process can help the athlete understand that supplement intake cannot reverse poor food choices and an inadequate diet, while a well-chosen diet can ensure maximal benefit from supplementation.
>
> D. Enette Larson-Meyer

Vitamins, minerals, and antioxidants that appear frequently in the scientific literature as potential ergogenic aids include sodium, vitamin D, calcium, iron, zinc, magnesium, and antioxidants [20]. Below, we review the evidence for their effectiveness.

Sodium

Sodium is the main electrolyte in the extracellular fluid and plasma. We obtain it primarily from table salt (sodium chloride, or NaCl) and processed foods (see Table 17.4). Adults are advised not to exceed a daily sodium intake of 2 g (equivalent to approximately 5 g of salt, since 39% of NaCl is Na), due to the association of excessive salt consumption with hypertension, although there is evidence that low sodium intake is also associated with cardiovascular events [21].

Athletes have increased demands of sodium if they experience large losses in the sweat during prolonged exercise (exceeding 2 hours), a situation known as "salty sweater" [20, 22]. Losses of other minerals, such as potassium, calcium, and magnesium, are much smaller. During a 90-min training session or game, football (soccer) players generally lose 3–5 g of salt [23]. A common adverse outcome in such cases is "heat cramping," a condition arising from the combination of sodium depletion, fluid loss, and muscle fatigue [24]. Another condition, spurred by the combination of sodium loss and excessive intake of fluids that contain no or little sodium (such as plain water), is hyponatremia (Greek for "low blood sodium"), a short-term sodium deficiency often manifested with nausea, vomiting, headache, fatigue, and muscle cramps [25].

Sodium replacement during exercise has been proposed to reduce the risk of hyponatremia and ameliorate exertional muscle cramping caused by a whole-body sodium deficit [11]. Additionally, sodium replacement is a pre-requisite for maintenance of water replacement and euhydration [26], as incorporating electrolytes into rehydration beverages improves the retention and distribution of ingested fluid. This aligns with rehydration objectives and contributes to optimizing performance. Typical electrolyte supplements contain 1.2–1.4 g/L of sodium and are recommended for consumption not only during but also before and after prolonged athletic activities [12].

As for hyponatremia, discontinuing fluid intake (in mild cases), using diuretics and consuming salty fluids (in cases of moderate severity), or intravenous administration of salt solution (in severe cases) constitute suggested therapeutic approaches [2, 25]. Another involvement of sodium in sports is through sodium phosphate, a salt that is accompanied by low certainty of evidence for ergogenic effects on aerobic capacity and peak power output [27, 28]. Finally, the efficacy of sodium supplementation in sports is demonstrated by the use of sodium bicarbonate as an ergogenic aid. There is high-quality of evidence that its alkaline character functions to neutralize the excessive acidity accompanying the anaerobic breakdown of carbohydrates during short maximal exercise and enhance the body's resistance to fatigue [12].

Vitamin D

Vitamin D exists in two main forms: cholecalciferol (vitamin D_3) and ergocalciferol (vitamin D_2). Cholecalciferol is primarily synthesized in the skin through exposure to sunlight. Additionally, it can be obtained in smaller amounts from certain foods, such as oily fish and fortified products. Ergocalciferol can be obtained through the UVB irradiation of ergosterol in plants and fungi, including mushrooms [29]. Vitamin D, whether derived from the skin or dietary sources, enters the bloodstream and, passing through the liver, is converted to 25-hydroxyvitamin D, or 25(OH)D. This re-enters the circulation and, in the kidneys, is converted to 1,25-dihydroxyvitamin D, or 1,25$(OH)_2$D, the biologically active form of the vitamin [30]. 1,25$(OH)_2$D facilitates the absorption of calcium and phosphorus from the intestine and kidneys, which makes it essential for development and bone health. Additionally, 1,25$(OH)_2$D is associated with skin development and possibly influences muscle function and performance [30].

Vitamin D adequacy in the body is better evaluated by measuring serum 25(OH)D, rather than 1,25$(OH)_2$D. The Institute of Medicine (IOM) declares vitamin D deficiency regarding bone health when serum 25(OH)D is below 12 ng/mL (30 nmol/L) and insufficiency at 12–20 ng/mL (30–50 nmol/L), while values above 20 ng/mL (50 nmol/L) are considered adequate [31, 32]. Values above 30 ng/mL (75 nmol/L) are not consistently associated with increased benefits, whereas caution may be warranted when serum 25(OH)D exceeds 50 ng/mL (125 nmol/L) [31]. The Endocrine Society sets the borders between deficiency, insufficiency, and adequacy higher, that is, at 20 and 30 ng/mL [33]. As a result, values between 20 and 30 ng/mL are considered adequate by one standard and insufficient by another, adding to the existing uncertainty about which vitamin D values are desirable.

Results from a meta-analysis revealed a 56% prevalence of vitamin D "inadequacy", defined as serum 25(OH)D $<$32 ng/mL, among athletes [34]. Because vitamin D participates in muscle function through its receptors, resulting in increased muscle contraction velocity and strength [35], many studies have investigated the physical aspect of athletes with vitamin D inadequacy and have highlighted a correlation between insufficient vitamin D levels and compromised physical performance [36]. Moreover, the existing data underscore the crucial role of adequate vitamin D in the adaptive response to intense exercise and in mitigating the risk of upper respiratory tract infections, skeletal muscle damage, stress fractures, elevated inflammatory markers, and impaired recovery following orthopedic injury, thus enhancing the athlete's capacity to train and compete effectively [20, 37–39]. Systematic reviews and meta-analyses conclude that vitamin D inadequacy can be restored by vitamin D supplementation, thus enhancing athletic performance and body composition [36, 40]. However, indiscriminate vitamin D supplementation is not advised, as there are potential compensatory responses to high doses, such as a decrease in parathyroid hormone [41]. The effectiveness of vitamin D supplementation on physical performance is questionable [11, 36, 39, 40] and probably contingent on the athletes' vitamin D status at the onset of supplementation [39, 41].

The EFSA Panel on Dietetic Products, Nutrition and Allergies refer that populations with mean baseline serum 25(OH)D concentration of 27 nmol/L or higher, show no support for an association between serum 25(OH)D concentration and physical performance in healthy older adults [42]. Additionally, a meta-analysis revealed that there is no statistically significant impact of vitamin D supplementation on enhancing maximum strength and power [43] and it is therefore not recommended to engage in indiscriminate vitamin D supplementation for athletes [37]. The findings from the literature underscore the absence of a unanimous agreement regarding the substantial correlation between low 25(OH)D levels and an elevated risk of bone fractures in the pediatric population. Moreover, there is a lack of evidence supporting the necessity of maintaining adequate vitamin D status for preventing nutritional rickets in pediatric cases [44]. Given the conflicting findings regarding the impact of vitamin D on injury risk and its potential benefits in the recovery process, it is essential to conduct additional studies with robust intervention designs. These studies will provide a more thorough assessment of the role of vitamin D supplementation in the recovery following sports injuries [39].

Calcium

Calcium is primarily linked to the formation and metabolic processes of osseous tissue. It also plays important roles in health and performance via its involvement in energy metabolism, skeletal, cardiac, and smooth muscle contraction, nervous transmission, intracellular signaling, and hormonal secretion [31]. A sound eating plan should be able to meet dietary calcium requirements. No definitive biomarker of calcium status is known. While a low bone mineral density might suggest chronic low calcium intake, it is essential to consider other important factors such as suboptimal vitamin D status and disordered eating [12]. Calcium supplementation may be required in cases of increased requirements (growth, menopause, amenorrhea, high sweat loss), coupled with inadequate intake, although evidence for this is limited. Likewise, evidence is lacking for an ergogenic effect of calcium supplementation.

Iron

Iron deficiency can appear in athletes due to a combination of factors, including compromised intestinal absorption, gastrointestinal bleeding, and menstrual blood loss [13]. Iron binds oxygen in hemoglobin, myoglobin, and cytochromes. This involvement in oxygen handling and aerobic energy production renders its deficiency a factor in reduced aerobic performance. A high intake of iron-rich foods may not always be an effective treatment for iron deficiency, and appropriate iron supplementation in athletes presenting with compromised iron stores is warranted [13, 20] Medical supervision of iron supplementation is needed to control for gastrointestinal distress and iron overload, which may result in excessive production of reactive oxygen and nitrogen species and organ dysfunction. There is no consensus about the optimal iron levels in athletes.

Zinc

As a cofactor of many enzymes and transcription factors, zinc is involved in metabolism, growth, and development. According to research findings, routine zinc supplementation is not recommended. However, findings from studies on the general population suggest that zinc supplementation may be effective in preventing and treating the common cold, a prevalent condition among athletes during transient immunodepression in the early post-exercise period. Thus, supplementation during periods of intense exercise could be advantageous, particularly in cases where there is a history of recurrent infections [45].

Magnesium

Magnesium is found primarily in muscle and bones. It is a cofactor of many enzymes, including those that use ATP as substrate or use nucleotides to synthesize DNA and RNA. Based on these important functions, magnesium supplementation is rather common among athletes. However, evidence for ergogenic effects is rather limited. A recent systematic review [46] located studies of mostly moderate and low quality that showed positive effects of magnesium supplementation on functional dynamic measures of muscle performance, such as countermovement jump, one-repetition maximum, and resistance to fatigue, but no effect on isokinetic or endurance performance.

Antioxidants

Antioxidant supplementation is a widely embraced approach by both the general population and athletes. Among the latter, antioxidant supplementation has gained popularity as a means to mitigate exercise-induced oxidative stress and muscle damage [47, 48], expedite the recovery process [49], and even enhance athletic performance [50]. This is based on the fact that prolonged, exhaustive exercise and immune system activation are associated with increased reactive oxygen and nitrogen species production, potentially leading to oxidative stress. Early studies on exercise and radical production emphasized the damaging effects of oxidants in muscles, but there is now ample evidence for beneficial biological effects of radicals, although high levels of reactive oxygen and nitrogen species promote muscle fatigue [51].

However, the impact of antioxidant supplementation on skeletal muscle adaptation to exercise training remains uncertain [47]. Apart from a lack of high-quality evidence for a beneficial effect of antioxidant supplementation on adaptations to training, a growing body of literature suggests that antioxidant supplementation might in fact hinder the signaling of critical adaptations, including muscle mitochondrial biogenesis, increased insulin sensitivity, and muscle hypertrophy [49]. Emerging evidence for antioxidant supplementation mitigating important exercise-induced adaptations contributes to the debate over the utility of antioxidant supplementation in athletes [45, 48].

Among exercisers, vitamins C and E stand out as the most commonly consumed vitamin supplements [48], although most of them meet the recommended dietary intakes for both vitamins through the diet [52]. There is moderate to low certainty of evidence for a protective effect of vitamin C and/or vitamin E supplementation against exercise-induced muscle soreness for up to four days [53, 54] and enhancement of physical performance [48].

Another antioxidant (considered by some a drug) is *N*-acetylcysteine, a derivative of the amino acid cysteine. There is moderate-quality evidence that *N*-acetylcysteine supplementation delays fatigue during high-intensity exercise by mitigating the accumulation of RONS, which interfere with muscle activity [55], and reduces muscle injury after fatiguing exercise [56]. Nevertheless, a meta-analysis concluded that *N*-acetylcysteine supplementation produces a range of effects on performance that range from beneficial to harmful [57]. Thus, whether *N*-acetylcysteine is ergogenic remains unclear.

In conclusion, the available evidence does not favor acute or chronic antioxidant supplementation for athletic performance [48, 53]. The majority of scientific research reinforces the notion that a well-balanced diet is able to meet potential heightened antioxidant needs athletes [11, 48, 50]. Nevertheless, the necessity is emerging for personalized approaches that take into account individual needs and deficiencies when prescribing antioxidant supplementation [58].

Other

There is little or no research on the potential ergogenic effects of other vitamins, minerals, or antioxidants, including some with purported benefits, such as folate, vitamin B_{12}, vitamin K, copper, selenium, and phytochemicals.

Vegan Athletes

A specific consideration is the adequacy of micronutrient intake by athletes following vegetarian diets. While the prevalence of vegetarianism in sports is not known, it is presumed to mirror the growing trend in the general population [59]. Studies indicate that, compared to omnivores, vegans run a higher risk for deficiencies in micronutrients that are predominantly found in animal-derived foods, such as vitamin B_{12}, riboflavin, iron, calcium, zinc, and vitamin D [11, 60]. Major sport organizations, including the International Olympic Committee, the Academy of Nutrition and Dietetics, Dietitians of Canada, and the American College of Sports Medicine advise vegan/vegetarian athletes to supplement their diets with vitamin B_{12} or consume fortified foods (e.g., fortified cereal or soy milk) to prevent cobalamin deficiency, leads to nerve damage and pernicious anemia [11, 12]. Yet, the influence of cobalamin deficiency on performance and adaptation to training remains unclear. Additionally, there is no evidence supporting vitamin B_{12} supplementation as being ergogenic [59].

Conclusion

While a balanced diet generally meets the nutritional needs of athletes, specific situations—such as intense training, nutrient deficiencies, or dietary restrictions—may necessitate targeted supplementation. However, the effectiveness of supplements, especially antioxidants, is not adequately supported by scientific evidence. Instead, supplementation may sometimes hinder adaptations to training and performance. Athletes should focus on a nutrient-dense diet and consider supplementation only when deficiencies are identified by qualified professionals, following the principle of "food first, but not always food only."

Athletes eating a well-balanced, nutrient-dense diet that meets their energy needs almost certainly do not require vitamin, mineral, or antioxidant supplementation.

Prior to considering the use of vitamin, mineral, or antioxidant supplementation, athletes should identify possible deficiencies through hematologic, biochemical, and nutritional assessment with the aid of qualified professionals.

Except for sodium supplementation during prolonged exercise, there is no high certainty of evidence that vitamin, mineral, or antioxidant supplementation enhances athletic performance or recovery.

References

1. Aronson JK. Defining 'nutraceuticals': neither nutritious nor pharmaceutical. Br J Clin Pharmacol. 2017;83(1):8–19. https://doi.org/10.1111/bcp.12935.
2. Mougios V. Exercise biochemistry. Human Kinetics Publishers Inc.; 2020.
3. Institute of Medicine. Dietary Reference Intakes: The Essential Guide to Nutrient Requirements. Washington, DC: The National Academies Press; 2006. p. 170–285.
4. European Food Safety Authority. Dietary reference values. Accessed 21 Nov 2023. https://www.efsa.europa.eu/en/topics/topic/dietary-reference-values
5. U.S. National Institutes of Health. Dietary Reference Intakes (DRIs): Recommended Dietary Allowances and Adequate Intakes, Vitamins. Accessed 21 Nov 2023. https://www.ncbi.nlm.nih.gov/books/NBK56068/table/summarytables.t2/?report=objectonly.
6. U.S. National Institutes of Health. Dietary Reference Intakes (DRIs): Recommended Dietary Allowances and Adequate Intakes, Elements. Accessed November 21, 2023. https://www.ncbi.nlm.nih.gov/books/NBK545442/table/appJ_tab3/?report=objectonly.
7. Castell LM, Stear SJ, Burke LM. Nutritional supplements in sport, exercise and health: an A-Z guide.; 2015. https://doi.org/10.4324/9781315772509.
8. Maughan RJ, Depiesse F, Geyer H. The use of dietary supplements by athletes. J Sports Sci. 2007;25(sup1):S103–13. https://doi.org/10.1080/02640410701607395.
9. Knapik JJ, Steelman RA, Hoedebecke SS, Austin KG, Farina EK, Lieberman HR. Prevalence of dietary supplement use by athletes: systematic review and meta-analysis. Sports Med. 2016;46(1):103–23. https://doi.org/10.1007/s40279-015-0387-7.
10. Beck KL, von Hurst PR, O'Brien WJ, Badenhorst CE. Micronutrients and athletic performance: a review. Food Chem Toxicol. 2021;158:112618. https://doi.org/10.1016/j.fct.2021.112618.
11. Thomas DT, Erdman KA, Burke LM. American College of Sports Medicine joint position statement. Nutrition and athletic performance. Med Sci Sports Exerc. 2016;48(3):543–68. https://doi.org/10.1249/MSS.0000000000000852.
12. Maughan RJ, Burke LM, Dvorak J, et al. IOC consensus statement: dietary supplements and the high-performance athlete. Br J Sports Med. 2018;52(7):439–55. https://doi.org/10.1136/bjsports-2018-099027.
13. Close GL, Kasper AM, Walsh NP, Maughan RJ. "Food First but Not Always Food Only": recommendations for using dietary supplements in sport. Int J Sport Nutr Exerc Metab. 2022;32(5):371–86. https://doi.org/10.1123/ijsnem.2021-0335.
14. Peeling P, Sim M, McKay AKA. Considerations for the consumption of vitamin and mineral supplements in athlete populations. Sports Med. 2023;53(S1):15–24. https://doi.org/10.1007/s40279-023-01875-4.
15. Greenhalgh T. How to read a paper. Getting your bearings (deciding what the paper is about). BMJ. 1997;315(7102):243–6. https://doi.org/10.1136/bmj.315.7102.243.
16. Schünemann HJ, Best D, Vist G, Oxman AD, GRADE Working Group. Letters, numbers, symbols and words: how to communicate grades of evidence and recommendations. CMAJ. 2003;169(7):677–80. http://www.ncbi.nlm.nih.gov/pubmed/14517128
17. Balshem H, Helfand M, Schünemann HJ, et al. GRADE guidelines: 3. Rating the quality of evidence. J Clin Epidemiol. 2011;64(4):401–6. https://doi.org/10.1016/j.jclinepi.2010.07.015.
18. Schünemann HJ, Brennan S, Akl EA, et al. The development methods of official GRADE articles and requirements for claiming the use of GRADE—a statement by the GRADE guidance group. J Clin Epidemiol. 2023;159:79–84. https://doi.org/10.1016/j.jclinepi.2023.05.010.
19. Larson-Meyer DE, Woolf K, Burke L. Assessment of nutrient status in athletes and the need for supplementation. Int J Sport Nutr Exerc Metab. 2018;28(2):139–58. https://doi.org/10.1123/ijsnem.2017-0338.
20. Castell LM, Stear SJ, Burke LM. Nutritional supplements in sport, exercise and health.
21. Mente A, O'Donnell M, Rangarajan S, et al. Associations of urinary sodium excretion with cardiovascular events in individuals with and without hypertension: a pooled analysis of data from four studies. Lancet. 2016;388(10043):465–75. https://doi.org/10.1016/S0140-6736(16)30467-6.

22. Shirreffs SM, Sawka MN. Fluid and electrolyte needs for training, competition, and recovery. J Sports Sci. 2011;29(sup1):S39–46. https://doi.org/10.1080/02640414.2011.614269.
23. Shirreffs SM, Sawka MN, Stone M. Water and electrolyte needs for football training and match-play. J Sports Sci. 2006;24(7):699–707. https://doi.org/10.1080/02640410500482677.
24. Eichner ER. The role of sodium in "Heat Cramping." Sports Med. 2007;37(4):368–70. https://doi.org/10.2165/00007256-200737040-00024.
25. Herfel R, Stone CK, Koury SI, Blake JJ. Iatrogenic acute hyponatraemia in a college athlete. Br J Sports Med. 1998;32(3):257–8. https://doi.org/10.1136/bjsm.32.3.257.
26. Shirreffs SM, Maughan RJ. Volume repletion after exercise-induced volume depletion in humans: replacement of water and sodium losses. Am J Physiol Renal Physiol. 1998;274(5):F868–75. https://doi.org/10.1152/ajprenal.1998.274.5.F868.
27. Buck CL, Wallman KE, Dawson B, Guelfi KJ. Sodium phosphate as an ergogenic aid. Sports Med. 2013;43(6):425–35. https://doi.org/10.1007/s40279-013-0042-0.
28. AIS Sports Supplement Framework. Australian Institute of Sport (AIS) position statement: supplements and sports foods in high performance sport. Published 2022. Accessed 19 Dec 2023. https://www.ais.gov.au/nutrition/supplements.
29. Holick MF. Sunlight and vitamin D for bone health and prevention of autoimmune diseases, cancers, and cardiovascular disease. Am J Clin Nutr. 2004;80(6):1678S–88S. https://doi.org/10.1093/ajcn/80.6.1678S.
30. Hossein-nezhad A, Holick MF. Vitamin D for health: a global perspective. Mayo Clin Proc. 2013;88(7):720–55. https://doi.org/10.1016/j.mayocp.2013.05.011.
31. Catharine RA, Institute of Medicine (U.S.). Committee to Review Dietary Reference Intakes for Vitamin D and Calcium. Dietary reference intakes : calcium, vitamin D. National Academies Press; 2011.
32. Ross AC, Manson JE, Abrams SA, et al. The 2011 report on dietary reference intakes for calcium and vitamin D from the Institute of Medicine: what clinicians need to know. J Clin Endocrinol Metab. 2011;96(1):53–8. https://doi.org/10.1210/jc.2010-2704.
33. Holick MF, Binkley NC, Bischoff-Ferrari HA, et al. Evaluation, treatment, and prevention of vitamin D deficiency: an Endocrine Society Clinical Practice Guideline. J Clin Endocrinol Metab. 2011;96(7):1911–30. https://doi.org/10.1210/jc.2011-0385.
34. Farrokhyar F, Tabasinejad R, Dao D, et al. Prevalence of vitamin D inadequacy in athletes: a systematic-review and meta-analysis. Sports Med. 2015;45(3):365–78. https://doi.org/10.1007/s40279-014-0267-6.
35. Ksiażek A, Zagrodna A, Słowińska-Lisowska M. Vitamin D, skeletal muscle function and athletic performance in athletes—a narrative review. Nutrients. 2019;11(8) https://doi.org/10.3390/nu11081800.
36. Selvan V, Ramachandran A, Mohamed A, Sidiq M. Can vitamin D supplementation influence the physical components and performance of athletes with deficiency? A systematic review. Physiother Quarter. 2023;31(2):7–13. https://doi.org/10.5114/pq.2023.125750.
37. Rawson ES, Miles MP, Larson-Meyer DE. Dietary supplements for health, adaptation, and recovery in athletes. Int J Sport Nutr Exerc Metab. 2018;28(2):188–99. https://doi.org/10.1123/ijsnem.2017-0340.
38. Crescioli C. Vitamin D, exercise, and immune health in athletes: a narrative review. Front Immunol. 2022:13. https://doi.org/10.3389/fimmu.2022.954994.
39. Giraldo-Vallejo JE, Cardona-Guzmán MÁ, Rodríguez-Alcivar EJ, et al. Nutritional strategies in the rehabilitation of musculoskeletal injuries in athletes: a systematic integrative review. Nutrients. 2023;15(4):819. https://doi.org/10.3390/nu15040819.
40. Zhang L, Quan M, Cao ZB. Effect of vitamin D supplementation on upper and lower limb muscle strength and muscle power in athletes: a meta-analysis. Sunderland C, ed. PLoS One. 2019;14(4):e0215826. https://doi.org/10.1371/journal.pone.0215826.
41. Owens DJ, Sharples AP, Polydorou I, et al. A systems-based investigation into vitamin D and skeletal muscle repair, regeneration, and hypertrophy. Am J Physiol Endocrinol Metab. 2015;309(12):E1019–31. https://doi.org/10.1152/ajpendo.00375.2015.

42. Bresson JL, Burlingame B, Dean T, et al. Dietary reference values for vitamin D. EFSA J. 2016;14(10):e04547. https://doi.org/10.2903/j.efsa.2016.4547.
43. Sist M, Zou L, Galloway SDR, Rodriguez-Sanchez N. Effects of vitamin D supplementation on maximal strength and power in athletes: a systematic review and meta-analysis of randomized controlled trials. Front Nutr. 2023:10. https://doi.org/10.3389/fnut.2023.1163313.
44. Armento A, Heronemus M, Truong D, Swanson C. Bone health in young athletes: a narrative review of the recent literature. Curr Osteoporos Rep. 2023;21(4):447–58. https://doi.org/10.1007/s11914-023-00796-5.
45. Bermon S, Castell LM, Calder PC, et al. Consensus Statement Immunonutrition and Exercise. Exerc Immunol Rev. 2017;23:8–50. http://www.ncbi.nlm.nih.gov/pubmed/28224969
46. Heffernan S, Horner K, De Vito G, Conway G. The role of mineral and trace element supplementation in exercise and athletic performance: a systematic review. Nutrients. 2019;11(3):696. https://doi.org/10.3390/nu11030696.
47. Kim J. Effect of high-dose vitamin C and E supplementation on muscle recovery and training adaptation: a mini review. Phys Act Nutr. 2023;27(2):8–12. https://doi.org/10.20463/pan.2023.0012.
48. Martinez-Ferran M, Sanchis-Gomar F, Lavie CJ, Lippi G, Pareja-Galeano H. Do antioxidant vitamins prevent exercise-induced muscle damage? A systematic review. Antioxidants. 2020;9(5):372. https://doi.org/10.3390/antiox9050372.
49. Merry TL, Ristow M. Do antioxidant supplements interfere with skeletal muscle adaptation to exercise training? J Physiol. 2016;594(18):5135–47. https://doi.org/10.1113/JP270654.
50. Higgins M, Izadi A, Kaviani M. Antioxidants and exercise performance: with a focus on vitamin E and C supplementation. Int J Environ Res Public Health. 2020;17(22):8452. https://doi.org/10.3390/ijerph17228452.
51. Powers SK, Radak Z, Ji LL. Exercise-induced oxidative stress: past, present and future. J Physiol. 2016;594(18):5081–92. https://doi.org/10.1113/JP270646.
52. Wardenaar F, Brinkmans N, Ceelen I, et al. Micronutrient intakes in 553 Dutch elite and sub-elite athletes: prevalence of low and high intakes in users and non-users of nutritional supplements. Nutrients. 2017;9(2):142. https://doi.org/10.3390/nu9020142.
53. Ranchordas MK, Rogerson D, Soltani H, Costello JT. Antioxidants for preventing and reducing muscle soreness after exercise. Cochrane Database Syst Rev. 2017;2017(12) https://doi.org/10.1002/14651858.CD009789.pub2.
54. Torre MF, Martinez-Ferran M, Vallecillo N, Jiménez SL, Romero-Morales C, Pareja-Galeano H. Supplementation with vitamins C and E and exercise-induced delayed-onset muscle soreness: a systematic review. Antioxidants. 2021;10(2):279. https://doi.org/10.3390/antiox10020279.
55. Cobley JN, McGlory C, Morton JP, Close GL. N-Acetylcysteine's attenuation of fatigue after repeated bouts of intermittent exercise: practical implications for tournament situations. Int J Sport Nutr Exerc Metab. 2011;21(6):451–61. https://doi.org/10.1123/ijsnem.21.6.451.
56. Pinheiro CHJ, Vitzel KF, Curi R. Effect of *N* -acetylcysteine on markers of skeletal muscle injury after fatiguing contractile activity. Scand J Med Sci Sports. 2012;22(1):24–33. https://doi.org/10.1111/j.1600-0838.2010.01143.x.
57. Rhodes K, Braakhuis A. Performance and side effects of supplementation with N-Acetylcysteine: a systematic review and meta-analysis. Sports Med. 2017;47(8):1619–36. https://doi.org/10.1007/s40279-017-0677-3.
58. Margaritelis NV. Personalized redox biology: designs and concepts. Free Radic Biol Med. 2023;208:112–25. https://doi.org/10.1016/j.freeradbiomed.2023.08.003.
59. West S, Monteyne AJ, van der Heijden I, Stephens FB, Wall BT. Nutritional considerations for the vegan athlete. Adv Nutr. 2023;14(4):774–95. https://doi.org/10.1016/j.advnut.2023.04.012.
60. Venderley AM, Campbell WW. Vegetarian Diets. Sports Med. 2006;36(4):293–305. https://doi.org/10.2165/00007256-200636040-00002.

Part III
Manual Therapies

Chapter 18
Effects of Anti-Inflammatory and Analgesic Drugs on Recovery and Performance in Sports

Glyn Howatson and Tommy Lundberg

Introduction

Analgesics, also known as "painkillers," are drugs that relieve pain without affecting consciousness or sensory perception (unlike anaesthetics). A variety of drugs have these properties, including paracetamol, nonsteroidal anti-inflammatory drugs (NSAIDs), and opioids (narcotics). While NSAIDs (Table 18.1) and paracetamol are available over the counter and are no longer considered doping agents, opioids are available by prescription and are banned in sports.

Inhibition of the activity of the cyclooxygenase enzyme family (COX), which promotes the formation of prostaglandins, is one of the main mechanisms by which over-the-counter analgesics exert their analgesic effect. Although paracetamol is not a NSAID, it is now generally accepted that it has a range of effects like NSAIDs and suppresses the activity of COX [1]. However, unlike NSAIDs, paracetamol also has antipyretic properties and a central site of action (e.g., in the central nervous system). The study of these effects is particularly important in an athletic context because the production of prostaglandins not only controls pain and inflammation but also influences the machinery that controls tissue remodelling during adaptive responses to exercise [2] (Fig. 18.1).

Professional athletes often use over-the-counter painkillers, but so do student and amateur athletes. Olympic athletes were found to be significantly more likely to use NSAIDs compared with control participants of the same age [3]. In addition, more than half of male and female soccer players who participated in the FIFA World Cup

G. Howatson (✉)
Northumbria University, Newcastle upon Tyne, UK
e-mail: glyn.howatson@northumbria.ac.uk

T. Lundberg
Karolinska Institute, Stockholm, Sweden

N. C. Apostolopoulos et al. (eds.), *Fundamentals of Recovery, Regeneration, and Adaptation to Exercise Stress: An Integrated Approach*,
https://doi.org/10.1007/978-3-031-44270-4_18

Table 18.1 Most common types of NSAIDs

Common NSAIDs
Ibuprofen
Naproxen
Diclofenac
Celecoxib
Mefenamic acid
Etoricoxib
Indomethacin
Acetylsalicylic acid (aspirin)

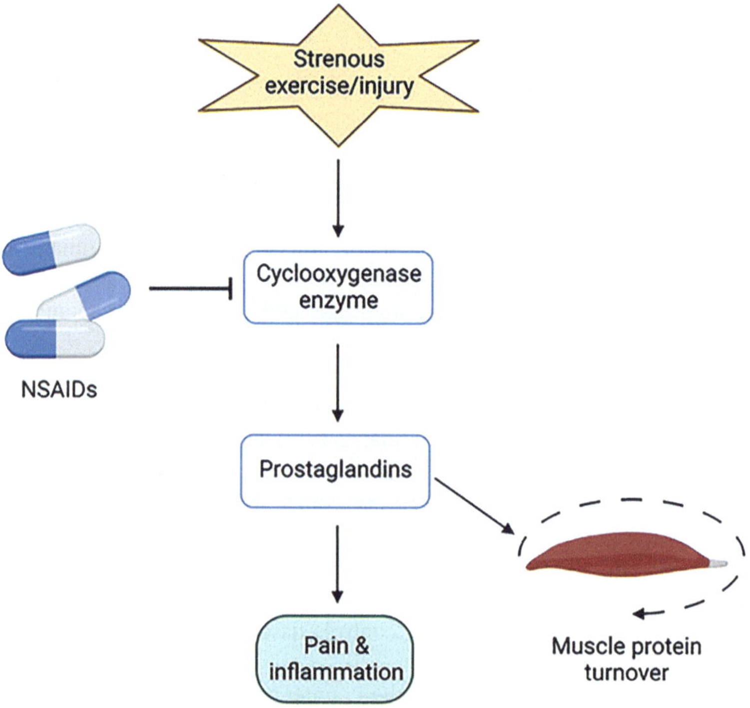

Fig. 18.1 Mechanisms of action of NSAIDs. Strenuous exercise or muscle injury leads to activation of the cyclooxygenase enzyme which promotes the formation of prostaglandins and the subsequent inflammatory response and (sometimes) pain. NSAIDs inhibit cyclooxygenase enzyme activity

took an NSAID at least once during the competition, and about seven players per national team did so before each match [4]. Other analgesics, such as paracetamol, were used far less frequently. College players, as well as elite athletes, triathletes, and marathon runners, were also said to take NSAIDs on a regular basis [5].

Although most sports organisations indicate that unwarranted medication should be avoided due to potential short- and long-term negative consequences, analgesics are still used extensively. NSAIDs are used by athletes to treat pain and inflammation caused by training, competition, or soft tissue injuries, as well as to give them a competitive edge and improve recovery. Numerous concerns have been raised regarding the impact of these medications on acute athletic performance and the impact on chronic training adaptations which are essential for long-term athletic development. This chapter will provide an overview of the current knowledge relating to the use of analgesics and NSAIDs in relation to their application to athletic performance. Specifically, the effects of these analogues on exercise performance, exercise recovery and recovery from injury are discussed.

Exercise Performance—Acute Effects of Analgesics and NSAIDs

An athlete's ability to tolerate pain may be a key element in high-intensity and/or longer distance athletic performance. The historic use of narcotic analgesics by athletes is supported by both anecdote and scientific data and has persisted long before the current era of doping. Tom Simpson, one of Britain's most successful cyclists, died on the slopes of Mont Ventoux during the 1967 Tour de France after using amphetamine and alcohol, which contributed to an inability to effectively regulate exercise tolerance and led to fatal heat exhaustion.

As for substances that are not banned, some pioneer studies examined the effects of acetylsalicylic acid (aspirin). Lisse et al. gave healthy male volunteers either a single dose of 650 mg aspirin or a placebo 30 minutes before a 3.2 km (2 mile) time trial in a double-blind crossover study [6]. They found no difference between the two groups. A few years later, Roi et al. conducted a single-blind, cross-over study to find out how aspirin affects maximal performance [7]. In this study, 18 young men underwent a maximal incremental bicycle ergometer test after receiving either 1000 mg of aspirin, 1000 mg of chewable buffered aspirin or a placebo. The results confirmed the earlier finding that aspirin dose had no effect on endurance performance on the measures of respiration, maximal oxygen uptake, heart rate or blood lactate.

In a more recent study, Mauger et al. had 13 skilled male cyclists complete a 10-mile (16.1 km) cycling time trial while taking paracetamol (3 × 500 mg) or a placebo [8]. In the paracetamol condition, participants cycled at a 4% higher average power output, had a higher heart rate and higher blood lactate levels, but there were no differences in the perception of effort. These results showed that pain perception influences athletic performance, and that paracetamol could improve athletic performance by reducing perceived exertion and increasing pain tolerance.

The same researchers also investigated whether paracetamol could enhance physical performance in the heat [9]. The results of the study showed that an acute dose of paracetamol (20 mg/kg lean body mass) enabled the participants—11 recreationally active men—to cycle an astonishing 4 minutes longer on average (+17%) in hot conditions (30 °C). Core, skin and body temperatures dropped noticeably, as did reported thermal stress. This suggests that paracetamol can reduce thermal stress during exercise and thus improve performance in hot environments.

Paracetamol has also been studied in relation to repeated sprint cycling. Nine recreationally active men completed 8 × 30 second sprints with 2 minutes of active rest in a randomised, cross-over design after taking either 1500 mg (3 × 500 mg) of paracetamol or a placebo [10]. After taking paracetamol, participants cycled with higher mean power in sprints 6, 7 and 8, and the relative decline in mean power was also lower. It was suggested that paracetamol improved performance by reducing self-perceived pain at a given work rate, allowing them to complete the exercise closer to their true performance limit.

It has been suggested that nociceptive afferent input and associated pain perception might alter both central and peripheral fatigue, contributing to the development of global neuromuscular fatigue. Morgan et al. asked 13 active men to perform 60 × 3 s of their maximal voluntary contractions followed by a 2 s passive recovery period to test this theory [11] following ingestion of 1000 mg of paracetamol or a placebo. They showed the average torque was higher after paracetamol than after the placebo suggesting that paracetamol use could improve repetitive maximal voluntary contractions by allowing better muscle activation and maintenance during exercise.

To provide a more quantitative summary of the results of taking paracetamol on performance, Grgic and Mikulic conducted a meta-analysis to examine the effects of paracetamol on endurance performance [12]. The meta-analysis included ten high-quality studies (n = 141 participants). All studies were randomised, double-blind and crossover designs. They showed placebo and paracetamol had similar effects on endurance performance. However, studies in which paracetamol was administered 45 to 60 minutes before exercise showed a small, but statistically significant ergogenic effect. Taking paracetamol increased performance to exhaustion in a subgroup analysis, but not time trial performance, which is more consistent with real-world athletic performance. It was concluded that paracetamol improves performance in endurance tests to exhaustion and when taken 45–60 minutes before exercise.

Overall, the research to date on acute physical performance in humans is certainly intriguing and suggests that analgesic drugs, in this case paracetamol, could improve both shorter- and longer-term performance parameters under certain circumstances through various physiological mechanisms related to the analgesic or antipyretic effects of these drugs. However, it is important to note that most of these studies were conducted with recreational exercisers and a meta-analysis found that the results on time trial performance were inconsistent. It is unclear whether highly trained elite athletes who are used to training at the highest level (both in terms of volume and intensity) also experience an increase in performance when using these drugs to enhance performance. In this context, it is worth mentioning a recent study that investigated the effect of caffeine and paracetamol on power output during a 6-minute cycling test in trained young men (average VO_{2max} 64 ml/kg/min) [13]. The results showed that paracetamol when co-ingested with caffeine or alone did not provide a performance improvement compared to caffeine during a 6-min cycling ergometer performance test. These results highlight the potential difference between recreational athletes and well-trained athletes, and probably also the difference between testing to exhaustion versus time trials.

The effectiveness of these interventions in a chronic training paradigm is also uncertain. In particular, one might wonder whether repeated use of pain medications could lead to improved tolerance to training loads and thus better adaptation. This is a fascinating but morally questionable topic, not least because of the potential ergogenic effect, but also because of the potentially harmful side effects of prolonged pharmacological use.

Recovery from Strenuous Exercise, Muscle Damage and Injury

Exercise-Induced Muscle Damage and the Use of NSAIDs for Recovery

This section focuses on the precipitation of exercise-induced muscle damage (EIMD) and how NSAIDs might be applied to combat the negative effects of muscle damage based on the currently available literature. Firstly, the magnitude of muscle damage can vary widely. In most cases, EIMD occurs after strenuous exercise that has an eccentric component, i.e., the muscle is under tension whilst actively lengthening. Simple examples include downhill running, rapid changes of direction and decelerations, repetitive submaximal eccentric loading over a long duration (such as a marathon or a fell run), heavy resistance training and so on. The magnitude of damage is often exacerbated when the exercise is unaccustomed and can be experienced by all people, including elite athletes. For example, the first spring venture out into the garden can result in muscle damage, which often manifests itself as stiff and sore muscles in the days after the event.

The nature of EIMD is biphasic, as it is initiated by mechanical loading, which can cause structural damage to muscle tissue and adjacent connective tissue, which is then followed by an inflammatory phase [14]. NSAIDs are thought to be most beneficial in the post-exercise period when inflammation has occurred [15]. The signs and symptoms of EIMD are generally quite consistent between individuals, but the magnitude of that response can vary greatly from person to person. The damage response also depends on the intensity and duration of the exercise, the speed of muscle contraction [16, 17] and the length of the muscle during contraction (longer muscle lengths tend to cause more damage [18]). The time course of these indices does not correlate strongly with each other, but they all persist for several days after the damaging bout of exercise [19]. For athletes, this usually manifests itself in an immediate loss of strength and power, increased muscle soreness that becomes apparent in the hours following exercise and persists for several days, and increased stiffness and swelling of the exercised limb, which negatively affects flexibility and range of motion. In addition, there is often an increase in intramuscular cytosolic and structural proteins in the circulating blood, such as creatine kinase (CK), myoglobin, troponin and myosin. Blood indices are indicative of the loss of cell membrane integrity and not necessarily of the magnitude of damage (unless levels are very high, which could indicate exercise-induced rhabdomyolysis). Often overlooked is the notion that reaction time and joint position sense can also be impaired over several days [20] and increase the risk of injury if not fully recovered.

From an athlete monitoring perspective, the most pragmatic indices to ascertain are function and soreness, as they are easily incorporated into daily monitoring, simple for science and medical teams to interpret, and relatively inexpensive to use. Considering that the negative effects of strenuous exercise can last for several days, and that the timing of competitions does not allow for full recovery from the previous competition or training, it is easy to understand why the prevalence of NSAID use is so high, especially among athletes and exercising populations who engage in exercise that could result in muscle damage [21–23].

The evidence for the use of NSAIDs in reducing the signs and symptoms of EIMD is inconclusive. This is complicated by different experimental designs, doses and the use of rodent and animal models. This section focuses on the use of NSAIDs in humans. The NSAIDs most commonly used in experimental models include ibuprofen, ketoprofen, naproxen and diclofenac, some of which are available over the counter and have been shown to be commonly used by athletes [23] to treat inflammation-related conditions, which include EIMD and muscle soreness [14].

Administration of NSAIDs has been used for prophylaxis and post-exercise treatment [24]. Prophylactic administration of 25 mg and 100 mg of ketoprofen before maximal eccentric exercise reduced muscle soreness by 19% and 10%, respectively, and improved muscle function by 9% and 16% [25]. In addition, taking 150 mg of diclofenac 2 weeks before and 2 weeks after stepping exercise showed a reduction in inflammation and a reduced appearance of CK. Similarly, Pizza et al. [26] showed that 2400 mg of ibuprofen reduced the appearance of CK in the blood. It is important to note, however, that these studies are of limited value as the athlete's muscle soreness was only slightly altered and there is limited evidence of a return of function.

Other studies examining physical performance showed unclear benefits. Tokmakidis et al. [27] administered 400 mg of ibuprofen every 8 hours for 48 hours after an eccentric exercise bout. Although there was less muscle soreness and CK on the day after the exercise bout, there were no effects on maximum strength, vertical jump performance or range of motion. Donnelly et al. [28] reported that a dose of 1200 mg ibuprofen given before a 45-minute downhill run, along with another dose of 600 mg every 6 hours for 72 hours after exercise, had no effect on muscle soreness or strength. Peterson et al. [29]. administered maximum over-the-counter doses of ibuprofen (1200 mg/day) and paracetamol (4000 mg/day) after a damaging exercise with no effect on the inflammatory response 24 hours after exercise.

Collectively, the current evidence for the use of NSAIDs to prevent and/or alleviate the signs and symptoms associated with EIMD is inconclusive. The use of higher doses of NSAIDs in future studies has been advocated by some [25], although caution is advised as habitual or chronic use of NSAIDs can have adverse effects leading to further medical complications. Furthermore, inflammation is the body's natural response to EIMD, and the use of pharmacological agents could inhibit protein synthesis, thereby attenuating the cycle of skeletal muscle damage, repair, and adaptation following eccentric exercise. Tissue protein synthesis following high-intensity eccentric exercise may be attenuated by the administration of over-the-counter doses of ibuprofen and paracetamol [30]. Although the longer-term effects of NSAIDs are underresearched, muscle regeneration and hypertrophy following NSAID administration has been demonstrated in animal models [31]. Given the effects of NSAIDs on EIMD, which are equivocal at best, the lack of evidence for return of performance and the potential negative long-term effects of administration, their use is not recommended as an effective strategy to prevent or treat the symptoms of EIMD.

NSAIDs and Recovery After Injury

Muscle, tendon, and ligament injuries can occur in a number of different situations, including explosive muscle actions, direct trauma to the muscles, or after long-term overuse. Therefore, it is difficult to extrapolate results from experimental models of muscle injury in which the role of NSAIDs has been studied to the rehabilitation of injuries that have occurred in real sports situations. It is also unusual for experimental studies to examine the entire rehabilitation process, i.e., from the first days after injury to full recovery of muscle function and return of the athlete to competition.

The muscle's own dormant stem cells, called satellite cells, play an important role in the muscle's repair process after injury. The satellite cells work together with other cell populations such as macrophages, endothelial cells and fibroblasts as part of the healing process [32, 33]. Several experimental models have shown that blocking the function of satellite cells negatively affects muscle repair processes [5, 34], and suppress collagen synthesis in tendon tissue [35]. This, of course, raises the obvious question of whether NSAIDs have a negative effect on muscle repair processes after muscle injury. Despite this, there are studies that have shown that NSAIDs can accelerate recovery and reduce loss of function during the initial phase of the rehabilitation process after muscle injury. In 2016, a team of researchers from Denmark investigated the effects of NSAIDs (1200 mg ibuprofen per day) on recovery from electrical stimulation-induced muscle injury in volunteers [36]. Contrary to the hypothesis of a negative effect, the results showed that NSAIDs accelerated satellite cell activation 2 days after injury and the total satellite cell pool was larger under NSAIDs than under placebo after 7 days. In addition, muscle function recovered better after 30 days under NSAIDs than under placebo. Although this study suggested a beneficial effect of NSAIDs, it is difficult to draw direct parallels with "real" muscle injuries that occur in the context of sports. In fact, there are several studies that have followed the muscle repair process after injury over a longer time. They have shown that what appears to be a positive effect at the beginning of the rehabilitation process is not necessarily maintained over time and may even be negative when NSAIDs are used [37].

Exercise-Induced Adaptations

Put simply, exercise is the stimulus for adaptation that is mediated by a cascade of complex cellular events that include (but are not limited to) the activation of inflammatory cells, oxidative stress, and satellite cells (Fig. 18.2). If NSAIDs are used to reduce inflammation, some of the adaptive stimuli could also be reduced, thereby attenuating the adaptive potential.

The ability of analgesics and NSAIDs to alter acute cellular responses to exercise is undoubted, as studies have shown effects on protein synthesis and the molecular response controlling muscle adaptation [2, 5]. The satellite cells already mentioned are not only important in the case of muscle damage, but also for the normal adaptation of the muscle to training, particularly following strength training. The satellite cells are thought to help increase the number of nuclei in the muscle fibres as the muscle gains size through training. As mentioned earlier, NSAIDs can inhibit satellite cells, which then has a negative effect on training adaptation. In addition, a 2002 study showed that the muscle's protein synthesis response, the process by which new muscle proteins are formed after exercise, can be inhibited by non-prescription NSAIDs [30].

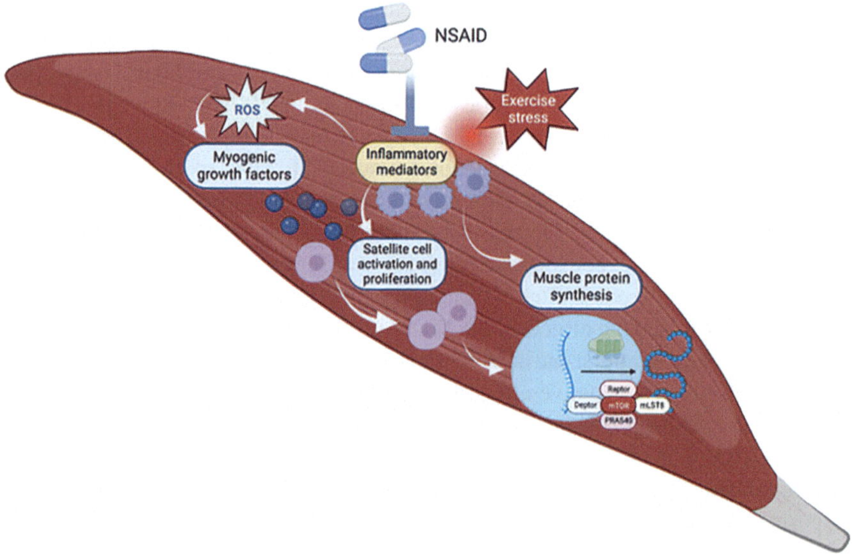

Fig. 18.2 Schematic representation of the exercise-induced response leading to increased protein synthesis and hence adaptative response to exercise

The effects of NSAIDs on acute exercise responses have little meaning for an athlete if the effects do not translate into altered long-term training adaptations. In investigating chronic training adaptations, three studies examined the effects on older people (60–80 years old) [38–40], while one study examined patients with knee osteoarthritis aged 50–70 years [41]. It is therefore highly questionable whether conclusions from these studies can be translated to training adaptations in athletes.

However, two studies have examined how NSAIDs affect specific training adaptations in young people. The first study found that taking 400 mg of ibuprofen immediately after each exercise session had no effect on muscle growth, strength gain or perception of muscle soreness after 6 weeks of strength training in young volunteers [42]. Six sets of biceps curls were performed twice per week during the training period. Medication was taken only on training days and totalled 800–1200 mg per week. The lack of effect on training adaptations in this study is likely due to the lower total dose compared to previous acute studies where the standard dose was the maximum over-the-counter dose of 1200 mg ibuprofen. However, it could be argued that the administration of a single dose taken near training sessions may reflect what many athletes would do in real life. One other point is that these participants are not well trained and therefore the potential for a positive adaptation is very large and so the response to a training stimulus will be profound, regardless of a modest dose of NSAIDs.

The second study involved young, moderately active men and women in a single-blind, randomised controlled trial [43]. During 8 weeks of resistance training of the knee extensor muscles, subjects were randomly assigned to either an experimental group receiving 1200 mg ibuprofen (n = 15) or a control group receiving a very low dose of 75 mg acetylsalicylic acid (n = 16). The results showed that quadriceps muscle volume was significantly reduced in the ibuprofen group compared to the control group receiving low-dose aspirin. The adverse effects of higher doses of ibuprofen were evident in some strength measurements, but not all. Overall, these results showed that higher doses of NSAIDs reduced the adaptations that younger people experience as a result of resistance training, which is arguably closer to an athletic population than the studies on older adults mentioned above.

In summary, there is insufficient information to draw firm conclusions, as only two studies have investigated how NSAIDs affect exercise adaptation in healthy young volunteers. However, it appears that higher doses of NSAIDs, but not lower doses, are able to reduce some training adaptations following resistance training. Although no studies have examined how paracetamol affects training adaptations in young people, a negative result might be expected given that paracetamol has been shown to have similar effects as NSAIDs on acute molecular processes. To date, no studies have been conducted on adaptations related to endurance training or in elite athletes. Given the high prevalence of use and the great interest in the possible attenuation of exercise-induced adaptation, there is an urgent need to understand the effects of analgesic use in well-trained cohorts participating in strength, endurance, and concurrent training paradigms.

Indications/Contraindications, Safety and Ethical Considerations

The use of pharmacological techniques, in this case to modulate athletic performance through pain-relieving drugs, is a highly controversial issue that raises ethical concerns about safety, possible side effects and problems with using drugs for competitive advantage, which is tantamount to doping. It also raises questions from a practical perspective. Anecdotal reports suggest that many marathon runners take painkillers before and during the race in the hope that it will mitigate the impact of the race. However, taking painkillers before the race does not protect against discomfort during or after the marathon [44].

At the FIFA World Cup, the reported use of NSAIDs by players was >30% every match and as high as 54% over the course of the tournament [22]. About a third of the NSAIDs prescribed were taken prophylactically, which is worrying given the lack of evidence to support such a strategy. The use of these drugs seems to be almost habitual, with similar trends in use at U17 and U20 World Cups [4].

In contact sports, the trends are even more worrying. In Collegiate American Football >95% of players had taken, or were using, NSAIDs. Prevalence was highest among linemen and line-backers, who are likely to have more contact than other positions [45]. In a recent survey of Division 1 Collegiate American Football players, 77% of the 226 players surveyed reported that they started taking NSAIDs in middle school [23]. The same paper showed that heavier players tended to use NSAIDs daily and had elevated resting blood pressure that was close to pre-hypertensive levels. It is not possible to know why these athletes were prescribed these drugs; some might be used to treat injuries for good reason, but it is very likely that many use them for acute pain management and to speed up recovery after a game and training. The obvious ethical and moral considerations of athletes habitually taking drugs at a young age is an emotive issue, and worryingly, in many cases the data simply does not support such a strategy.

Overall, the use of analgesics to treat exercise-induced pain appears to have relatively few benefits and a significant risk of unfavourable side effects, especially when higher doses and long duration are used. Although the available data suggest a reduced exercise response, there is evidence that paracetamol might improve performance under some circumstances. As a result, there are publications advocating the inclusion of paracetamol in the WADA Doping List [46]. It is difficult to draw a clear line on what substances should be allowed and those that should be banned, as there are so many other drugs and substances that can have a pain-relieving effect. This is especially true when one considers that there could well be a medical justification for the use of certain substances.

Regardless of the doping issue, it seems logical that athletes who use painkillers are more likely to sustain injuries and tissue damage. NSAIDs can cause more severe bleeding after trauma as they affect platelet function and aggregation. This is a real concern for athletes participating in impact sports or other activities where the likelihood of contact is high, thereby increasing the risk of acute trauma or soft tissue injury. Although the exact effect of the timing of NSAID use on muscle repair is uncertain, athletes could return to play or competition too soon if the pain response is continuously suppressed, increasing the risk of injury or recurrence of previous injuries.

It is also clear that pain-relieving drugs can have other side effects. Paracetamol is associated with liver toxicity [47]. Taking NSAIDs can increase the risk of bleeding and gastrointestinal problems. For example, there is a 3–5 fold increased risk of gastric ulcers, intestinal perforation, constipation and bleeding when taking NSAIDs [48]. In addition, the risk of longer-term cardiovascular and cerebrovascular events is increased by one third [49]. The true extent of the potential side effects and problems arising from the use of painkillers and NSAIDs is not yet fully known, but there is evidence in other areas of research of the potential issues that could be problematic for our athletes. In the US Army, NSAID use during basic combat training is high and the risk of stress fractures is increased more than fivefold [50]. Although not directly applicable to athletes, these data illustrate the problematic nature of NSAIDs in people undergoing strenuous training, which is not dissimilar to a regime that athletes might be exposed to.

Lastly, NSAIDs can increase the overall mortality rate and the risk of heart attacks, sudden cardiac arrests and other cardiac events. The dangers vary between painkillers; for example, the increased risk of death appears to be much greater with diclofenac than with aspirin or paracetamol. Although the overall risk is still very low, in absolute terms in healthy people, the risk increases with higher doses and/or with prolonged use of these pharmaceutical interventions. While there are few absolute contraindications to taking NSAIDs, for some conditions it is advisable to consult a doctor (Table 18.2).

What We Know and Don't Know—Future Directions for Research

Despite the large number of studies that have been conducted so far, there are still many unanswered questions and contradictory results. Much of the lack of agreement is due to differences in exercise, the population studied, and the dose of the

Table 18.2 Conditions that could be potential contraindications to use NSAIDs

Special considerations
Are pregnant or trying for a baby
Have had an allergic reaction to NSAIDs in the past
Have had stomach ulcers in the past
Have had problems with heart, liver, kidneys, or blood pressure
Are taking other medicines

drug. With so many variations, it is no wonder that there are no definitive recommendations. There is an urgent need to understand how analgesic and anti-inflammatory drugs can alter athletic performance, recovery, and training adaptations. Consequently, athletes have yet to receive a clear message on how to use these substances while minimising side effects.

Although some recent research suggested that taking paracetamol for individual sporting events increased performance, there is no evidence that this also applies to competitive performance in highly trained athletes. Studies examining the effect of analgesics on applied athletic performance in elite athletes are needed to determine transferability to populations of athletes. In addition, it is important to recognise that little is known about the specific processes by which analgesics or anti-inflammatory drugs influence the adaptive training response, particularly in highly trained men and women. As can be seen from this research, the dose and duration of drug treatment are likely to be important factors. Another important factor is that very little is known about the interaction of these analogues with female biology, particularly with oral contraceptives and the menstrual cycle, where inflammatory indices may be elevated. As female participation continues to increase and is recognised at the highest levels, this is an area of high research priority to provide athletes, coaches, and practitioners with an evidence base for informed interventions.

Given the alarmingly high use of analgesic and anti-inflammatory drugs in professional sport, we recommend greater awareness of the potentially harmful effects of these substances on health. It is also important to emphasise that the use of NSAIDs and possibly even paracetamol can impair chronic training adaptations and hinder longer-term athletic development. As there is currently some circumstantial evidence to suggest that these drugs might enhance athletic performance in certain circumstances, it is important that we continue to explore how analgesic medications affect performance and training adaptations in elite athletes, and that the moral dilemmas associated with such activities in sport are explored publicly and transparently. Given that over-the-counter painkillers work well when used for their intended medical purpose (i.e., to treat acute pain symptoms), there is definitely still a place and appropriate time for elite athletes to use these therapies. However, before using painkillers and NSAIDs frequently or habitually in the context of training or competition, all athletes should carefully consider the risks and benefits with their coaches, physicians, and other professional staff (Fig. 18.3).

Conclusions

Non-steroidal antiinflammatories and analgesics are commonly used interventions to provide performance enhancement and also to facilitate accelerated recovery in athletic populations. The research suggested there might be small benefits from using analgesics, but the use of analgesics and non-steroidal anti-inflammatories is far from conclusive. There are some suggestions that markers of damage and recovery might be reduced although it seems there is little or no effect on the return of

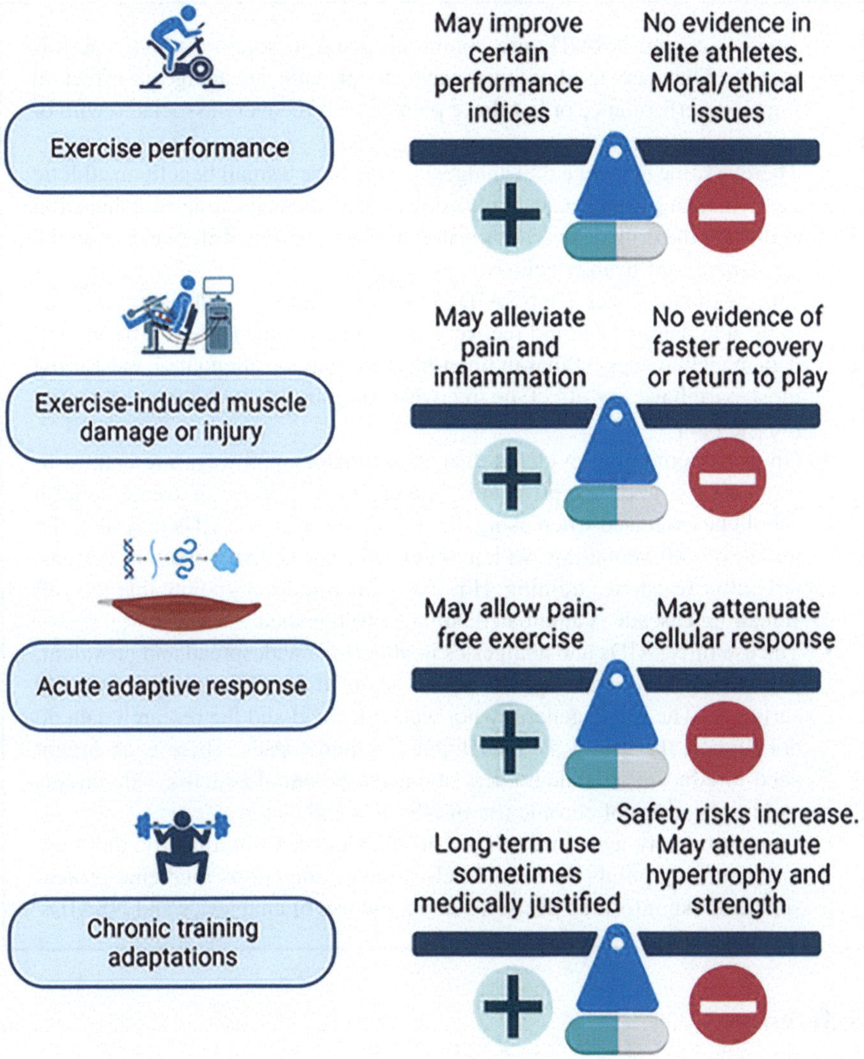

Fig. 18.3 Cost-benefit considerations for the use of analgesic and anti-inflammatory drugs in sport

muscle function or in reducing muscle soreness following strenuose exercise. Given the potential importance of the inflammatory pathway in developing skeletal muscle adaptations, particularly resistance training, non-steroid anti-inflammatories might not be prudent to use, particularly when training is targeting an adaptative response. Notwithstanding, it is important to note that these interventions do not extinguish the inflammatory cascade, rather they reduce the inflammatory response. Athletes, coaches, and practitioners should be mindful of using these analogues in light of the limited literature and the modest benefit which should be considered amongst any cost to the athlete.

- Analgesics and NSAIDs are commonly used to support athletic performance. They are used in two ways: to provide an analgesic effect to improve performance or to relieve pain and aid recovery associated with or following performance or training
- There is some evidence that analgesics may have a small benefit on athletic performance proxies, especially in tests that measure time to exhaustion and when the drug dose is administered 45–60 minutes before exercise, but the data are far from conclusive
- The use of analgesics and NSAIDs as an intervention to reduce the response to muscle damage and aid recovery is extremely inconsistent. The limited data available suggest that indirect markers may be attenuated, but for the most part have no effect on recovery of muscle function or reducing muscle pain
- Given the emerging evidence that inflammatory pathways are central to promoting a good adaptive environment post-exercise, extreme caution should be exercised when using these interventions. NSAIDs may alter the milieu of cell signalling, which could reduce adaptive potential, particularly after resistance training. However, it is important to note that the cell signalling cascade is attenuated but not extinguished
- The use of NSAIDs and analgesics in athletes is widespread and prevalent, particularly in contact sports, which seem to become habitual from an early age. The use is generally not well-informed and the research data do not support the use of such analogues in many cases. There is an urgent need to educate all stakeholders about the potential benefits, side effects and implications of chronic use of NSAIDs and analgesics
- There are many gaps in the body of knowledge. Consequently, there are many starting points that could help athletes and sports medicine professionals make informed decisions about the use of analgesics and NSAIDs

References

1. Graham GG, Davies MJ, Day RO, Mohamudally A, Scott KF. The modern pharmacology of paracetamol: therapeutic actions, mechanism of action, metabolism, toxicity and recent pharmacological findings. Inflammopharmacology. 2013;21(3):201–32. https://doi.org/10.1007/s10787-013-0172-x.
2. Trappe TA, Liu SZ. Effects of prostaglandins and COX-inhibiting drugs on skeletal muscle adaptations to exercise. J Appl Physiol 1985. 2013;115(6):909–19. https://doi.org/10.1152/japplphysiol.00061.2013.
3. Corrigan B, Kazlauskas R. Medication use in athletes selected for doping control at the Sydney Olympics (2000). Clin J Sport Med. 2003;13(1):33–40.
4. Tscholl PM, Vaso M, Weber A, Dvorak J. High prevalence of medication use in professional football tournaments including the World Cups between 2002 and 2014: a narrative review with a focus on NSAIDs. Br J Sports Med. 2015;49(9):580–2. https://doi.org/10.1136/bjsports-2015-094784.

5. Lundberg TR, Howatson G. Analgesic and anti-inflammatory drugs in sports: implications for exercise performance and training adaptations. Scand J Med Sci Sports. 2018;28(11) https://doi.org/10.1111/sms.13275.
6. Lisse JR, MacDonald K, Thurmond-Anderle ME, Fuchs JE Jr. A double-blind, placebo-controlled study of acetylsalicylic acid (ASA) in trained runners. J Sports Med Phys Fit. 1991;31(4):561–4.
7. Roi GS, Garagiola U, Verza P, et al. Aspirin does not affect exercise performance. Int J Sports Med. 1994;15(5):224–7. https://doi.org/10.1055/s-2007-1021050.
8. Mauger AR, Jones AM, Williams CA. Influence of acetaminophen on performance during time trial cycling. J Appl Physiol 1985. 2010;108(1):98–104. https://doi.org/10.1152/japplphysiol.00761.2009.
9. Mauger AR, Taylor L, Harding C, Wright B, Foster J, Castle PC. Acute acetaminophen (paracetamol) ingestion improves time to exhaustion during exercise in the heat. Exp Physiol. 2014;99(1):164–71. https://doi.org/10.1113/expphysiol.2013.075275.
10. Foster J, Taylor L, Chrismas BCR, Watkins SL, Mauger AR. The influence of acetaminophen on repeated sprint cycling performance. Eur J Appl Physiol. 2014;114(1):41–8. https://doi.org/10.1007/s00421-013-2746-0.
11. Morgan PT, Bowtell JL, Vanhatalo A, Jones AM, Bailey SJ. Acute acetaminophen ingestion improves performance and muscle activation during maximal intermittent knee extensor exercise. Eur J Appl Physiol. 2018;118(3):595–605. https://doi.org/10.1007/s00421-017-3794-7.
12. Grgic J, Mikulic P. Effects of paracetamol (acetaminophen) ingestion on endurance performance: a systematic review and meta-analysis. Sports Basel Switz. 2021;9(9):126. https://doi.org/10.3390/sports9090126.
13. Jessen S, Eibye K, Christensen PM, Hostrup M, Bangsbo J. No additive effect of acetaminophen when co-ingested with caffeine on cycling performance in well-trained young men. J Appl Physiol Bethesda Md 1985. 2021;131(1):238–49. https://doi.org/10.1152/japplphysiol.00108.2021.
14. Howatson G, van Someren KA. The prevention and treatment of exercise-induced muscle damage. Sports Med. 2008;38(6):483–503. https://doi.org/10.2165/00007256-200838060-00004.
15. Schoenfeld BJ. The use of nonsteroidal anti-inflammatory drugs for exercise-induced muscle damage: implications for skeletal muscle development. Sports Med. 2012;42(12):1017–28. https://doi.org/10.2165/11635190-000000000-00000.
16. Chapman D, Newton M, Sacco P, Nosaka K. Greater muscle damage induced by fast versus slow velocity eccentric exercise. Int J Sports Med. 2006;27(8):591–8. https://doi.org/10.1055/s-2005-865920.
17. Chapman DW, Newton M, McGuigan M, Nosaka K. Effect of lengthening contraction velocity on muscle damage of the elbow flexors. Med Sci Sports Exerc. 2008;40(5):926–33. https://doi.org/10.1249/MSS.0b013e318168c82d.
18. Nosaka K, Sakamoto K. Effect of elbow joint angle on the magnitude of muscle damage to the elbow flexors. Med Sci Sports Exerc. 2001;33(1):22–9. https://doi.org/10.1097/00005768-200101000-00005.
19. Nosaka K, Newton M, Sacco P. Delayed-onset muscle soreness does not reflect the magnitude of eccentric exercise-induced muscle damage. Scand J Med Sci Sports. 2002;12(6):337–46. https://doi.org/10.1034/j.1600-0838.2002.10178.x.
20. Paschalis V, Nikolaidis MG, Theodorou AA, Giakas G, Jamurtas AZ, Koutedakis Y. Eccentric exercise affects the upper limbs more than the lower limbs in position sense and reaction angle. J Sports Sci. 2010;28(1):33–43. https://doi.org/10.1080/02640410903334764.
21. Warner DC, Schnepf G, Barrett MS, Dian D, Swigonski NL. Prevalence, attitudes, and behaviors related to the use of nonsteroidal anti-inflammatory drugs (NSAIDs) in student athletes. J Adolesc Health. 2002;30(3):150–3.
22. Vaso M, Weber A, Tscholl PM, Junge A, Dvorak J. Use and abuse of medication during 2014 FIFA World Cup Brazil: a retrospective survey. BMJ Open. 2015;5(9):e007608. https://doi.org/10.1136/bmjopen-2015-007608.

23. Tso J, Hollowed C, Liu C, et al. Nonsteroidal anti-inflammatory drugs and cardiovascular risk in American football. Med Sci Sports Exerc. 2020;52(12):2522–8. https://doi.org/10.1249/MSS.0000000000002404.
24. Baldwin LA. Use of nonsteroidal anti-inflammatory drugs following exercise-induced muscle injury. Sports Med. 2003;33(3):177–85.
25. Sayers SP, Knight CA, Clarkson PM, Van Wegen EH, Kamen G. Effect of ketoprofen on muscle function and sEMG activity after eccentric exercise. Med Sci Sports Exerc. 2001;33(5):702–10. https://doi.org/10.1097/00005768-200105000-00004.
26. Pizza FX, Cavender D, Stockard A, Baylies H, Beighle A. Anti-inflammatory doses of ibuprofen: effect on neutrophils and exercise-induced muscle injury. Int J Sports Med. 1999;20(2):98–102. https://doi.org/10.1055/s-2007-971100.
27. Tokmakidis SP, Kokkinidis EA, Smilios I, Douda H. The effects of ibuprofen on delayed muscle soreness and muscular performance after eccentric exercise. J Strength Cond Res. 2003;17(1):53–9. https://doi.org/10.1519/1533-4287(2003)017<0053:teoiod>2.0.co;2.
28. Donnelly AE, Maughan RJ, Whiting PH. Effects of ibuprofen on exercise-induced muscle soreness and indices of muscle damage. Br J Sports Med. 1990;24(3):191–5. https://doi.org/10.1136/bjsm.24.3.191.
29. Peterson JM, Trappe TA, Mylona E, et al. Ibuprofen and acetaminophen: effect on muscle inflammation after eccentric exercise. Med Sci Sports Exerc. 2003;35(6):892–6. https://doi.org/10.1249/01.MSS.0000069917.51742.98.
30. Trappe TA, White F, Lambert CP, Cesar D, Hellerstein M, Evans WJ. Effect of ibuprofen and acetaminophen on postexercise muscle protein synthesis. Am J Physiol Endocrinol Metab. 2002;282(3):E551–6. https://doi.org/10.1152/ajpendo.00352.2001.
31. Soltow QA, Betters JL, Sellman JE, Lira VA, Long JH, Criswell DS. Ibuprofen inhibits skeletal muscle hypertrophy in rats. Med Sci Sports Exerc. 2006;38(5):840–6. https://doi.org/10.1249/01.mss.0000218142.98704.66.
32. Franco I, Fernandez-Gonzalo R, Vrtačnik P, Lundberg TR, Eriksson M, Gustafsson T. Healthy skeletal muscle aging: the role of satellite cells, somatic mutations and exercise. In: International review of cell and molecular biology, vol. 346; 2019. p. 157–200. https://doi.org/10.1016/bs.ircmb.2019.03.003.
33. Bazgir B, Fathi R, Valojerdi MR, Mozdziak P, Asgari A. Satellite cells contribution to exercise mediated muscle hypertrophy and repair. Cell J. 2016;18(4) https://doi.org/10.22074/CELLJ.2016.4714.
34. Mikkelsen UR, Langberg H, Helmark IC, et al. Local NSAID infusion inhibits satellite cell proliferation in human skeletal muscle after eccentric exercise. J Appl Physiol 1985. 2009;107(5):1600–11. https://doi.org/10.1152/japplphysiol.00707.2009.
35. Christensen B, Dandanell S, Kjaer M, Langberg H. Effect of anti-inflammatory medication on the running-induced rise in patella tendon collagen synthesis in humans. J Appl Physiol Bethesda Md 1985. 2011;110(1):137–41. https://doi.org/10.1152/japplphysiol.00942.2010.
36. Mackey AL, Rasmussen LK, Kadi F, et al. Activation of satellite cells and the regeneration of human skeletal muscle are expedited by ingestion of nonsteroidal anti-inflammatory medication. FASEB J. 2016;30(6):2266–81. https://doi.org/10.1096/fj.201500198R.
37. Mackey AL, Mikkelsen UR, Magnusson SP, Kjaer M. Rehabilitation of muscle after injury - the role of anti-inflammatory drugs. Scand J Med Sci Sports. 2012;22(4):e8–14. https://doi.org/10.1111/j.1600-0838.2012.01463.x.
38. Trappe TA, Carroll CC, Dickinson JM, et al. Influence of acetaminophen and ibuprofen on skeletal muscle adaptations to resistance exercise in older adults. Am J Physiol Regul Integr Comp Physiol. 2011;300(3):R655–62. https://doi.org/10.1152/ajpregu.00611.2010.
39. Jankowski CM, Shea K, Barry DW, et al. Timing of Ibuprofen use and musculoskeletal adaptations to exercise training in older adults. Bone Rep. 2015;1:1–8. https://doi.org/10.1016/j.bonr.2014.10.003.

40. Jankowski CM, Gozansky WS, MacLean PS, et al. N-acetyl-4-aminophenol and musculoskeletal adaptations to resistance exercise training. Eur J Appl Physiol. 2013;113(5):1127–36. https://doi.org/10.1007/s00421-012-2529-z.
41. Petersen SG, Beyer N, Hansen M, et al. Nonsteroidal anti-inflammatory drug or glucosamine reduced pain and improved muscle strength with resistance training in a randomized controlled trial of knee osteoarthritis patients. Arch Phys Med Rehabil. 2011;92(8):1185–93. https://doi.org/10.1016/j.apmr.2011.03.009.
42. Krentz JR, Quest B, Farthing JP, Quest DW, Chilibeck PD. The effects of ibuprofen on muscle hypertrophy, strength, and soreness during resistance training. Appl Physiol Nutr Metab. 2008;33(3):470–5. https://doi.org/10.1139/H08-019.
43. Lilja M, Mandić M, Apró W, et al. High doses of anti-inflammatory drugs compromise muscle strength and hypertrophic adaptations to resistance training in young adults. Acta Physiol. 2018;222(2) https://doi.org/10.1111/apha.12948.
44. Kuster M, Renner B, Oppel P, et al. Consumption of analgesics before a marathon and the incidence of cardiovascular, gastrointestinal and renal problems: a cohort study. BMJ Open. 2013;3(4):1–9. https://doi.org/10.1136/bmjopen-2012-002090.
45. Holmes N, Cronholm PF, Duffy AJ 3rd, Webner D. Nonsteroidal anti-inflammatory drug use in collegiate football players. Clin J Sport Med. 2013;23(4):283–6. https://doi.org/10.1097/JSM.0b013e318286d0fa.
46. Lippi G, Sanchis-Gomar F. Acetaminophen and sport performance: doping or what? Eur J Appl Physiol. 2014;114(4):881–2. https://doi.org/10.1007/s00421-014-2852-7.
47. Ryder SD, Beckingham IJ. ABC of diseases of liver, pancreas, and biliary system. Other causes of parenchymal liver disease. BMJ. 2001;322(7281):290–2.
48. Castellsague J, Riera-Guardia N, Calingaert B, et al. Individual NSAIDs and upper gastrointestinal complications: a systematic review and meta-analysis of observational studies (the SOS project). Drug Saf. 2012;35(12):1127–46. https://doi.org/10.2165/11633470-000000000-00000.
49. Coxib and traditional NSAID Trialists' (CNT) Collaboration, Bhala N, Emberson J, et al. Vascular and upper gastrointestinal effects of non-steroidal anti-inflammatory drugs: meta-analyses of individual participant data from randomised trials. Lancet Lond Engl. 2013;382(9894):769–79. https://doi.org/10.1016/S0140-6736(13)60900-9.
50. Hughes JM, McKinnon CJ, Taylor KM, et al. Nonsteroidal anti-inflammatory drug prescriptions are associated with increased stress fracture diagnosis in the US Army population. J Bone Miner Res. 2019;34(3):429–36. https://doi.org/10.1002/jbmr.3616.

Chapter 19
Stretching to Improve Muscle Performance, Recovery, and Health

Olyvia Donti and Gregory C. Bogdanis

Introduction

Stretching of human skeletal muscles is defined as a series of manipulations of joint positions aiming to increase muscle-tendon and tissue extensibility and maximal joint range of motion (flexibility) [1, 2]. Increased joint range of motion (ROM) following long-term stretching interventions may be explained by increased stretch tolerance [1, 3] and/or changes in tissue mechanical properties [4, 5] while some recent studies also report changes in muscle morphology [6, 7]. Exercise or training sessions generally begin with a warm-up, which is followed by the main workout, and end with a cool-down phase that includes a progressive reduction of effort and intensity, as well as stretching [8, 9].The use of stretching in the warm-up phase has been extensively examined in the last decades, with conflicting results regarding its positive or negative effects [10–12]. However, recent studies show that a combination of static and dynamic stretching with general and specific short dynamic efforts should be an integral part of a warm-up aiming to optimize performance [13, 14]. The use of stretching in the post-exercise phase has also been examined in several studies, while stretching is included in the guidelines of the American College of Sports Medicine [8] and the American Heart Association [15] to promote recovery. The main aims of stretching exercises applied after training include the re-establishment of ROM and the reduction of muscle tightness and soreness, as well as the improvement of flexibility of prime mover muscles, and other secondary muscles involved in the general musculoskeletal health [8, 16, 17].

O. Donti · G. C. Bogdanis (✉)
Physical Education and Sport Science, National and Kapodistrian University of Athens, Athens, Greece
e-mail: gregbogdanis@gmail.com

N. C. Apostolopoulos et al. (eds.), *Fundamentals of Recovery, Regeneration, and Adaptation to Exercise Stress: An Integrated Approach*,
https://doi.org/10.1007/978-3-031-44270-4_19

Several different post-exercise stretching methods are typically used in athletic training, such as static stretching, passive stretching, and proprioceptive neuromuscular facilitation (PNF) stretching [2, 18]. Notably, despite the wide adoption of stretching in post-exercise and rehabilitation protocols, evidence on the effectiveness of stretching on short-term recovery is heterogenous and contradicting, with a number of studies indicating that post-exercise stretching is not effective for improving recovery [19, 20]. Moreover, several studies reported that stretching may be ineffective in relieving perceived muscle pain or in reducing DOMS [21, 22]. However, before reaching a conclusion regarding the short-term effects of post-exercise stretching, one has to consider several factors influencing the effectiveness of stretching as a recovery mode, such as the characteristics of the preceding exercise, the level of muscle fatigue and/or muscle damage [23], the type and modality of stretching interventions [8], the training level, the level of muscle strength of the populations examined [16], and the combination of stretching with other recovery techniques, among others [8, 24, 25]. Short-term effects of post-exercise stretching on recovery should always be examined in relation to longer-term adaptations [8].

This chapter will provide an overview of the current knowledge relating to the use and effectiveness of stretching performed before and after training with the aim to improve performance and to aid recovery and muscle regeneration. Specifically, stretching types, mechanisms underpinning stretch-induced increases in range of motion and their potential ability to enhance short-term (≤1-h post-intervention) and delayed recovery (at 24, 48, and 72 h) following strenuous exercise will be discussed. Evidence for the use of stretching to reduce injury incidence, alleviate pain and reduce inflammation will also be discussed as well as the effect of stretching training on muscle and connective tissue remodelling.

Types of Stretching

There are several different types of stretching incorporated in a pre- competition and pre-activity warm-up, as well as during the post-exercise phase [2, 18]:

Types of Stretching

- Static stretching
- Dynamic stretching
- Ballistic stretching
- Proprioceptive neuromuscular facilitation (PNF) stretching.
- Foam rolling and roller massage (when used as means of increasing joint ROM)

Static Stretching

This form of stretching involves the lengthening of muscle at different intensities ranging from a mild stretch sensation to the point of discomfort (POD) [2, 26] and then holding that lengthened position for a prescribed time (usually 15–60 s) [18, 26]. Static stretching may be performed passively, when the muscle is lengthened by an external force (i.e., another person or equipment) or actively, when the individual is self-stretching their muscles by moving and holding the joint(s) to the stretched position by contracting the antagonist muscles [18]. Static stretching is commonly used in sports, fitness, health, and rehabilitation environments, and is generally considered an effective method for increasing joint ROM [27, 28], improving performance [29], reducing pain [30] and decreasing the likelihood of sports-related muscle injuries [31]. Static stretching is also used during the warm-up phase before training or competition [11, 12]. However, several studies suggested that static stretching may impair force and power output if performed *immediately* before intense activities [32, 33]. The possibility of static stretching performed during warm-up, to impair subsequent performance has made some practitioners avoid or significantly limit its use during the warm-up for competition in the last two decades. However, recent evidence shows that it may be considered as an integral part of the warm-up, and this will be discussed later in the Chapter [12, 29, 34].

Dynamic Stretching

Dynamic stretching involves the performance of a relatively slow and controlled movement through the full ROM of the active joint(s) [35, 36] by contracting the agonist and relaxing the antagonist muscle(s) which are lengthened. Specifically, the contraction of the agonist muscles during the deliberate slow movement results in the relaxation of the lengthened muscles due to the principle of reciprocal inhibition [37]. In contrast, if the movement of the limbs is fast, then the stretch reflex is activated and the muscles under stretching are momentarily contracting [37]. The controlled movement of the limb back to the neutral position is executed using an eccentric contraction of the agonist muscles [18]. It is suggested that dynamic stretching is an effective method for increasing joint ROM [29, 36, 38] and improving performance [29, 36, 39], while its effects on recovery and in reducing the incidence of sport-related injuries have not been examined. Dynamic stretching differs from ballistic stretching in that the latter typically involves higher velocity movements with bouncing actions at the end of the ROM [29].

Ballistic Stretching

This type of stretching describes the action of moving a joint at or very close to its maximum ROM with a fast motion by executing a rapid swing of a body part or the torso without decelerating at the end of ROM ("ballistic" movement) [40]. This ballistic movement is typically a bouncing rhythmic motion using the momentum of a swinging body segment to lengthen the muscle [26, 41]. The rapid change in tension, that increases with the magnitude and rate of the stretch, can activate the stretch reflex resulting in a contraction of the muscle under stretch, and may exert a rapid strain on the involved tissues, which increases the risk of injury especially in individuals unaccustomed to this type of stretching [41–43]. Nevertheless, ballistic stretching has been used to increase joint ROM [42] and to acutely improve athletic performance [44, 45], while no study has examined its effect on recovery or injury risk [8].

Proprioceptive Neuromuscular Facilitation (PNF)

PNF combines static stretching and isometric contractions in a cyclical pattern [46]. Two of the more common techniques are the contract relax (CR) and contract relax antagonist contract (CRAC) techniques [47]. The CR method includes moving the joint to a stretched position, followed immediately by a short (3–15 s) isometric contraction of the stretched muscle with submaximal (20–50%) or near maximal intensity while the muscles are at a long muscle length. Afterwards, with the stretched muscles relaxed, the joint is moved to a position of further stretching and the cycle is repeated 2–6 times [47, 48]. The CR technique induces the Golgi tendon organ reflex, which causes relaxation of the muscles when tension is high. Compared to the CR technique, the CRAC technique involves an additional contraction of the agonist muscles (i.e., those moving the joint in the direction of the stretch), which induces relaxation of the stretched muscles due to the reciprocal inhibition reflex [49]. Research indicates that both the CR and CRAC techniques are effective for improving and maintaining ROM, for increasing muscular strength and power, and for increasing athletic performance [48, 50]. In clinical settings, PNF is used by therapists to restore functional ROM, increase strength in patients who have sustained soft tissue damage or have received invasive surgeries and to enhance recovery from strenuous activities [47, 51]. Despite its efficacy in increasing ROM, PNF stretching is not commonly used in athletic warm-up and post-exercise routines, possibly because there is a requirement for partner assistance, it may be uncomfortable or painful, and muscle contractions performed at highly stretched muscle lengths may induce muscle fatigue or even muscle damage [51, 52], especially with individuals who are weak and/or unaccustomed to flexibility training.

Foam Rolling and Roller Massage

More recently, foam rolling and rolling massage have become popular techniques to reduce muscle tightness and increase joint ROM [53]. Foam rolling is a form of self-massage in which the targeted muscles are compressed utilizing a rolling device with the aim to relax and to slightly stretch them in a massage-like motion [54]. Common rolling tools include foam rollers and various types of roller massage bars, sticks or small balls in various sizes and foam densities. With foam rollers, athletes commonly use their bodyweight to apply pressure to the soft tissues during the rolling motion, while roller massagers are applied by the individual's upper extremities to exert pressure to the target muscles [55, 56]. These rolling motions exert pressure and generate friction between the soft tissue and the foam rolling device, resembling the motions of massage [56].

Besides increasing ROM, foam rolling, and roller massage have also been shown to attenuate DOMS after intense exercise and to decrease latent trigger point sensitivity [57]. The physiological mechanisms underpinning these effects remain unclear, although some evidence suggests modified corticospinal excitability, enhanced fascial layer sliding and increases in arterial perfusion after intervention [55]. It is generally suggested that rolling is a "myofascial release" technique, a term describing a therapeutic intervention for releasing soft tissue from areas of abnormally tight fascia [53, 58]. However, evidence is insufficient to support that the primary mechanism underlying rolling and other similar devices is the release of myofascial restrictions [53, 58, 59].

- Static stretching is more effective than dynamic and ballistic stretching to increase joint ROM in the short and long-term and easy to apply
- Dynamic stretching is also effective in increasing joint ROM albeit to a lesser extent than static stretching
- Dynamic stretching may include functional and sport-specific movements that help increase muscle temperature and decrease muscle stiffness
- Ballistic stretching involves high velocity movements with bouncing actions at the end ROM which may increase injury risk in unaccustomed to stretching training participants
- Foam rolling is a form of self-massage purposing to relax and to slightly stretch muscles in a massage-like motion

Main Aims of Stretching Training

Stretching is primarily used to induce acute or long-term increases in joint ROM or flexibility [2, 18]. The simplest definition of flexibility is the ROM available in a joint or group of joints [60] and from a functional perspective it implies the ability to move a joint or a related series of joints through an unrestricted, pain-free ROM [61]. This has implications on both short- and long-term performance, due to changes in architectural and mechanical parameters of the muscle-tendon unit [29, 62]. Stretching is also used as an injury prevention modality [63, 64], and as a post-exercise intervention aiming to reduce muscle soreness and pain [65, 66]. Finally, stretching has also been used to enhance recovery [67], while there are some indications for muscle and connective tissue regeneration, such as increased fascicle length (possibly through the addition of sarcomeres in series and fascia remodeling [6, 68, 69].

Main Aims of Stretching Training

- to increase joint ROM in the short and long term
- to prepare the muscle-tendon unit for athletic performance (i.e., transiently increase muscle extensibility)
- to aid in injury prevention
- to promote recovery and reduce delayed onset muscle soreness (DOMS)
- to reduce muscle pain and muscle imbalances in various musculoskeletal conditions and assist rehabilitation following injury or disease
- to induce muscle and connective tissue (i.e., fascia) remodeling

Stretching to Increase Joint Range of Motion

Stretching to increase joint ROM takes soft tissue structures beyond their resting length by imposing unaccustomed and relatively high tensile forces on muscles, tendons, fascia, and other structures [2, 16]. This involves taking the joint beyond a comfortable ROM, thus inducing subsequent transient reductions in muscle and tendon stiffness, as well as neural adaptations that cause an increased tolerance to the imposed stretch [16]. Improvement of stretch tolerance requires practice in uncomfortable and/or extreme ROM positions [16, 70]. Stretching discomfort is difficult to quantify and relates to stretching intensity, duration and pain tolerance [71]. However, when stretching is used for recovery, high levels of discomfort and/or pain are in contrast with the aim of this approach and are therefore avoided. Thus, for recovery purposes, an individual should be directed to perform movements that involve low or no pain [16]. Some studies investigating low-intensity stretching (e.g. microstretching®) compared to static stretching [72], as well as moderate- and high intensity stretching [73, 74], found that ROM improved with both low and high-intensity stretching, but high-intensity static stretching also caused inflammation (i.e. increase in hsCRP levels), slower recovery times and a decrease in muscle function (eccentric and isometric peak torque) compared with low intensity stretching [74, 75]. The definition and importance of low-intensity stretching (microStretching®) for recovery and regeneration is further discussed in another Chapter 20. *The Genesis and Evolution of Low-Intensity Gentle Stretching—microStretching®* of this book, by Apostolopoulos, N.C.

All stretching types (static, dynamic, ballistic, and PNF) induce acute and long-term increases in ROM [28, 31, 40, 46, 55]. Some studies reported that dynamic and PNF stretching provide either similar or greater ROM increases than static stretching [76–78]. However, a recent meta-analysis found that chronic static stretching training provided greater flexibility gains than PNF and ballistic stretching [79]. Recently, it was also reported that foam rolling training induces small increases in ROM only when longer duration training interventions (>4 weeks) were used, with responses being muscle or joint specific (i.e., increases were reported when foam rolling was applied on quadriceps and hamstring muscles but not on triceps surae) [55].

It is important to note that long-term (12–26 weeks) unilateral static stretching interventions can induce contra-lateral ROM increases [3, 7, 62, 80]. The contralateral response to stretching may be attributed to central neural adaptations (i.e., increased pain tolerance [80–82]); similar to cross-education effects reported in strength training studies [83]. In addition, passive static stretching of a muscle group has also been found to induce small to moderate increases in passive ROM in non-stretched joints of other muscle groups (i.e., stretching of leg muscles affecting shoulder ROM), with the duration of the stretching being the most important variable influencing these non-local effects [84].

Mechanisms Underpinning Stretch Induced Changes in Range of Motion

Various theories have been proposed to explain acute and short-lasting (3- to 8-week) increases in muscle extensibility observed after stretching. Skeletal muscles comprise contractile tissue woven together by fibrous connective tissue that gradually blends into tendons [85]. During stretching, the ability of a muscle to extend over a predetermined endpoint, defined as muscle extensibility, increases [1, 86]. In human studies, viscoelastic deformation and recovery have been tested on hamstring and ankle plantar-flexor muscles [87–89]. The results suggest viscoelastic deformation as a potential mechanism for short lasting increases in muscle length and extensibility [1, 5]. However, this length increase is transient, and its magnitude and duration are dependent upon the duration, intensity, and type of stretching applied [1, 4]. Several studies suggest that stretching improves ROM mainly by enhancing stretch tolerance, i.e., the ability of the individual to tolerate the sensory discomfort caused by stretching [1, 5, 13, 86]. In addition, recent evidence points out that there may also be changes in muscle architecture, i.e., an increase in fascicle length, due to long term intense stretching (to or above the point of discomfort), especially in young individuals [6, 7, 62, 68].

Decreased Tissue Stiffness

Changes in ROM have been attributed to changes in the passive properties of the muscle-tendon unit [90–92]. Muscle-tendon unit stiffness is used to assess the changes in the elastic properties of the muscle-tendon unit after static stretching [1, 5] and is calculated from the slope of the torque-angle curve during passive joint movement [93]. There is a significant association between changes in ROM and muscle-tendon unit stiffness (MTS) after static stretching [93, 94].

The Torque-Joint Angle Curve
Torque generation during a movement depends on muscle length, which is, in turn, modified by joint angle. Therefore, maximum torque generated during parts of the full range of motion may differ by several-fold, depending on muscle length. The graph below shows the torque-angle curve for knee extension, measured by isokinetic dynamometry at an angular velocity of 60°/s.

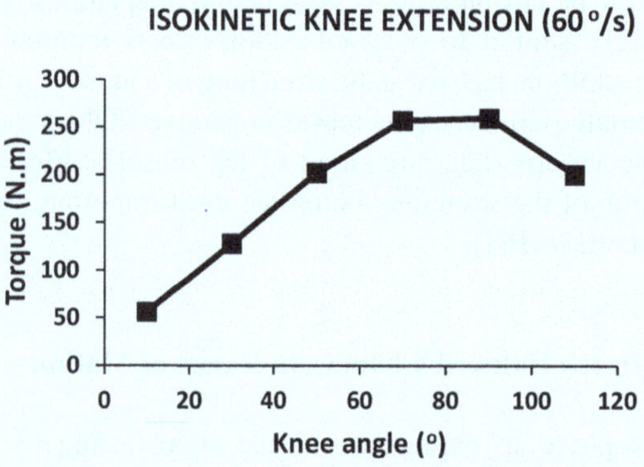

Acute static stretching reduces muscle-tendon stiffness in a time dependent manner, i.e., a longer stretch decreases muscle-tendon stiffness to a greater extent [90]. In contrast, long-term stretching training (for 8 weeks or less), including all types of stretching (i.e., static, dynamic, and proprioceptive neuromuscular facilitation stretching), does not seem to change muscle-tendon stiffness [4]. Recently, it was also found that longer-term static stretching training (3–12 weeks) has no significant effect on muscle-tendon stiffness, with the same reported for tendon stiffness after 3–6 weeks of static stretching [86, 92]. In contrast, Andrade et al. [68] and Longo et al. [3] reported a significant decrease in the muscle stiffness of the plantar flexors after 12 weeks of static stretching, with Moltubakk et al. [80] observing a significant decrease in passive torque at any given angle in the plantar flexors after 24 weeks of static stretching. Passive stiffness reflects the extensibility of the muscle-tendon unit [1]. Although the effect of static stretching for a longer period (≥ 12 weeks) may reduce muscle-tendon stiffness [80], this needs to be further examined.

Improved Stretch Tolerance

A sensory theory has been proposed to explain increases in ROM following stretching, suggesting that increases in muscle extensibility observed after a single stretching session and after short-term stretching programs (3- to 8-week) are due to modified sensation of pain [1, 4]. To what extent this adaptation is a peripheral or central phenomenon, or a combination of both, remains to be established [95]. Stretch tolerance may be influenced by nociceptive nerve endings, mechanoreceptors, or proprioceptors [95]. Alternatively, stretching may change some other aspects of the sensory neural pathways [96]. For example, afferent input from muscles and joints during stretch may interfere with signals from nociceptive fibers, subsequently inhibiting an individual's perception of pain [95, 96]. In addition, changes in stretch tolerance may be psychologically mediated [97]. It is possible that participants anticipate the positive effects of stretch and, therefore, their perception of the stretch discomfort is dampened [98]. Prior experiences or motivation to stretch, and elevated confidence from positive stretch benefits are all potential psychological contributors explaining the participants' altered perception of the discomfort and willingness to tolerate greater stretch over time [97]. However, more research needs to be conducted to substantiate this viewpoint.

Muscle and Fascia Remodeling

Muscle structure can be described by quantifying its architectural parameters, namely, fascicle length and angle, muscle thickness and cross-sectional area [99]. Mechanical loading induced by muscle stretching triggers alterations in cellular signaling and gene expression, which may modify the physiological, structural, and contractile properties of muscle fibers [100, 101]. Animal studies have shown significant increases in fascicle length, possibly through sarcomerogenesis in series, after long periods of immobilization in a stretched position [102, 103]. Cross-sectional studies *in humans* observed greater fascicle length at rest and greater fascicle elongation during stretching in flexibility trained compared to untrained subjects [104, 105] and the same was found in flexibility trained children [106]. Recent stretching interventions in humans found that static stretching induces trivial increases in fascicle length at rest and small increases in fascicle length during stretching (fascicle extensibility) [7, 62, 68, 69]. Increases in fascicle length at rest have been found after 6–12 weeks of stretching training [7, 62, 68], while increases in fascicle length during stretching may appear earlier, i.e., after 3–4 weeks of static stretching training [82].

It may be postulated that muscle strain induced by high-volume of intense static stretching may trigger mechanotransduction signaling pathways associated with changes in muscle thickness [107]. Stretching intensity modifies these responses, with maximal intensity stretches promoting higher acute inflammatory responses compared to submaximal stretching [74]. Behm and Kibele [108] reported that stretches held at the maximal point of discomfort induced muscle damage and delayed onset muscle soreness, similar to that observed following eccentric exercise [108]. Thus, the level of muscle stress, muscle damage and muscle inflammation due to intense high-volume stretching may induce some degree of muscle hypertrophy [109]. In addition, the time during which a muscle is held under intense static stretching is related to ischemia attributed to vascular occlusion, which in turn induces changes in metabolic sensors that regulate muscle growth [110]. It has been shown, mainly in animal studies, that chronic stretching to long muscle lengths, such as in the case of casting of a limb, induces changes in fascicle length by sarcomere addition, in order to reduce passive tension and to maintain optimal actin-myosin overlap [102, 103]. However, immobilization also causes muscle atrophy, and this must be taken into account when examining the adaptations to this type of passive stretching.

The characteristics of the stretching protocols (total stretching duration/volume and intensity) are key elements to induce morphological changes of the muscles. Longer-term stretching interventions (>12 weeks) [7, 68], overloaded static stretching [6] and high-intensity and/or long-duration stretching bouts [3] have been found to increase fascicle length at rest and during stretching (fascicle extensibility), muscle thickness and muscle cross-sectional area. However, fascicle angle remains unaffected from static stretching training [3, 7, 62]. It is important to note that it remains largely undetermined how these changes may affect the mechanics of muscle contraction, i.e., whether contraction velocity, total work, metabolic cost, etc. are affected when fibers are lengthened by sarcomere addition in series.

Stretching may also affect the fascia system by removing or decreasing fascial restrictions which occur in response to inactivity, injury, inflammation and disease [111]. Skeletal muscles and fascia form a network of multidirectional myofascial continuity [112]. The fascia system represents a tensional network that surrounds, interweaves between and interpenetrates all organs, muscles bones and nerve fibers, providing the body with a functional structure that connects almost all body systems [112]. The fascial system includes adipose tissue, adventitia, neurovascular sheaths, aponeuroses, deep and superficial fasciae, dermis, epineurium, joint capsules, ligaments, membranes, meninges, myofascial expansions, periostea, retinacula, septa, tendons (including the endotendon, peritendon, epitendon and paratendon), visceral fasciae, and all the intramuscular and intermuscular connective tissues, including endomysium/perimysium/epimysium [113].

Imaging studies in humans have shown that intermuscular and extramuscular fascial tissues also provide a pathway for force transmission, dependent on the mechanical properties of myofascial tissue linkages [114, 115]. Myofascial tissue that is stiffer or more compliant than normal has been shown to influence the magnitude of intermuscular force transmission [112, 115]. Several mechanisms, acting individually or in complex, may trigger stiffness increases of the connective tissue. In contrast, if the connective tissues are loaded properly, the inherent networking cells (fibroblasts), adapt their matrix remodelling activity so that the tissue architecture is modified according to mechanical loading of daily demands [111, 112]. High doses of tensile stress stimulate the excretion of pro-inflammatory messenger substances by the fibroblasts. Applied in the right context and with an appropriate dosage, this can be interpreted as the start of a healthy repair and adaptation processes. For example, fascial hydration changes occur in response to mechanical stimuli such as stretching exercises [112]. This is of importance because alterations in the water content are directly related to connective tissue stiffness [116]. Nevertheless, research is still limited on the effect of stretching interventions targeting fascia tissue or muscles morphologically connected through fascia tissue such as adductor and rectus abdominis muscles [117].

- Acute increases in joint ROM are due to improved stretch tolerance and a transient decrease in tissue stiffness (depending on stretch duration and intensity)
- Long-term increases in joint ROM are also due to improved stretch tolerance while evidence is limited and conflicting on the chronic effect of stretching on muscle stiffness
- Long-term increases in joint ROM are associated with changes in muscle morphology, i.e., muscle fascicle length and anatomical cross-sectional area (depending on stretch duration and intensity)

Stretching to Prepare the Muscle-Tendon Unit for Athletic Performance

All types of stretching have been used in pre-activity routines to acutely enhance performance [2, 10, 11]. Static stretching has been traditionally performed in warm-up routines, with the aim to transiently increase joint ROM [118, 119] and improve performance, especially in sports requiring the ability to move through a large ROM [34]. Several earlier studies suggested that long duration and high intensity static stretching could acutely impair muscle power and strength if performed immediately before intense exercise [10–12]. Due to the widespread belief that static stretching may impair performance, a considerable number of practitioners had limited or even abandoned its use during the warm-up and replaced it with dynamic stretching. Nevertheless, an increasing body of evidence shows that performance decrements following static stretching are observed only when long-duration (>30 or >45 s per muscle group) and high intensity static stretching is performed, or when the subsequent activity is performed immediately after stretching [12, 13, 34]. In contrast, static stretching increases ROM and may be performed without concern for performance reduction, especially when stretching duration is kept around 15–20 s per muscle group [120]. In fact, sprint, and agility performance, as well as muscle power have been shown to improve rather than decrease following short-duration static stretching, especially when it was performed intermittently, and by stretching-accustomed athletes, such as gymnasts [34, 120]. Notably, even if longer duration static stretching is performed, any small reduction in performance may be reversed by performing a few repetitions of an activation exercise, such as 3–5 jumps [34, 121]. Thus, short-duration and moderate-intensity static stretching may be an integral part of warm-up, with no adverse but rather positive effects on performance [13, 34, 120, 121]. Dynamic and PNF stretching may also acutely increase subsequent performance possibly because of the similarity of dynamic stretching movements with *sport-specific movement patterns* [122]. Dynamic stretching also increases *muscle temperature* [123], *and may increase central drive*, which all contribute to enhanced muscle power [124, 125]. In practical terms, it is suggested that longer-duration PNF stretching may be performed a long time period before (e.g., >10 min) while dynamic stretching may be performed closer to task performance [126].

The main long-term effect of stretching training is the increase in flexibility, an important component of physical fitness and health [127]. Enhanced ROM increases the distance over which muscle force is applied or absorbed and allows athletes to demonstrate strength through the ROM typically used in their sport [105, 128]. Also, static stretching induces a moderate improvement of muscle force (around 2%) at longer muscle lengths, which may be beneficial for both performance and injury prevention purposes [12, 105]. The importance of flexibility for performance depends on the determinants of performance in each sport. In some sports like artistic, rhythmic gymnastics and ballet, flexibility is probably the most important performance determinant [129, 130], while in other sports like swimming and weight-lifting flexibility demands are joint specific (i.e., shoulders, ankle, wrist) [131–133]. Furthermore, there are sports like marathon running that enhanced flexibility may not be an important parameter for performance. Nevertheless, the interactions between increased flexibility and human locomotion have imperative consequences in both athletic and therapeutic settings [134].

Importantly, high volume and high intensity static stretching training in humans has the potential to induce increases in fascicle length, muscle thickness and muscle cross sectional area, which may have a direct effect on muscle mechanical function (e.g., force – length relationship) [7, 62, 68]. Some studies indicate that increased fascicle length may shift the optimal muscle length for force production to longer lengths and may widen the entire force-length relationship [62, 105], but this remains to be verified for stretching training interventions. In addition, future research should examine the effect of longitudinal fascicle growth following static stretching on velocity of contraction during shortening [68], as well as on the torque-angle relationship which may improve muscle power, especially at specific muscle lengths and joint positions [68, 105, 135]. These adaptations may be useful not only for athletes, but also in patients and other inactive individuals [135].

- Performance decrements following static stretching are observed only when long-duration (>45 s per muscle group) and high intensity static stretching is performed or when the subsequent activity is performed immediately after stretching
- Static stretching increases ROM and may be performed without concern for performance reduction, especially when stretching duration is kept around 15–20 s
- Dynamic stretching also increases muscle temperature and may increase central drive, which both contribute to enhanced muscle power
- Static stretching within a warm-up that includes additional post-stretching dynamic stretching and activity is recommended for reducing muscle injuries and increasing joint ROM with insignificant effects on subsequent athletic performance

Stretching to Prevent Injury

Evidence for the effectiveness of acute and chronic stretching for improving ROM is extensive [10, 11]. Although enhanced flexibility can positively impact sports performance as well as daily living activities, much less is known about the direct effects of stretching training on injury prevention [31]. The acute and long-term effects of stretching on injury prevention have been largely misinterpreted for many reasons (e.g., type of injury examined, mechanism of injury incidence, etc.), but mostly due to the study design approaches implemented.

For example, in the . study of Pope et al. [136], 1093 army recruits were examined during 12 weeks of military training. The authors found a significant correlation between dorsiflexion ROM and injury incidence. Poor flexibility was associated with a 2.5 times higher risk for injuries compared with average dorsiflexion ROM and 8 times higher risk compared with people with a high level of flexibility. However, the stretch-training intervention that was applied for 12 weeks had no significant effects on the incidence of these injuries, and the authors concluded that flexibility training does not reduce the risk of injury. However, in that study injuries were caused by quite intense military training, and this makes it difficult to establish a link between flexibility training and injury risk. In addition, 12 weeks of stretching training may be a short period to improve or correct motor patterns and induce morphological changes in the muscles. Many studies since then examined the effects of stretching on injury prevention and reported conflicting results [64, 138–140]. Some studies suggested that stretching may induce a reduction in injuries through muscle tendon elongation [137] while others reported that stretching before an event will not reduce injuries [138]. The confusion, apart from not discriminating between acute and long-term stretch benefits, resides in whether the reviews are examining all-injury risk versus muscle-tendon unit injury risk. Lower body fractures, meniscal lesions, or other injuries cannot be prevented by stretching during warm-up or with flexibility training in general. In the randomized trial of Jamtvedt et al. [140],

2337 physically active adults were examined before and after 12 weeks of stretching training. It was found that stretching did not induce clinically important reductions in all-injury risk, but it reduced the risk of experiencing soreness. In that study, stretching reduced the risk of injuries to muscles, ligaments, and tendons, with an incidence rate of 0.66 injuries per person-year in the stretching group and 0.88 injuries per person-year in the control group [140]. In line with these results other studies also reported a lower incidence rate (16.7%) in the flexibility trained compared with the non-trained groups, citing an average reduction of injuries by 32% following long-term stretching training [141]. Due to the difficulty of separating out the effects of stretching from the remaining training content, more long-term studies are needed to determine the effectiveness of regular stretching in maintaining muscle-tendon health during intense training [142].

Flexibility for musculoskeletal health is particularly important for young athletes who are subjected to intense training [128]. High training and competition volumes which are maintained for inappropriately long time, combined with incorrect movement patterns, low ROM and muscle tightness may have a negative impact on muscle health [132]. Improved flexibility through regular stretching training, not only enhances motor competence in developing children [143], but may also reduce overuse injury risk due to ROM deficits in child and adolescent athletes [132, 144]. Notably, ROM deficits and imbalances during the early sports years may lead to long-term health problems into adulthood [144, 145].

One must keep in mind that the acute effects of stretching on ROM and muscle mechanical properties (i.e., reduction of muscle stiffness) are transient, which means that warm-up stretching can contribute to acute injury prevention only to a small degree [64]. In contrast, long-term stretching may improve the energy-absorbing capabilities of the muscle-tendon unit by increasing muscle-tendon extensibility, thus preventing musculotendinous injuries especially those caused by high loads associated with rapid force absorption such as sprinting and agility drills [146, 147]. It is not known if the increased fascicle extensibility following stretching training reflects changes in-series elastic (i.e., the muscle internal aponeuroses, the structural protein "titin", the elastic elements in the cross-bridges aponeurosis) or contractile elements (i.e., sarcomeres) [148]. Also, it remains undetermined how these changes may affect the mechanics of muscle contraction, and the storage and release of elastic energy [147, 148], which may all reduce the risk of muscle injuries.

In addition, long-term stretching training increases joint ROM and reduces tissue resistance to stretch, as well as the resistance to the intended movement, thereby improving movement efficiency [18]. Stretching training may induce a shift in the active muscle force-length relationship toward longer length thus, improving force and power output at longer muscle lengths [105]. Since some injuries occur close to the end of the range of motion with the muscle in a lengthened state, this shift of the force-length relationship could play a role in reducing such injuries [7, 12, 140, 141]. Interestingly, early studies have shown that the athletes most at risk of a hamstring injury are those with a previous hamstring injury history [149]. Previously injured individuals usually show a shift of the hamstrings force-length curve towards shorter lengths, and this highlights the important role of interventions aiming to reverse this negative adaptation [149]. Stretching interventions have demonstrated improvements in the viscosity and compliance of the tendon, which may be important for lowering muscle stiffness and contribute to injury prevention in sports requiring high intensity stretch–shortening cycles [138, 150].

Poor flexibility can lead to poor posture, agonist-antagonist muscle imbalances and altered motor patterns that can hamper motor skills performance, muscle synergies and balance, increasing the mechanical load on the joints [12, 150, 151]. Muscles which are shortened demonstrate atrophy and become weaker, leading to muscle imbalances [31, 63, 64]. A more compliant muscle and tendon system through enhanced ROM are better able to absorb disruptive perturbations thereby lowering the metabolic cost of movement [12, 31, 152], but again, this needs to be further examined. A combination of flexibility training, education in correct movement patterns, and muscle strengthening might be needed to return the body to proper function and reduce the possibility of pain, and re-injury [31].

- The acute effects of stretching on range of motion and muscle mechanical properties (i.e., muscle stiffness) are transient, thus, warm-up stretching can contribute to acute injury prevention only to a small degree
- *In contrast*, long-term stretching improves the energy-absorbing capabilities of the muscle-tendon unit, thus preventing musculotendinous injuries especially those caused by high loads associated with rapid force absorption
- Improved flexibility through regular stretching training in child and adolescent athletes, enhances motor competence and also reduces injury risk
- Poor flexibility can lead to poor posture, agonist-antagonist muscle imbalances and altered motor patterns that can hamper motor skills performance, muscle synergies and balance, increasing the mechanical load on the joints and eventually the risk of injury

Stretching to Aid Recovery and to Reduce Delayed Onset Muscle Soreness

Recovery is defined as the process of returning to a normal (homeostatic) or new improved health and performance state (psychological, metabolic, neural) after fatigue caused by exercise [153]. The increasing physical demands of athletic training and competition has exacerbated the physical and mental load placed on athletes [154]. By accelerating recovery, the athlete can train at the same or higher level of intensity, duration, frequency, or volume [153]. Recovery time between successive trainings and competitions may be insufficient to allow athletes to fully recover, leading to fatigue which may increase the likelihood of under-performance, non-functional overreaching, injury, and illness [153, 155].

Post-exercise stretching is commonly used among athletes for improving the recovery of strength and ROM, and for reducing the level of delayed onset muscle soreness (DOMS) in the next few days [8, 149, 156]. The value of using stretching to improve recovery may be demonstrated when it is used after a demanding and intense exercise that causes fatigue and muscle damage. Static stretching, active-passive stretching [157–159] and PNF stretching [19, 157] have been used for recovery purposes while no study used dynamic or ballistic stretching [8]. However, to date, evidence of beneficial effects of stretching on the recovery of performance and muscle soreness is limited and conflicting. A recent systematic review identified only eleven studies that used post-exercise stretching aiming to improve short-term (i.e., until ≤1-h post recovery) and delayed recovery (i.e., 24, 48, and 72 h post recovery) of strength, ROM and DOMS, in adults of different training statuses [8]. The studies included in that systematic review examined recovery following strength training exercise protocols that induced muscle soreness by emphasizing the eccentric component [20, 160–163]. The results of these studies showed no effect of post-exercise stretching protocols on strength [19, 162, 163] or vertical jump performance [158], with one study reporting that maximum isometric force was significantly higher after cycling than after stretching or passive recovery [19]. Both, static and PNF stretching did not enhance ROM immediately post-exercise [157]. In another study, it was found that ROM decreased significantly following a high intensity functional training session with no difference between the static stretching, the foam rolling or the control group, and the same was found 24 h after the intervention [164]. However, there is evidence that ROM may be higher at 24, 48, and 72 h post stretching intervention when compared to cold water immersion or to a control group [158].

The acute effects of stretching on delayed onset muscle soreness are also conflicting. At this point, it should be stressed that muscle damage and the subsequent delayed-onset muscle soreness are induced by either unaccustomed exercise or by high intensity or long duration eccentric (i.e., lengthening) muscle contractions [165, 166]. Muscle damage is characterized by ultrastructural alterations in muscle tissue which are accompanied by several clinical signs and symptoms (e.g., reduced muscle strength and range of motion, increased muscle soreness and swelling, efflux of myocellular proteins) [166]. The time course of recovery following exercise-induced muscle damage depends on the extent of initial muscle damage, which in turn is influenced by the intensity and duration of exercise, joint angle and thus muscle length employed, as well as the muscle groups used during exercise [166]. Torres et al. [163] examined the effect of stretching immediately after and 24, 48, and 72 h post eccentric exercise on maximal concentric peak torque muscle soreness, muscle stiffness, and plasma creatine kinase and reported that a single bout of muscle stretching performed immediately after exercise did not influence the levels of the main markers of exercise-induced muscle damage.

Interestingly, in the study of Torres, examining the effect of stretching at 24, 48, and 72 h post eccentric exercise on muscle soreness, muscle stiffness, and plasma creatine kinase it was reported that stretching repeated at 24 h post eccentric exercise may have favorable effects on muscle stiffness [163]. Overall, there is no hard evidence that acute stretching enhances recovery of strength, ROM, and DOMS in comparison to passive recovery or to other recovery modalities, such as cycling and cold-water immersion [8]. Even for secondary outcomes, such as blood lactate and serum creatine kinase, no strong case can be made for stretching accelerating or improving recovery [8]. However, evidence is limited (only 11 studies), heterogenous, and overall confidence in cumulative evidence is very low [8]. It is also possible that the level of fatigue induced by the exercise protocols was too high, as most studies included populations not engaged in regular physical activity, and thereby less well-adapted to the acute effects of fatiguing exercise [8].

There is some interesting evidence regarding the effects of myofascial rolling (i.e., foam rolling and roller massage) on the process of recovery. A systematic review reported that myofascial rolling for at least 90 seconds per muscle group may be the minimal duration to achieve a short-term reduction in pain and soreness in the lower limbs, with no upper limit found [167]. However, results were insufficient to provide a conclusive recommendation for impacting acute ROM [57]. Another study compared static stretching to foam rolling as a recovery means immediately after and at 24 h post a bout of high-intensity functional training [164] and found that ROM decreased significantly immediately post exercise with no difference between static stretching, foam rolling or control. In contrast, a recent systematic review examining the effect of foam rolling on range of motion, recovery and markers of athletic performance, reported that foam rolling increased range of motion immediately post exercise, and appeared to be useful for recovery from exercise-induced muscle damage [168]. Collectively, evidence suggests that rolling techniques may help attenuate decrements in muscle performance and DOMS after intensive exercise [57, 168, 169], however, further research to substantiate this notion is warranted.

Effects of Stretching Intensity on Recovery

One important aspect of stretching manipulations aiming to enhance recovery, and to reduce muscle tightness and soreness is the intensity of stretching. Stretching aims to elongate soft tissue structures within the current boundaries of tissue extensibility [16]. However, the different intensity that should be used for stretching to improve flexibility and for stretching to reduce delayed-onset muscle soreness (DOMS) should be noted. This is because post-exercise stretching to improve flexibility usually puts muscles and tendons under relatively high and in some cases unaccustomed tensile stresses until a high degree of discomfort is reached [170]. However, depending on the previous activity, muscles may be fatigued, and stretching musculotendinous tissue to the point of discomfort during the post-exercise phase may lead to tissue strains or even in muscle damage if the tensile strength and extensibility of the tissue is compromised by fatigue [12, 16]. Thus, stretching using pain-free motions with low intensity, such as microstretching, may be typically suggested as a recovery modality [171]. However, current evidence on the acute effects of stretching on delayed-onset muscle soreness, indicates that it cannot significantly affect the magnitude or the time-course of DOMS following muscle damaging exercise [172]. Nevertheless, the effects of mild stretching on DOMS should be further examined, as there is a study reporting reduced DOMS when low intensity stretching is performed after muscle damaging exercise [17]. Also, several practitioners, clinicians and athletes have reported potential benefits of low intensity stretching to the recovery process, which are supported by indirect research evidence [171]. Moreover, low intensity stretching also results in lower tissue inflammation, as shown by a comparison of high and low intensity stretching [75] and may thus be more beneficial for recovery compared with high intensity stretching [17].

Another way to view the positive effects of stretching on the reduction of muscle soreness is by looking at its long-term effects. Studies that employed several weeks of stretching training report decreases in both DOMS symptoms and general soreness [173, 174]. In a randomized, controlled study, participants (N = 2125) self-reported soreness outcomes [140]. It was found that performing static stretching for 12 weeks both before and after distance running reduced the soreness in the legs, buttocks, and back compared to a non-stretch control group [140]. Furthermore, 8 weeks of static or PNF hamstrings stretching training before a single session of DOMS-inducing exercise reduced pain severity compared to a non-stretching control group with no difference between static stretching and PNF [174]. In contrast, some studies failed to detect any effect of stretching training on DOMS [24, 175]. For example, Lund et al. [175] assessed the extent of quadriceps soreness from a series of eccentric contractions followed by a 7-day stretching protocol and concluded that 7 days of quadriceps stretching after a DOMS-inducing exercise did not reduce pain. Another study showed a similar lack of pain reduction when groups were assigned to either dynamic or static calf stretching for 5 days after a DOMS-inducing exercise [24]. Compared to other long-term stretching studies it seems that static stretching is not effective in reducing DOMS when only a single bout of stretching or short-term stretching interventions (5–7 days) are used. However, static stretching for more than 3 weeks seems to attenuate DOMS [30], while low intensity of stretching is considered as a "key characteristic" of its effectiveness during recovery [74, 171].

Stretch-induced reductions following stretching training in DOMS may be attributed to increased pain tolerance or alterations in extracellular matrix properties, providing protection against tissue damage [86, 176]. Furthermore, while individual muscle-tendon unit fibers must have sufficient stiffness to accommodate and positively utilize the high strains imposed during sprint running, jumping, and other activities, there is also a need for architectural adaptations and responses such as muscle and tendon fascicle rotation and translation (sliding) [12, 31]. Reduced musculotendinous stiffness following stretching may, in some cases, result in a higher energy-absorbing capacity to store and release a large amount of elastic energy during subsequent, intense stretch-shortening cycle type of activities, which, in turn, may reduce DOMS and enhance recovery following intense eccentric exercise [152].

- Evidence for the acute effect of stretching (static, PNF) on recovery parameters (i.e, muscle strength, ROM, jumping ability) is limited. However, longer-term stretching interventions showed beneficial effects of stretching on recovery following strenuous exercise
- Static stretching is not effective in reducing DOMS when only a single bout of stretching or short-term stretching interventions (5–7 days) are used
- However, static stretching for more than 3 weeks seems to attenuate DOMS, while low intensity of stretching is considered as a "key characteristic" of its effectiveness during recovery

Stretching for Musculoskeletal Pain Relief and for Rehabilitation Following Injury or Disease

Musculoskeletal pain is a common ailment in modern society [30]. The World Health Organization (WHO) reports that approximately two billion people worldwide have musculoskeletal conditions, and, among musculoskeletal disorders, low back pain causes the highest burden on society with a prevalence of 568 million people [177]. Neck pain is also a growing disorder due to the increase in the number of office and computer workers. Although musculoskeletal conditions contribute substantially to disability worldwide, the effect of stretching on pain modulation and the role of flexibility in the prevention and treatment of musculoskeletal disorders has not been largely examined [30, 177]. Few studies that implemented acute stretching protocols (static and PNF stretching) have reported small to large magnitude decreases in quadriceps and anterior knee pain [178] as well as reductions in headache [179]. Other studies examining the acute effects of stretching, combined it with other modalities such as manual therapy, mobilization and massage and reported positive outcomes [180–182]. Long-term stretching interventions report consistent alleviation of pain, which is not limited to the time immediately after stretching but instead may last longer, depending on the frequency and duration of the stretching training program [30]. These studies demonstrated reductions in pain for a variety of joints and muscles, which has mainly been attributed to an increased pain tolerance. The most prevalent stretching program for pain reduction has been that prescribed for nonspecific low-back pain [30, 182]. The positive effect of stretching on low-back pain depends on the intensity, duration and location of stretch [30].

Poor adductor flexibility has been associated with groin pain and injury [183]. Groin pain is an umbrella term that includes distinct pathological processes (potentially overlapping) and multifactorial causes [184–186]. Stretching has therefore been hypothesized to aid in recovery from groin pain. However, despite the clinical narrative supporting the use of stretching for the recovery or improvement of groin pain, trials specifically assessing the effects of stretching only on recovery of groin pain or injury in athletes are missing, and therefore a clear cause–effect relationship cannot be established [187, 188]. Nevertheless, stretching as part of a multimodal exercise or therapy program (i.e., stretching in combination with other interventions) is effective in alleviating groin pain, showing the multifaceted nature of rehabilitation after a sports injury [186].

The proposed mechanisms through which stretching may alleviate pain symptoms and accelerate recovery include improvements in joint range of motion (ROM) [183, 186, 189], and reduction in muscle stiffness [183]. In healthy adults, static stretching has been shown to increase pressure pain thresholds and tolerance to stretch thus, altering the sensation or perception of discomfort [96, 190]. Other mechanisms underlying acute and chronic pain reduction have been proposed to be related to gate control theory, diffuse noxious inhibitory control, myofascial meridians/connective points, and reflex-induced increases in parasympathetic nervous activity [30, 31]. Recently it was suggested that foam rolling and roller massage may also help alleviate pain and decrease latent trigger point sensitivity [55, 57, 59]. It is important to note that while foam rolling can affect the myofascial tissues, other mechanisms such as rolling-induced global effects on muscle tone from changes in the parasympathetic system, reflexes and pain tolerance may have a large influence on improvements in range of motion and pain sensitivity [59].

One of the aims of rehabilitation from an injury of the muscle-tendon unit is to ensure a return of ROM within the constraints of tissue healing [16]. Unfortunately, the studies that used only stretching for recovery purposes applied an extremely wide range of stretching intensities, precluding any valid conclusions to be drawn [8, 16]. However, stretching is widely used to ensure proper alignment of collagen in the previously injured muscle tissue and to improve ROM and, most importantly, strength and power across the force length relationship [12, 18, 105]. Regarding rehabilitation purposes, it is commonly accepted that stretching is of great importance in treating strain injuries and it can also contribute to the effectiveness of the rest rehabilitation program [191, 192]. The use of the systematic stretching in combination or without core stability exercises, also enhances the functional gains of chronic low-back pain patients compared to the control groups and helps to alleviate pain [193, 194].

- Long-term stretching interventions report consistent alleviation of pain, which is not limited to the time immediately after stretching but instead may last longer, depending on the frequency and duration of the stretching training program
- Stretching is widely used to ensure proper alignment of collagen in the previously injured muscle tissue and to improve ROM and, most importantly, strength and power across the force length relationship
- Increased range of motion is important for correct body posture and movement patterns thus, optimizing muscle synergies and mechanical loading on muscle and joint structures

Stretching and Muscle Regeneration

Regeneration may be considered as the process of remodeling and repairing muscle, connective tissue and other cells of the body following damage, inflammation or oxidative stress caused by exercise [195, 196]. Regeneration usually results in improved structure and function of the cells involved (e.g., muscle and connective tissue) and reduces the level of damage when a similar exercise stimulus is employed in subsequent sessions [195, 196].

Human skeletal muscle responds to mechanical loading by adapting its structure, namely, fascicle length and angle, muscle thickness and cross-sectional area [165, 197]. As mentioned before, mechanical loading induced by muscle stretching may induce an increase in the number of sarcomeres in series and induce "longitudinal fascicle growth" [147]. Recent studies examining longitudinal fascicle growth following eccentric training, reported increases in the length and number of in-series sarcomeres [147]. However, evidence on the effects of long-term stretching on sarcomere addition comes only from animal studies using casting protocols [102, 198], and studies in humans performing real-life stretching protocols, in terms of stretching intensity and duration, are still lacking. In theory, if a muscle is placed in an overstretched position for a sufficient duration and under a sufficient stretch intensity, then the number of sarcomeres in series is increased to maintain optimal actin-myosin overlap, and to reduce the passive tension caused by the stretch [165, 199]. This longitudinal muscle fascicle growth may shift the optimal muscle length for active force production to longer muscle lengths and more extended joint positions [147, 199]. This adaptation may also broaden the torque-angle relationship's plateau region, and the ability to generate higher forces during shortening [147]. Nevertheless, evidence that longitudinal fascicle growth induced by static stretch training can shift the passive and active length-tension relationship in humans is only indirect, with data coming from comparisons between stretch-trained and untrained groups of individuals, such as between rhythmic gymnasts and controls [105].

Furthermore, there are indications that the fascia system may respond to high stretching loads by modifying its matrix remodeling activity so that the tissue architecture responds better to mechanical loading [111]. High doses of tensile stress may also stimulate the excretion of pro-inflammatory messenger substances enhancing repair and adaptation processes [112, 200]. The extracellular matrix turnover of muscle and tendon is also influenced by mechanical loading. For example, blood flow, oxygen demand, and the level of collagen synthesis and matrix metalloproteinases increase with mechanical loading [201]. These changes may modify the biomechanical properties of the tissue (i.e., viscoelastic characteristics) as well as the structural properties of the in collagen (i.e., cross-sectional area) and contribute to altered resistance to loading and enhanced tolerance to strenuous exercise [201]. A recent review suggested that applying gentle continuous stretching to myofascial connective tissue can improve the regulation of oxidative stress, the expression of pro- and anti-inflammatory genes and mediators, and enhance remodeling of the extracellular matrix, expressed by changes in collagen and matrix metalloproteinases levels in tissues subjected to stretching [202]. There is also evidence in healthy humans showing that acute and chronic stretching training induces angiogenesis, by increasing the microvascular endothelial function and minimalizing capillary diameter thus, increasing hypoxia-inducible factor 1-alpha (HIF-1alpha) and vascular endothelial growth factor A (VEGF-A) expression, which promote regeneration [203]. However, most of these suggestions regarding connective tissue and extracellular matrix adaptations remain to be determined in humans performing stretching protocols.

A summary on the acute and long-term effects of stretching related to muscle-tendon performance and neuromuscular function recovery is presented in page 26

- When a muscle is placed in an overstretched position for a sufficient duration and under a sufficient stretch intensity, then the number of sarcomeres in series is increased to maintain optimal actin-myosin overlap, and to reduce the passive tension caused by the stretch
- The longitudinal muscle fascicle growth may shift the optimal muscle length for active force production to longer muscle lengths and more extended joint positions, which broadens the torque-angle relationship's plateau region, and the ability to generate higher forces during shortening
- The fascia system may respond to high stretching loads by modifying its matrix remodeling activity so that the tissue architecture responds better to mechanical loading

> The wind does not break a tree that bends
>
> African proverb

Summary of effects of stretching related to muscle-tendon performance and neuromuscular function recovery

Acute effects	Long-term effects
Re-established or increased range of motion (ROM)	Increased range of motion (ROM)
No effect on muscle strength	Increased muscle strength (especially with PNF stretching) Increased range of force application and absorption
Temporary shift in optimal muscle length for force production	Shift in the optimal muscle length for force production
Unaffected movement efficiency	Improved movement efficiency and body posture
Pain relief	Pain relief (musculoskeletal disorders)
Decreased muscle stiffness	No changes in muscle stiffness

Acute effects	Long-term effects
Muscle relaxation	Muscle relaxation
Decreased DOMS (contradicting evidence)	Decreased DOOMS (promising evidence)
Increased fascicle length (extensibility) during stretching (transient effect)	Changes in muscle architecture (increases in fascicle length at rest and cross-sectional area. Improved muscle mechanical and contractile function)
No acute effect on injury risk	Decreased risk for muscle and tendon injury -especially at long muscle lengths

References

1. Weppler C, Magnusson S. Increasing muscle extensibility: a matter of increasing length or modifying sensation? Phys Ther. 2010;90:438–49.
2. Alter MJ. Science of flexibility. 3rd ed. Champain: Human Kinetics; 2004. p. 17–373.
3. Longo S, Cè E, Bisconti AV, et al. The effects of 12 weeks of static stretch training on the functional, mechanical, and architectural characteristics of the triceps surae muscle–tendon complex. Eur J Appl Physiol. 2021;121:1743–58.
4. Freitas SR, Mendes B, Le Sant G, Andrade RJ, Nordez A, Milanovic Z. Can chronic stretching change the muscle-tendon mechanical properties? A review. Scand J Med Sci Sports. 2018;28(3):794–806. https://doi.org/10.1111/sms.12957.
5. Magnusson SP. Passive properties of human skeletal muscle during stretch maneuvers. Scand J Med Sci Sports. 1998;8(2):65–77. https://doi.org/10.1111/j.1600-0838.1998.tb00171.x.
6. Simpson CL, Kim BDH, Bourcet MR, Jones GR, Jakobi JM. Stretch training induces unequal adaptation in muscle fascicles and thickness in medial and lateral gastrocnemii. Scand J Med Sci Sports. 2017;27(12):1597–604. https://doi.org/10.1111/sms.12822.
7. Panidi I, Bogdanis GC, Terzis G, et al. Muscle architectural and functional adaptations following 12-weeks of stretching in adolescent female athletes. Front Physiol. 2021;12:701338. https://doi.org/10.3389/fphys.2021.701338.
8. Afonso J, Clemente FM, Nakamura FY, et al. The effectiveness of post-exercise stretching in short-term and delayed recovery of strength, range of motion and delayed onset muscle soreness: a systematic review and meta-analysis of randomized controlled trials. Front Physiol. 2021:553. https://doi.org/10.3389/fphys.2021.677581.
9. ACSM. In: Bayles MP, Swank AM, editors. ACSM's guidelines for exercise testing and prescription. 10th ed. Philadelphia: Wolters Kluwer; 2018.
10. Behm DG, Chaouachi A. A review of the acute effects of static and dynamic stretching on performance. Eur J Appl Physiol. 2011;111:2633–51. https://doi.org/10.1007/s00421-011-1879-2.
11. Kay AD, Blazevich AJ. Effect of acute static stretch on maximal muscle performance: a systematic review. Med Sci Sport Exer. 2012;44(1):154–64. https://doi.org/10.1249/MSS.0b013e318225cb27.
12. Behm DG, Blazevich AJ, Kay AD, McHugh M. Acute effects of muscle stretching on physical performance, range of motion, and injury incidence in healthy active individuals: a systematic review. Appl Physiol Nutr Metab. 2016;41(1):1–11. https://doi.org/10.1139/apnm-2015-0235.
13. Blazevich AJ, Gill ND, Kvorning T, et al. No effect of muscle stretching within a full, dynamic warm-up on athletic performance. Med Sci Sports Exerc. 2018;50(6):1258–66. https://doi.org/10.1249/MSS.0000000000001539.

14. Opplert J, Babault N. Acute effects of dynamic stretching on muscle flexibility and performance: an analysis of the current literature. Sports Med. 2018;48:299–325. https://doi.org/10.1007/s40279-017-0797-9.
15. American Heart Association. Guidelines and statements, 2020. Available online at: https://professional.heart.org/en/guidelines-and-statements/guidelines-and-statements-search. Assessed 13 June 2023.
16. Sands WA, McNeal JR, Murray SR, et al. Stretching and its effects on recovery: a review. Strength Cond. 2013;35(5):30–6. https://doi.org/10.1519/SSC.0000000000000004.
17. Apostolopoulos NC, Lahart IM, Plyley MJ, et al. The effects of different passive static stretching intensities on recovery from unaccustomed eccentric exercise—a randomized controlled trial. Appl Physiol Nutr Metab. 2018;43:806–15. https://doi.org/10.1139/apnm-2017-0841.
18. Behm DG. The science and physiology of flexibility and stretching: implications and applications in sport performance and health. Routledge; 2018.
19. Mika A, Mika P, Fernhall B, Unnithan VB. Comparison of recovery strategies on muscle performance after fatiguing exercise. Am J Phys Med Rehabil. 2007;86:474–81. https://doi.org/10.1097/PHM.0b013e31805b7c79.
20. Cè E, Limonta E, Maggioni MA, Rampichini S, Veicsteinas A, Esposito F. Stretching and deep and superficial massage do not influence blood lactate levels after heavy-intensity cycle exercise. J Sports Sci. 2013;31:856–66. https://doi.org/10.1080/02640414.2012.753158.
21. Cheung K, Hume P, Maxwell L. Delayed onset muscle soreness: treatment strategies and performance factors. Sports Med. 2003;33:145–64. https://doi.org/10.2165/00007256-200333020-00005.
22. Wessel J, Wan A. Effect of stretching on the intensity of delayed-onset muscle soreness. Clin J Sport Med. 1994;4:83–7. https://doi.org/10.1097/00042752-199404000-00003.
23. Dupuy O, Douzi W, Theurot D, Bosquet L, Dugué B. An evidence-based approach for choosing post-exercise recovery techniques to reduce markers of muscle damage, soreness, fatigue, and inflammation: a systematic review with meta-analysis. Front Physiol. 2018;9:403. https://doi.org/10.3389/fphys.2018.00403.
24. Xie Y, Feng B, Chen K, Andersen LL, Page P, Wang Y. The efficacy of dynamic contract-relax stretching on delayed-onset muscle soreness among healthy individuals: a randomized clinical trial. Clin J Sport Med. 2018;28(1):28–36. https://doi.org/10.1097/JSM.0000000000000442.
25. West AD, Cooke MB, LaBounty PM, Byars AG, Greenwood M. Effects of G-trainer, cycle ergometry, and stretching on physiological and psychological recovery from endurance exercise. J Strength Cond Res. 2014;28:3453–61. https://doi.org/10.1519/JSC.0000000000000577.
26. Bandy WD, Irion JM, Briggler M. The effect of time and frequency of static stretching on flexibility of the hamstring muscles. J Orthop Sport Phys. 1994;77:1090–6. https://doi.org/10.1093/ptj/77.10.1090.
27. Haff GG. Roundtable discussion: flexibility training. Strength Cond J. 2006;28(2):64–85.
28. Konrad A, Alizadeh S, Daneshjoo A, et al. Chronic effects of stretching on range of motion with consideration of potential moderating variables: a systematic review with meta-analysis. JSHS. 2023;8:S2095-2546(23)00057-1. https://doi.org/10.1016/j.jshs.2023.06.002.
29. Kokkonen J, Nelson AG, Eldredge C, Winchester JB. Chronic static stretching improves exercise performance. Med Sci Sports Exerc. 2007;39(10):1825–31. https://doi.org/10.1249/mss.0b013e3181238a2b.
30. Diz JBM, de Souza JRL, Leopoldino AAO, Oliveira VC. Exercise, especially combined stretching and strengthening exercise, reduces myofascial pain: a systematic review. J Physiother. 2017;63(1):17–22. https://doi.org/10.1016/j.jphys.2016.11.008.
31. Behm DG, Kay AD, Trajano GS, Alizadeh S, Blazevich AJ. Effects of stretching on injury risk reduction and balance. J Clin Exerc Physiol. 2021;10(3):106–16. https://doi.org/10.31189/2165-6193-10.3.106.
32. Winchester JB, Nelson AG, Landin D, Young MA, Schexnayder IC. Static stretching impairs sprint performance in collegiate track and field athletes. J Strength Cond Res. 2008;22(1):13–9. https://doi.org/10.1519/JSC.0b013e31815ef202.

33. Young WB, Elias G, Power J. Effects of static stretching volume and intensity on plantar flexor explosive force production and range of motion. J Sport Med Phys Fit. 2006;46(3):403–11.
34. Donti O, Tsolakis C, Bogdanis GC. Effects of baseline levels of flexibility and vertical jump ability on performance following different volumes of static stretching and potentiating exercises in elite gymnasts. J Sports Sci Med. 2014;13(1):105–13.
35. Fletcher IM. The effect of different dynamic stretch velocities on jump performance. Eur J Appl Physiol. 2010;109:491–8. https://doi.org/10.1007/s00421-010-1386-x.
36. Konrad A, Močnik R, Nakamura M, Sudi K, Tilp M. The impact of a single stretching session on running performance and running economy: a scoping review. Front Physiol. 2021;11:1768. https://doi.org/10.3389/fphys.2020.630282.
37. Kido A, Tanaka N, Stein RB. Spinal reciprocal inhibition in human locomotion. J Appl Physiol. 2004;96(5):1969–77. https://doi.org/10.1152/japplphysiol.01060.2003.
38. Bacurau RF, Monteiro GA, Ugrinowitsch C, Tricoli V, Cabral LF, Aoki MS. Acute effect of a ballistic and a static stretching exercise bout on flexibility and maximal strength. J Streng Cond Res. 2009;23:304–8. https://doi.org/10.1519/JSC.0b013e3181874d55.
39. Nelson AG, Kokkonen J. Acute ballistic muscle stretching inhibits maximal strength performance. Res Q Exerc Sport. 2001;72:415–9. https://doi.org/10.1080/02701367.2001.10608978.
40. Konrad A, Tilp M. Effects of ballistic stretching training on the properties of human muscle and tendon structures. J Appl Physiol. 2014;117(1):29–35. https://doi.org/10.1152/japplphysiol.00195.2014.
41. Mahieu NN, McNair P, De Muynck M, et al. Effect of static and ballistic stretching on the muscle-tendon tissue properties. Med Sci Sports Exerc. 2007;39(3):494–501. https://doi.org/10.1249/01.mss.0000247004.40212.f7.
42. Shrier I, Gossal K. Myths and truths of stretching: individualized recommendations for healthy muscles. Phys Sportsmed. 2000;28(8):57–63. https://doi.org/10.3810/psm.2000.08.1159.
43. Page P. Current concepts in muscle stretching for exercise and rehabilitation. Int J Sports Phys Ther. 2012;7(1):109–19.
44. Samuel MN, Holcomb WR, Guadagnoli MA, Rubley MD, Wallmann H. Acute effects of static and ballistic stretching on measures of strength and power. J Strength Cond Res. 2008;22(5):1422–8. https://doi.org/10.1519/JSC.0b013e318181a314.
45. Jaggers JR, Swank AM, Frost KL, Lee CD. The acute effects of dynamic and ballistic stretching on vertical jump height, force, and power. J Strength Cond Res. 2008;22(6):1844–9. https://doi.org/10.1519/JSC.0b013e3181854a3d.
46. Konrad A, Gad M, Tilp MJ. Effect of PNF stretching training on the properties of human muscle and tendon structures. Scand J Med Sci Sports. 2015;25(3):346–55. https://doi.org/10.1111/sms.12228.
47. Sharman MJ, Cresswell AG, Riek S. Proprioceptive neuromuscular facilitation stretching mechanisms and clinical implications. Sports Med. 2006;36:929–39. https://doi.org/10.2165/00007256-200636110-00002.
48. Hindle K, Whitcomb T, Briggs W, Hong J. Proprioceptive neuromuscular facilitation (PNF): its mechanisms and effects on range of motion and muscular function. J Hum Kinet. 2012;31:105–13. https://doi.org/10.2478/v10078-012-0011-y.
49. Etnyre BR, Abraham LD. H-reflex during static stretching and two variations of proprioceptive neuromuscular facilitation techniques. Electroencephalogr Clin Neurophysiol. 1986;63(2):174–9. https://doi.org/10.1016/0013-4694(86)90010-6.
50. Feland JB, Myrer JW, Merrill RM. Acute changes in hamstring flexibility: PNF versus static stretch in senior athletes. Phys Ther Sport. 2001;2:186–93. https://doi.org/10.1054/ptsp.2001.0076.
51. Marek SM, Cramer JT, Fincher AL. Clinical studies - acute effects of static and proprioceptive neuromuscular facilitation stretching on muscle strength and power output. J Ath Train. 2005;40(2):94.
52. Butterfield TA, Herzog W. The magnitude of muscle strain does not influence serial sarcomere number adaptations following eccentric exercise. Pflugers Arch. 2006;451:688–700. https://doi.org/10.1007/s00424-005-1503-6.

53. Behm DG, Wilke J. Do self-myofascial release devices release myofascia? Rolling mechanisms: a narrative review. Sports Med. 2019;49(8):1173–81. https://doi.org/10.1007/s40279-019-01149-y.
54. Peacock CA, Krein DD, Silver TA, Sanders GJ, Von Carlowitz KPA. An acute bout of self-myofascial release in the form of foam rolling improves performance testing. Int J Exerc Sci. 2014;7(3):202.
55. Konrad A, Nakamura M, Tilp M, Donti O, Behm DG. Foam rolling training effects on range of motion: a systematic review and meta-analysis. Sports Med. 2022;52(10):2523–35. https://doi.org/10.1007/s40279-022-01699-8.
56. Beardsley C, Škarabot J. Effects of self-myofascial release: a systematic review. J Bodyw Mov Ther. 2015;19(4):747–58. https://doi.org/10.1016/j.jbmt.2015.08.007.
57. Cheatham SW, Kolber MJ, Cain M, Lee M. The effects of self-myofascial release using a foam roll or roller massager on joint range of motion, muscle recovery, and performance: a systematic review. Int J Sports Phys Ther. 2015;10(6):827.
58. Wilke J, Krause F, Vogt L, Banzer W. What is evidence-based about myofascial chains: a systematic review. Arch Phys Med Rehabil. 2016;97(3):454–61. https://doi.org/10.1016/j.apmr.2015.07.023.
59. Wilke J, Müller AL, Giesche F, Power G, Ahmedi H, Behm DG. Acute effects of foam rolling on range of motion in healthy adults: a systematic review with multilevel meta-analysis. Sports Med. 2020;50(2):387–402. https://doi.org/10.1007/s40279-019-01205-7.
60. De Vries HA. Physiology of exercise: for physical education and athletics. 4th ed. Dubuque: Wm. C; 1986.
61. Armiger P, Martyn MA. Stretching for functional flexibility. Philadelphia: Lippincott Williams & Wilkins; 2010.
62. Panidi I, Donti O, Konrad A. Muscle architecture adaptations to static stretching training: a systematic review with meta-analysis. Sports Med Int Open. 2023;9(1):1–27. https://doi.org/10.1186/s40798-023-00591-7.
63. Gleim GW, McHugh MP. Flexibility and its effects on sports injury and performance. Sports Med. 1997;24:289–99. https://doi.org/10.2165/00007256-199724050-00001.
64. McHugh MP, Cosgrave CH. To stretch or not to stretch: the role of stretching in injury prevention and performance. Scand J Med Sci Spor. 2010;20:169–81. https://doi.org/10.1111/j.1600-0838.2009.01058.x.
65. Safran MR, Seaber AV, Garrett WE. Warm-up and muscular injury prevention an update. Sports Med. 1989;8:239–49. https://doi.org/10.2165/00007256-198908040-00004.
66. High DM, Howley ET, Franks BD. The effects of static stretching and warm-up on prevention of delayed-onset muscle soreness. Res Q Exerc Sport. 1989;60(4):357–61. https://doi.org/10.1080/02701367.1989.10607463.
67. Torres R, Carvalho P, Duarte JA. Effects of a static stretching program on clinical and biochemical markers of muscle damage induced by eccentric exercise. Port J Sport Sci. 2005;5:274–87.
68. Andrade RJ, Freitas SR, Hug F, et al. Chronic effects of muscle and nerve-directed stretching on tissue mechanics. J Appl Physiol. 2020;129(5):1011–23. https://doi.org/10.1152/japplphysiol.00239.2019.
69. Freitas SR, Mil-Homens P. Effect of 8-week high-intensity stretching training on biceps femoris architecture. J Strength Cond Res. 2015;29(6):1737–40. https://doi.org/10.1519/JSC.0000000000000800.
70. Hutton RS. Neuromuscular basis of stretching exercises. In: Komi PV, editor. Strength and power in sport. Oxford, UK: Blackwell Scientific; pp; 1992. p. 29–38.
71. Borg G. Borg's perceived exertion and pain scales. Champaign: Human Kinetics; 1998.
72. Wyon M, Felton L, Galloway S. A comparison of two stretching modalities on lower-limb range of motion measurements in recreational dancers. J Strength Cond Res. 2009;23(7):2144–8. https://doi.org/10.1519/JSC.0b013e3181b3e198.
73. Wyon MA, Smith A, Koutedakis YA. Comparison of strength and stretch interventions on active and passive ranges of movement in dancers: a randomized controlled trial. J Strength Cond Res. 2013;27(11):3053–9. https://doi.org/10.1519/JSC.0b013e31828a4842.

74. Apostolopoulos N, Metsios GS, Flouris AD, Koutedakis Y, Wyon MA. The relevance of stretch intensity and position—a systematic review. Front Psychol. 2015;6:1128. https://doi.org/10.3389/fpsyg.2015.01128.
75. Apostolopoulos N, Metsios GS, Taunton J, Koutedakis Y, Wyon M. Acute inflammation response to stretching: a randomised trial. Ital J Sports Rehabil Posturol. 2015;2:368–81.
76. O'Hora J, Cartwright A, Wade CD, Hough AD, Shum GL. Efficacy of static stretching and proprioceptive neuromuscular facilitation stretch on hamstrings length after a single session. J Streng Cond Res. 2011;25:1586–91. https://doi.org/10.1519/JSC.0b013e3181df7f98.
77. Perrier ET, Pavol MJ, Hoffman MA. The acute effects of a warm-up including static or dynamic stretching on countermovement jump height, reaction time, and flexibility. J Streng Cond Res. 2011;25:1925–31. https://doi.org/10.1519/JSC.0b013e3181e73959.
78. Amiri-Khorasani M, Abu Osman NA, Yusof A. Acute effect of static and dynamic stretching on hip dynamic range of motion during instep kicking in professional soccer players. J Streng Cond Res. 2011;25:1647–52. https://doi.org/10.1519/JSC.0b013e3181db9f41.
79. Thomas E, Bianco A, Paoli A, Palma A. The relation between stretching typology and stretching duration: the effects on range of motion. Int J Sports Med. 2018;39:243–54. https://doi.org/10.1055/s-0044-101146.
80. Moltubakk MM, Villars FO, Magulas MM, et al. Altered triceps Surae muscle-tendon unit properties after six months of static stretching. Med Sci Sports Exerc. 2021;53(9):1975–86. https://doi.org/10.1249/MSS.0000000000002671.
81. Trajano GS, Seitz LB, Nosaka K, Blazevich AJ. Passive muscle stretching impairs rapid force production and neuromuscular function in human plantar flexors. Eur J Appl Physiol. 2019;119:2673–84. https://doi.org/10.1007/s00421-019-04244-0.
82. Blazevich AJ, Cannavan D, Waugh CM, Fath F, Miller SC, Kay AD. Neuromuscular factors influencing the maximum stretch limit of the human plantar flexors. J Appl Physiol. 2012;113:1446–55. https://doi.org/10.1152/japplphysiol.00882.2012.
83. Manca A, Dragone D, Dvir Z, Deriu F. Cross-education of muscular strength following unilateral resistance training: a meta-analysis. Eur J Appl Physiol. 2017;117:2335–54. https://doi.org/10.1007/s00421-017-3720-z.
84. Behm DG, Alizadeh S, Anvar SH, Drury B, Granacher U, Moran J. Non-local acute passive stretching effects on range of motion in healthy adults: a systematic review with meta-analysis. Sports Med. 2021;51:945–59. https://doi.org/10.1007/s40279-020-01422-5.
85. Hollinshead WH, Rosse CM. Textbook of anatomy. 4th ed. Baltimore: Lippincott Williams & Wilkins; 1985.
86. Blazevich AJ, Cannavan D, Waugh CM, et al. Range of motion, neuromechanical, and architectural adaptations to plantar flexor stretch training in humans. J Appl Physiol. 2014;117:452–62. https://doi.org/10.1152/japplphysiol.00204.2014.
87. De Weijer VC, Gorniak GC, Shamus E. The effect of static stretch and warm-up exercise on hamstring length over the course of 24 hours. J Orthop Sports Phys Ther. 2003;33:727–33. https://doi.org/10.2519/jospt.2003.33.12.727.
88. Chan SP, Hong Y, Robinson PD. Flexibility and passive resistance of the hamstrings of young adults using two different static stretching protocols. Scand J Med Sci Sports. 2001;11:81–6. https://doi.org/10.1034/j.1600-0838.2001.011002081.x.
89. Willy RW, Kyle BA, Moore SA, Chleboun GS. Effect of cessation and resumption of static hamstring muscle stretching on joint range of motion. J Orthop Sports PhysTher. 2001;31:138–44. https://doi.org/10.2519/jospt.2001.31.3.138.
90. Takeuchi K, Sato S, Kiyono R, et al. High-intensity static stretching in quadriceps is affected more by its intensity than its duration. Front Physiol. 2021;12:709655. https://doi.org/10.3389/fphys.2021.709655.
91. Gajdosik RL, Vander Linden DW, Williams AK. Influence of age on length and passive elastic stiffness characteristics of the calf muscle-tendon unit of women. Phys Ther. 1999;79(9):827–38.
92. Konrad A, Tilp M. Increased range of motion after static stretching is not due to changes in muscle and tendon structures. Clin Biomech. 2014;29(6):636–42. https://doi.org/10.1016/j.clinbiomech.2014.04.013.

93. Nordez A, Cornu C, McNair P. Acute effects of static stretching on passive stiffness of the hamstring muscles calculated using different mathematical models. Clin Biomech. 2006;21(7):755–60. https://doi.org/10.1016/j.clinbiomech.2006.03.005.
94. Guissard N, Duchateau J. Effect of static stretch training on neural and mechanical properties of the human plantar-flexor muscles. Muscle Nerve. 2004;29(2):248–55. https://doi.org/10.1002/mus.10549.
95. Folpp H, Deall S, Harvey LA, Gwinn T. Can apparent increases in muscle extensibility with regular stretch be explained by changes in tolerance to stretch? Aust J Physiother. 2006;52(1):45–50. https://doi.org/10.1016/s0004-9514(06)70061-7.
96. Magnusson SP, Simonsen EB, Aagaard P, Sørensen H, Kjaer M. A mechanism for altered flexibility in human skeletal muscle. J Physiol. 1996;497(1):291–8. https://doi.org/10.1113/jphysiol.1996.sp021768.
97. Behm DG, Kay AD, Trajano GS, Alizadeh S, Blazevich AJ. Effects of acute and chronic stretching on pain control. J Clin Exerc Physiol. 2021;10(4):150–9.
98. Law RY, Harvey LA, Nicholas MK, Tonkin L, De Sousa M, Finniss DG. Stretch exercises increase tolerance to stretch in patients with chronic musculoskeletal pain: a randomized controlled trial. Phys Ther. 2009;89(10):1016–26. https://doi.org/10.2522/ptj.20090056.
99. Lieber RL, Fridén J. Functional and clinical significance of skeletal muscle architecture. Muscle Nerve. 2000;23:1647–66. https://doi.org/10.1002/1097-4598(200011)23:11<1647::aid-mus1>3.0.co;2-m.
100. Ferraro E, Giammarioli AM, Chiandotto S, Spoletini I, Rosano G. Exercise-induced skeletal muscle remodeling and metabolic adaptation: redox signaling and role of autophagy. Antioxid Redox Signal. 2014;21:154–76. https://doi.org/10.1089/ars.2013.5773.
101. Franchi MV, Atherton PJ, Reeves ND, et al. Architectural, functional and molecular responses to concentric and eccentric loading in human skeletal muscle. Acta Physiol (Oxf). 2014;210(3):642–54. https://doi.org/10.1111/apha.12225.
102. Tabary JC, Tabary C, Tardieu C, Tardieu G, Goldspink G. Physiological and structural changes in the cat's soleus muscle due to immobilization at different lengths by plaster casts. J Physiol. 1972;224:231–44. https://doi.org/10.1113/jphysiol.1972.sp009891.
103. Cox VM, Williams PE, Wright H, et al. Growth induced by incremental static stretch in adult rabbit latissimus dorsi muscle. Exp Physiol. 2000;85(2):193–202.
104. Donti O, Panidis I, Terzis G, Bogdanis GC. Gastrocnemius Medialis architectural properties at rest and during stretching in female athletes with different flexibility training background. Sports. 2019;7(2):39. https://doi.org/10.3390/sports7020039.
105. Moltubakk MM, Magulas MM, Villars FO, Seynnes OR, Bojsen-Møller J. Specialized properties of the triceps surae muscle-tendon unit in professional ballet dancers. Scand J Med Sci Sports. 2018;28(9):2023–34. https://doi.org/10.1111/sms.13207.
106. Panidi I, Bogdanis GC, Gaspari V, et al. Gastrocnemius Medialis architectural properties in flexibility trained and not trained child female athletes: a pilot study. Sports. 2020;8(3):29. https://doi.org/10.3390/sports8030029.
107. Mohamad NI, Nosaka K, Cronin J. Maximizing hypertrophy: possible contribution of stretching in the interest rest period. Strength Cond J. 2011;33:81–7. https://doi.org/10.1519/SSC.0b013e3181fe7164.
108. Behm DG, Kibele A. Effects of differing intensities of static stretching on jump performance. Eur J App Physiol. 2007;101:587–94.
109. Tidball JG, Dorshkind K, Wehling-Henricks M. Shared signaling systems in myeloid cell-mediated muscle regeneration. Development. 2014;141:1184–96. https://doi.org/10.1242/dev.098285.
110. McCully KK. The influence of passive stretch on muscle oxygen saturation. In: Takahashi E, Bruley D, editors. Advances in experimental medicine and biology. Boston: Springer; 2010. p. 317–22.
111. Wilke J, Hespanhol L, Behrens M. Is it all about the fascia? A systematic review and meta-analysis of the prevalence of extramuscular connective tissue lesions in muscle strain injury. Orthop J Sports Med. 2019;7(12):2325967119888500. https://doi.org/10.1177/2325967119888500.

112. Wilke J, Schleip R, Yucesoy CA, Banzer W. Not merely a protective packing organ? A review of fascia and its force transmission capacity. J Appl Physiol. 2018;124(1):234–44. https://doi.org/10.1152/japplphysiol.00565.2017.
113. Adstrum S, Hedley G, Schleip R, Stecco C, Yucesoy CA. Defining the fascial system. J Bodyw Mov Ther. 2017;21(1):173–7. https://doi.org/10.1016/j.jbmt.2016.11.003.
114. Huijing PA. Intra-, extra-and intermuscular myofascial force transmision of synergists and antagonists: effects of muscle length as well as relative position. J Mech Med Biol. 2002;2(03n04):405–19. https://doi.org/10.1142/S0219519402000496.
115. Bernabei M, Van Dieen JH, Maas H. Altered mechanical interaction between rat plantar flexors due to changes in intermuscular connectivity. Scand J Med Sci Sports. 2017;27(2):177–87. https://doi.org/10.1111/sms.12644.
116. Haut TL, Haut RC. The state of tissue hydration determines the strain-rate-sensitive stiffness of human patellar tendon. J Biomech. 1997;30(1):79–81. https://doi.org/10.1016/s0021-9290(96)00108-x.
117. Karlsson MK, Dahan R, Magnusson H, Nyquist F, Rosengren BE. Soccer and gender effect of groin pain. Dtsch Z Sportmed. 2014;65:38–42. https://doi.org/10.5960/dzsm.2013.095.
118. Power K, Behm D, Cahill F, Carroll M, Young W. An acute bout of static stretching: effects on force and jumping performance. Med Sci Sports Exerc. 2004;36:1389–96. https://doi.org/10.1249/01.mss.0000135775.51937.53.
119. Reid D, McNair PJ. Passive force, angle, and stiffness changes after stretching of hamstring muscles. Med Sci Sports Exerc. 2004;36(11):1944–8. https://doi.org/10.1249/01.mss.0000145462.36207.20.
120. Avloniti A, Chatzinikolaou A, Fatouros I, et al. The acute effects of static stretching on speed and agility performance depend on stretch duration and conditioning level. J Strength Cond Res. 2016;30(10):2767–73. https://doi.org/10.1519/JSC.0000000000000568.
121. Tsolakis C, Bogdanis GC. Acute effects of two different warm-up protocols on flexibility and lower limb explosive performance in male and female high level athletes. J Sports Sci Med. 2012;11(4):669.
122. Behm DG, Sale DG. Velocity specificity of resistance training. Sports Med. 1993;15:74–388. https://doi.org/10.2165/00007256-199315060-00003.
123. Fletcher IM, Jones B. The effect of different warm-up stretch protocols on 20 meter sprint performance in trained rugby union players. J Strength Cond Res. 2004;18(4):885–8. https://doi.org/10.1519/14493.1.
124. Barbosa GM, Trajano GS, Dantas GA, Silva BR, Vieira WHB. Chronic effects of static and dynamic stretching on hamstrings eccentric strength and functional performance: a randomized controlled trial. J Strength Cond Res. 2020;34(7):2031–9. https://doi.org/10.1519/JSC.0000000000003080.
125. Behm DG, Kay AD, Trajano GS, Blazevich AJ. Mechanisms underlying performance impairments following prolonged static stretching without a comprehensive warm-up. Eur J App Physiol. 2021;121:67–94. https://doi.org/10.1007/s00421-020-04538-8.
126. Konrad A, Stafilidis S, Tilp M. Effects of acute static, ballistic, and PNF stretching exercise on the muscle and tendon tissue properties. Scand J Med Sci Sports. 2017;27(10):1070–80. https://doi.org/10.1111/sms.12725.
127. Magnusson P, Renström P. The European College of Sports Sciences Position statement: the role of stretching exercises in sports. Eur J Sport Sci. 2006;6(2):87–91. https://doi.org/10.1080/17461390600617865.
128. Donti O, Konrad A, Panidi I, Dinas PC, Bogdanis GC. Is there a "Window of Opportunity" for flexibility development in youth? A systematic review with meta-analysis. Sports Med Open. 2002;8(1):88. https://doi.org/10.1186/s40798-022-00476-1.
129. Donti O, Bogdanis GC, Kritikou M, Donti A, Theodorakou K. The relative contribution of physical fitness to the technical execution score in youth rhythmic gymnastics. J Hum Kinet. 2016;51:143. https://doi.org/10.1515/hukin-2015-0183.
130. Sands WA, McNeal JR, Penitente G, et al. Stretching the spines of gymnasts: a review. Sports Med. 2016;46:315–27. https://doi.org/10.1007/s40279-015-0424-6.

131. Beach ML, Whitney SL, Dickoff-Hoffman SA. Relationship of shoulder flexibility, strength, and endurance to shoulder pain in competitive swimmers. J Orthop Sports Phys Ther. 1992;16(6):262–8. https://doi.org/10.2519/jospt.1992.16.6.262.
132. Cejudo A, Sánchez-Castillo S, Sainz de Baranda P, Gámez JC, Santonja-Medina F. Low range of shoulders horizontal abduction predisposes for shoulder pain in competitive young swimmers. Front Psychol. 2019;10:478. https://doi.org/10.3389/fpsyg.2019.00478.
133. Kruszewski M, Kruszewski A, Tabęcki R, Mierzejewski B, Pągowski Ł. Range of motion in selected joints in relation to sports performance and technique effectiveness in weightlifting. Pol J Sport Tour. 2022;29(1):9–13. https://doi.org/10.2478/pjst-2022-0002.
134. Zügel M, Maganaris CN, Wilke J, et al. Fascial tissue research in sports medicine: from molecules to tissue adaptation, injury and diagnostics: consensus statement. Br J Sports Med. 2018;52:1497. https://doi.org/10.1136/bjsports-2018-099308.
135. Warneke K, Brinkmann A, Hillebrecht M, Schiemann S. Influence of long-lasting static stretching on maximal strength, muscle thickness and flexibility. Front Physiol. 2022;13:878955. https://doi.org/10.3389/fphys.2022.878955.
136. Pope R, Herbert R, Kirwan J. Effects of ankle dorsiflexion range and preexercise calf muscle stretching on injury risk in army recruits. Aust J Physiother. 1998;44:165–77. https://doi.org/10.1016/s0004-9514(14)60376-7.
137. Woods K, Bishop P, Jones E. Warm-up and stretching in the prevention of muscular injury. Sports Med. 2007;37:1089–99. https://doi.org/10.2165/00007256-200737120-00006.
138. Witvrouw E, Mahieu DL, McNair P. Stretching and injury prevention: an obscure relationship. Sports Med. 2004;34:443–9. https://doi.org/10.2165/00007256-200434070-00003.
139. Shrier I. Stretching before exercise does not reduce the risk of local muscle injury: a critical review of the clinical and basic science literature. Clin J Sport Med. 1999;9(4):221–7. https://doi.org/10.1097/00042752-199910000-00007.
140. Jamtvedt G, Herbert RD, Flottorp S, et al. A pragmatic randomised trial of stretching before and after physical activity to prevent injury and soreness. Br J Sports Med. 2010;44(14):1002–9. https://doi.org/10.1136/bjsm.2009.062232.
141. Hartig DE, Henderson JM. Increasing hamstring flexibility decreases lower extremity overuse injuries in military basic trainees. Am J Sports Med. 1999;27(2):173–6. https://doi.org/10.1177/03635465990270021001.
142. Emery CA, Meeuwisse WH. The effectiveness of a neuromuscular prevention strategy to reduce injuries in youth soccer: a cluster-randomised controlled trial. Br J Sports Med. 2010;44(8):555–62. https://doi.org/10.1136/bjsm.2010.074377.
143. Chagas DDV, Barnett LM. Adolescents' flexibility can affect motor competence: the pathway from health-related physical fitness to motor competence. Percept Mot Skills. 2023;130(1):94–111. https://doi.org/10.1177/00315125221128638.
144. Nguyen AD, Zuk EF, Baellow AL, Pfile KR, DiStefano LJ, Boling MC. Longitudinal changes in hip strength and range of motion in female youth soccer players: implications for ACL injury, a pilot study. J Sport Rehabil. 2017;26(5):358–64. https://doi.org/10.1123/jsr.2015-0197.
145. Madden C, Putukian M, McCarty E, Young C. Netter's sports medicine E-book. Elsevier Health Sciences; 2013.
146. Webber S, Kriellaars D. Neuromuscular factors contributing to in vivo eccentric moment generation. J Appl Physiol. 1997;1985(83):40–5. https://doi.org/10.1152/jappl.1997.83.1.40.
147. Hinks A, Franchi MV, Power GA. The influence of longitudinal muscle fascicle growth on mechanical function. J Appl Physiol. 2022;133(1):87–103. https://doi.org/10.1152/japplphysiol.00114.2022.
148. Herzog W. The problem with skeletal muscle series elasticity. BMC Biomed Eng. 2019;1(1):1–14. https://doi.org/10.1186/s42490-019-0031-y.
149. Popp JK, Bellar DM, Hoover DL, et al. Pre-and post-activity stretching practices of collegiate athletic trainers in the United States. J Strength Cond Res. 2017;31(9):2347–54. https://doi.org/10.1519/JSC.0000000000000890.
150. Pickering Rodriguez EC, Watsford ML, Bower RG, Murphy AJ. The relationship between lower body stiffness and injury incidence in female netballers. Sports Biomech. 2017;16(3):361–73. https://doi.org/10.1080/14763141.2017.1319970.

151. Panoutsakopoulos V, Bassa E. Countermovement jump performance is related to ankle flexibility and knee extensors torque in female adolescent volleyball athletes. J Funct Morphol Kinesiol. 2023;8(2):76. https://doi.org/10.3390/jfmk8020076.
152. Witvrouw E, Mahieu N, Roosen P, McNair P. The role of stretching in tendon injuries. Br J Sports Med. 2007;41(4):224–6. https://doi.org/10.1136/bjsm.2006.034165.
153. Kellmann M, Bertollo M, Bosquet L, et al. Recovery and performance in sport: consensus statement. Int J Sports Physiol Perform. 2018;13:240–5. https://doi.org/10.1123/ijspp.2017-0759.
154. Ekstrand J, Spreco A, Davison M. Elite football teams that do not have a winter break lose on average 303 player-days more per season to injuries than those teams that do: a comparison among 35 professional European teams. Br J Sports Med. 2018;53:9506. https://doi.org/10.1136/bjsports-2018-099506.
155. Bengtsson H, Ekstrand J, Hägglund M. Muscle injury rates in professional football increase with fixture congestion: an 11-year follow-up of the UEFA champions league injury study. Br J Sports Med. 2013;47:743–7. https://doi.org/10.1136/bjsports-2013-092383.
156. Judge LW, Petersen JC, Bellar DM, et al. An examination of preactivity and postactivity stretching practices of crosscountry and track and field distance coaches. J Strength Cond Res. 2013;27(9):2456–64. https://doi.org/10.1519/JSC.0b013e318257703c.
157. McGrath RP, Whitehead JR, Caine DJ. The effects of proprioceptive neuromuscular facilitation stretching on post-exercise delayed onset muscle soreness in young adults. Int J Exerc Sci. 2014;7:14–21.
158. Muanjai P, Namsawang J. Effects of stretching and cold-water immersion on functional signs of muscle soreness following plyometric training. J Phys Educ Sport. 2015;15:128–35. https://doi.org/10.7752/jpes.2015.01021.
159. César EP, Júnior CSR, Francisco RN. Effects of 2 intersection strategies for physical recovery in Jiu-Jitsu athletes. Int J Sports Physiol Perform. 2021;16:1–6. https://doi.org/10.1123/ijspp.2019-0701.
160. Cooke MB, Nix CM, Greenwood LD, Greenwood MC. No differences between alter G-trainer and active and passive recovery strategies on isokinetic strength, systemic oxidative stress and perceived muscle soreness after exercise-induced muscle damage. J Streng Cond Res. 2018;32:736–47. https://doi.org/10.1519/JSC.0000000000001750.
161. Kokkinidis E, Tsamourtas A, Buckenmeyer P, Machairidou M. The effect of static stretching and cryotherapy on the recovery of delayed muscle soreness. Exerc Soc J Sport Sci. 1998;1998(19):45–53.
162. Torres R, Carvalho P, Duarte JA. Effects of a static stretching program on clinical and biochemical markers of muscle damage induced by eccentric exercise. Rev Port Ciên Desp. 2005;5:274–87. https://doi.org/10.5628/rpcd.05.03.274.
163. Torres R, Pinho F, Duarte JA, Cabri JMH. Effect of single bout versus repeated bouts of stretching on muscle recovery following eccentric exercise. J Sci Med Sport. 2013;16:583–8. https://doi.org/10.1016/j.jsams.2013.01.002.
164. De Oliveira F, Paz GA, Corrêa Neto VG, et al. Effects of different recovery modalities on delayed onset muscle soreness, recovery perceptions, and performance following a Bout of high-intensity functional training. Int J Environ Res Public Health. 2023;20(4):3461. https://doi.org/10.3390/ijerph20043461.
165. Franchi MV, Reeves ND, Narici MV. Skeletal muscle remodeling in response to eccentric vs. concentric loading: morphological, molecular, and metabolic adaptations. Front Physiol. 2017;8:447. https://doi.org/10.3389/fphys.2017.00447.
166. Peake JM, Neubauer O, Della Gatta PA, Nosaka K. Muscle damage and inflammation during recovery from exercise. J Appl Physiol. 2017;122(3):559–70. https://doi.org/10.1152/japplphysiol.00971.2016.
167. Hughes GA, Ramer LM. Duration of myofascial rolling for optimal recovery, range of motion, and performance: a systematic review of the literature. Int J Sports Phys Ther. 2019;14(6):845.

168. Skinner B, Moss R, Hammond L. A systematic review and meta-analysis of the effects of foam rolling on range of motion, recovery and markers of athletic performance. J Bodyw Mov Ther. 2020;24(3):105–22.
169. MacDonald GZ. Foam rolling as a recovery tool following an intense bout of physical activity (Doctoral dissertation, Memorial University of Newfoundland), 2013.
170. Bandy WD, Irion JM. The effect of time on static stretch on the flexibility of the hamstring muscles. Phys Ther. 1994;74(9):845–50. https://doi.org/10.1093/ptj/74.9.845.
171. Apostolopoulos N. microStretching-A practical approach for recovery and regeneration. New Stud Åthl. 2010;25:81–97.
172. Herbert RD, de Noronha M, Kamper SJ. Stretching to prevent or reduce muscle soreness after exercise. Cochrane Database Syst Rev. 2011;(7) https://doi.org/10.1002/14651858.CD004577.pub3.
173. Kay AD, Rubley B, Talbot C, Mina M, Baross AW, Blazevich AJ. Stretch imposed on active muscle elicits positive adaptations in strain risk factors and exercise-induced muscle damage. Scand J Med Sci Sports. 2018;28(11):2299–309. https://doi.org/10.1111/sms.13251.
174. Chen CH, Nosaka K, Chen HL, Lin MJ, Tseng KW, Chen TC. Effects of flexibility training on eccentric exercise induced muscle damage. Med Sci Sports Exerc. 2011;43(3):491–500. https://doi.org/10.1249/MSS.0b013e3181f315ad.
175. Lund H, Vestergaard-Poulsen P, Kanstrup I-L, Sejrsen P. The effect of passive stretching on delayed onset muscle soreness, and other detrimental effects following eccentric exercise. Scand J Med Sci Sports. 1998;8(4):216–21. https://doi.org/10.1111/j.1600-0838.1998.tb00195.x.
176. Blazevich AJ. Adaptations in the passive mechanical properties of skeletal muscle to altered patterns of use. J Appl Physiol. 2019;126(5):1483–91. https://doi.org/10.1152/japplphysiol.00700.2018.
177. World Health Organization WH. Musculoskeletal conditions. Updated 2021. Accessed 16 Feb 2021. https://www.who.int/news-room/fact-sheets/detail/musculoskeletal-conditions.
178. Valenza MC, Torres-Sánchez I, Cabrera-Martos I, ValenzaDemet G, Cano-Cappellacci M. Acute effects of contract-relax stretching vs. tens in young subjects with anterior knee pain: a randomized controlled trial. J Strength Cond Res. 2016;30(8):2271–8. https://doi.org/10.1097/JSC.0000000000000215.
179. Lin LY, Wang RH. Effectiveness of a neck stretching intervention on nurses' primary headaches. Workplace Health Saf. 2015;63(3):100–6. https://doi.org/10.1177/2165079915571355.
180. Camargo PR, Alburquerque-Sendín F, Avila MA, Haik MN, Vieira A, Salvini TF. Effects of stretching and strengthening exercises, with and without manual therapy, on scapular kinematics, function, and pain in individuals with shoulder impingement: a randomized controlled trial. J Orthop Sports Phys Ther. 2015;45(12):984–97. https://doi.org/10.2519/jospt.2015.5939.
181. Gutiérrez-Espinoza H, Araya-Quintanilla F, Gutiérrez-Monclus R, Ríos-Riquelme M, Álvarez-Bueno C, MartínezVizcaino V, Cavero-Redondo I. Does pectoralis minor stretching provide additional benefit over an exercise program in participants with subacromial pain syndrome? A randomized controlled trial. Musculoskelet Sci Pract. 2019;44:102052. https://doi.org/10.1016/j.msksp.2019.102052.
182. Cleland JA, Childs JD, Palmer JA, Eberhart S. Slump stretching in the management of non-radicular low back pain: a pilot clinical trial. Man Ther. 2006;11(4):279–86. https://doi.org/10.1016/j.math.2005.07.002.
183. Ramazzina I, Bernazzoli B, Braghieri V, Costantino C. Groin pain in athletes and non-interventional rehabilitative treatment: a systematic review. J Sports Med Phys Fit. 2019;59:1001–10. https://doi.org/10.23736/S0022-4707.18.08879-5.
184. Jansen JA, Mens JM, Backx FJ, Kolfschoten N, Stam HJ. Treatment of longstanding groin pain in athletes: a systematic review. Scand J Med Sci Sports. 2008;18:263–74. https://doi.org/10.1111/j.1600-0838.2008.00790.x.
185. King E, Ward J, Small L, Falvey E, Franklyn-Miller A. Athletic groin pain: a systematic review and meta-analysis of surgical versus physical therapy rehabilitation outcomes. Br J Sports Med. 2015;49:1447–51. https://doi.org/10.1136/bjsports-2014-093715.

186. Machotka Z, Kumar S, Perraton LG. A systematic review of the literature on the effectiveness of exercise therapy for groin pain in athletes. Sports Med Arthrosc Rehabil. 2009;1:1–10. https://doi.org/10.1186/1758-2555-1-5.
187. Serner A, Van Eijck CH, Beumer BR, Hölmich P, Weir A, DeVos RJ. Study quality on groin injury management remain slow: a systematic review on treatment of groin pain in athletes. Br J Sports Med. 2015;49(12):813. https://doi.org/10.1136/bjsports-2014-094256.
188. Afonso J, Claudino JG, Fonseca H, et al. Stretching for recovery from groin pain or injury in athletes: a critical and systematic review. J Funct Morphol Kinesiol. 2021;6(3):73. https://doi.org/10.3390/jfmk6030073.
189. Kachingwe AF, Grech S. Proposed algorithm for the management of athletes with athletic pubalgia (sports hernia): a case series. J Orthop Sports Phys Ther. 2008;38:768–81. https://doi.org/10.2519/jospt.2008.2846.
190. Magnusson SP, Aagard P, Simonsen E, Bojsen-Møller F. A biomechanical evaluation of cyclic and static stretch in human skeletal muscle. Int J Sports Med. 1998;19(5):310–6. https://doi.org/10.1055/s-2007-971923.
191. Malliaropoulos N, Papalexandris S, Papalada A, Papacostas E. The role of stretching in rehabilitation of hamstring injuries: 80 athletes follow-up. Med Sci Sports Exerc. 2004;36(5):756–9. https://doi.org/10.1249/01.mss.0000126393.20025.5e.
192. Kim G, Kim H, Kim WK, Kim J. Effect of stretching-based rehabilitation on pain, flexibility and muscle strength in dancers with hamstring injury: a single-blind, prospective, randomized clinical trial. J Sports Med Phys Fitness. 2017;58(9):1287–95. https://doi.org/10.23736/S0022-4707.17.07554-5.
193. Chan EWM, Adnan R, Azmi R. Effectiveness of core stability training and dynamic stretching in rehabilitation of chronic low back pain patient. Malays J Mov Health Exerc. 2019;8(1):1–13. https://doi.org/10.15282/mohe.v8i1.210.
194. Khalil TM, Asfour SS, Martinez LM, Waly SM, Rosomoff RS, Rosomoff HL. Stretching in the rehabilitation of low-back pain patients. Spine. 1992;17(3):311–7. https://doi.org/10.1097/00007632-199203000-00012.
195. Musarò A. The basis of muscle regeneration. AdvBiol. 2014; https://doi.org/10.1155/2014/612471.
196. Chargé SB, Rudnicki MA. Cellular and molecular regulation of muscle regeneration. Physiol Rev. 2004;84(1):209–38. https://doi.org/10.1152/physrev.00019.2003.
197. Bénard MR, Harlaar J, Becher JG, Huijing PA, Jaspers RT. Effects of growth on geometry of gastrocnemius muscle in children: a three-dimensional ultrasound analysis. J Anat. 2011;219:388–402. https://doi.org/10.1111/j.1469-7580.2011.01402.x.
198. Goldspink G, Tabary C, Tabary JC, Tardieu C, Tardieu G. Effect of denervation on the adaptation of sarcomere number and muscle extensibility to the functional length of the muscle. J Physiol. 1974;236:733–42. https://doi.org/10.1113/jphysiol.1974.sp010463.
199. Pincheira PA, Boswell MA, Franchi MV, Delp SL, Lichtwark GA. Biceps femoris long head sarcomere and fascicle length adaptations after 3 weeks of eccentric exercise training. J Sport Health Sci. 2022;11(1):43–9. https://doi.org/10.1016/j.jshs.2021.09.002.
200. Berrueta L, Muskaj I, Olenich S, et al. Stretching impacts inflammation resolution in connective tissue. J Cell Physiol. 2016;231(7):1621–7. https://doi.org/10.1002/jcp.25263.
201. Kjaer M, Langberg H, Miller BF, et al. Metabolic activity and collagen turnover in human tendon in response to physical activity. J Musculoskelet Neuronal Interact. 2005;5(1):41–52.
202. Król M, Kupnicka P, Bosiacki M, Chlubek D. Mechanisms underlying anti-inflammatory and anti-cancer properties of stretching—a review. Int J Mol Sci. 2022;23(17):10127. https://doi.org/10.3390/ijms231710127.
203. Høier B, Rufener N, Bojsen- Møller J, Bangsbo J, Hellsten Y. The effect of passive movement training on angiogenic factors and capillary growth in human skeletal muscle. J Physiol. 2010;588:3833–45. https://doi.org/10.1113/jphysiol.2010.190439.

Chapter 20
The Genesis and Evolution of Low-Intensity Gentle Stretching—microStretching®

Nikos C. Apostolopoulos

Introduction

> "The task is not so much to see what no one yet has seen, but to think what nobody yet has thought about that which everybody sees".
>
> –Arthur Schopenhauer (1788–1860)

"Movement is what is closest to the soul". With this phrase, the Greek philosopher, Aristotle (384–322 BCE) identified the critical nature of 'kinesis' in the life and nature of humans. Kinesis is defined as an undirected response to stimulation of the muscles of the body [64], which enables us to move about our immediate environment, to explore nature, to obtain sustenance, and to procreate. To survive, organisms must effectively respond to the challenge of maintaining their physiological integrity and preserve homeostasis, in the face of an ever-changing environment [1, 165].

From an evolutionary perspective, movement is intimately linked to the genesis of the nervous system, and plays a crucial role in its maintenance and adaptation [56]. For instance, the sensing of the position of the body in space (proprioception) depends on how muscle tension and joint position affect the mechanical stress and strain of the embedded somatosensory neurons and the ability of these neurons to generate electrical signals encoding the mechanical inputs [102, 172, 173]. Proprioception plays a crucial role in regulating cellular, tissue, and organ function [102].

N. C. Apostolopoulos (✉)
Founder microStretching® and Stretch Therapy™, Toronto, ON, Canada
e-mail: nikos.apostolopoulos@microstretching.com

N. C. Apostolopoulos et al. (eds.), *Fundamentals of Recovery, Regeneration, and Adaptation to Exercise Stress: An Integrated Approach*,
https://doi.org/10.1007/978-3-031-44270-4_20

Through movement we develop an appreciation of actions of the body and learn how to control and fine tune the information sent to the brain for processing, allowing us to learn, to develop skills, and to appreciate the body and oneself [101, 144]. According to Marleau-Ponty, the body is a material agent of all our perceptual acts, and it is in our perception that we understand ourselves in terms of our body and its movement(s) [79], and where consciousness and reality occupy the same conceptual space [8]. Specific sensory receptors, classified by the type of stimulus to which they respond—mechanoreceptors (sensing pressure), nociceptors (sensing pain), thermoreceptors (sensing temperature), and chemoreceptors (sensing chemicals), provide updates on changes in status. In addition, three somatic sensors provide information concerning the body relative to itself and to the environment—interoceptors, exteroceptors, and proprioceptors, provide updated information concerning the internal body, the external body relative to the environment, and the body relative to movement in time and space (see Box 1 below). These neologisms and their first usage appeared in a book entitled, "The Integrative Action of The Nervous System" by British neurophysiologist, Sir Charles S Sherrington [195].

The three somatic sensors provide information about the state of the body and its surroundings. Although many aspects make them particularly unique and distinct from one another, they work collectively, not exclusively, as was previously thought [40]. This viewpoint is important because it allows for the establishment of novel concepts, and in the creation of new ways of supporting, understanding, and articulating experiences, thereby shaping future somatic practices, including therapeutic actions.

Interoception — a collection of senses providing conscious and subconscious information about the internal state of the body, including both emotional (feelings, pain, psychological sense of well-being, etc.) and somatic (visceral, vasomotor, etc.) state. Interoception is the overall process by which the nervous system (central and autonomic) senses, interprets, and integrates signals originating from within the body [42].

Exteroception — the process whereby the nervous system senses, interprets, and integrates signals from the external environment. This information is gathered through the primary senses of the body (vision, audition, olfaction, taste, and somatosensation). Exteroception enables the body to form an estimate about continuously evolving external states of the environment across both conscious and subconscious levels [165].

Proprioception — allows for the perception of the location, movement, and action of parts of the body. It is the process by which the nervous system interprets the concept of joint position and movement, muscle force, and effort [207]. Proprioception arises from signals of the afferent sensory receptors located within the muscle (muscle spindles), the tendon (Golgi tendon organs), the joints (Ruffini endings), and the skin (Pacinian, Ruffini, Meissner's corpuscles, and Merkel's disks) [44]. Proprioception allows for the spatial orientation of the body without visual cues.

My thoughts on stretching begin with the premise that stretching is a process with a dynamic interface, which influences various components of the body from the nano-, the micro-, and the macro- perspective [7]. Research has observed that the mechanical impact of stretching has a direct effect on inflammation-regulation mechanisms within connective tissue [10, 23, 119, 168]. As a practicing therapist, experienced scientist, and developer of Stretch Therapy™ and the therapeutic techniques of microStretching, the response(s) of the body to physical effort and the etiology of both injury and recovery, I am interested in trying to learn everything I can concerning the causes and effects of acute and chronic adverse events, and how to use this knowledge in my practice to achieve a better outcome by improving my technique, incorporating the latest research finding, and continually asking questions. Prior to embarking on the studies for my PhD, years of anecdotal evidence afforded me the opportunity to formulate ideas and hypotheses concerning the response(s) of the human body to the practice of stretching. I became interested in the impact of the magnitude of muscle stretch on injury prevention and recovery—in particular, the potential positive effects of low-intensity, gentle stretching (microStretching) on muscle, and muscle function. Focused on mitigating the effects of chronic pain and inflammation, my observations over a number of years were instrumental in defining specific principles and procedures that have come to characterise the practice of microStretching (see below).

A systematic literature review on both stretch intensity (magnitude of the stretch) and body position assumed during stretching, whether loaded—the body supports its entire weight, or unloaded—body mass all or somewhat supported (i.e., sitting, lying down, etc.), concluded that body position may influence the magnitude of the force generated prior to and during the stretch, potentially altering the response of the muscle and tendon tissue(s) [11].

It can be argued that the mechanical act of stretching forms a dynamic juncture between the three somatic sensations (interoception, exteroception, and proprioception), with inflammation and the inflammatory response significantly altering the sensitivity of the musculoskeletal tissue(s), especially if it is associated with tissue damage.

Overall, the evidence and its analysis indicate that stretching should be conceived as a regulatory mechanism, which may play a principal role in the re-establishment of the body's homeostasis from injury, disease, and ameliorating inflammation and the inflammatory response [23, 119, 219].

A "Sketchy" History of Stretching

> "O, let him pass. He hates him that would upon the *rack* of this tough world *stretch* him out longer ..."
>
> *William Shakespeare*. 'King Lear' (1605–1606), act 5, sc. 3, I. 314

While no artifacts in terms of written word, stone carvings, paintings, etc., have been found that illustrate stretching per se, it seems reasonable to assume that as long as humans have, hunted, participated in warfare, or been involved in social-cultural activities (sport, dance, gymnastics, etc.), some form of stretching must have been practiced.

Prior to the dawn of human civilisation (4000 and 3000 B.C.E), humans lived a semi-nomadic existence as hunters and gatherers, following herds of animals to hunt, gathering edible plants, residing in caves or simple huts, using basic stone and bone tools [106], and living within seasonal environmental constraints [74]. With their survival depending on their ability to move about, whether it be to provide food for the family, or to their household, or to a new region, they needed to develop a number of physical abilities, e.g., running, leaping, throwing objects, etc. [63]. Realising the importance of and the maintenance of some form of functionality and to develop an efficiency of movement, it is likely that they attempted to improve their physical abilities and talents to facilitate their fighting and survival skills. Inevitably, their strenuous physical lifestyle led to various discomforts and injuries, which required the development of various remedial activities; however, it is believed that diagnosing, treating, and preventing diseases were spiritual in nature [63, 83, 212]. Given such a setting, it seems extremely likely that some form of stretching or similar activity was practiced, especially if prehistoric man observed the act of pandiculation amongst the other mammals.

Pandiculation (Latin: Pandiculare = stretching): the involuntary stretching of the soft tissues of most animal species, including humans. It is associated with transitions between cyclic biological behaviors, especially the sleep-wake rhythm [224]. When accompanied by yawning it is referred to as the stretch-yawning syndrome [24].

Pandiculation is an old, almost ubiquitous behavior occurring in a similar form and circumstance across a wide spectrum of species [24]. It has been associated with the maintenance of arousal and attention and is involved in setting and maintaining both the central nervous system and the locomotor system, allowing animals to perceive environmental stimuli, and to respond to them with appropriate motor actions [13, 15, 24, 223]. The arousal associated with vigilance and attentiveness can be measured by motor activity and physiological variables, such as muscle tension, heart rate, etc., [15], and the act of stretching may be a necessary form of pre-activity which is needed to ensure maximum fitness and efficiency for hunting, gathering, attacking, or escaping. With the emergence of the "city state" denoting the beginning of civilisation, the rise of agriculture and trade brought economic stability and the wide-spread availability of food, as well as an increased time for recreational and cultural pursuits [215]. The recorded history of this time is a rich mosaic of people, places, events, and social forces, entwined with religion, education, and government activities, defining the arts, sports, and the pastimes of different cultures [87].

The ancient Egyptians, Greek city states (Athens, Sparta, Thebes, etc.), and Romans engaged in many sport activities as part of their education and recreation—many aimed at encouraging courageous competitiveness [45]. Within these cultures, children were instructed in running and leaping, wrestling, throwing the javelin and discus, dancing (form of a military drill), militaristic training (staged battles or brawls), boxing, swimming, ball games, swinging and lifting weights, and formation movement [63, 87, 89, 100]. According to the ancient Greek philosopher, Aristotle: 'expertise in letters and drawing [while] useful for life and having many uses, gymnastics [contribute] to courage' [183]. The ideal Athenian citizen was a balanced combination of soldier, athlete, artist, statesman, and philosopher [58, 73], and the model Spartan, possessed heightened physical prowess in the form of enduring physical pain and fostering the strength and endurance required for being a warrior. Young children were encouraged to develop strength through sports [100, 113, 134].

It is not surprising that the high level of involvement in physical education, sport, and other activities necessitated a sound foundation in medicine [143]. The first known study of what was termed "therapeutic gymnastics" was conducted by Herodicus (5th century BCE), the "father of sports medicine" [70]. A Greek physician and *paidotribes* (a formal tutor in sport wrestling and gymnastics), Herodicus was the first to suggest that bad health was the result of an imbalance between diet and physical activity, and that patients should minimise bed rest, and start moving as part of recovery and rehabilitation, which is now a commonly prescribed practice in contemporary medicine [70].

Another Greek physician, Hippocrates (c. 460–380 BCE), the "father of scientific medicine", developed many of the principles concerned with the treatment and management of athletic injuries, including proper setting of bones, that may have occurred in Greek wrestling and pankration [136].

Hippocrates was credited with the creation of the Hippocratic bench (Scamnum) (Fig. 20.1) [220]. The bench was an apparatus used to reduce the fractures and dislocation of the lower limb which occurred often in wrestling matches and other competitive games during classical times [169]. Hippocrates considered that a sound knowledge of the structure and function of the spine was an essential requisite for understanding disease [38]. He used the bench as part of his management of spinal deformities, and the treatment of spinal injuries. The combination of applying simultaneous traction (tensile force = stretching) of spine and focal pressure (Fig. 20.2) was an approach used by Hippocrates for correcting spinal deformities and fractures [109].

The Hippocratic bench is considered the forerunner of the traction devices commonly used today in orthopaedics and neurosurgery. Other "applications" of this concept include the sinister piece of equipment used to torture individuals during the inquisition of Medieval times [27], i.e., the so-called torture 'rack' (Fig. 20.3). Both devices used forceful stretching of the body through the application of

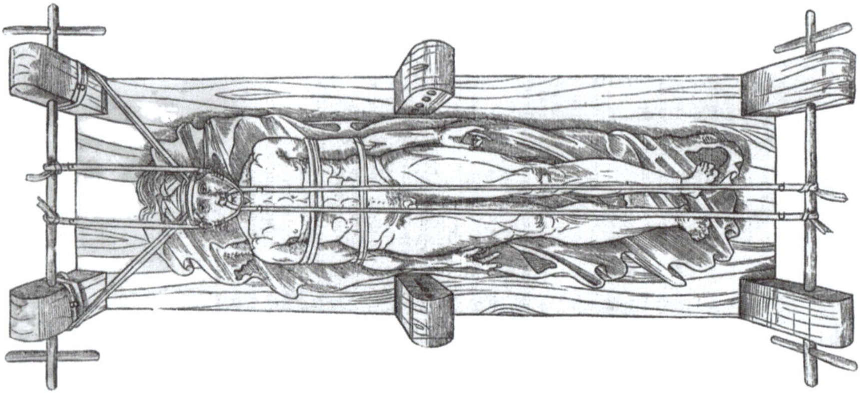

Fig. 20.1 "With the patient lying upon the scamnum either prone, supine or on the side, the dislocated limb is placed uppermost, whilst the fixed portion of the joint is beneath. The limb is extended by placing the patient on the scamnum with the pulleys on each side, to which the cords are tied. The cords can be tightened by twisting them like a winepress, though care must be taken in doing so not to tear the muscles or the tendons [169]". "Let the patient with a dislocated hip be ***stretched*** upon the scamnum like that upon which we ***stretch*** those who have dislocated their spine"….[169] (G. Guidi. The Corrections of Dislocations. Wellcome Collections, London)

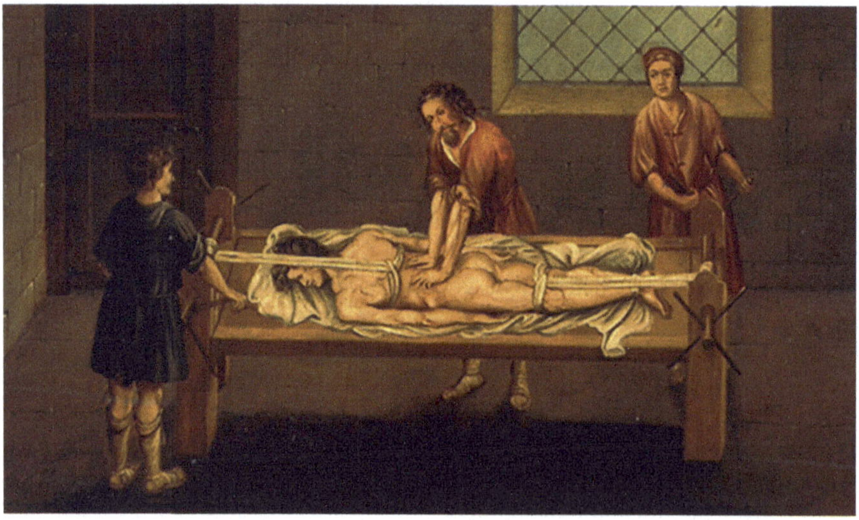

Fig. 20.2 Directions for surgeons straightening curvatures of the spine: "The bench is to be covered with clothing or anything else which is soft, and the patient is to be stoved with vapor if necessary, or bathed with much hot water, and then he is to be ***stretched*** along the scamnum on his face, with his arms laid alongside and bound to the body…."[169] (The use of the Scamnum Hippocrates. Oil painting by H. M. Hayman, 1923. Wellcome Collections, London)

Fig. 20.3 Image of the sixteenth century torture rack, invented by John Holland, Duke of Exeter (Wellcome Collection, London). The photo is of the rack at the Tower of London (Original image by David Bjorgen Creative Commons Attribution-ShareAlike)

increasing tension. While both devices provided a gradual increase in pull, with the ability of holding a sustained stretch at a desired point, the similarities end there.

With the rack, the victim was tied down in the supine position, and using a cranking device, the ropes fixed to the victim's wrists and ankles were gradually tightened, often at a snail's pace, resulting in the body being stretched beyond comprehension. The increased pressure placed on the shoulders, arms, legs, back and hips, eventually resulted in the limbs being dislocated and even ripped from their sockets [71, 75]. In addition, the ability of holding the stretch at a desired point allowed the inquisitor to simultaneously introduce other methods of torture, such as pulling out the victim's nails, burning with hot candles, or driving spikes into their spine [33].

Summary

Although, the history of stretching is quite sketchy and scarce, one thing is certain—its use is as varied as the comparison between the Hippocratic bench, used for treatment, and its progeny, the 'rack,' used for torture. This dichotomy led to envisioning the difference as a Jekyll and Hyde condition—see the "NO PAIN vs. PAIN" story in the inset below.

NO PAIN vs PAIN — a Jekyll and Hyde story of opposing forces concerning recovery, regeneration, and healing.

In many societies, the use of symbols and signals are the means by which one defines and interprets what exists, and to what extent things exist. In Jekyll and Hyde, we see the juxtaposition of good vs evil. As the story goes, Dr. Henry Jekyll, a respected, successful doctor, is aware of the duplicity of the life he lives, as expressed by his evil counterpart, Mr. Edward Hyde. Jekyll and Hyde are two equipotent, coexistent, and eternally opposed components that make up the "normal" individual [197].

Pain, a well-known body signal, represents a relative measure of one's physical, mental, and emotional state at any point in time, as well as providing a measure of progress toward wholeness. Many individuals have been led to believe that "pain" and "discomfort" are a "normal" part of the healing process, a rite of passage so to speak, which must be endured, similar to Dr. Jekyll enduring his evil counterpart, Mr. Hyde. However, after treating many individuals with microStretching, the discourse often expressed is the disbelief that something so gentle, associated with "no pain", has a profound effect on the body.

The dichotomy illustrated above is central to the practice and evolution of stretching. To a significant extent, the use of pain or discomfort continues to be a target level with both practitioners and personal trainers subscribing to the concept of "*no pain, no gain*" mantra. Although this has merit concerning the physical training of the body and its subsequent adaptation, for the proper recovery, regeneration, and healing of the body from training and trauma, this precept needs to be replaced with a new tenet—the "*greatest gain can be made without incurring pain*", for pain is the signal that informs us that something is amiss (Fig. 20.4).

According to the International Association for the Study of Pain (IASP), pain is defined as: "an unpleasant sensory and emotional experience associated with, or resembling that associated with, actual or potential tissue damage" [158]. Pain is a predictable consequence of trauma affiliated with significant physiological, emotional, and mental consequences [39]. It is an aspect of interoception (sense of the physiological condition of the body) and a specific behavioral motivation [46]. Research in the field of psychoneuroimmunology suggests that trauma creates problems through the dysregulation of the inflammatory response. [104]. Therefore, the sensation of pain during stretching needs to be properly assessed, for similar to the rack, albeit to a reduced extent, the pain and discomfort caused by "over-stretching" will be affiliated with an inflammatory response.

Fig. 20.4 Dr. Jekyll & Mr. Hyde (Modified image copyright © Dr. Nikos C. Apostolopoulos. All rights reserved)

The Cutting of the Gordian Knot—The Evolution and Future of Stretching Practice (microStretching)

> The Gordian knot refers to a highly intricate and complex knot tied by king Gordius of Phrygia (8th Century BCE). The problem was solved by Alexander the Great who instead of trying to untie the knot, simply cut the knot in 333 BCE with his sword. By cutting the knot, and not spending time trying to untie it, Alexander's approach symbolised the ability "to think outside of the box," and suggests that sometimes the best answer to a complex problem is the simplest one...

Stretching is considered a movement in which an external and/or internal force is applied to the musculoskeletal cells and tissues—in particular the muscle-tendon complex [77, 114, 129, 225]. Through sensory inputs from mechanoreceptors (tension/no tension), nociceptors (pain/no pain), and the proprioceptors (muscle spindles, Golgi tendon organs, joint, ligament, and cutaneous receptors), one can assume that a real time representation of the state of the body is created during stretching.

The sense of tension or force exerted by the extrafusal muscle fibres during stretching is perceived by the specialised mechanosensory organs of proprioception, the muscle spindles and the Golgi tendon organs. These sensory organs differ in morphology, location, measured output, and effect [25, 91, 147]. Collectively, they share the ability to assess changes in mechanical conditions, namely in muscle length (muscle spindles) or in actively generated force (Golgi tendon organs) [14, 171]. Because the muscle spindles and Golgi tendon organs share a common ability of sensing their biomechanical environment, they initiate a neural response in specialised muscle sensory afferent fibres, modulating the local muscle tension by forming muscle spindle and Golgi tendon organ reflex arcs [34].

The muscle sensory afferent neurons, referred to as proprioceptive afferents, are located in the dorsal root ganglion. These afferents provide information about muscle length (muscle spindle) and muscle tension (Golgi tendon organ) to the central nervous system [41, 173]. It should be stressed that the muscle spindle is a stretch detector sensing how much and how fast a muscle is lengthened and shortened [138]. Embedded within the belly of the extrafusal muscle fibres the muscle spindle is innervated peripherally by a primary Type Ia proprioceptive afferent neuron with several secondary Type II proprioceptive afferent neurons (Table 20.1) [155]. In contrast, the Golgi tendon organ shows little activation with muscle stretch but is highly sensitive to the contraction of the motor units that are in series with the Golgi tendon organ they innervate [51, 91, 155].

The Golgi Tendon organ is primarily concerned with the sense of effort and response to changes in muscle load [91]. Located at the myotendinous junction, it is innervated by a single Type Ib afferent neuron. This afferent neuron consists of a similar activation threshold and conduction velocity as the Type Ia afferent neuron of the muscle spindle, exhibiting dynamic sensitivity and the ability to signal rapid changes in contractile force (Table 20.1) [51].

Centrally, both the Type Ia and Type Ib afferent neurons, which have their cell bodies in the dorsal root ganglion, project into the spinal cord. The muscle spindle Type Ia afferent neuron directly excites alpha motor neurons that innervate the same

Table 20.1 Afferent sensory neurons

Group		Diameter (μm)	Conduction velocity (m/s)	Function		Common term
Myelinated	*Type I*	12–20	70–120	Largest afferent	Muscle spindle	*Type Ia* primary afferent
					Golgi Tendon Organ	*Type Ib* afferent
	Type II	5–12	30–70	Touch, pressure, muscle spindle (nuclear chain fibre) [afferent]		*Type II* secondary afferent
	Type III	3–5	12–30	Some pain and visceral receptors		*δ*
Type IV [unmyelinated]		0.4–1.2	0.5–2.0	Pain, temperature, mechanoreception		*C*—Fibres

and synergist muscles, while recruiting inhibitory neurons that innervate the antagonist motor pools [214]. This direct synapse with the alpha motor neuron forms the central pathway of the simple myotatic stretch reflex [41]. In contrast, the Golgi tendon organ Type Ib afferent sensory neuron directly synapse onto excitatory and inhibitory interneurons, ultimately inhibiting alpha motor neurons of the same muscle and exciting that of the antagonist muscle [214]. In addition, the involvement of the proprioceptor afferent neurons in the brainstem and spinal reflexes, contribute to the body schema [95], suggesting that the comprehension of the body during stretching can be both conscious and subconscious. Because it is beyond the scope of this chapter to present in detail the structures and intricacies of both the muscle spindles and the Golgi tendon organs, as well as the structure and relationship of both the afferent sensory neurons, the interneurons, and the efferent motor neurons, the reader is directed to the chapter entitled, "*Proprioception—The Sixth Sense: Understanding Our Relationship with the Internal and External environment*", by Apostolopoulos, N.C.

During stretching, the mechanical force conveyed to the cells and tissues is associated with a specific magnitude (i.e., how much) and rate (i.e., change with respect to time). With cells and tissues experiencing more than one form of mechanical stress, in essence a complex combination of tension, compression and shear stress [17] (Fig. 20.5), an understanding of how the intensity of a stimulus may interrupt their structure and function, suggests the stimulus's relevance to the proper recovery, regeneration, and adaptation of the cells and tissues. A properly imposed stress will minimise any stress induced damage, help increase the tolerance to such damage, as well as reducing the effect of inflammation and the inflammatory response [7, 105, 126]. Therefore, knowledge of the mechanical properties of the various tissues, both at the macroscopic (muscle, tendons, and myotendon unit) and microscopic (sarcomere, myofibrillar proteins, extracellular matrix, etc.) levels, their stress-strain characteristics, and history of previous insults are pivotal to the proper design of a stretching protocol [7].

Stretching is a mechanotransduction process, since a mechanical force is applied to the muscle, and in response, there is a cascade of biochemical signals, which are

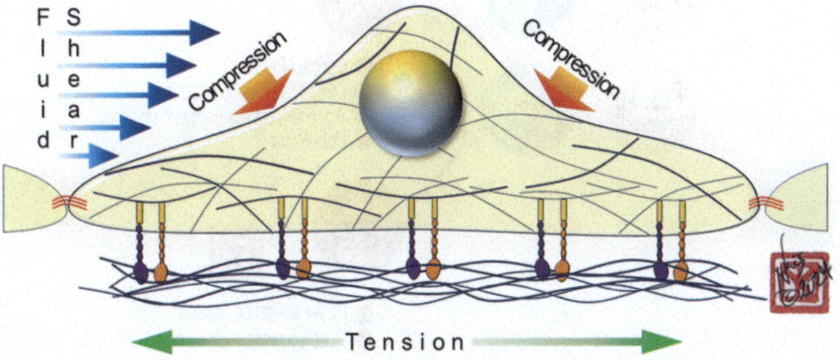

Fig. 20.5 Three basic mechanical stresses (Compression, Tension, and Fluid Shear) (Published with kind permission of copyright © Dr. Nikos C. Apostolopoulos 2024. All rights reserved)

integrated to affect an appropriate response to the perturbation of tissue and cellular homeostasis [7]. In turn, the biochemical signal expressed in response to a mechanical stress is responsible for regulating a large variety of biological processes at different levels of resolution [14]. The information concerning both the magnitude and the rate of the imposed load stress (i.e., stretching) transferred and exchanged at both the macro- and microscopic levels, occurring via inter- and intracellular pathways, is important for determining the influence of the fundamental mechanisms of stretching on the homeostasis and the proper function of the cells and tissues. In this way, the body is relatively quick in responding immediately to resolve any trauma or injury, and in initiating longer term adaptations to prevent similar disturbances in the future [11].

Methods of Stretching

While stretching can be achieved through active (self-stretch) and passive (with a partner or therapist) means [157, 225], there are three basic methods of stretching: static, active (ballistic and dynamic), and proprioceptive neuromuscular facilitation [20, 194], with the common denominator being the placement of an external load on the muscles and joint(s) to achieve a lengthening impetus being placed on a target muscle [7] (Fig. 20.6).

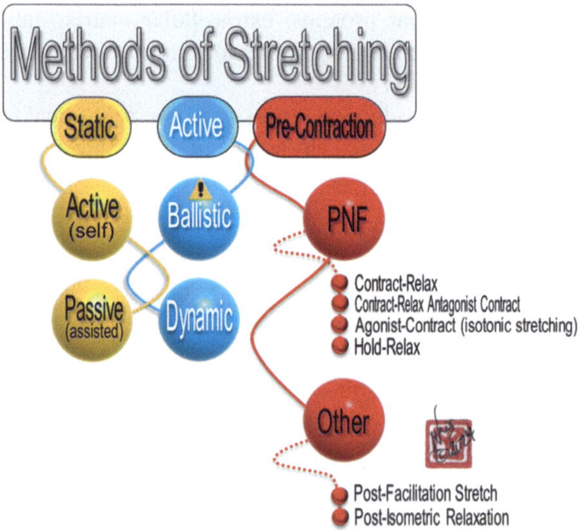

Fig. 20.6 Methods of stretching (Published with kind permission of copyright © Dr. Nikos C. Apostolopoulos 2024. All rights reserved)

Static Stretching

Static stretching is used to improve joint mobility and range of motion, and to prevent contracture. Static stretching involves the movement of a limb to its end range of motion, whereupon at this position the stretch is held anywhere between 15 and 60 s, at the point of discomfort and/or pain [16, 20, 61, 188, 200]. Static stretching can be performed passively (partner or therapist) or actively (self-stretching). Increases in range of motion are attributed to changes in the length and stiffness of the myotendon unit of the limb being stretched [20, 170]. However, a controversy exists as to whether the change is due to a decrease in myotendon unit stiffness [233][233], or simply an increase in the tolerance to the pain evoked by the stretching [123, 130].

During static stretching, the inverse myotatic reflex, mediated by the Golgi tendon organ, is activated in response to the prolonged stretch, which produces a relaxation of the muscle, and therefore, allows for an increase in the range of motion [61] [61]. Other benefits associated with static stretching include: a reduction in pain and injury [189, 200], a decrease in soreness [12, 82], and an improved performance [238]. However, systematic reviews assessing the significance of static stretching in the prevention or risk of sports injury have reported no reduction in injury rates [31, 199] nor has a change in the muscle-tendon mechanical properties been found [65]. Controversy continues concerning whether static stretching improves performance, reduces pain and soreness, or increases the range of motion, and whether any changes reflect an increase in stretch tolerance, a decrease in myotendon unit stiffness, or some other factor. However, from a mechanical point of view, it is clear that stretching exerts a tensile force on the muscle, which depending on the magnitude and duration of the stretch, may elicit pain and/or discomfort, and potentially cause tissue damage.

Active—Ballistic Stretching

To begin, the joint is passively moved to its maximum range of motion, whereupon a dynamic "ballistic" movement, in the form of a rhythmic bouncing motion, using momentum of a swinging body segment to lengthen the muscle, is applied at the end range of motion [16, 115, 132]. This momentum places the muscle under tension, activating the stretch reflex, leading to the contraction of the already stretched muscle. The increased tension is a byproduct of an amplified magnitude and rate of the stretch and is believed to be responsible for a potential strain or rupture of the muscle, suggesting that ballistic stretching is disadvantageous for improving range of motion [3, 157, 196]. Papers that describe ballistic stretching as being damaging provide a clinical commentary [157], a non-systematic literature review [196], and a specific guideline [3]. In contrast, articles that support the use of ballistic stretching, include randomised clinical trials [122, 149], which report that on its own, or in combination with other techniques, ballistic stretching is an appropriate approach for training and rehabilitation [132]. As with static stretching, the controversy concerning the efficacy of ballistic stretching illustrates the need for further robust research.

Active—Dynamic Stretching

Introduced as a better alternative to static stretching for increasing range of motion, dynamic stretching involves moving a limb through its full range of motion to its end range several times in a slow and deliberate manner from the neutral position of the joint [16, 150, 157]. If limbs are moved too quickly, the tendency to swing the limb exists, activating the stretch reflex at the endpoint of the movement during lengthening of the muscle [150]. Use of slow deliberate movement to the neutral position is executed with an eccentric contraction of the antagonist muscle relaxing the lengthened muscle because of reciprocal inhibition, in other words the muscle is reflexively inhibited thereby increasing range of motion [150].

Proprioceptive Neuromuscular Facilitation

Developed at the Kabat-Kaiser institute (1946–1956) by neurophysiologist Herman Kabat and physical therapist Margaret "Maggie" Knott, the proprioceptive neuromuscular facilitation method was aimed at treating individuals with a variety of neurological conditions (e.g., polio, stroke) to improve muscle elasticity [112, 222]. This method of stretching has since been used to treat numerous musculoskeletal conditions.

Proprioceptive neuromuscular facilitation is believed to have a positive effect on both active and passive range of motion [84]. Currently, two techniques, collectively referred to as proprioceptive neuromuscular facilitation procedures are utilised: (a) contract-relax, and (b) contract-relax-antagonist-contract [60, 84, 157]. Each procedure uses a volitional contraction to increase range of motion. The rationale behind contract-relax is to use successive maximal excitations of motor neurons during volitional contractions to reflexively promote their subsequent inhibition [99]. With the contract-relax-antagonist-contract procedure, a concentric contraction of the opposing (antagonist) muscle after the contraction and relaxation of the target muscle, is used to reduce the activation of the target muscle through reciprocal inhibition [84, 99, 112, 193, 222]. An example of a paired target muscle and opposing muscle would be the quadriceps and hamstring muscles.

It is believed that proprioceptive neuromuscular facilitation results in a greater increase in range of motion compared to static stretching [130, 131, 188], with the effects lasting 90 min or more [67, 178]. Changes in range of motion are associated with a decreased passive resistive torque (i.e., resistance to stretch) at the same joint angle prior to the intervention of this stretching method [103, 116]. Differences in the torque-angle curve of passive resistive torque indicate a decrease in the soft tissue component of the muscle (muscle or tendon stiffness) [178]. However, one study has suggested that increases in range of motion are due to an increase in stretch tolerance, since the increased range of motion and concomitant higher passive resistive torque values were observed at the maximal joint position [151]. Similar to other methods of stretching, additional research is needed to determine the efficacy of this method, and to ascertain the extent and etiology of the response(s) in terms of muscle physiology.

microStretching

Objective

This section outlines the merits of using a low-intensity gentle stretching regime (microStretching) aimed at developing an increased ease of motion, while moderating and modulating the inflammation and tissue damage associated with more strenuous stretching procedures. A decrease in inflammation allows for improved recovery and regeneration of the muscle and connective tissue. A sample of active (self-stretching) microStretching exercises is presented to illustrate the principles and variables associated with the technique. By the end of this section, the reader will have a working knowledge of the attributes and principles of microStretching.

Working Definitions

Stretch intensity is a subjective sensation felt during the stretch, ranging from light — a slight pulling, to discomfort and pain. It is assessed via a rating scale (numerical, visual analogue, or soreness) (see Fig. 20.8) [11].

Stretch position concerns the nature of the support provided to the muscle during the execution of the stretching exercise. The issue here is whether the muscle is being supported during stretching (i.e., the loading of the muscle). The loading is different when the body position is stable and balanced with a broad base of support (lying down vs standing) compared to a muscle being stretched while helping to support the body during the duration of the stretch (lying down vs standing hamstring stretch) [1, 11].

A systematic literature review, referencing 152 studies, aimed to examine two key elements of stretching—stretch intensity and stretch position, in terms of their significance to range of motion, delayed onset muscle soreness, pain, and inflammatory response(s) in different populations (athletes, clinical, elderly, and general) [11]. While most studies reported duration and frequency, only a few reported intensity. One reason for this is that duration and frequency are easily quantifiable (i.e.,

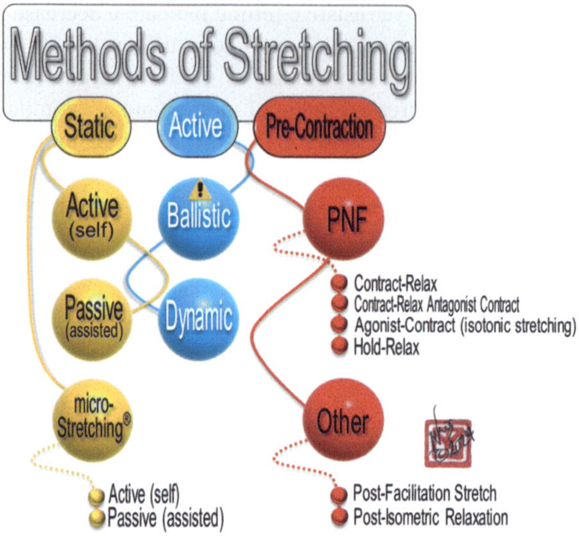

Fig. 20.7 Methods of stretching—microStretching (Published with kind permission of copyright © Dr. Nikos C. Apostolopoulos 2024. All rights reserved)

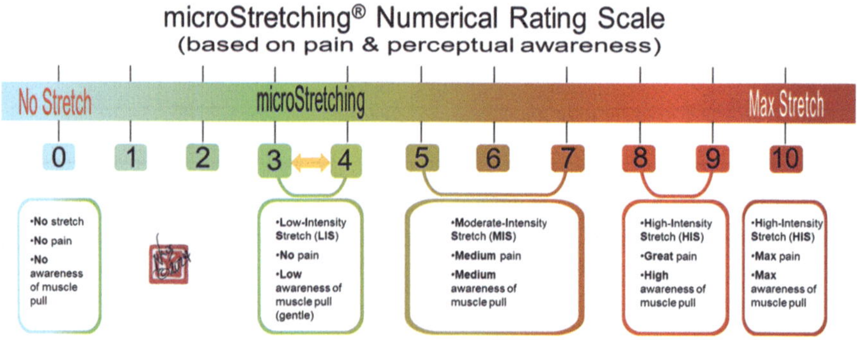

Fig. 20.8 Numerical Rating Scale (NRS) categorising intensity of microStretching (Published with kind permission of copyright © Dr. Nikos C. Apostolopoulos 2024. All rights reserved)

how long to hold, and how often to repeat the stretch) and more easily controlled than intensity. Intensity is more difficult to measure, and often involves a measure that depends on the meaning and interpretation of an individual's experience [12, 47, 62]. As previously mentioned, the position assumed by the body during the stretch, whether loaded or unloaded, may influence the intensity of the stretch [11]. In other words, if the muscles of the body are very tense and not relaxed while in the stretch position, their ability to deform is diminished, and places a large demand on the musculoskeletal tissue [52]. In addition, if during stretching, the muscle is exposed to a high-tension load (i.e., an intense stretch), the high mechanical force is associated with a physical disruption of the muscle fibres, resulting in tissue damage, inflammation, and scar tissue [50, 90, 152].

Studies referencing both stretch intensity and stretch position as part of the experimental protocol more often revealed a benefit from the stretching [177, 227, 236]. This finding may have been the result of a decreased static loading [216], and/or a decrease in the stretch reflexes engaged to protect damaged passive structures, or local instability caused by injuries [176, 204].

To address the issue of stretch intensity and stretch position, microStretching was developed as a static stretching method best defined as a "low-intensity stretching" technique (Fig. 20.7). It is based on a realisation of the importance of the magnitude of the internal forces generated during stretching, and therefore, minimises the loading accordingly. Similar to other methods of stretching (static, active, and proprioceptive neuromuscular facilitation), microStretching consists of both active (self-stretching) and passive (therapist or partner assisted) approaches. Although increases in active and passive range of motion have been observed with microStretching [236], this is a secondary outcome. A greater importance is placed on the physiological and psychological changes associated with training and tissue damage. In particular, the timely resolution of inflammation, which is important to achieving the restoration of tissue homeostasis, and a return to a "normal" pre-inflamed state [19, 72, 153].

Therefore, microStretching aims, to decrease neuromuscular tension, activating the relaxation response, and reducing inflammation [5, 6].

Irritation of muscle and connective tissue as a result of insufficient recovery, or further tissue damage through overzealous non-training activities, often causes secondary damage before proper resolution has occurred, resulting in the ongoing release of pro-inflammatory cells (neutrophils, monocytes, and macrophages) [124, 154, 235]. These actions prevent adequate recovery, regeneration, and adaptation of the body [166, 228]. It has been observed that physiological and psychological changes influence the neuroendocrine, autonomic nervous, and immune systems, collectively referred to as the mediators of adaptation [43, 142]. With human regenerative powers having limits, as witnessed by the presence of scar tissue, weakened function, and diminished mobility [174], therapeutic interventions should consist of approaches that stimulate the body's regenerative capacity, while modulating and resolving the inflammatory response [7].

To create a better understanding of the development of, and the principles behind microStretching, a look at the attributes of training may be of help. The variables that comprise the adaptive stimulus of training, namely, intensity (how much), duration (how long), frequency (how often), and volume (how many). These are the key variables used to prescribe training. In addition, the observed response to training provides feedback concerning how one might/should alter the stimulus to ensure proper recovery, regeneration, and adaptation. In general, the amount of change needed depends on the initial intensity of the stimulus [4]. However, it is important to note that the same stimulus may result in a response of differing magnitude in different individuals, i.e., a "one size fits all" approach does not work, whether it be the activity stimulus or the rest/recovery period, and therefore, the programme for each individual must be unique [139].

Applying these thoughts to pain. An individual learns about the perception of pain through their own experience [48]. The awareness of the sensation/feeling of pain occurs in a response to an applied stimulus, which can be described essentially by the same set of variables as presented above for training, i.e., a minimum stimulation is needed for detection of the stimulus, i.e., pain, which can be described by its intensity, duration, and frequency.

In the case of stretching, the process begins with the application of a tensile force [129, 225] along the axis of the muscle. The cells (muscle fibres) will respond to the applied forces by sending out signals conveying complex three-dimensional information to indicate detailed information of the current stress level and status of the cells [55, 94, 110, 184, 226]. With cells and tissues in a state of a tensile pre-stress (a level of isometric tension), any mechanical perturbation of the force will alter the environment [88]. Once an external load is applied, the structural elements of the muscle move relative to one another, changing both their orientation and spacing—until a new equilibrium configuration is attained [203]. The changes of orientation and spacing of discrete elements within the muscle suggest that proper stretching is a means of enabling, yet controlling, changes in tissue that has been traumatised or injured [203].

Research in the clinical development of pharmacological agents identifies a dose-response relation between the amount of a prescribed drug and its effectiveness on an illness or disease. This process allows for the prescription of a most effective dose [179]. A similar approach was used in the development of microStretching. Through years of clinical evidence, based on my own experience following a severe injury and resulting in spinal fusion (Cervical Discs 5 & 6), I was able to find the range of an effective stretch intensity that produced a "just noticeable difference", i.e., a threshold intensity that did not elicit pain and discomfort, but was adequate enough to alleviate the pain, and produce improvement. The technique has now been used on individuals at the clinic, who have presented with inflammatory conditions, including fibromyalgia, rheumatoid and osteoarthritis, post-surgery, over-training, knee, low back pain, etc.

Since one's perception of pain is a dynamic experience [167], the position and support of the body (i.e., stretch position) was important in generating a decrease in pain intensity and sensitivity. This initial step is critical as it acts to provide positivity early on and allows the individual to relax and focus during the treatment. Being in a relaxed state ensures an active participation, but more importantly educates them in terms of sensing the proper intensity of the stretch and allows them to replicate the sensation when performing their self-stretches (active) at home.

Research on the effect of stretch intensity has shown through the measurement of blood biomarkers for inflammation and the inflammatory response that high-intensity stretching was associated with inflammation. In contrast, low-intensity stretching (microStretching) was responsible for resolving inflammation, decreasing muscle soreness, and increasing muscle function [7, 9, 12, 152]. The extent of muscle tissue disruption is determined by the combination of the intensity and severity of the exercise [9, 59, 211]. The prescription of stretching, specifically its magnitude, enhances resolution of inflammation of muscle and connective tissue [7, 23, 90, 152, 168]. The proper application of the stimulus attributes—intensity (how much), duration (how long), frequency (how often), volume (how many), body position (stability, balance, and control—SBC™) [1, 11, 52, 173, 236], and stretch sequencing (SS™), are fundamental to the prescription of microStretching—see below.

And Now the Answers to the Critical Issues

Intensity (How Much)

For stretching, intensity is defined as the magnitude of force or torque applied to the joint and muscle during a stretching exercise [90]. The answer to 'how much' concerns the elongation of the muscle for a given rate of stretch [239], with too much force during stretching linked to inflammation and fibrosis [35, 140]. Intensity can also be defined in terms of feeling, i.e., the perception associated with a certain amount of exertion, a psychophysical response, an internal representation associated with a given magnitude of the stimulus, which evokes a conscious interpretation within the mind [98, 128, 205]. In this way, the values associated with any intensity of stretching can be expressed by the individual, and later can be adjusted to produce a match to the stimulus. In short, intensity is a subjective response unique to individuals. It is this subjective nature that explains why stretch intensity remains under researched [11].

Research has shown that a high-intensity passive static stretch at an intensity level of 80% of a maximum perceived stretch produced inflammation [7, 152]. Another study which examined stretches from 30 to 90% of a maximum perceived stretch reported no inflammation at the 30% level of stretching, but inflammation occurred at the 90% level [7].

Maximal stretching is associated with pain and promotes a higher acute inflammatory response compared to submaximal stretching [152]. Behm and Kibele [21] observed that stretches held at a maximal point of discomfort, as well as submaximal stretching at a level greater than the 50% level of discomfort, were responsible for muscle damage, similar to what is seen with eccentric induced muscle damage associated with delayed onset muscle soreness [21].

The intensity of microStretching is aimed at 30–40% of maximum perceived stretch (Fig. 20.8). The working hypothesis to support this position is primarily concerned with the functioning of muscle spindles and their response to higher levels of stretch. The Golgi tendon organs are much less sensitive to passive tension [91, 148], and require changes in muscle length approaching the physiological limit to be activated [85].

Embedded within the extrafusal muscle fibres, the muscle spindle is the most frequently found sense organ in skeletal muscle [118]. This fusiform shaped capsule consists of three intrafusal muscle fibres (nuclear bag$_1$ fibre (*dynamic*), nuclear bag$_2$ fibre (*static*), and the nuclear chain fibre), with each being aligned in parallel to the larger extrafusal muscle fibres. The intrafusal muscle fibres inform the central nervous system about changes in the velocity of stretching and the length of individual muscles [118].

The sensory innervation of the intrafusal muscle fibres consists of two types of afferent neurons, a single Type Ia neuron and a variable number of Type II neurons (Table 20.1). The former innervates all three intrafusal muscle fibres at the non-contractile equatorial region, forming a primary ending characteristic of an "annulospiral morphology" [210]. Terminations of the Type II afferent, referred to as secondary endings, have a characteristic "flower-spray morphology" predominantly found at the juxtaequatorial region of the intrafusal nuclear bag$_2$ fibre (*static*) and intrafusal nuclear chain fibre, where they intermingle with motor terminals (Fig. 20.9) [18, 86, 159, 187, 191]. Because of their distinct endings on the intrafusal muscle fibres, the Type Ia afferent encodes a stretch-evoked stimuli differently from the Type II afferent [202].

The Type Ia afferent displays a supralinear summation of action potentials when activated during a passive muscle stretch, and a pause in firing when the muscle is shortened [53], which makes the Type Ia afferent better suited for encoding stretch-velocity or dynamic stretch [137]. The Type II afferent, with a firing rate that is linear with muscle length, is better suited for relaying information about static stretching, whether the muscle is passively stretched, held at a new length, or shortened [86, 137, 202]. Although the receptor potential and firing rates of the sensory afferent neurons are proportional to muscle length and velocity, these responses can be modulated by higher centres in the brain, which suggest voluntary control of movement (i.e., stretch position), as well as passive limb displacement by external forces [69].

Efferent innervation of the contractile polar regions of the intrafusal muscle fibres primarily occurs by the gamma (fusimotor) and the beta (skeleto-fusimotor) motor neurons [76]. Activation of these motor neurons helps to maintain the sensitivity of the muscle spindle during the contraction of the extrafusal muscle fibres [120, 121]. In addition, along with the primary motor neuron of the extrafusal fibers, the alpha (skeleto-motor) motor neuron and the beta motor neuron innervates the extrafusal muscle fibres (Fig. 20.9).

Please note that presenting details of the structures and working of the muscle spindle and the afferent sensory and efferent motor neurons of the peripheral nervous system are beyond the scope of this chapter; however, the reader is directed to the chapter in this book entitled, "*Proprioception – The Sixth Sense: Understanding Our Relationship with the Internal and External environment*" by Apostolopoulos, N.C., as well as the article by Catala, M and Kubis, N. referenced below.[1]

When the body is in a relaxed, supported, and comfortable position, it is expected that the skeletal muscles (agonists, synergists, and antagonists) assume a resting muscle (myofascial) tone. In this instance, the tone is the result of the intrinsic molecular viscoelastic properties of muscle, i.e., devoid of neural activity, and undetectable by electromyography [37, 135]. This tone has been described as the 'tension in a relaxed muscle' or 'the resistance felt by the examiner during the passive

[1] CATALA, M. and KUBIS, N. 2013. Chapter 3. Gross anatomy and development of the peripheral nervous system. *In:* SAID, G. and KRARUP, C. (eds.). *Handbook of Clinical Neurology.* New York: Elsevier.

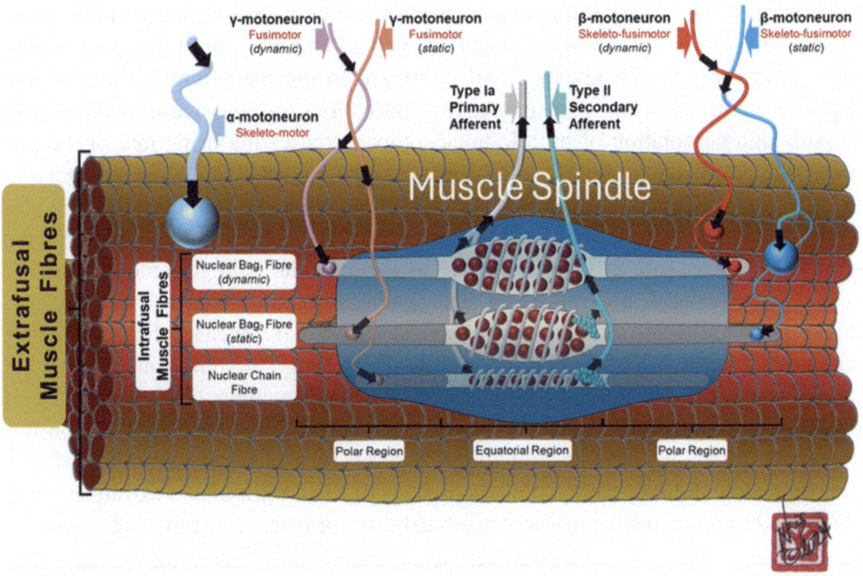

Fig. 20.9 Schematic diagram showing the relationship of the sensory afferent neurons, the efferent motor neurons, and the muscle spindle and intrafusal muscle fibres (Published with kind permission of copyright © Dr. Nikos C. Apostolopoulos 2024. All rights reserved)

stretching of a joint when the muscles are at rest' [37, 78], and differs from the muscle tone generated by the activation of the stretch reflex.[2]

In a completely relaxed individual, activation of the gamma motor neuron of the muscle spindle can be triggered by changes in behavioral, as well as environmental conditions (i.e., high- versus low-intensity stretching) [180]. This expressed gamma motor neuronal activity at the level of the muscle spindle further supports the concept that the sensitivity of the gamma motor neuron is controlled by the central nervous system independent of muscle activity. In other words, individuals can mentally effect the discharge of the gamma motor neuron by comprehending the intensity of the stimulus (i.e., stretch) [181].

The intrafusal nuclear bag fibres traverse the fusiform capsule at both ends, inserting and terminating at different distances on the extrafusal endomysium [190]. The nuclear chain intrafusal fibres are fixed at both ends within the capsule by elastic fibres [186] (Fig. 20.9). Along the distal portions of the muscle spindle, the spindle and non-muscle tissues are continuous [133]. In combination with the transmembrane connectors, intracellular filaments, and extracellular matrix components, a complete attachment of the muscle spindle to the extrafusal muscle arises [133]. This anatomical configuration creates a complete synchronisation between the intrafusal and extrafusal muscle fibres. This is important functionally since a major component of the passive stretch of a relaxed muscle fibre is the non-linear elastic segments of the extracellular tissues [30], the contractile (e.g. actin-myosin forces) and non-contractile units (e.g., titin), and the ancillary firing rate of the afferent neurons [29].

[2] For review—GANGULY, L.; KULSHRESHTHA, D.; ALMOTIRI, M. and JOG, M. 2021. Muscle tone physiology and abnormalities. *Toxins* 13 (article 282): 1–20.

Experiments have observed that muscle force-related variables define a unique transformation between muscle mechanical events and spindle firing rates, with sarcomere level cross-bridge kinetics, and is critical to understanding muscle spindle dependence on sensorimotor responses to perturbations (i.e., passive stretching) [28].

Although stimulation of muscle nociceptors increases the firing rate of the primary and secondary muscle spindles innervating the muscle [127], the presence of the Type III and Type IV afferents (see box below), in close proximity to the annulospiral structures, suggests that the majority of the pain receptors are found within the muscle spindles [160, 161]. Irrespective of whether the pain is acute or chronic, peripheral or central, nociceptive or neuropathic, the underlying origin of pain is inflammation and the inflammatory response [156]. The release of inflammatory mediators produces an acute discharge in the sensory afferents innervating the injured or inflamed tissue [237]. With high-intensity stretching, there is a release of inflammatory mediators [7, 152], making it conceivable that the pain and discomfort associated with high-intensity stretching [7, 152] is in response to disruptions of both the contractile and non-contractile components of the intrafusal and extrafusal muscle fibres. However, further research needs to be performed to confirm this hypothesis.

Type III and IV Fibres

The Type III (thin-myelinated) and Type IV (non-myelinated, aka Type C-fibres) afferents are known to be excited by substances produced or liberated during muscle pain, inflammation and ischemia, and sustained static muscle contractions [97]. Representing the largest group of sensory fibres leaving skeletal muscle, their axons are characterised by their small size and slow conduction velocity [96] (See Table 20.1). Type IV is present in every tissue of the muscle with the exception of capillaries [1, 2]. Within muscle the Type III afferents are sensitive to mechanical (stretch) and chemical change [80], and the Type IV are high-threshold mechanoreceptors, and are very sensitive to pain producing substances [2]. With the capsulated part of the muscle spindle being isolated from the extrafusal tissue, tight junctions between the capsular sheet cells act as a barrier to the diffusion of substances into the periaxial space of the capsule [160]. However, some leakage does happen in the pole regions through the open end of each capsular sleeve where the intrafusal fibres of both nuclear bag fibres exit [160]. In addition, the capillaries supplying the intramuscular nerves of the intrafusal muscle fibres do not leak compared to the extrafusal muscle fibres [107]. The thick membrane and tight sleeves of the muscle spindle with the blood/nervous barrier predisposes the unmyelinated Type III- and Type IV-afferent fibres to an accumulation of inflammatory agents over the intrafusal sensory terminals [66]. A sustained efferent activation of the muscle spindle causes a high concentration of inflammatory and contraction metabolites activating and sensitising the Type III- and Type IV-afferents [160]. Their activation in response to intrafusal inflammation suggests that muscle spindles also act as pain receptors in both health and muscle pain syndromes [161] (Fig. 20.10).

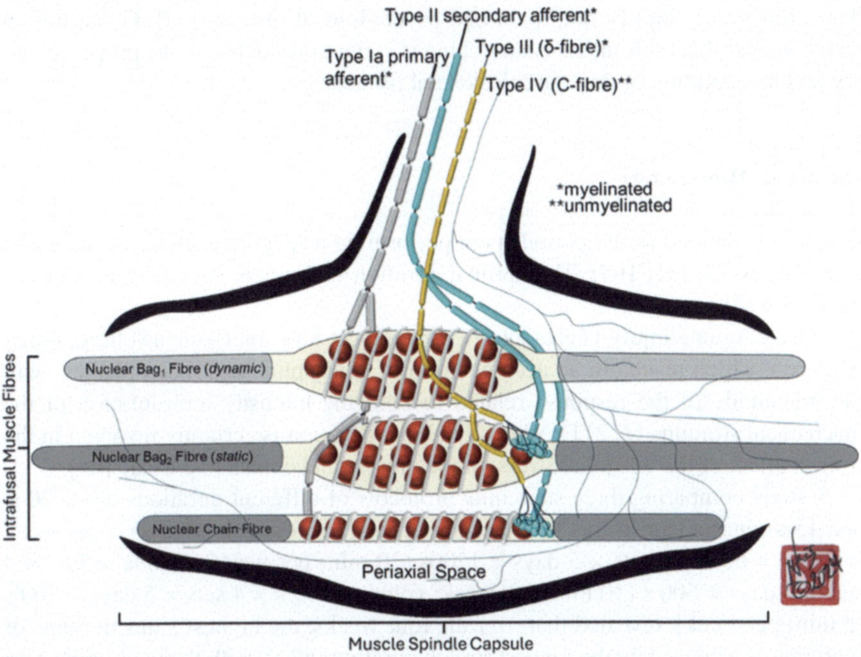

Fig. 20.10 A schema illustaring the sensory affernet neuron innervation of the muscle spindle (Published with kind permission of copyright © Dr. Nikos C. Apostolopoulos 2024. All rights reserved)

The sympathetic nervous system is highly involved in the generation of pain and influencing the intensity of the inflammatory processes [92, 93]. The sympathetic fibres of the sympathetic nervous system penetrate and innervate the muscle spindle capsule directly [175]. Adrenergic terminals are distributed within the lamellar layers inside the spindle capsule (within the periaxial space, adjacent to sensory endings), and have a 'neuroeffective association' with the nuclear bag and chain intrafusal muscle fibres in the polar regions [32, 182].

Direct sympathetic innervation of human muscle spindle fibres are of crucial importance for understanding the enhanced sympathetic activity during stress and chronic muscle pain syndromes [97, 162]. In addition, as mentioned above, individuals can mentally influence the discharge of the gamma motor neuron independent of muscle activity, suggesting the importance of the central nervous system [181].

Selective and differential control of the sensitivity of the muscle spindle by the central nervous system affords an awareness of the intensity of the stretch, and a potential role in the adaptation response. Adaptation of the whole muscle sensory apparatus could mediate an awareness of the state of the body [181] in the presence or absence of pain. Growing evidence suggests that pain conditions, especially those of a chronic nature, are associated with physical and psychological triggers that initiate and amplify both pain and psychological distress [54]. Given this, it would appear that high-intensity stretching is counterproductive to the proper recovery and regeneration of the musculoskeletal tissue.

Duration (How Long)

Duration is defined as the period spent performing a specific exercise or the entire exercise session [81, 164]. The optimal duration with microStretching is approximately 60 s [6].

Muscle trauma/injury is characterised by a local reaction (pain, swelling, bruising, etc.), which results in an accumulation of neutrophils and macrophages, with the magnitude of the response related to both the intensity and duration of the mechanical stimulus [7, 211]. With stretching, duration is certainly involved in the overall effect of the tensional stress on the muscle and connective tissue [68].

A study comparing three stretching protocols of different durations (60 s, 30 s, and 15 s) on the hamstrings of older individuals, for four sets, five days per week [volume = 60 s × 4 sets × 5 days = 1200s (20 min) per week; volume = 30 s × 4 sets × 5 days = 600 s (10 min) per week; volume = 15 s × 4 sets × 5 days = 300 s (5 min) per week], revealed that over the four weeks, the highest gains in range of motion was achieved by the longest stretch duration (60 s), with the gains persisting a further 4 weeks [62]. In addition, although this is anecdotal evidence, at the microStretching clinic, it was observed that if a stretch was held greater than 60 s (e.g., 120–180 s), patients experienced muscle "tightness". In this instance, patients modified the protocol on their own, believing that holding the stretch for longer (180 s) and doing less repetitions would be more beneficial and at the same time decrease the amount of time needed to stretch. Although the intensity of the stretch was low (3 or 4 out of 10—see Fig. 20.8), the feeling of tightness may be attributed to a slight decrease in blood flow because of the longitudinal stretching of the fibres and vessels, or perhaps due to a falling oxygen saturation due to the prolonged nature of the stretch and decreased blood flow. A study investigating the influence of a continuous passive stretch on muscle oxygen saturation observed a large decrease in oxygen saturation approaching levels comparable to complete ischemia [141]. As well, the size of the decrease in oxygen saturation is dependent on the location (proximal vs distal) within the muscle. A study reported that during plantar flexion exercise, the distal portion of the gastrocnemius generated high intramuscular pressures related to a greater pennation angle, suggesting that the tightness experienced may be due to increased intramuscular pressures affecting blood flow [146]. One study observed that stretching reduced capillary diameters, which would increase vascular resistance and reduce blood flow [111]. Another study suggested that during passive stretching, the incompressible intracellular fluid within the muscle fibres constrains the geometry of the intramuscular connective tissue, which alters the mechanical behaviour of the muscle [198].

Frequency (How Often)

Frequency is described as the number of activity sessions per time period (i.e., day) [108]. It has been suggested that the repeated processing and integration of a stimulus over a period of time is important to the learning or mastery of a new skill, with the manipulation of the amount of repetition allowing for a greater potential of implementing and learning a new behaviour [221]. On the other hand, it has been suggested that repetitive exposure to a stimulus predisposes the body to trauma and injury, and results in inflammation. With repeated exposure, a compensatory change in motor control is developed, which is focused on restoring homeostasis and/or providing greater protection [201]. Often, pain or discomfort associated with injury or trauma causes an involuntary shifting of some of the load from the affected limb and joint towards the sound limb. When this is established as a habit, an adaptive sensorimotor response is triggered [57] in which an automatic neuromuscular response becomes linked to the compensatory motor pattern. This new patterned response results in a repetitive strain to the unaffected side, creating a 'habit' [57, 234]. In order to reverse this 'habit', the tissue associated with the initial injury must be repaired, and the altered neuromuscular pathway established as part of the compensatory shift must be replaced by the creation of a new motor pattern [26]. In this regard, repetition of a motor pattern hones the motor system to master an alternate strategy of movement, with the greatest changes to the muscle-tendon unit, which occurs within the first four sessions of an exercise [208, 209].

Based on anecdotal evidence from the microStretching clinic, and the suggestion made by Taylor et al., the recommendation for active (self-microStretching) is the performance of three repetitions of a stretch per muscle group. For example, if a person is microStretching their left calf, they will hold the stretch for 60 s at an intensity level of 30 to 40% of a maximum perceived stretch, followed by a stretch in the same format for the right calf, with this sequence being repeated two more times for a total of 6 min (3 min per side time two).

Volume (How Many)

The number of sessions across a distinct period of time (e.g., a week) describes volume, which plays an important role in the restoration of function for injured, damaged, or overused muscle and connective tissue [81]. Mathematically, it is characterised by the other three attributes of a stimulus (intensity, duration, and frequency) and is considered when referencing a proper dose-response relation [125]. Similar to training recommendations, where the various combinations of load and volume are used to target specific physiological adaptations through activation of distinct cellular and molecular events, volume allows the therapist to target specific mechano-biological adaptations [185, 213]. The prescription of volume helps to reinforce the mechanical changes needed to alter the structural and cellular behaviour of the tissue associated with an adverse tension, which will enhance blood supply and neural conductivity as well as reduce inflammation [6].

Experimental evidence suggests that task persistence increases when pain or tension is less intense [36, 206, 218]. In microStretching, the gentle nature of the low-intensity active stretching promotes adaptation yet does not activate the pain pathway(s), which encourages and motivates individuals to engage in regular activity. The recommendation is to have the person stretch daily.

Body (Stretch) Position (Stability, Balance, and Control-SBC™)

Body (stretch) position is the first of the two vital principles of microStretching. It refers to a step in the protocol aimed at minimising the potential for a muscle contraction during a stretch. Proper body positioning allows for greater comfort for the individual and relaxation of the targeted muscle group. It decreases the tension within the surrounding musculature since the body is in an unloaded and supported position during stretching. Supporting the body decreases any muscle tone associated with a compensation shift due to injury and over-protection [26].

Supporting of the body is managed with the use of a therapy bed (plinth), chair, wall, floor, or other aid. By supporting the body, it is easier to place the joint(s) in different angles since the muscles are in a limited or non-contracted state. A study comparing a hamstring stretch in the standing position to a supine (lying-down) position observed that when in the supine position, the isolation of the hamstring muscle was more comfortable, and promoted a relaxation response [1].

The concept of stability, balance, and control for working on the soleus and gastrocnemius muscles is demonstrated in Fig. 20.10. In Fig. (A), the body position is not supported, and the muscles will be active in terms of providing balance, meaning that the calf muscles are activated isometrically. In this case, the loading of the muscles during the stretch will be confounded by the ongoing muscle activation. A more efficient and effective way to stretch both the soleus (B) and gastrocnemius (C) muscles would be the use of a seated position. This places the body in a stable, balanced, and controlled position. For instructions for stretches B and C see Appendix below (Fig. 20.11)

Stretch Sequence (SS™)

Stretch sequencing is the second vital principle of microStretching. Each stretch sequence is done in an established order of assigned microStretching stretches based on the origin and insertion of the muscle groups and their overlap (Fig. 20.12). In addition, the particular ordering is used to address the issue of the potential generation of any muscular tension of the body during stretching (Fig. 20.13) [6].

For instance, if an individual is suffering from hamstring issues (i.e., tightness, pain, spasms, etc.), the therapist would assign stretches for the calf, glutes, and hamstring regions, but not necessarily in that order. Stretch sequencing does not necessarily need to follow the order of muscles of a closed or open kinetic chain (i.e., calf ➔ hamstring ➔ glutes).

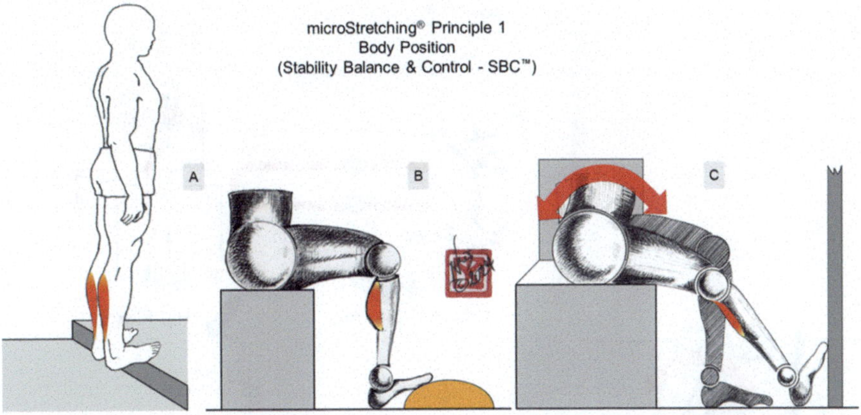

Fig. 20.11 Traditional positioning (**a**) versus body positioning for stability, balance, and control (**b** and **c**) (Published with kind permission of copyright © Dr. Nikos C. Apostolopoulos 2024. All rights reserved)

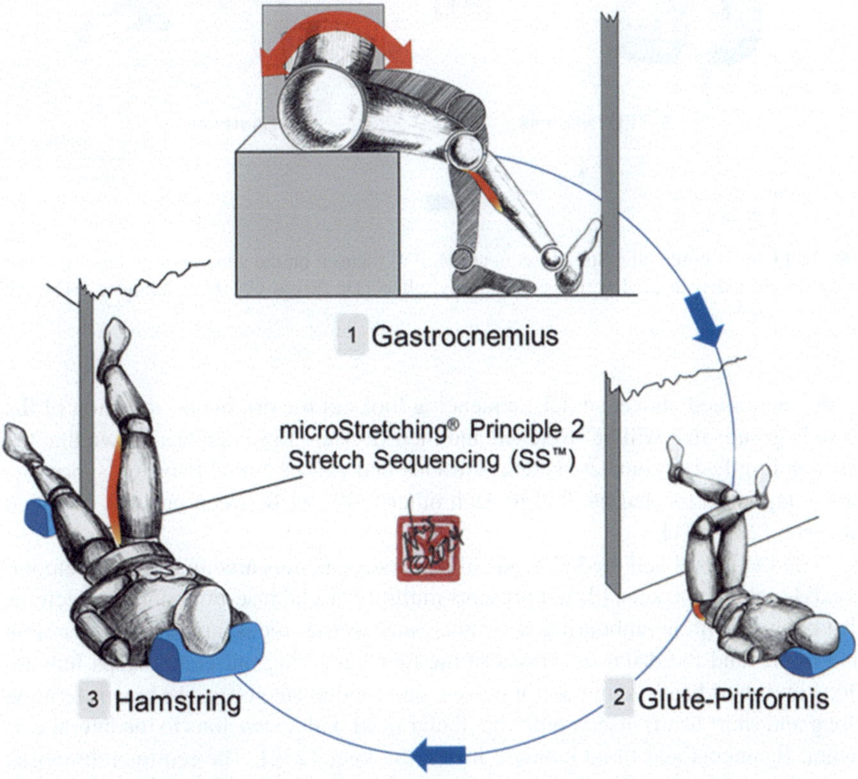

Fig. 20.12 microStretching stretch sequence (SS™) of the hamstring muscle group (Published with kind permission of copyright © Dr. Nikos C. Apostolopoulos 2024. All rights reserved)

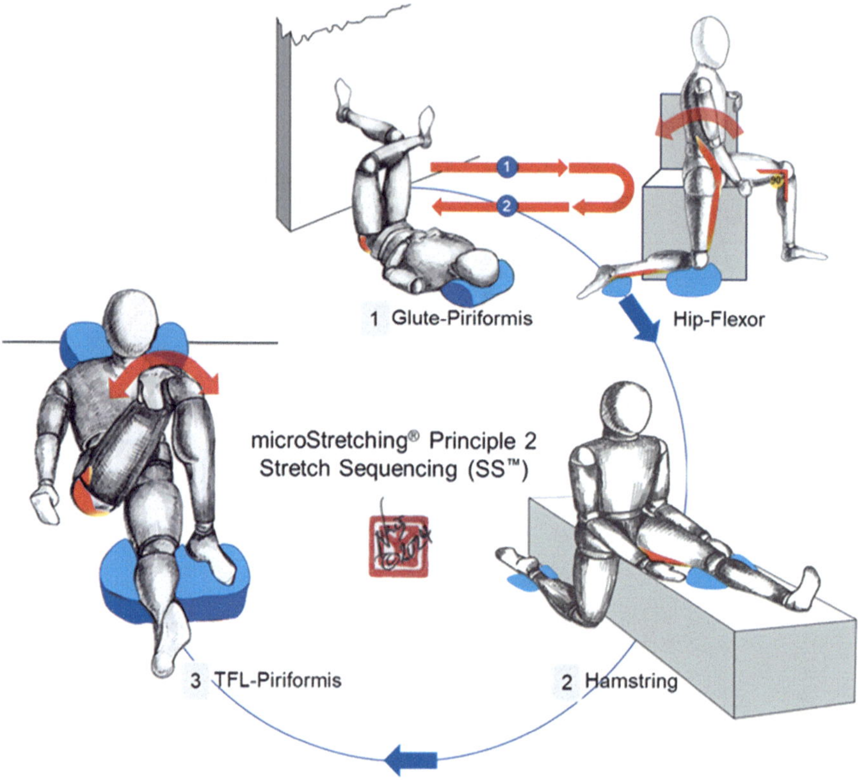

Fig. 20.13 microStretching Stretch sequence (SS™) focused on the generation of tension in the body (Published with kind permission of copyright © Dr. Nikos C. Apostolopoulos 2024. All rights reserved)

As mentioned above, stretch sequencing looks at the origin and insertion of the muscle groups that will be stretched and their overlap. Research has shown that the surrounding fascia connect skeletal muscles and can be found between synergists and antagonists located parallel to each other [240], or between muscles arranged in-series [229–231].

With the lateral and medial heads of the gastrocnemius arising from the femoral condyles above the knee [22], it presents multiple fascial insertions and connections that connect with neighbouring structures, such as the soleus, arising from the tibia and fibula, and the distal insertions of the four hamstring muscles (biceps femoris (long and short head), semimembranosus, semitendinosus) [49]. The biceps femoris (long and short head) inserts onto the fibular head with extensions to the lateral collateral ligaments and tibial condyle below the knee [232]. The semimembranosus inserts into the posteromedial tibia, with the semitendinosus joining the sartorius and gracilis tendons forming a common attachment proximal to the medial tibia [117, 232]. The proximal hamstring complex tendons (biceps femoris (longhead), semimembranosus, semitendinosus) originate in close proximity to one another on the posterior surface of the ischial tuberosity [192], and are separated from the gluteus maximus by a bursa [217]. The deep surface of the gluteus maximus overlaps and covers a number of structures, including the gluteus medius muscles, pelvic bones, as well as the proximal attachments of the hamstring muscles [145, 163, 232]. A stretch of the calf region (Fig. 20.12-1) will both directly and indirectly stretch and relax the distal insertions of the hamstring muscle group. Similarly, stretching the glute-piriformis (Fig. 20.12-2) will direct and indirect stretch and relax the proximal hamstring complex. With both the origin and insertion of the hamstring muscles being stretched before performing a hamstring stretch (Fig. 20.12-3), a more effective stretch of the hamstring muscle group is ensured.

The stretch protocol below (Fig. 20.13) highlights the other important aspect of the stretch sequence, i.e., addressing the generation of any muscular tension of the body during stretching. Here, the patient has been assigned a stretching protocol targeting the glute-piriformis (Fig. 20.13-1), hamstring (Fig. 20.13-2), and TFL-piriformis (Fig. 20.13-3) muscles, in order to alleviate tightness in the glute region. However, during the first stretch (glute-piriformis), the patient experiences a pinching sensation in the anterior region of the hip (in this case the left hip). This is associated with pain and discomfort preventing full relaxation of the patient. It also suggests that the left hip flexor is engaged instead of the intended glute-piriformis area. Based on this observation, the therapist should alter the stretch sequence by having the patient stretch the hip-flexor muscle group first (bilaterally). Once this is relaxed, the patient would then continue stretching their glute-piriformis muscle, followed by the rest of the microStretching protocol.

Summary

microStretching® is a static stretching method; however, unlike other forms of stretching, it is primarily a recovery and regeneration method focused on decreasing neuromuscular tension, activating the relaxation response, and mitigating inflammation and the inflammatory response—an increase in the range of motion is a secondary outcome. The systematic approach used in microStretching utilises specific values for the programme attributes: intensity, duration, frequency, and volume. In addition, microStretching follows two guiding principles: (1) body (stretch) position (stability, balance, and control) and (2) stretch sequencing. Two important considerations for programme development include: proper stretch intensity and body position. A stretch associated with pain and discomfort has been shown to be indicative of additional tissue damage as shown by an increased inflammatory response. Using a lower stretch intensity and placing the body in a position minimising any muscle contraction (isometric, isotonic, and/or concentric) during stretching are essential for designing and implementing a proper therapeutic stretching protocol for the recovery, regeneration, adaptation, and restoration of the tissues of the body.

1. What is the mechanism of low-intensity gentle stretching?
2. Can low-intensity gentle stretching enhance performance?
3. In the time course of recovery from injury where does low-intensity gentle stretching fit in?
4. How effective is low-intensity gentle stretching in preventing injury or re-injury?
5. What place does low-intensity gentle stretching have for return-to-play following injury?
6. Can low-intensity gentle stretching be introduced 24 h post-knee surgery?
7. What role does low-intensity gentle stretching have in mitigating or resolving inflammation?
8. Does low-intensity gentle stretching activate the vagus nerve and engage the relaxation response?
9. Can low-intensity gentle stretching be used as a pre-sleep routine to facilitate a deeper sleep for proper recovery and regeneration?
10. Can the principle of stability, balance, and control (SBC™) be used as a means of assessing imbalances in the body due to overtraining or injury?

Appendix: Examples of microStretching® Technique[3,4,5] (Self-microStretches)

Key Points:

All stretches are done *bilaterally* adhering to the attributes and principles of microStretching as described below:

Attributes & principles		Descriptions
Attributes	Intensity (how much)	30 to 40% of maximum perceived stretch (p. 13–17)
	Duration (how long)	60 s (p. 17–18)
	Frequency (how often)	3× per muscle group, once per day (p. 18)
	Volume (how many)	Daily, preferable before sleep (p. 18)

[3] Apostolopoulos, N.C. (2023). microStretching®—Level One Manual.
[4] Apostolopoulos, N.C. (2023). microStretching®—Level Two Manual.
[5] Apostolopoulos, N.C. (2023). microStretching®—Coach and Fitness Professional Manual.

Attributes & principles		Descriptions
Principles	1. Body (stretch) Position (Stability, Balance, & Control—SBC™)	Supporting the body during the stretch eliminating the potential of any muscle contractions (isometric, isotonic, or concentric) (p. 19, Fig. 20.11)
	2. Stretch Sequencing (SS™)	It establishes a particular order of assigned stretches based on the origin and insertion of the muscle groups and their overlap (p. 19–20, Fig. 20.12), as well as addressing the generation of any muscular tension of the body during stretching (p. 20, Fig. 20.13). If another muscle group is activated first as opposed to the target muscle group, it must be addressed before proceeding to the target muscle group.

A. Soleus microStretch

Target Muscle(s)

- *Primary*: Soleus [1]
- Secondary: Flexor Hallucis Longus, Digitorum Longus

Description of microStretch

- Sitting squarely on chair with feet shoulder width apart
- Ball of foot on edge of an object [2]
- Position lower leg with heel placed on ground [3]
- Change angle of ankle or knee to perceive a low-intensity gentle stretch in the soleus muscle
- Prevent knees from falling inwards [4]

Key Points

- Keep heel planted on the ground throughout the duration of the stretch [3]
- If tension is felt in the opposing muscle in the front of the shin (Tibialis Anterior) versus the calf (Soleus [1]), then the front of the shin [5] needs to be addressed first by performing a Tibialis Anterior stretch. Once addressed, resume stretching the Soleus muscle [1]
- Hold low-intensity gentle stretch for 60 s and then switch sides. Repeat this cycle two more times for a total of 6 min (3 min per side) (Fig. 20.14)

B. Gastrocnemius microStretch

Target Muscle(s)

- *Primary*: Gastrocnemius [1]
- Secondary: Plantaris, Popliteus, Hamstring (Biceps Femoris (Short & Long Head), Semi-Membranosus & -Tendinosus), Flexor Hallucis Longus, Digitorum Longus

Description of microStretch

- Sitting squarely on chair with feet shoulder width apart

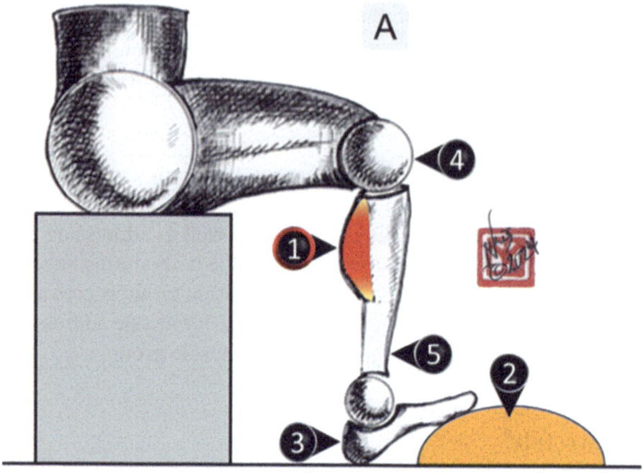

Fig. 20.14 Soleus microStretch (Published with kind permission of copyright © Nikos C. Apostolopoulos 2024. All rights reserved)

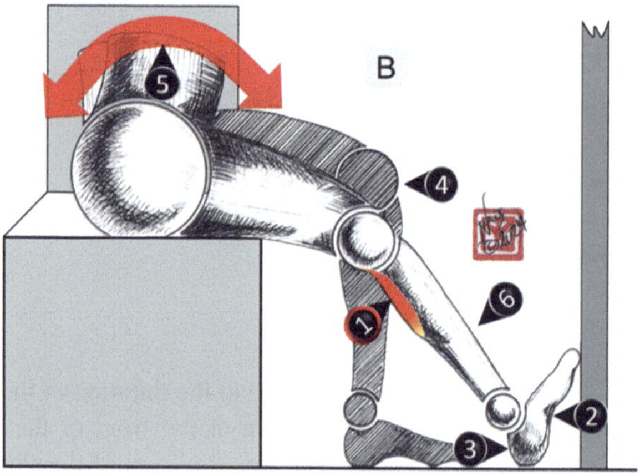

Fig. 20.15 Gastrocnemius microStretch (Published with kind permission of copyright © Nikos C. Apostolopoulos 2024. All rights reserved)

- Extend the right lower leg from the knee towards the wall
- Gently place the right foot and toes on the wall [2]
- Keep the heel of the right foot planted on the ground [3]
- Ensure left knee is bent at 90 degrees and prevent it from falling inwards [4]
- By flexing or extending the lower back and hip [5] perceive a low-intensity gentle stretch in the upper calf region near the back of the knee [1]

Key Points

- Place right foot toes on the wall [2]
- Keep right heel planted on the ground [3]
- If tension is felt in the opposing muscle in the front of the shin (Tibialis Anterior) versus the calf (Gastrocnemius [1]), then the front of the shin [6] needs to be addressed first by performing a Tibialis Anterior stretch. Once addressed, resume stretching the Gastrocnemius muscle [1]

- Hold low-intensity gentle stretch for 60 s and then switch sides. Repeat this cycle two more times for a total of 6 min (3 min per side) (Fig. 20.15)

C. Glute-Piriformis microStretch

Target Muscle(s)

- *Primary*: Piriformis, Gluteus-Minimus, -Medius, -Maximus [1]
- Secondary: Superior & Inferior Gemelli, Obturator Externus, Quadratus Femoris

Description of microStretch

- Place both feet on the wall shoulder width apart
- Keep hips, pelvis, and lower back on the ground [2]
- Pillow underneath the head ensures a flat back [3]
- Cross left ankle past the right knee joint [4]

Key Points

- Not necessary to have right knee at 90 degrees [5]
- Important to feel stretch only in the left gluteal region [1]
- If tension is felt in the anterior left hip, it may indicate a tight left Hip-Flexor muscle [6]. This needs to be addressed first before proceeding with the Glute-Piriformis stretch [1]. Once addressed, resume stretching the Glute-Piriformis muscles [1]
- Hold low-intensity gentle stretch for 60 s and then switch sides. Repeat this cycle two more times for a total of 6 min (3 min per side) (Fig. 20.16)

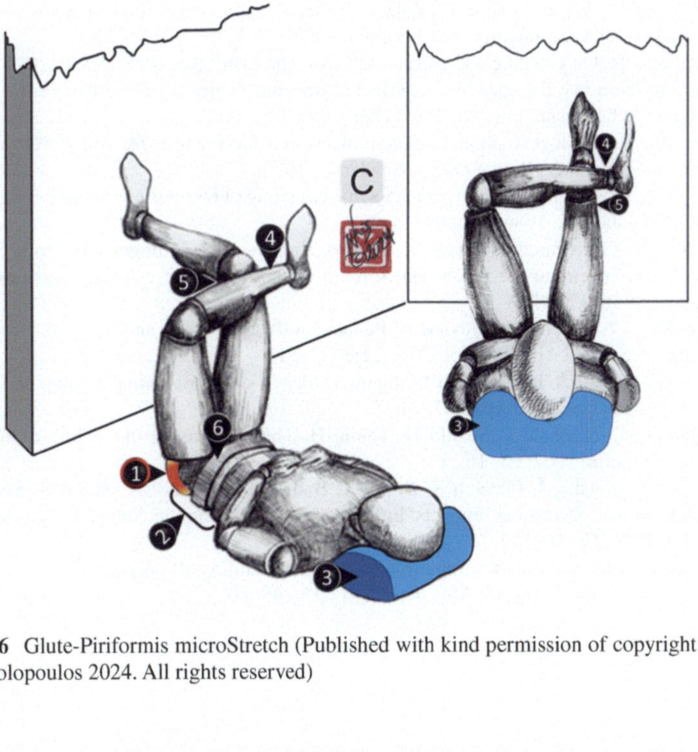

Fig. 20.16 Glute-Piriformis microStretch (Published with kind permission of copyright © Nikos C. Apostolopoulos 2024. All rights reserved)

References

1. Abdel-Aziem AA, Draz AH, Mossad DM, Abdelraou OR. Effect of body position and type of stretching on hamstring muscles flexibility. Int J Med Res Health Sci. 2013;2:399–406.
2. Abrahams VC. Group III and IV receptors of skeletal muscle. Can J Physiol Pharmacol. 1986;64:509–14.
3. ACSM. ACSM's guidelines for exercise testing and prescription. Baltimore: Lippincot Williams Wilkins; 2006.
4. Alvarado S, Kanter-Braem B, Manz K, Masciopinto P, Mckenna E, Nelson D, Williams C, Korek K. Sensation and perception: a unit lesson plan for high school psychology teachers. National Standards for High School Psychology Curricula. 2011:1–46.
5. Apostolopoulos N. Microstretching – a new recovery and regeneration technique. New Stud Athl. 2004;19:46–54.
6. Apostolopoulos N. microStretching – a practical approach for recovery and regeneration. New Stud Athl. 2010;25:81–97.
7. Apostolopoulos NC. Stretch intensity and the inflammatory response: a paradigm shift. Cham: Springer; 2018.
8. Apostolopoulos NC. What is movement? Int J Kinesiol Sport Sci. 2019;7:9–11.
9. Apostolopoulos N, Metsios GS, Nevill A, Koutedakis Y, Wyon M. Stretch intensity vs. inflammation: a dose-dependent association? Int J Kinesiol Sport Sci. 2015a;3:27–31.
10. Apostolopoulos N, Metsios GS, Taunton J, Koutedakis Y, Wyon M. Acute inflammation response to stretching: a randomised trial. Ital J Sport Rehabil Posturol. 2015b;2:368–81.
11. Apostolopoulos NC, Metsios GS, Flouris AD, Koutedakis Y, Wyon MA. The relevance of stretch intensity and position – a systematic review. Front Psychol. 2015c;6:1–25.
12. Apostolopoulos NC, Lahart IM, Plyley MJ, Taunton J, Nevill AM, Koutedakis Y, Wyon M, Metsios GS. The effects of different passive static stretching intensities on recovery from unaccustomed eccentric exercise – a randomized controlled trial. Appl Physiol Nutr Metab. 2018;43:806–15.
13. Askenasy JJ. Is yawninig an arousal defense reflex? J Psychol. 1989;123:609–21.
14. Assaraf E, Blecher R, Heinemann-Yesushalmi L, Krief S, Vinestock RC, Biton IE, Brumfeld V, Rotkopf R, Avisar E, Agar G, Zelzer E. Piezo2 expressed in proprioceptive neurons is essential for skeletal integrity. Nat Commum. 2020;11:1–15.
15. Baenninger R. On yawning and its functions. Psychon Bull Rev. 1997;4:198–207.
16. Bandy W, Irion JM, Briggler M. The effect of time and frequency of static stretching on flexibility of hte hamstring muscles. Phys Ther. 1997;77:1090–6.
17. Banes AJ. Mechanical strain and the mammaian cell. In: Frangos JA, editor. Physical forces and the mammalian cell. San Diego: Academic; 1993.
18. Banks RW. Observations on the primary sensory ending of tenuissimus muscle spindles in the cat. Cell Tissue Res. 1986;246:309–13.
19. Bannenberg GL, Chiang N, Ariel A, Arita M, Tjonahen E, Gotlinger KH, Hong S, Serhan CN. Molecular circuits of resolution: formation and actions of resolvins and protectins. J Immunol. 2005;174:4345–55.
20. Behm DG, Chaouachi A. A review of the acute effects of static and dynamic stretching on performance. Eur J Appl Physiol. 2011;11:2633–51.
21. Behm DG, Kibele A. Effects of differing intensities of static stretching on jump performance. Eur J Appl Physiol. 2007;101:587–94.
22. Benjamin M, Theobald P, Suzuki D, Toumi H. The anatomy of the Achilles tendon. The Achilles Tendon. 2007;3:5–16.
23. Berrueta L, Muskaj I, Olenich S, Butler T, Badger GJ, Colas RA, Spite M, Serhan CN, Langevin HM. Stretching impacts inflammation resolution in connective tissue. J Cell Physiol. 2016;231:1621–7.
24. Bertolucci LF. Pandiculation: nature's way of maintaining the functional integrity of the myofascial system? J Bodyw Mov Ther. 2011;15:268–80.

25. Bewick GS, Banks RW. Mechanotransduction in the muscle spindle. Pflugers Arch - Eur J Physiol. 2014;467:175–90.
26. Bhatia D. Muscular load compensation patterns in patients with knee osteo-arthritis. Smart healthcare for disease diagnosis and prevention. Elsevier; 2020.
27. Bishop C. Stretching the truth?: The'rack'in Anglo-Saxon England. J Aust Early Mediev Assoc. 2013;9:89–111.
28. Blum KP, D'Incamps BL, Zytnicki D, Ting LH. Force enconding in muscle spindles during stretch of passive muscle. PLoS ONE Comput Biol. 2017;13:1–24.
29. Blum KP, Nardelli P, Cope TC, Ting LH. Elastic tissue forces mask muscle fiber forces underlying muscle spindle Ia afferent firing rates in stretch of relaxed rat muscle. J Exp Biol. 2019;222:1–6.
30. Blum KP, Campbell KS, Horslen BC, Nardelli P, Housley SN, Cope TC, Ting LH. Diverse and complex muscle spindle afferent firing properties emerge from multiscale muscle mechanics. elife. 2020;9:1–32.
31. Bo Lauersen J, Bertelsen DM, Bo Andersen L. The effectiveness of exercise interventions to prevent sports injuries: a systematic review an dmeta-analysis of randomised controlled trials. Br J Sports Med. 2014;48:871–7.
32. Bombardi C, Grandis A, Chiocchetti R, Bortolami R, Johansson H, Lucchi ML. Immunohistochemical localization of alpha$_{1a}$-adrenorecepros in muscle spindles of rabbit masseter muscle. Tissue Cell. 2006;38:121–5.
33. Borg A. Torture and punishment. London: HM Stationary Office; 1975.
34. Bornstein B, Konstantin N, Alessandro C, Tresch MC, Zelzer E. More than movement: the proprioceptive system as a new regulator of musculoskeletal biology. Curr Opin Physiol. 2021;20:77–89.
35. Brand PW. Clinical mechanics of the hand. Hand rehabilitation in occupational therapy. Routledge; 2012.
36. Buhle J, Wager TD. Performance-dependent inhibition of pain by an executive working memory task. Pain. 2010;149:19–26.
37. Campbell WW, Barohn RJ. Muscle Tone. DeJong's the neurologic examination. 7th ed. Philadelphia: Lippincott Williams & Wilkins; 2019.
38. Capps E, Page T, Rouse W. Hippocrates: on joints. Hippocrates: the loeb classical library. 1927;3:200–397.
39. Carr DB, Goudas LC. Acute pain. Lancet. 1999;353:2051–8.
40. Ceunen E, Vlaeyen JW, Van Diest I. On the origin of interoception. Front Psychol. 2016;7:743.
41. Chen HHSH, Arber S, Frank E. Development of the monosynaptic stretch reflex circuit. Curr Opin Neurobiol. 2003;13:96–102.
42. Chen WG, Schloesser D, Arensdorf AM, Simmons JM, Cui C, Valentino R, Gnadt JW, Nielsen L, St. Hillaire-Clarke C, Spruance V, Horowitz TS, Vallejo YF, Langevin HM. The emerging science of interoception: Sensing, integrating, interpreting, and regualting signals within the self. Trends Neurosci. 2021;44:3–16.
43. Chikanza IC, Grossman AB. Reciprocal interactions between the neuroendocrine adn immune systems during inflammation. Rheum Dis Clinics N Am. 2000;26:693–711.
44. Cobo R, Garcia-Piqueras J, Cobo J, Vega JA. The human cutaneous sensory corpuscles: an update. J Clin Med. 2021;10:1–12.
45. Cornell T. On war and games in the ancient world. In: Cornell TJ, Allen TB, editors. War and games. Suffolk: The Boydell Press; 2002.
46. Craig AD. A new view of pain as a homeostatic emotion. Trends Neurosci. 2003;26:303–7.
47. Creswell JW. Research design: qualitative, quantitative, and mixed method approaches. Thousand Oaks: Sage; 2009.
48. Crofford LJ. Chronic pain: where the body meets the brain. Trans Am Clin Climatol Assoc. 2015;126:167–83.

49. Cruz-Montecinos C, Blanche AG, Sanchex DL, Cerda M, Sanzana-Cuche R, Cuesta-Vargas A. In vivo relationship between pelvis motion and deep fascia displacement of the medial gastrocnemius: anatomical and functional implications. J Anat. 2015;227:665–72.
50. Cutlip RG, Baker BA, Hollander M, Ensey J. Injury and adaptive mechanisms in skeletal muscle. J Electromyogr Kinesiol. 2009;19:358–72.
51. Davies P, Petit J, Scott JJ. The dynamic response of Golgi tendon organs to tetanic contraction of in-series motor units. Brain Res. 1995;690:82–91.
52. Decoster LC, Scanlon RL, Horn KD, Cleland J. Standing and supine hamstring stretching are equally effective. J Athl Train. 2004;39:330–4.
53. De-Doncker L, Piquet F, Petit J, Falempin M. Characterization of spindle afferents in rat soleus muscle using ramp-and-hold and sinusoidal stretches. J Neurophysiol. 2003;89:442–9.
54. Diatchenko L, Fillingim RB, Smith SB, Maixner W. The phenotypic and genetic signatures of common musculoskeletal pain conditions. Nat Rev Rheumatol. 2013;9:340–50.
55. Dufort CC, Paszek MJ, Weaver VM. Balancing forces: architectural control of mechanotransduction. Nat Rev Mol Cell Biol. 2011;12:308–19.
56. Dzwiza-Ohlsen EN, Kempermann G. The embodied mind in motion: a neuroscientific and philosophical perspective on prevention and therapy of dementia. Front Psychol. 2023;14:1–12.
57. Egbert MD, Barandiaran XE. Modeling habits as self-sustaining patterns of sensorimotor behavior. Front Hum Neurosci. 2014;8:1–15.
58. El-Harami J. Entertainment and recreation in the classical world-tourism products. J Manag Sustain. 2015;5:168–78.
59. Eming SA, Wynn TA, Martin P. Inflammation and metabolism in tissue repar and regeneration. Science. 2017;356:1026–30.
60. Entyre BR, Abraham LD. H-reflex during static stretching and two variations of proprioceptive neuromuscular facilitation techniques. Electroencephalogr Clin Neurophysiol. 1986;63:174–9.
61. Feland JB, Myrer JW, Merrill RM. Acute changes in hamstring flexibility: PNF versus static stretch in senior athletes. Phys Ther Sport. 2001a;2:186–93.
62. Feland JB, Myrer JW, Schulthies S, Fellingham G, Measom G. The effect of duration of stretching of the hamstring muscle group for increasing range of motion in people aged 65 or older. Phys Ther. 2001b;81:1110–7.
63. Frantzopoulou A, Douka S, Kaimakamis V, Matsaridis A, Terzoglou M. Acrobatic gymnastics in Greece from ancient times to the present day. Stud Phys Cult Tour. 2011;18:337–42.
64. Fraser A. Pandiculation: the comparative phenomenon of systematic stretching. Appl Anim Behav Sci. 1989;23:263–8.
65. Freitas SR, Mendes B, Le Sant G, Andrade RJ, Nordez A, Milanovic Z. Can chronic stretching change the muscle-tendon mechanical properties. J Med Sci Sport. 2017;28:794–806.
66. Fukami Y. Studies of capsule space of cat muscle spindles. J Physiol. 1986;376:281–97.
67. Funk DC, Swank AM, Mikla BM, Fagan TA, Farr BK. Impact of prior exercise on hamstring flexibility: a comparison of proprioceptive neuromusclar facilitation and static stretching. Nat Strength Cond Assoc. 2003;17:489–92.
68. Gajdosik RL. Passive extensibility of skeletal muscle: review of thr eliterature with clinical implications. Clin Biomech. 2001;16:87–101.
69. Gardner EP, Johnson KO. The somatosensory system: receptors and central pathways. In: Kandel ER, Schwartz JH, Jessell TM, Siegelbaum SA, Hudspects AJ, editors. Principles of neural science. 5th ed. New York: McGraw-Hill Education; 2014.
70. Georgoulis AD, Kiapidou I-S, Velogianni L, Stergiou N, Boland A. Herodicus, the father of sports medicine. Knee Surg Sports Traumatol Arthrosc. 2007;15:315–8.
71. Giacomazzo B (2022) The sickening history of the torture rack, the medieval torture device that stretched victims' limbs until they dislocated [Online]. Available: https://allthatsinteresting.com/rack-torture. Accessed 26.02.2023 2023

72. Gilroy DW, Lawrence T, Perretti M, Rossi AG. Inflammatory resolution: new opportunities for drug discovery. Nat Rev Drug Discov. 2004;3:401–16.
73. Golden M. Children and childhood in classical Athens. JHU Press; 2015.
74. Gowdy J. Hunter-gatherers and the mythology of the market. In: The Cambridge encyclopedia of hunters and gatherers; 1999. p. 391–8.
75. Grabianowski E (2021) 10 Medieval torture devices [Online]. Available: https://history.howstuffworks.com/10-medieval-torture-devices.htm. Accessed 26.02.2023 2023
76. Grill SE, Rymer WZ. Reflex actions of muscle afferents on fusimotor innervation in decerebrated cats: an assessment of beta contributions. Exp Brain Res. 1985;59:282–95.
77. Guissard N, Duchateau J. Neural aspects of muscle stretching. Exerc Sport Sci Rev. 2006;34:154–8.
78. Gurfinkel V, Cacciatore TW, Cordo P, Horak F, Nutt J, Skoss R. Postural muscle tone in the body axis of health humans. J Neurophysiol. 2006;96:2678–87.
79. Hanson M. Phenomenology of perception. Honors, College of Saint Benedict & Saint John's University; 1993.
80. Hayes SG, Kindig AE, Kaufman MP. Comparison between the effect of static contraction and tendon stretch on the discharge of group III and IV muscle afferents. J Appl Physiol. 2005;99:1891–6.
81. Herold F, Müller P, Gronwald T, Müller NG. Dose–response matters!–a perspective on the exercise prescription in exercise–cognition research. Front Psychol. 2019;10:2338.
82. High DM, Howley ET, Franks BD. The effects of static stretching and warm-up on prevention of delayed onset muscle soreness. Res Quar Exec Sport. 1989;60:357–61.
83. Hill L. Exercise science and sport medicine: an ancient account. Russ J Phys Educ Sport. 2019;14:93–7.
84. Hindle KB, Whitcomb TJ, Briggs WO, Hong J. Proprioceptive neuromuscular facilitation (PNF): its mechanisms and effects on range of motion and muscular function. J Hum Kinet. 2012;31:105–13.
85. Houk JC, Singer JJ, Goldman MR. Adequate stimulus for tendon organs with observations on mechanics of ankle joint. J Neurophysiol. 1970;34:1051–65.
86. Hunt CC. Mammalian muscle spindle: peripheral mechanisms. Physiol Rev. 1990;70:643–63.
87. Hurd A, Anderson DM, Mainieri T. Kraus' recreation and leisure in modern society. Jones & Bartlett Learning; 2021.
88. Ingber DE. Mechanobiology and diseases of mechanotransduction. Ann Med. 2003;35:564–77.
89. International Olympic Academy, I. 18th International session on Olympic studies for postgraduate students: 50 years of Olympic education. Athens: Eptalofos S.A; 2012.
90. Jacobs CA, Sciacia AD. Factors that influence the efficacy of stretching programs for patients with hypomobility. Sport Health. 2011;3:520–3.
91. Jami L. Golgi tendon organs in mammalian skeletal muscle: functional properties and central actions. Physiol Rev. 1992;72:623–66.
92. Janig W. Complex regional pain syndrome. In: Robertson D, Biaggioni I, Burnstock G, Low PA, Paton JFR, editors. Primer on the autonomic nervous system. Academic; 2012.
93. Janig W, Baron R. Complex regional pain syndrome: mystery explained? Lancet Neurol. 2003;2:687–97.
94. Janmey PA, Miller RT. Mechanisms of mechanical signaling in development and disease. J Cell Sci. 2011;124:9–18.
95. Jayasinghe SAL, Sarlegna FR, Schiedt RA, Sainburg RL. Somatosensory deafferentation reveals laterlized roles of proprioception in feedback and adaptive feedforward control of movement and posture. Curr Opin Physiol. 2021;19:141–7.
96. Johansson H, Sojka P. Pathophysiological mechanisms involved in genesis and spread of muscular tension in occupational muscle pain and in chronic musculoskeletal pain syndromes: a hypothesis. Med Hypotheses. 1991;35:196–203.

97. Johansson H, Sjolander P, Djupsjobacka M, Bergenheim M, Pedersen J. Pathophysiological mechanisms behind work-related muscle pain syndromes. Am J Ind Med. 1999:104–6.
98. Johnson KO, Hsiao SS, Yoshioka T. Neural coding and the basic law of psychophysics. Neuroscientist. 2002;8:111–21.
99. Kabat H. Studies on neuromuscular dysfunction: XIII. New concepts and techniques of neuromuscular re-education for paralysis. Permanente Found Med Bull. 1950;8:121–43.
100. Kamen D. The life cycle in archaic Greece. In: Shapiro H, editor. The Cambridge companion to archaic Greece. New York: Cambridge University Press; 2007.
101. Kassab R, Alexandre F. Integration of exteroceptive and interoceptive information within the hippocampus: a computational study. Front Syst Neurosci. 2015;9:87.
102. Katta S, Krieg M, Goodman MB. Feeling force: physical and physiological principles enabling sensory mechanotransduction. Annu Rev Cell Dev Biol. 2015;31:347–71.
103. Kay AD, Husbands-Beaseley J, Blazevich AJ. Effects of contract-relax, static stretching, and isometric contractions on muscle-tendon mechanics. Med Sci Sports Exerc. 2015;47:2181–90.
104. Kendall-Tackett K. Psychological trauma and physical health: a psychoneuroimmunology approach to etiology of negative health effects and possible interventions. Psychol Trauma Theory Res Pract Policy. 2009;1:35.
105. Kendroud S, Fitzgerald LA, Murray I, Hanna A. Physiology, nociceptive pathways. In: StatPearls [Internet]. StatPearls Publishing; 2021.
106. Kennedy L. The Prehistoric Ages: how humans lived before written records [Online]. History Channel; 2019. Accessed 16.02.2023 2023
107. Kennedy WR, Yoon KS. Permeability of muscle spindle capillaries and capsule. Muscle Nerve. 1979;2:101–6.
108. Kesaniemi YA, Danforth E, Jensen MD, Kopelman PG, Lefèbvre P, Reeder BA. Dose-response issues concerning physical activity and health: an evidence-based symposium. Med Sci Sports Exerc. 2001;33:S351–8.
109. Khan MJ, Srinivasan VM, Jea AH. The history of bracing for scoliosis. Clin Pediatr. 2016;55:320–5.
110. Kim JB. Three-dimensional tissue culture models in cancer biology. Semin Cancer Biol. 2005;15:365–77.
111. Kindig CA, Poole DC. Sarcomeere length-induced alterations of capillary hemodynamicss in rat spinotrapezius muscle: vasoacrtive vs passive control. Microvasc Res. 2001;61:64–74.
112. Knott M, Voss D. Proprioceptive neuromusclar facilitation: patterns and techniques. 2nd ed. New York: Harper & Row; 1968.
113. Knotterus JD, Berry PE. Spartan society: structual ritualization in an ancient social system. Humboldt J Soc Relat. 2002;271:1–41.
114. Knudson D. The biomechanics of stretching. J Exer Sci Physiotherapy. 2006;2:3–12.
115. Konrad A, Tilp M. Effects of ballistic stretchiung training on the properties of human muscle and tendon structures. J Appl Physiol. 2014;436:735–41.
116. Konrad A, Stafilidis S, Tilp M. Effects of acute static, ballistic, and PNF stretching exercise on the muscle and tendon tissue properties. Scand J Med Sci Sports. 2017;27:1070–80.
117. Koulouris G, Connell D. Evaluation of the hamstring muscle complex following acute injury. Skeletal Radiol. 2003;10:582–9.
118. Kroger S, Watkins B (2021) Muscle spindle function in healthy and diseased muscle 11
119. Krol M, Kupnicka P, Bosiacki M, Chlubek D. Mechanisms underlying anti-inflammatory and anti-cancer properties of stretching – a review. Int J Mol Sci. 2020;23:1–21.
120. Kucera J, Walro JM, Reichler J. Innervation of develioping intrafusal muscle fibers in the rat. Am J Anat. 1988a;183:344–58.
121. Kucera J, Walro JM, Reichler J. Motor and sensory innervation of muscle spindles in the neonatal rat. Anat Embryol (Berl). 1988b;177:427–36.
122. Kumar CKK, Chakrabarty S. A comparative study of static stretching versus ballisic stretching on gthe flexibility of the hamstring muscles of athletes. Br J Sports Med. 2010;

123. Law RYW, Harvey LA, Nicholas MK, Tonkin I, Desousa M, Finniss DG. Stretch exercises increase tolerance to stretch in patients with chronic musculoskeletal paon: a randomized controlled trial. Phys Ther. 2009;89:1016–26.
124. Lech M, Anders H-J. Macrophages and fibrosis: how resident and infiltrating mononuclear phagocytes orchestrate all phases of tissue injury and repair. Biochimica et Biophysica Acta (BBA)-Mol Basis Dis. 2013;1832:989–97.
125. Lee I-M, Skerrett PJ. Physical activity and all-cause mortality: what is the dose-response relation? Med Sci Sports Exerc. 2001;33:S459–71.
126. Liu H, Hu J, Zheng Q, Feng X, Zhan F, Wang X, Xu G, Hua F. Piezo1 channels as force sensors in mechanical force-related chronic inflammation. Front Immunol. 2022;13:11.
127. Lund JP, Sadeghi S, Athanassiadis T, Salas NC, Auclair F, Thivierge B, Arsenault I, Rompre P, Westberg K-G, Kolta A. Assessment of the potential role of muscle spindle mechanoreceptor afferents in chronic muscle pain in the rat masseter muscle. PLoS One. 2010;5:1–21.
128. Mackay DM. Psychophysics of perceived intensity: a theoretical basis for Fechner's and Stevens' laws. Science. 1963;139:1213–6.
129. Maffulli N, King J, Helms P. Training in élite young athletes (the Training of Young Athletes (TOYA) Study): injuries, flexibility and isometric strength. Br J Sports Med. 1994;28:123–36.
130. Magnusson SP, Simonsen E, Aagaard P, Sorensen H, Kjaer M. A mechanism for altered flexibility in human skeletal muscle. J Physiol Lond. 1996a;497:291–8.
131. Magnusson SP, Simonsen E, Dyhre-Poulsen P, Aagaard P, Mohr T, Kjaer M. Viscoelastic stress relaxation during static stretc in human skeletal muscle in the absence of EMG activity. Med Sci Sports Exerc. 1996b;6:323–8.
132. Mahieu NN, Mcnair P, De Muynck M, Stevens V, Blanckaert I, Smits N, Witvrouw E. Effect of static and ballistic stretching on the muscle–tendon tissue properties. Detection and Approach of Intrinsic Risk Factors for Achilles Tendinopathy. 2007;75
133. Maier A. Extracellular matrix and transmembrane linkages at the termination of intrafusal fibers and the outer capsule in chicken muscle spindles. J Morphol. 1996;228:335–46.
134. Marrou H. A history of educaton in antiquity. Madison: University of Wisconsin Press; 1982.
135. Masi AT, Hannon JC. Human resting muscle tone (HRMT): narrative introduction and modern concepts. J Bodyw Mov Ther. 2008;12:320–32.
136. Masterson D. The ancient Greek origins of sports medicine. Br J Sports Med. 1976;10:196–202.
137. Matthews PB. Evolving views on the internal operation and functional role of the muscle spindle. J Physiol. 1981;320:1–30.
138. Matthews PB. Where anatomy led, physiology followed: a survey of our developing understanding of the muscle spindle, what it dows and how it works. J Anat. 2015;227:104–14.
139. Matthews WJ, Stewart N, Wearden JH. Stimulus intensity and the perception of duration. J Exp Psychol Hum Percept Perform. 2011;37:303–34.
140. Mcclure PW, Blackburn LG, Dusold C. The use of splints in the treatment of joint stiffness: biologic rationale and an algorithm for making clinical decisions. Phys Ther. 1994;74:1101–7.
141. Mccully KK. The influence of passive stretch on muscle oxygen saturation. In: Oxygen transport to tissue XXXI. Springer; 2010. p. 317–22.
142. Mcewen BS. Interacting mediators of allostasis and allostatic load: towards an understanding of resilience in aging. Metab Clin Exp. 2003;52:10–6.
143. Mcintosh PC. Landmarks in the history of physical education. Routledge; 2013.
144. Meehan E, Carter B. Moving with pain: what principles from somatic practices can offer to people living with chronic pain. Front Psychol. 2021;11(620):381.
145. Miller SL, Gill J, Webb GR. The proximal origin of the hamstrings and surrounding anatomy encountered during repair: a cadaveric study. JBJS. 2007;89:44–8.
146. Miura H, Mccully K, Nioka S, Chance B. Relationship between muscle architectural features and oxygenation status determined by near infrared device. Eur J Appl Physiol. 2004;91:273–8.
147. Moore JC. The golgi tendon organ: a review and update. Am J Occupa Ther. 1984;38:227–36.

148. Moore M. Golgi tendon organs neuroscience update with relevance to stretching and proprioception in dancers. J Dance Med Sci. 2007;11:85–92.
149. Morcelli MH, Oliveira JMCA, Navega MT. Comparison of static, ballistic and contract-relax stretching in hamstring muscle. Fisioterapia e Pesquisa. 2013;20:244–9.
150. Murphy DR. A criticall ook at static stretching: are we doing our patients harm. Chiro Sport Med. 1991;5:67–70.
151. Nakamura M, Ikezoe T, Tokugawa T, Ichihashi N. Acute effects of stertching on passive properties of human gastrocnemius muscle-tendon unit: analysis of differences between holdf-relax and static stretching. J Sport Rehabil. 2015;24:286–92.
152. Nogueira CJ, Da Silva AG, Brandão PP, Vale RG, Apostolopoulos NC, Dantas EH. Acute inflammatory responses induced by flexibility training with two different intensities. Motricidade. 2022;18:246–53.
153. Norling LV, Spite M, Yang R, Flower RJ, Perretti M, Serhan CN. Cutting edge: humanized nano-proresolving medicines mimic inflammation-resolution and enhance wound healing. J Immunol. 2011;186:5543–7.
154. Oishi Y, Manabe I. Macrophages in inflammation, repair and regeneration. Int Immunol. 2018;30:511–28.
155. Oliver KM, Florez-Paz DM, Badea TC, Mentis GZ, Menon V, De Nooij JC. Molecular correlates of muscle spindle and golgi tendon organ afferents. Nat Commun. 2021;12:1–19.
156. Omoigui S. The biochemical origin of pain – proposing a new law of pain: the origin of all pain is inflammation and the inflammatory response. Part 1 of 3 – A unifying law of pain. Med Hypotheses. 2007;69:70–82.
157. Page P. Current concepts in muscle stretching for exercise and rehabilitation. IJSPT. 2012;7:109–19.
158. PAIN-IASP, I. A. F. T. S. O (2021) IASP Announces revised definition of pain [Online]. Available: https://www.iasp-pain.org/publications/iasp-news/iasp-announces-revised-definition-of-pain/. Accessed 28 Mar 2023
159. Partanen JV. Ephaptic transmission from type ii afferents to static γ and β efferents causes comples repetitive discharge: an hypotheses. Muscle Nerve. 2016;53:508–12.
160. Partanen JV, Ojala TA, Arokoski JPA. Myofascial syndrome and pain: a neurophysiological approach. Pathophysiology. 2010;17:19–28.
161. Partanen JV, Lajunen H-R, Liljandeer SK. Muscle spindles as pain receptors. BMJ Neurol Open. 2023;5:1–4.
162. Passatore M, Roatta S. Influence of sympathetic nervous system on senosorimotor funciton: whiplash associated disorders (WAD) as a model. Eur J Appl Physiol. 2006;98:423–49.
163. Pérez-Bellmunt A, Miguel-Pérez M, Brugué MB, Cabús JB, Casals M, Martinoli C, Kuisma R. An anatomical and histological study of the structures surrounding the proximal attachment of the hamstring muscles. Man Ther. 2015;20:445–50.
164. Pesce C. Shifting the focus from quantitative to qualitative exercise characteristics in exercise and cognition research. J Sport Exer Psychol. 2012;34:766–86.
165. Petzchner FH, Garfinkel SN, Paulus MP, Koch C, Khalsa SS. Computational models of interoception and body regulation. Trends Neurosci. 2021;44:63–76.
166. Philippou A, Maridaki M, Theos A, Koutsilieris M. Cytokines in muscle damage. Adv Clin Chem. 2012;58:49–97.
167. Pinto EA, Van Ryckeghem DM, Meulders A, Torta DM, Claus A, Van Damme S. Motor action changes pain perception: a sensory attenuation paradigm in the context of pain. Pain. 2021;162:2060–9.
168. Pizza FX, Koh TJ, Mcgregor SJ, Brooks SV. Muscle inflammatory cells after passive stretches, isometric contractions, and lengthening contractions. J Appl Physiol. 2002;92:1873–8.
169. Power DA. The Scamnum Hippocratis. Proc R Soc Med. 1925;18:15–7.
170. Power K, Behm DG, Farrel C, Carroll M, Young W. An acute bout of static stretching: effects on force and jumping performance. Med Sci Sports Exerc. 2004;36:1389–96.

171. Prochazka A. Proprioception: clinical relevance and neurophysiology. Curr Opin Physiol. 2021;23(100):440.
172. Proske U, Gandevia SC. The kinesthetic senses. J Physiol. 2009;587:4139–46.
173. Proske U, Gandevia SC. The proprioceptive senses: their roles in signaling body shape, body position and movement, and muscle force. Physiol Rev. 2012;
174. Purnell BA, Hines PJ. Repair and regeneration. American Association for the Advancement of Science; 2017.
175. Radovanovic D, Peikert K, Lindstrom M, Domellöf FP. Sympathetic innervation of human muscle spindle. J Anat. 2015;226:542–8.
176. Raub JA. Psychophysiologic effects of Hatha Yoga on musculoskeletal and cardiopulmonary function: a literature review. J Altern Complement Med. 2002;8:797–812.
177. Rees S, Murphy A, Watsford M, Mclachlan K, Coutts A. Effects of proprioceptive neuromusclar facilitation stretching on stiffness and force-producing characteristics of the ankle of inactive women. J Strength Cond Res. 2007;21:572–7.
178. Reiner M, Tilp M, Guilhem G, Morales-Artacho A, Nakamura M, Konrad A. Effects of a single proprioceptive neuromuscular facilitation stretching exercise with and without post-stretching activation on the muscle function and mechanical properties of the plantar flexor muscles. Front Physiol. 2021;12(732):654.
179. Rhea MR, Alvar BA, Burkett LN, Ball SD. A meta-analysis to determine the dose response for strength development. Med Sci Sports Exerc. 2003;35:456–64.
180. Ribot E, Roll JP, Vedel J-P. Efferent discharges recorded from single skeletomotor and fusimotor fibres in man. J Physiol. 1986;375:251–68.
181. Ribot-Ciscar E, Rossi-Durand C, Roll J-P. Increased muscle spindle sensitivity to movement during reinforcement manoeuvres in relaxed huam subjects. J Physiol. 2000;523:271–82.
182. Roatta S, Windhorst U, Ljubisavljevic M, Johansson H, Passatore M. Sympathetic modualtion of muscle spindle afferent sensitivity to stretch in rabbit jaw closing muscles. J Physiol. 2002;540:237–48.
183. Robertson EA. Growing up Greek: the differing journeys through childhood in ancient Athens and Sparta. Master, Stellenbosch University; 2018.
184. Ross TD, Coon BG, Yun S, Baeyens N, Tanaka K, Ouyang M, Schwartz MA. Integrins in mechanotransduction. Curr Opin Cell Biol. 2013;25:613–8.
185. Ruimerman R, Hilbers P, Van Rietbergen B, Huiskes R. A theoretical framework for strain-related trabecular bone maintenance and adaptation. J Biomech. 2005;38:931–41.
186. Rumpelt H-J, Schmalbruch H. Zur morphologie der bauelemente von muskelspindeln bei mensch und ratte. Z Zellforsch Mikrosk Anat. 1969;102:601–30.
187. Rumsey JW, Das M, Bhalkikar A, Stancescu M, Hickman JJ. Tissue engineering the mechanosensory circuit of the stretch reflex arc: sensory neuron innervation of intrafusal muscle fibers. Biomaterials. 2010;31:8218–27.
188. Sady SP, Wortman M, Blanke D. Flexibility training: ballistic, static or proprioceptive neuromuscular facilitation. Arch Phys Med Rehabil. 1982;63:261–3.
189. Safran M, Seaber A, Garrett W. Warm-up and muscular injury prevention: an update. Sport Med. 1989;8:239–49.
190. Sahgal V, Subramani V, Sahgal S. Attachments of human intrafusal fibers. Brain Res Bull. 1987;19:411–20.
191. Schroder JM, Bodden H, Hamacher A, Verres C. Scanning electron microscopy of teased intrafusal muscle fibers from rat muscle spindles. Muscle Nerve. 1989;12:221–32.
192. Schroeter S, Heiss R, Hammer CM, Grim C, Engelhardt M, Hotfiel T. Diagnosis of proximal hamstring injuries. Sports Orthop Traumatol. 2022;38:47–57.
193. Sharman MJ, Cresswell AG, Riek S. Proprioceptive neuromuscular facilitation stretching: mechanisms and clinical implications. Sports Med. 2006;36:929–39.
194. Shellock FG, Prentice WE. Warming-up and stretching for improved physcial performance and prevention of sports related injuries. Sports Med. 1985;2:267–78.

195. Sherrington CS. The integrative action of the nervous system. New Haven: Yale University Press; 1906.
196. Shrier I, Gossal K. Myths and truths of stretching. Phys Sportsmed. 2000;28:1–10.
197. Singh SM, Chakrabarti S. A study in dualism: the strange case of Dr. Jekyll and Mr. Hyde. Indian J Psychiatry. 2008;50:221.
198. Sleboda DA, Roberts TJ. Incompressible fluid plays a mechanical roel in teh development of passive muscle tension. Biol Lett. 2017;13:1–5.
199. Small K, Mcnaughton L, Matthews M. A systematic review into the efficacy of static stretching as part of a warm-up for the prevention of exercise-related injury. Res Sport Med. 2008;16:213–31.
200. Smith CA. The warm-up procedure: to stretch or not to stretch. A brief review. JOSPT. 1994;19:12–7.
201. Solomonow M. Neuromuscular manifestations of viscoelastic tissue degradation following high and low risk repetitive lumbar flexion. J Electromyogr Kinesiol. 2012;22:155–75.
202. Sonner MJ, Walters MC, Ladle DR. Analysis of proprioceptive sensory innervation of the mouse soleus: a whole-mount musscle approach. PLoS One. 2017;12(1):1–18.
203. Stamenovic D, Fredberg JJ, Wang N, Butler JP, Ingber DE. A microstructural approach to cytoskeletal mechanics based on tensegrity. J Theor Biol. 1996;181:125–36.
204. Sterling M, Jull G, Wright A. The effect of musculoskeletal pain on motor activity and control. J Pain. 2001;2:135–45.
205. Stevens SS. Issues in psychophysical measurement. Psychol Rev. 1971;78:426.
206. Tabor A, Van Ryckeghem DM, Hasenbring MI. Pain unstuck: the role of action and motivation. Clin J Pain. 2020;36:143–9.
207. Taylor JL. Proprioception. In: Squire LR, editor. Encyclopedia of neuroscience. Academic; 2009.
208. Taylor DC, Dalton JD Jr, Seaber AV, Garrett WE Jr. Viscoelastic properties of muscle-tendon units: the biomechanical effects of stretching. Am J Sports Med. 1990;18:300–9.
209. Taylor DC, Brooks DE, Ryan JB. Viscoelastic characteristics of muscle: passive stretching versus muscular contractions. Med Sci Sports Exerc. 1997;29:1619–24.
210. Thornell LE, Carlsson L, Eriksson PO, Liu JX, Österlund C, Stål P, Pedrosa-Domellöf F. Fibre typing of intrafusal fibres. J Anat. 2015;227:136–56.
211. Tidball JG. Inflammatory processes in muscle injury and repair. Am J Phys Regul Integr Comp Phys. 2005;288:R345–53.
212. Tipton C. Historical perspective: the antiquity of exercise, exercise physiology and the exercise prescription for health. In: Simopoulos AP, editor. Nutrition and fitness: cultural, genetic and metabolic aspects. Shanghai: Karger Publishers; 2008.
213. Toigo M, Boutellier U. New fundamental resistance exercise determinants of molecular and cellular muscle adaptations. Eur J Appl Physiol. 2006;97:643–63.
214. Tuthill JC, Azim E. Proprioception. Curr Biol. 2018;28:R194–203.
215. UNESCO (1996) History of humanity: scientific and cultural development, v. II: From the third millenium to teh seventh century B.C.y.
216. Vallath N. Perspectives on yoga inputs in the management of chronic pain. Indian J Palliat Care. 2010;16:1.
217. Van Mieghem IM, Boets A, Sciot R, Van Breuseghem I. Ischiogluteal bursitis: an uncommon type of bursitis. Skeletal Radiol. 2004;33:413–6.
218. Van Ryckeghem DM, Rost S, Kissi A, Vögele C, Crombez G. Task interference and distraction efficacy in patients with fibromyalgia: an experimental investigation. Pain. 2018;159:1119–26.
219. Varvogli L, Darviri C. Stress management techniques: evidence-based procedures that reduce stress and promote health. Health Sci J. 2011;5:74.
220. Vasiliadis ES, Grivas TB, Kaspiris A. Historical overview of spinal deformities in ancient Greece. Scoliosis. 2009;4:1–13.

221. Vleugels LW, Swinnen SP, Hardwick RM. Skill acquisition is enhanced by reducing trial-to-trial repetition. J Neurophysiol. 2020;123:1460–71.
222. Voss DE. Proprioceptive neuromuscular facilitation. Am J Phys Med Rehabil. 1967;46:838–98.
223. Walusinski O. Yawning: unsuspected avenue for a better understanding of arousal and interoception. Med Hypotheses. 2006;67:6–14.
224. Walusinski O, Meenakshisundaram R, Thirumalaikolundusubramanian P, Diwakar S, Dhanalakshmi G. Yawning: Comparative study of knowledge and beliefs, popular and medica. The mystery of yawning in physiology and disease. Karger Publishers; 2010.
225. Weerapong P, Hume P, Kolt G. Stretching: mechanisms and benefits for sport performance and injury prevention. Phys Ther Rev. 2004;9:189–206.
226. Wells RG. Tissue mechanics and fibrosis. BioChim et BioPhys Acta. 2013;1832:884–90.
227. Wicke J, Gainey K, Figueroa M. A comparison of self-administered proprioceptive neuromuscular facilitation to static stretching on range of motion and flexibility. J Strength Cond Res. 2014;28:168–72.
228. Widgerow AD. Cellular resolution of inflammation – catabasis. Wound Repair Regen. 2012;20:2–7.
229. Wilke J, Krause F. Myofascial chains of the upper limb: a systematic review of anatomical studies. Clin Anat. 2019;32:934–40.
230. Wilke J, Krause F, Vogt L, Banzer W. What is evidence-based about myofascial chains: a systematic review. Arch Phys Med Rehabil. 2016;97:454–61.
231. Wilke J, Debelle H, Tenberg S, Dilley A, Maganaris C. Ankle motion is associated with soft tissue displacement in the dorsal thigh: an in vivo investigation suggesting myofascial force transmission across the knee joint. Front Physiol. 2020;11:1–7.
232. Williams PL, Bannister LH, Berry MM, Colins P, Dyson M, Dussek JE, Ferguson MW. Gray's anatomy: the anatomical basis of medicine and surgery. New York: Churchill Livingstone; 1995.
233. Wilson G, Wood G, Elliott B. The relationship between stiffness of the musculature and static flexibility: an alternative explanation for the occurrence of muscular injury. Int J Sports Med. 1991;12:403–7.
234. Wise CH. Orthopaedic manual physical therapy: from art to evidence. Philadelphia: FA Davis Company; 2015.
235. Wynn TA, Vannella KM. Macrophages in tissue repair, regeneration, and fibrosis. Immunity. 2016;44:450–62.
236. Wyon M, Felton L, Galloway S. A comparison of two stretching modalities on lower-limb range of motion measurements in recreational dancers. J Strength Cond Res. 2009;23:2144–8.
237. Yaksh TL. Mediators and mechanisms of inflammatory pain. In: Serhan CN, Ward PA, Gilroy DW, editors. Fundamentals of inflammation. Cambridge: Cambridge University Press; 2010.
238. Young W, Behm DG. Should static stretching be used during warm-up or for strength and power activities? Strength Cond J. 2002;24:33–7.
239. Young W, Elias G, Power J. Effects of static stretching volume and intensity on plantar flexor explosive force production and range of motion. J Sports Med Phys Fitness. 2006;46:403.
240. Yucesoy CA. Epimuscular myofascial force transmission implies novel principles for muscular mechanics. Exerc Sport Sci Rev. 2010;38:128–34.

Chapter 21
Healing Champions: The Power of Physical Therapy in Athlete Recovery and Performance

Tugce Kalaycioglu and Gul Baltaci

Introduction

As the art and science of neuromusculoskeletal care evolved, it has become increasingly clear that manual therapy techniques (i.e., sports physical therapy) are essential for the proper management of patients with any neuromusculoskeletal issues. Most sports activities place a high physical demand on the athlete's body, which can cause tissue overload. The intense exercise stress on the body may lead to postural and biomechanical changes, soft tissue inflammation, as well as myofascial and muscle problems. Soft tissue pain may affect and limit the range of motion (ROM) about the joint, along with a decrease in strength flexibility and performance [1]. Delayed-onset muscle soreness (DOMS), characterized by pain and tenderness after 22–72 hours post exercise occurs after an overload activity (i.e., eccentric, or plyometric exercise) of the tissue, triggering an inflammatory response. The adaptation to this overload is associated with lactic acid accumulations, pain, local edema, tightness of the soft tissue structure, spasms and painful trigger points of muscles, altering the body's biomechanics and kinematics [2]. Overloading is responsible for a decrease in performance and functionality, thereby increasing the risk of injury. Studies have indicated that massage 24–96 hours after intense exercises may reduce the symptoms of DOMS such as pain, fatigue, and improve perceived recovery [1, 3].

T. Kalaycioglu (✉)
Private Clinic, Point Therapy, Ankara, Turkey
e-mail: pt.tugcekalaycioglu@gmail.com

G. Baltaci
Private Clinic, Point Therapy, Ankara, Turkey

Faculty of Health Sciences, Department of Physiotherapy and Rehabilitation,
Istanbul Atlas University, Istanbul, Turkey
e-mail: ygul.baltaci@gmail.com

N. C. Apostolopoulos et al. (eds.), *Fundamentals of Recovery, Regeneration, and Adaptation to Exercise Stress: An Integrated Approach*,
https://doi.org/10.1007/978-3-031-44270-4_21

Manual therapy techniques are useful methods for rehabilitation and the recovery process for soft tissue abnormalities, arthrokinematic dysfunctions, functional disorders, lymphatic or vascular problems and pain [4]. Physical therapy, a manual technique, increases the oxygen supply of the tissues, accelerates the healing, helps to remove lactic acid and other metabolic waste from the affected and injured site [1]. This chapter aims to highlight the importance of physical therapy for recovery and regeneration by referring to soft tissue mobilization, passive range of motion (PROM) flexibility, joint mobilization, and taping.

Soft Tissue

Therapeutic techniques have been used to prevent exercise stress reactions and treat the exercise induced pathologies related to overuse, delayed onset muscle soreness (DOMS), and biomechanics deficit such as iliotibial band syndrome effect on knee biomechanics [1]. The most popular soft tissue recovery technique is massage however, in recent decades physical therapy has increased in popularity. Soft tissue physical therapy techniques for recovery include trigger point or myofascial release, positional release therapy, manual lymphatic drainage, cupping, and acupressure. These approaches are applied directly by the physiotherapist using their hands or specific tools. In sports, these treatments aid in recovery by releasing the myofascial and connective tissue, decrease pain, improve functional movement and joint range of motion (ROM), and reduce the effect of DOMS [1].

1. **Trigger Point Techniques**

Trigger points are sensitive painful spots. They are a palpable tight band of skeletal muscle fibers generally formed after a sudden overload or repetitive physical activities of the muscle. Motor (tightness, functional disorders of muscle) and sensorial (pain and tenderness) abnormalities and autonomic phenomena (anxiety, depression) can be accompanied with trigger point(s) formation [5].

Overuse syndrome and repetitive strain injury related trigger points are the most common amongst athletes, with athletes complaining about weakness and pain [6]. They negatively affect the biomechanics of the body, influencing performance, as well as being associated with chronic inflammation, sleep disorders, and psychological issues. Trigger points are characterized by a twitch response that can be stimulated mechanically (palpation, ischemic pressure, or dry needle insertion). The twitch response is an electrical activity found within the muscle fiber band containing the trigger point. It is a spinal reflex arc [5, 7]. Trigger point therapy consists of two approaches used to deactivate these sensitive points: non-invasive and invasive (see below).

Reflex Arc

The reflex arc is a pathway in our body allowing for quick and automatic responses to mechanical stimuli. It involves different parts, like sensory receptors, neurons, the spinal cord, and muscles.

In the case of pain, when we experience frequent or intense pain, it can trigger the reflex arc. Specialized pain receptors detect the pain and send signals through sensory neurons to the spinal cord. Instead of the pain signal fading, it keeps activating the reflex arc, causing the pain to persist.

At the same time, the reflex arc can cause a feedback loop, where signals from the sensory neurons repeatedly trigger muscle contractions, leading to muscle spasms. These spasms add to the ongoing pain creating a cycle of discomfort.

Sometimes, this reflex arc can continue for a long time, even for years or decades. However, introducing a new stimulus can disrupt it. For example, when a needle is used, it stimulates different sensory receptors and pathways, diverting attention away from the original pain. This diversion can help relax the affected muscle and provide relief from the persistent pain associated with the reflex arc.

Non-invasive

This is very painful consisting of stretching, ischemic compression, and electrotherapy [6]. How it works can be explained by the following mechanisms. Compression of the trigger point occurs with the use of stretching creating an ischemic state by decreasing the local perfusion of the trigger point. When the compression is released, hyperemia occurs aiding in the removal of inflammatory metabolites. This non-invasive approach desensitizes the nerve endings during compression of the trigger point [5, 6].

Invasive Trigger Point Technique

Unlike the previous process, the invasive technique makes use of needles to cause a release of the trigger point(s), by either dry needling or local injections. Dry needling compared to injections is relatively painless. When the trigger point is stimulated by dry needling, a twitch is observed suggesting the release of the muscle fiber associated with the trigger point. Its effect can last for several hours, with the patient continuing to feel a twitch of the muscle fiber post treatment, eventually subsiding. Invasive trigger point release is suggested to patients twice a week to once every 2 weeks [5, 6].

Local injection can include various procedures with local anesthetics, corticosteroids, and botulinum toxin [8]. Once the needle is inside the trigger point, a small amount (0.2 mL) of anesthetic is injected. After that, the needle is pulled back to the subcutaneous tissue level. This process of needling and injecting is repeated in each direction until the local twitch response is no longer observed or the muscle tightness disappears [9]. Hong suggested that patients who received either lidocaine injections or dry needling for trigger points experienced almost complete pain relief right after the procedure if local twitch responses were elicited [10]. The difference between local injection and dry needling is that local injections are applied by doctors, whereas in many countries, physiotherapists and doctors are authorized to apply dry needling.

2. **Positional Release Therapy**

Positional release therapy, known by its parent term strain-counterstrain, is a therapeutic technique that uses a position of comfort of the body, its appendages, and its tissues to resolve somatic dysfunction. Somatic dysfunction is defined as a disturbance in the sensory or proprioceptive system that can elicit lesions of the musculoskeletal system including osseous and soft tissue [11]. In spinal and appendicular strains, injuries, or lesions distortion from the normal anatomical posture or position is often evident. This eccentric state is often relieved by placing the joint of the patient in an exaggerated degree of the deformity or distortion found during an examination [12, 13]. This distorted position is held for 90 to 120 seconds, with a spontaneous release often occurring [14]. When one link in the chain is unkinked, others nearby untangle, producing profound pain relief [15].

Discovery of Positional Release Therapy—The Lucky Accident

Dr. Lawrence Jones, an osteopathic physician, is credited with the discovery of the therapy in the early 1950s. He initially called it positional release technique later coining the term strain-counterstrain. Jones described his clinical discovery as "a lucky accident and nothing more" [1]. After Jones failed to help a patient with severe back pain, the patient said that his greatest challenge was sleeping at night and that if he could find a comfortable position, he might get relief. Jones assisted the patient into various positions and discovered that a fetal position provided the greatest pain reduction. He left him in this position while he examined another patient. Upon his return, the patient arose without pain for the first time in 4 months. Jones didn't understand how placing a patient in a position of comfort for a short period of time could provide complete cessation of unrelenting pain after so many traditional therapies had failed. He then experimented with patient positioning with moderate success. Three years later he accidentally discovered that treatment of anterior pelvic tender points often relieved posterior pelvic pain. Based on this observation, Jones believed that tender points were the result of a counterstrain mechanism: If a tissue is abruptly strained, the opposing tissue (antagonist) is counterstrained in its attempt to stabilize against the straining force, resulting in the production of antagonist tender points preventing the agonist strained tissue from fully healing [16].

Essentially, this positional release therapy is the opposite of stretching. For example, if a patient has a tight, tender area on the calf, the clinician will traditionally dorsiflex the foot to stretch the calf to reduce the tightness and pain [15].

Unfortunately, this might lead to muscle guarding and increased pain. Using the same example, a clinician who employs positional release technique would place the tender point in the position of greatest comfort (plantar flexion), shortening the muscle or tissue to relax them. Being a gentle and passive technique, positional release therapy has been advocated for the treatment of acute, subacute, and chronic somatic dysfunction in people of all ages [17].

Tender points, in contrast to myofascial trigger points, are not associated with hyper-irritable bands of tissue, but are discrete areas of tissue tenderness that can occur anywhere in the body [15, 18]. Myofascial trigger points are hyper-irritable nodules of knotted muscular tissue that often entrap nerves and local vessels causing pain, inflammation, and loss of function [6].

The art lies in finding the specific direction in which a painful joint can be moved too that will release muscular tension as well as relieve pain (the position of ease or comfort). When passively placed in such a position, inhibition of painful stimuli results, and the range of motion of the area will usually be significantly enhanced [19].

3. **Myofascial Release Techniques**

Fatigue, both peripheral and central, is important when considering the proper recovery, regeneration, and adaptation of the athlete to exercise stress. Peripheral fatigue can occur during many specific conditions such as: exercise type, ROM, muscle length, and force production. The structure responsible for all these are the muscles [20].

Muscles and soft tissues are enveloped by fascial tissue, a varied and complex organ that is particularly significant in supporting and providing cohesion to the body structures [21]. Fascia consists of elastic fibers running in different directions within each layer, with varying arrangements and thickness depending on the body part, surface, and gender. In normal conditions, visceral fasciae are flexible, allowing mobility while also providing stability. However, the fascia can transmit forces and cause pain due to the presence of free nerve endings and receptors. In addition, injuries, inflammation, or infections can cause the fascia to restrict the motion of organs. Because of its supportive role, fascia stabilizes the plexuses of the body helping to determine proper movement and postural balance. Functional correct biomechanics movements and body postural balance depend on healthy fascia which is a supportive and stabilized plexus all around the body [21–23].

Myofascial release therapy is based on the idea that the fascia can become tight and restricted due to injury, inflammation, or prolonged periods of inactivity. Heightened fascial tightness is associated with symptoms that can occur away from the site of injury, suggesting that use of traditional localized recovery techniques provide poor or a temporary solution. This can cause pain, stiffness, and reduced range of motion in the affected area [24, 25]. Therefore, myofascial therapy is focused on treating the area of whole area supported by the fascia. It consists of gentle, sustained pressure and stretching to the affected areas, with the aim of restoring mobility, reducing pain, and improving overall function. By applying gentle pressure and stretching to the fascia, myofascial release therapy aims to release tension and restore the fascia's natural elasticity. This therapy can be used to treat a variety of conditions, including chronic pain such as headaches, fibromyalgia, back and neck pain, and sports injuries [26, 27]. It is typically performed by a trained physical therapist, and may be combined with other treatments, such as stretching, exercise, and hot or cold therapy, to achieve the best results. For injuries that involve myofascial tension or restriction, it is important for the treatment and rehabilitation program to specifically target this aspect to promote complete recovery and optimal function. This may involve incorporating myofascial release therapy or other techniques that aim to release tension in the fascia and restore its natural elasticity. By addressing the myofascial component of the injury, the rehabilitation program can help reduce pain, improve range of motion, and enhance overall function [26]. There are many different techniques for applying myofascial release therapy.

Fatigue

Fatigue is defined as the decrease in the ability of a muscle to generate force that occurs as a result of exercise. It can be classified into two categories: peripheral and central. Peripheral fatigue is typically attributed to alterations occurring at or below the neuromuscular junction. In contrast, central fatigue is defined as a progressive reduction in voluntary muscle activation during exercise, mediated by various intrinsic motoneuronal, spinal, and supraspinal factors [28].

Peripheral fatigue depends on multi-factors such as the type of movement (open versus closed chain), involvement of one side or both sides of the body (unilateral versus bilateral), types of muscle activity, muscle length, range of motion, rate of force development, sustained versus intermittent effort, and maximal versus submaximal efforts [21]. It occurs from the motor neuron to the muscle [29].

Unlike peripheral fatigue, central fatigue originates from the cerebrum (brain) and extends to the motor neuron, and is associated with task failure [21, 28–31].

3.1 Instrument-assisted Soft Tissue Mobilization (IASTM)

Instrument-assisted soft tissue mobilization (IASTM) is a therapeutic technique that is rooted in the principles of soft tissue mobilization first introduced by James Cyriax [32]. IASTM is commonly utilized to break down myofascial adhesions in the body's connective tissue. These techniques are employed to achieve a range of goals, such as enhancing blood flow to tissues, reducing the risk of blood clotting, minimizing swelling, alleviating discomfort related to muscle spasms, increasing flexibility, elevating skin and muscle temperature, and lessening the effects of delayed onset muscle soreness [28].

The Graston and ERGON techniques are among the various IASTM techniques available. The ERGON technique is an innovative approach to therapy that employs static and dynamic manipulations of the body's soft tissues, using specialized clinical equipment to treat pathological conditions. This technique is a recently developed therapeutic approach that has been backed by the latest research evidence suggesting that it can significantly enhance sports rehabilitation. It enables therapists to improve the range of motion, alleviate pain, enhance athletic performance, raise tissue temperature, and enhance blood circulation within just a few sessions [1] (Fig. 21.1).

The Graston technique involves the use of stainless-steel instruments, potentially including up to six variations (Fig. 21.2). According to some evidence, this technique may increase blood flow to treated areas by promoting changes in collagen fibers thereby improving healing [28]. A 10-minute treatment has been observed to increase skin temperature in the targeted area, indicating enhanced blood flow and the delivery of nutrients and oxygen to the tissues [28].

3.2 Foam Roller (Self- Myofascial Release Technique)

In recent years, self-myofascial release (SMR) has emerged as a popular technique for addressing myofascial restrictions and restoring normal soft-tissue

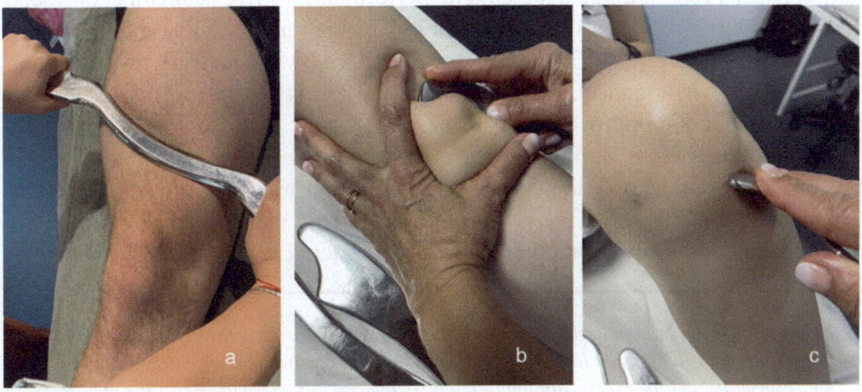

Fig. 21.1 Applications of IASTM Techniques

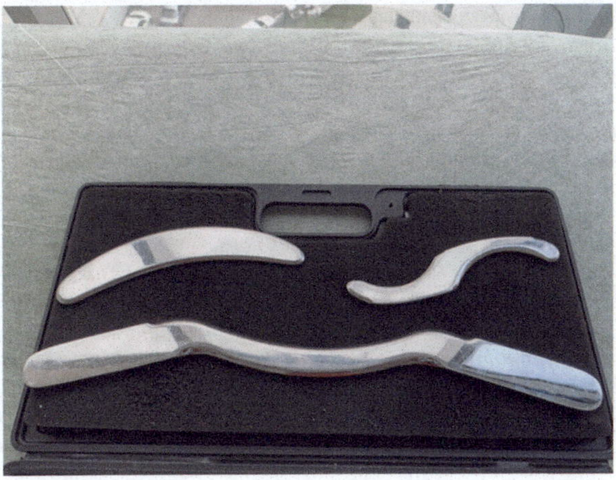

Fig. 21.2 IASTM tools

extensibility. It is often used in conjunction with traditional approaches to treating soft-tissue injuries and recovery. Typically, SMR involves using a myofascial roller to apply pressure to the affected soft tissue, with the patient using their body weight to vary the intensity and target specific areas of the body. By adjusting their body position, patients can isolate injured areas and effectively treat restrictions in the soft tissue [29].

Foam roller, an SMR tool, has many benefits for athletes in terms of flexibility, performance, and recovery. Stiffness or tightness is a frequent cause of workout-related injuries. If a muscle is unable to extend fully but is compelled to do so, there is a possibility for injury or minor damage, which may cause soreness [29]. According to a literature review published by the American College of Sports Medicine, foam rolling seems to improve flexibility prior to exercise and leads to reduced soreness and fatigue after exercise. Also, improved muscle flexibility results in better joint mobility and movement [33]. These findings indicate that foam rolling plays a role in influencing athletic performance [33]. Cheatham et al. found that foam rolling is a useful technique for enhancing joint mobility and improving overall performance [34]. Furthermore, another study observed that foam rolling prior to basic performance tests, including heavy weightlifting, agility drills, and jumping, resulted in improved performance [35].

Foam rolling before, during, and after a workout can enhance blood flow, facilitate efficient muscle function, initiate recovery, and alleviate soreness (see box below). Additionally, it has the potential to aid in pain management by promoting motion and temporarily distracting the nervous system [29].

As a professional handball player who has been involved in sports for years, I was introduced to foam rolling quite late. In the last 5–6 years, it has become a routine exercise for me, especially before every training session. It has greatly contributed to my recovery process during injuries.

I can say that it is a basic warm-up and cool-down aid that most athletes should use. In a short period of time (10–12 minutes) before training, it can prepare you for the workout and release the stiffness in your muscles before starting.

Initially, it was my biggest aid with a fat-pad injury in my left foot. Before and after every training session, I used the small roller to prepare my feet with small circular movements. I also did exercises with duo-roller on my upper and lower calves during knee problems. I had problems with my piriformis in my back, but I minimized it with rollers. I strongly recommend it. Although I am no longer plagued with major injuries, I continue to do roller exercises before and after training. With my goal being to minimize my injuries, and I believe that I have achieved this with regular use of the foam roller. I believe it is a great tool for helping with muscle spasms and tensions.

At first, I didn't know how to use it, but with the help of a physiotherapist and a little research, I learned how to apply it to my muscles over time. With just 10–15 minutes of roller exercises per day, an athlete can minimize their injuries and maximize their performance during training and matches. I highly recommend it to every athlete regardless of age or sport.

Serpil Iskenderoglu
Turkish National Handball Player

3.3 Fascial Mobilization and Manipulation Techniques

Fascial Manipulation® (FM®) is a hands-on therapeutic approach developed by Luigi Stecco, which focuses on the deep muscular fascia - a fibrous, collagenous tissue that connects all the body's connective tissues [36]. The FM® methodology is centered around a comprehensive movement analysis system that facilitates the evaluation of fascial mobility in three cardinal planes (sagittal, frontal, and transverse) of movement, treating it as an integrated three-dimensional entity. This approach allows for the identification of regional influences that may arise from the intricate connections between the musculoskeletal and fascial systems [21–23, 36]. One study which investigated sport injury prevention in individuals with chronic ankle instability demonstrated significant improvement in patient outcomes over a 6-month treatment period. The one-year follow-up revealed a complete absence of reported symptoms. The application of the FM® technique resulted in improved range of motion, reduced symptomatology, and proved helpful in preventing injury in the study group [37].

Another manual therapy technique involving the assessment and treatment of musculoskeletal pain and dysfunction associated with fascia is the Fascial Distortion Model (FDM). It was developed by American osteopath Stephen Typaldos in the 1990s, and it is based on the principle that there are six specific types of fascial distortions, each with its own characteristic symptoms and treatment techniques (see box below).

The Six Types of Fascial Distortions

1. Trigger band: A localized area of fascial tension that can cause pain and restricted movement
2. Continuum: A linear pattern of fascial tension that can extend across multiple structures
3. Herniation: A bulging or protruding area of fascia that can compress underlying structures
4. Cylinder: A circular pattern of fascial tension that can cause pain and limited movement
5. Tectonic fixation: A distortion of the fascia that can cause instability and abnormal joint motion
6. Folding: A distortion of the fascia that can cause an area of tissue to fold in on itself, resulting in pain and limited movement

Fascial distortion model practitioners use a hands-on approach to assess and identify these types of fascial distortions and then use specific manual techniques to release the fascial tension and restore normal movement and function. It has gained in popularity in recent years, and there is ongoing research to evaluate its effectiveness in treating musculoskeletal pain and dysfunction [38, 39].

Fascial mobilization and manipulation techniques, which are frequently employed for the purpose of treating injuries in athletes, can also be utilized in the post-performance recovery phase allowing for the potential resolution for both mechanical and functional issues such as muscle and fascial dysfunctions [40] especially cylindrical technique, which is a form of a fascial mobilization method of FDM utilized by athletes to increase their joint range of motion after a performance.

My patient, a 28-year-old basketball player, presented to me with a persistent shoulder pain. He had trouble lifting his shoulder above his head and reported significant pain during these movements. Previous treatments included prolonged sessions of physical therapy and multiple injections. Based on the patient's history and symptoms, I suspected a possible association with a fascial dysfunction. During the examination, the patient indicated a distinct line of pain using his finger in front of the shoulder joint, along with a generalized pain in the shoulder when lowering his arm, which he attempted to alleviate by massaging the affected area. The pain initially began as a distinct line sensation but gradually extended throughout the shoulder with continued training. Prior to preparing the treatment plan, I discussed the intended approach and potential outcomes with the patient.

I initiated the treatment by focusing on what I believed to be a cylindrical distortion related to fascial dysfunction. The therapy was quite painful, and the patient expressed occasional discomfort during the procedure. However, after 4-5 minutes, when we assessed the shoulder movements, a significant portion of the radiating pain had subsided, and he reported a reduction in the sharp pain experienced while lowering his arm. Only pain along a line extending in front of the shoulder persisted. Subsequently, I performed the treatment for the suspected trigger band distortion, which resulted in the athlete experiencing almost no pain while lifting his arm. Nonetheless, there remained a limitation in motion without pain only during the final angles of elevation. To address this, I proceeded with cervical herniated trigger point treatment, the athlete was able to perform complete shoulder elevation without any pain.

We continued the treatment program with weekly sessions, and after three treatments, the patient was completely symptom-free and was advised to engage in exercise.

– Atilla Cagatay Sezik, PhD candidate, PT, FDM practitioner

4. **Manual Lymphatic Drainage Therapy**

After performance, athletes may face several challenges depending on the intensity and duration of the event. These challenges include fatigue caused by the accumulation of metabolic byproducts and waste materials, edema and hematomas resulting from excessive strain on the body's anatomical structures, as well as muscle soreness and increased muscle tone. These physiological adaptations can result in a significant reduction in physical properties such as strength, endurance, elasticity, and proprioception. Passive rehabilitation methods, such as post-competition massage or manual lymphatic drainage therapy, have been shown to be effective in promoting recovery and reducing some of these physiological adaptations [41].

Manual lymphatic drainage therapy can be integrated into rehabilitation paradigms by both medical practitioners and allied health clinicians to treat somatic dysfunctions and pathologies. The theoretical foundation for utilizing these forms of manual therapy rests on the following principles: firstly, the stimulation of the lymphatic system by increasing lymph circulation; secondly, the acceleration of the removal of biochemical waste products from body tissues; thirdly, the enhancement of body fluid dynamics to facilitate edema reduction; and fourthly, the reduction of sympathetic nervous system responses and the concurrent increase in parasympathetic nervous tone, resulting in a state of non-stressed body framework [41, 42].

Manual lymphatic drainage can help muscle recovery and restore homeostasis after exercise by promoting increased blood flow to the muscles. Additionally, this therapy has been shown to decrease fatigue and muscle soreness by reducing the concentration of lactate in the muscles. This is achieved through improved blood flow to the muscles, which facilitates increased oxidation and the removal of harmful metabolites [43]. Overall, this form of manual therapy can have a positive impact on post-exercise recovery and can aid in reducing the negative effects of strenuous physical activity on the body [41].

5. **Cupping**

Cupping therapy is a treatment modality that has gained in popularity in sports over the last decades due to its ease of application, and minimal adverse effects. Its use can be traced back thousands of years in traditional Chinese medicine for the treatment of blood disorders, pain relief, inflammatory conditions, physical relaxation, and overall physical and mental well-being [1]. The technique involves placing a dome-shaped cup made of glass, bamboo, or plastic onto the skin and creating a negative pressure within the cup either by applying heat directly or through an air pump device. Cupping can be applied through a variety of techniques. Wet cupping involves making a skin incision before placing the cup, which draws blood, and is more commonly associated with alternative medicine. On the other hand, dry cupping requires only the placement of the cup, which is left in place (static cupping) or moved along the skin with lubricants (moving cupping). The negative pressure inside the cup is believed to lift and separate tissue, promoting the release of interfaces between soft tissues such as skin, fascia, neural tissues, muscles, ligaments, and tendons, resulting in a massage-like effect [44].

The primary mechanism of cupping therapy revolves around vascular and blood adaptations, which involve decompression of injured areas and elimination of inflammatory agents and blood. Consequently, local blood and lymphatic circulation increase, leading to enhanced oxygenation and tissue metabolism. The observed decrease in pain is attributed to the reduced irritation of pain receptors caused by a decline in the concentration of edema and hematomas in tissues. Furthermore, local vasodilation can result in heightened parasympathetic activity and muscle relaxation. It is noteworthy that no harmful effects of cupping therapy have been established through research [45].

Moving cupping has been observed to be beneficial for treating conditions such as Iliotibial Band and Priformis syndrome [46]. This therapy is believed to enhance range of movement (ROM) by addressing pain, deep scar tissues, adhesions, and swelling in the muscles [44, 46]. The normal ROM may be hindered due to various factors such as joint capsule, musculotendinous tissue, scar and fibrous tissue, muscle spasm, skin, and fascial restrictions, which can lead to decreased flexibility, strength, endurance, motor coordination, and pain. With physiotherapists aiming to improve or maintain ROM to promote the overall health and quality of life of patients, moving cupping may prove to be a valuable tool. For instance, if a patient is suffering from a hamstring issue, cupping may aid in its recovery by facilitating an increase in ROM, thereby potentially decreasing the risk of injury, enhancing athletic performance, and preventing musculoskeletal injuries. Moving cupping is employed as an aggressive method for mitigating swelling and hematomas according to static application. The use of an emollient is necessary during this massage technique to enable smooth movement of the cups for 15 minutes [44].

During the static application of cupping therapy, suction cups are placed on the skin to create a negative pressure for duration of 5–15 minutes, depending on the individual's tolerance. Emollients are not used in this procedure. The patient may experience initial superficial discomfort, which is replaced by a sensation of deep heat. The use of cups of various sizes allows the application of cupping therapy to all areas of the body [1, 45].

6. **Acupressure**

   Acupressure is a form of alternative medicine that utilizes the same pressure points and meridians as acupuncture. However, instead of the use of needles, acupressure applies finger pressure to specific points on the body. It is believed to stimulate the body's natural self-healing mechanisms and promote balance within the body's energy systems. Acupressure can be used as a healthcare and self-care technique to maintain and improve overall health and wellbeing, potentially preventing the onset of diseases and disorders [47].

Passive Range of Motion (PROM)

Normal ROM is crucial for proper function of the body. When ROM is compromised, the body compensates by increasing movement at other segments, which can lead to hypermobility and eventually pain and laxity. Assessing ROM, especially passively, is important for the health of the joints [48].

Passive range of motion (PROM) is a technique used for both assessment and treatment, where the therapist moves the bony segments around a joint axis without the patient's contribution. This ROM is usually slightly greater than active range of motion (AROM) due to the small amount of end-range joint play motion that can be created with overpressure. It assesses the joint's ability to move through its normal range of motion without muscle restriction and is distinct from assessments of muscle flexibility or joint arthrokinematics. Identifying the limiting structure through end-feel is important when PROM is restricted in order to determine appropriate treatment [48].

In case of restriction of the athletes after performance, the therapist holds the end range of movement for 10–20 seconds and repeats it 3–5 times during the treatment [48] (Table 21.1).

Table 21.1 Types of End Feel

End feel	Sensation
Soft Tissue	Soft, yielding
Muscle/Tendon	Firm, slight give, elastic
Capsule	Very firm
Muscle Spasms	Abrupt stop, hard
Edema	Soft, boggy
Bone	Hard, unyielding
Empty	Pain before resistance
Springy	Rebound

Flexibility

Flexibility refers to the length of muscles, and it should not be mistaken for joint PROM. It is evaluated separately and forms a critical aspect of a comprehensive musculoskeletal examination. In clinical practice, two-joint or multi-joint muscles are typically more prone to tightness or injury [48].

Static stretching is a common technique purported to improve ROM within sporting and rehabilitation settings and it is a passive lengthening of a muscle which is then sustained [49]. Some believe that increases in ROM results from an increase in stretch tolerance and a decrease in passive stiffness of the myo-tendon unit [49], with others believing that the increases is due to sarcomerogenesis [50, 51]. As myofibers comprise serially aligned sarcomeres, the longitudinal growth of the muscle fascicle occur via an increase in serial sarcomere number referring to the process of sarcomerogenesis [52].

Completed before exercise, the effects on performance are unclear, with some studies observing positive effects and others impaired performance [53, 54]. As a result, the European College of Sports Sciences and the American College of Sports Medicine [55] do not recommend the use of static stretching and instead promote dynamic stretching. Despite these recommendations, static stretching is still used in warm-ups and promoted by coaches [49, 56, 57].

A passive and active stretching technique focused on recovery and regeneration is microstretching®. A natural byproduct of this low intensity gentle stretching is the increase in both active and passive range of motion [58]. microStretching, developed by Dr. N. C. Apostolopoulos, is often used as a part of a rehabilitation or injury prevention program, and can be performed by numerous therapists practicing manual therapy techniques (physiotherapists, massage therapists, chiropractors etc.). Similar to training, it subscribes to the use of the attributes that define and characterize a stimulus, intensity (how much), duration (how long), frequency (how often), and volume (how many). The intensity of the stretch is held between a 30–40% of maximum perceived stretch on a numerical rating scale anchored between 0% (no stretch) to 100% (pain and discomfort). The stretch is held for 60 seconds, with each stretch being repeated three times per muscle group once per day for a minimum of 5 days per week [59–62]. microStretching is further discussed in another Chapter

20. *The Genesis and Evolution of Low-Intensity Gentle Stretching—microStretching®* of this book by, Apostolopoulos, N.C.

Good ROM, or flexibility around a joint, is important for the performance and activities of daily living as it allows full usage of the functional range. There are also suggestions that poor flexibility [63] and higher myo-tendon unit stiffness [64] are associated with a greater risk of muscular injury. This occurs because demands in energy absorption and release may rapidly exceed the stiffness capacity of the myo-tendon unit [65]. As a result, there is a need to balance the associated benefits of increasing ROM to decrease injury risk, against reducing performance from a level of stiffness that contributes to force production during dynamic movements where the myo-tendon unit stiffness is important. Furthermore, this balance is made following limited research on the intensity of stretching as a variable [66].

Research indicates that a 30-second passive, static stretch is more effective than a 15-second stretch, and additional gains are minimal with an extra 30 seconds (60 seconds in total) [67]. In a study investigating the duration of stretching in the elderly, observed that a 60 second stretch was superior to both a 30 and 15 second for increases in the ROM of the hamstring muscle group [68]. A 60 second stretch produced greater gains in ROM (60-second stretch =2.4° per week, 30-second stretch = 1.3° per seek 15-second stretch = 0.6° per week), which persisted longer than the gains in any other group [68]. Combining soft tissue mobilization with stretching can enhance its effectiveness. Dynamic stretching, which involves actively moving into a position that stretches the muscle, may not be as effective as static stretching in increasing flexibility but may be more effective in enhancing athletic performance.

Joint Mobilization

Joint mobility refers to the passive and involuntary movements that occur between two joint surfaces during movement, known as arthrokinematic motion. These movements are essential for normal osteokinematic motion and involve two specific types of passive movements: traction/separation and translatoric gliding, referring to all points of the body moving uniformly in the same direction. These movements are necessary to prevent compression and/or subluxation of a joint, as outlined by Kaltenborn et al. in 1999 [69, 70]. A potential cause of joint play reduction is capsular tightness, which obstructs the typical arthrokinematic gliding in the joint. To identify this condition, it is possible to evaluate the end-feel during passive range of motion and joint play [69].

Joint mobilization aims to enhance the arthrokinematic range of motion, promote optimal joint alignment and tracking, and reduce muscle spasm and pain by triggering articular sensory receptors. It was shown that joint mobilization produces mechanical effects on the joint capsule that align with enhanced range of motion. Some practitioners indicated that the main benefit of joint mobilization is the improvement of the sensorimotor system [69]. Also, it was concluded that joint mobilization can effectively improve the range of joint motion and stimulate the recovery of proprioception.

In light of these advantages, joint mobilization techniques represent one of the modalities employed during the recovery phase of athletes. Also, joint mobilization has been shown to effectively enhance ankle joint range of motion and balance during the recovery process in athletes with chronic ankle instability. Additionally, research suggests that in recovery process mobilization with movement can be more effective in improving ankle dorsiflexion range of motion (as assessed through the weight-bearing lunge test) than joint mobilization [71].

Arthrokinematics and Osteokinematics

Arthrokinematics and osteokinematics are two concepts that differentiate between different aspects of joint movement.

Osteokinematics refers to the gross movements of bones at joints, such as flexion, extension, abduction, adduction, internal rotation, and external rotation. These terms describe the movements that occur around a center of rotation, known as the joint axis.

Arthrokinematics, on the other hand, encompasses the small amplitude motions of the bones specifically at the joint surface. These movements are essential for maintaining the long-term integrity of the joint. Joint surface movements are often referred to as joint play motions or component motions.

Joint surfaces move with respect to one another by simultaneously (1) rolling, (2) gliding, and (3) spinning. This discussion focuses on how joint surfaces roll and glide with respect to one another, and largely ignores the spin component [72].

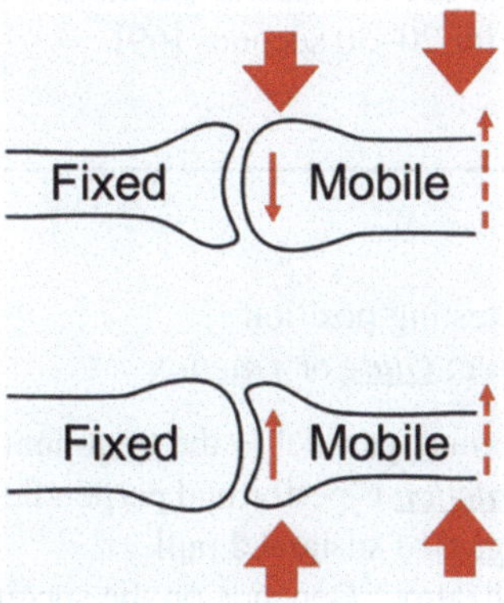

Fig. 21.3 The convex-concave rules of the joints [54, 73]

Glide and Roll

In order to enhance range of motion and reduce pain, joint mobilization frequently employs the techniques of bony segment rolling and gliding. Joint surfaces are rounded, causing movement to occur around multiple axes at once. Movement can occur in different planes (sagittal, frontal, or transverse), but it's not just one plane. When the convex and concave joint surfaces move on each other, there are rolling and sliding movements. If the joint surfaces match well, then there is more sliding, but if they don't match as well, then there is more rolling. Which way the joint moves depends on which surface is moving.

It is important to understand the convex-concave rule for proper use of joint mobilization. When a concave surface moves on a fixed convex surface, the joint will move in the same direction as the bone. On the other hand, when a convex surface moves on a fixed concave surface, the joint will move in the opposite direction of the bone (Fig. 21.3). To summarize, the therapist will move a bone with a convex joint surface in the opposite direction of the restricted bone movement, and a bone with a concave joint surface in the same direction as the restricted bone movement [69].

Traction and Compression

Traction is a technique that involves a sustained force to separate the joint surfaces at a right angle to the treatment plane. It has a global effect on the joint capsule, unlike glides that target a specific aspect of the capsule. Grade I and II tractions are commonly used for pain relief, while grade III is employed for stretching the surrounding tissues [69]. Kaltenborn recommends using intermittent traction with 10-second holds and a rest between each repetition. However, other authors suggest maintaining the traction for 20–30 seconds [69].

Joint Mobilization

- Position of joint: Resting position
- Type of mobilization: *Glide* or *Traction*
  - A *glide* is performed parallel to the treatment plane using graded oscillations, while *traction* is performed perpendicular to the treatment plane and involves a graded sustained pull.
- Direction of mobilization: Depends on the specific restriction of the joint.
  - *Glide* technique involves applying force parallel to the treatment plane with grades I–IV, whereas *traction* involves applying force at vertical angles to the treatment plane with grades I–III

- Grade of mobilization: Depends on the goal of the treatment
  - *Glide* technique: Grade I and II glides are used for pain. Grade III and IV *glides* are used to stimulate mechanoreceptors and stimulate proprioceptive fibers.
  - *Traction* technique: Graded I mobilization involves minimal joint separation and is used to eliminate compressive forces. Grade II mobilization involves taking up slack in the joint, during which the ligaments and capsule becomes taut. Grade III involves taking up slack and applying more traction force, which stretches the surrounding tissues [73]

Taping

Taping, which has become part of the physiotherapy toolbox for rehabilitation has been proven to be more effective than orthosis or splinting [74–76]. Similar to splinting, it provides support while allowing for the movement of the joint and by resting of the tissue. Taping also allows for the reduction of pain, edema, and an increase in functionality [74].

Taping is an indispensable treatment modality for the prevention and rehabilitation of injuries [73, 77], helping individuals to return to sports and daily activities [78]. The techniques used in the clinic I have been applying and teaching for more than 20 years.

There are 3 different taping techniques applied by physiotherapists in the clinic. The advantages and disadvantages of all three are shown in Table 21.2.

Table 21.2 Comparison of Taping Techniques

Parameter	Athletic [79–84]	Mcconnell [85–88]	Kinesio [30, 89–91]
Latex-No	x	X	√
Elastic-rigid	x	x	√
Skin breath	x	x	√
ROM	x	x	√
Long-use	x-4 hour	x-18 hour	√-72 hour
Resistance to water	x	x	√
Increase circulation	x	x	√
Use clinical application	x	x	√
No material before taping	x	x	√

1. Athletic Taping is a technique that is applied after an injury, and it is beneficial for the occurrence of homeostasis of the body [92]. In addition, it is used to prevent sports injuries. Functioning as a splint, it limits the range of motion of sprained joint(s), prevents edema, and as a supportive protective tool may prevent further injuries from occurring during recovery, stabilization and immobilization of the tissues, and joints. Although athletic taping, may play a vital role after injuries or phases of return to sports, its use may curb performance [93, 94].
2. McConnell patellar taping is a well-known and popular application especially used in patellofemoral pain syndrome [88, 95–97]. However, clinical evidence about the success of this application is still unclear. The clarity of evidence is narrowed because of technical differences in insufficient randomized controlled studies and parameters.
3. Kinesio taping technique is designed to have a lifting effect on the skin and allow for air circulation between the skin and the external environment, similar to the structural properties and flexibility of human skin [98, 99], without limiting joint movements [100]. It is a treatment method that can be combined with other modalities such as cold application, hydrotherapy, manual therapy, electrical stimulation, and acupuncture [100]. The tapes, which maintain their elasticity for 3–4 days, consist of polymer elastic fibers wrapped in 100% cotton fibers. While the bands are elongated up to 55–60% of their current state, they do not show a transverse stretching feature [100–103].

1. Athletic Taping

The Benefits of Athletic Taping

Prevention of Injuries

Taping is known as one of the strategies to prevent injuries in contact sports by reducing the possibility of extrinsic factors [104] including poor or inappropriate equipment regarding the bats, balls, racquets, nets, sticks, clubs, and other items integral to the game or training session, clothing and footwear in especially basketball, running, tennis, hockey, and wrestling and involving a forceful impact with an opponent or implement (such as a hockey stick or ball) [72, 77, 78, 105, 106]. It can also prevent overuse syndromes of wrist and ankle in some sports like football or gymnastics [75, 107]. The seriousness of injuries can be reduced by taping. Taping may be superior to rigid or semi-rigid orthotics in preventing some injuries [75, 79, 108, 109].

Prophylactic taping is often used in amateur and professional sports. Additionally, as stated above, taping decreases formation and seriousness of injuries [76, 80, 81, 105]. With taping decreasing the risk of foot and ankle injuries during functional rehabilitation and return to sports it may be a preferred preventive strategy [85] (Fig. 21.4a–g).

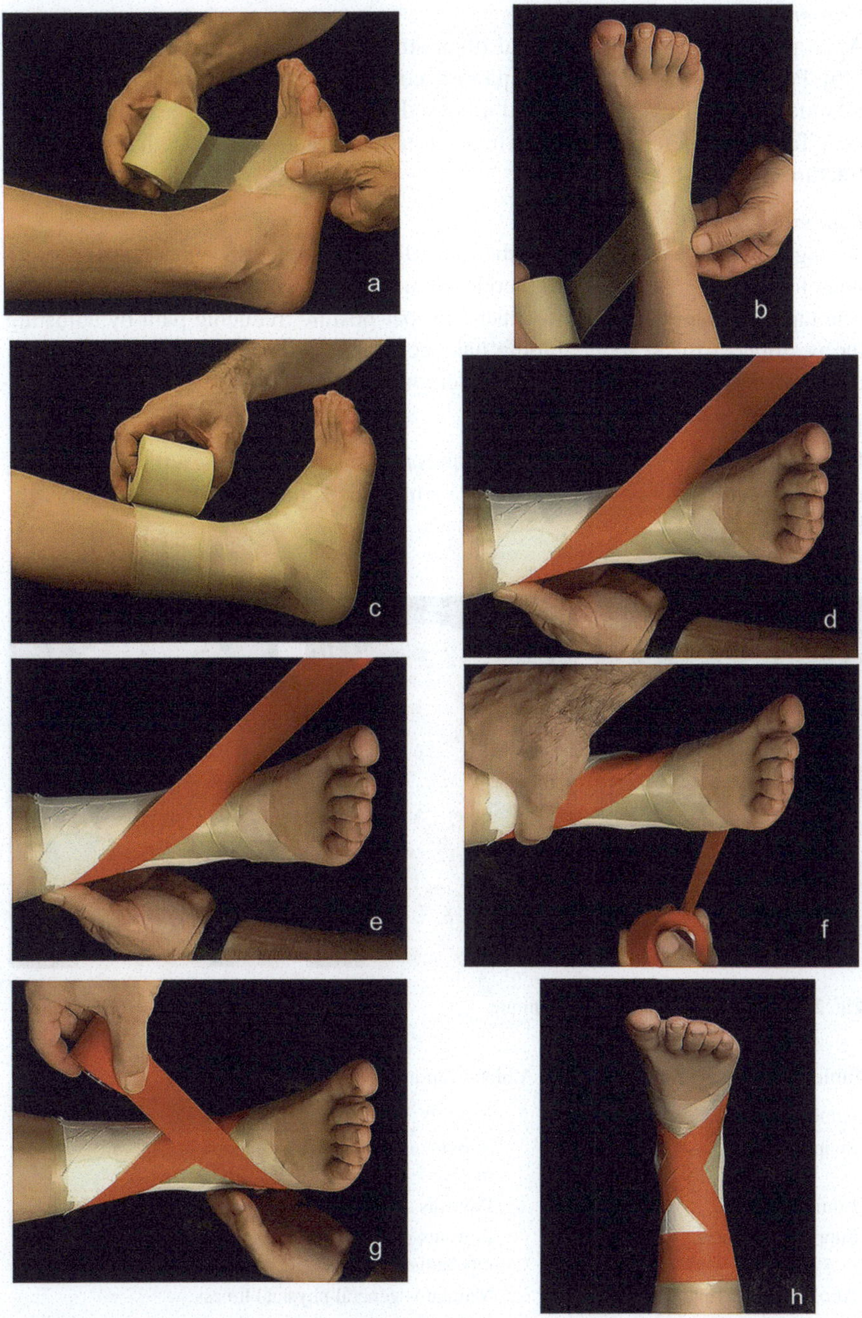

Fig. 21.4 Athletic tape application to ankle prevention

Conservative Treatment
Appropriate taping reduces medial tibial stress syndrome or shin-split symptoms [86]. Patellofemoral taping techniques are considered effective among conservative treatments [89] (Fig. 21.5a–c), and it is used as an option for finger and wrist injuries. Taping is preferred as the primary treatment option in Jones fracture or rib fractures [81, 90–92].

Pain management
Taping may play a crucial role in the reduction of pain. It may support nerve structures found along a neural path thereby reducing tension and pain [93]. The retraction taping technique is used to ensure proper posture, reducing pain by adjusting the position of the scapula. Taping has been suggested as an effective method in conservative treatment, such as patellofemoral pain syndrome [94, 95].

Proprioception
It has been suggested that taping techniques may aid in increasing ankle proprioception and position sense [96]. Table 21.3 summarizes the aims and benefits of athletic taping.

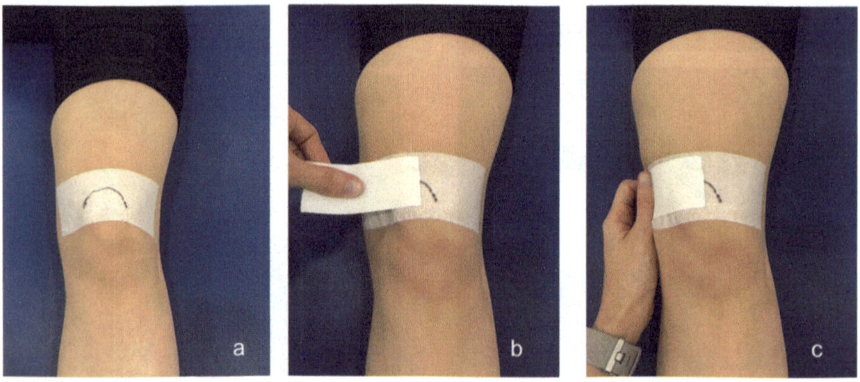

Fig. 21.5 Patellofemoral taping technique

Table 21.3 The aims and benefits of Athletic Taping

Aims	Benefits
Supports injured tissue	Stimulates circulation with painless joint range of motion
Limits activity	Prevents edema increase
Supports and increases recovery	Prevents injury deterioration, compensatory mechanisms, tissue atrophy
Allows painful functional motion	Maintains general physical fitness (Strength, endurance etc.) after injury
Allows optimal and sustainable motion by protecting the area	Regaining/ Maintaining abilities lost from inhibitor factors (ache, the fear of re-injury)
Reduces pain	Provides ease of use like orthosis

Taping and Orthosis

Taping provides protection and support of an injured area allowing for optimal functional motion and the reduction of pain. It is frequently used in sport rehabilitation, allowing for return to activity by providing rapid recovery. It prevents re-injury, the increase of injury symptoms, or the emergence of undesired adaptations (muscle activation patterns) [97–99]. Similar to taping, advantages and disadvantages of orthosis exist. Existing studies indicate that use of orthosis is better and effective on the recovery of injured tissues compared to taping [100, 101]. However, some orthoses are not suitable, especially in contact sports, in terms of their placement on the human body and because of their physical features (generally metal, rigid plastic or carbon fiber ones) [102, 105, 108]. Table 21.4 explains differences between orthosis and taping.

Best Time for Taping

Taping for sports rehabilitation can be applied and used during an acute period of injury, daily life, and to aid with therapeutic exercises [90]. It should be applied in a manner that lasts for the duration of the sport and removed later. When taping, it is important to not negatively impede any anatomical structures such as vascular and nerve tissues. Feedback should always be received from the athlete, making sure that circulation is not hampered. When there is a need for longer periods, suitable materials and techniques should be chosen [98, 100].

The athlete should also be instructed to be aware of any adverse effects and to remove the tape to prevent tissue damage. Many studies conducted on the mechanical effect of taping indicated that there is some movement restriction after exercise, with some researchers observing that ankle taping rapidly loses its effect of support (within 30–40 minutes) during exercise [72, 80, 81, 85, 100, 105].

Table 21.4 Differences between orthosis and taping [75, 76, 79–81, 85, 97–101, 107–109]

Orthosis	Taping
No need of an expert to apply, can be applied by athlete	Applied by experts
Reusable	Looks thinner than orthosis
Non allergic	More adaptive for different anatomic locations
Adjustable	Appropriate for all sports
Cost-efficient	
Could be inappropriate for some sports	

2. Kinesio Taping

Kinesio taping technique uses a special tape made with an acrylic, adhesive, colored elastic cotton that can extend up to 140% of its actual length. The band, which can stretch 55–60% of the resting length in the longitudinal direction, can adapt to the stretch and relaxation of the skin, imitating the thickness and flexibility of the skin. It is similar to the epidermis layer of the skin in terms of its thickness and weight and can adhere to the epidermis from between 24 hours and 3 days depending on the application technique. Kinesio taping is widely used by physiotherapists, to help reduce pain, swelling and edema, increase muscle activity and function, regulate spasticity and proprioceptive input [103, 110].

Its mechanism creates a positional stimulation by sending a signal to the central nervous system stimulating the mechanoreceptors through the skin depending on the degree of tension applied [111, 112]. It helps to [103]:

(a) Correct the alignment of the fascia tissue, create more space for circulation by removing the fascia and skin, subcutaneous soft tissues on the painful
(b) Create sensory stimulation to limit or increase the movement
(c) To reduce edema by directing exudate to lymph tracts

Kinesio-Taping generally acts on 5 basic physiological systems:

Skin

With several pathological conditions being associated with increased pressure on blood vessels, lymph tracts, and nociceptors in the dermis, use of kinesio-tape on the epidermis creates convulsions (folds) in the tissue lifting the epidermis. The space created by the kinesio-tape in the dermis reduces the effect of pressure on the blood and lymphatic vessels, diminishes stimulation of the nociceptors thereby regulating the homeostasis of the skin [113–115].

Fascia

During embryological development of the human body, the dermis and fascia are differentiated from the mesoderm. With fascia consisting of both deep and superficial layers, use of Kinesio-taping will influence the fascia and the muscle it surrounds [116–119]. For example, in quadriceps or hamstring strain, fascial integrity is impaired, reducing mobility of the affected muscle. Kinesio-tape ensures a smoothness of the fascia allowing for easier contraction of the muscle thereby accelerating its healing [119–121].

Muscle

With the application of Kinesio taping, it is possible to establish the appropriate length/stretch ratio of the muscle tissue. This length/stretch ratio assures the formation of optimal actin-myosin cross-bridging during muscle contraction and thus allows an optimal force release of the muscle. This characteristic of kinesio-tape, which is not found with other taping techniques, accelerates the healing process, as well as preserves the anatomical integrity of the muscles [122–124]. Figure 21.6 shows m. longissimus cervicis (10–20% tension of tape) and contralateral ligament (100% tension of tape) taping techniques after whiplash injury [103, 125, 126].

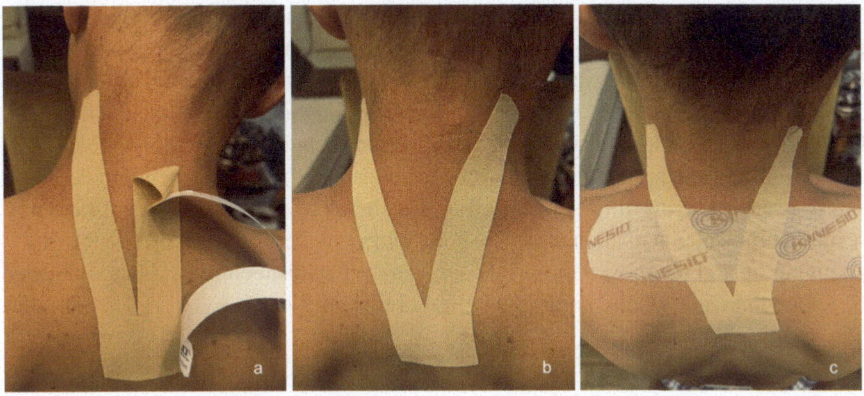

Fig. 21.6 Kinesio taping for m.longissimus cervicis (**a**, **b**) and for contralateral ligament (**c**)

Joint

Correct joint biomechanics is possible with Kinesio taping. With ligaments and tendons of a joint affected by injury, as a result of imbalance between the agonists and antagonists muscle groups about a joint, kinesio-taping corrects for this imbalance, ensuring proper function [127–129], and a reduction in pain [130–133]. One important effect of Kinesio taping of the joint is the increase in kinesthetic awareness during movement, helping correct any mechanically damaged structure [134–136].

Circulatory/Lymphatic Systems

Application of Kinesio-tape on the skin increases interstitial lymphatic fluid flow providing better interstitial fluid exchange, reducing edema and pain in the injured area as well as decreasing tissue temperature. By creating convulsions, as discussed above in the skin section, Kinesio-Tape opens low pressure channels in the obstructed areas affecting the epidermis and dermis, thereby reducing pain and re-establishing homeostasis [137–139]. In addition, by increasing circulation, it reduces fatigue, by enabling the fast removal of waste materials accumulated in the interstitial gap after exercise. This is extremely beneficial for acute situations or in all pathologies where lymphatic fluid flow is impaired [137–139].

The contraindications for Kinesio taping are on areas of active malignancy, cellulite or skin irritation, open wounds, and deep vein thrombosis [103]. Caution should be exercised in cases such as diabetes, kidney diseases and congestive heart failure, on baroreceptors in the neck in the carotid artery, on fragile or healing skin, and during pregnancy [103].

Robust scientific data on the effect mechanisms and effectiveness of taping techniques with regards to increases in muscle strength are lacking and the quality of evidence is low. According to some studies, by supporting the muscle tissue around the joint, this not only allows for better joint stability and proper efficient movement, but aids in the strengthening of muscle [140, 141]. In addition, by reducing any external pressure on muscles, ligaments, tendons, and nerves, a decrease in tension increases the proper proprioception sensation of these tissues.

The literature [142, 143] on Kinesio taping suggests many uses for musculoskeletal injuries, improvement in sports performance, the treatment of pediatric structural issues (i.e., Osgood Schlaters), and the prevention of injuries [111–134, 144, 145]. Within Türkiye [146, 147], many clinical applications have looked at the combination of kinesio taping with other therapeutic techniques [113–116, 137, 138]. As a result of these applications, it has been suggested that taping techniques can be safely used for many injuries as well as for their prevention. Since injuries may result in biomechanical changes to the body, physical therapists need to be informed about taping and the most appropriate application for this technique. It should never be applied in cases of any contraindication or discomfort.

Conclusion

In summary, physical therapy is crucial for athletes' recovery, regeneration, and adaptation to exercise stress. Physical therapists use a range of techniques to help athletes recover from intense exercise by reducing pain, inflammation, and promoting tissue healing. They assist in restoring damaged tissues, improving their strength and mobility. Physical therapy helps athletes adapt to the demands of their sports by designing personalized exercise programs that enhance their physical abilities. In addition, physical therapy plays a significant role in preventing injuries by identifying and addressing imbalances and weaknesses that may increase the risk of getting hurt. Physical therapy takes a holistic approach, considering athletes' unique needs and working together with other healthcare professionals and coaches. Overall, physical therapy is vital for athletes to reach their full potential, stay injury-free, and excel in their sports.

1. How effective is physical therapy for an athlete's recovery?
2. How can you apply Myofascial Release Techniques effectively?
3. How effective are the Six Fascial Distortion Models (FDM) for the proper recovery and regeneration of the body?
4. What are some of the benefits of manual lymphatic drainage therapy for athletes post-performance, and how does it contribute to their recovery process?
5. What are some of the key benefits of the use of cupping therapy in sports medicine following injury?
6. Would grade III traction by typically used to stretch the surrounding tissues in joint therapy?
7. Is the use of orthosis generally considered more effective than taping for the recovery of injured tissues?
8. How effective and beneficial is kinesio taping for the recovery of the athlete from injury?

References

1. Mylonas K, Angelopoulos P, Tsepis E, Billis E, Fousekis K. Soft-tissue techniques in sports injuries prevention and rehabilitation. Contemporary advances in sports science. IntechOpen; 2021.
2. Micklewright D. The effect of soft tissue release on delayed onset muscle soreness: a pilot study. Phys Ther Sport. 2009;10(1):19–24.
3. PM T, Shoemaker JK. Effleurage massage, muscle blood flow and long-term post-exercise strength recovery. Int J Sports Med. 1995;16:478–83.
4. Zink C, Van Dyke JB. Canine sports medicine and rehabilitation. Wiley; 2018.
5. Frontera WR, Silver JK, Rizzo TD. Essentials of physical medicine and rehabilitation e-book. Elsevier Health Sciences; 2018.
6. Travell JG, Simons DG. Myofascial pain and dysfunction: the trigger point manual. Lippincott Williams & Wilkins; 1992.
7. Bentley S. Exercise-induced muscle cramp: proposed mechanisms and management. Sports Med. 1996;21:409–20.
8. Wong CS, Wong SH. A new look at trigger point injections. Anesthesiology research and practice; 2012.
9. Alvarez DJ, Rockwell PG. Trigger points: diagnosis and management. Am Fam Physician. 2002;65(4):653.
10. Hong C-Z. Lidocaine injection versus dry needling to myofascial trigger point. The importance of the local twitch response. Am J Phys Med Rehabil. 1994;73(4):256–63.
11. Korr IM. Proprioceptors and somatic dysfunction. J Am Osteopath Assoc. 1975;74:638–50.
12. Korr IM. The neural basis of the osteopathic lesion. J Am Osteopath Assoc. 1947;47(4):191–8.
13. IM K. The somatic approach to the disease process. J Am Osteopath Assoc. 1951;51(4):201–5.
14. Korr IM. Somatic dysfunction, osteopathic manipulative treatment, and the nervous system: a few facts, some theories, many questions. J Am Osteopath Assoc. 1986;86(2):97–102.
15. Speicher TE, Draper DO. Top 10 positional-release therapy techniques to break the chain of pain, part 1. 2006.
16. Jones LH, Kusunose R, Goering E. Jones strain-counterstrain. Jones Strain Counterstrain Incorporated Boise; 1995.
17. Huguenin LK. Myofascial trigger points: the current evidence. Phys Ther Sport. 2004;5(1):2–12.
18. Speicher T. Clinical guide to positional release therapy. Human Kinetics; 2016.
19. Chatow L, Delany J. Ch 9, Summary of modalities. In: Clinical application of neuromuscular techniques. Edinburgh: Churchill Livingstone; 2002. p. 201–8.
20. Sands WA, Apostolopoulos N, Kavanaugh AA, Stone MH. Recovery-adaptation. Streng Conditi J. 2016;38(6):10–26.
21. Stecco C, Schleip R. A fascia and the fascial system. J Bodyw Mov Ther. 2016;20(1):139–40.
22. Stecco L, Stecco C. Fascial Manipulation. Piccin: Practical Part; 2012.
23. Stecco A, Stern R, Fantoni I, De Caro R, Stecco C. Fascial disorders: implications for treatment. PM&R. 2016;8(2):161–8.
24. Kumka M, Bonar J. Fascia: a morphological description and classification system based on a literature review. J Canad Chiropractic Assoc. 2012;56(3):179.
25. Schleip R, Jäger H, Klingler W. What is 'fascia'? A review of different nomenclatures. J Bodywork Move Ther. 2012;16(4):496–502.
26. Mauntel TC, Clark MA, Padua DA. Effectiveness of myofascial release therapies on physical performance measurements: a systematic review. Athlet Train Sports Health Care. 2014;6(4):189–96.
27. McKenney K, Elder AS, Elder C, Hutchins A. Myofascial release as a treatment for orthopaedic conditions: a systematic review. J Athlet Train. 2013;48(4):522–7.

28. Portillo-Soto A, Eberman LE, Demchak TJ, Peebles C. Comparison of blood flow changes with soft tissue mobilization and massage therapy. J Altern Complement Med. 2014;20(12):932–6.
29. Stull K. Complete guide to foam rolling. Human Kinetics; 2017.
30. Boerio D, Jubeau M, Zory R, Maffiuletti NA. Central and peripheral fatigue after electrostimulation-induced resistance exercise. Med Sci Sports Exerc. 2005;37(6):973–8.
31. Williams C, Ratel S. Human muscle fatigue. Routledge; 2009.
32. Baker RT, Nasypany A, Seegmiller JG, Baker JG. Instrument-assisted soft tissue mobilization treatment for tissue extensibility dysfunction. Int J Athlet Ther Train. 2013;18(5):16–21.
33. Schroeder AN, Best TM. Is self myofascial release an effective preexercise and recovery strategy? A literature review. Curr Sports Med Rep. 2015;14(3):200–8.
34. Cheatham SW, Kolber MJ, Cain M, Lee M. The effects of self-myofascial release using a foam roll or roller massager on joint range of motion, muscle recovery, and performance: a systematic review. Int J Sports Phys Ther. 2015;10(6):827.
35. Peacock CA, Krein DD, Silver TA, Sanders GJ, Von Carlowitz K-PA. An acute bout of self-myofascial release in the form of foam rolling improves performance testing. Int J Exerc Sci. 2014;7(3):202.
36. Stecco A, Gesi M, Stecco C, Stern R. Fascial components of the myofascial pain syndrome. Curr Pain Headache Rep. 2013;17:1–10.
37. Brandolini S, Lugaresi G, Santagata A, Ermolao A, Zaccaria M, Marchand AM, et al. Sport injury prevention in individuals with chronic ankle instability: fascial Manipulation® versus control group: a randomized controlled trial. J Bodyw Mov Ther. 2019;23(2):316–23.
38. Thalhamer C. A fundamental critique of the fascial distortion model and its application in clinical practice. J Bodyw Mov Ther. 2018;22(1):112–7.
39. Typaldos S. Introducing the fascial distortion model. AAO J. 1994;4(2):14–8.
40. Lee JK, Kim CM. Comparison of maximum isometric strength, proprioceptive, dynamic balance, and maximum angle by applying the fascial distortion model to chronic ankle instability subjects. J Korean Phys Ther. 2021;33(5):224–30.
41. Vairo GL, Miller SJ, Rier NCI, Uckley WI. Systematic review of efficacy for manual lymphatic drainage techniques in sports medicine and rehabilitation: an evidence-based practice approach. J Manual Manipul Ther. 2009;17(3):80E–9E.
42. Schillinger A, Koenig D, Haefele C, Vogt S, Heinrich L, Aust A, et al. Effect of manual lymph drainage on the course of serum levels of muscle enzymes after treadmill exercise. Am J Phys Med Rehabil. 2006;85(6):516–20.
43. Korosec BJ. Manual lymphatic drainage therapy. Home Health Care Manag Pract. 2004;16(6):499–511.
44. Murray D, Clarkson C. Effects of moving cupping therapy on hip and knee range of movement and knee flexion power: a preliminary investigation. J Manual Manipul Ther. 2019;27(5):287–94.
45. Chirali I. Cupping therapy. Traditional Chinese medicine E-book Churchill Livinstone. Philadelphia: Elsevier; 2007.
46. Jam B. Tissue distraction release with movement (TDR-WM) to replace tissue compression techniques. J Novel Physiother. 2015;5(4):77–97.
47. Young J. Acupressure: simple steps to health: discover your body's powerpoints for health and relaxation. HarperCollins UK; 2013.
48. Coates JC. Manual therapy. In: Canine sports medicine and rehabilitation; 2018. p. 120–35.
49. Magnusson S, Simonsen E, Aagaard P, Gleim G, McHugh M, Kjaer M. Viscoelastic response to repeated static stretching in the human hamstring muscle. Scand J Med Sci Sports. 1995;5(6):342–7.
50. Zöllner AM, Abilez OJ, Böl M, Kuhl E. Stretching skeletal muscle: chronic muscle lengthening through sarcomerogenesis. 2012.
51. Salvini TF, Coutinho EL, Russo TL, DeLuca C. One-minute bouts of passive stretching after immobilization increase sarcomerogenesis in rat soleus muscle. J Morphol Sci. 2017;23(3):0.

52. Hinks A, Franchi MV, Power GA. The influence of longitudinal muscle fascicle growth on mechanical function. J Appl Physiol. 2022;133(1):87–103.
53. Young WB. The use of static stretching in warm-up for training and competition. Int J Sports Physiol Perform. 2007;2(2):212–6.
54. Behm DG, Chaouachi A. A review of the acute effects of static and dynamic stretching on performance. Eur J Appl Physiol. 2011;111:2633–51.
55. Garber CE, Blissmer B, Deschenes MR, Franklin BA, Lamonte MJ, Lee I-M, et al. American College of Sports Medicine position stand. Quantity and quality of exercise for developing and maintaining cardiorespiratory, musculoskeletal, and neuromotor fitness in apparently healthy adults: guidance for prescribing exercise. Med Sci Sports Exerc. 2011;43(7):1334–59.
56. Simenz CJ, Dugan CA, Ebben WP. Strength and conditioning practices of National Basketball Association strength and conditioning coaches. J Strength Cond Res. 2005;19(3):495–504.
57. Judge LW, Petersen JC, Bellar DM, Craig BW, Wanless EA, Benner M, et al. An examination of preactivity and postactivity stretching practices of crosscountry and track and field distance coaches. J Strength Cond Res. 2013;27(9):2456–64.
58. Wyon M, Felton L, Galloway S. A comparison of two stretching modalities on lower-limb range of motion measurements in recreational dancers. J Strength Cond Res. 2009;23(7):2144–8.
59. Apostolopoulos N. Microstretching®: a new recovery regeneration technique. New Stud Athlet. 2004;19(4):47–56.
60. Apostolopoulos N. microStretching-A practical approach for recovery and regeneration. New Stud Athlet. 2010;25:81–97.
61. Apostolopoulos NC. Stretch intensity and the inflammatory response: a paradigm shift. Springer; 2018.
62. Popp JK, Bellar DM, Hoover DL, Craig BW, Leitzelar BN, Wanless EA, et al. Pre-and post-activity stretching practices of collegiate athletic trainers in the United States. J Strength Cond Res. 2017;31(9):2347–54.
63. Witvrouw E, Danneels L, Asselman P, D'Have T, Cambier D. Muscle flexibility as a risk factor for developing muscle injuries in male professional soccer players: a prospective study. Am J Sports Med. 2003;31(1):41–6.
64. Watsford ML, Murphy AJ, McLachlan KA, Bryant AL, Cameron ML, Crossley KM, et al. A prospective study of the relationship between lower body stiffness and hamstring injury in professional Australian rules footballers. Am J Sports Med. 2010;38(10):2058–64.
65. Lorimer AV, Hume PA. Stiffness as a risk factor for Achilles tendon injury in running athletes. Sports Med. 2016;46:1921–38.
66. Apostolopoulos N, Metsios GS, Flouris AD, Koutedakis Y, Wyon MA. The relevance of stretch intensity and position—a systematic review. Front Psychol. 2015;6:1128.
67. Bandy WD, Irion JM. The effect of time on static stretch on the flexibility of the hamstring muscles. Phys Ther. 1994;74(9):845–50.
68. Feland JB, Myrer JW, Schulthies SS, Fellingham GW, Measom GW. The effect of duration of stretching of the hamstring muscle group for increasing range of motion in people aged 65 years or older. Phys Ther. 2001;81(5):1110–7.
69. Kaltenborn FM. Manual mobilization of the joints; the Kaltenborn method of joint examination and treatment. The extremities; 1999. p. 21–8.
70. Kaltenborn FM, Evjenth O, Morgan D. Manual mobilization of the extremity joints: basic examination and treatment techniques. Oslo: Olaf Norlis; 1989.
71. Yin Y, Wang J, Wu C, Sun J. Joint mobilization treatment for chronic ankle instability function recovery: a meta-analysis. Chinese J Tissue Eng Res. 2022;26(33):5407.
72. Bullock-Saxton J, Janda V, Bullock M. The influence of ankle sprain injury on muscle activation during hip extension. Int J Sports Med. 1994;15(06):330–4.
73. Logan CA, Bhashyam AR, Tisosky AJ, Haber DB, Jorgensen A, Roy A, et al. Systematic review of the effect of taping techniques on patellofemoral pain syndrome. Sports Health. 2017;9(5):456–61.
74. Baquie P. Taping: general principles. Aust Fam Physician. 2002;31(2):155–7.

75. Greene TA, Hillman SK. Comparison of support provided by a semirigid orthosis and adhesive ankle taping before, during, and after exercise. Am J Sports Med. 1990;18(5):498–506.
76. Hughes LY, Stetts DM. A comparison of ankle taping and a semirigid support. Phys Sportsmed. 1983;11(4):99–103.
77. Hewetson TJ, Austin K, Gwynn-Brett K, Marshall S. An illustrated guide to taping techniques: principles and practice. Elsevier Health Sciences; 2009.
78. Birrer RB, Poole B. Taping of sports injuries: review of a basic skill: general principles, specifics for the ankle. J Musculoskeletal Med. 2004;21(4):197–205.
79. Callaghan MJ. Role of ankle taping and bracing in the athlete. Br J Sports Med. 1997;31(2):102–8.
80. Olmsted LC, Vela LI, Denegar CR, Hertel J. Prophylactic ankle taping and bracing: a numbers-needed-to-treat and cost-benefit analysis. J Athl Train. 2004;39(1):95.
81. Pederson TS, Ricard MD, Merrill G, Schulthies SS, Allsen PE. The effects of spatting and ankle taping on inversion before and after exercise. J Athl Train. 1997;32(1):29.
82. Alt W, Lohrer H, Gollhofer A. Functional properties of adhesive ankle taping: neuromuscular and mechanical effects before and after exercise. Foot Ankle Int. 1999;20(4):238–45.
83. Aminaka N, Gribble PA. A systematic review of the effects of therapeutic taping on patellofemoral pain syndrome. Database of Abstracts of Reviews of Effects (DARE): quality-assessed reviews [Internet], 2005.
84. Lohrer H, Alt W, Gollhofer A. Neuromuscular properties and functional aspects of taped ankles. Am J Sports Med. 1999;27(1):69–75.
85. Wilkerson GB. Biomechanical and neuromuscular effects of ankle taping and bracing. J Athl Train. 2002;37(4):436.
86. Ugalde V, Batt ME, Chir M. Shin splints: current theories and treatment. Crit Rev Phys Rehabil Med. 2001;13(2-3):37.
87. Araújo CGA, Macedo CSG, Ferreira D, Shigaki L, da Silva RA. Mcconnell's patellar taping does not alter knee and hip muscle activation differences during proprioceptive exercises: a randomized placebo-controlled trial in women with patellofemoral pain syndrome. J Electromyogr Kinesiol. 2016;31:72–80.
88. Clifford A, Dillon S, Hartigan K, O'Leary H, Constantinou M. The effects of McConnell patellofemoral joint and tibial internal rotation limitation taping techniques in people with Patellofemoral pain syndrome. Gait Posture. 2020;82:266–72.
89. Sisk D, Fredericson M. Taping, bracing, and injection treatment for Patellofemoral pain and patellar tendinopathy. Curr Rev Musculoskelet Med. 2020;13:537–44.
90. Manfroy PP, Ashton-Miller JA, Wojtys EM. The effect of exercise, prewrap, and athletic tape on the maximal active and passive ankle resistance to ankle inversion. Am J Sports Med. 1997;25(2):156–63.
91. Specchiulli F, Cofano R. A comparison of surgical and conservative treatment in ankle ligament tears. Thorofare: SLACK Incorporated; 2001. p. 686–8.
92. Trégouët P, Merland F, Horodyski M. A comparison of the effects of ankle taping styles on biomechanics during ankle inversion. Ann Phys Rehabil Med. 2013;56(2):113–22.
93. Edmonds DW, McConnell J, Ebert JR, Ackland TR, Donnelly CJ. Biomechanical, neuromuscular and knee pain effects following therapeutic knee taping among patients with knee osteoarthritis during walking gait. Clin Biomech. 2016;39:38–43.
94. Leibbrandt DC, Louw QA. The use of McConnell taping to correct abnormal biomechanics and muscle activation patterns in subjects with anterior knee pain: a systematic review. J Phys Ther Sci. 2015;27(7):2395–404.
95. Moss CL, Gorton B, Deters S. A comparison of prescribed rigid orthotic devices and athletic taping support used to modify pronation in runners. J Sport Rehabil. 1993;2(3):179–88.
96. Simoneau GG, Degner RM, Kramper CA, Kittleson KH. Changes in ankle joint proprioception resulting from strips of athletic tape applied over the skin. J Athl Train. 1997;32(2):141.
97. Lau KK-L, Cheng KC-C. Effectiveness of taping on functional performance in elite athletes: a systematic review. J Biomech. 2019;90:16–23.

98. Seah R, Mani-Babu S. Managing ankle sprains in primary care: what is best practice? A systematic review of the last 10 years of evidence. Br Med Bull. 2011;97(1):105–35.
99. Hatch RL, Hacking S. Evaluation and management of toe fractures. Am Fam Physician. 2003;68(12):2413–8.
100. Handoll HH, Rowe BH, Quinn KM, de Bie R, Cochrane Bone J, Group MT. Interventions for preventing ankle ligament injuries. Cochrane Database Syst Rev. 1996;2010(1)
101. Verhagen EA, van der Beek AJ, van Mechelen W. The effect of tape, braces and shoes on ankle range of motion. Sports Med. 2001;31:667–77.
102. Doodkorte RJ, Vercoulen TF, Roth AK, de Bie RA, Willems PC. Instrumentation techniques to prevent proximal junctional kyphosis and proximal junctional failure in adult spinal deformity correction—a systematic review of biomechanical studies. Spine J. 2021;21(5):842–54.
103. Kase K. Clinical therapeutic applications of the Kinesio (! R) taping method. Albuquerque, 2003.
104. Parkkari J, Kujala UM, Kannus P. Is it possible to prevent sports injuries? Review of controlled clinical trials and recommendations for future work. Sports Med. 2001;31:985–95.
105. Cordova ML, Ingersoll CD, Palmieri RM. Efficacy of prophylactic ankle support: an experimental perspective. J Athl Train. 2002;37(4):446.
106. Cordova ML, Ingersoll CD, LeBlanc MJ. Influence of ankle support on joint range of motion before and after exercise: a meta-analysis. J Orthop Sports Phys Ther. 2000;30(4):170–82.
107. Jessee AD, Gourley MM, Valovich McLeod TC. Bracing and taping techniques and patellofemoral pain syndrome. J Athl Train. 2012;47(3):358–9.
108. Gross MT, Bradshaw MK, Ventry LC, Weller KH. Comparison of support provided by ankle taping and semirigid orthosis. J Orthop Sports Phys Ther. 1987;9(1):33–9.
109. Canelon MF. Silicone rubber splinting for athletic hand and wrist injuries. J Hand Ther. 1995;8(4):252–7.
110. Stockheimer KR, Baltacı G, Forrester GG, Frassine S, Wolkenberg A. Research requires deep knowledge of the modality to be tested. J Physiother. 2016;62(2):118.
111. Kirmizigil B, Chauchat JR, Yalciner O, Iyigun G, Angin E, Baltaci G. The effectiveness of kinesio taping in recovering from delayed onset muscle soreness: a crossover study. J Sport Rehabil. 2019;29(4):385–93.
112. Saltan A, Baltaci G, Ankarali H. Does Kinesio® taping improve balance and functional performance in older adults? A pilot study. J Sports Med Phys Fitness. 2018;59(8):1346–52.
113. Orhan C, Kaya Kara O, Kaya S, Akbayrak T, Kerem Gunel M, Baltaci G. The effects of connective tissue manipulation and Kinesio Taping on chronic constipation in children with cerebral palsy: a randomized controlled trial. Disabil Rehabil. 2018;40(1):10–20.
114. Eraslan L, Yuce D, Erbilici A, Baltaci G. Does Kinesiotaping improve pain and functionality in patients with newly diagnosed lateral epicondylitis? Knee Surg Sports Traumatol Arthrosc. 2018;26:938–45.
115. Pekyavas NO, Baltaci G. Short-term effects of high-intensity laser therapy, manual therapy, and Kinesio taping in patients with subacromial impingement syndrome. Lasers Med Sci. 2016;31:1133–41.
116. Aytar A, Ozunlu N, Surenkok O, Baltacı G, Oztop P, Karatas M. Initial effects of kinesio® taping in patients with patellofemoral pain syndrome: a randomized, double-blind study. Isokinet Exerc Sci. 2011;19(2):135–42.
117. Hung M-H, Chen H-Y, Chang Y-C, Chiu C-W, Chang H-Y. Effects of the direction of Kinesio taping on sensation and postural control before and after muscle fatigue in healthy athletes. Sci Rep. 2023;13(1):1282.
118. Annino G, Alashram A, Romagnoli C, Balducci E, De Paolis M, Manzi V, et al. Acute effects of Kinesio taping on functional performance in healthy soccer players: a randomized, controlled crossover trial. J Funct Morphol Kinesiol. 2022;8(1):2.
119. Rana A, Tyagi D, Alghadir AH, Khan M. Effects of Kinesio taping on calf muscle fatigue in college female athletes: a randomized controlled trial. Medicine. 2022;101(43):e31004-e.

120. Malhotra D, Sharma S, Chachra A, Dhingra M, Alghadir AH, Nuhmani S, et al. The time-based effects of Kinesio Taping on acute-onset muscle soreness and calf muscle extensibility among endurance athletes: a randomized cross-over trial. J Clin Med. 2022;11(20):5996.
121. Özmen T, Koparal SS, Karataş Ö, Eser F, Özkurt B, Gafuroğlu TÜ. Comparison of the clinical and sonographic effects of ultrasound therapy, extracorporealshock wave therapy, and Kinesio taping in lateral epicondylitis. Turkish J Med Sci. 2021;51(1):76–83.
122. Kiseljak D, Medved V. The effects of Kinesio Taping® on muscle interplay within the Lumbo–Pelvic–Hip complex: a randomized placebo-controlled trial. Sports. 2023;11(3):70.
123. Li R, Qin R, Tan Y, Liu H, Wang K, Cheng L. Effect of kinesio taping intervention on the muscle strength and balance of college basketball players with functional ankle instability. Front Physiol. 2023;14:496.
124. Kalaycioglu T, Apostolopoulos NC, Goldere S, Duger T, Baltaci G. Effect of a core stabilization training program on performance of ballet and modern dancers. J Strength Cond Res. 2020;34(4):1166–75.
125. Shih H-S, Chen S-S, Cheng S-C, Chang H-W, Wu P-R, Yang J-S, et al. Effects of Kinesio taping and exercise on forward head posture. J Back Musculoskelet Rehabil. 2017;30(4):725–33.
126. Alahmari KA, Reddy RS, Tedla JS, Samuel PS, Kakaraparthi VN, Rengaramanujam K, et al. The effect of Kinesio taping on cervical proprioception in athletes with mechanical neck pain—a placebo-controlled trial. BMC Musculoskelet Disord. 2020;21(1):1–9.
127. Bicici S, Karatas N, Baltaci G. Effect of athletic taping and kinesiotaping® on measurements of functional performance in basketball players with chronic inversion ankle sprains. Int J Sports Phys Ther. 2012;7(2):154.
128. Khalili SM, Barati AH, Oliveira R, Nobari H. Effect of combined balance exercises and Kinesio Taping on balance, postural stability, and severity of ankle instability in female athletes with functional ankle instability. Life. 2022;12(2):178.
129. Biz C, Nicoletti P, Tomasin M, Bragazzi NL, Di Rubbo G, Ruggieri P. Is kinesio taping effective for sport performance and ankle function of athletes with chronic ankle instability (CAI)? A systematic review and meta-analysis. Medicina. 2022;58(5):620.
130. Lee JH, Rhim HC, Jang K-M. Effect of adding Kinesio Taping to exercise therapy in the treatment of patellofemoral pain syndrome. Medicina. 2023;59(4):754.
131. Nelson NL. Kinesio taping for chronic low back pain: a systematic review. J Bodyw Mov Ther. 2016;20(3):672–81.
132. Senbursa G, Pekyavas NO, Baltaci G. Comparison of physiotherapy approaches in low back pain: a randomized controlled trial. Korean J Family Med. 2021;42(2):96.
133. Sheng Y, Duan Z, Qu Q, Chen W, Yu B. Kinesio taping in treatment of chronic non-specific low back pain: a systematic review and meta-analysis. J Rehabil Med. 2019;51(10):734–40.
134. Kochman M, Kasprzak M, Kielar A. ACL reconstruction: which additional physiotherapy interventions improve early-stage rehabilitation? A systematic review. Int J Environ Res Public Health. 2022;19(23):15893.
135. Baltaci G, Ozunlu Pekyavas N, Atay OA. Short-time effect of sterile kinesio tape applied during anterior cruciate ligament reconstruction on edema, pain and range of motion. Res Sports Med. 2021;31:1–12.
136. Labianca L, Andreozzi V, Princi G, Princi AA, Calderaro C, Guzzini M, et al. The effectiveness of Kinesio Taping in improving pain and edema during early rehabilitation after Anterior Cruciate Ligament reconstruction: a prospective, randomized, control study. Acta Bio Medica: Atenei Parmensis. 2022;92(6):e2021336.
137. Trofa DP, Obana KK, Herndon CL, Noticewala MS, Parisien RL, Popkin CA, et al. The evidence for common nonsurgical modalities in sports medicine, part 1: kinesio tape, sports massage therapy, and acupuncture. JAAOS Glob Res Rev. 2020;4(1):e1900104.
138. Eraslan L, Baltaci G. Effectiveness of kinesio taping and night splinting along with physical therapy intervention on symptoms and functionality: a double-blind randomized controlled trial for moderate carpal tunnel syndrome. Am J Phys Med Rehabil. 2023;10:1097.

139. Naci B, Ozyilmaz S, Aygutalp N, Demir R, Baltaci G, Yigit Z. Effects of Kinesio Taping and compression stockings on pain, edema, functional capacity and quality of life in patients with chronic venous disease: a randomized controlled trial. Clin Rehabil. 2020;34(6):783–93.
140. Alabbad M, Muaidi Q. The effect of the Kinesio Tape on the muscle power performance of elite weightlifters. J Bodyw Mov Ther. 2021;27:26–33.
141. Sheikhi B, Letafatkar A, Marchetti PH, Eftekhari F, Wallace BJ, Maselli F, et al. Effects of Kinesio Taping on Tuck jump performance in competitive male athletes. Int J Sports Med. 2023;44:516.
142. Tran L, Makram AM, Makram OM, Elfaituri MK, Morsy S, Ghozy S, et al. Efficacy of kinesio taping compared to other treatment modalities in musculoskeletal disorders: a systematic review and meta-analysis. Res Sports Med. 2021;31:1–24.
143. Williams S, Whatman C, Hume PA, Sheerin K. Kinesio taping in treatment and prevention of sports injuries: a meta-analysis of the evidence for its effectiveness. Sports Med. 2012;42:153–64.
144. Gianola S, Iannicelli V, Fascio E, Andreano A, Li L, Valsecchi M, et al. Kinesio taping for rotator cu disease. status and date: New, published in. 2021(8).
145. Song W, Yang Y. Effect of Kinesio taping on delayed-onset muscle soreness in elite athletes. J Sports Med Phys Fitness. 2021;62(5):667–72.
146. Tekin D, Agopyan A, Baltaci G. Balance training in modern dancers: proprioceptive-neuromuscular training vs kinesio taping. Med Probl Perform Art. 2018;33(3):156–65.
147. Yildiz S, Pamuk U, Baltaci G, Yucesoy CA. Effects of Kinesio Taping on muscle contractile properties: assessment using tensiomyography. J Sport Rehabil. 2021;31(3):263–70.

Chapter 22
Acupuncture: It's Relationship to Recovery, Regeneration and Function

Colin Harris and Richard Hazel

Introduction

History of Acupuncture

Acupuncture is an ancient treatment that can trace its routes back to 2500–3000 years ago in China. The first classical medical text in Traditional Chinese Medicine (TCM), the Huandi Neijing contains two sections, the Su Wen and the Ling Shu and its composition is placed around the late Warring States of Sun Tzu, 475–221 BC, through to the Han dynasty around 206–220 BC. The Ling Shu is the first medical text that describes acupuncture treatment principles and techniques and outlines specific medical conditions and their acupuncture point prescriptions for treatment. In Traditional Chinese Medicine theory, acupuncture modulates the flow of Qi (energy) and Xue (blood) in the body through various mapped out meridians which result in re-establishing a homeostasis in the organs based on certain laws of yin-yang and the five elements. Insufficiency or blockage of energy along the meridians leads to stagnation of the flow of blood and restriction in the tissues causing tension and pain.

One of the descriptions used frequently in the Ling Shu is the sensation a patient should feel from the insertion of the acupuncture needle for the treatment to be effective. These classically described sensations are feelings of dull and achy, heaviness, and fullness and is called "De Qi" which is translated to simply big qi.

C. Harris (✉)
Vital Balance Therapy, Coquitlam, BC, Canada
e-mail: vitalbalancetherapy@gmail.com

R. Hazel
Range of Motion Acupuncture, Buffalo, NY, USA
e-mail: info@richhazel.com

N. C. Apostolopoulos et al. (eds.), *Fundamentals of Recovery, Regeneration, and Adaptation to Exercise Stress: An Integrated Approach*,
https://doi.org/10.1007/978-3-031-44270-4_22

However, over time medical science has been able to correlate these sensations and movement of what is typically referred to as energy with the activation of specific neural fibers. Corradino states that this De Qi sensation occurs with the stimulation of the A-delta fibers by the acupuncture needles and the reaction that takes place that transmits that sensation to the brain [1].

The History of Orthopedic Acupuncture and the Use of Motor Points

Much of today's modern application of acupuncture from an orthopedic perspective can be traced back to five physicians who pioneered research into physical medicine, rehabilitation, myofascial pain.

Duchenne De Boulogne 1806–1875 French Neurologist
Mapped motor points and explored the effects of electrical stimulation on diseased nerves and muscles.
https://www.britannica.com/biography/Duchenne-de-Boulogne

Dr. Janet Travell, MD

Dr. Janet Travell, MD (1901–1997) was one of the most influential doctors in the world of orthopedic acupuncture because of her contributions and research on treating myofascial pain. In 1983, Dr. Travell and Dr. David Simons co-authored Myofascial Pain and Dysfunction: The Trigger Point Manual. Dr. Travell used a hypodermic needle to inject procaine or lidocaine to deactivate the active (painful) trigger points that present as tight or knotted bands of tissue [2].

Dr. Karel Lewit, MD

Dr. Karel Lewit, MD (1916–2014), co-founder of The Prague School of Rehabilitation, was familiar with the work of Janet Travell and proposed that the analgesic effect of trigger point injections was due to the needle and not the injected anesthetic. In 1979 Dr. Lewit published his study of "The Needle Effect" where he demonstrated that pain could be effectively treated with acupuncture needles as long as the source of the pain was reached by the needle. This method was referred to as *Dry Needling* [3].

Dr. Joseph Wong, MD

In the late 1960s Dr. Joseph Wong (1931–2017) a Canadian physician, specializing in pain and physical medicine and rehabilitation, used his training in acupuncture to treat pain. Dr. Wong was aware of motor points (the muscle's zone of innervation) used for electromyography and felt that using acupuncture to needle these zones of innervation could be physically therapeutic. Dr. Wong experimented with needling deeper, into motor points, trigger points, myotendinous junctions, ligaments and near nerve plexuses and produced excellent results in the treatment of pain. Dr. Wong called his approach "neuro-anatomical acupuncture" [4, 5].

Dr. C. Chan Gunn, MD

In the 1970s Dr. C Chan Gunn MD, founder and president of the Institute for the Study and Treatment of Pain (iSTOP) in Vancouver, Canada, began treating chronic pain using acupuncture needles to stimulate motor points in order to release shortened muscles. Dr. Gunn treated the specific muscles that were causing joint pain in addition to the related spinal segments. It is a form of dry needling called Intramuscular Stimulation (IMS) that is based on Canon's Law of Denervation Supersensitivity. Dr. Gunn was recognized in 1979 for his paper "Prespondylosis" and Some Pain Syndromes Following Denervation Supersensitivity [6, 7].

> I was treated by Dr. Chan Gunn in the late 80's following chronic pain and tension as a result of a whiplash injury in a motor vehicle accident. This was before I became an acupuncturist, but I found this method of combined trigger point needling, the use of motor points and the needling of corresponding spinal segments incredibly effective in releasing chronically tight muscles.
>
> Colin Harris MSN, DOMP, DTCM, RN, R.Ac, C.SMA

Dr. Alejandro Elloreaga Claraco, MD

Dr. Alejandro Elorriaga Claraco MD, Sports Medicine Specialist (Spain), is Program Director and Assistant Clinical Professor at the McMaster University Department of Anaesthesia in Hamilton, Canada, and probably one of the most influential educators in the field of orthopedic acupuncture. Dr. Elorriaga Claraco started developing his Neurofunctional Acupuncture in 1995. He needled spinal segments, nerve plexes, motor, and trigger points as well as system regulatory (distal) acupuncture points to treat injury and pain. Treatment with electric stimulation acupuncture is done primarily using motor points of the muscles, but there are other methods used at spinal segments to influence the autonomic nervous system for blood perfusion and pain relief [8].

The styles of acupuncture described above use a biomedical approach and distinguish themselves from the traditional acupuncture approach. This more contemporary acupuncture style involves insertion of needles in neuro-reactive sites of the body for the purpose of inducing central, autonomic, or motor and sensory neuromodulatory responses.

What's in a Name: Dry Needling, Acupuncture, Intramuscular Stimulation

Acupuncture literally translates to needle penetration (Acu = needle; puncture = penetration). Although most traditional acupuncturists use the terminology of moving "qi" or energy in their practice, most acupuncture trials do not claim to move qi along meridians or channels. Many research studies of acupuncture have used Western medical diagnoses and not Traditional Chinese Medicine. The dry needling technique developed by Dr. Janet Travell was a term used to describe the use of stimulation using empty syringe needles as opposed to "wet" needling from injection therapy. Dry needling is sometimes referred to as trigger point needling. A trigger point is a tight band or knotted area of muscle which is highly sensitive and often painful to the touch. These areas often cause referred pain to another part of the body. Dry needling involves the insertion of needles into those tight or tender points causing a twitch response resulting in a release of that tight tissue (Fig. 22.1).

Although dry needling may seem like a relatively new concept, that has been recently adopted and promoted by therapists other than acupuncturists, it is really rooted in the ancient acupuncture techniques. The dry needling technique was described as needling "ashi" points or pain points when pressed and is referred to in old texts such as the Yellow Emperor's Inner Classic, and Chapter 13 of the Ling Shu or Sun Si-Miao in the Tang Dynasty. It involved needling those "ashi" points whether they were acupoints or not [9]. There are significant similarities between

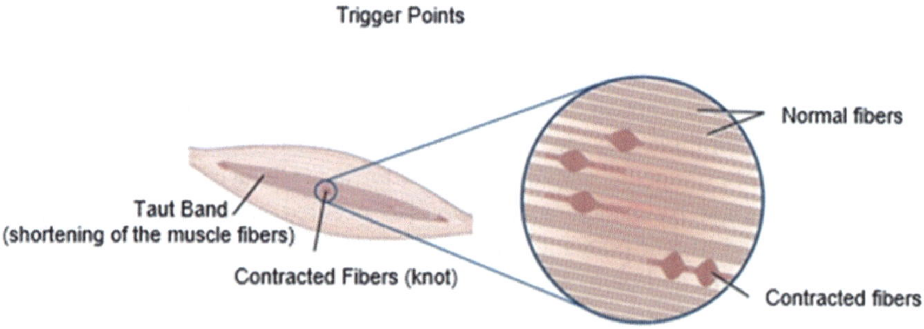

Fig. 22.1 Depiction of taut bands in muscle fibers

the description of "ashi" points and Travell's definition of myofascial trigger point [10].

Another model using acupuncture for musculoskeletal pain was Dr. Chan Gunn's intramuscular stimulation. Dr. Gunn's technique involved releasing shortened muscles caused by denervated tissues with particular focus on shortening of the paraspinal muscles. Dr. Gunn believed that shortened mutifidi muscles led to disc compression and narrowing of the intervertebral foramina resulting in peripheral neuropathy. Gunn used the more descriptive term intramuscular stimulation to distinguish his approach from other techniques of acupuncture [11, 12].

Many of these techniques emerged from the exploration of the role of tender or trigger points as a cause of muscular pain. All of these techniques have been used in one form or another in acupuncture for thousands of years and the use of these techniques are shared across a variety of disciplines today.

Acupuncture Treatment

Motor Point Acupuncture

Beyond the use of dry needling techniques for trigger points, one of the main stays of the orthopedic acupuncturist is the use of motor points. Motor point acupuncture is a form of acupuncture that targets the specific points on the body that control the function of individual muscles. The muscle motor point's electrophysiological definition is an area in the belly of the muscle where the motor nerve pierces that tissue and the location of the terminal end of that nerve or motor-end plate [13]. It is the stimulation of this motor point that culminates in contraction of the skeletal muscle (Fig. 22.2).

The earliest mapping of motor points occurred through electromyography and was subsequently supplemented by cadaver dissections. The tests conducted by

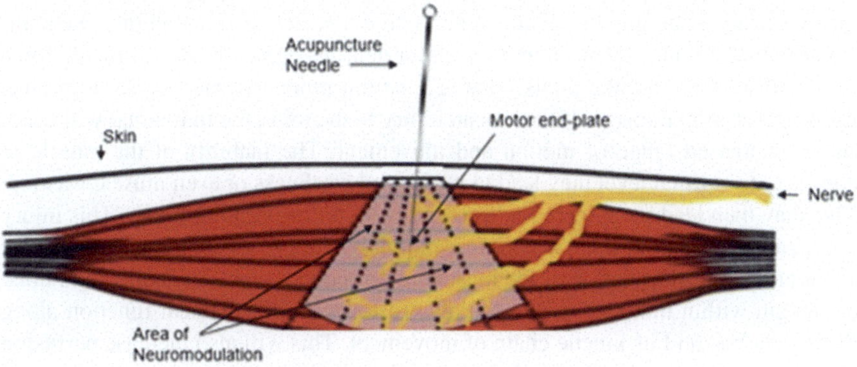

Fig. 22.2 Needling the muscle motor point

electromyographers enabled physicians to evaluate the motor unit, in other words the functional status of the nerves and the skeletal muscle fibers. The tests consisted of determining the electrical activity of the muscle with the use of a transcutaneously placed intramuscular electrode.

The motor unit is a functional unit composed of the following structures:

- The Anterior horn cell
- Its axon cylinder
- Terminal branches
- The neuromuscular junction
- The muscle fiber

One of the commonly used guides for the location of motor points is the Anatomical Guide for the Electromyographer: The Limbs and Trunk by Aldo O. Perotto, MD. (2011), (5th ed.). The first edition was published in 1975. The current edition is divided into sections organized by anatomical regions that includes innervations and attachments of muscles of the arms, and hands, the shoulder girdle, the legs and feet, the pelvis and hip, the perineal and paraspinal regions and the abdominal wall, intercostals and diaphragm [14].

Efficacy of Acupuncture Stimulation of Motor Points and Neuroactive Sites

Inflammation or swelling and direct or repetitive trauma can cause changes in the Golgi tendon organs of the muscles. This resulting trauma, pain, or neurogenic inflammation can cause motor inhibition by modifying the afferent information within the muscle spindles [15]. When muscle groups are inhibited, they do not receive an adequate impulse from the nerve to the muscle to contract due to chemical or physical trauma [16]. Inhibitory motor neurons decrease the contractile force of the muscles around the joints. This is a natural protective mechanism to protect from further joint damage. If there is an injury to the joint, the individual will experience decreased range of motion and movement. The inability of the muscle to contract at a normal level may lead to prolonged weakness or even muscle wasting. This may then lead to an increased susceptibility to further joint injury. This injury cycle can perpetuate if there is no intervention to address the dysfunction [17].

Acupuncture stimulation of the muscle motor point restores activation and muscle length within that motor unit, but also restores bio-mechanical function along that fascia line and its kinetic chain of movement. This will also increase perfusion in ischemic tissue along that pathway [18]. In other words insertion of the needle into the motor point with the application of some electrical stimulation will cause the muscle to contract and relax. That contraction will reset the muscle so that is more elastic and flexible which will quickly restore joint function.

> Unfortunately, most current acupuncture research efforts are focusing on the applications of acupuncture for the treatment of pain, while in my opinion the use of acupuncture as a segmental and regional sensory-motor regulatory intervention could play in the future a fundamental role in the area of treatment of movement disorders.
>
> Dr. Elorriaga Claraco
> Contemporary Medical Acupuncture Program
> McMaster University, Ontario, CAN

The most effective neuro-reactive sites for acupuncture stimulation are the sensory and motor innervation associated areas of the musculoskeletal system. These are typically located in the muscle belly where motor points are found in addition to ligaments, the periosteum, neurovascular bundles, and muscle-tendon junctions [19, 20]. Dr. Elorriaga Claraco states that stimulation of the relevant sensory and motor somatic receptors can restore normal sensory-motor activity in the treated segment in movement disorders [19, 20].

In 1980, 56 patients who failed to respond to traditional therapy for chronic low back pain for at least 12 weeks duration were entered into a randomized clinical trial. The purpose was to compare a regimen with and without needling of muscle motor points. Before the trial, all patients had undergone a standard therapy regimen of physiotherapy, remedial exercises, and occupational therapy for 8 weeks without improvement. The study consisted of 29 study subjects and 27 control patients. All patients continued with the standard therapy however the 29 study subjects also received needling at the muscle motor points for an average of 7.9 treatments. Patients were assessed at time of discharge, 12 weeks post and 27 weeks post discharge. The group that was treated with needling was significantly better than the control group and 18 of the 29 study subjects had returned to their original work, with only four returning to work from the control group [21].

In a 2015 study published in the European Journal of Integrative Medicine, the authors compared the effects of motor point acupuncture in the cases of shoulder abduction dysfunction, with standard acupuncture treatment of acupoints on meridians in the shoulder area. The results from the 40 patients showed that motor point acupuncture therapy was more effective on shoulder abduction dysfunction than standard acupuncture therapy as measured by physical functional ability [15].

Acupuncture and Electroacupuncture

Acupuncture involves the manual manipulation of a filiform needle by methods such as twisting or thrusting at different speeds. In manual acupuncture Langevin [22, 23] hypothesized that such motions lead to distortion of the connective tissues, stimulating the mechanoreceptors which create neural signals that correspond to the

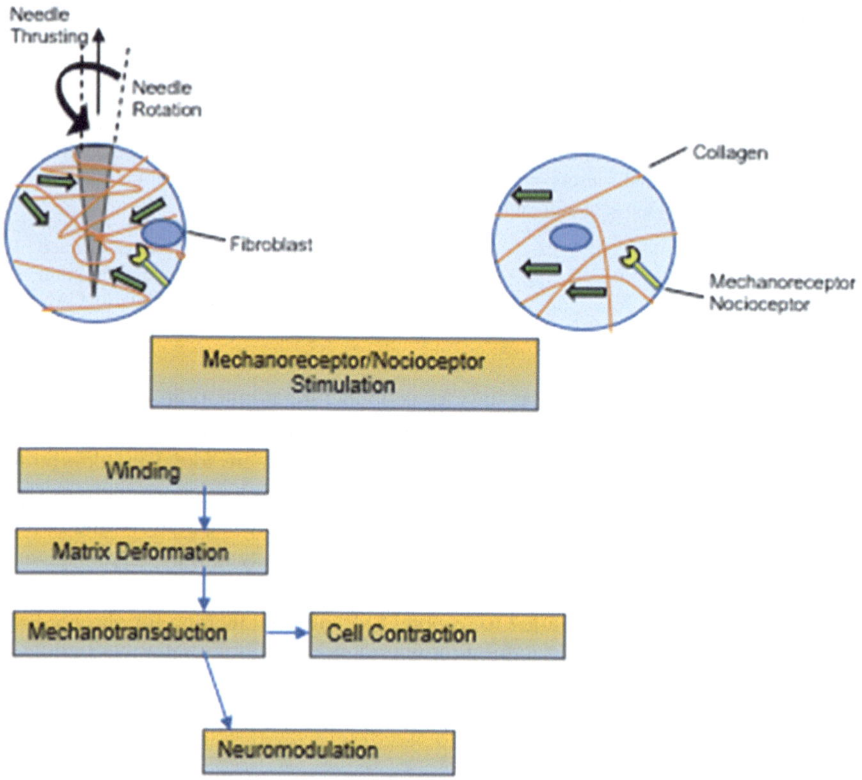

Fig. 22.3 Neuromodulation via acupuncture

historical sensations of "De Qi". The needle grasp sensation is likely the winding of the tissue around the needle during needle rotation and thrusting, and the subsequent mechanical coupling of the needle and connective tissue. This process of manipulation transmits a mechanical signal to the connective tissues called mechanotransduction [24] (Fig. 22.3).

Electroacupuncture can also be used to stimulate the needles inserted into tissues or acupoints. Electroacupuncture is the application of a direct current to the acupuncture needle with electrical impulses at different frequencies, stimulating different fibers to target different tissue responses. With this technique, impulses pass between two needles with the current delivered neuromodulating a greater surface area. Electroacupuncture has the ability to activate mechanoreceptors sending signals to the brain to relevant nuclei that modulate pain sensations [25].

Electroacupuncture has been shown to activate the nervous system by both alleviating sensory and affective inflammatory pain in ranges from 2 to 10 Hz. It blocks pain by activating a variety of chemicals in the body. Electroacupuncture techniques have been shown to target specific endogenous opioid receptors for pain relief and reduce proinflammatory cytokines (tumor necrosis factor-alpha (TNF-α), interleukin-1 beta (IL-1β), and interleukin-6(IL-6)) peripherally and in the spinal cord [24]. As an example, 2 Hz millicurrent can release beta-endorphins, 4 Hz targets enkephalin release via Delta receptors and 100 Hz targets Kappa receptors in the spinal cord for dynorphin release [26].

Various manipulation techniques of the acupuncture needles produce a location reaction response in the tissues. This feedback has been analyzed to include red and white bloods cells, prostaglandins, histamine, amino acids, nerve growth factor and calcitonin gene-related peptide (CGRP), a potent natural vasodilator. CGRP encourages local congregation of these elements which can help to explain part of acupuncture's effectiveness for local needling for inflammation [1].

> Acupuncture and electrostimulation are key allies in the minimization of trophic changes and the acceleration of soft tissue healing.
>
> –Dr. Anthony Lombardi, DC, Hamilton Back Clinic
> Former senior instructor
> Contemporary Medical Acupuncture Program
> McMaster University, Ontario, CAN
> https://www.chiroeco.com/trophic-changes-musculoskeletal/

Inflammation, Pain, and Acupuncture

Inflammation is the body's natural response to injuries. However, prolonged inflammation can result in muscle weakness and atrophy in the area surrounding the injury. Continued inflammation in the musculoskeletal system generated at the site of an original injury typically releases inflammatory substances such as Substance P, Calcitonic Gerne Related Peptide and Neurokinin. Prolonged inflammation can lead to the accumulation of fibrous tissue that become taut muscle bands and trigger points. These trophic changes lead to compressed blood vessels and decreased blood flow [27]. Clinically these tight bands are often found over injured muscles because of changes in blood circulation and motor inhibition. This can also lead to compensatory patterns in the body as it attempts to remain dynamically functional. Acupuncture may have a role to play as an effective treatment in reversing the stimulus of these knots and tight bands.

> Dr. Vladimir Janda MD (1928–2002)—Department of Rehabilitation and Medicine
>
> - Pain and Inflammation are symptoms not the problem
> - Often the site of pain is not the cause of pain
> - Altered peripheral input due to pain leads to changes in muscle activation, causing faulty movement patterns
>
> http://www.jandaapproach.com/2012/11/02/timeless-vladimir-janda-quotes-and-concepts/

Langevin [22] found that manual manipulation of acupuncture needles resulted in microtrauma around the needle resulting in the release of various proinflammatory mediators. There is significant research that demonstrates the body can drive inflammation via autonomic dysregulation through overresponse of the immune system [28]. Some research has suggested that acupuncture may play a role in downregulating the immune system via the autonomic nervous system [29]. This study showed that acupuncture with electrical stimulation can increase dopamine production. Dopamine is a neurotransmitter that inhibits inflammatory cytokine (Interleukin-6) production [26].

In addition to the role of proinflammatory mediators, acupuncture has the ability to increase blood flow to affected areas, improving the removal of waste products and the delivery of oxygen and nutrients to the tissues. Studies have demonstrated that acupuncture can alter skin temperature suggesting changes in local circulation. Secondly, this effect suggests that this may create an effect on the development of persistent pain [30]. The effect of acupuncture on local Nitric Oxide (NO) levels was examined in a randomized, double-blind, crossover study. Acupuncture was performed on the following points on four of the twelve meridians, Li4, Pc6, Pc8 Lu6 and Ht5 (Fig. 22.4).

The concentration of nitrate and nitrite in plasma was analyzed from the area treated to determine NO formation. Nitrate and nitrite are the end-products of metabolism of NO and their levels are reliable indicators of the formation of

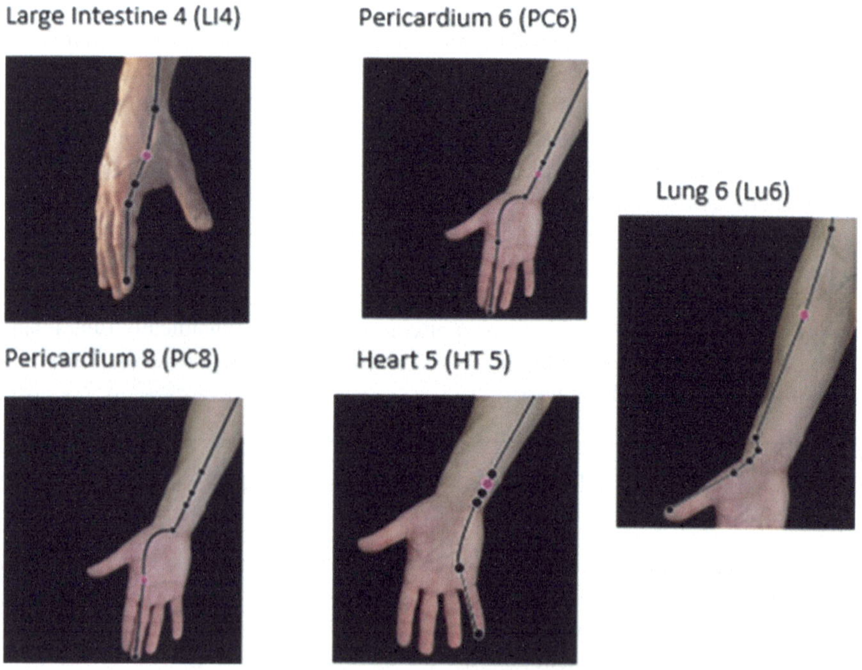

Fig. 22.4 Acupuncture point locations

NO. This information was confirmed with NO-hemoglobin levels in the same samples. Local circulation was examined with a laser tissue-blood flow meter. Meter readings were obtained before acupuncture and at 5 and 60 minutes after the acupuncture needles were removed. The study found that acupuncture increased the plasma concentrations of nitrates and nitrites and increased blood flow in the treated region. This is consistent with a previous study showing an increase in skin temperature in treated regions [30].

Similar findings of increased circulation were found in another study that measured the muscle blood volume and oxygenation of the trapezius muscle when stimulated with acupuncture. Results of the study found oxygenation and blood volume increased in the area stimulated with acupuncture but not in the distal region of the same muscle [31]. Because ischemia is an important aspect in the development of certain types of pain and its persistence, the increase in circulation resulting from the increase in NO levels could play a role in the use of acupuncture for the purpose of pain relief [30].

> Every day I am more amazed at the profound effects a tiny little needle can have on the nervous system. Kinetic chains activate and become more coordinated, pain reduces, and inflammation is reduced. Truly a wonder.
>
> Joshua Margolis, LAc, DACM, DOMTP, Farmacopia, Santa Rosa, CA
> Former faculty – Orthopedic Acupuncture and Pain Management
> Acupuncture and Integrative Medicine College. Berkely CA

Delayed Onset Muscle Soreness (DOMS) and Acupuncture

Delayed onset muscle soreness is a benign exercise-induced myogenic condition that develops after unaccustomed vigorous physical activity consisting of symptoms of localized muscle soreness and tenderness and pain with movement. It usually peaks after a period of 12–24 hours slowly resolving within 4–7 days [31]. Although the most likely cause of DOMS is microtrauma to the muscles and inflammation, other theories have been proposed involving the production of neurotrophic factors in muscle fibers/satellite cells [31]. DOMS is not necessarily experienced during rest but rather in response to movement. DOMS does not produce increased intramuscular pressure at rest. However, small increases in intramuscular pressure may take place during movement, increasing pain [32]. Therefore, DOMS may discourage an individual's participation in further exercise or any rehabilitative therapy.

A study in 2008 investigated the efficacy of acupuncture on the symptoms of DOMS. It determined that the insertion of acupuncture needles in tender points had an analgesic effect and may reduce discomfort and improve muscle activation [33]. There were significant differences between the control and tender point group's pain scores immediately after treatment and up to 3 days post exercise [34]. Reducing discomfort from DOMS may accelerate recovery and has considerable interest for the performance of athletes.

Safety and Effectiveness of Acupuncture

A systemic literature search of PubMed and EBSCO databases was performed for randomized controlled trials between 2000 and 2020 for the efficacy of the treatment of acupuncture for nonspecific chronic low back pain. A meta-analysis concluded that acupuncture was a safe and effective method in reducing pain in adults with nonspecific low back pain. In another review of 24 research papers examining the outcomes of immunohistochemical results, electrophysiological results and motor behaviour scores, all but one of those studies favoured acupuncture and electroacupuncture for the recovery and regeneration of nerves [35]. Acupuncture is a safe treatment with few side effects and has been demonstrated as a cost-effective treatment modality on its own or as a complement to other interventions [36]. Many professional sports teams in a variety of sports now have acupuncturists on staff for not only the treatment of injuries but for prevention purposes and improved athletic performance [37].

Summary

There is minimal research out there in the acupuncture community related to the use of motor points and their treatment of movement disorders, postural imbalances, and recovery and rehabilitation. Most acupuncture research tends to focus on the management of pain and the use of acupuncture points on the meridians or based on TCM pattern pathology. However, there are many acupuncturists with an interest in musculoskeletal disorders or athletic performance that are embracing the use of motor points in their clinical practice and getting great results. Further research needs to take place to support the use of a more orthopedic acupuncture style for the purposes of recovery, regeneration, adaptation to exercise stress and improved functional movement. In that way, the profession can bring scientific evidence to the table to substantiate what is happening anecdotally with this type of therapy.

> If you want fast and excellent results treating sports injuries, overuse injuries, chronic pain, flexibility, and mobility, sometimes going against tradition is what is best for your patient. I do not assess or treat using TCM theory. My clinical practice consists of the use of motor points and trigger points, and I am not going to apologize for getting great results or for sharing what I've learned from Western Medicine using acupuncture.
>
> Dr. Richard Hazel DAc., LAc, New York
> Acupuncture Outsider – Podcast

Taut bands that are not active trigger points and do not cause pain can still restrict joint mobility and contribute to dysfunction.

1. What is the cause of latent trigger points if there is not a high level of calcitonin gene related peptide (cgrp) substance?
2. Is there a high level of acetylcholine?
3. Are the taut bands still driven by segmental sensitization or are they due to chronic hypertonicity like posture?
4. Which develops first muscle trigger points or related spinal segmental sensitization?
5. Does the answer to this question above change treatment protocols?

Electrical stimulation of a needle in a motor point seems to improve muscle elasticity, contractility and extensibility and improves joint mobility.

6. Is it due to improved muscle elasticity alone or is there a fascial consideration or a gamma motor nerve/muscle spindle (muscle tone) consideration?

Dr. Chan Gunn used motor points to demonstrate that low back pain could be treated with motor points of the affected segments such as the glutes, hamstrings or gastrocnemius, thereby calming those segments.

7. Using electroacupuncture, what is the differentiation of improvement between neuromodulation via the spinal segment versus functional improvement of the associated joints in the affected segments?

References

1. Corradino ML. Neuropuncture: solidifying Acupuncture's authenticity. J Acupunct Integr Med. 2019. jaimonline.org/neuropuncture-solidifying-acupunctures-authenticity.
2. Wilson VP. Janet G. Travell, MD: a daughter's recollection. Tex Heart Inst J. 2003;30(1):8–12.
3. Lewit K. The needle effect in the relief of myofascial pain. Pain. 1979;6(1):83–90. https://doi.org/10.1016/0304-3959(79)90142-8.
4. Ronda. Dr. Joseph Wong. Acupuncture Canada, 10 Nov. 2017. acupuncturecanada.org/blog/dr-joseph-wong.
5. Wong JY. A manual of neuro-anatomical acupuncture: musculo-skeletal disorders. Toronto Pain and Stress Clinic Incorporated; 1999.
6. The Institute for the Study and Treatment of Pain – Dr. Chit Chan Gunn. istop.wildapricot.org/Dr.Chit-Chan-Gunn.
7. Gunn CC. 'Prespondylosis' and some pain syndromes following denervation supersensitivity. Spine. Ovid Technologies (Wolters Kluwer Health). 1980;5(2):185–92. https://doi.org/10.1097/00007632-198003000-00014.
8. Dr. Alejandro Elorriaga Claraco – Program Director – McMaster University Contemporary Medical Acupuncture Program. mcmasteracupuncture.com/meet-our-faculty/dr-alejandro-elorriaga-claraco-program-director.
9. Jiang S, Zhao J. The historical source of "trigger points": classical Ashi points. World J Acupunct Moxibustion (WJAM). 2016;26(2):11–4.

10. Legge D. A history of dry needling. J Musculoskelet Pain. 2014;22(3):301–7. https://doi.org/10.3109/10582452.2014.883041.
11. Cheng KJ. Neurobiological mechanisms of acupuncture for some common illnesses: a clinician's perspective. J Acupunct Meridian Stud. 2014;7(3):105–14. https://doi.org/10.1016/j.jams.2013.07.008.
12. Kalichman L, Vulfsons S. Dry needling in the management of musculoskeletal pain. J Am Board Fam Med. 2010;23(5):640–6. https://doi.org/10.3122/jabfm.2010.05.090296.
13. Lee JH, Kim HW, Im S, et al. Motor entry point acupuncture entry points and terminal intramuscular nerve endings of the musculocutaneous nerve to biceps and brachialis muscles. Surg Radiol Anat. 2010;32(3):213–20.
14. Delagi EF, Perotto A. Anatomical guide for the electromyographer: the limbs and trunk. 5th ed. Thomas; 2011.
15. Wang ZJ, Xing YL, Gao X, et al. Motor entry point acupuncture compared with the standard acupuncture for treatment of shoulder abduction dysfunction after stroke: a randomized clinical trial. Eur J Integr Med. 2015;7(Supplement 1):26. https://doi.org/10.1016/j.eujim.2015.09.067.
16. Le Pera D, Graven-Nielsen T, Valeriani M, et al. Inhibition of motor system excitability at cortical and spinal level by tonic muscle pain. Clin Neurophysiol. 2001;112(9):1633–41. https://doi.org/10.1016/s1388-2457(01)00631-9.
17. Hopkins T, Ingersoll CD. Arthrogenic muscle inhibition: a limiting factor in joint rehabilitation. J Sport Rehabil. 2000;9:135–59.
18. Nils J, Daenen L, Cras P. Nociception affects motor output: a review on sensory-motor interaction with focus on clinical implications. Clin J Pain. 2012;28(2):175–81.
19. Elorriaga CA. Electrical stimulation of acupuncture needles: an overview (part 1). Acu-Press. 2004;5(1):3–4.
20. Elorriaga Claraco A, Fargas-Babjak A. Neuromodulation, an emerging acupuncture paradigm: implications. Acu-Press. 2003;4(1):4.
21. Gunn CC, Millerand WE, Little AS, Mason KE. Dry needling of muscle motor points for chronic low-Back pain. Spine. 1980;5(3):279–91.
22. Langevin HM, Churchill DL, Cipolla MJ. Mechanical signaling through connective tissue: a mechanism for the therapeutic effect of acupuncture. FASEB J. Federation of American Societies for Experimental Biology. 2001;15(12):2275–82. https://doi.org/10.1096/fj.01-0015hyp.
23. Israel E, Churchill DG, Junru W, et al. Evidence of connective tissue involvement in acupuncture. FASEB J. Federation of American Societies for Experimental Biology. 2002;16(8):872–4. https://doi.org/10.1096/fj.01-0925fje.
24. Zhang R, Lao L, Ren K, Berman BM. Mechanisms of acupuncture-electroacupuncture on persistent pain. Anesthesiology. 2014;120(2):482–503.
25. Cao X. Scientific bases of acupuncture analgesia. Acupunct Electro-ther Res. Cognizant. 2001;27(1):1–14. https://doi.org/10.3727/036012902816026103.
26. Hennessey S. Can acupuncture bio-hack the autonomic nervous system and down-regulate inflammation? J Acupunct Integr Med. 2019. jaimonline.org/can-acupuncture-bio-hack-the-autonomic-nervous-system-and-down-regulate-inflammation
27. Devor M. Unexplained peculiarities of the dorsal root ganglion. Pain. 1999;82:S27–35. https://doi.org/10.1016/s0304-3959(99)00135-9.
28. Pongratz G, Straub RH. The sympathetic nervous response in inflammation. Arthritis Res Ther. Springer Science+Business Media. 2014;16(6):1–12. https://doi.org/10.1186/s13075-014-0504-2.
29. Rosas R, et al. Dopamine mediates vagal modulation of the immune system by electroacupuncture. Nat Med. Nature Portfolio. 2014;20(3):291–5. https://doi.org/10.1038/nm.3479.
30. Tsuchiya M, Sato EF, Inoue M, Asada A. Acupuncture enhances generation of nitric oxide, and increases local circulation. Anesth Analg. 2007;104(2):301–7. https://doi.org/10.1213/01.ane.0000230622.16367.fb.

31. Ohkubo M, Hamaoka T, Niwayama M, et al. Local increase in trapezius muscle oxygenation during and after acupuncture. Dyn Med. 2009;8(1):1–8. https://doi.org/10.1186/1476-5918-8-2.
32. Smith LL. Acute inflammation: the underlying mechanism in delayed onset muscle soreness? Med Sci Sports Exerc. 1990;23(5):542–9.
33. Parisotto Antonassi D, Rodacki CLN, Lodovico A, Ugrinowitsch C, Rodacki ALF. Immediate effects of acupuncture on force and delayed onset of muscle soreness. Med Acupunct. 2020;33(3):203–11. https://doi.org/10.1089/acu.2020.1491.
34. Itoh K, Ochi H, Kitakoji H. Effects of tender point acupuncture on delayed onset muscle soreness (DOMS) – a pragmatic trial. Chin Med. 2008;3(1):14.
35. Asano H, Plonka D, Weeger J. Effectiveness of acupuncture for nonspecific chronic low back pain: a systematic review and meta-analysis. Med Acupunct. 2022;34(2):96–106. https://doi.org/10.1089/acu.2021.0057.
36. Lund I, Lundeberg T. Mechanisms of acupuncture. Acupunct Relat Ther. 2016;4(4):26–30. https://doi.org/10.1016/j.arthe.2016.12.001.
37. Yucel E. NFL teams are tackling injuries with acupuncture and cupping. Medill Reports Chicago, 1 Apr. 2020. news.medill.northwestern.edu/chicago/nfl-teams-are-tackling-injuries-with-acupuncture-and-cupping.

Chapter 23
Neuromusculoskeletal Complaints Related to the Spine and Peripheral Joints

Sean Abdulla, Simon Wang, Eric St-Onge, and Karin Hammerich

Introduction

Chiropractic Principles and Practice

Chiropractic practice, as defined by the College of Chiropractors in Ontario [1], Canada, is the assessment of conditions related to the spine, nervous system, and joints and the diagnosis, prevention, and treatment, primarily by adjustment of:

- dysfunctions or disorders arising from the structures or functions of the spine and the effects of those dysfunctions or disorders on the nervous system; and
- dysfunctions or disorders arising from the structures or functions of the joints.

Chiropractic, as an accepted healthcare profession, is relatively young, existing for over 100 years. However, the history of manual therapy dates to the time of Hippocrates. Thanks to both research and education, chiropractic has been recognized for its expertise in spinal manipulative therapy (SMT). Commensurate with technological advances in science and research, the attitudinal views of chiropractic are changing [2].

S. Abdulla (✉) · S. Wang · E. St-Onge
Department of Clinical Education, Canadian Memorial Chiropractic College, Toronto, ON, Canada
e-mail: drseanabdulla@gmail.com

K. Hammerich
Department of Clinical Education, Canadian Memorial Chiropractic College, Toronto, ON, Canada

Department of Chiropractic Principles and Practice, Canadian Memorial Chiropractic College, Toronto, Canada

N. C. Apostolopoulos et al. (eds.), *Fundamentals of Recovery, Regeneration, and Adaptation to Exercise Stress: An Integrated Approach*,
https://doi.org/10.1007/978-3-031-44270-4_23

Chiropractors are equipped with a large toolbox to manage patients and guide clinical practice. Often, they employ a multimodal approach encompassing clinical experience, patient preferences and beliefs, and best available evidence combined with various interventions in a program of care [3, 4]. Manual therapy techniques, specifically SMT, extremity manipulative therapy (EMT), and joint mobilization continue to be the primary choice of intervention despite a broad spectrum of techniques [5–8]. Many theories have been proposed to explain clinical phenomena resulting from a chiropractic manipulation, and not one theory can explain all the neurophysiological effects. However, many scientists in related fields of biomechanics and neurophysiology have made discoveries that explore the results of chiropractic treatment.

> ...evidence-based Chiropractic care is a dynamic approach that integrates best-available evidence, clinical expertise, and patient preference, which, combined is crucial in delivering safe and effective patient-centred Chiropractic care [160].
>
> – Dr. Keshena Malik
> (Ontario Chiropractic Association (OCA)
> Evidence-Based Framework Advisory Council (EBFAC)

Chapter Objectives

This chapter will highlight manual therapy approaches commonly utilized by chiropractors. The aim is to provide a framework that will assist practitioners when treating the spine and peripheral joints. A review of the current state of the literature focused on the efficacy of spinal manipulation for treating spine pain and propose the potential therapeutic neurophysiological mechanisms will be presented. The chapter will conclude by discussing the effects of spinal SMT on sports-related performance outcomes and new research perspectives. The following are the specific learning objectives:

- Describe Spinal Manipulative Therapy (SMT), mobilization, and traction techniques
- Propose mechanisms for the therapeutic benefits of spinal manipulation
- Review the clinical efficacy of spinal manipulation for the treatment of spine pain
- Examine the effect of spinal manipulation and performance outcomes
- Explore new research perspectives

What Is Spinal Manipulative Therapy (SMT)?

This therapy is defined as "a high-velocity, low-amplitude (HVLA) dynamic force applied to a specific location of the spine for therapeutic reasons (i.e., induce motion, reduce pain and improve function) resulting in a mechanical deformation of the spine and surrounding tissues" [9]. The HVLA technique is also referred to as a thrust [10]. The audible pop sound that often accompanies SMT is termed a cavitation or audible release [11, 12]. It is hypothesized that the cavitation is the "formation or release of gas from the synovial fluid within the joint caused by separation of joint surfaces resulting in an intracapsular reduction in pressure" [13]. An audible pop after delivery of the HVLA manipulation may indicate a successful treatment, however, there are no improved therapeutic pain outcomes reported with or without cavitation [14]. HVLA manual therapies may be applied to the spine through manual techniques, mechanically assisted devices, and impulse hammers [15].

Chiropractic manipulation techniques have essentially remained the same over time with slight variations in procedure discovered by clinicians wanting to render those techniques more amenable to certain categories of patient care (Fig. 23.1). These named chiropractic techniques or approaches have been developed and practiced using a systematic approach guided by theoretical frameworks [16]. When SMT is indicated, the choice of method (manipulation and/or mobilization) used will depend on the chiropractors' training, patient preference, clinical assessment findings, and the region of application of the manipulation; with the emphasis placed on mitigating the risk of adverse events [17].

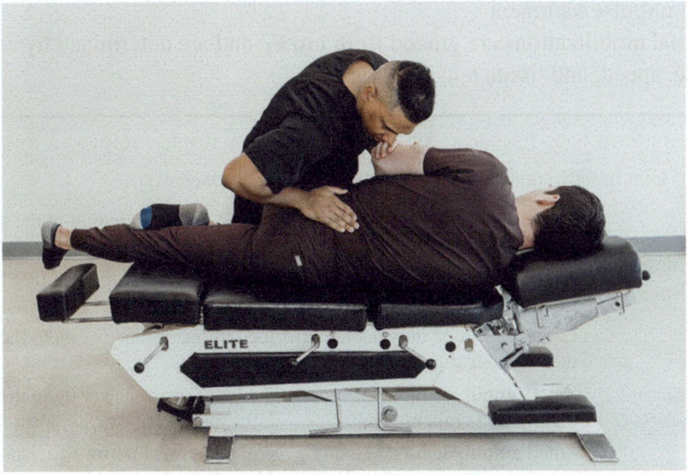

Fig. 23.1 Side posture Sacro-Iliac (SI) joint manipulation indicated for SI joint pain

What Are Spinal Joint Mobilizations?

Joint mobilizations are adjunctive techniques performed to improve joint mobility and reduce pain in the spine and extremities [18, 19]. Mobilizations are a series of rhythmic oscillatory movements graded I to IV relative to amplitude, speed, and tissue tension (Table 23.1) [20, 21]. The graded oscillatory movements are hypothesized to stimulate mechanoreceptors in order to modulate pain and improve proprioceptive function [16]. In contrast to SMT that apply a HVLA thrust to the spinal tissues and structures, spinal mobilizations employ a low velocity, and varied amplitude of passive therapeutic motion resulting in intrinsic joint movement [20]. The grade and direction of force application (rotation, flexion, extension, or lateral flexion) of the mobilization performed will vary depending on the region of application, available range of motion within the tissue and joints of the spine, and the desired clinical effect [16].

Although it is not the objective to achieve a cavitation, it is possible to occur when performing spinal mobilizations. When performing a spinal mobilization technique, the general principle is to mobilize below pain thresholds to avoid muscle guarding, progressing through the strengths of applied force (grades of mobilizations) [22, 23]. This is followed up by re-assessment ("check before, re-recheck after") for joint movement, muscle tonicity, and/or pain [24]. SMT should be considered if graded mobilizations are tolerated or when clinically indicated [23]. Aside from oscillatory techniques, mobilizations can also be performed as a progressive stretch or as a sustained continuous stretch (Table 23.2) [16].

- SMT is a high-velocity, low amplitude (HVLA) dynamic force applied to specific location of the spine
- HVLA is applied by manual techniques, mechanically assisted devices, and impulse hammers
- Spinal mobilizations are graded from I to IV and are determined by amplitude, speed, and tissue tension

Table 23.1 Joint mobilization strength of grading (Maitland Concepts)

Grade I	A small-amplitude movement performed at the beginning of the available range.
Grade II	A large-amplitude movement performed within a resistance free part of the available range.
Grade III	Large amplitude movement performed into resistance or up to the limit of the available range.
Grade IV	Small amplitude movement performed into resistance or up to the limit of the available range.
Grade V (Manipulation)	Small amplitude high velocity general movement performed usually, but not always, at the end of the available range.

Adapted directly from Hengeveld et al. [21]

Table 23.2 Types of graded mobilization techniques

Mobilization technique	Grade	Procedure	Indication for use
Oscillatory	I-IV	Cyclic, rhythmic, low-velocity passive motion applied. Alternating pressure (on and off) is delivered at different parts of the available range.	Decrease pain and increase range of motion. Activate mechanoreceptors and improve proprioceptive function.
Progressive Stretch	I-IV	Successive short-amplitude, spring-type pressures or a series of short-amplitude stretch movements. The pressure or stretch is imparted at progressive increments of the range.	Improve mechanical or soft tissue restriction or both.
Continuous Stretch	I-IV	Sustained, gradually increased stretch or pressure without interruption. Maintaining a stretch or pressure throughout the mobilization procedure is recommended when immediate tissue feedback is desired.	Improving extensibility of the periarticular soft tissues about a spinal joint via collagen fiber realignment and viscosity. Stretch adaptively shortened, fibrosed periarticular soft tissue structures.

Adapted directly from Bergmann and Peterson [16] and Hengeveld et al. [21]

Spinal Traction

Traction therapy is the application of mechanical or manual force to separate the vertebra of the spine [25]. The passive translational force is applied perpendicular to the joint surface, resulting in the separation of the bony surfaces [16]. Regarded as a safe technique, traction is a frequently used technique for pain and radiculopathy related to the neck or low back [25]. The therapeutic efficacy of traction may result from decreased mechanical pressure on the intervertebral disc and affected nerve roots, promoting intracapsular joint movement, the expansion of the intervertebral foramen size, the stretching of soft tissues, and pain modulation [25–29] (Table 23.3).

Recent meta-analysis and systematic reviews regarding the effectiveness of traction for cervical or lumbar spine radiculopathy is equivocal, with minimal clinical value reported longterm [30–32]. This is in line with best practice guidelines that do not recommend traction for low back pain with or without the presence of radicular symptoms [33].

Table 23.3 Types for applied traction, Kaltenborn's Techniques

Kaltenborn's techniques	Procedure	Indication for use
Grade I: Loosen	Small amplitude movement applied to offset compressive forces about the joint. No appreciable joint separation noted.	May be used for pain relief.
Grade II: Taking up the slack	Varied amount movement to tighten the surrounding tissues of the joint. The amplitude of movement may be impacted if joint pathology exists.	May be used to inhibit pain and promote joint mobility.
Grade III: Stretch	Additional force to induce a stretch on the capsule and surrounding tissues traversing the joint. The amplitude of movement may be impacted if joint pathology exists.	May be used to inhibit pain, stretch tissues, and increase joint mobility.

Adapted from Bergmann and Peterson [16], Kaltenborn et al. [24], Heiser et al. [159]

Spinal Manipulation

Potential Mechanisms for the Therapeutic Effects

Before discussing the science of SMT, scientific evidence supports two different goals. The first category of evidence supports clinical treatment plans with the second providing an understanding of clinical treatments and the mechanisms of how they work. The first type comes in the form of clinical practice guidelines (CPG), meta-analyses, systematic reviews, and randomized clinical trials (RCT). Although case control studies and reports may be considered part of the first group, they should not be relied on to guide treatment choices. Instead, the purpose of case control studies and case reports is to guide the design of future RCTs informing CPGs. This second type of evidence primarily comes in the form of non-clinical randomized controlled trials, such as basic science experiments and animal model experiments. The risk of confusing these two forms of evidence can result in supporting treatment plans with the wrong type of science. Most of this section will report on the second type which is concerned with understanding how SMT works.

What Are the Mechanisms That Improve Clinical Symptoms and Function?

Mechanotransduction

Mechanotransduction occurs when living cells convert mechanical stimuli into electrical and chemical signals [34]. Mechanotransduction is currently the only overarching theory that accounts for all the affects that have been reported clinically after SMT. When tissue cells are mechanically stretched or compressed, mechanoreceptors are stimulated initiating a cascade of neurophysiological pathways. These

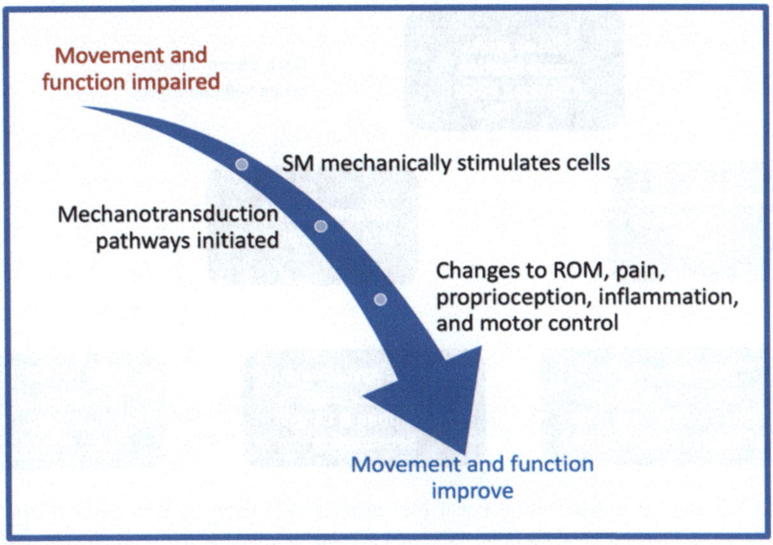

Fig. 23.2 Mechanotransduction pathway to effects of spinal manipulation

pathways signal the body and brain to make changes that are temporarily beneficial, especially for someone suffering with pain and disability (Fig. 23.2).

SMT has been demonstrated to cause a bombardment of muscle spindle and Golgi Tendon Organ (GTO) afferent signals (similar to other mechanoreceptors) to the spinal cord, followed by an immediate efferent burst of EMG activity [35–38] and a drastic reduction of afferent signal for a short period [8, 35, 39, 40]. This observation is thought to be responsible for the H-reflex responses to local muscles [40], and decreased reactivity to painful stimulus [41–44].

The Dysfunctional Movement Model

This model proposed by Jay Triano [45], suggests the therapeutic mechanisms of high velocity low amplitude SMT are due to mechanotransduction signals (Fig. 23.3).

Triano describes the theory as follows:

1. Normal movement patterns become disrupted by a single or multiple overload event resulting in abnormal mechanical stress within the tissue(s). These stresses, if left unresolved, are transformed through mechanotransduction processes into disease states.
2. Normal motion consists of a rhythmic sequence of feed-forward and feed-back control systems that provide smooth function and result in local tissue stresses consistent with the task being attempted.
3. Abnormal movement consists of dysrhythmic motion that concentrates stress, inconsistent with the task demands, in one or more tissues leading to nociceptive bombardment and/or inflammatory cascades disrupting the feed-forward and feed-back loops.

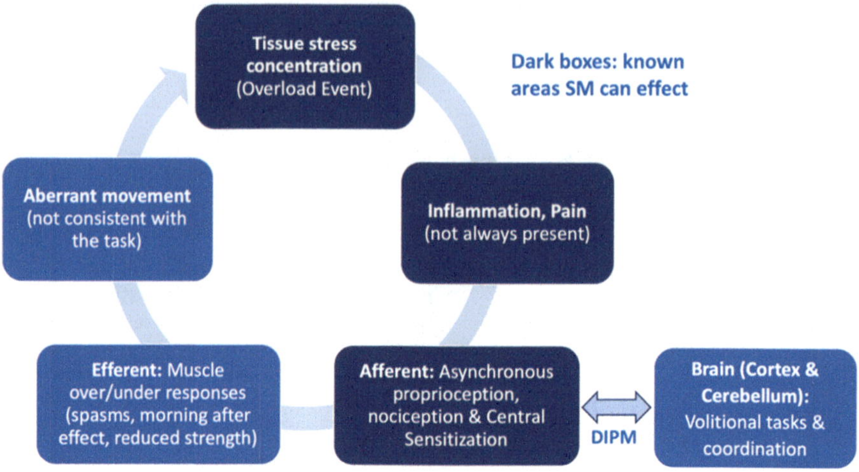

Fig. 23.3 Triano Modified Dysfunctional Movement Model (adapted with permission). This model proposes that abnormal movement that can concentrate stress in tissues leading to nociceptive bombardment and / or inflammatory cascades that disrupt the feed-forward and feed-back communication in the nervous system. These processes, if left unresolved, can perpetuate local stress, which repeats the cycle. Dark boxes represent locations in the model where spinal manipulation may be able to break this cycle, through restoration of normal movement. DIPM (Descending Inhibitory Pain Mechanism)

4. These processes, if left unresolved, engage the mechanotransduction of local stress into abnormal structure (e.g., swelling, fibrosis, etc.) and ultimately disease.

Specific Findings Supporting the Dysfunctional Movement Model

Randoll et al. [46] categorized three mechanisms of back pain relief from SMT that rely on decreasing the temporal summation (progressive buildup) of pain:

1. Segmental (spinal) processes may decrease pain perception through inhibition of the dorsal horn neurons' activity.
2. Descending pathways originating from various brain structures, including the brainstem, these may be activated by SMT and subsequently modulate the spinal neurons' activity (known as descending inhibitory pain mechanisms, DIPM).
3. Nonspecific cerebral processes (expectations, placebo) that modulate pain perception without a spinal involvement could decrease nociceptive activity and pain perception.

Further support for Triano's model is also provided by Pickar and Bolton suggesting that high velocity low amplitude SMT may take advantage of two signaling characteristics of the nervous system: [8]

1. Inherent high frequency signaling properties of dynamically-sensitive primary afferent neurons (from mechanoreceptors) and
2. Response properties of post-synaptic (spinal) neurons

Research findings to Support the Above Theories on SMT

1. Restoration of normal movement is supported by research that observed increased and improved motion. Multiple studies have demonstrated an increased global ROM [47–55] and gapping of facet joints [56–60] with the possibility that the quality of motion becomes smoother [61].
2. Alterations of how the central nervous system perceives pain. People with neck and low back pain (LBP) are reported to have more sensitivity to pressure (decreased pain pressure thresholds (PPT)), increased subjective pain sensitivity, and increased inflammatory markers (e.g., TNF α, Chemotactic cytokines (CCL2, CCL3), NO) [51, 62]. How SMT creates changes to these outcomes through neurophysiological mechanisms of pain reduction has been compiled in a review by Gevers-Montoro et al. [63] The authors provide an excellent summary on the current state of the evidence explaining changes seen after SMT (e.g. changes to pain related brain activity, DIPM, inhibition of central sensitization and temporal summation, changes to inflammatory markers, reactive oxygen species (ROS), PPT and pain sensitivity). Preliminary data suggests SMT may alter the way the brain perceives pain [64] and adapts to pain [65]. The following subsections provide details on the different mechanisms studied:
   (a) Decreased pain perception. Local and distal hypoalgesia (increased PPT) via descending inhibitory pain mechanism have been demonstrated to utilize the noradrenergic, serotonergic, and adenosine systems [41, 42, 51, 66–70].
   (b) Changes to blood inflammatory and pain **markers** have also been shown, with changes (primarily decreases) in cytokine/chemokines reported, although some marker response differs in acute versus chronic pain populations [62, 71]. A systematic review and meta-analysis has shown that there is some evidence that spinal manipulation can alter substance P, neurotensin, oxytocin, and various cytokine levels (e.g., TNF, IL-1, 2 and 6), and may influence cortisol levels [66, 72, 73].
   (c) Reactive Oxygen Species (ROS) have the ability to sensitize tissues that have been under stress from acute injury [74]. An animal study with immobilization-induced allodynia, demonstrated the prevention of plasmatic ROS increases with instrument SMT [75].
3. Decreased resting muscle reactivity. Reduced erector spinae muscle activity has been reported following SMT during standing full lumbar flexion, where activation levels are closer to the normal flexion relaxation response [76, 77], but not all studies have observed this change [78]. Muscle reactivity to painful pressure on the spinous processes has been reduced in some individuals [43]. Interestingly, relaxation of pelvic floor after SMT was seen with an increase in pelvic inlet area in women in their second trimester of pregnancy [79].
4. Increased voluntary muscle activation. Few studies have demonstrated that back muscle activation [80], leg muscle strength [81–83], jaw bite strength [84] and hand grip strength [85] can be increased during voluntary contraction. One possible reason for this is that the increase in strength is due to a decrease in muscle

inhibition caused by pain. Studies on participants with neck and knee pain demonstrated increased biceps and quadriceps muscle activation after SM [51, 86, 87], other studies unable to reproduce findings of increased muscle activity [88–90]. Furthermore, not all the studies that have shown an increase in strength following SMT were done with pain patients, as several studies included participants not in pain on the day of their participation [81, 82, 84]. This suggests increased muscle activation originated from more neural drive/signal and not only from less inhibition. Interestingly, the increase in muscle strength following spinal manipulation appears possible in populations with cortical brain damage, such as in a chronic stroke population [83]. Improved feed forward muscle activation latency (faster reaction) of core musculature (transversus abdominus/internal oblique) during trunk perturbation, following SMT have been reported [91].

5. Decreased spinal stiffness has been observed, with several authors suggesting that this may be as a result of a viscoelastic tissue response to mechanical forces of SMT or due to neuromuscular changes [92, 93]. This data supports the idea that SMT can affect local tissue stress concentrations.
6. Changes in proprioception has been demonstrated with improved cervical [94], lumbar [95] and elbow joint positioning [96]. Furthermore, ankle joint position sense was shown to significantly improve, along with an improvement in the ability to take a faster step after 3 months of chiropractic care [97].
7. Changes in sensorimotor integration, cortical excitability, and improved fine motor training. Multiple studies have observed that spinal manipulation alters various measures such as sensorimotor integration, central inhibition, and cortical excitability [65, 81–83, 98–106]. These integrative and processing changes are related to the way we learn new motor skills, suggesting that SMT may alter motor skill learning [59] and in particular motor skill retention [107].
8. Changes in the pathway between the cerebellum and primary somatosensory and motor cortex. SMT in individuals with recurrent neck pain has been observed to affect the excitability of the pathway between the cerebellum and both the primary somatosensory (S1) [107] and primary motor cortex (M1) [99, 108]. Healthy individuals normally decrease the inhibition in cerebellum-M1 pathway when learning a novel task, but those with recurrent neck pain remained inhibited [99, 108]. The group that received SMT prior to performing the motor acquisition regained the ability to disinhibit this pathway [108].

The above findings suggest that SMT creates a temporary "window of opportunity", to retrain impaired motor function, promote desensitization and increase tissue growth, supporting the positive pragmatic long-term results of SMT [69]. Empowering people to re-engage with their normal activities of daily living provides the stimulus to tissues needed for repair (Davis's law) [109]. This idea supports the use of SMT for the improvement of function and the reduction of disability [78, 110–115].

> Davis's law states that soft tissue will model along the lines of stress [109].

Non-specific Cerebral Responses

Although the above models on the mechanisms of SMT provide a good theoretical basis, it is important to recognize a growing area of research that influences the entire system and perceived patient outcomes, the contextual factors of the encounter with a health professional like a chiropractor [116]. For example, the history and exam alone have been shown to produce therapeutic effects in those with LBP [117]. However, SMT has been suggested to have greater effects on pain sensitivity than expectations of receiving treatment alone [89, 90, 118, 119] and may improve psychological outcomes compared to verbal interventions [120]. This will no doubt be an important research area to revisit in the future to maximize the clinical benefits to patients.

- The Dysfunctional Movement Model proposes therapeutic mechanisms of spinal manipulation. This model is supported by research
- Mechanotransduction is currently the only overarching explanation for the various clinical responses from SMT
- The impact of contextual factors on the therapeutic effects of SMT must be considered

Clinical Implications and Current Evidence: SMT for the Treatment of Spinal Pain

Pain related to the spine is one of the leading causes of disability worldwide [121], affecting over 11% of the global population at any given time, with 80% of people report experiencing an episode of neck or low back pain at least once during their lifetime [4, 122, 123]. The economic burden remains high, due to work absenteeism, limitations in daily activities, medical expenses, and changes in productivity [4, 122, 123].

SMT is considered an effective treatment for neuromusculoskeletal conditions. It is generally regarded as safe, however the risk associated with SMT remains unclear, partially due to a gap in reporting [124]. Those being treated with SMT usually report low back, neck, and head pain [5], although extensive research supports the use of SMT for the treatment of musculoskeletal pain related to the spine. SMT, when combined with a multimodal care program (education, reassurance, self-management strategies and exercise) is regarded as a first line treatment for spine pain (cervical, thoracic and low back) and cervicogenic related headaches [3, 4, 110, 111, 113, 114, 123, 125, 126].

Previous research has recommended thoracic SMT for mechanical cervico-thoracic spine and chest wall pain [54, 127, 128], yet current systematic reviews do not endorse thoracic spine manipulation citing the lack of quality evidence. More research needs to be conducted focused on the clinical effectiveness of SMT for the treatment of thoracic spine pain [129].

Adverse events following SMT and mobilization are common, with symptoms being typically mild and transient with muscles soreness, stiffness and headaches being the most commonly reported [130, 131]. Though rare, serious complications include pathological fractures, neurological deficits, vertebral artery dissection and death [23, 130]. The debate persists regarding the safety of cervical spine SMT and vertebrobasilar (VBA) stroke [132]. Although an association with cervical spinal manipulation and cervicocerebral artery dissection exists, no causal link has been established. Cassidy et al., reports that risk of VBA stroke associated with chiropractic care is no different than seeing a primary care physician [133]. The clinical symptom of cervicocerebral artery dissection includes neck pain and headaches [134]. However, consideration should be given for ruling out serious pathologies for patients seeking chiropractic care for neck pain and headaches. Therefore, a comprehensive history, physical examination should be performed to determine a pre-existing pathology, or whether the origin of pain is vascular or mechanical in nature [23, 130]. Although VBA stroke is rare, and difficult to diagnose, efforts to mitigate risk of adverse events should be taken when subjects present with neck or head pain and/or neurological deficits to the medical provider [131, 135].

- SMT is a safe and effective for the treatment of neck and low back pain
- Therapeutic benefits of thoracic SMT have been previously documented, continued research to determine the clinical efficacy is warranted
- SMT is regularly offered in a multimodal care approach (exercise, education, soft tissue therapy, modalities) for spine pain
- Adverse events post-SMT are benign and short-term

SMT and Performance Related Measures

As discussed earlier, SMT is an effective intervention for the treatment of neuromusculoskeletal complaints and spinal pain with the potential neurophysiological therapeutic effects linked to SMT being supported by a wealth of evidence [136]. However, little is known about SMT as it relates to performance enhancement.

> *As long as I see the Chiropractor, I feel like I'm one step ahead of the game.*
>
> –Tom Brady

Athletes and individuals seek chiropractic care to improve athletic performance [137]. Anecdotal evidence suggests that athletic performance can be improved with SMT, as witnessed from chiropractic and athlete testimonials and theoretical frameworks [138–140]. A survey conducted on treatment interventions and outcomes report 76% of Royal College of Chiropractors Sports Sciences Canada (RCCSS(C)) Fellows believe they could cause direct and specific improvements in an athlete's sports performance with use of chiropractic treatment techniques the most popular interventions utilized being SMT/mobilization and extremity manual therapy (EMT) [139]. Current approaches to studying the impact of Chiropractic care on athletic performance have focused on the immediate effects of SMT on biomechanical, physiological and range of motion measures of performance [137, 138, 141]. The effects of SMT on performance-based outcomes "report minimal to no immediate effects in running and cycling sprint times, grip strength, free throw accuracy, golf ball driving distance, and hip flexor flexibility" [142]. These equivocal findings may be related to the methodological study design and the performance outcomes being measured [138]. One of the challenges remains the ability to best measure performance-related metrics, and to discern whether they lead to meaningful sports-related outcomes [140]. A recent study by Corso et al. [142] has attempted to investigate underlying neuromuscular mechanisms contributing to athletic performance post SMT in soccer players. The authors tested the feasibility of their research design, taking into consideration some of the methodological pitfalls and study limitations of previous research: recruitment, sample size, scheduling, collection of outcome measures, control group, temporal intervals of outcomes, and self-reported subject feedback and expectations [142]. Findings align with previous studies favoring SMT for performance-based outcomes and support continued research and larger scale studies investigating neuromuscular mechanisms underlying the effects of spinal manipulation for performance enhancement [142, 143].

Most chiropractic care and performance-based research investigate SMT as a single intervention, yet athletic performance is multi-dimensional being technical, tactical, physiological, psychological, and sociological [137, 144, 145]. Chiropractic care is rooted in traditional characteristics of the biopsychosocial model of care. Aside from physical interventions, chiropractors traditionally connect with patients providing psychological and social support [146]. The entire clinical encounter, which may include the combined influences of physical interventions (SMT/Mobilizations, EMT, Soft tissue Therapy), the psychological effects of receiving treatment, and the provision of contextualized care may have contributed to the athletes perception of improved performance according to Lee et al. [137] Given these complexities, practical approaches are required to investigate the many facets of chiropractic care for performance enhancement [137].

Sports performance is defined as a combination of specific physical routines or procedures performed by someone who is trained or skilled in a physical activity and influenced by physiological, psychological, and sociocultural factors [138].

- Practitioners credit improved athletic performance with therapeutic treatments, however, this was not substantiated by research previously
- Improved performance post-SMT has been reported, however the evidence is weak
- Higher quality research investigating SMT and performance must consider the patients context (contextualized care)
- Chiropractic care is rooted in traditional characteristics of the biopsychosocial model of care

Extraspinal or Extremity Manipulations and Mobilizations

Chiropractors are considered specialists in the assessment and treatment of conditions related to the spine and nervous system. Accredited Chiropractic institutions teach students how to assess, diagnose and treat disorders and dysfunctions arising from the structures of the peripheral joints [1]. High-velocity low amplitude (HVLA) manipulation and mobilizations may be applied to the peripheral joints, much like the spine. This is commonly referred to as "extraspinal" or "extremity" manipulation. The joints of the extremities are an ensemble of hinge, ball-and-socket, saddle, and condylar. Depending on the amount of axis available to the joint, specific manipulative procedures are available to the practitioner. For example, the hip joint with its multiple axis of motion will have more "manipulative options" than a hinge joint (i.e., elbow).

Proper joint biomechanics consists of the optimal interaction between osseus structures, the articular surface of the joint, the surrounding ligaments, and muscles [147]. Events such as injuries, surgeries and muscle fatigue may interfere with their proper functioning of causing to aberrant joint biomechanics [147, 148]. Aberrant joint biomechanics are associated with reduced range of motion, increased pain, and possibly early osteoarthritic changes, which may promote further issues within the joint [149]. Aberrant joint biomechanics may lead to compensatory joint biomechanics within the kinetic chain [150, 151]. Recognizing abnormal joint biomechanics (structure or function) and the impact on the kinetic chain is important when performing a clinical assessment to guide treatment, such as EMT. EMT can play a role in restoring range of motion of said joint which may help to restore optimal function across the kinetic chain [152, 153].

EMT can be a useful component in a patient's rehabilitation program. In individuals with chronic ankle instability and chronic lateral ankle sprains, EMT significantly improved ankle dorsiflexion, dynamic balance, and function [152, 153] (Fig. 23.4). Elderly patients with hip osteoarthritis receive immediate improvement

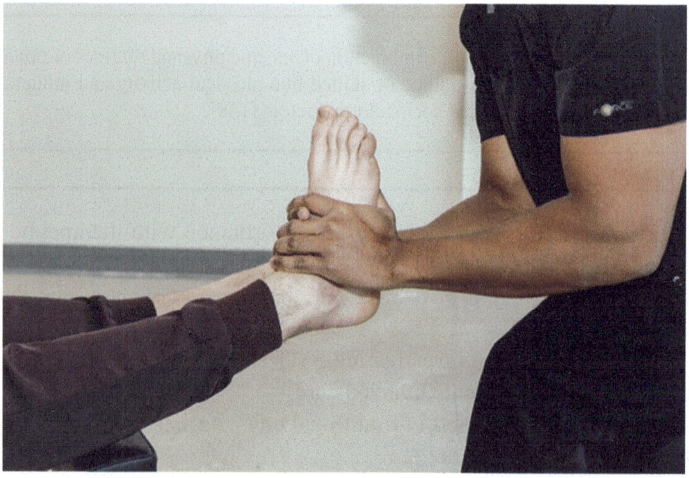

Fig. 23.4 Ankle mortise joint mobilization. Long-axis distraction is applied to the tibiotalar joint

of pain and range of motion [154]. EMT may play a role in athletic performance. As aforementioned, motor tasks demand proper joint arthrokinematics spanning across the kinetic chain to deliver optimal movement. In tasks such as the squat, an increase in ankle dorsiflexion can aide in achieving a deeper squat [155].

Although research pertaining to the neurophysiological effects of EMT continues to grow, more needs to be done. Preliminary studies observe an increase of corticospinal motor activity 30 minutes post-talocrural joint manipulation, as demonstrated by increased excitability of the tibialis anterior muscle [156]. In contrast, corticospinal excitability decreased in the mobilization group. A possible explanation for this finding may be due to the sustained passive tension to the joint and surrounding structures as this finding can be seen after bouts of passive stretching. It is also suggested that this 30-minute time frame post manipulation may serve as an optimal period to enforce movement and skill acquisition as corticospinal motor activity is increased following motor learning tasks.

An increase in corticospinal motor activity may help improve postural sway. In a randomized control trial, participants received a series of upper and lower extremity manipulations. Postural sway was improved in the frontal plane in participants that received non-specific long-axis distraction manipulation in the lower extremity. When administered in the upper extremity, participants exhibited improvements in the sagittal plane [157]. Further research is needed to quantify a possible real-world application.

Lower extremity manipulation may affect a subjects' center of pressure [158]. In a randomized, controlled crossover pilot study, individuals stood on a force plate with eyes closed for 3 minutes before and after standing on an incline board with or without joint manipulation. Center of pressure and overall movement were found to be diminished in the intervention group during the latter sections of the 3-minute observation. The increased afferent communication from the joints post-manipulation to the dorsospino-cerebellar tract may explain how postural control was affected [158].

- EMT can play a role in restoring range of motion of the joints of the extremities which may help to restore optimal function across the kinetic chain.
- EMT research pertaining to neurophysiological mechanisms are not as numerous compared to SMT research.
- Current research suggests that EMT may have positive effects on balance and temporary increase in corticospinal motor activity post-EMT. Research in this domain is ongoing.

Continued Research

There is extensive chiropractic research providing insight on the therapeutic benefits of SMT and EMT, yet the pursuit to better understand the biomechanical and neurophysiological effects remain. Continued high quality research surrounding

SMT as it pertains to safety, efficacy, athletic performance, contextual factors (psychosocial), clinical prediction models, and neurological mechanisms is required.

Conclusion

The rise in clinical research in the field of Chiropractic and spinal manipulative therapy (SMT) is encouraging. Attitudes towards chiropractic are generally positive. Chiropractors are regarded as neuromusculoskeletal experts, treating spine pain and peripheral joint injuries. The therapeutic mechanisms of SMT, with respect to recovery and regeneration can be understood by the current findings that support the process of mechanotransduction, central nervous system modulations, and contextual contributions. Current research supports the use of manipulative therapy for the treatment of spinal pain and peripheral joints. However, continued research is needed to better comprehend the fundamental neurophysiological mechanisms that surround spinal and peripheral joint manipulation. Although there was an attempt to present the current evidence regarding SMT and EMT, research is ever evolving, and this paper may not have captured all the research in this space. As new research emerges, practitioners must continue to apply evidence to clinical practice.

Chiropractic care is acknowledged as a safe, non-invasive therapy for treating spine related pain, headaches and neuromusculoskeletal injuries. Manipulative therapy, whether it is applied to the spine or extremities, is one of many methods used by chiropractors and other health care professionals to help improve pain and function in patients. (OCA's EBFAC) [160].

1. What mechanisms and neurophysiological effects are responsible for the positive clinical outcomes reported in the literature?
2. How can we best study performance related metrics, and to discern whether they lead to meaningful sports related outcomes with SMT?
3. What protocols can be developed to ensure that SMT matches the unique circumstance of a clinical situation (i.e. stage of recovery, chronicity)?
4. What is the effectiveness of combining SMT, strength, and stretching exercises for improving athlete performance post lower back injury?
5. With chiropractic care rooted in traditional characteristics of the biopsychosocial model of care what effect does SMT have on PTSD?
6. What are the paired effects of SMT and EMT on improvement of proprioception of the lower leg in the elderly?

References

1. Chiropractic scope of practice; 2018. Accessed 26 Jan 2023. https://cco.on.ca/members-of-the-public/scope-of-practice-and-authorized-acts/.
2. Busse JW, Pallapothu S, Vinh B, et al. Attitudes towards chiropractic: a repeated cross-sectional survey of Canadian family physicians. BMC Fam Pract. 2021;22(1) https://doi.org/10.1186/s12875-021-01535-4.
3. Bussières AE, Stewart G, Al-Zoubi F, et al. The treatment of neck pain–associated disorders and whiplash-associated disorders: a clinical practice guideline. J Manip Physiol Ther. 2016;39(8):523–564.e27. https://doi.org/10.1016/j.jmpt.2016.08.007.
4. Sutton DA, Côté P, Wong JJ, et al. Is multimodal care effective for the management of patients with whiplash-associated disorders or neck pain and associated disorders? A systematic review by the Ontario Protocol for Traffic Injury Management (OPTIMa) Collaboration. Spine J. 2016;16(12):1541–65. https://doi.org/10.1016/j.spinee.2014.06.019.
5. Hurwitz EL. Epidemiology: spinal manipulation utilization. J Electromyogr Kinesiol. 2012;22(5):648–54. https://doi.org/10.1016/j.jelekin.2012.01.006.
6. LeFebvre R, Peterson D, Haas M. Evidence-based practice and Chiropractic care. J Evid Based Complement Altern Med. 2013;18(1):75–9. https://doi.org/10.1177/2156587212458435.
7. Tuttle AH, Shurtleff D. Spinal manipulation: what you need to know. https://www.nccih.nih.gov/.
8. Pickar JG, Bolton PS. Spinal manipulative therapy and somatosensory activation. J Electromyogr Kinesiol. 2012;22(5):785–94. https://doi.org/10.1016/j.jelekin.2012.01.015.
9. Funabashi M, Nougarou F, Descarreaux M, Prasad N, Kawchuk G. Influence of spinal manipulative therapy force magnitude and application site on spinal tissue loading: a biomechanical robotic serial dissection study in Porcine motion segments. J Manip Physiol Ther. 2017;40(6):387–96. https://doi.org/10.1016/j.jmpt.2017.05.003.
10. LaPelusa A, Bordoni B. High velocity low amplitude manipulation techniques. https://www.ncbi.nlm.nih.gov/.
11. Cramer GD, Kim R, Raju PK, et al. Quantification of cavitation and gapping of lumbar zygapophyseal joints during spinal manipulative therapy. J Manip Physiol Ther. 2012;35(8):614–21. https://doi.org/10.1016/j.jmpt.2012.06.007.
12. Kaltenborn FM, Kaltenborn TB, Vollowitz E. Manual mobilization of the joints. Joint examination and basic treatment, vol. 3. 1st ed. Norli; 2008.
13. Sillevis R, Cleland J. Immediate effects of the audible pop from a thoracic spine thrust manipulation on the autonomic nervous system and pain: a secondary analysis of a randomized clinical trial. J Manip Physiol Ther. 2011;34(1):37–45. https://doi.org/10.1016/j.jmpt.2010.11.007.
14. Moorman AC, Newell D. Impact of audible pops associated with spinal manipulation on perceived pain: a systematic review. Chiropr Man Therap. 2022;30(1) https://doi.org/10.1186/s12998-022-00454-0.
15. Triano JJ. Biomechanics of spinal manipulative therapy, vol. 1; 2001.
16. Bergmann TF, Peterson DH. Chiropractic technique: principles and procedures. 3rd ed. Elsevier; 2011.
17. de las Fernandez Penas C, Cleland J, Dommerholt J, editors. Manual therapy for musculoskeletal pain syndromes an evidence- and clinical-informed approach. 1st ed. Churchill Livingstone; 2015.
18. Skirven TM, Osterman AL, Fedorczyk JM, Amadio PC. Rehabilitation of the hand and upper extremity, vol. 1. 6th ed. Elsevier; 2011.
19. Hooks TR. Physical rehabilitation of the injured athlete. 4th ed. Elsevier; 2012.
20. Slater H, Fernandez de las Penas C. In: Fernandez-de-las-Penas C, Cleland JA, Huijbregts PA, editors. Neck and arm pain syndromes. Elsevier; 2011.
21. Hengeveld E, Banks K. In: Wells P, editor. Maitland's peripheral manipulation. 4th ed. Elsevier; 2005. www.elsevierhealth.com.

22. Mulligan EP. Principles of joint mobilization continuing ED. Published online 2001.
23. Puentedura EJ. In: Fernández-de-las- Peñas C, Cleland JA, Dommerholt J, editors. Manual therapy for musculoskeletal pain syndromes an evidence-and clinical-informed approach, vol. 1. 1st ed. Elsevier; 2016.
24. Kaltenborn FM, Evjenth O, Kaltenborn TB, Morgan D, Vollowitz E. Manual mobilization of the joints: the spine, vol. 2. 4th ed. Norli, Universitetsgaten; 2003.
25. Catapang G. Pain procedures in clinical practice. 3rd ed. Elsevier; 2011.
26. Romeo A, Vanti C, Boldrini V, et al. Cervical radiculopathy: effectiveness of adding traction to physical therapy-a systematic review and meta-analysis of randomized controlled trials, vol. 98; 2018. https://academic.oup.com/ptj.
27. Choi J, Lee S. Influences of spinal decompression therapy and general traction therapy on the pain, disability, and straight leg raising of patients with intervertebral disc herniation.
28. Pellecchia GI. Lumbar traction: a review of the literature; 1994. www.jospt.org.
29. Batavia M. Contraindications in physical rehabilitation. Elsevier; 2006.
30. Vanti C, Turone L, Panizzolo A, Guccione AA, Bertozzi L, Pillastrini P. Vertical traction for lumbar radiculopathy: a systematic review. Arch Physiother. 2021;11(1) https://doi.org/10.1186/s40945-021-00102-5.
31. Alrwaily M, Almutiri M, Schneider M. Assessment of variability in traction interventions for patients with low back pain: a systematic review. Chiropr Man Therap. 2018;26(1) https://doi.org/10.1186/s12998-018-0205-z.
32. Vanti C, Panizzolo A, Turone L, et al. Effectiveness of mechanical traction for lumbar radiculopathy: a systematic review and meta-analysis. Phys Ther. 2021;101(3) https://doi.org/10.1093/ptj/pzaa231.
33. Low back pain and Sciatica in over 16s: assessment and management NICE guideline; 2016. www.nice.org.uk/guidance/ng59.
34. Ingber DE. Mechanobiology and diseases of mechanotransduction. Ann Med. 2003;25:546–77.
35. Pickar JG. Neurophysiological effects of spinal manipulation, vol. 2; 2002.
36. Electromyographic responses of back and limb muscles associated with spinal manipulative therapy.
37. Synzons BP, Herzog W, Leonard T, Nguyen H. Reflex responses associated with activator treatment. Vol 23; 2000.
38. Herzog W, Scheele D, Conway PJ. Electromyographic responses of back and limb muscles associated with spinal manipulative therapy. Spine (Phila Pa 1976). 1999;24(2):146–52.
39. Sung PS, Kang YM, Pickar JG. Effect of spinal manipulation duration on low threshold Mechanoreceptors in Lumbar Paraspinal muscles a preliminary report. Vol 30.
40. Reed WR, Cao DY, Long CR, Kawchuk GN, Pickar JG. Relationship between biomechanical characteristics of spinal manipulation and neural responses in an animal model: effect of linear control of thrust displacement versus force, thrust amplitude, thrust duration, and thrust rate. Evid Based Complement Alternat Med. 2013;2013 https://doi.org/10.1155/2013/492039.
41. Grayson JE, Barton T, Cabot PJ, Souvlis T. Spinal manual therapy produces rapid onset analgesia in a rodent model. Man Ther. 2012;17(4):292–7. https://doi.org/10.1016/j.math.2012.02.004.
42. Martins DF, Mazzardo-Martins L, Cidral-Filho FJ, Stramosk J, Santos ARS. Ankle joint mobilization affects postoperative pain through peripheral and central adenosine A1 receptors. Phys Ther. 2013;93(3):401–12. https://doi.org/10.2522/ptj.20120226.
43. Lehman GJ, Mcgill SM. Spinal manipulation causes variable spine kinematic and trunk muscle electromyographic responses. www.elsevier.com/locate/clinbiomech.
44. Yeo HK, Wright A. Hypoalgesic effect of a passive accessory mobilisation technique in patients with lateral ankle pain. Man Ther. 2011;16(4):373–7. https://doi.org/10.1016/j.math.2011.01.001.
45. Triano J. Foundations of Chiropractic practice: updating ourselves. 2016; Lecture.

46. Randoll C, Gagnon-Normandin V, Tessier J, et al. The mechanism of back pain relief by spinal manipulation relies on decreased temporal summation of pain. Neuroscience. 2017;349:220–8. https://doi.org/10.1016/j.neuroscience.2017.03.006.
47. Anderst WJ, Gale T, LeVasseur C, Raj S, Gongaware K, Schneider M. Intervertebral kinematics of the cervical spine before, during, and after high-velocity low-amplitude manipulation. Spine J. 2018;18(12):2333–42. https://doi.org/10.1016/j.spinee.2018.07.026.
48. Hanney WJ, Puentedura EJ, Kolber MJ, Liu X, Pabian PS, Cheatham SW. The immediate effects of manual stretching and cervicothoracic junction manipulation on cervical range of motion and upper trapezius pressure pain thresholds. J Back Musculoskelet Rehabil. 2017;30(5):1005–13. https://doi.org/10.3233/BMR-169573.
49. Ditcharles S, Yiou E, Delafontaine A, Hamaoui A. Short-term effects of thoracic spine manipulation on the biomechanical organisation of gait initiation: a randomized pilot study. Front Hum Neurosci. 2017:11. https://doi.org/10.3389/fnhum.2017.00343.
50. Galindez-Ibarbengoetxea X, Setuain I, Andersen LL, et al. Effects of cervical high-velocity low-amplitude techniques on range of motion, strength performance, and cardiovascular outcomes: a review. J Altern Complement Med. 2017;23(9):667–75. https://doi.org/10.1089/acm.2017.0002.
51. Suter E, Mcmorland G. Brief report decrease in elbow flexor inhibition after cervical spine manipulation in patients with chronic neck pain. www.elsevier.com/locate/clinbiomech.
52. Whittingham W, Nilsson N. Active range of motion in the cervical spine increases after spinal manipulation (toggle recoil). J Manip Physiol Ther. 2001;24(9):552–5. https://doi.org/10.1067/mmt.2001.118979.
53. Gavin D. The effect of joint manipulation techniques on active range of motion in the midthoracic spine of asymptomatic subjects. J Manual Manipul Ther. 1999;7(3):114–22. https://doi.org/10.1179/106698199790811672.
54. Krauss J, Creighton D, Jonathan DE, Podlewska-Ely J. The immediate effects of upper thoracic translatoric spinal manipulation on cervical pain and range of motion a randomized clinical trial. J Man Manip Ther. 2008;12(2):93–9.
55. Nilsson N, Christensen HW, Hartvigsen J. Lasting changes in passive range motion after spinal manipulation: a randomized, blind, controlled trial. J Manip Physiol Ther. 1996;19(3):165–8.
56. Cramer GD, Ross K, Pocius J, et al. Evaluating the relationship among cavitation, zygapophyseal joint gapping, and spinal manipulation: an exploratory case series. J Manip Physiol Ther. 2011;34(1):2–14. https://doi.org/10.1016/j.jmpt.2010.11.008.
57. Cramer GD, Gregerson DM, Knudsen JT, Hubbard BB, Ustas LM, Cantu JA. The effects of side-posture positioning and spinal adjusting on the Lumbar Z joints a randomized controlled trial with sixty-four subjects. Number. 2002;27:2459–66. https://doi.org/10.1097/01.BRS.0000031267.86149.E3.
58. Cramer GD, Ross JK, Raju PK, et al. Distribution of cavitations as identified with accelerometry during lumbar spinal manipulation. J Manip Physiol Ther. 2011;34(9):572–83. https://doi.org/10.1016/j.jmpt.2011.05.015.
59. Cramer GD, Cambron J, Cantu JA, et al. Magnetic resonance imaging zygapophyseal joint space changes (gapping) in low back pain patients following spinal manipulation and side-posture positioning: a randomized controlled mechanisms trial with blinding. J Manip Physiol Ther. 2013;36(4):203–17. https://doi.org/10.1016/j.jmpt.2013.04.003.
60. Cramer GD, Tuck NR Jr, Fonda SD, Schliesser JS, Patel P. Effects of side-positioning and side-posture adjusting on the Lumbar zygapophysial joints as evaluated by magnetic resonance imaging: a before and after study with randomization. J Manip Physiol Ther. 2000;23(6):380–94.
61. Mieritz RM, Hartvigsen J, Boyle E, Jakobsen MD, Aagaard P, Bronfort G. Lumbar motion changes in chronic low back pain patients: a secondary analysis of data from a randomized clinical trial. Spine J. 2014;14(11):2618–27. https://doi.org/10.1016/j.spinee.2014.02.038.
62. Teodorczyk-Injeyan JA, Triano JJ, McGregor M, Woodhouse L, Injeyan HS. Elevated production of inflammatory mediators including nociceptive chemokines in patients with neck

pain: a cross-sectional evaluation. J Manip Physiol Ther. 2011;34(8):498–505. https://doi.org/10.1016/j.jmpt.2011.08.010.
63. Gevers-Montoro C, Provencher B, Descarreaux M, Ortega de Mues A, Piché M. Neurophysiological mechanisms of chiropractic spinal manipulation for spine pain. Eur J Pain. 2021;25(7):1429–48. https://doi.org/10.1002/ejp.1773.
64. Weber KA, Wager TD, Mackey S, Elliott JM, Liu WC, Sparks CL. Evidence for decreased Neurologic Pain Signature activation following thoracic spinal manipulation in healthy volunteers and participants with neck pain. Neuroimage Clin. 2019:24. https://doi.org/10.1016/j.nicl.2019.102042.
65. Navid MS, Lelic D, Niazi IK, et al. The effects of chiropractic spinal manipulation on central processing of tonic pain - a pilot study using standardized low-resolution brain electromagnetic tomography (sLORETA). Sci Rep. 2019;9(1) https://doi.org/10.1038/s41598-019-42984-3.
66. Molina-Ortega F, Lomas-Vega R, Hita-Contreras F, et al. Immediate effects of spinal manipulation on nitric oxide, substance P and pain perception. Man Ther. 2014;19(5):411–7. https://doi.org/10.1016/j.math.2014.02.007.
67. Skyba DA, Radhakrishnan R, Rohlwing JJ, Wright A, Sluka KA. Joint manipulation reduces hyperalgesia by activation of monoamine receptors but not opioid or GABA receptors in the spinal cord. Pain. 2003;106(1–2):159–68. https://doi.org/10.1016/S0304-3959(03)00320-8.
68. Gorrell LM, Beath K, Engel RM. Manual and instrument applied cervical manipulation for mechanical neck pain: a randomized controlled trial. J Manip Physiol Ther. 2016;39(5):319–29. https://doi.org/10.1016/j.jmpt.2016.03.003.
69. Saava C, Giakas G, Efstatiou M. The role of the descending inhibitory pain mechanism in musculoskeletal pain following high-velocity, low amplitude thrust manipulation: a review of the literature. J Back Musculoskelet Rehabil. 2014;24(4):377–82.
70. Song JX, Gan Q, Cao JL, Wang ZB, Rupert RL. Spinal manipulation reduces pain and Hyperalgesia after Lumbar intervertebral Forament inflammation in the rat. J Manip Physiol Ther. 2006;29(1):5–13.
71. Teodorczyk-Injeyan JA, McGregor M, Triano JJ, Injeyan SH. Elevated production of nociceptive CC Chemokines and sE-Selectin in patients with low back pain and the effects of spinal manipulation. Clin J Pain. 2018;34(1):68–75. https://doi.org/10.1097/AJP.0000000000000507.
72. Kovanur Sampath K, Mani R, Cotter JD, Tumilty S. Measureable changes in the neuroendocrinal mechanism following spinal manipulation. Med Hypotheses. 2015;85(6):819–24. https://doi.org/10.1016/j.mehy.2015.10.003.
73. Plaza-Manzano G, Molina F, Lomas-Vega R, Martínez- Amat A, Achalandabaso A, Hita-Contreras F. Changes in biochemical markers of pain perception and stress response after spinal manipulation. J Orthop Sports Phys Ther. 2014;44(4):231–9. https://doi.org/10.2519/jospt.2014.4996.
74. Westlund KN, Kochukov MY, Lu Y, Mcnearney TA. Impact of central and peripheral TRPV1 and ROS levels on proinflammatory mediators and nociceptive behavior; 2010. http://www.molecularpain.com/content/6/1/46.
75. Duarte FCK, Kolberg C, Riffel APK, Souza JA, Belló-Klein A, Partata WA. Spinal manipulation therapy improves tactile Allodynia and peripheral nerve functionality and modulates blood oxidative stress markers in rats exposed to knee-joint immobilization. J Manip Physiol Ther. 2019;42(6):385–98. https://doi.org/10.1016/j.jmpt.2018.11.023.
76. Lalanne K, Lafond D, Descarreaux M. Modulation of the flexion-relaxation response by spinal manipulative therapy: a control group study. J Manip Physiol Ther. 2009;32(3):203–9. https://doi.org/10.1016/j.jmpt.2009.02.010.
77. Bicalho E, Palma Setti JA, Macagnan J, Rivas Cano JL, Manffra EF. Immediate effects of a high-velocity spine manipulation in paraspinal muscles activity of nonspecific chronic low-back pain subjects. Man Ther. 2010;15(5):469–75. https://doi.org/10.1016/j.math.2010.03.012.

78. Xia T, Long CR, Vining RD, et al. Association of lumbar spine stiffness and flexion-relaxation phenomenon with patient-reported outcomes in adults with chronic low back pain - a single-arm clinical trial investigating the effects of thrust spinal manipulation. BMC Complement Altern Med. 2017;17(1) https://doi.org/10.1186/s12906-017-1821-1.
79. Haavik H, Murphy BA, Kruger J. Effect of spinal manipulation on pelvic floor functional changes in pregnant and nonpregnant women: a preliminary study. J Manip Physiol Ther. 2016;39(5):339–47. https://doi.org/10.1016/j.jmpt.2016.04.004.
80. Keller TS, Colloca CJ. Mechanical force spinal manipulation increases trunk muscle strength assessed by electromyography: a comparative clinical trial. J Manip Physiol Ther. 2000;23(9):585–95.
81. Niazi IK, Türker KS, Flavel S, Kinget M, Duehr J, Haavik H. Changes in H-reflex and V-waves following spinal manipulation. Exp Brain Res. 2015;233(4):1165–73. https://doi.org/10.1007/s00221-014-4193-5.
82. Christiansen TL, Niazi IK, Holt K, et al. The effects of a single session of spinal manipulation on strength and cortical drive in athletes. Eur J Appl Physiol. 2018;118(4):737–49. https://doi.org/10.1007/s00421-018-3799-x.
83. Holt K, Niazi IK, Nedergaard RW, et al. The effects of a single session of chiropractic care on strength, cortical drive, and spinal excitability in stroke patients. Sci Rep. 2019;9(1) https://doi.org/10.1038/s41598-019-39577-5.
84. Haavik H, Özyurt MG, Niazi IK, et al. Chiropractic manipulation increases maximal bite force in healthy individuals. Brain Sci. 2018;8(5) https://doi.org/10.3390/brainsci8050076.
85. Botelho MB, Andrade BB. Effect of cervical spine manipulative therapy on judo athletes' grip strength. J Manip Physiol Ther. 2012;35(1):38–44. https://doi.org/10.1016/j.jmpt.2011.09.005.
86. Suter E, McMorland G, Herzog W, Bray R. Decrease in quadriceps inhibition after Sacroiliac joint manipulation in patients with anterior knee pain. J Manip Physiol Ther. 1999;22(3):149–53.
87. Suter E, McMorland G, Herzog W, Bray R. Conservative lower back treatment reduces inhibition in knee-extensor muscles. J Manip Physiol Ther. 2000;23(2):76–80.
88. Crowell MS, Wofford NH. Lumbopelvic manipulation in patients with patellofemoral pain syndrome. J Manual Manipul Ther. 2012;20(3):113–20. https://doi.org/10.1179/2042618612Y.0000000002.
89. Grindstaff TL, Hertel J. "Lumbopelvic joint manipulation on quadriceps activation of individuals with patellofemoral pain syndrome" in the Article 'Abstracts: Accepted Platform Presentations AAOMPT. Vol 2012. Maney Publishing; 2009. http://www.maney.com.uk/journals/jmtandwww.ingentaconnect.com/content/maney/jmt.
90. Bautista-Aguirre F, Oliva-Pascual-Vaca A, Heredia-Rizo AM, Bosca-Gandia JJ, Ricard F, Rodriguez-Blanco C. Effect of cervical vs. thoracic spinal manipulation on peripheral neural features and grip strength in subjects with chronic mechanical neck pain - a randomized controlled trial. Eur J Phys Rehabil Med. 2017;53(3):333–41.
91. Marshall P, Murphy B. The effect of Sacroiliac joint manipulation on feed-forward activation times of the deep abdominal musculature. J Manip Physiol Ther. 2006;29(3):196–202. https://doi.org/10.1016/j.jmpt.2006.01.010.
92. Wong AYL, Parent EC, Dhillon SS, Prasad N, Kawchuk GN. Do participants with low back pain who respond to spinal manipulative therapy differ biomechanically from nonresponders, untreated controls or asymptomatic controls? Spine (Phila Pa 1976). 2015;40(17):1329–37. https://doi.org/10.1097/BRS.0000000000000981.
93. Edgecombe TL, Kawchuk GN, Long CR, Pickar JG. The effect of application site of spinal manipulative therapy (SMT) on spinal stiffness. Spine J. 2015;15(6):1332–8. https://doi.org/10.1016/j.spinee.2013.07.480.
94. Gong W. Effects of cervical joint manipulation on joint position sense of normal adults.
95. Gong W. The influence of lumbar joint mobilization on joint position sense in normal adults.

96. Haavik H, Murphy B. Subclinical neck pain and the effects of cervical manipulation on elbow joint position sense. J Manip Physiol Ther. 2011;34(2):88–97. https://doi.org/10.1016/j.jmpt.2010.12.009.
97. Holt KR, Haavik H, Lee ACL, Murphy B, Elley CR. Effectiveness of Chiropractic care to improve sensorimotor function associated with falls risk in older people: a randomized controlled trial. J Manip Physiol Ther. 2016;39(4):267–78. https://doi.org/10.1016/j.jmpt.2016.02.003.
98. Taylor HH, Murphy B. The effects of spinal manipulation on central integration of dual somatosensory input observed after motor training: a crossover study. J Manip Physiol Ther. 2010;33(4):261–72. https://doi.org/10.1016/j.jmpt.2010.03.004.
99. Daligadu J, Haavik H, Yielder PC, Baarbe J, Murphy B. Alterations in cortical and cerebellar motor processing in subclinical neck pain patients following spinal manipulation. J Manip Physiol Ther. 2013;36(8):527–37. https://doi.org/10.1016/j.jmpt.2013.08.003.
100. Taylor HH, Murphy B. Altered sensorimotor integration with cervical spine manipulation. J Manip Physiol Ther. 2008;31(2):115–26. https://doi.org/10.1016/j.jmpt.2007.12.011.
101. Haavik H, Murphy B. The role of spinal manipulation in addressing disordered sensorimotor integration and altered motor control. J Electromyogr Kinesiol. 2012;22(5):768–76. https://doi.org/10.1016/j.jelekin.2012.02.012.
102. Lelic D, Niazi IK, Holt K, et al. Manipulation of dysfunctional spinal joints affects sensorimotor integration in the prefrontal cortex: a brain source localization study. Neural Plast. 2016;2016 https://doi.org/10.1155/2016/3704964.
103. Haavik H, Niazi IK, Jochumsen M, Sherwin D, Flavel S, Türker KS. Impact of spinal manipulation on cortical drive to upper and lower limb muscles. Brain Sci. 2017;7(1) https://doi.org/10.3390/brainsci7010002.
104. Haavik H, Niazi IK, Holt K, Murphy B. Effects of 12 weeks of Chiropractic care on central integration of dual somatosensory input in chronic pain patients: a preliminary study. J Manip Physiol Ther. 2017;40(3):127–38. https://doi.org/10.1016/j.jmpt.2016.10.002.
105. Haavik H, Niazi IK, Jochumsen M, et al. Chiropractic spinal manipulation alters TMS induced I-wave excitability and shortens the cortical silent period. J Electromyogr Kinesiol. 2018;42:24–35. https://doi.org/10.1016/j.jelekin.2018.06.010.
106. Taylor Haavik H, Murphy B. Altered central integration of dual somatosensory input following cervical spine manipulation. J Manip Physiol Ther. 2000;33(3):178–88.
107. Andrew D, Yielder P, Haavik H, Murphy B. The effects of subclinical neck pain on sensorimotor integration following a complex motor pursuit task. Exp Brain Res. 2018;236(1):1–11. https://doi.org/10.1007/s00221-017-5103-4.
108. Baarbé JK, Yielder P, Haavik H, Holmes MWR, Murphy BA. Subclinical recurrent neck pain and its treatment impacts motor training-induced plasticity of the cerebellum and motor cortex. PLoS One. 2018;13(2) https://doi.org/10.1371/journal.pone.0193413.
109. Myofascial adhesions, Davis's Law & relative flexibility. :6–7. Accessed 28 Jan 2023. nasm.org/ces.
110. Coulter ID, Crawford C, Hurwitz EL, et al. Manipulation and mobilization for treating chronic low back pain: a systematic review and meta-analysis. Spine J. 2018;18(5):866–79. https://doi.org/10.1016/j.spinee.2018.01.013.
111. Bronfort G, Evans R, Anderson AV, Svendsen KH, Bracha Y, Grimm RH. Spinal manipulation, medication, or home exercise with advice for acute and subacute neck pain a randomized trial; 2012. www.annals.org.
112. Goertz CM, Pohlman KA, Vining RD, Brantingham JW, Long CR. Patient-centered outcomes of high-velocity, low-amplitude spinal manipulation for low back pain: a systematic review. J Electromyogr Kinesiol. 2012;22(5):670–91. https://doi.org/10.1016/j.jelekin.2012.03.006.
113. Bussières AE, Stewart G, Al-Zoubi F, et al. Spinal manipulative therapy and other conservative treatments for low back pain: a guideline from the Canadian chiropractic guideline initiative. J Manip Physiol Ther. 2018;41(4):265–93. https://doi.org/10.1016/j.jmpt.2017.12.004.

114. Côté P, Wong JJ, Sutton D, et al. Management of neck pain and associated disorders: a clinical practice guideline from the Ontario Protocol for Traffic Injury Management (OPTIMa) Collaboration. Eur Spine J. 2016;25(7):2000–22. https://doi.org/10.1007/s00586-016-4467-7.
115. Hondras MA, Long CR, Cao Y, Rowell RM, Meeker WC. A randomized controlled trial comparing 2 types of spinal manipulation and minimal conservative medical care for adults 55 years and older with subacute or chronic low back pain. J Manip Physiol Ther. 2009;32(5):330.
116. Sherriff B, Clark C, Killingback C, Newell D. Impact of contextual factors on patient outcomes following conservative low back pain treatment: systematic review. Chiropr Man Therap. 2022;30(1):20.
117. Louw A, Goldrick S, Bernstetter A, et al. Evaluation is treatment for low back pain. J Manual Manipul Ther. 2021;29(1):4–13. https://doi.org/10.1080/10669817.2020.1730056.
118. Bialosky JE, George SZ, Horn ME, Price DD, Staud R, Robinson ME. Spinal manipulative therapy-specific changes in pain sensitivity in individuals with low back pain (NCT01168999). J Pain. 2014;15(2):136–48. https://doi.org/10.1016/j.jpain.2013.10.005.
119. Bialosky JE, Bishop MD, Robinson ME, Barabas JA, George SZ. The influence of expectation on spinal manipulation induced hypoalgesia: an experimental study in normal subjects. BMC Musculoskelet Disord. 2008:9. https://doi.org/10.1186/1471-2474-9-19.
120. Williams NH, Hendry M, Lewis R, Russell I, Westmoreland A, Wilkinson C. Psychological response in spinal manipulation (PRISM): a systematic review of psychological outcomes in randomised controlled trials. Complement Ther Med. 2007;15(4):271–83. https://doi.org/10.1016/j.ctim.2007.01.008.
121. Hartvigsen J, Hancock MJ, Kongsted A, et al. What low back pain is and why we need to pay attention. Lancet. 2018;391(10137):2356–67. https://doi.org/10.1016/S0140-6736(18)30480-X.
122. Gevers-Montoro C, Provencher B, Descarreaux M, Ortega de Mues A, Piché M. Clinical effectiveness and efficacy of Chiropractic spinal manipulation for spine pain. Front Pain Res. 2021:2. https://doi.org/10.3389/fpain.2021.765921.
123. Wong JJ, Côté P, Sutton DA, et al. Clinical practice guidelines for the noninvasive management of low back pain: a systematic review by the Ontario Protocol for Traffic Injury Management (OPTIMa) Collaboration. Eur J Pain. 2017;21(2):201–16. https://doi.org/10.1002/ejp.931.
124. Nielsen SM, Tarp S, Christensen R, Bliddal H, Klokker L, Henriksen M. The risk associated with spinal manipulation: an overview of reviews. Syst Rev. 2017;6(1) https://doi.org/10.1186/s13643-017-0458-y.
125. Oliveira CB, Maher CG, Pinto RZ, et al. Clinical practice guidelines for the management of non-specific low back pain in primary care: an updated overview. Eur Spine J. 2018;27(11):2791–803. https://doi.org/10.1007/s00586-018-5673-2.
126. Côté P, Yu H, Shearer HM, et al. Non-pharmacological management of persistent headaches associated with neck pain: a clinical practice guideline from the Ontario protocol for traffic injury management (OPTIMa) collaboration. Eur J Pain. 2019;23(6):1051–70. https://doi.org/10.1002/ejp.1374.
127. Vanti C, Ferrari S, Morsillo F, Tosarelli D, Pillastrini P. Manual therapy for non-specific thoracic pain in adults: review of the literature. J Back Musculoskelet Rehabil. 2008;21(3):143–52. https://doi.org/10.3233/BMR-2008-21301.
128. National Health and Medical Research Council (Australia), Australian Acute Musculoskeletal Pain Guidelines Group. Evidence-based management of acute musculoskeletal pain. National Health and Medical Research Council; 2000.
129. Southerst D, Marchand AA, Côté P, et al. The effectiveness of noninvasive interventions for musculoskeletal thoracic spine and chest wall pain: a systematic review by the Ontario protocol for traffic injury management (OPTIMa) Collaboration. J Manip Physiol Ther. 2015;38(7):521–31. https://doi.org/10.1016/j.jmpt.2015.06.001.
130. Swait G, Finch R. What are the risks of manual treatment of the spine? A scoping review for clinicians. Chiropr Man Therap. 2017:25. https://doi.org/10.1186/s12998-017-0168-5.

131. Funabashi M, Pohlman KA, Goldsworthy R, et al. Beliefs, perceptions and practices of chiropractors and patients about mitigation strategies for benign adverse events after spinal manipulation therapy. Chiropr Man Therap. 2020;28(1) https://doi.org/10.1186/s12998-020-00336-3.
132. Whedon JM, Song Y, Mackenzie TA, Phillips RB, Lukovits TG, Lurie JD. Risk of stroke after chiropractic spinal manipulation in Medicare B beneficiaries aged 66 to 99 years with neck pain. J Manip Physiol Ther. 2015;38(2):93–101. https://doi.org/10.1016/j.jmpt.2014.12.001.
133. David Cassidy J, Boyle E, Côté P, et al. Risk of vertebrobasilar stroke and Chiropractic care results of a population-based case-control and case-crossover study. Vol 33.
134. Wang Y, Cheng W, Lian Y, et al. Characteristics and relative factors of headache caused by cervicocerebral artery dissection. J Neurol. 2019;266(2):298–305. https://doi.org/10.1007/s00415-018-9111-5.
135. Paanalahti K, Holm LW, Nordin M, Asker M, Lyander J, Skillgate E. Adverse events after manual therapy among patients seeking care for neck and/or back pain: a randomized controlled trial. BMC Musculoskelet Disord. 2014;15(1) https://doi.org/10.1186/1471-2474-15-77.
136. Gyer G, Michael J, Inklebarger J, Tedla JS. Spinal manipulation therapy: is it all about the brain? A current review of the neurophysiological effects of manipulation. J Integr Med. 2019;17(5):328–37. https://doi.org/10.1016/j.joim.2019.05.004.
137. Lee A, Abdulla S, Stuber K, et al. What can Chiropractors do to support athletic performance? a focused ethnographic qualitative study of sports Chiropractic clinical encounters; 2023.
138. Botelho MB, Alvarenga BAP, Molina N, Ribas M, Baptista AF. Spinal manipulative therapy and sports performance enhancement: a systematic review. J Manip Physiol Ther. 2017;40(7):535–43. https://doi.org/10.1016/j.jmpt.2017.03.014.
139. Miners AL, deGraauw C. A survey of Fellows in the College of Chiropractic Sports Sciences (Canada): their intervention practices and intended therapeutic outcomes when treating athletes. J Can Chiropr Assoc. 2010;54(4):282–92.
140. Miners AL. Chiropractic treatment and the enhancement of sport performance: a narrative literature review. J Can Chiropr Assoc. 2010;54(4):210–21.
141. Corso M, Mior SA, Batley S, et al. The effects of spinal manipulation on performance-related outcomes in healthy asymptomatic adult population: a systematic review of best evidence. Chiropr Man Therap. 2019;27(1) https://doi.org/10.1186/s12998-019-0246-y.
142. Corso M, Liang C, Tran S, Howitt S, Srbely J, Mior SA. The immediate effect of spinal manipulation on ball velocity and neuromuscular function during an instep kick in former varsity soccer players: a feasibility study; 2020. www.nsca.com.
143. Deutschmann KC, Jones AD, Korporaal CM. A non-randomised experimental feasibility study into the immediate effect of three different spinal manipulative protocols on kicking speed performance in soccer players. Chiropr Man Therap. 2015;23(1) https://doi.org/10.1186/s12998-014-0046-3.
144. Burns L, Weissensteiner JR, Cohen M. Lifestyles and mindsets of Olympic, Paralympic and world champions: is an integrated approach the key to elite performance? Br J Sports Med. 2019;53(13):818–24. https://doi.org/10.1136/bjsports-2018-099217.
145. Bangsbo J. Performance in sports - with specific emphasis on the effect of intensified training. Scand J Med Sci Sports. 2015;25:88–99. https://doi.org/10.1111/sms.12605.
146. Gliedt JA, Schneider MJ, Evans MW, King J, Eubanks JE. The biopsychosocial model and chiropractic: a commentary with recommendations for the chiropractic profession. Chiropr Man Therap. 2017;25(1) https://doi.org/10.1186/s12998-017-0147-x.
147. Hassebrock JD, Gulbrandsen MT, Asprey WL, Makovicka JL, Chhabra A. Knee ligament anatomy and biomechanics; 2020. www.sportsmedarthro.com.
148. Royer PJ, Kane EJ, Parks KE, et al. Fluoroscopic assessment of rotator cuff fatigue on glenohumeral arthrokinematics in shoulder impingement syndrome. J Shoulder Elb Surg. 2009;18(6):968–75. https://doi.org/10.1016/j.jse.2009.03.002.

149. Taruc-Uy R, Lynch SA. Diagnosis and treatment of osteoarthritis. Prim Care. 2013;40(4):821–36.
150. Taylor JB, Wright ES, Waxman JP, Schmitz RJ, Groves JD, Shultz SJ. Ankle Dorsiflexion affects hip and knee biomechanics during landing. Sports Health. 2022;14(3):328–35. https://doi.org/10.1177/19417381211019683.
151. Pierannunzii L. Pelvic posture and kinematics in femoroacetabular impingement: a systematic review. J Orthop Traumatol. 2017;18(3):187–96. https://doi.org/10.1007/s10195-016-0439-2.
152. Weerasekara I, Osmotherly P, Snodgrass S, Marquez J, de Zoete R, Rivett DA. Clinical benefits of joint mobilization on ankle sprains: a systematic review and meta-analysis. Arch Phys Med Rehabil. 2018;99(7):1395–1412.e5. https://doi.org/10.1016/j.apmr.2017.07.019.
153. Kim H, Moon S. Effect of joint mobilization in individuals with chronic ankle instability: a systematic review and meta-analysis. J Funct Morphol Kinesiol. 2022;7(3):66.
154. Beselga C, Neto F, Alburquerque-Sendín F, Hall T, Oliveira-Campelo N. Immediate effects of hip mobilization with movement in patients with hip osteoarthritis: a randomised controlled trial. Man Ther. 2016;22:80–5. https://doi.org/10.1016/j.math.2015.10.007.
155. Gomes J, Neto T, Vaz JR, Schoenfeld BJ, Freitas SR. Is there a relationship between back squat depth, ankle flexibility, and Achilles tendon stiffness? Sports Biomech. 2022;21(7):782–95. https://doi.org/10.1080/14763141.2019.1690569.
156. Fisher BE, Piraino A, Lee YY, et al. The effect of velocity of joint mobilization on corticospinal excitability in individuals with history of ankle sprain the effect of velocity of joint mobilization on corticospinal excitability.; 2016. www.jospt.org.
157. Malaya CA, Haworth J, Pohlman KA, Powell C, Smith DL. Impact of extremity manipulation on postural sway characteristics: a preliminary, randomized crossover study. J Manip Physiol Ther. 2020;43(5):457–68. https://doi.org/10.1016/j.jmpt.2019.02.014.
158. Schueren S, Hunger H, Pham H, Smith DL, Layne C, Malaya CA. Immediate effect of lower extremity joint manipulation on a lower extremity somatosensory illusion: a randomized, controlled crossover clinical pilot study. Front Hum Neurosci. 2022;8(16):1011997.
159. Heiser R, O'Brien VH, Schwartz DA. The use of joint mobilization to improve clinical outcomes in hand therapy: a systematic review of the literature. J Hand Ther. 2013;26(4):297–311. https://doi.org/10.1016/j.jht.2013.07.004.
160. Best available Evidence Care. https://chiropractic.on.ca/chiropractic-care/attain-care/chiropractic-treatments/.

Chapter 24
Manual Soft-Tissue Manipulation and Its Use in Recovery, Regeneration, and Adaptation to Exercise Stress

Seth Forman and Kika Mela

Introduction

- **The Story of the Patron**
- **A team approach**
- **Chapter objectives**
- **Understand the impact of MSTM on the body**

The Story of the Patron

An Athletic Trainer (ATC), Massage Therapist (LMT), and a Physical Therapist (PT) walk into a bar. One of the patrons noticed and declared he would give $1000 to whomever could get rid of his shoulder pain. A chiropractor and a Muscle Activation Techniques™ Specialist (MAT) were having lunch together and decided to join the discussion. The ATC suggested cupping and stim while the LMT said to release the spiral fascial line. The PT insisted on rotator cuff exercises as the chiropractor argued to first release the scalenes and then adjust the thoracic area. The MAT Specialist shook her head and recommended that it could be a muscle imbalance in his opposite hip or foot. After a heated debate, the group decided that in the best interest of the gentleman, they should work together and split the $1000. The ATC would start with stim, the LMT would release the fascia before the

S. Forman
Athletic Development Training, Inc., West Milford, NJ, USA

K. Mela (✉)
Mela Therapeutics, Inc., Aventura, FL, USA
e-mail: kika.mela@comcast.net

N. C. Apostolopoulos et al. (eds.), *Fundamentals of Recovery, Regeneration, and Adaptation to Exercise Stress: An Integrated Approach*,
https://doi.org/10.1007/978-3-031-44270-4_24

chiropractor did the adjustment. The MAT Specialist would follow by activating the necessary muscle groups and the PT would finish off the session with corrective exercises. Happy with their decision, the group looked for the patron to share. They searched the bar, and he was nowhere to be found. When they approached the bartender to see where the patron had gone, the bartender told them he left about 30 minutes ago.

"What about the $1000?" the chiropractor asked.

"Oh, he gave it to me after he finished his drinks." the bartender said.

"Why would he do that? We hadn't provided the best treatment plan for him yet" the PT insisted.

The bartender smiled as he explained. "He said he didn't have any more pain in his shoulder thanks to me and left".

"Well, what did you do to him?" the LMT demanded.

"I gave him three shots of Tequila. He won't be feeling anything for the rest of the day!"

There are several points to this story. The first is that when it comes to dealing with the body, there are as many solutions as there are practitioners. The second is that the best plan is the one where everyone works together for the betterment of the individual. The third is that the individual often cares more about the results than the method, regardless of whether it is long lasting or healthy.

Professor Sidney Dekker, PhD, an expert in complexity and systems thinking, says, "You cannot reduce a complex system to one of its parts, because if you do, you will tragically and dramatically oversimplify things." By taking a certain perspective, you obscure some things and make others highly visible. The outcome of a system is never produced by one single part; rather, the outcome is produced via the interactions of all the parts [1].

> *...everyone needs habits of mind that allow them to dance across disciplines* [2].

A Team Approach

The Story of The Patron in the bar identifies numerous practitioners trying to determine the best treatment plan for the patron. The approach offered by each specialist is different, yet they could all be successful. How can this be? In practice, treating a client is complex with many paths to achieving the same goal. Using an integrative or multi-disciplinary approach provides more angles to attack issues from. You can begin to *think critically* about complexity, resulting in deeper questions about seemingly unrelated events that could have a significant impact on appropriate treatment approaches and training outcomes.

> *I had a client who had seen numerous doctors concerning a pain in her right mid-thoracic region. The treatment she received was not helping and it was recommended she try exercise. After 8-weeks she reported a subjective decrease in pain by 50% and an improved*

quality of sleep. Unfortunately, we hit a wall in her progress. I convinced her to see a doctor in our network who had helped other clients. After some bloodwork, they believed she had an undiagnosed tick born infection that manifested itself in that tissue, so they treated her accordingly. Amazingly, the pain went away! Having this doctor in our network provided a perspective we would have never considered based on our backgrounds and provided insight into each individual client so we could maximize their recovery and performance.

–Seth Forman, M.Ed., LMT, CSCS

- Could research determine the ideal combination of treatment options?
- Does any treatment option reduce or enhance the efficacy of another when combined?

Chapter Objectives

The purpose of this chapter is to discuss our interpretation of the research and its application to improve treatment approaches using manual soft-tissue manipulation (MSTM). We will explore how MSTM stimulates changes in the body, with the purpose of enhancing recovery, regeneration, and adaptation to exercise. We have found that understanding the tissues and changes that can be made provides practitioners with an understanding of what techniques can help promote recovery, regeneration, and adaptation to improve movement for clients. For athletes, the presence of any limitation to movement, or recovery time from injury are of vital importance, especially during their season. Long recovery times do not align with game schedules, and methods that can speed up the recovery process and result in mid-to-long-term resolution of the issue(s) are needed.

Understand the Impact of MSTM on the Body

Most of the research reviewed either investigated differences between two methods, or efficacy of a specific method. From the research perspective we will discuss how MSTM affects the body through traditional and non-traditional techniques. Additional perspective will be offered from our experiences and those of the professionals we interviewed. It is important to note that the body of existing literature would take multiple lifetimes to review, apply, and ultimately understand. This journey of research and practice is never ending, it requires a relentless pursuit to push our knowledge and understanding forward. In the words of Neuroscientist Antonio Damasio "*the road to discovery is twisted, to say the least* [3]. Those of us who study, and practice certainly know this to be true. Ask different practitioners why they use MSTM, and you may receive responses that range from releasing tight tissue to improving range of motion (ROM) to reducing pain.

One of the current topics in manual therapy is the role of the fascial system in pain and performance. This chapter also summarizes some of the new insights into the adaptive nature of the fascial system and how it can help us understand the effects of MSTM techniques. There is still much that is unknown about the central nervous system (CNS), mechanoreceptors, nociceptors, and potential changes within tissue (muscle and connective) in response to MSTM.

Manual Soft-Tissue Manipulation (MSTM)

- **Definition of MSTM**
- **Why MSTM and How Does It Work?**
- **How is it used in Recovery, Regeneration, and the Adaptation to Exercise**

What Is MSTM?

Since humans have had hands, it has been instinctive to rub on the body to relieve pain, effect ROM, and/or improve function. Manipulating the soft tissues of the body is commonly used by many members of an athlete's medical team. The team can include massage therapists, physical therapists, ATCs, MAT Specialists, chiropractors, some coaches, and even administered by the individual to themselves. Techniques applied vary by practitioner and the philosophy on how they believe the body responds. Manual Soft Tissue Manipulation (MSTM) is the application of force to the soft tissues of a body with the intention of improving tissue extensibility, induce relaxation, modulate pain, and reduce inflammation [4]. Massage can be defined as "*a set of manipulations and/or maneuvers performed with the hands from a specialized therapist and applied to the human body with preventive, hygienic, therapeutic, rehabilitating, and psychological objectives in a methodical, orderly, and reasoned manner*" [5]. Excluding joint manipulation techniques, massage therapy can be used to describe MSTM interchangeably, though some higher-level therapists may protest the term. "More complex definitions can be found for different purposes (e.g. muscle, sports performance, rehabilitation), and various types of massage therapies are available" [6].

Why MSTM and How Does It Work?

There are numerous reasons why a person might apply soft tissue work. Most often, MSTM is applied to reduce pain, improve ROM, and enhance function or performance [6–18]. Condensed down, it's basically to help the tissue effectively contract and manage loads. That's it! If it hurts to lift your arm overhead, then the body is unable to appropriately create or distribute the forces necessary to do it effectively, fully, and/or pain free. MSTM is applied as a solution to achieve this by releasing tissue, reducing inflammation, restructuring connective tissue with the purpose of improving ROM, improving function, and thus reducing pain.

What happens when a therapist feels a "tissue release" while applying MSTM? Has the structure of the tissue changed? Is the change permanent? Research has shown that to make a permanent change to a structure such as the iliotibial (IT) band in a short amount of time, a minimum force of 60 kg creating a 3–8% elongation of tissue is needed [7]. Unfortunately, this is accompanied by tearing of the tissue resulting in inflammation. To prevent tearing and inflammation, a lesser force that only stretches the tissue 1–1.5% is recommended but needs to be applied for ONE HOUR to be effective [18]. Spending an hour on a dense area of connective tissue such as the IT band or thoracolumbar fascia isn't practical. If the lighter loading is within the elastic limits of the tissue, it may even result in a gradual return to the original resting length [7]. So how are the reported changes in tonus and increases in ROM achieved? To understand what could be happening (at a basic level,) we need to briefly explore the sensory system and touch. It should also be noted that more research is needed to understand how structures such as the superficial fascia, and muscle fascia respond to MSTM.

Tactile stimulation of the tissues creates a cascade of events relayed throughout the CNS and back to the tissues involved [13]. The underlying sensory receptors embedded in the tissue detect changes like pressure, pain, temperature, etc. so that the body can respond to it appropriately. These receptors include Pacini, Golgi organs, Ruffini endings, muscle spindles, and interstitial muscle receptors [18]. Typically, muscle nerves have close to 3 times the number of sensory fibers than motor fibers. The majority belong to the smaller diameter types III and IV interstitial muscle receptors which act mainly as mechanoreceptors, and which are also found profusely in the fascia. There are over 250 million nerve endings found in the fascia which makes it the largest sensory organ in the body [19–21]. Schleip summarized four feedback loops to explain the palpable tissue response or "release" following manual tissue manipulation [18]. These feedback loops overlap and can be viewed in Fig. 24.1. All the feedback loops are associated with the initial MSTM which stimulates the underlying mechanoreceptors and in turn stimulate the nervous system. After receptor stimulation, each loop takes a slightly different route which is summarized below [18].

The first loop is the Intrafascial Circulation Loop. Here the receptors respond with stimulation of the autonomic nervous system (ANS) resulting in the alteration of fluid dynamics. The targeted local tissue's viscosity and blood supply is thought to be affected enough to possibly be perceived by the therapist applying the work. The second is the Hypothalamus-Loop, here stimulation of interstitial mechanoreceptors with deep pressure triggers the vagus nerve to increase vagal tone which results in a palpable relaxation response throughout the body. The third loop is the Fascial Contraction Loop, this loop uses the premise that the fascia holds smooth muscle cells which may be involved in contraction of the fascia independent and

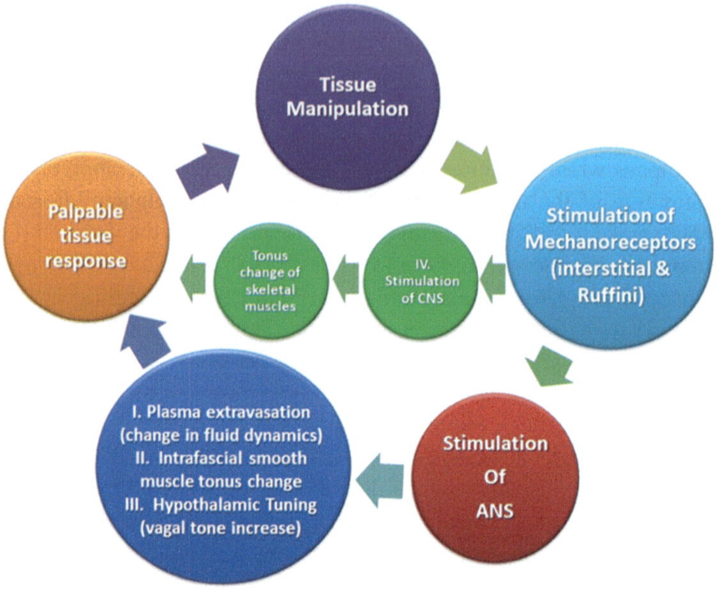

Fig. 24.1 The Four Feedback Loops of how tissues respond to MSTM. I: Intrafascial Circulation Loop, II: Hypothalamus Loop, III: Fascial Contraction Loop, and IV the CNS Loop

perhaps complementary to the muscular system. It is thought that the regulation of the tone of the fascia is achieved by the sympathetic nervous system along with CO2 levels and can be manipulated by manual stimulation of the intrafascial mechanoreceptors for short term fascial plasticity. Stimulation of the mechanoreceptors triggers the ANS, which then alters the tone of intrafascial smooth muscle and possibly elicits a change in tone of the tissue being palpated. The last loop is the Central Nervous System Loop, which states that the stimulation of the intrafascial mechanoreceptors affect proprioceptive input to the CNS and can alter the firing rate of the associated skeletal muscle motor units. This lowered firing rate and resulting change in tonus may be what the practitioner is feeling as a "release" [18].

MSTM in Recovery, Regeneration, and Adaptation to Exercise

Massage or MSTM has been a method assumed to enhance the recovery and regeneration of muscles after intense exercise though research is conflicted on its effectiveness [9, 11]. As put forth at the beginning of this book, the emphasis of recovery is the restoration of the body from any physical, mental, or emotional catalyst which enables the athlete to train at the required intensity or complete the required load at the next training session. *"Recovery from exercise implies a progressive loss of functional capacity that can be reversed"* [22]. Regeneration refers to the repair of the tissue at the cellular level, specifically the biochemical and cellular responses involved with such mechanisms as inflammation, stem cells, and fibroblast

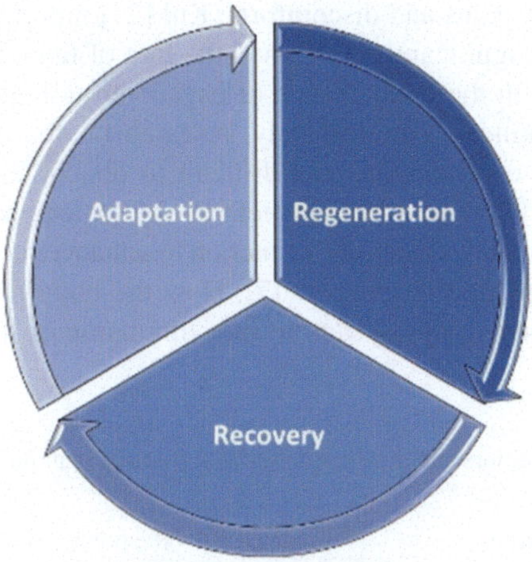

Fig. 24.2 The constant interplay between Recovery, Regeneration, and Adaptation

formation etc. Adaptation is the ability of these tissues to respond to an internal or external stimulus. Recovery, regeneration, and adaptation to exercise within the body is closely linked and difficult to separate as each require the plasticity of all tissues involved (see Fig. 24.2) [23].

Competition within sports takes a physical, emotional, and mental toll on an athlete and may be helped through MSTM [9, 11, 24]. Mental as well as physical fatigue limits high level performance and curtails the daily activities of those with a limited capacity [12]. Enhanced recovery and regeneration are often goals to improve performance. In his experience, renowned physical therapist, athletic trainer and strength and conditioning coach Dr. Charlie Weingroff feels that (manual) therapy is best used to accelerate acute recovery so you can train harder the next day [25]. To him, therapy can be used to mitigate the impact of acidosis, hypoxia, pain, and thermal changes. If the therapy can aid in putting the tissues in a more adaptable state the next day, then the athlete can get more intense days, weeks, etc. with less repercussions. This idea is also supported and put into practice by strength and conditioning coach Pete Bommarito of Bommarito Performance. Bommarito feels that soft tissue work is a critical component in his methodology for getting the most out of his athletes [26].

> *Almost any touch can have an effect on Oxytocin, Serotonin, and a decrease in creatine kinase which is associated with pain. ...you can spread peanut butter on someone, and it can have an effect. There is no such thing as a placebo, you just measured the wrong thing...*
>
> –Dr. Charlie Weingroff

The addition of MSTM is assumed to act as a supporter in motion therapy and special therapeutic exercises to control skeletal-muscle pains [27]. Increasing range of motion and elimination of muscular spasms is considered to be effective in decreasing temporary pains and discomforts. Kul [27] reported that the superficial stroking of massage significantly decreased the rate of tenderness and the pain of patients and resulted in the maintenance of psychophysiological parameters up to three days after treatment. Unfortunately, research on the effects of soft tissue manipulation can be contradictory and difficult to obtain. In practice, soft tissue manipulation is used to create biomechanical, physiological, neurological, and psychophysiological benefits to the body's function to enhance recovery, rehabilitation, and compliment adaptation to exercise [7]. How the body responds to touch is a direct response by the nervous system to the stimulation of the sensory receptors. (See Table 24.1).

Table 24.1 Mechanoreceptors and their responses to soft-tissue manipulation [18, 21]

Receptor	Found In	Stimulated By	Responds With	Soft Tissue Manipulation
Golgi Receptors G. End Organs G. Tendon Organs	Dense Connective Tissue Ligaments Myotendinous junctions Joint capsules	GEOs to Slow active stretch GTOs to muscular contractions	Reduced tone in muscle fibers Feedback during active contractions	Soft tissue manipulation while tissue is contracting Yoga Slow active stretching
Pacini & Paciniform Type II	Tendinous sites Deep spinal ligaments Muscular fasciae Aponeurosis Joint capsules Medial & Lateral knee joint	Rapid pressure Vibrations	Proprioception for control of movement	Tapotement Cross fiber friction
Ruffini Type II	Outer layer of joint capsules Dura mater Ligaments of peripheral joints Deep fascia of the hand Ant/post ligamentous & capsular structures of the knee	Long term pressure Tangential forces Lateral stretch	Lowering of sympathetic nervous system activity	Sustained myofascial tension/pressure Craniosacral therapies*
Interstitial Muscle Receptors Type III & IV	Dense connective tissues	Light touch Tension Pressure Pain Rapid as well as sustained pressure	CNS regulation of blood flow Changes heart rate, blood pressure, respiration, etc. Stimulation of Type III can ↓↑ blood pressure Plasma extravasation	Light touch Effleurage Petrissage Myofascial tension/pressure

Biomechanically, it is thought to reduce tissue adhesions, increase muscular compliance, increase ROM at a joint, and reduce both active and passive tissue resistance [13, 28]. Stecco and Schliep have found that it is possible to alter collagen expression via applying mechanical pressure on tissues [18, 21].

Physiologically it is thought to increase the temperature and blood flow of the skin or affected tissues [13]. It is unknown if an increase in temperature or blood flow will have a positive impact on performance, but it is surmised that increased circulation will aid in the recovery and rehabilitation process. Improved waste removal and enhanced delivery of proteins and other nutrients needed for muscle repair can be a factor in reducing swelling, reducing pain, decreasing ischemia, and thus reducing soreness associated with delayed onset muscle soreness (DOMS) [9]. Using laser Doppler flowmetry (LDF) and reflection photoplethysmography (PPG), Monteiro Rodrigues et al., observed that the application of effleurage with compression for 10 minutes to one lower limb increased both the perfusion and skin microcirculation bilaterally regardless of stroke direction [5].

Psychophysiological changes may occur with reduction in anxiety and enhanced relaxation. Touch has been found to stimulate the Parasympathetic Nervous System (PNS) which may increase relaxation hormones and reduce stress hormones [6, 9, 12, 16, 18]. Weinberg et al. [24] inferred that massage could have a positive effect on athletic performance by reducing tension, anger, fatigue, depression, anxiety, and confusion. He found a positive relationship between massage and profile of mood states (POMS) and between massage and the Adjective Checklist POMS subscales of tension, depression, anger, fatigue, and confusion. Mental or emotional recovery may also be influenced by soft tissue work [12, 24].

The neurological effect of MSTM is multi-layered and complex. An application of force creates a neurological response with a frequently validated response being the engagement of the ANS [18]. Increased parasympathetic activity is most often noted with the reduction of stress hormones, increase in endorphins, reduction of blood pressure and reduced heart rate [6].

One desired aspect of recovery is the reduction of DOMS post exercise. DOMS is the painful condition created after heavily intense, exhaustive, or unaccustomed exercise [9, 12, 29]. DOMS occurs between 24 and 72 hours post exercise and can result in chronic pain, diminished muscle function, soreness, loss of strength, reduced ROM, increased muscle tightness, increased metabolic rate, and altered athletic performance [9]. The painful condition could be a result of the exercise-induced damage to the sarcolemma and the muscle cell membrane releasing biochemical markers such as lactate dehydrogenase (LDH), creatine kinase (CK), aspartate aminotransferase (AST), and myoglobin (Mb) [14]. Creatine kinase is a particularly large molecule (80,000 Da) which cannot penetrate back into the bloodstream following muscular damage. Instead, it must pass from the intercellular fluid into the lymphatic system where it is then transferred to general circulation via the thoracic lymph nodes [14]. Serum CK levels tend to increase 24 hours following exercise and may persist for more than 4 days [14, 30, 31]. Recovering from this muscle damage is important and MSTM is one of the ways that is used to accelerate recovery.

Current research is changing the perspective of the causation of DOMS and therefore how we should look at prevention and reduction of it using MSTM. Wilke outlined several of these changes in their 2021 article which is summarized here [29].

- Lactic acid levels as a source of pain were disproved many years ago as the lactate levels return to baseline shortly after exercise is finished
- CK levels were not shown to correlate with the pain of DOMS upon palpation or movement.
- Muscle damage is seemingly less of a cause as activation of the mechanosensitive type-III afferents due to structural lesions is conceivable for initial feelings of pain but not for the delayed or lingering sensations of pain
- Free radicals may not contribute to causation as production increases 24–48 hours after maximum DOMS sensations occur
- Intramuscular swelling from the gradual rise of inflammation can exert pressure on the type-III afferents causing the delayed and sustained activation of them

While the mechanisms related to DOMS are still controversial there is research showing damage in the deep fascia potentially playing a significant role in the pain associated with DOMS [29]. Connective tissues are rich in mechanoreceptors, and nociceptors compared with muscle tissue, it is plausible that our experiential association of muscle pain is more connective tissue oriented.

> *Although theories explaining the pathogenesis of DOMS have long focused on the skeletal muscle, there is accumulating evidence suggesting a prominent role of the collagenous connective tissue. Based on the available literature, strain forces associated with eccentric contraction may cause micro-ruptures and inflammation of the deep fascia. As experimental research clearly demonstrates that fascia is more pain-sensitive than muscle following chemical, thermal, electrical, and mechanical irritation, we propose that delayed onset soft tissue stiffness (DOSS) is a more precise descriptor of the post-exercise phenomenon...*
>
> –Jan Wilke and Michael Behringer [29]

For recovery purposes, it has been suggested that in some instances petrissage, effleurage, tapotement, and cross fiber friction were able to reduce the effects of DOMS [6, 14]. Bakar Y, Coknaz H, Karli U, et al. found that Manual Lymphatic Drainage (MLD), a massage technique that follows the flow of the lymphatic pathways, applied immediately after exercise reduced connective tissue or muscle damage [14]. It was reported that when MLD was applied for 30 minutes there was a faster fall in lactate, CK, myoglobin, muscular enzymes of LDH, and a continued decrease in CK 24-hours post exercise. This study suggests that MLD contributed to alleviation of muscular damage and enhanced the regenerative process of muscle cells after submaximal exercise.

If You Can't Contract, You Can't Adapt…

Adaptation to exercise requires not only the plasticity of the involved tissues, but the proper transference of forces generated by muscular contractions through healthy collagen structures [23]. This point was captured well during our interview with Eric Owens (co-founder of Delos Therapy) when he said, "*If you can't contract, you can't adapt.*" [32] This concept also seems to align well with the limited research on the role of massage in sports performance. In a 2008 literature review on the subject, most articles highlighting successful outcomes had applied a combination of interventions, including postural education, strengthening exercises, and stretching [11]. Recent research has followed the same model by including additional exercise programs and treatment modalities such as heat or cryotherapy in conjunction with MSTM treatments [16, 33]. There may be a correlation between MSTM and temporary restoration of muscle and connective tissue function, or reduction of perceived pain during movement [34, 35]. During this temporary reduction in pain or enhanced function lies an opportunity to improve function of the tissue via muscular contraction, namely isometric and eccentric actions depending upon the pathology being treated [35–39].

> *We know from our clinical experience fibrotic muscle, full of gristle, is stiff, lacks ROM and doesn't adapt to exercise very well. This fibrotic change, simply due to the aging process, requires multi-directional pressure to separate fascicles that are stuck together three-dimensionally. The result is reversing the aging process, leading to pain free, fluid movement and tissue that is very adaptable to a variety of loads and exercise.*
>
> –Eric Owens, co-founder of Delos Therapy

If most of the reasons for MSTM are about improving functionality and reducing discomfort, is there a best way to achieve this? Unfortunately, there is no concise answer. As mentioned previously, any touch can elicit a neurological response. Weingroff mentioned that if the application of short, intermittent MSTM induced a change in function, then your body's tissue could already do that [25]. The body had not created a long-term physical change and therefore there was not a structural but a neurological alteration to the tissue with the manual intervention. This is positive in that if the function has been enhanced and pain lessened, then the application of force is possibly more tolerable, and longer lasting solutions may be probable, and adaptation enhanced.

In practice, Dr. Weingroff has found that if the assessment shows normal joint function, then there may be a motor function issue. When the joints can all do what they want, any manual therapy can be used to minimize the shock and resistance phase to the adaptation process [25]. He uses manual therapy to accelerate adaptation and uses the training process (applying force) for changing tissue. Therapy is used to accelerate acute recovery so you can train harder the next day.

So, what caused the body to make the change in the first place? Why did your body feel this limitation in range of motion or altered pattern of movement was the best option? What would be the long-term repercussions of compensation for injury? Most importantly, how can we assess and then correct the body's solution with lasting results using MSTM? Unfortunately, the research is not consistent enough to answer these questions satisfactorily. The experts interviewed for this chapter were unanimous in their opinion that for long term corrections, exercise or targeted force needed to go through the tissues for adaptations to occur [19, 25, 32, 41]. McGill found in his research that mobility without stability can lead to vulnerability and injury [42]. MSTM is used to enhance the ability of tissue to handle force, but what happens if the muscle can't contract efficiently? What happens if motor unit recruitment or intra and/or intermuscular coordination is inefficient? Will this affect the force distribution and therefore appropriate remodeling of the tissue? What can be done?

When a muscle exerts a force, and that force is transferred through connective tissue to bone to create movement around an axis of rotation, it doesn't work in isolation [43]. To perform a particular action, the CNS must not only activate the correct muscles, but it must also orchestrate a tuning between muscles to prevent unwanted motion [43]. Within each involved muscle during a motion, there is intramuscular coordination of the fibers that are affected by speed, direction, and magnitude of the force applied [42]. There is also intermuscular coordination between the muscle groups to where each muscle may have the same goal of joint motion, but between them may be different functions [42]. All the muscles activated during a motion can simultaneously have multiple roles ranging from stabilizing, balancing, protecting, preventing motion, or even stabilizing or preventing motion at other joints [42]. Most muscles can create movement in multiple planes of motion so more than one muscle or group of muscles must be involved to prevent unwanted motion. During hip flexion, the group of muscles that flex the hip can also create hip internal rotation, adduction, abduction, and hip external rotation [44]. Understanding this orchestrated engagement of coactivation between muscles can help guide treatment approaches [45].

The endless possibility of muscular recruitment also means that similar motion patterns can be achieved with different motor patterns [46]. According to Kandel et al., this is how compensation patterns begin [43]. If one muscle isn't up to doing the action, then others could take over and help [47]. While it is great that your body will find a way to do an action or an exercise, it is perhaps not so great when it's consistently repeated over time. The muscles that helped might not be the best choice or designed to handle that action for long periods of time or under heavy loads. During videofluoroscopy of a dead lift, McGill discovered that a minor motor control error causing the slight rotation of a spinal joint put undue stress on tissues and resulted in an injury [45]. Poor motor control reducing muscular reaction-time or causing deficient co-activation is a factor of decreased joint stability [48].

Increase in speed of a motion also creates undesired motion of other body parts which is another issue that the CNS must control. This can be seen in running where at the beginning of the swing phase the hip flexes rapidly. Rapid hip flexion (acceleration of the thigh in a forward direction) will create rapid passive knee flexion which then needs to be opposed by knee extensors [49]. During the latter part of the phase, the tibia swings forward from momentum so the hamstrings activate to decelerate the motion and prevent hyperextension of the knee prior to ground contact [49, 43]. It could be concluded that one characteristic of skilled performance is the ability to accomplish a task with the most efficient and effective activation of muscles that span associated joints, which can be referred to as tuning (timing of activation between muscles).

Improved contraction or coordination of muscles or muscle groups is not always what is first thought of when deciding to apply MSTM. As mentioned previously, releasing tight tissues, or creating fascial changes while improving mood are common goals when working with soft tissue. However, there are other methods which aim to stimulate the tissue or to enhance contraction options and thus improve function. Muscle Activation Techniques™ (MAT), Muscle Energy Techniques (MET), Muscle Firing Patterns (MFP), and NeuroKinetic Therapy™ (NKT) are a few specific techniques designed to enhance the contraction of muscles, muscle groups, or movement patterns [10, 33, 47, 50–52]. These types of applications are discussed later in the chapter.

- Why does the body feel a limitation in range of motion or altered pattern of movement and what is the best option for improvement?
- What would be the long-term repercussions of compensation for injury and how do we know?
- How can we assess and then correct the body's solution with lasting results using MSTM?
- What happens if motor unit recruitment or intra and/or intermuscular coordination is inefficient?
- Will this affect the force distribution and therefore appropriate remodeling of the tissue? What can be done?

How Is MSTM Applied?

- **Techniques used**
- **The 3Ds**
- **Is deeper work better?**

Manual soft tissue manipulation or massage is done with commonly used hand to skin techniques that apply force into the tissue for the purpose of tissue relaxation, stimulation, or improved muscular contraction. These techniques include classic massage strokes such as effleurage, petrissage, friction, and tapotement, myofascial release techniques, various forms of dynamic soft tissue mobilization (DSTM), and tissue contraction methods [8, 11].

Effleurage consists of long, gliding strokes applied with gentle pressure through open palms or with the thumbs. It is used to apply lotion, transition between strokes, and can have a relaxation effect [6, 8, 11].

Petrissage uses the palms, fingers, knuckles, and/or forearms to lift and spread the tissue with a kneading or wringing effect [8]. This squeezing of the tissues is applied to mobilize deep muscle and subcutaneous tissue, assist venous return, break up adhesions, and release tension in the tissues [6, 8].

Friction is comprised of short, quick circular or linear movements of the palms, fingers, or thumbs. For broader application the palms can be used but friction is mostly applied more precisely to the tendons or ligaments to reduce muscle spasms, break up adhesions, and draw heat to the tissues [8].

Tapotement is rapid and rhythmic striking of the tissues using fingertips or the edge of the hands [8]. It is used to stimulate the tissues through the direct mechanical force or by a reflex action [6, 8]. Extremely rapid tapotement with very light pressure and short burst of duration is considered vibration and is thought to improve circulation, relieve muscle tension, and decrease stress [8].

Dynamic Soft Tissue Mobilization (DSTM) consists of the combination of classic massage strokes with passive or active lengthening of the tissue. For example, effleurage is applied both longitudinally and across the muscle fibers as the tissue is elongated passively, actively or while the patient eccentrically contracts the tissue [11].

These classic massage strokes vary in how the practitioner engages the tissues: how much pressure is applied, what direction the force is moving, and how long the stroke lasts.

> *You will never know exactly what it is that made changes: Is it the skin? Is it the nerves? Is it the muscles? The tactile therapy of Kinesio tape? Even a subjective response counts if they feel different.*
>
> –Dr. Charlie Weingroff

The 3 D's: Depth, Direction, and Duration

The nervous system response to MSTM is dependent on the Depth, Direction, and Duration of force applied (Fig. 24.3). Our review of the literature has found little guidance in the precise application of the 3Ds to create a desired effect as it is difficult to measure. We have not come across adequate and consistent definitions quantifying the 3Ds in soft-tissue manipulation. Therefore, for the purpose of this chapter we will define them as follows.

Depth: How physically deep into the underlying tissues the touch reaches. Superficial applies to the epidermis down to the deep fascia. Deep fascia is a layer of fascia located deep to the fascia profundus and superficial to the epimysium. Deep applies from the epidermis to the perimysium. However, based on

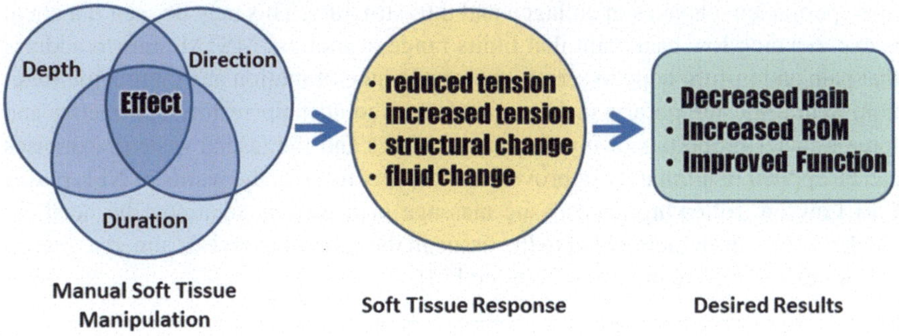

Fig. 24.3 All MSTM techniques are a combination of Direction, Depth, and Duration which can stimulate mechanoreceptors and/or structural changes

ultrasound imaging of the application of force into tissue, even a very light touch can be seen to reach all the way to the perimysium [53].

Duration: how long the applied force lasts in seconds or minutes. Touch can be intermittent or sustained [25].

Direction: the line of force that is applied into the tissues

With the fiber: in the same direction as the muscle or fascial fibers run

Against the fiber: perpendicular to the muscle or fascial fibers direction

Oblique to the fibers: any direction between parallel and perpendicular to the fibers

In practice, there are infinite variations on how to combine the 3Ds of MSTM with the most important considerations being the goal and is it appropriate. Too much time on the tissue or too much force could also have a negative effect on the tissue and be inappropriate. For depth, one therapist's moderate pressure could be considered extremely heavy to another, or the pressure too light to be effective for the client. Time on the tissue could last milliseconds or minutes depending on whether stimulation resulting in activation or inhibition is desired. Most applications of massage strokes or manual therapy on a particular tissue last between a few seconds to 1.5 minutes [18]. Rarely is a practitioner encouraged to linger longer on the tissue continuously. As mentioned earlier, this amount of time is too short to make a structural change to dense connective tissue [18]. The direction and/or force application will vary depending on how you wish to affect the structure or fluid dynamics. A connective tissue loaded rapidly will act stiffly and deform less than a tissue that is loaded slowly [7].

Dr. Charlie Weingroff has found that to make permanent changes to collagen realignment, application of force needs to be unidirectional and long duration. He feels that if pressure has a latent period where you come off the tissue, it's main value would be as a sensorimotor experience. Intermittent application of force can change tension of contractile tissue, affect the ANS to engage the PNS or stimulate the central nervous system (CNS). He described how pin and stretch (DSTM) may result in an increase in ROM, but without changing tissue structure. If you do something immediately afterwards such as a muscular contraction, you may start to make more permanent changes in collagen and thus structure. This may be seen if a client is experiencing low back pain that limits range of motion. MSTM aims to address that pain and in-turn hope to see an improved range of motion and contractile function. While the temporary suspension of pain could help restore contractile and force transfer properties of myofascial structures and the fascial system, exercises can be applied to ultimately improve function over time. In the words of NFL player Carl Lawson, following deep tissue massage or a Rolfing treatment he needs to "lock it in" by introducing load to the tissue in the form of exercises aimed to recruit motor units that were previously altered [54].

- What would be the ideal combination of depth, direction, and duration to achieve a particular result in the tissues?
- What is the ideal amount of treatment for a particular tissue?
- Is there a point where it no longer helps but hinders?
  - If so, what is that threshold and what occurs in the tissues?

Is Deeper Work Better?

Based on our interviews, those who have been doing manual therapy the longest seem to have all come to the same conclusion, despite extremely different routes. They have found that applying aggressive pressure is not necessary for creating changes or for the recipient to experience the sensation of "going deep". Those interviewed have found that "going deep" safely and effectively is more about specificity and not about the amount of force applied [41]. They are also in agreement that any long-lasting changes to the tissues are best achieved with the combined effort of MSTM and application of force through the tissue with muscular contractions.

In our personal experience, "deeper is better" has been a mantra for many massage therapists and manual practitioners. We have heard stories of how much pain they have inflicted on a client in the name of therapy or how proud patients are of the suffering they can endure at the hands of their therapist, all in the name of healing. In our interview with Dr. Robert Schleip, he said that in the past, he and his colleagues encouraged deeper work to effect fascial change but has recently acknowledged that current research is suggesting a lighter approach may be more beneficial. His group is finding that a lighter touch is more effective in removing the fluid stuck around the epimysium which manifests as localized tissue pain and may be more beneficial for the reduction of DOMS [19]. Mark Perkins states that the mechanoreceptors responsible for neurologically reducing tension in the tissue are more superficial, so bone crushing pressure is not necessary [41].

According to Perkins, too deep and too painful can be more stimulatory than inhibitory [41]. He feels that since the mechanoreceptors are more superficial, then it is not necessary to push deeper into the tissue to create a response. If painful, the depth can cause the body to resist the invasion and push back by increasing tension, commonly referred to as "guarding". He has found in practice that deeper work can be harmful by overstimulating the system or by creating too much change in the function. He recounted an experience with a client where the deep work she desired overstimulated her system and resulted in her being sick for two days! Another time he had a baseball pitcher with a stiff throwing arm and spent a lot of time releasing the subscapularis and surrounding tissues. The next day, the pitcher was unable to throw well and refused to have that kind of work done at that intensity ever again. With these experiences as examples, Perkins feels that there is a level of adaptation that can be exceeded, and a good practitioner finds the line between too much and not enough [41].

As a high-performance consultant and having been involved in three World Cup Soccer teams, Dawn Scott has also found that in practice working too deep or aggressively in the tissue can have a negative effect [55]. Trained as a massage therapist herself, she is a firm believer in massage but prefers flushing of the tissues with a lighter touch to promote recovery and to improve sleep. She recounted a time where a massage therapist worked too aggressively on an athlete and contributed to the worsening of an injury.

In the massage industry, deeper pressure is often considered the best way to "release" the tissue and restore it to a preferred resting length. This thought process has permeated the industry and is rarely challenged. The deeper work can reduce tension within a tissue due to engagement of the ANS, but at what cost? Is the tissue now too inhibited to function and stabilize? Is the joint stability now compromised? Is the client going to have an inflammatory response that causes pain the next day? Was the deep work necessary?

- If deep work is inhibitory, how long does the inhibition last?
- Does the inhibition enhance or reduce performance when loading the tissues?
- Could this inhibition contribute to injury?

Other Manual Soft-Tissue Manipulation Techniques

- **Improving Contraction**
- **Changing Tissue Structure**
  - **Myofascial release therapy**
  - **Fascial manipulation**

We discussed the basic palpation techniques used to elicit a neurological response of the tissues for release, increased ROM, and for a relaxation response. The results of manual technique combinations rely on the skillful application of forces (3Ds) to specific anatomical locations on the body, as well as the pathology at hand [7]. There are also specific styles or techniques that have been developed for assessing and addressing the problem to create a desired result outside of just relaxation of the tissue. In practice there are various techniques used with the goal of improving muscular contraction or changing collagenous structure.

Improving Muscular Contraction

Muscle Activation Techniques™ (MAT) is a method developed by Greg Roskopf which addresses the muscle system's role in chronic pain and injury [50, 56]. During our interview, he spoke in depth about the science and methodology of MAT. For him, MAT recognizes that anytime there is stress, trauma or overuse, the resultant inflammation alters the communication between the nervous system and the muscle system. The core philosophy of MAT is that muscle tightness is secondary to muscle weakness. From an MAT perspective, when muscles can't contract efficiently, they can't shorten effectively and therefore the opposing muscles are not inhibited to allow the motion and thus may present as being "tight". MAT looks at what's weak vs what's tight which is contrary to most soft tissue modalities.

The conventional thought process of most modalities is to find the limitations in range of motion and address or "release" the tight tissues preventing the motion. MAT looks for limitations in ROM as an ***indicator*** of neuromuscular inefficiencies or instability with a lack of communication between the nervous system and the muscle system (interrupted alpha-gamma coactivation). Once the most limited pain-free motion is determined, manual muscle testing is used to test the appropriate muscles in shortened ranges to determine if the muscles that are supposed to contract can contract on demand. Muscles that test weak are stimulated via specific stimulatory palpation or isometric contractions in the shortened range. Following the stimulation, the muscles are retested for verification of activation and ROM is reassessed. Appropriate follow up exercises are then recommended to continue the enhancement of the muscle's communication between the nervous system and the muscle system. A central belief of MAT is that during an exercise or movement, the engaged muscles will get stronger, but the weak muscles will become detrained and possibly weaker. Activation of the weaker muscles will not only include them in the desired movements for training, but also reduce tightness, joint instability, pain, and dysfunction.

In practice, improvement of muscle contractions has shown to positively affect other ranges of motion and pain not directly addressed. For example: activation of the big toe musculature to increase the toe's ROM has at times corresponded with an increase in shoulder function and reduction of shoulder pain. Theoretically, if the toe's limitation created compensations going up the kinetic chain to the shoulder, then improving the toe function resulted in the body no longer needing the compensations and thus the shoulder was no longer compromised.

–Kika Mela

Muscle Firing Patterns (MFP) is the sequential order of muscular recruitment to produce a coordinated movement of agonists, synergists, and antagonists [52]. The work of Vladimir Janda explores the foundation for this method [47]. In this model, the ideal sequence produces an efficient movement of acceleration, deceleration, and stability [52]. Abnormal firing occurs when a synergist or stabilizer is the dominant muscle which results in the movement becoming less efficient, and the joint position may be strained or compromised [47, 52]. In a simplified explanation, the sequence will activate in order of prime mover, stabilizer, then synergist. According to Fritz, when a stabilizer is required to move the limb (acceleration) and/or control the movement (deceleration), it can become short and tight [52]. For example, in hip flexion the psoas major would be the prime mover. If it becomes short and tight, it may decrease the function of the hip extensors, mainly the gluteus maximus. This could result in less contribution of the gluteus maximus during hip extension and increased contribution of other muscles such as the hamstrings or low back muscles. The same action could be created using a different motor pattern [46]. MFP resolves the misfiring by either inhibition of the overactive muscle and/or manual stimulation of the under active muscle [47, 52].

Current understanding of coordination of movement may eliminate the simple agonist, antagonist, synergist labels. MFP attempts to manipulate certain perceived inefficient motor patterns. For instance, standing up from a chair could either be knee extensor dominant, or hip extensor dominant [46]. MFP goal may be to encourage greater contribution of the glutes during the action after discovering a reliance on the hamstrings or low back for the action of hip extension [52].

Neurokinetic Therapy is a technique that utilizes the inhibition of tightened muscles to change movement patterns and reprogram the motor control center (MCC) [51]. By using manual muscle testing progressively from core to distal muscles, it identifies weakened agonist muscles that have been reciprocally inhibited by tight antagonist muscles. The antagonists of the weakened muscles are then inhibited by the therapist's preferred release techniques. A retest of the weakened muscles should show an increase in strength and therefore a reprogramming of the MCC.

Muscle Energy Techniques

Two osteopathic physicians, Fred Mitchell, Sr., and Fred Mitchell, Jr., are credited with developing muscle energy techniques (MET) with the intent to "*treat soft tissue, mobilize joints, stretch tight muscles and fascia, reduce pain, and improve circulation and lymphatic drainage. MET are defined as a manual treatment in which a patient produces a contraction in a precisely controlled position and direction against a counterforce applied by a manual therapist*" [57]. There are two important details highlighting the difference between MET and proprioceptive neuromuscular facilitation (PNF). First, MET is usually performed with the application of lower forces (≤20% of maximum voluntary contraction or MVC) compared with PNF (≥25% MVC). Second, "*the contraction during MET is performed at the initial barrier of tissue resistance, rather than at the end (ROM) of a joint"*, as is the case when applying PNF stretching [57].

MET application in research differs in 4 key aspects (*see* Table 24.2), presenting a limitation in assessing specific variable control required to accrue evidence for benefits from specific applications of this technique.

A 2019 systematic review of MET [57] identified two applications and their potential roles in increasing ROM, decreasing pain, and restoring function. The review revealed that positive clinical outcomes can be achieved with the use of MET in both asymptomatic and symptomatic populations. It appears that MET applied with an average of 4 repetitions with 5 second isometric contractions at 20% perceived MVC regardless of the duration of stretch, or relaxation between contractions, increases ROM in asymptomatic populations. Also identified in the literature is an upper end of perceived MVC at 75% applied in asymptomatic groups (aiming to increase hamstrings extensibility), and patients with chronic lateral epicondylitis. MET provided positive patient outcomes in these groups as well. It's possible that both the low and high intensity contraction acts on mechanoreceptors altering stretch perception, and pain.

It seems MET is effective in reducing pain, increasing ROM, and restoring function in patients experiencing pain without the presence of any observable pathology. It also appears to be an effective strategy to create an opportunity for application of exercises with the goal of "locking in" the acute improvement in ROM and elimination of pain (the goal of MSTM described earlier in this chapter) [57, 58].

In an article published in 2021 [58], MET and ischemic compression (IC) were compared in decreasing the presence of latent trigger points (LTrPs) and increasing dorsiflexion ROM. LTrPs are thought to cause muscle dysfunction while not causing pain during times of stress or activity, they do not present referred pain observed in active trigger points (ATrPs), which could be an important diagnostic factor in deciding what type of treatment to use. The MET protocol used 7 second isometric holds of 20% MVC at the barrier of dorsiflexion followed by a 60 second stretch held at the end of the new ROM created, then repeated three times. For the IC group, "*a LTrP was identified, and a gradual pressure was applied until the participant felt a 7/10 level of discomfort on the visual analogue scale*" [58] (VAS). Pressure was held for 30 seconds on one LTrP, three separate trigger points in the gastrocnemius were treated. Both groups participated in a total of three treatments over a ten-day period.

Table 24.2 Variables that effect MET outcomes

# of Repetitions	How many times the contraction and/or stretch is applied
% of MVC	What % of the maximum voluntary contraction of the tissue is required
Duration of stretch	How long is the stretch held
Duration of relaxation	How long is the duration of the relaxation between repetitions

MET was significantly more effective when compared with ischemic compression for reducing LTrPs in the gastrocnemius and increasing dorsiflexion, for both acute and midterm treatment assessments [58]. These results suggest the potential of MET to address LTrPs in a more time efficient way than techniques that require direct application of force on each LTrP. Time to manually treat appropriate tissue pathologies is essential in MSTM so that exercises can be applied to reinforce tissue changes for optimal function. More research is warranted to understand what combinations of MSTM and exercises match up to achieve such results.

> *The hypoalgesic effects of MET... is explained by the inhibitory Golgi tendon reflex, activated during the isometric contraction that in turn leads to the reflex relaxation of the muscles. Also, the muscle and joint mechanoreceptors were activated leading to sympatho-excitation evoked by somatic afferents and localized activation of the periaqueductal gray matter. This plays a role in the descending modulation of pain. On application of MET, there may be a reduction in proinflammatory cytokines and it may also desensitize the peripheral nociceptors* [58].

- Is there a way to measure if MSTM improves motor unit recruitment?
- Can it change intra and/or intermuscular coordination?
- If MSTM affects the force distribution, would this affect appropriate remodeling of the tissue?
- How can this be measured?

Changing Tissue Structure

Connective tissue structure is not static but is constantly adapting to stresses such as compression, tension, and shear loading [16, 23, 38, 59, 60]. Research thus far has to do with adaptations of tendon and ligament structures in-vitro to these stresses. In-vivo studies have also been conducted to verify in-vitro observations in both rat and human subjects [23, 38, 39, 61]. Observation of connective tissue responses to different exercise protocols could prove helpful in further understanding how to manipulate depth, direction, and duration to promote a change in structure.

Myofascial Release Therapy (MRT)

"*Myofascial Release Therapy (MRT) is a manual application of a low-load and long-duration stretch to the myofascial complex, which is intended to restore optimal length, decrease pain, and improve function*" [16]. Low-load and long-duration stretch could affect tissue structure as the target of treatment versus MET which targets mechanoreceptors to modulate pain and restore movement.

MRT protocols have shown morphological tissue changes both in-vivo via ultrasound (US) [63] and in-vitro cellular adaptations [60, 62]. MRT could be an effective treatment option when muscular contraction, and joint range of motion causes pain, eliminating the use of contraction in combination with treatment like that used during MET.

In-vitro bioengineered tendon (BET) exposed to repetitive motion strain (RMS) as seen in overuse injuries, such as those in sports, respond favorably to MRT protocols indicating that its use could promote improved recovery leading to higher quality training sessions [62]. Differences in strain direction have an effect on cell growth, morphology, interleukin 1, 6, and 8 secretion, fibroblast morphology, and cellular proliferation [62]. BET responds to various magnitudes of directional strain and duration with different outcomes. For example, a magnitude resulting in 12% increase in tendon elongation may upregulate collagen synthesis, however it can also develop negative structural changes, increase wound area, and increase inflammatory markers [62]. Strain magnitudes of 3–6% applied with longer duration (2–5 minutes) significantly reduced wound area [62].

Recall our earlier discussion "is applying deeper pressure better"? These results suggest that dose dependent and prophylactic MRT may potentially regulate inflammation and wound healing responses in patients. Additionally, MRT could be used to improve contractile function in muscle following RMS, promoting its use as a recovery method where high volume training cycles are implemented. "*…MRT may modify the sensitivity of muscle to the contractile actions of acetylcholine in vivo, as well as play a direct role in new muscle growth*" [62].

Morphological tissue changes resulting from 4 minutes MRT has shown improved gliding function and flexibility of the deep fascia and the vastus lateralis muscle, whereas 10 or 20 minutes of heat application was unable to change the gliding function and flexibility of muscle and fascia [63]. Possible mechanisms for these results include "*changes in the extracellular matrix (eg, collagen synthesis, secretion, and architecture) and gene activation that might result from MRT applied for longer than 2 minutes*" [63]. It's possible that low-load, and long-duration MRT could be highly beneficial in promoting recovery, regeneration, and adaptation to exercise stress, and may have significant tissue remodeling capabilities in the presence of tendinopathy and other conditions showing altered directional orientation of fascial structures. What effect could MRT have on tendinopathy and other connective tissue disorders accompanied by pathological stiffness, or fibrotic tissue?

Fascial Manipulation (FM)

Another method of MSTM gaining attention is Fascial Manipulation (FM). In short, the technique itself "*involves deep friction over specific points mapped by Luigi Stecco named Centre of Coordination (CC) and Centre of Fusion (CF)*" [64]. FM research has shown it to be a method that is effective in treating chronic aspecific low back pain (CALBP), fascial fibrosis, chronic ankle instability (CAI), shoulder issues, and as a preventative method for injury [28, 64–67]. FM presents a deep understanding of the fascial system and through the selection of precise points to treat, it aims to restore the gliding of underlying tissue layers [65–67]. It is of interest to understand if the FM reasoning process could be used to accurately identify points to be treated in combination with other MSTM techniques based on the findings.

Combining FM with specific physiotherapy protocols showed that one session of FM and one session of manual therapy per week for 4 weeks resulted in faster recovery times compared to manual therapy twice per week [65]. Reduction in pain and disability following one treatment of FM was equivalent to 4 sessions of manual therapy. Additionally, at one-month and three-month follow-up the combined FM/MT group showed sustained improvements compared to the manual therapy only group [65].

In a recently published study, FM applied to semi-professional footballers (soccer players) with chronic ankle instability (CAI) revealed results that warrant the attention of all practitioners who use MSTM in their practice [64]. Three sessions of FM lasting 45 minutes one time per week over three weeks was provided for the study group. The control group in this study suffered five injuries resulting in interruption of competitive activities ranging between 6 days to the entire season. The study group experienced no injuries, allowing them to compete through the entire season with increased physical abilities according to the follow-up evaluations. "*Three treatment sessions of 45 minutes showed long-term improvements maintained at the 6 months' follow-up and prevented the interruption of sports activities at one year* [64].

Discoveries by the Stecco family (Luigi, Carla, and Antonio) in anatomy, physiology, fascia along with the development of FM is revolutionizing our understanding of the effects of MSTM.

> *For over 14 years I would use Mark Perkin's expertise in FM to help us mitigate the effects of our field-based and weightroom protocols to promote faster recovery making it possible to keep a highly intensive schedule and achieve maximal performance results.*
>
> –Seth Forman, M.Ed., LMT, CSCS

Understanding Connective Tissue and Its Response to Loading

- **Introduction to fascia**
- **Adaptive nature of connective tissue and its response to loading**
- **Stress shielding**
- **Stress relaxation**

Science has just scratched the surface in understanding the effects of MSTM. An area of study that has moved the needle in recent years is connective tissue research.

Introduction to Fascia

As with many other terms, fascia has been a term of debate without an agreed upon definition among the vast specialists researching and working in medical and physical sciences [68].

The Fascia Research Society established a Fascia Nomenclature Committee in 2013 to provide some foundational clarity, so we can all be working from the same definitions and help push the conversation forward [69].

Here are the two current definitions of fascia.

- A fascia is a sheath, a sheet, or any other dissectible aggregations of connective tissue that forms beneath the skin to attach, enclose, and separate muscles and other internal organs [69].
- The Fascial System consists of the three-dimensional continuum of soft, collagen-containing, loose, and dense fibrous connective tissues that permeate the body. It incorporates elements such as adipose tissue, adventitia and neurovascular sheaths, aponeuroses, deep and superficial fascia, epineurium, joint capsules, ligaments, membranes, meninges, myofascial expansions, periostea, retinacula, septa, tendons, visceral fascia, and all the intramuscular and intermuscular connective tissues, including endo-, peri-, and epimysium. The fascial system surrounds, interweaves between, and interpenetrates all organs, muscles, bones, and nerve fibers, endowing the body with a functional structure and providing an environment that enables all body systems to operate in an integrated manner [69].

> There are two different usages of the term fascia. While the first one is recommended for the description of histological details and of small-scale topographical relationships, the second usage is recommended for the description of larger functional properties, such as muscular force transmission, fibrotic processes, or sensory capacities (such as proprio-, intero-, or nociception) within the fascial net [69].

Adaptive Nature of Connective Tissue and Its Response to Loading

Collagen has a unique adaptation response to load by optimizing its structure and function based on the stresses it experiences [38, 61, 62, 70, 71]. It can change its genetic expressions, shape of fibrils, number of proteoglycans, types of enzymatic or non-enzymatic cross-links, and the components of its extracellular matrix (ECM). Adaptation in collagen function is not limited to one type of connective tissue as it has been observed across the fascial system in tendon, ligament, cartilage, and even bone [23, 72] Over the past two decades imaging technology and biochemical discoveries have led to advancements in our understanding of collagen throughout the connective tissue network and research has only begun to unravel the mystery of this underappreciated full body system [73].

> Several professionals interviewed all expressed similar thoughts that to make lasting changes in a tissue or range of motion, specifically applied external load needs to be directed through the tissue after the manual manipulation of the tissue.

Tensile, compressive, and shear loading of the same genetic collagen cells all produce completely different and fascinating responses [61]. For example, placed under a tensile load as experienced by tendons, the connective tissue increases production of tendon collagen [40]. Tendon collagen (types I and III) are the most prevalent in mammalian striated muscle and dominates tensile regions of the fascial system [69]. In tensile regions, collagen fibers become long and line up along the direction of tension allowing the fibers to efficiently transfer forces from end to end. Placing the same tensile loaded collagen under a compressive load results in increased production of joint collagen (Collagen type II) which makes up articular structures and cartilage [40]. The compressive load also results in the rounding of collagen cells and the recruitment of large, water filled proteoglycans to increase compressive strength of the tissue [40].

Shear force occurs when there is a combination of compression and sliding between structures (ex. the retinacula) and results in three distinctive tissue reactions. First, a large influx of proteoglycans to address compression. Second, the creation of oval or somewhat directionally oriented cells (along the direction of shear). Third, the production of lubricin along the top edge of the tissue to hold water along the edge. The lubricin minimizes friction between structures that need to slide and glide past one another [40].

It doesn't take long when you are looking at the screen of an ultrasound machine during muscular contraction to understand the importance of shear function for optimal interactions between fascial layers. Shear relationships are evident throughout the body between superficial fascia and deep fascia, between deep fascia and epimysium, and between retinacula and tendon [40]. Increased cohesion between two connective tissue layers could result in the disruption of function by impeding movement and possibly causing pain. This may be modulated by the ECM [74].

Connective Tissue Response to Duration of Load

Connective tissues are responsive to load for 10 cumulative minutes (this 10-minute period includes rest periods) from the onset of exercise before becoming non-responsive for the next 6 hours [40]. This discovery has led to connective tissue training protocols that have been shown to increase collagen synthesis and has provided a useful alternative to treat tendinopathies [59, 61]. Greater collagen synthesis has been demonstrated with nutritional strategies as well [71]. For example, gelatin and vitamin C taken together one hour prior to exercise has shown significant increases in collagen synthesis, which could be beneficial in decreasing the risk of tendon/ligament injury, and potentially accelerate return to play [59].

- Would it be possible to increase collagen synthesis in tissue that is under passive tension from an external force such as MSTM or stretching (passive and/or active)?

Connective Tissue Response to Unloading

As connective tissue responds to loading, it also adapts to unloading such as bed rest, being placed in a cast, extreme cases of sedentary behaviors, or immobilization. This lack of motion leads to fascial thickening, scarring, and disorganization of the extracellular matrix which is a result of increasing collagen fibrils, non-directional collagen orientation, and non-enzymatic cross-linking [74, 77]. One study showed that unloading of the dominant leg (as if injured) for 23 days resulted in quadriceps atrophy of 10%, a 21% reduction of MVC torque, a 16.7% reduction in electromechanical delay (EMD), a 16.9% reduction of tendon stiffness, and yet showed no difference in central activation ratio, which is *"the ability to activate the available muscle mass."* [70, 76].

Could MSTM following immobilization be beneficial to reestablish tensile force adaptation in tendons? Perhaps a low-level sustained isometric, stretching, or various manual techniques could reduce the alterations in structure, reorganize the ECM, and improve the function of the muscle, muscle fascia, and tendon, without causing damage to the healing tissue. Understanding of the complex interactions between muscle and the various connective tissue structures is essential for practitioners applying MSTM so that the combination of techniques applied, and exercise interventions prescribed are appropriate for the circumstances. Further research is required to understand effective strategies that accompany this type of situation.

In practice, applying MSTM to connective tissues is used to aid in the "realignment" of the fibers [32]. Perhaps combining the manual therapy with the application of force through the tissues will help to enhance or maintain any changes made in fiber alignment.

Stress Shielding

Stress shielding describes a protective mechanism of force transfer in the presence of a central core tendinopathy, or other scar formation within tensile regions of connective tissue and occurs during changes in tendon length while under resistive load [40, 77, 78]. In stress shielding, healthy portions of tendon transfer force *around* the scar, resulting in radial compression of the scar. For example, when jump roping or performing leg extensions, the healthy collagen fibers of the patellar tendon will continuously compress any scarring within the tendon. While this strategy allows an individual to function, it does little to help the scarred tissue adapt for tensile function. The radial compression of a scar within a tendon result in changes that help resist compressive load such as the attraction of large proteoglycans to draw water into the tissue, an increase in collagen type II (versus collagen I and III found in tensile regions), and the rounding of collagen cells [40]. This structure does not belong in a tensile region (tendon).

Stress Relaxation

Stress relaxation refers to structural changes in connective tissue in the presence of a central core tendinopathy, and possibly fibrosis [40]. During an isometric hold, healthy tendinous tissue will reach a point where it fatigues causing the tensile stresses to be redirected through the scar or tendinopathy. This forces the scar tissue to adapt by changing its features from connective tissue under compression to connective tissue under tension [40, 75]. This adaptation literally heals scarring within tensile regions of connective tissue over time, realigning the collagen fibers [40, 72].

An example of a stress relaxation protocol would include 4 sets of one 30 second isometric effort around 75% maximum effort with 2 minutes rest between sets [61]. Time of the isometric hold can start with a tolerable time, which could be 10 seconds or less, and build up to 30 seconds, or more (optimal time has not been agreed upon, and likely differs for most people). If the central core tendinopathy is in the patellar tendon the isometric tension should be focused on that region, a good exercise example for this pathology would be a mid-range isometric hold on a leg extension machine with the pin at the bottom of the weight stack. Research providing imagery shows the patellar tendon following 18 months of an applied stress relaxation protocol and produced positive results [72].

Efforts at 70% maximal voluntary contraction (MVC) for 5 sets lasting 45 seconds with 2 minutes rest also showed great results, following the isometric intervention, participants eliminated pain during a single leg descending squat, with the elimination of pain remaining for 45 minutes following the intervention [78]. Participants increased their maximal voluntary isometric contraction (MVIC) torque immediately after the isometric protocol which also held for 45 minutes later. Perhaps the most interesting thing is that the isometric protocol released short interval intracortical inhibition (SICI), which is basically an inhibitory pathway resulting in reduced numbers of action potentials (lower force production). The decrease in intracortical inhibition was associated with pain reduction indicating its potential role in pain and motor unit recruitment.

Since isometrics do not cause prolonged fatigue or delayed onset muscle soreness (DOMS/DOSS) at the same degree as eccentric exercise, this isometric protocol can be applied daily (as well as multiple times per day separated by at least 6 hours) [40]. An in-vivo study on this approach was applied 5 days per week for 50 days. Following the 50 days, ultrasound showed a "filling in" of collagen where the tendinopathy existed along with more directionally oriented fibers and an increased amount of collagen, thickening the tendon, which resulted in a more resilient tendon, reduction of pain, and an increase in performance [40].

For healthy tissue, exercises such as jump rope, sprints, change of direction activities, and reactive drills that aim to bridge the gap between training and competing are great ways to develop robust connective tissue structures. The combination of Vitamin C and gelatin one hour prior to activity has shown up to a 50% increase in collagen synthesis [71]. It would be beneficial to understand if this nutritional intervention also yields greater collagen synthesis following MSTM interventions. Nonetheless, practitioners can use this information to apply effective protocols depending on the client's needs.

- What is the force profile of tissues post-MSTM at various depths and duration of application?
- Can nutrition be combined with manual therapy to achieve greater results?

Benefits of Contraction by Type

- **Isometric**
- **Eccentric**
- **Concentric**
- **Plyometric**

Heavy and slow eccentric loads create beneficial enzymatic cross-links between collagen fibers along the direction of applied force [39]. This adaptation is important to promote smooth transfer of force through the connective tissue matrix from muscle fascia to tendon to periosteum where applied force results in movement [77]. Eccentric loading can play an essential role in maintaining a healthy balance of muscle fascia and tendinous stiffness during the in-season because high force high speed movements promote greater stiffness in tendinous structures [79], this is discussed in more detail in the plyometric section.

Isometric

During isometric loading, beneficial enzymatic cross-links are formed similar to that described during eccentric loading. While eccentric loading results in a greater amount of these beneficial cross-links, it also causes more muscle fatigue and soreness [40], therefore choosing between eccentric and isometric applications depends on timing in respect to a variety of variables, one being time to competition. As discussed earlier isometric protocols (stress relaxation) can be applied to treat tendinopathy. Isometrics can also be used as a preventative strategy during high-volume plyometric programs, like those we see during the weeks leading up to the preseason. Plyometrics can result in a level of tendon stiffness that increases the potential for soft-tissue injuries [79]. Therefore, it is important to apply exercise strategies that maintain smooth force transfer as well as maintain an appropriate amount of tendon stiffness.

Eccentric

Heavy eccentric loads and isometric exercises can achieve this goal. Since eccentric work can cause severe DOMS, it should be applied appropriately based on the competition schedule [80]. Isometrics on the other hand do not cause the same levels of fatigue and DOMS as eccentric training and can have a place in recovery, regeneration, and preparation for competition.

Concentric

While concentric action is arguably the most focused contractile type during a resistance training exercise, connective tissues do not have the level of enzymatic cross-link adaptation that comes with eccentric and isometric loading [39]. Concentric actions are important within training for additional reasons, one highly valuable reason includes central nervous system (CNS) signaling to optimize intramuscular tuning, which is the timing of motor unit stimulation to optimize high speed high force movements such as sprinting and jumping [81].

- Can isometrics applied during muscle energy techniques (MET), muscle activation techniques (MAT), or active/passive lengthening techniques play a role in promoting optimal function in a similar way as isometric and eccentric loading?
- Could these therapies that combine a manual technique with contraction be promoting faster recovery?

Plyometric

The purpose of plyometrics is to absorb and redirect force as fast as possible. A high velocity eccentric load (usually the combination of body weight and gravity) is absorbed quickly, this absorption of force ends when the amount of counterforce is equal to the load (amortization phase or acceptance of force), finally movement in the desired direction occurs, we can refer to this as the redirection phase. Pete Bommarito describes the actions during any plyometric activity as follows:

1. Absorb force
2. Accept force
3. Redirect force

Activities such as sprinting, bounding, depth drop-jumps, and continuous jump squats qualify as plyometrics. During these rapid movements, muscles hold a quasi-isometric action while tendons stretch and recoil quickly to generate fast movement [84, 85]. If a high volume of plyometrics is programmed into a training phase, there is a risk. Any force greater than what the muscle-tendon complex can handle (together as one structure) will result in a soft-tissue injury to the myofascial and/or tendinous structures. Therefore, a combination of loading approaches is essential to maintain an appropriate balance between connective tissue stiffness and compliance.

Tendon adaptation to plyometric activities is to become like a steal rod [79]. Since the tendon needs to become a steel rod to quickly and efficiently transfer the force produced by the muscle, ultimately increasing power, where does MSTM fit in? Is it possible that inhibition of the muscle or reduced tendon stiffness through massage or stretching may be counterproductive? Or at a certain point is it complementary? One study found that stretching interventions used to enhance flexibility tend to decrease economy [82], while massage prior to activity reduced the maximal voluntary contraction force (MVC) of the quadriceps and hamstrings by 9% [83]. This is important to understand when working with athletes because the isometric applications we can add within our treatments may act as a potential way to reset mechanoreceptors, resulting in optimal intra and intermuscular tuning. Perhaps this is a key component to enhanced performance via recovery? (Table 24.3)

How do we verify if inhibition is productive or not? If it feels subjectively looser, is it just relaxed or temporarily weaker? Will that impede or enhance performance? If weaker, do other muscles have to take on more load for the task and if so, at what cost?

Table 24.3 The types of muscle contractions and their benefits [39, 40, 77, 79, 82, 85]

Type	What it is	Benefits of
Eccentric	Tissue is lengthened while being loaded	Heavy and slow loads create collagen crosslinks along direction of loading for smooth force transfer
Isometric	Tissue is loaded without change in length or angle of joint	Creates collagen crosslinks along direction of loading with less damage or fatigue to tissue Soft tissue injury prevention during heavy plyometrics
Concentric	Tissue shortens while being loaded	CNS signaling optimization of intramuscular stimulation
Plyometric	Rapid loading of tissue	Enables tissue to absorb, accept, and redirect force

Putting It All Together

Internationally known Strength and Conditioning Coach Pete Bommarito has long embraced a multi-disciplinary approach to training his professional athletes. Manual Therapy plays a significant role in his programming as he feels that healthy athletes in an aligned state focused on training can execute max power. His data driven field and weight room sessions are guided by the medical reports to train the athletes as aggressively and safely as possible for optimal long-term results. He uses an array of medical disciplines to help athletes recover from injury, help increase ROM, and then stabilize the new ranges through strategic exercises to achieve proper joint alignments. If aggressive manual work has shut down the CNS, he feels that it is imperative to follow up with re-assessment to see if the therapy was too much and then re-activate before heavy training [26].

Based on how an athlete looks and sounds on the field he can tell if their performance is where it needs to be and how effective the medical team's treatments are. In addition to physical therapy and chiropractic work, he stresses the use of manual therapy for recovery and MAT for looking outside of the painful area to figure out the cause of the issue.

Renowned therapist Sandy Fritz has been practicing massage since 1979 and writes some of the most detailed massage textbooks in the industry [86]. During our interview she explained that when providing manual work there is only a mechanical push or pull and what changes is time, depth, and how much drag is applied [86]. For her, the strain you create on the tissue is what triggers an adaptation such as enhancing the gliding of the fascial layers or effecting fluid dynamics. Longer duration massage therapy has a unique ability to positively affect the whole body while shorter duration or "spot work" is more symptom management. She tends to prefer the full body session as it allows a therapist to move through all the tissues deliberately and within a normal range to create a sensory input for the whole body and effect several interconnecting systems. She described "spot work" for a painful area as an opportunity to create a counter-irritation or hyperalgesia to exaggerate a painful symptom with the goal to override or interrupt the CNS and perhaps force the tissue to now engage in a new way [86].

When dealing with a particular issue, Fritz feels that getting stuck in one approach or mindset will limit you so suggests not fixating on the tool. Instead, focus on the results and what tools you can use. She prefers to start with the results you want, base the treatment on that goal, use the least complicated way to achieve that goal, and stay within your scope of practice. She has found that most athletes want to sleep and recover, not do more work on the table [86].

Both Bommarito and Fritz summarize effective strategies in positively effecting recovery, regeneration, and adaptation to exercise by incorporating MSTM. In practice, we have seen where the application of MSTM can cause a physiological imbalance or a negative response and set an athlete back or the same work can keep a different athlete performing at a high level while adapting to the training and potentially reducing the risk to injury. The patient or athlete on the table is unique and requires manipulation of MSTM variables that apply to their needs and desired goals.

There is an art and science to all things: the technical application of science, and the artful application of intuition using a combination of techniques for the best results. Science and research do not cover everything. Some practitioners are great at the technical. Some are great at the "art". The best practitioners are able to do both. We hope to have provided information for readers to advance in both aspects.

References

1. Bubbs M. Peak : the new science of athletic performance that is revolutionizing sports. Chelsea Green Publishing; 2019.
2. Epstein D. Range : why generalists triumph in a specialized world. Riverhead Books; 2019.
3. Damasio A. Feeling and knowing: making minds conscious. Robinson; 2022:4.
4. Donatelli R, Al E. Physical therapy of the shoulder. 5th ed. Elsevier/Churchill Livingstone; 2012. p. 305.
5. Monteiro Rodrigues L, Rocha C, Ferreira HT, Silva HN. Lower limb massage in humans increases local perfusion and impacts systemic hemodynamics. J Appl Physiol. 2020;128(5):1217–26. https://doi.org/10.1152/japplphysiol.00437.2019.
6. Weerapong P, Hume PA, Kolt GS. The mechanisms of massage and effects on performance, muscle recovery and injury prevention. Sports Med (Auckland, NZ). 2005;35(3):235–56. https://doi.org/10.2165/00007256-200535030-00004.
7. Threlkeld AJ. The effects of manual therapy on connective tissue. Phys Ther. 1992;72(12):893–902. https://doi.org/10.1093/ptj/72.12.893.
8. Shryer D. Beauty of the basics. Mass Ther J. 2022;61(4):14–9.
9. Gasibat Q, Suwehli W. Determining the benefits of massage mechanisms: a review of literature. Rehabil Sci. 2017;2(3):58–67. https://doi.org/10.11648/j.rs.20170203.12.
10. Akram H, Bashir S, Zia A, Noor R, Shakeel A. Comparison of muscle energy technique and mobilization with movement to reduce pain and improve functional status in subjects with anterior innominate ilio-sacral dysfunction. J Bodyw Mov Ther. Published online December 2022. https://doi.org/10.1016/j.jbmt.2022.11.003.
11. Brummitt J. The role of massage in sports performance and rehabilitation: current evidence and future direction. N Am J Sports Phys Ther. 2008;3(1):7–21.
12. Cafarelli E, Flint F. The role of massage in preparation for and recovery from exercise. Sports Med. 1992;14(1):1–9. https://doi.org/10.2165/00007256-199214010-00001.

13. Bialosky JE, Bishop MD, Price DD, Robinson ME, George SZ. The mechanisms of manual therapy in the treatment of musculoskeletal pain: a comprehensive model. Man Ther. 2009;14(5):531–8. https://doi.org/10.1016/j.math.2008.09.001.
14. Bakar Y, Coknaz H, Karlı Ü, Semsek Ö, Serın E, Pala ÖO. Effect of manual lymph drainage on removal of blood lactate after submaximal exercise. J Phys Ther Sci. 2015;27(11):3387–91. https://doi.org/10.1589/jpts.27.3387.
15. Hilbert JE. The effects of massage on delayed onset muscle soreness. Br J Sports Med. 2003;37(1):72–5. https://doi.org/10.1136/bjsm.37.1.72.
16. Khan ZK, Ahmed SI, Baig AAM, Farooqui WA. Effect of post-isometric relaxation versus myofascial release therapy on pain, functional disability, rom and qol in the management of non-specific neck pain: a randomized controlled trial. BMC Musculoskelet Disord. 2022;23(1) https://doi.org/10.1186/s12891-022-05516-1.
17. Shah JP, Thaker N, Heimur J, Aredo JV, Sikdar S, Gerber L. Myofascial trigger points then and now: a historical and scientific perspective. PM&R. 2015;7(7):746–61. https://doi.org/10.1016/j.pmrj.2015.01.024.
18. Schleip R. Fascial plasticity—a new neurobiological explanation: part 1. J Bodyw Mov Ther. 2003;7(1):11–9. https://doi.org/10.1016/s1360-8592(02)00067-0.
19. Schleip R. Zoom Interview with Dr. Schleip on Fascia. Published online June 23, 2022.
20. Schleip R. The Fascial Network our richest sensory organ. Massage & Bodywork. Published online September 2022, pp. 40–51.
21. Schleip R. Fascial plasticity—a new neurobiological explanation part 2. J Bodyw Mov Ther. 2003;7(2):104–16. https://doi.org/10.1016/s1360-8592(02)00076-1.
22. Bergin M, Tucker K, Vicenzino B, Hodges PW. "Taking action" to reduce pain—has interpretation of the motor adaptation to pain been too simplistic?. Lee Y, ed. PLoS One. 2021;16(12):e0260715. https://doi.org/10.1371/journal.pone.0260715.
23. Smeets JSJ, Horstman AMH, Vles GF, et al. Protein synthesis rates of muscle, tendon, ligament, cartilage, and bone tissue in vivo in humans. Markofski MM, ed. PLoS One. 2019;14(11):e0224745. https://doi.org/10.1371/journal.pone.0224745.
24. Weinberg R, Jackson A, Kolodny K. The relationship of massage and exercise to mood enhancement. Sport Psychol. 1988;2(3):202–11. https://doi.org/10.1123/tsp.2.3.202.
25. Weingroff C. Interview on MSTM with Dr. Charlie Weingroff. Published online June 8, 2022.
26. Bommarito P. Interview with Pete Bommarito on MSTM and training. Published online August 16, 2022.
27. Kul A, Ugur M. Comparison of the efficacy of conventional physical therapy modalities and Kinesio taping treatments in shoulder impingement syndrome. Eurasia J Med. 2019;51(2):138–43. https://doi.org/10.5152/eurasianjmed.2018.17421.
28. Ercole B, Antonio S, Julie Ann D, Stecco C. How much time is required to modify a fascial fibrosis? J Bodyw Mov Ther. 2010;14(4):318–25. https://doi.org/10.1016/j.jbmt.2010.04.006.
29. Wilke J, Behringer M. Is "Delayed Onset Muscle Soreness" a false friend? The potential implication of the fascial connective tissue in post-exercise discomfort. Int J Mol Sci. 2021;22(17):9482. https://doi.org/10.3390/ijms22179482.
30. Hortobágyi T, Denahan T. Variability in Creatine Kinase: methodological, exercise, and clinically related factors. Int J Sports Med. 1989;10(02):69–80. https://doi.org/10.1055/s-2007-1024878.
31. Smith LL, Keating MN, Holbert D, et al. The effects of athletic massage on delayed onset muscle soreness, Creatine Kinase, and neutrophil count: a preliminary report. J Orthop Sports Phys Ther. 1994;19(2):93–9. https://doi.org/10.2519/jospt.1994.19.2.93.
32. Owens E. Interview with Eric Owens. Published online September 22, 2022.
33. Phadke A, Bedekar N, Shyam A, Sancheti P. Effect of muscle energy technique and static stretching on pain and functional disability in patients with mechanical neck pain: a randomized controlled trial. Hong Kong Physiother J. 2016;35:5–11. https://doi.org/10.1016/j.hkpj.2015.12.002.

34. Stanek J, Sullivan T, Davis S. Comparison of compressive myofascial release and the Graston technique for improving ankle-dorsiflexion range of motion. J Athl Train. 2018;53(2):160–7. https://doi.org/10.4085/1062-6050-386-16.
35. Budini F, Rafolt D, Christova M, Gallasch E, Tilp M. The recovery of muscle spindle sensitivity following stretching is promoted by isometric but not by dynamic muscle contractions. Front Physiol. 2020:11. https://doi.org/10.3389/fphys.2020.00905.
36. Carano A. Effects of continuous and intermittent forces on human fibroblasts in vitro. Eur J Orthodont. 1996;18(1):19–26. https://doi.org/10.1093/ejo/18.1.19.
37. Majima T, Yasuda K, Tsuchida T, et al. Stress shielding of patellar tendon: effect on small-diameter collagenfibrils in a rabbit model. J Orthop Sci. 2003;8(6):836–41. https://doi.org/10.1007/s00776-003-0707-x.
38. Moore DR, Phillips SM, Babraj JA, Smith K, Rennie MJ. Myofibrillar and collagen protein synthesis in human skeletal muscle in young men after maximal shortening and lengthening contractions. Am J Physiol Endocrinol Metabol. 2005;288(6):E1153–9. https://doi.org/10.1152/ajpendo.00387.2004.
39. Heinemeier KM, Olesen JL, Haddad F, et al. Expression of collagen and related growth factors in rat tendon and skeletal muscle in response to specific contraction types. J Physiol. 2007;582(3):1303–16. https://doi.org/10.1113/jphysiol.2007.127639.
40. Keith B. Loading & nutrition to improve connective tissue function.
41. Perkins M. Google Meet interview with Mark Perkins on massage and myofascial therapies. Published online June 14, 2022.
42. Mcgill S. Ultimate back fitness and performance. Backfitpro Inc; 2017. p. 12–29.
43. Kandel ER. In: Schwartz JH, Jessell TM, Siegalbaum SA, Hudspeth AJ, editors. Principles of neural science. 5th ed. Mcgraw-Hill Medical; 2012. p. 475–497, 784–785.
44. Kapandji A. Physiology of the joints - Volume 2 : the lower limb. Handspring Publishing Lim; 2019.
45. McGill SM. Low back stability: from formal description to issues for performance and rehabilitation. Exerc Sport Sci Rev. 2001;29(1):26–31. https://doi.org/10.1097/00003677-200101000-00006.
46. McGill SM, Grenier S, Kavcic N, Cholewicki J. Coordination of muscle activity to assure stability of the lumbar spine. J Electromyogr Kinesiol. 2003;13(4):353–9. https://doi.org/10.1016/s1050-6411(03)00043-9.
47. Janda V. Janda Compendium. Distributed By Optp; 1997. p. 105–45.
48. Liebenson C. Rehabilitation of the spine : a practitioner's manual. 2nd ed. Lippincott Williams & Wilkins; 2007. p. 46.
49. Novacheck TF. The biomechanics of running. Gait Posture. 1998;7(1):77–95. https://doi.org/10.1016/s0966-6362(97)00038-6.
50. Roskopf G. Interview with Greg Roskopf on MAT in relation to MSTM. Published online June 21, 2022.
51. Weinstock D. NeuroKinetic therapy : an innovative approach to manual muscle testing. North Atlantic Books; 2010. p. xvi–ix.
52. Fritz S. Sports & exercise massage : Comprehensive care in athletics, fitness & rehabilitation. 2nd ed. Elsevier Mosby; 2013.
53. Harrington Dr M. Interview with Mark Harrington, DO. Published online August 16, 2022.
54. Lawson C. Interview with NFL DE Carl Lawson on MSTM. Published online August 22, 2022.
55. Scott D. Interview with Dawn Scott on massage and soccer. Published online November 10, 2022.
56. Roskopf G. Jumpstart into MAT. Muscle activation techniques; 2021. p. 5–8.
57. Thomas E, Cavallaro AR, Mani D, Bianco A, Palma A. The efficacy of muscle energy techniques in symptomatic and asymptomatic subjects: a systematic review. Chiroprac Manual Ther. 2019;27(1) https://doi.org/10.1186/s12998-019-0258-7.

58. Clarke J, Allen L. The effect of muscle energy techniques on latent trigger points of the gastrocnemius muscle. The Sport Journal. Published online February 12, 2021. https://doi.org/1543-9518.
59. Paxton JZ, Hagerty P, Andrick JJ, Baar K. Optimizing an intermittent stretch paradigm using ERK1/2 phosphorylation results in increased collagen synthesis in engineered ligaments. Tissue Eng Part A. 2012;18(3–4):277–84. https://doi.org/10.1089/ten.tea.2011.0336.
60. Herchenhan A, Uhlenbrock F, Eliasson P, et al. Lysyl oxidase activity is required for ordered collagen Fibrillogenesis by tendon cells. J Biol Chem. 2015;290(26):16440–50. https://doi.org/10.1074/jbc.m115.641670.
61. Steffen D, Mienaltowski MJ, Baar K. Scleraxis and collagen I expression increase following pilot isometric loading experiments in a rodent model of patellar tendinopathy. Matrix Biol. 2022;109:34–48. https://doi.org/10.1016/j.matbio.2022.03.006.
62. Zein-Hammoud M, Standley PR. Modeled osteopathic manipulative treatments: a review of their in vitro effects on fibroblast tissue preparations. J Osteopath Med. 2015;115(8):490–502. https://doi.org/10.7556/jaoa.2015.103.
63. Ichikawa K, Takei H, Usa H, Mitomo S, Ogawa D. Comparative analysis of ultrasound changes in the vastus lateralis muscle following myofascial release and thermotherapy: a pilot study. J Bodyw Mov Ther. 2015;19(2):327–36. https://doi.org/10.1016/j.jbmt.2014.11.018.
64. Brandolini S, Lugaresi G, Santagata A, et al. Sport injury prevention in individuals with chronic ankle instability: Fascial Manipulation® versus control group: a randomized controlled trial. J Bodyw Mov Ther. 2019;23(2):316–23. https://doi.org/10.1016/j.jbmt.2019.01.001.
65. Branchini M, Lopopolo F, Andreoli E, Loreti I, Marchand AM, Stecco A. Fascial Manipulation® for chronic aspecific low back pain: a single blinded randomized controlled trial. F1000Res. 2015;4:1208. Published 2015 Nov 3. https://doi.org/10.12688/f1000research.6890.2.
66. Day JA, Stecco C, Stecco A. Application of Fascial Manipulation© technique in chronic shoulder pain—anatomical basis and clinical implications. J Bodyw Mov Ther. 2009;13(2):128–35. https://doi.org/10.1016/j.jbmt.2008.04.044.
67. Pintucci M, Simis M, Imamura M, et al. Successful treatment of rotator cuff tear using Fascial Manipulation ® in a stroke patient. J Bodyw Mov Ther. 2017;21(3):653–7. https://doi.org/10.1016/j.jbmt.2016.12.007.
68. Wittgenstein L, Georgallides A. Tractatus Logico - Philosophicus. Ekdóseis Íambos; 2016.
69. Schleip R. Fascia: the tensional network of the human body : the science and clinical applications in manual and movement therapy. Churchill Livingstone/Elsevier; 2012.
70. Myrick KM, Voss A, Feinn RS, Martin T, Mele BM, Garbalosa JC. Effects of season long participation on ACL volume in female intercollegiate soccer athletes. J Exp Orthop. 2019;6(1) https://doi.org/10.1186/s40634-019-0182-8.
71. Shaw G, Lee-Barthel A, Ross ML, Wang B, Baar K. Vitamin C–enriched gelatin supplementation before intermittent activity augments collagen synthesis. Am J Clin Nutr. 2016;105(1):136–43. https://doi.org/10.3945/ajcn.116.138594.
72. Baar K. Stress relaxation and targeted nutrition to treat patellar tendinopathy. Int J Sport Nutr Exerc Metab. 2019;29(4):1–5. https://doi.org/10.1123/ijsnem.2018-0231.
73. Myers TW. Anatomy trains: myofascial meridians for manual and movement therapists. 4th ed. Elsevier Health Sciences; 2020. p. 5.
74. Slimani L, Micol D, Amat J, et al. The worsening of tibialis anterior muscle atrophy during recovery post-immobilization correlates with enhanced connective tissue area, proteolysis, and apoptosis. Am J Physiol Endocrinol Metabol. 2012;303(11):E1335–47. https://doi.org/10.1152/ajpendo.00379.2012.
75. Baar K. Minimizing injury and maximizing return to play: lessons from engineered ligaments. Sports Med. 2017;47(S1):5–11. https://doi.org/10.1007/s40279-017-0719-x.
76. Scott WB, Oursler KK, Katzel LI, Ryan AS, Russ DW. Central activation, muscle performance, and physical function in men infected with human immunodeficiency virus. Muscle Nerve. 2007;36(3):374–83. https://doi.org/10.1002/mus.20832.

77. Delos Institute: "Strength Training and Fascia" course (April 17, 2021). www.youtube.com. Accessed April 23, 2023. https://www.youtube.com/watch?v=rbfA-sF_MqA
78. Rio E, Kidgell D, Purdam C, et al. Isometric exercise induces analgesia and reduces inhibition in patellar tendinopathy. Br J Sports Med. 2015;49(19):1277–83. https://doi.org/10.1136/bjsports-2014-094386.
79. Barr K. Strength training to improve connective tissue health and performance. Presented at: 2022 PFSCCA education series; 2022. Accessed 2022. https://pfscca.com/2022-pfscca-education-series/.
80. Jones DA, Newham DJ, Torgan C. Mechanical influences on long-lasting human muscle fatigue and delayed-onset pain. J Physiol. 1989;412(1):415–27. Accessed April 18, 2023. https://doi.org/10.1113/jphysiol.1989.sp017624.
81. Carrasco DI, Delp MD, Ray CA. Effect of concentric and eccentric muscle actions on muscle sympathetic nerve activity. J Appl Physiol. 1999;86(2):558–63. https://doi.org/10.1152/jappl.1999.86.2.558.
82. Hughes DC, Ellefsen S, Baar K. Adaptations to endurance and strength training. Cold Spring Harb Perspect Med. 2017;8(6):a029769. https://doi.org/10.1101/cshperspect.a029769.
83. Wiktorsson-Möller M, Oberg B, Ekstrand J, Gillquist J. Effects of warming up, massage, and stretching on range of motion and muscle strength in the lower extremity. Am J Sports Med. 1983;11(4):249–52. https://doi.org/10.1177/036354658301100412.
84. Kurokawa S, Fukunaga T, Fukashiro S. Behavior of fascicles and tendinous structures of human gastrocnemius during vertical jumping. J Appl Physiol (1985). 2001;90(4):1349–58. https://doi.org/10.1152/jappl.2001.90.4.1349.
85. Reeves ND, Narici MV. Behavior of human muscle fascicles during shortening and lengthening contractions in vivo. J Appl Physiol. 2003;95(3):1090–6. https://doi.org/10.1152/japplphysiol.01046.2002.
86. Fritz S. Interview with Sandy Fritz on massage therapy. Published online August 30, 2022.

Part IV
Machine Based Therapies

Chapter 25
In the Pursuit of Excellence—The Importance of Electrical Stimulation Modalities for Athletics

Amir Pakravan

Introduction

Scientists' fascination with electricity and the potential healing properties of electrical stimulation goes back centuries. Ancient Greeks are believed to have used electricity from torpedo fish in reducing childbirth pain and there is evidence of early Roman physicians using electrical characteristics of torpedo fish in managing gout and headaches [1, 2].

Such ongoing experiments eventually led to the development of electrophysiology and electrochemistry by late eighteenth century when in 1790 for the first time Luigi Galvani reproduced movement in frog's leg muscles by connecting electricity [2]. It was later in mid-nineteenth century that Michael Faraday demonstrated movement by electrical stimulation of nerves [2, 3].

Since then and in particular from the 1960s onwards, with significant advancements in electrophysiology in combination with medical technology and data science, use of electrical stimulation has become an integral part of neural and musculoskeletal rehabilitation.

Neuromuscular Electrical Stimulation (NMES) or transcutaneous application of controlled electrical currents to muscles and nerves is now recognized as a safe non-invasive method of producing muscle activity which is widely used in rehabilitation as well as improving muscle bulk, strength, and function, and reducing edema and promoting recovery.

A. Pakravan (✉)
Centre for Sports and Exercise Medicine, William Harvey Research Institute, School of Medicine and Dentistry, Queen Mary University of London, London, UK

School of Allied Health and Science, University of Suffolk, Ipswich, UK

Super League Basketball, Canon Medical Arena, Sheffield, UK
e-mail: Amir_pak@yahoo.com

N. C. Apostolopoulos et al. (eds.), *Fundamentals of Recovery, Regeneration, and Adaptation to Exercise Stress: An Integrated Approach*,
https://doi.org/10.1007/978-3-031-44270-4_25

NMES Physiology and Mechanism of Action

The NMES procedure essentially involves passing electrical currents with specific pre-determined characteristics through electrodes that are directly or through a coupling medium attached onto the skin over the target area. Numerous studies have looked at physiological effects of NMES and the mechanism through which clinical effects can be achieved. A basic knowledge of muscle activation mechanism can help with better understanding of NMES mechanism of action.

Muscle movement is controlled by the central nervous system through a complex network of afferent and efferent nerves. At the muscle level, these effects are exerted by motor units. A motor unit is made up of a muscle fibre and the axon of a peripheral motor neuron in the ventral horn of the spinal cord. This is responsible for stimulating the contraction of the muscle fibre. Axon branches of a single motor neuron may reach multiple muscle fibres in the same unit. Depending on the size, speed or specific function of a muscle, the size and number of motor units may vary. An individual muscle may be controlled by hundreds to thousands of motor units provided by multiple motor neurons which work in harmony and are collectively known as a motor pool or motor nucleus. Different muscles that produce the same force vector receive a common synaptic input from central nervous system, albeit independent from each other [4–6].

Neural activation of muscle fibres happens through an electrical flux travelling along the nerve precipitated by changes in the neuron cell membrane voltage, known as the Action Potential. The action potential is controlled by cell membrane permeability to and the availability of different ions or electrically charged atoms in the intracellular and extracellular environment [4, 5, 7]. The cellular interface between each motor neuron axon branch and each individual muscle fibre is known as a motor endplate. It is at this neuromuscular junction that the neural action potential, or electrical flux, is translated into post-synaptic or muscle fibre depolarization and action potential which in turn will cause the release of calcium ions and the cross-bridging of actin and myosin filaments leading to muscle fibre contraction [4, 6, 8].

Through a similar mechanism, the electrical current of appropriate characteristics delivered by NMES peripherally depolarizes the target motor neuron axon initiating an action potential causing muscle contraction. A healthy motor neuron and motor unit are requirements for success of a peripheral stimulation [4, 9].

It is, however, important to recognise fundamental differences between muscle contraction produced by NMES and voluntary central nervous system driven muscle contraction (See Table 25.1 below). Physiologically voluntary muscle activation is driven by tasks aimed at producing a specific amount of force where different motor units are recruited at different times with the smallest units first. This enables the activating of precise force vectors required for specific tasks whilst allowing other unused motor units time to rest. In NMES a large group of motor units are stimulated synchronously and in reverse order to the physiological pattern, with largest motor units recruited first. This is because large diameter axons have a lower

Table 25.1 Fundamental differences between muscle contraction produced by NMES and voluntary central nervous system contraction (physiological)

NMES muscle contraction	Voluntary physiological muscle contraction
Not task related	Task driven
Multiple motor units at the same time	Different motor units at different times
Largest / closest motor units first	Smallest required motor units first

activation threshold and respond more easily to electrical stimulation. These larger units are primarily associated with type II fast twitch fibres that are more fatigable, accounting for neuromuscular fatigue and discomfort [4, 9, 10].

Depending on tissue impedance, the motor units activated are generally those closest to the electrodes. This spatial recruitment limitation means that at intensities commonly used in NMES, only about 30–40% of muscle cross-sectional area is activated. Therefore, to reach deeper or smaller motor units, higher intensity is needed which may not be comfortably tolerated [4, 9, 10].

These differences explain why NMES in general appears to be more effective in maintaining and recovering muscle bulk, strength, and function after a period of reduced mobility and disuse as opposed to healthy active subjects [4, 9].

Muscle activation in addition to commonly recognized effects on mobility, function, posture, and edema, can have a more systemic impact on general health . Skeletal muscles release a special kind of cytokine or signal transmitting peptide, known as a Myokine. This is produced by the myocytes in response to activation. Myokines have known autocrine or local effects on muscle metabolism and adaptive changes in muscle tissue [11, 12]. These adaptations and improvements in muscle function following NMES are also observed in the homologous muscle contralateral to the muscle being treated [9, 13].

There is growing evidence that myokines released by skeletal muscle contractions impact other organs and tissues such as the liver, brain, and adipose tissue through paracrine and endocrine signals similar to hormones. In fact, the skeletal muscle system is increasingly recognized as the body's largest secretory organ, with more than 600 myokines currently identified, although only a handful have been thoroughly investigated. This is an emerging and exciting field in muscle research which can explain how muscle contraction induced by NMES enables a regulatory effect on whole body metabolism and management of metabolic disorders such as diabetes and possibly cancer [11, 12, 14, 15].

Practical and Clinical Applications of NMES

Use of NMES in rehabilitation and sport settings is increasingly common. There is ample evidence to support observable changes in skeletal muscle undergoing a medium to long term treatment programme. These changes include increased muscle bulk or hypertrophy, primarily due to increased muscle fibre type II distribution, increased vascularity, triggering specific genes expression, and neural adaptations such as new axonal sprouting and central nervous system activity collectively enabling an increased muscle force and enhanced function [9, 16, 17].

Clinical Applications of NMES

- Improve muscle bulk and strength
- Improve mobility and function
- Improve athletic performance
- Rehabilitation after injury or surgery
- Prehabilitation
- Prevent/control Metabolic disorders

Most common applications of NMES are in muscle strengthening, improving muscle function, motor re-education in neurological impairment, assisting in the maintenance and improvement in joint range of motion and research [4, 9]. Depending on the clinical setting, NMES can be used as both a rehabilitation or prehabilitation modality and an adjunct to strength and functional training. The procedure has successfully been used in both pre and post orthopedic surgery to improve surgical outcome and accelerate strength and functional recovery. This is of particular interest during periods of reduced mobility and subsequent muscle atrophy [4, 9].

NMES in combination with routine rehabilitation protocols can enhance post-surgical recovery in major knee, hip, and abdominal operations by accelerating gains in muscle bulk, strength, and function and reducing time spent in hospital as well as post-operative pain and medium to long term care dependency [18–21]. Similarly, use of NMES as part of pre-operative prehabilitation for total knee replacement has been shown to expedite a better return to normal activities after operation [22].

More recently application of NMES in prehabilitation before cardiothoracic surgery has shown promising results as a safe and effective option for post-operative improvement in functional capacity and maintaining strength [23, 24]. Equally, it has been found as a useful addition to prehabilitation in cancer patients undergoing acute treatments or those unable or for various reasons not allowed to actively exercise [25, 26]. This helps improve their functional and endurance capacity, independence, and quality of life.

This modality is recognized as a feasible, valid, and safe option for fighting sarcopenia, age-related loss of muscle bulk, and muscle atrophy in the elderly population with tangible improvements in functional ability and quality of life [27].

In sport and performance, it is widely believed that the use of NMES in the healthy and athletic population, in combination with active voluntary exercise, can lead to measurable improvements in strength and endurance for athletic performance. Improvements have been seen in muscle torque, sprint and jump performance, maximal oxygen consumption, and ventilatory anaerobic threshold [28–31]. Interestingly, despite the higher baseline values for elite and professional athletes, significant benefits in function and performance can be acquired with the use of NMES over a number of weeks [32–34]. Application of this modality in warm up and post-exercise recovery has also been investigated with promising results [35, 36].

NMES Treatment Protocol Variables

To ensure optimum results, minimal side effects, and higher compliance with NMES, careful consideration should be given to the specific indication, appropriate subject selection, and the relevant variable parameters for electrical stimulation.

The electrode is typically positioned on a perceived motor point, a specific point in the muscle belly near the motor neuron entry point. Muscle stimulation over the motor point is easier, requiring lower intensity, and is better tolerated. Body charts illustrating specific motor points for various muscles are currently available, helping to remove any guesswork from the practitioner [4, 37] (Fig. 25.1).

However, it is important to realise motor points do not always clearly correlate with neuron entry points, and there may be differences between each person which may require a trial-and-error approach to best identify the optimum electrode position [4, 38]. Precise positioning of motor points is possible by use of special low frequency and low amplitude pen electrodes [37, 38]. Using this technique, it has

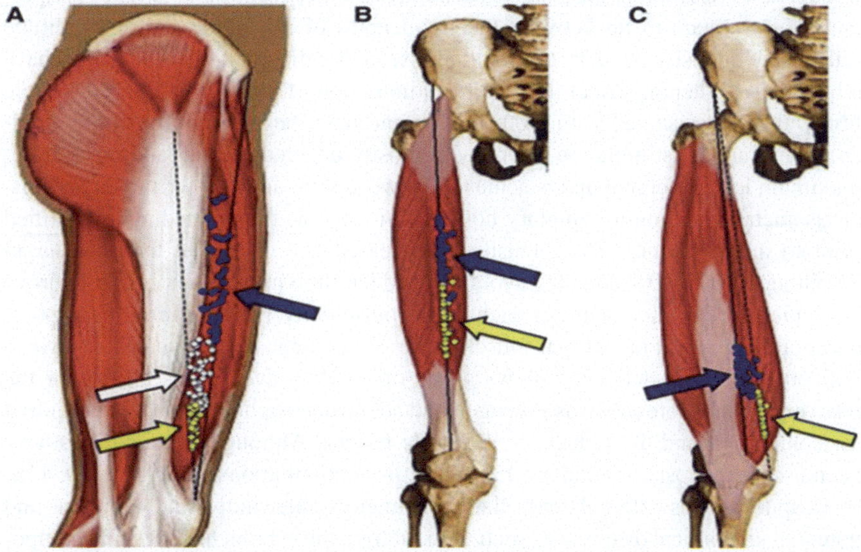

Fig. 25.1 Position of the motor points for the quadriceps muscles in 53 healthy subjects. The arrows indicate the average positions of the motor points along the respective reference lines (continuous solid black = proximal motor point; dashed black line = central and distal motor point). (**a**) Vastus Lateralis (**b**) Rectus Femoris and (**c**) Vastus Medialis (See Ref. [39])

been demonstrated that individual muscles may have different motor points [39]. The finding suggests non-synchronous multi-channel activation of muscles is likely to recruit different fibres within the muscle and delay fatigue [39].

NMES electrode numbers, size, shape, and placement orientation relative to the target muscle can influence the intensity required, comfort, and compliance of the patient. It is possible to recruit more motor units by increasing the number of electrodes delivering the stimulus [9]. There are various sizes and shapes of electrodes available. They are positioned longitudinally or transversely to the muscle orientation. Although optimum electrode placement near the motor point is probably still the most important factor, larger electrode sizes appropriate for the target muscle size, placed longitudinally, are generally believed to reduce stimulation density, hence associated with less discomfort and better compliance [40, 41, 42].

Stimulation Current Density
This is a measure of the amount of electrical current over specific surface area needed to trigger a muscle contraction and is calculated as mA/cm^2. Lower current density is associated with more comfort and potentially better compliance [41].

Other machine related NMES parameters include pulse frequency, amplitude, and duration, and duty cycle. Pulse frequency is directly related to achievable torque, however, finding the balance between higher frequency and fatigue is important. This appears to lie between the frequencies of 30–50 Hz. Pulse amplitude reflects the intensity of each stimulation pulse, and although higher intensities have a better strengthening effect through the stimulation of a larger number of muscle fibres, this is associated with more fatigue and pain, negatively affecting compliance. In practice, stimulation intensity is mostly decided by trial and error at the maximum level tolerated and is achieved between 40% and 60% of the target muscle isometric maximum voluntary contraction force as determined and quantified prior to the procedure. The intensity is increased and decreased from the target amplitude over 5–10 seconds to avoid any sudden limb movements and to improve convenience. The time span for each single pulse delivered is referred to as pulse duration or width. For optimal muscle fibre recruitment this is typically between 300–600 μs. Pulse delivery follows an on-off or duty cycle. This allows for the recovery of substrates such as glycogen and adenosine triphosphate (ATP) required for contraction and the reduction of muscle fatigue. Although various ratios have been used, the most common are 1:3 or 1:5, from stimulation to rest [4, 9, 10, 42].

Despite various external controllable parameters, inter-individual variations and potential anatomical differences such as in motor neuron branching or body composition are important and less controllable determinants of differences in individual responses to NMES intervention [9].

Transcutaneous Electrical Nerve Stimulation (TENS)

TENS is another electrical stimulation procedure primarily used as a pain management modality. By definition, TENS refers to any method of stimulating the nerve through transcutaneous electrical current. However, this method only stimulates afferent peripheral nerves and unlike NMES, is not associated with motor neuron stimulation or muscle activation and therefore is not used for muscle strengthening or maintenance [10, 43, 44].

Similar to NMES, the delivery mechanism is through the use of electrodes attached onto intact skin. The machine is often a small hand-held battery powered unit with easy instructions for self-administration. Mechanism of action is through triggering an action potential but only for pain management through afferent nerves [43–45].

Currently three types of TENS exist. Each one is used for specific and different applications. Conventional TENS stimulates large diameter, low threshold non-noxious (A-beta) afferent nerves with use of a high frequency 50–100 Hz. Low intensity, small width 50–200 μs pulses produce a comfortable paraesthesia through reducing nociceptor cell activity at a segmental level in central nervous system. This can effectively control pain in dermatomal distribution [43–45].

The second application is referred to as Acupuncture-like TENS (AL-TENS), because of its characteristic sensation. During AL-TENS, The stimulation pulse is of a low frequency 2–4 Hz, with a higher intensity, longer duration 100–400 μs. This activates smaller diameter, high threshold sensory nerves (A-delta) that do not respond to conventional settings [10, 43, 45].

Intense TENS, the third type of application, uses a higher frequency of 200 Hz. This method is primarily used for analgesia associated with painful minor procedures (i.e. suture removal). The signal delivered is of a high intensity that can be tolerated by the patient. This delivers a counter irritant impulse by stimulating A-delta afferent nerves [43, 45].

Ample evidence exists for the effectiveness of TENS as a safe non-invasive option in both acute and chronic settings. TENS allows for better management of musculoskeletal, nociceptive, and neuropathic pain as well as pain associated with inflammatory aetiology [43–48].

Precautions and Contra-Indications to Electrical Stimulation

Electrical stimulation modalities are considered safe when used for the correct indication in the appropriately selected candidates. However, certain limitations exist such as convenience, discomfort, fatigue, and potential practicalities of the procedure. Although the majority of these can be addressed through modifications in controllable parameters already discussed, personal experiences and final outcome are also partly determined by uncontrollable individual factors.

Although intensities used in these modalities fall within a safe spectrum, and the flux remains localized to the distance between the two electrodes, an electrical current can pass through the whole body via highly ionised bodily fluids, suggesting that appropriate precautions need to be considered [4]. Use of electrical stimulation is contraindicated in unstable cardiac conditions or with those who have a cardiac pacemaker, and, in individuals who are at risk of acute bleeding or thromboembolism, and skin lesions where the electrodes should be applied. NMES should not be applied over an area of bony fracture especially if considered unstable. It should be avoided over specific areas such as carotid sinus, larynx, and through the chest wall [4, 49, 50]. Precaution is advised during pregnancy to avoid application of electrodes onto abdomen or lumbar spine [4]. Despite previous concerns regarding use of electrical stimulation in cancer, there is growing evidence for the application of both NMES and TENS in cancer pain control and improved function and quality of life [25, 26, 51].

Conclusion

Therapeutic applications of electrical stimulation have long been a subject of human curiosity. With the advancement of science and technology, there is growing prospect for evidence-based utilization of modalities such as NMES and TENS in the optimization of functional and performance abilities across the spectrum of health from the frail elderly population to elite high-performance athletes.

Appropriate use of these options can enhance recovery from injuries, post-surgical, or after any period of immobility. Electrical stimulation modalities have an important role in improving the quality of life in people suffering with chronic or debilitating conditions or undergoing aggressive treatment. It aids in better management of pain and increased functional capabilities.

This is a growing field in performance science with ongoing research into various current and future applications of different electrical stimulation modalities.

1. How can we deliver a more personalized NMES for optimum effect based on specific application, specific target area, and individual variations?
2. What protocols can be developed to ensure that NMES matches the unique circumstance of a clinical situation (i.e. stage of recovery, level of fatigue)?
3. The effectiveness of NMES for enhanced recovery for shoulder subluxation in stroke rehabilitation?
4. The effectiveness of pre- and post-op NMES compared to other modalities for enhanced recovery after total hip replacement surgery?
5. The use of NMES for speeding up the return-to-play following knee joint injury and surgery?
6. The effectiveness of combining NMES, strength, and stretching exercises for improving lower limb pain and function in athletes post Achilles tendon tear?

References

1. Martin RA. “Electric Rays”. Reef Quest Centre for Shark Research. Archived from the original on 15 October 2008. Retrieved 12 October 2008.
2. Dolhem R. Histoire de l'électrostimulation en médecine et en rééducation [The history of electrostimulation in rehabilitation medicine]. Ann Readapt Med Phys. 2008;51(6):427–31.
3. Cambridge NA. Electrical apparatus used in medicine before 1900. Proc R Soc Med. 1997;70(9):635–41.
4. Reed B. The physiology of neuromuscular electrical stimulation. Pediatr Phys Ther. 1997;9(3):96–102.
5. Farina D, Negro F, Muceli S, Enoka RM. Principles of motor unit physiology evolve with advances in technology. Physiology (Bethesda). 2016;31(2):83–94.
6. Heckman CJ, Enoka RM. Physiology of the motor neuron and the motor unit. In: Handbook of clinical neurophysiology, vol. 4; 2004. p. 119–47.
7. Grider MH, Jessu R, Kabir R. Physiology, action potential. [Updated 2022 May 15]. In: StatPearls [Internet]. Treasure Island: StatPearls Publishing; 2022.
8. Mihailoff GA, Haines DE. Chapter 24 - Motor system I: peripheral sensory, brainstem, and spinal influence on anterior horn neurons. In: Fundamental neuroscience for basic and clinical applications. 5th ed; 2018. p. 346–59.
9. Maffiuletti NA. Physiological and methodological considerations for the use of neuromuscular electrical stimulation. Eur J Appl Physiol. 2010;110(2):223–34.
10. Doucet BM, Lam A, Griffin L. Neuromuscular electrical stimulation for skeletal muscle function. Yale J Biol Med. 2012;85(2):201–15.
11. Lee JH, Jun HS. Role of Myokines in regulating skeletal muscle mass and function. Front Physiol. 2019;10:42.
12. Carson BP. The potential role of contraction-induced Myokines in the regulation of metabolic function for the prevention and treatment of type 2 diabetes. Front Endocrinol (Lausanne). 2017;8:97.
13. Cattagni T, Lepers R, Maffiuletti NA. Effects of neuromuscular electrical stimulation on contralateral quadriceps function. J Electromyogr Kinesiol. 2018 Feb;38:111–8.
14. Whitham M, Febbraio MA. The ever-expanding myokinome: discovery challenges and therapeutic implications. Nat Rev Drug Discov. 2016;15(10):719–29.
15. Sanchis-Gomar F, Lopez-Lopez S, Romero-Morales C, Maffulli N, Lippi G, Pareja-Galeano H. Neuromuscular electrical stimulation: a new therapeutic option for chronic diseases based on contraction-induced Myokine secretion. Front Physiol. 2019 Nov;28(10):1463.
16. Guo Y, Phillips EB, Atherton PJ, Piasecki M. Molecular and neural adaptations to neuromuscular electrical stimulation; implications for ageing muscle. Mech Ageing Dev. 2021;193:111402.
17. Mancinelli R, Toniolo L, Di Filippo ES, Doria C, Marrone M, Maroni CR, Verratti V, Bondi D, Maccatrozzo L, Pietrangelo T, Fulle S. Neuromuscular electrical stimulation induces skeletal muscle fiber remodeling and specific gene expression profile in healthy elderly. Front Physiol. 2019;10:1459.
18. Klika AK, Yakubek G, Piuzzi N, Calabrese G, Barsoum WK, Higuera CA. Neuromuscular electrical stimulation use after total knee arthroplasty improves early return to function: a randomized trial. J Knee Surg. 2022;35(1):104–11.
19. Peng L, Wang K, Zeng Y, Wu Y, Si H, Shen B. Effect of neuromuscular electrical stimulation after total knee arthroplasty: a systematic review and meta-analysis of randomized controlled trials. Front Med (Lausanne). 2021;8:779019.
20. Zhao Y, Zhang B, Wang Y, Liu J. Effectiveness of neuromuscular electrical stimulation for enhanced recovery after total hip replacement surgery: a randomized controlled trial. Geriatr Orthop Surg Rehabil. 2022;13:21514593221129528.
21. Hardy EJ, Hatt J, Doleman B, Smart TF, Piasecki M, Lund JN, Phillips BE. Post-operative electrical muscle stimulation attenuates loss of muscle mass and function following major

abdominal surgery in older adults: a split body randomised control trial. Age Ageing. 2022;51(10):afac234.
22. Walls RJ, McHugh G, O'Gorman DJ, et al. Effects of preoperative neuromuscular electrical stimulation on quadriceps strength and functional recovery in total knee arthroplasty. A pilot study. BMC Musculoskelet Disord. 2010;11:119.
23. Sumin AN, Oleinik PA, Bezdenezhnykh AV, Bezdenezhnykh NA. Prehabilitation in cardiovascular surgery: the effect of neuromuscular electrical stimulation (randomized clinical trial). Int J Environ Res Public Health. 2023;20(3):2678.
24. Oleinik P, Sumin AN, Bezdenezhnykh AV. Effects of the use neuromuscular electrical stimulation of the lower extremities in the prehabilitation of patients in a before cardiac surgery. Eur J Prevent Cardiol. 2021;28(1):zwab061.365.
25. Crevenna R. From neuromuscular electrical stimulation and biofeedback-assisted exercise up to triathlon competitions—regular physical activity for cancer patients in Austria. Eur Rev Aging Phys Act. 2013;10:53–5.
26. Crevenna R, Palma S, Licht T. Cancer prehabilitation—a short review. Memo Mag Eur Med Oncol. 2021;14:39–43.
27. Jandova T, Narici MV, Steffl M, Bondi D, D'Amico M, Pavlu D, Verratti V, Fulle S, Pietrangelo T. Muscle hypertrophy and architectural changes in response to eight-week neuromuscular electrical stimulation training in healthy older people. Life (Basel). 2020;10(9):184.
28. Gondin J, Cozzone PJ, Bendahan D. Is high-frequency neuromuscular electrical stimulation a suitable tool for muscle performance improvement in both healthy humans and athletes? Eur J Appl Physiol. 2011;111(10):2473–87.
29. Hasan S, Kandasamy G, Alyahya D, Alonazi A, Jamal A, Iqbal A, Unnikrishnan R, Muthusamy H. Effect of plyometric training and neuromuscular electrical stimulation assisted strength training on muscular, sprint, and functional performances in collegiate male football players. PeerJ. 2022;10:e13588.
30. Thomé GR, Costa RA, Marquezi ML, Aparecido JML, Durigan JLQ, Amorim CF, Liebano RE. Effects of neuromuscular electrical stimulation on torque and performance in recreational distance runners: a randomized controlled trial. J Bodyw Mov Ther. 28:369.
31. Veldman MP, Gondin J, Place N, Maffiuletti NA. Effects of neuromuscular electrical stimulation training on endurance performance. Front Physiol. 2016;7:544.
32. Filipovic A, Kleinöder H, Dörmann U, Mester J. Electromyostimulation—a systematic review of the effects of different electromyostimulation methods on selected strength parameters in trained and elite athletes. J Strength Cond Res. 2012;26(9):2600–14.
33. Babault N, Cometti G, Bernardin M, Pousson M, Chatard JC. Effects of electromyostimulation training on muscle strength and power of elite rugby players. J Strength Cond Res. 2007 May;21(2):431–7.
34. Kim J, Park J, Yang J, Kim Y, Kim I, Shim H, Jang C, Kim M, Kim M, Lee B. Effects of 8-week electromyostimulation training on upper-limb muscle activity and respiratory gas analysis in athletes with disabilities. Int J Environ Res Public Health. 2022;20(1):299.
35. Wicke J, Chianchiano B, Garner S, Cola JL. Neuromuscular electric stimulation as an alternative to dynamic warm-up for anaerobic power activities. Int J Sports Exerc Med. 2022;8(2):219.
36. Malone JK, Blake C, Caulfield BM. Neuromuscular electrical stimulation during recovery from exercise: a systematic review. J Strength Cond Res. 2014;28(9):2478–506.
37. Gobbo M, Maffiuletti NA, Orizio C, Minetto MA. Muscle motor point identification is essential for optimizing neuromuscular electrical stimulation use. J Neuroeng Rehabil. 2014;11:17.
38. Franz A, Klaas J, Schumann M, Frankewitsch T, Filler TJ, Behringer M. Anatomical versus functional motor points of selected upper body muscles. Muscle Nerve. 2018;57(3):460–5.
39. Botter A, Oprandi G, Lanfranco F, Allasia S, Maffiuletti NA, Minetto MA. Atlas of the muscle motor points for the lower limb: implications for electrical stimulation procedures and electrode positioning. Eur J Appl Physiol. 2011;111(10):2461–71.
40. Forrester B, Petrofsky JS. Effect of electrode size, shape, and placement during electrical stimulation. J Appl Res. 2004;4(2):346.

41. Flodin J, Juthberg R, Ackermann PW. Effects of electrode size and placement on comfort and efficiency during low-intensity neuromuscular electrical stimulation of quadriceps, hamstrings and gluteal muscles. BMC Sports Sci Med Rehabil. 2022;14(1):11.
42. Glaviano NR, Saliba S. Can the use of neuromuscular electrical stimulation be improved to optimize quadriceps strengthening? Sports Health. 2016;8(1):79-85.
43. Johnson M. Transcutaneous electrical nerve stimulation: mechanisms. Clin Appl Evid Rev Pain. 2007;1(1):7–11.
44. Johnson M. Transcutaneous Electrical Nerve Stimulation (TENS): research to support practice. Oxford University Press; 2014.
45. Teoli D, An J. Transcutaneous electrical nerve stimulation. StatPearls Publishing LLC. Last Update: October 31, 2022.
46. Grover CA, McKernan MP, Close RJH. Transcutaneous Electrical Nerve Stimulation (TENS) in the emergency Department for Pain Relief: a preliminary study of feasibility and efficacy. West J Emerg Med. 2018 Sep;19(5):872–6.
47. Báez-Suárez A, Martín-Castillo E, García-Andújar J, García-Hernández JÁ, Quintana-Montesdeoca MP, Loro-Ferrer JF. Evaluation of different doses of transcutaneous nerve stimulation for pain relief during labour: a randomized controlled trial. Trials. 2018;19(1):652.
48. Almeida TCDC, Figueiredo FWDS, Barbosa Filho VC, de Abreu LC, Fonseca FLA, Adami F. Effects of transcutaneous electrical nerve stimulation (TENS) on proinflammatory cytokines: protocol for systematic review. Syst Rev. 2017;6(1):139.
49. ELECTROPHYSICAL AGENTS - contraindications and precautions: an evidence-based approach to clinical decision making in physical therapy. Physiother Can. 2010 Fall;62(5):1–80.
50. Lago AF, de Oliveira AS, de Souza HCD, da Silva JS, Basile-Filho A, Gastaldi AC. The effects of physical therapy with neuromuscular electrical stimulation in patients with septic shock: study protocol for a randomized cross-over design. Medicine (Baltimore). 2018;97(6):e9736.
51. Bennett MI, Johnson MI, Brown SR, Radford H, Brown JM, Searle RD. Feasibility study of Transcutaneous Electrical Nerve Stimulation (TENS) for cancer bone pain. J Pain. 2010;11(4):351–9.

Chapter 26
Pulsed Electromagnetic Field Therapy

Siddharth M. Agrawal

Introduction

Definition

Pulsed electromagnetic field (PEMF) therapy is the application of pulsating magnetic fields generated by electromagnets having particular frequencies, waveforms and intensities, using a specially designed apparatus [1]. PEMF devices are designed in a variety of configurations that allow for targeted treatment areas of the body or provide whole body stimulation with electromagnetic fields. When applied optimally, it can enhance the body's oxygen levels and improve neurological functions by triggering several cellular mechanisms (electroporation, increase in calcium, reduces pro-inflammatory cytokines, etc.) [2].

> Electroporation occurs when electric charges create temporary pores in the cell membrane thereby increasing cell membrane permeability to ions and macromolecules such as DNA, RNA, drugs, and proteins

A study suggests that treatment with characteristic pulsed electromagnetic fields appears to have disease-modifying and fitness boosting effects [3].

S. M. Agrawal (✉)
Digital Wellness Directory, PEMF-devices.com, Pune, India
e-mail: info@digitalwellness.directory; psysid@gmail.com

N. C. Apostolopoulos et al. (eds.), *Fundamentals of Recovery, Regeneration, and Adaptation to Exercise Stress: An Integrated Approach*,
https://doi.org/10.1007/978-3-031-44270-4_26

How Is PEMF Applied?

Pulsed electromagnetic fields (PEMFs) are applied by using an apparatus that usually has two main components—a control unit and applicators. The applicators contain the electromagnets designed to non-invasively treat target areas, with whole-body PEMF mats also commonly used. The control unit provides the electric signal that drives the electromagnets to produce the PEMFs. In most systems, the control unit will offer different programs for treatment frequency, intensity, and duration. The applicators contain electromagnets and are usually fashioned in form of mats, pads, helmets, or rings.

A PEMF treatment session begins once the applicators have been positioned on or worn by the user and the PEMF device has been configured to produce the desired pulsed electromagnetic field. A session could last anywhere between 5 min to several hours, depending on the treatment parameters discussed in the section below entitled, "Determining the Treatment Parameters". A course of PEMF treatment sessions could vary from multiple times per day to weekly or monthly applications. Those who own their own PEMF devices use PEMF regularly, although no research is available which investigates the effects of PEMF therapy when used for several years. Contraindications exist and should be studied before application. You can find some of the known contraindications in the section entitled, "Contraindications" (below).

Cellular Mechanisms and Areas of Application

Pulsed electromagnetic fields have a number of complex systemic effects on various cells in the body. Pulsating magnetic fields stimulate movement of electrolytes and ions in the tissues and fluids of the body resulting in a vast array of micro-electric, chemical, and mechanical effects in the tissues of the body [2, 4]. They not only increase oxygen uptake in the mitochondria, with less lactate production, but also cause changes in the genetic structures [5]. By understanding some of the basic actions of PEMFs on various cell bodies, research on application for diseases and sports medicine becomes possible.

Bioavailability of charged ions such as calcium, potassium, sodium, chlorine, etc. in the cell membrane is a crucial element in healing and regeneration. Various membrane ion channels need to be triggered to enable healing processes after injury [6].

PEMFs induce the process of electroporation (see above in definition section). PEMFs penetrate the tissue and affect cellular energy by depositing charges near the membrane protein binding sites in red blood cells [7]. This allows for better nutrient up-take and increased cellular ion transport. The primary action of the PEMFs is on the physical level of the plasma membrane of the cells. It has a direct action on the voltage-gated calcium channels resulting in an increase in calcium ion availability and transport in the cell plasm. This triggers biochemical pathways that include enzymatic actions such as production of nitric oxide [8].

Nitric oxide is an important molecule that is produced by almost all cells in the body. It helps in vasodilation or keeping the blood vessels supple and strong. It has a beneficial effect on both the immune and the circulatory systems [9]. For athletes, it improves physical performance [10]. In 2018, scientists from the Institute of Anatomy in Germany conducted a study to investigate the cellular mechanism of PEMF and found that it can trigger voltage-gated calcium channels and may help in production of nitric oxide. They concluded that correct use of PEMF can lead to reduced inflammation and improve cellular resilience [2].

Anti-inflammatory Effects

PEMF exhibits anti-inflammatory effects by modulating the expression of various cellular information processing networks and signalling cascades restoring them to homeostatic production levels [11].

Pro-inflammatory cytokines are produced predominantly by activated macrophages and are involved in the up-regulation of inflammatory reactions. Some of these pro-inflammatory cytokines such as interleukins (IL-1β, IL-6), and tumour necrosis factor-α (TNF-α) are involved in the process of pathological pain [12]. Application of PEMFs have been found to reduce the release of TNF-α, IL-1β, IL-6 and IL-8 [13]. PEMFs upregulate the expression of transforming growth factor (TGF) and bone morphogenetic proteins (BMPs) to increase cell proliferation, osteogenesis, and chondrogenesis. PEMFs inhibit pro-inflammatory cytokines through adenosine receptors A2A and A3 BMPs [14].

> Adenosine Receptors: Adenosine is most commonly recognized with the molecule adenosine triphosphate (ATP) and is used by the body for metabolism. Adenosine initiates its biological effects via adenosine receptors. A2A is known to increase internal adenosine production, resulting in activation of protein kinase A (PKA) and phosphorylation of the cyclic response element binding protein (CREB) [15]. A3 is known to have anti-inflammatory effects [16].

Reduces Oxidative Stress

Several peer-reviewed studies suggest that PEMFs stimulate cell mitochondria, resulting in enhanced cell respiration/oxidative phosphorylation including the Krebs cycle or citric acid cycle, whereby increased enzyme levels are known to be associated with improved oxygen metabolism with less oxidative waste, known as reactive oxygen species (ROS) [13, 17, 18]. PEMFs induce human cryptochrome to modulate intracellular reactive oxygen species [19]. The implications of reducing oxidative stress are manifold. It not only helps against fighting diseases [20] but also improves athletic performance by stimulating the Krebs cycle and increased production of ATP [21].

> Krebs cycle is a part of cellular respiration process by which nutrients are broken down to release energy, which is stored in the form of ATP with the subsequent release of waste products.

In 2017, scientists from Germany and China collaborated to study the effects of extremely low frequency (ELF) PEMF on anti-oxidative defence mechanisms in human osteoblasts. They concluded that PEMF treatment may be used as an adjunct to conventional therapy supporting bone formation under oxidative stress conditions such as during fracture healing [22].

Boosts Regeneration

Mesenchymal stem cells (MSCs) are stromal cells that have the ability to self-renew and possess the capability of multilineage differentiation. MSCs are found in a variety of tissues, such as umbilical cord, endometrial polyps, menses blood, bone marrow and adipose tissue. Studies show that these cells can differentiate to form cells with the characteristics of adipose, cartilage, and bone cells and form human bone tissue after transplantation [23]. In a study from Korea, it was observed that electromagnetic fields (EMFs) had a protective effect against oxygen and glucose deprivation (OGD)-induced human mesenchymal stem cell (MSC) death. They found PEMF to protect the MSCs from OGD-induced cell death. The underlying mechanisms resulted in reduction of intracellular levels of calcium and ROS/RNS [24].

Common Areas of Application

The cellular mechanisms of PEMFs have widespread effects on human physiology, with PEMF therapy being applicable for a host of neurological and physiological disorders [25]. It has been observed to be beneficial for nerve regeneration [26], enhanced muscle repair [27], enhanced mental clarity [28], and reduction in fatigue [29]. PEMF therapy has been studied to improve outcomes of spinal fusion [30].

In Eastern Europe, over the past 60 years, PEMF therapy has been used to treat pain and oedema in soft tissue. Most high-quality meta-analysis', including this one from The Central European Journal of Medicine, recommend further trials and continued research on intensities, duration, and frequencies to discover more properties of PEMFs [31].

Within a clinical setting pulsed electromagnetic fields have been applied to treat the following:

- Pain management [32]
- Musculoskeletal chronic pain [33]
- Delayed or non-union fractures [34]
- Osteoporosis [35]
- Tendon disorders [36]
- Fibromyalgia symptoms [37]
- Multiple Sclerosis [38]
- Nerve regeneration [39] and Angiogenesis [25]
- Depression and anxiety [40]
- Parkinson's and Alzheimer's disease [41]
- Traumatic brain injury [42]

Other areas currently under research include cancer [43], epilepsy [44], eye disorders [45], and stroke rehabilitation [46].

Determining the Treatment Parameters

Like any other treatment, PEMF therapy can be experimented with clinically, by referencing published clinical trials and studies. Treatment parameters are important to achieve similar outcomes. Understanding the waveform of the electromagnetic field will help the practitioner get closer to choosing the right equipment and settings to achieve similar outcomes. To administer PEMF treatment, the following parameters defining the waveform of the electromagnetic field are considered:

1. Frequency (measured in Hz)
2. Amplitude/intensity (measured in Gauss (G) or milliTesla (mT) or Tesla(T))
3. Pulse Rise—Time (measured in nanoseconds)
4. Pulse Shape (Square, Sine, Triangular, etc.) (Fig. 26.1)

A combination of these parameters defines the waveform and consequently the techniques, protocols, and effects of treatment. Application techniques come in various terminologies based on how the electromagnetic field is applied. For example, high-intensity (1+ T) techniques such as transcranial magnetic stimulation (TMS) is used to explore brain functioning, while repetitive TMS (rTMS) is used to induce changes in brain activity that can last beyond the stimulation period [47].

Another factor would be the area covered by the field. The size/area of the field would be determined by the body part being treated. For example, a ring applicator would work well for a shoulder as one would be able to surround the shoulder joint with a PEMF field. A pillow or helmet applicator would be suitable for treating the head, a whole-body PEMF mat would be the obvious choice for treating the

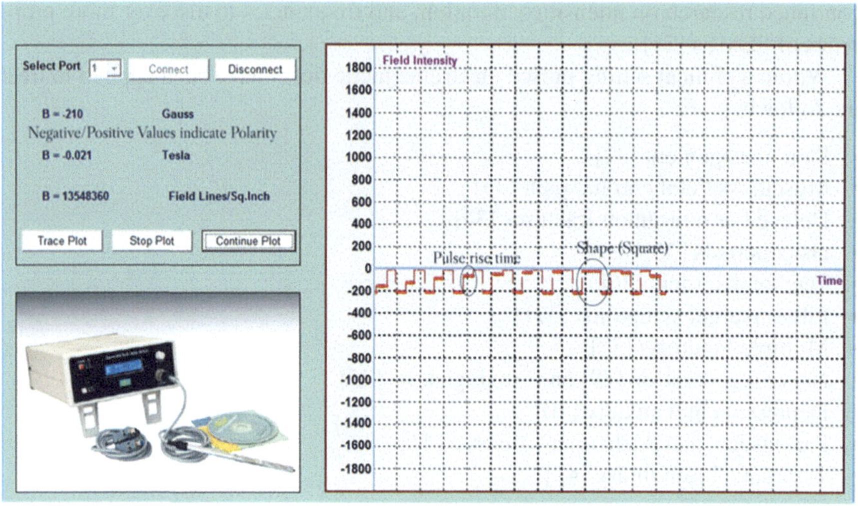

Fig. 26.1 Screenshot showing PEMF characteristics measured using a Tesla/Gauss meter with a Hall Effect sensor. The frequency, intensity, shape, and rise-time of the pulsed electromagnetic field can be measured this way. (Screenshot Courtesy: Scientech Technologies Pvt Ltd)

entire-body. The target tissue or area of application in the body also matters. Brain conditions such as Parkinson's disease may respond best if the PEMF application is made to a specific area of the brain. A functional neuroimaging study of patients with Parkinson's disease revealed that repetitive Transcranial Magnetic Stimulation increased the concentration of endogenous dopamine in the ipsilateral striatum [48]. Stimulation of the left dorsolateral prefrontal cortex (DLPFC) with 5–25 Hz rTMS increases ipsilateral dopamine release in Brodmann areas 25/12 and 32 as well as in Brodmann area 11, which is in the medial orbitofrontal cortex [49]. However, no significant changes were observed during stimulation of the right DLPFC. This suggests that it is necessary for the field to reach the right area of the brain for the desired effects on brain performance (Fig. 26.2).

Besides these parameters, the length of the treatment would also affect the outcomes which varies depending on the parameters above. For example, some athletes may use PEMF regularly up to 7 days a week for several years, with others sporadically using PEMF before and after events or workouts. The long-term and short-term effects of these applications would also vary depending on the intensity and frequency of the PEMF [50].

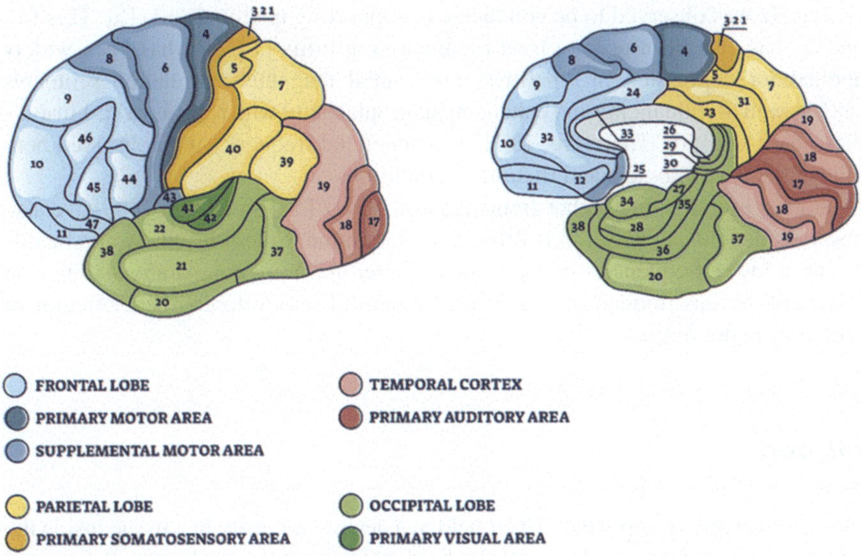

Fig. 26.2 Brodmann areas

Frequency

Most PEMF research utilizes lower frequencies, usually below 100 Hz [51]. Several meta-analysis studies compare the effects of different frequencies and intensities [51–53]. However, the lack of the availability of the comparative factors published in these trials, the non-disclosure of the parameters, and the treatment time of these various studies suggests the further need for comprehensive robust research.

Certain frequencies at certain intensities would suit some cells better than others [51] with some frequencies and intensity combinations working better for different molecular mechanisms and cell types. Neuronal cells for example display most energetic responses at 10 Hz, with the response dropping by 50% when moving to 5 and 15 Hz [5].

Schumann resonances are a naturally occurring geomagnetic field of the Earth. All biology has evolved in their presence. The Schumann resonances ranging between 1 and 20 Hz stabilize brain function and affect the brainwaves [53]. The dominant frequencies of the brainwaves are related to cognition and awareness states such as sleeping and level of alertness. Further, imbalanced brainwaves (faster or slower) are seen in a variety of mental and brain disorders [54]. PEMF therapy was found to be beneficial in the Schumann range, and not just for brain disorders. A study, utilizing 7.83 Hz, considered the fundamental Schumann resonance frequency, observed that at this frequency the growth of cancer cells was inhibited [55]. In another study, exposure to the Schumann resonance field was associated with a reduction in stress as well as exhibiting cardioprotective effects [56].

Ten Hz was observed to be conducive in supporting neuronal cells [5]. This frequency has also been used to treat rheumatoid arthritis [14, 57], has been widely applied for depression in repetitive transcranial magnetic stimulation protocols [58], as well as influencing and enhancing neuroplasticity with regards to Parkinson's disease effects [50]. Therefore, 10 Hz is a beneficial frequency with observed benefits in both neurological and rheumatic conditions.

Studies investigating higher frequencies at 50 Hz [59] and 75 Hz [60] have also observed benefits, however, it is difficult to determine if these results would be different at lower frequencies or intensities. Therefore, more comparative studies on effects of various frequency ranges will be helpful in developing better treatment protocols in the future.

Intensity

In order to properly apply the PEMF field at a desired intensity in a tissue inside the body, an appropriate calculation of the field intensity at the applicator surface and distance from the applicator needs to be made. This is measured with use of a gauss or tesla meter with a Hall Effect sensor. The intensity and distance have an inverse square relationship, so the further away a PEMF applicator is from the target tissue, the lower amount of its source intensity is available to the tissue. According to Coulomb's Inverse Square Law of Magnetism we learn that the intensity of a magnetic field drops by the square of the distance.

> A Hall Effect sensor detects the presence of magnetic fields. The Hall Effect PEMF product developers use measurement and mapping devices utilizing data from Hall Effect Sensors. These devices record the magnetic field frequency, intensity, and time (see Fig. 26.1).
>
> The Hall Effect is the principle by which a measurable voltage is obtained when a current carrying conductor, or a semiconductor, is exposed to a perpendicular magnetic field. The voltage is measured at the right angle to the path of the current.

A NASA report in 2008 observed that low intensity (0.01–0.1 Gauss) 10 Hz PEMF modulates more than 10,000 genes in neuronal cells. "Specific genes such as human germline oligomeric matrix protein, prostaglandin endoperoxide synthase 2, early growth response protein 1, and insulin-like growth factor binding protein 3 precursor are highly up-regulated. Keratin Type II cytoskeletal 7, mitotic kinesin like protein 1, transcription factor 6 like 1, mitotic feedback 26 control protein, and cellular retinoic acid binding protein are down-regulated. These two sets are only an example from the approximately 320 gene changes expressed as a consequence of exposure to the field." As a result of these gene modulation effects, PEMF exposure can significantly increase the cell count and viability of neuronal cells [5]. This study observed that even at low intensities, PEMF can have gene modifying effects *in-vitro*.

Medium to high-intensity pulsed electromagnetic fields are generally considered to be between the 0.3 and 2 T range. With the penetration depth of high-intensity PEMF being much higher, it is possible to deliver higher energy fields deeper into the body, inducing a higher physiological response. These ranges are frequently used in Transcranial Magnetic Stimulation (TMS) treatments [61].

Other Waveform Characteristics

Pulse dynamics such as the rise-time and shape of the waveform are relevant as charged ions in cellular membranes are influenced by these characteristics. The rise-time measured in seconds would be the speed at which the electromagnetic field is introduced to the cell membrane. The shape of the pulse would affect the direction of the magnetic field as well as the energy delivered to the target tissue(s). Square waves transfer the highest energy with the fastest rise times, while sinusoidal waves provide a gentle but more consistent and penetrative field. With many PEMF devices one is able to switch the current direction in order to deliver multiphasic fields [62].

Presently, much of PEMF use is in the experimental domain. There are some parameters such as frequencies and intensities that are commonly used, however further optimization of treatment parameters specific to individuals and conditions is always plausible. With the technology being investigated for a long time, some treatment frequencies such as 10 Hz are widely now accepted to be regenerative as confirmed in several studies and trials [14, 57, 58]. However, there is still a lot of work remaining in devising treatment parameters with more comparative studies needed. For practical applications, the treatment parameters currently used, and based on existing research studies and clinical evidence, are suggested for the recovery, regeneration, and adaptation of the athletes. The aim of the subsequent sections below is to provide a starting point for researchers and practitioners to apply PEMF therapy in sports medicine fields.

Sports Physical Therapy and Kinesiology

Sports physiotherapists and kinesiologists regularly advise athletes on injury prevention to mitigate injury risk. With injury risks varying from sport to sport, PEMF therapy has tremendous use as a tool for rehabilitation medicine with great benefits observed after orthopaedic surgery and neuro-rehabilitation [42, 46, 63–65]. As these disciplines aim to enhance bone, joint and muscle recovery, and improve flexibility endurance and energy, sports physiotherapists and kinesiologists may consider adopting PEMF therapy for use with sports associated with frequent injuries. With PEMF, kinesiologists may see an improvement in movement and movement patterns in athletes, with physiotherapists possibly noticing improved bone, muscle and soft tissue health thereby aiding in a rapid and better recovery from sports injuries.

Musculoskeletal Healing

PEMF therapy has been observed to have an accelerated improvement in bone fracture healing [66] and the ability to reduce neuroinflammation [42]. In addition, flexibility might improve in the short-term due to the effects of PEMFs on the softening of the collagen associated with scars or adhesions and its ability to enhance circulation [67].

PEMF promotes osteogenesis, since it has a positive effect on bone and cartilage regeneration [68, 69]. Another common condition amongst athletes is arthritis, with the prevalence of osteoarthritis reported to be much higher in this population [70]. This study observed that the incident and prevalence of osteoarthritis in lower limbs was higher amongst former elite athletes compared to the general population [71]. Synovitis plays an important role in the pathogenesis of arthritis, which is closely related to the inflammation and pain of the joint. The synovial membrane is a connective tissue on the inner surface of the synovial joint capsule playing an important role in the maintenance of joint homeostasis [72]. In 2020, scientists from the Army Medical University of China conducted a study on the anti-inflammatory effects in of PEMFs on synovitis. They concluded that PEMF noninvasively exhibits the anti-inflammatory effect on synovitis via the upregulation of the efferocytosis in macrophages in the p38 MAP kinase pathway (p38MAPK). In cell biology, efferocytosis is the process by which apoptotic cells are removed by phagocytic cells [71]. Recent studies show that p38MAPK activation affects the control of cytoskeleton remodelling [73] responsible for maintaining homeostasis during acute dynamic stretching of muscles [74].

Articular cartilage injury is prevalent in football players as a result of chronic joint stress or acute traumatic injuries which often progresses to a painful impairment of joint function limiting sports participation [75]. A chapter published in *The Patellofemoral Joint* observed positive results with the use of PEMF in active patients presenting with symptomatic cartilage lesions involving the patellofemoral joint. Both *In-vitro* and *in-vivo* studies demonstrated that PEMFs have the ability to influence cartilage metabolism through pro-anabolic and anti-catabolic activities. In addition, clinical studies demonstrated the efficacy of PEMFs in patients presenting with symptomatic knee cartilage lesions [76].

Tendinopathies

Tendinopathy, which is increasing in prevalence, accounts for a substantial part of all sports injuries and occupational disorders [77]. Recently, more than a dozen scientists from the *PMU Spinal Cord Injury & Tissue Regeneration Center* in Austria provided some insight on how PEMF therapy can be optimized for the non-invasive therapy of tendinopathies at the cellular level [78].

One of the first clinical trials of PEMF therapy for treatment of rotator cuff tendinitis was conducted in 1984, with the scientists concluding that PEMF may be useful in treatment of severe and persistent rotator cuff tendinitis and other chronic tendon lesions [79]. In a randomized clinical trial study at the University of California in 2015, application of pulsed electromagnetic fields after rotator cuff surgery was found to be safe, reducing postoperative pain, analgesic use, and stiffness in the short term [80].

Delayed Onset Muscle Soreness

In 2016, a randomized clinical trial studied the effects of PEMF therapy on delayed onset muscle soreness (DOMS)-related variables in elbow flexors at 24, 48, and 72 h after DOMS induction exercise. The results indicated that PEMF may be useful as a modality to reduce DOMS symptoms with its application found to be more effective than the sham treatment in reducing the physiological symptoms associated with DOMS [81].

Therefore, as discussed above, use of PEMF as a tool by sports physical therapists and kinesiologists may prove to be a useful modality in the treatment of and regeneration of neurons [82], muscles, bones, joints, and cartilages. Further experimentation with PEMF and its impact in these tissues will be helpful in establishing new injury prevention and recovery strategies and protocols.

Sports Performance Training

Athletic and sports trainers provide treatments and protocols for preventing, diagnosing, and treating musculoskeletal injuries and diseases in athletes. They also provide athletes with emergency treatment and implement rehabilitation programs in case of sports injuries. The PEMF treatment protocols for musculoskeletal injury recovery would be similar to those employed by sports physiotherapists, however as coaches are also involved with injury prevention, they need treatments that can improve performance. PEMF therapy has interesting applications in this field due its effects on the cardiovascular system and tissue reengineering.

High level athletes have great compensation abilities for being physiologically efficient at maintaining homeostasis or balancing performance levels through allostatic mechanisms.

Allostasis is the efficient regulation to satisfy the body's needs before they arise by budgeting those needed resources such as oxygen, insulin etc., as opposed to homeostasis, in which the goal is a steady state.

The ability to compensate in response to physical stressors suggests that athletes have the potential of preventing further injury. Applying PEMF therapy can be an effective rapidly acting homeostasis strategy aiding in the maintenance of a better regulatory condition to avoid injury. One goal of training is to enable the athlete to perform longer and better, while maintaining the body's homeostasis. The cardiovascular system plays an important role in maintaining homeostasis during exercise and physically demanding sports. To accommodate the increased metabolic activity in skeletal muscle, the cardiovascular system must properly control the transport of oxygen and carbon dioxide, as well as help to buffer the pH level of active tissues. Proper β-adrenergic signalling is indispensable for modulating heart rate. In 2016, Italian scientists conducted a study on low-frequency PEMF and β-adrenergic signalling. They found that PEMF therapy reduced cardiac β-adrenergic response at molecular, functional and adaptative levels [83]. In *in-vivo* and *in-vitro* models, PEMF was found to stimulate processes critical for angiogenesis (formation of new blood vessels) [84].

Applied sports technologists can measure the effectiveness of PEMF treatment for sports performance using heart rate variability tests. Heart rate variability parameters are relevant in the analysis of stress that the body experiences during training providing insights into physiological recovery after training [85]. Heart rate variability testing can have a great value in evaluating sports performance when analysed as a trend over time. Balancing the autonomic responses for optimum performance depends on the recovery-stress ratio and the type of training dose [85].

In 2007, scientists from Austria conducted a clinical trial to study the effects of PEMF therapy on heart rate variability. They studied the effects of 50 Gauss or 0.005 Tesla PEMFs after physical exercise. They concluded that the PEMF treatment improved heart rate variability and accelerated recovery after exercise [86].

Besides improving heart rate variability efficiency, PEMF therapy improves blood circulation [87] and, therefore, tissue oxygenation, allowing for more efficient muscle activity and faster recovery after muscle activity [88].

Pulsed electromagnetic fields can also improve production of ATP in tissues by between 100% and 600% [89]. This This study also showed the potential of PEMF therapy for reducing the effects of jet lag. It also shows that PEMF treatment for ATP enhancement is not "one size fits all", individual reactions will vary. However, the average boost in ATP levels was more than 100% across two groups (healthy British and Chinese volunteers).

In addition to supporting heart health in athletic training, the use of PEMF treatment in conjunction with athletic training has been observed to enhance the anabolic effects on the bone and cartilage matrix and its anti-catabolic effects of inflammation in the joint environment [90, 91]. These qualities, in addition to the positive effects on homeostasis, blood circulation, bone joint and cartilage preservation [87] suggests that coaches should consider the use of PEMF therapy as a viable performance enhancement tool.

Similar to any other treatment protocols, performance testing while using PEMF therapy can provide additional optimization ideas derived from practical applications. This data would be crucial to understand the combined neurological and physical recovery effects of PEMF is having on the performance of the athlete and enable sports performance coaches to implement it in practice. Brain injury prevention and treatment with PEMF is also possible, this will be discussed in the subsequent section on Neurology.

Body Building and Strength Training

Strength training and body building focus on increasing power by optimizing muscle mass and hypertrophy and reducing loss of bone density. A proper training schedule, nutrition and supplementation are important aspects of strength training [92]. PEMF therapy can be a useful adjunct as it can help improve muscle function [93], bone density [94] and tissue oxygenation [95]. These actions can result in improvement of performance and a decrease in recovery time after workouts and injuries.

Power sports such as weight lifting, and American style football seem to be associated with higher systemic blood pressure levels [96]. Several studies reported that athletes participating in weightlifting, water sports and American style football have higher blood pressure levels and are more likely to suffer from prehypertension or hypertension compared to other studies on endurance athletes [97, 98]. Pulsed electromagnetic field therapy is known to improve blood pressure and vascular function in hypertensive individuals [99].

> Celebrity trainer and professional body builder Kris Gethin reveals, "I have used PEMF for many years and I also suggest it to my clients which include high performing celebrities and professional athletes. I have found that ATP energy is restored much faster so I can push my workouts harder with less rest periods and recovery is much quicker through increased oxygenation and heat shock protein production. I have had stem cell procedures several times and using PEMF during these times have helped further mobilize stem cells for a much more efficacious treatment.

In a recent review, various studies have demonstrated the positive effects of PEMF of 0.2–5.0 mT and 15–75 Hz on the proliferation and differentiation of Mesenchymal stem cells (MSCs) at various treatment times [100].

PEMFs help muscles work harder, longer and recover faster. Muscle strength increased according to a study using PEMF therapy for Parkinson's disease functional force improvement [101].

All physical sports demand excellent strength training, and it's especially important for weight lifters and bodybuilders. Using PEMF can help to regain joint strength, leading to quicker recovery, and improved performance [102].

Sports Psychology and Neurology

Adaptation to physical activity and training is a complex physiological process with greater to normal stress on the body, being interspersed with sufficient recovery periods. As a rule of thumb, it is assumed that athletes can adapt to training, however the degree of adaptation depends on many factors of which one of them is mental performance [103]. In professional sports, the focus shifts towards delivering high level performance in a competitive environment from simply enjoying the sport or the health benefits. High-level athletes can experience psychological effects that are different than the recreational athlete [104]. Coaches and sports psychologists alike regularly use sports psychology principles to motivate and improve performance of athletes. Studies suggest that PEMF therapy may be used as an adjunct modality for improving mental health, sleep, and the neuroplasticity of the brain tissues in sports [105–107].

Mental Health

In a double-blind pilot study utilizing a low field magnetic stimulation at an intensity level of less than 30 Gauss, the authors observed that the stimulation of the brain with the PEMF field surrounding the brain produced a mood-enhancing effect that was beneficial for the treatment of treatment-resistant depression [105]. Thus, PEMFs can be used for professional athletes who have regular stress factors such as hectic work schedules, frequent travels, and the competitiveness and politics of sports.

Sleep

Adequate sleep is an essential for efficient and effective body function, with sleep deprivation being common amongst professional sports impacting performance negatively [106, 108]. PEMFs have been found through brain entrainment effects to improve sleep and hence optimize mental and cognitive functioning. Regular application of PEMF therapy for insomnia has been shown to improve sleep [106]. With deep sleep being very important for sports training, affecting the physical and mental performance of the athlete. The use of PEMFs may prove to be another tool for better performance.

Neurological Conditions

In 1997, Dr. Sandyk, a well-known PEMF researcher, who has done a considerable amount of work in various modes of electromagnetic field therapy for neurological conditions, published a case study of a successful resolution of sleep paralysis by weak electromagnetic field in a patient with multiple sclerosis. He concluded that resolution of sleep paralysis in this patient, by alternating current pulsed applications of EMFs, was related to the enhancement of melatonin circadian rhythms and cerebral serotoninergic neurotransmission [107].

In a study from Laurentian University, Canada, it was found that 30 min of weak PEMFs once per week for 5 weeks reduced phobias and improved symptoms in treatment-resistant depression in traumatic brain injury (TBI) patients [109]. In addition, PEMF therapy was observed to improve the brain function in Parkinson's disease, Alzheimer's disease [110], and stroke [111].

Concussions

With concussions being a regular occurrence in several sports (i.e., football, ice hockey, rugby, etc.), PEMF may prove to be a great tool in mitigating the effects of concussions and improving recovery. In 1999, scientists from Israel conducted a study on applying slow-rate or low-frequency magnetic stimulation to 64 patients with a moderate head injury. They observed and concluded that magnetic field stimulation was beneficial and that it should be considered as a complementary therapy [112]. In another study investigating traumatic brain injury and the use of PEMF, it was concluded that PEMF resulted in a reduction in inflammation of the brain [42].

In a study conducted in 2016, a reduction of chronic pain was observed in individuals suffering from mild TBI related headaches with the use of repetitive transcranial magnetic stimulation. With no side-effects being seen the authors suggested further randomized clinical trials [113]. In support of further use of rTMS for treatment of brain injuries, a study from Johns Hopkins University endorsed the application of magnetic stimulation for brain injuries and reported that low frequencies are advocated since no negative side-effects were observed [113]. As we can observe, PEMFs not only balance brain chemistry, but they also help stabilize and improve neurological health, which combined can produce an improvement in reaction time and performance. Standing balance tests have shown magnetic fields decrease sway, indicating enhanced homeostatic neurologic muscular control mechanisms [114]. PEMFs may prove to be beneficial for helping to improve the proprioception of the concussed athlete.

Comparison with Transcranial Direct Current Stimulation (tDCS)

"Brain zapping" using tDCS electrotherapy applications has shown positive results [115]. However, the effects of tDCS and transcranial pulse current stimulation (tPCS) are limited by the flow of current along the path of least resistance. Neither can they penetrate and affect the brain tissues with their original waveforms because of this natural resistance. Therefore, electric stimulation produces the effects on the brain through higher resistance and therefore requires higher voltage. With PEMF therapy application being *contact-less* because it is electromagnetic, this allows for deeper penetration and stimulation of the brain using lower intensity, providing both an excitatory stimulation and also a clearer waveform or signal allowing for a more harmonic treatment [51]. PEMF may prove to be a better method than tDCS [116] for optimizing neurotransmitters, allowing for faster recovery from brain injury and for a better mental and physical performance.

Summary

Often times, athletes spend long hours training, practicing, and competing with little work-life balance, which can decrease their quality of life. Using PEMF therapy can have meaningful benefits when applied in sports psychology and neurology by enhancing mental performance and an improved brain function leading to happier, better rested and better performing athletes.

Sports Nutrition

Optimized diet and nutrition is prescribed for athletes to achieve peak performance in training and competition [117]. In athletes who have to undertake heavy-training, negative physiological changes occur in the immune system along with increased inflammation and oxidative stress [118]. PEMF may prove to be a useful tool for sports nutritionists working with elite athletes, due to its effects on reducing inflammation [11], reducing oxidative stress [43], and supporting the immune system [119].

Sports medicine suggests that oxygen greatly enhances the body's absorption of vitamins, minerals, amino acids, proteins and other important nutrients [120]. Oxygen improves brain function and memory [121]. However, it's widely known that in high doses oxygen is toxic [122]. Thus, there is the need for the correct catalyst which raises the cell membrane potential itself.

Although oxygen is not recognised as a nutrient in the context of nutritional science and in relation to whole-body physiology, it is frequently described as such in studies at a cellular level [123, 124]. The central role of oxygen as a nutrient is in mitochondrial respiration, acting as an electron acceptor thereby enabling ATP to be formed through oxidative phosphorylation. This process is fundamental to aerobic organisms, with the oxidation of glucose and fatty acids requiring the continuous provision of oxygen. Several core metabolic pathways are central to mitochondrial oxidative phosphorylation—glycolysis, glycogenolysis, lipolysis, and the tricarboxylic acid (Krebs) cycle.

There are differences between cell types according to the level of respiration and oxidative phosphorylation that is functionally required. White adipocytes have fewer mitochondria which contain limited cristae, with most of the volume of these cells being due to the lipid droplet [125]. Brown adipocytes, in marked contrast, contain large numbers of mitochondria with a highly developed and dense cristae structure, especially in rodents adapted to cold environments when maximum non-shivering thermogenesis is required [125]. In these circumstances, brown fat mitochondria utilise substantial amounts of oxygen in order to sustain the oxidation of fatty acids and other substrates at high rates, with ATP synthesis being bypassed through a proton leakage pathway regulated by uncoupling protein-1 (UCP1) [126].

Adaptogen can be any substances or agents that promote normal functioning of the body by being able to do both, activate or inhibit, depending on the imbalance that needs correction. Adaptogenic herbs are commonly used for stress reduction

PEMF therapy may work in synergy with nutritional programs, appearing to qualify as an adaptogen and an anabolic when paired with rigorous physical training. In a research article, it was concluded that PEMF treatment was capable of inducing metabolic adaptations independently of exercise. Metabolic adaptations refer to the changes in the body which allow it to utilise energy more effectively and efficiently. It was found that PEMF treatment accentuates adipose browning and insulin-sensitivity in response to exercise. UCP1 transcript levels increased in inguinal white adipose tissue, but not interscapular brown adipose tissue, following 6 weeks PEMF intervention (10 min/week) in conjunction with exhaustive exercise [127].

High-altitude exposures can reduce splanchnic perfusion (blood supply to the gastrointestinal tract, liver, spleen, and pancreas) and blood oxygen levels causing hypoxic and oxidative stress. These stressors might injure the intestinal barrier leading to consequences such as bacterial translocation and local/systemic inflammatory responses [128]. In 2019, scientists from China conducted a study on the effects of PEMF therapy for high-altitude stress ulcer (HSU) in rats. They found that low-frequency PEMFs can penetrate stomach tissues to relieve the symptoms of high-altitude stress ulcer and promote the regeneration of disturbed tissues [129]. Therefore, PEMF can be studied further as a useful modality to improve high-altitude acclimatization in athletes and also shows protective effects on the gastric and digestive system.

Nutritionists also pay attention to bone health in athletes as bones are nutritionally modulated tissues. In 2022, scientists from Italy published a research review on the effects of PEMF therapy in bone healing in animal models. They concluded that PEMF proved useful in bone injury and in conditions of bone metabolic impairment (such as osteoporosis induced by diabetes or corticosteroids), as well as in the prevention of osteoporosis from disuse [130].

Based on these studies and others, sports nutritionists and researchers can consider applying PEMF therapy as an adjunct treatment to support nutritional programs due to anti-inflammatory and oxidative stress reduction effects and improvements in digestive system and bone health. Further research can be carried out to deduce the effects of various PEMF configurations for best results with particular nutrients to devise even more effective nutritional strategies for athletes.

Contraindications

Numerous clinical trials [131–134] and meta-analyses [33, 135] report no relevant adverse effects from the use of PEMF therapy.

Use of PEMF is contraindicated in the case of pregnancy and individuals with pacemaker or other electronic implants. It's contraindicated for pregnancy as it affects the growth rate of the foetus. While it's not allowed for patients with pace makers and electronic implants as the electromagnetic field can interfere with the functioning of the electronic circuitry [33].

In some cases, PEMF therapy is not recommended for malignancies since it increases circulation, it can potentially promote tumour growth due to improved metabolism of tumour cells, and should be discontinued during administration of chemotherapy, but has proven to be useful during recovery [136]. Research is ongoing on PEMF for cancer and treatment methods are being devised which are showing benefits [137, 138].

Some people may experience skin rash and aggravated pain [139]. Nausea, fatigue, agitation, headaches, dizziness or vertigo, and warm/cold sensations might manifest temporarily [140]. Changing or optimizing the parameters of the PEMF used in the treatment may prevent some of these effects from manifesting as specific frequencies and amplitudes will affect the target tissues and cells differently [51, 137].

Conclusion

Current knowledge and clinical evidence on PEMF therapy suggests potential applications in several sports medicine disciplines such as physiotherapy, athletic training, and nutrition. Knowledge about the body's electromagnetic field with regards to sports training and rehabilitation science, may provide another avenue of pursuit, especially if the concern is to improve sports performance and decrease the effects of inflammation and the inflammatory response.

Good sports training and rehabilitation science should include knowledge about our body's electromagnetic functions, and this knowledge can be used to provide improvement in sports performance.

With rehabilitation and injury mitigation being paired with faster recovery and sharper mental performance, the field of PEMF may prove to be an exciting area for sports medicine. However, further studies and collaborative work between researchers and practitioners of PEMF are needed to ask the questions and formulate the research as to the benefit of PEMF for the improved recovery, regeneration, and adaptation of the athlete to the training stimulus.

1. How effective is PEMF in injury prevention and improving sportsperformance?
2. What inherent mechanisms of PEMF aid in the recovery, regeneration,and adaptation of the body to exercise?
3. The effectiveness of use of PEMF for neurological conditions (i.e.,concussions), and oxidative stress in exercise?
4. The usefulness of PEMF pre or post sports training for enhancingperformance, speed of recovery, and healing of connective tissue?
5. The effectiveness of PEMF in combination with sports physiotherapyfor treatment of sports injuries (i.e., muscle strains, sprains, tears)?
6. Use of PEMF for ameliorating inflammation and the inflammatoryresponse?
7. Does PEMF improve the quality of sleep for the athletepre-competition?
8. What parameters of PEMF are important to consider for the treatment ofsports injuries?

> Applying pulsed electromagnetic field therapy for sports performance, sports medicine and physical therapy involves selection of treatment parameters that will influence the outcome. The treatment parameters include the characteristics of the pulsed magnetic field (intensity, waveform, frequency, etc.), the regimen of treatment sessions (duration of sessions, number of sessions, timing of sessions (before exercise, after exercise, sleep-time, etc.), and the duration of use (short-term/long-term). Due to the large number of variables in the treatment parameters, a practitioner would require an in-depth knowledge of clinical trials and research review articles to devise treatment protocols for the desired outcome. Secondly, they would need easy access to a range of PEMF therapy equipment that can allow them to administer PEMF with greater success.

References

1. Gaynor JS, Hagberg S, Gurfein BT. Veterinary applications of pulsed electromagnetic field therapy. Res Vet Sci. 2018;119:1–8. https://doi.org/10.1016/j.rvsc.2018.05.005.
2. Funk RH. Coupling of pulsed electromagnetic fields (PEMF) therapy to molecular grounds of the cell. Am J Transl Res. 2018;10(5):1260–72. https://www.ncbi.nlm.nih.gov/pubmed/29887943
3. Bassett CA. Fundamental and practical aspects of therapeutic uses of pulsed electromagnetic fields (PEMFs). Crit Rev Biomed Eng. 1989;17(5):451–529. https://www.ncbi.nlm.nih.gov/pubmed/2686932
4. Kim CH, Wheatley-Guy CM, Stewart GM, Yeo D, Shen WK, Johnson BD. The impact of pulsed electromagnetic field therapy on blood pressure and circulating nitric oxide levels: a double blind, randomized study in subjects with metabolic syndrome. Blood Press. 2020;29(1):47–54. https://doi.org/10.1080/08037051.2019.1649591.
5. Goodwin TJ, Lyndon B. Johnson Space Center, Houston, Texas 77058. Physiological and molecular genetic effects of time-varying electromagnetic fields on human neuronal cells. Published online 2003. https://ntrs.nasa.gov/api/citations/20030075722/downloads/20030075722.pdf
6. Tyler SEB. Nature's electric potential: a systematic review of the role of bioelectricity in wound healing and regenerative processes in animals, humans, and plants. Front Physiol. 2017;8:627. https://doi.org/10.3389/fphys.2017.00627.
7. Loginov VA. Change in charge of the erythrocyte membrane during exposure to a pulsed electromagnetic field. Biofizika. 1991;36(4):614–20. https://www.ncbi.nlm.nih.gov/pubmed/1793745
8. Salais V, Lopez CA , Perzabal CT, Journals R. Mechanisms of action and effects of pulsed electromagnetic fields (PEMF) in medicine. Published August 24, 2020. Accessed 25 May 2023. https://respubjournals.com/medical-research-surgery/Mechanisms-of-Action-And-Effects-of-Pulsed-Electromagnetic-Fields-PEMF-in-Medicine.php.
9. Lundberg JO, Weitzberg E. Nitric oxide signaling in health and disease. Cell. 2022;185(16):2853–78. https://doi.org/10.1016/j.cell.2022.06.010.
10. Oral O. Nitric oxide and its role in exercise physiology. J Sports Med Phys Fitness. 2021;61(9):1208–11. https://doi.org/10.23736/S0022-4707.21.11640-8.

11. Ross CL, Zhou Y, McCall CE, Soker S, Criswell TL. The use of pulsed electromagnetic field to modulate inflammation and improve tissue regeneration: a review. Bioelectricity. 2019;1(4):247–59. https://doi.org/10.1089/bioe.2019.0026.
12. Zhang JM, An J. Cytokines, inflammation, and pain. Int Anesthesiol Clin. 2007;45(2):27–37. https://doi.org/10.1097/AIA.0b013e318034194e.
13. Vincenzi F, Ravani A, Pasquini S, et al. Pulsed electromagnetic field exposure reduces hypoxia and inflammation damage in neuron-like and microglial cells. J Cell Physiol. 2017;232(5):1200–8. https://doi.org/10.1002/jcp.25606.
14. Iwasa K, Reddi AH. Pulsed electromagnetic fields and tissue engineering of the joints. Tissue Eng Part B Rev. 2018;24(2):144–54. https://doi.org/10.1089/ten.TEB.2017.0294.
15. Sheth S, Brito R, Mukherjea D, Rybak L, Ramkumar V. Adenosine receptors: expression, function and regulation. Int J Mol Sci. 2014;15(2):2024–52. https://doi.org/10.3390/ijms15022024.
16. Gessi S, Merighi S, Varani K, Leung E, Mac Lennan S, Borea PA. The A3 adenosine receptor: an enigmatic player in cell biology. Pharmacol Ther. 2008;117(1):123–40. https://doi.org/10.1016/j.pharmthera.2007.09.002.
17. Hollenberg AM, Huber A, Smith CO, Eliseev RA. Electromagnetic stimulation increases mitochondrial function in osteogenic cells and promotes bone fracture repair. Sci Rep. 2021;11(1):19114. https://doi.org/10.1038/s41598-021-98625-1.
18. Woo SH, Kim B, Kim SH, Jung BC, Lee Y, Kim YS. Pulsed electromagnetic field potentiates etoposide-induced MCF-7 cell death. BMB Rep. 2022;55(3):148–53. https://doi.org/10.5483/bmbrep.2022.55.3.119.
19. Sherrard RM, Morellini N, Jourdan N, et al. Low-intensity electromagnetic fields induce human cryptochrome to modulate intracellular reactive oxygen species. PLoS Biol. 2018;16(10):e2006229. https://doi.org/10.1371/journal.pbio.2006229.
20. Maurya PK, Dua K, editors. Role of oxidative stress in pathophysiology of diseases. Singapore: Springer; 2020. https://doi.org/10.1007/978-981-15-1568-2.
21. Hadžović-Džuvo A, Valjevac A, Lepara O, Pjanić S, Hadžimuratović A, Mekić A. Oxidative stress status in elite athletes engaged in different sport disciplines. Bosn J Basic Med Sci. 2014;14(2):56–62. https://doi.org/10.17305/bjbms.2014.2262.
22. Ehnert S, Fentz AK, Schreiner A, et al. Extremely low frequency pulsed electromagnetic fields cause antioxidative defense mechanisms in human osteoblasts via induction of •O2- and H2O2. Sci Rep. 2017;7(1):14544. https://doi.org/10.1038/s41598-017-14983-9.
23. Ding DC, Shyu WC, Lin SZ. Mesenchymal stem cells. Cell Transplant. 2011;20(1):5–14. https://doi.org/10.3727/096368910X.
24. Hyeok Jung J, Young KJ. Electromagnetic field (10 Hz, 1 mT) protects mesenchymal stem cells from oxygen-glucose deprivation-induced cell death by reducing intracellular Ca^{2+} and reactive oxygen species. J Appl Biomed. 2017;15(2):112–8. https://doi.org/10.1016/j.jab.2016.11.003.
25. Peng L, Fu C, Wang L, et al. The effect of pulsed electromagnetic fields on angiogenesis. Bioelectromagnetics. 2021;42(3):250–8. https://doi.org/10.1002/bem.22330.
26. Sisken BF, Kanje M, Lundborg G, Kurtz W. Pulsed electromagnetic fields stimulate nerve regeneration in vitro and in vivo. Restor Neurol Neurosci. 1990;1(3):303–9. https://doi.org/10.3233/RNN-1990-13419.
27. Ong MTY, Man GCW, Lau LCM, et al. Effect of pulsed electromagnetic field as an intervention for patients with quadriceps weakness after anterior cruciate ligament reconstruction: a double-blinded, randomized-controlled trial. Trials. 2022;23(1):771. https://doi.org/10.1186/s13063-022-06674-2.
28. Hattapoğlu E, Batmaz İ, Dilek B, Karakoç M, Em S, Çevik R. Efficiency of pulsed electromagnetic fields on pain, disability, anxiety, depression, and quality of life in patients with cervical disc herniation: a randomized controlled study. Turk J Med Sci. 2019;49(4):1095–101. https://doi.org/10.3906/sag-1901-65.

29. Granja-Domínguez A, Hochsprung A, Luque-Moreno C, et al. Effects of pulsed electromagnetic field therapy on fatigue, walking performance, depression, and quality of life in adults with multiple sclerosis: a randomized placebo-controlled trial. Braz J Phys Ther. 2022;26(5):100449. https://doi.org/10.1016/j.bjpt.2022.100449.
30. Bose B. Outcomes after posterolateral lumbar fusion with instrumentation in patients treated with adjunctive pulsed electromagnetic field stimulation. Adv Ther. 2001;18(1):12–20. https://doi.org/10.1007/bf02850247.
31. Wagner B, Steiner M, Markovic L, Crevenna R. Successful application of pulsed electromagnetic fields in a patient with post-COVID-19 fatigue: a case report. Wien Med Wochenschr. 2022;172(9–10):227–32. https://doi.org/10.1007/s10354-021-00901-2.
32. Yang X, He H, Ye W, Perry TA, He C. Effects of pulsed electromagnetic field therapy on pain, stiffness, physical function, and quality of life in patients with osteoarthritis: a systematic review and meta-analysis of randomized placebo-controlled trials. Phys Ther. 2020;100(7):1118–31. https://doi.org/10.1093/ptj/pzaa054.
33. Paolucci T, Pezzi L, Centra AM, Giannandrea N, Bellomo RG, Saggini R. Electromagnetic field therapy: a rehabilitative perspective in the management of musculoskeletal pain – a systematic review. J Pain Res. 2020;13:1385–400. https://doi.org/10.2147/JPR.S231778.
34. Punt BJ, den Hoed PT, Fontijne WPJ. Pulsed electromagnetic fields in the treatment of nonunion. Eur J Orthop Surg Traumatol. 2008;18(2):127–33. https://doi.org/10.1007/s00590-007-0271-8.
35. Ebid A, El-Boshy M, El-Shamy S, Thabet A, Abedalla M, Ali T. Long-term effect of full-body pulsed electromagnetic field and exercise protocol in the treatment of men with osteopenia or osteoporosis: a randomized placebo-controlled trial. F1000Res. 2021;10:649. https://doi.org/10.12688/f1000research.54519.3.
36. Accessed 26 June 2023. https://www.thelancet.com/journals/lancet/article/PIIS0140-6736(84)92219-0/fulltext.
37. Sutbeyaz ST, Sezer N, Koseoglu F, Kibar S. Low-frequency pulsed electromagnetic field therapy in fibromyalgia: a randomized, double-blind, sham-controlled clinical study. Clin J Pain. 2009;25(8):722–8. https://doi.org/10.1097/AJP.0b013e3181a68a6c.
38. Lappin MS, Lawrie FW, Richards TL, Kramer ED. Effects of a pulsed electromagnetic therapy on multiple sclerosis fatigue and quality of life: a double-blind, placebo controlled trial. Altern Ther Health Med. 2003;9(4):38–48. https://www.ncbi.nlm.nih.gov/pubmed/12868251
39. Ross CL, Syed I, Smith TL, Harrison BS. The regenerative effects of electromagnetic field on spinal cord injury. Electromagn Biol Med. 2017;36(1):74–87. https://doi.org/10.3109/15368378.2016.1160408.
40. van Belkum SM, Bosker FJ, Kortekaas R, Beersma DGM, Schoevers RA. Treatment of depression with low-strength transcranial pulsed electromagnetic fields: a mechanistic point of view. Prog Neuro-Psychopharmacol Biol Psychiatry. 2016;71:137–43. https://doi.org/10.1016/j.pnpbp.2016.07.006.
41. Pennisi G, Ferri R, Lanza G, et al. Transcranial magnetic stimulation in Alzheimer's disease: a neurophysiological marker of cortical hyperexcitability. J Neural Transm (Vienna). 2011;118(4):587–98. https://doi.org/10.1007/s00702-010-0554-9.
42. Rasouli J, Lekhraj R, White NM, et al. Attenuation of interleukin-1beta by pulsed electromagnetic fields after traumatic brain injury. Neurosci Lett. 2012;519(1):4–8. https://doi.org/10.1016/j.neulet.2012.03.089.
43. Vadalà M, Morales-Medina JC, Vallelunga A, Palmieri B, Laurino C, Iannitti T. Mechanisms and therapeutic effectiveness of pulsed electromagnetic field therapy in oncology. Cancer Med. 2016;5(11):3128–39. https://doi.org/10.1002/cam4.861.
44. Guo Y, Liu Y, Wang X. Electromagnetic activity: a possible player in epilepsy. Acta Epileptol. 2020;2(1) https://doi.org/10.1186/s42494-020-00019-9.
45. Tsisel'skiĭ IV, Kashintseva LT, Skrinnik AV. The effect of a pulsed electromagnetic field on the hemodynamics of eyes with glaucoma. Oftalmol Zh. 1990;3:154–7. https://www.ncbi.nlm.nih.gov/pubmed/2255478

46. Kubis N. Non-invasive brain stimulation to enhance post-stroke recovery. Front Neural Circuits. 2016;10:56. https://doi.org/10.3389/fncir.2016.00056.
47. Klomjai W, Katz R, Lackmy-Vallée A. Basic principles of transcranial magnetic stimulation (TMS) and repetitive TMS (rTMS). Ann Phys Rehabil Med. 2015;58(4):208–13. https://doi.org/10.1016/j.rehab.2015.05.005.
48. Strafella AP, Paus T, Barrett J, Dagher A. Repetitive transcranial magnetic stimulation of the human prefrontal cortex induces dopamine release in the caudate nucleus. J Neurosci. 2001;21(15):RC157. https://doi.org/10.1523/jneurosci.21-15-j0003.2001.
49. Cho SS, Strafella AP. rTMS of the left dorsolateral prefrontal cortex modulates dopamine release in the ipsilateral anterior cingulate cortex and orbitofrontal cortex. PLoS One. 2009;4(8):e6725. https://doi.org/10.1371/journal.pone.0006725.
50. Chervyakov AV, Chernyavsky AY, Sinitsyn DO, Piradov MA. Possible mechanisms underlying the therapeutic effects of transcranial magnetic stimulation. Front Hum Neurosci. 2015;9:303. https://doi.org/10.3389/fnhum.2015.00303.
51. Mansourian M, Shanei A. Evaluation of pulsed electromagnetic field effects: a systematic review and meta-analysis on highlights of two decades of research in vitro studies. Biomed Res Int. 2021;2021:6647497. https://doi.org/10.1155/2021/6647497.
52. Tong J, Chen Z, Sun G, et al. The efficacy of pulsed electromagnetic fields on pain, stiffness, and physical function in osteoarthritis: a systematic review and meta-analysis. Pain Res Manag. 2022;2022:9939891. https://doi.org/10.1155/2022/9939891.
53. Saroka KS, Vares DE, Persinger MA. Similar spectral power densities within the Schumann resonance and a large population of quantitative electroencephalographic profiles: supportive evidence for Koenig and Pobachenko. PLoS One. 2016;11(1):e0146595. https://doi.org/10.1371/journal.pone.0146595.
54. Roohi-Azizi M, Azimi L, Heysieattalab S, Aamidfar M. Changes of the brain's bioelectrical activity in cognition, consciousness, and some mental disorders. Med J Islam Repub Iran. 2017;31:53–312. https://doi.org/10.14196/mjiri.31.53.
55. Tang JY, Yeh TW, Huang YT, Wang MH, Jang LS. Effects of extremely low-frequency electromagnetic fields on B16F10 cancer cells. Electromagn Biol Med. 2019;38(2):149–57. https://doi.org/10.1080/15368378.2019.1591438.
56. Elhalel G, Price C, Fixler D, Shainberg A. Cardioprotection from stress conditions by weak magnetic fields in the Schumann resonance band. Sci Rep. 2019;9(1):1645. https://doi.org/10.1038/s41598-018-36341-z.
57. Hong JE, Lee CG, Hwang S, et al. Pulsed electromagnetic field (PEMF) treatment ameliorates Murine model of collagen-induced arthritis. Int J Mol Sci. 2023;24(2):1137. https://doi.org/10.3390/ijms24021137.
58. Sonmez AI, Camsari DD, Nandakumar AL, et al. Accelerated TMS for depression: a systematic review and meta-analysis. Psychiatry Res. 2019;273:770–81. https://doi.org/10.1016/j.psychres.2018.12.041.
59. Zhang X, Liu X, Pan L, Lee I. Magnetic fields at extremely low-frequency (50 Hz, 0.8 mT) can induce the uptake of intracellular calcium levels in osteoblasts. Biochem Biophys Res Commun. 2010;396(3):662–6. https://doi.org/10.1016/j.bbrc.2010.04.154.
60. Fini M, Giavaresi G, Torricelli P, et al. Pulsed electromagnetic fields reduce knee osteoarthritic lesion progression in the aged Dunkin Hartley Guinea pig. J Orthop Res. 2005;23(4):899–908. https://doi.org/10.1016/j.orthres.2005.01.008.
61. Markovic L, Wagner B, Crevenna R. Effects of pulsed electromagnetic field therapy on outcomes associated with osteoarthritis: a systematic review of systematic reviews. Wien Klin Wochenschr. 2022;134(11–12):425–33. https://doi.org/10.1007/s00508-022-02020-3.
62. Goetz SM, Truong CN, Gerhofer MG, Peterchev AV, Herzog HG, Weyh T. Analysis and optimization of pulse dynamics for magnetic stimulation. PLoS One. 2013;8(3):e55771. https://doi.org/10.1371/journal.pone.0055771.

63. Lullini G, Cammisa E, Setti S, Sassoli I, Zaffagnini S, Marcheggiani Muccioli GM. Role of pulsed electromagnetic fields after joint replacements. World J Orthop. 2020;11(6):285–93. https://doi.org/10.5312/wjo.v11.i6.285.
64. Fiani B, Kondilis A, Runnels J, Rippe P, Davati C. Pulsed electromagnetic field stimulators efficacy for noninvasive bone growth in spine surgery. J Korean Neurosurg Soc. 2021;64(4):486–94. https://doi.org/10.3340/jkns.2020.0269.
65. Marks RA. Spine fusion for discogenic low back pain: outcomes in patients treated with or without pulsed electromagnetic field stimulation. Adv Ther. 2000;17(2):57–67. https://doi.org/10.1007/bf02854838.
66. Miyamoto H, Sawaji Y, Iwaki T, et al. Intermittent pulsed electromagnetic field stimulation activates the mTOR pathway and stimulates the proliferation of osteoblast-like cells. Bioelectromagnetics. 2019;40(6):412–21. https://doi.org/10.1002/bem.22207.
67. Farndale RW, Murray JC. Pulsed electromagnetic fields promote collagen production in bone marrow fibroblasts via athermal mechanisms. Calcif Tissue Int. 1985;37(2):178–82. https://doi.org/10.1007/bf02554838.
68. Sollazzo V, Massari L, Caruso A, Mattei MD, Pezzetti F. Effects of low-frequency pulsed electromagnetic fields on human osteoblast-like CellsIn wtro. Electro- Magnetobiol. 1996;15(1):75–83. https://doi.org/10.3109/15368379609016167.
69. Traina GC, Fontanesi G, Costa P, et al. Effect of electromagnetic stimulation on patients suffering from non-union. A retrospective study with a control group. J Bioelectron. 1991;10(1–2):101–17. https://doi.org/10.3109/15368379109031402.
70. Gouttebarge V, Inklaar H, Backx F, Kerkhoffs G. Prevalence of osteoarthritis in former elite athletes: a systematic overview of the recent literature. Rheumatol Int. 2015;35(3):405–18. https://doi.org/10.1007/s00296-014-3093-0.
71. Ouyang J, Zhang B, Kuang L, et al. Pulsed electromagnetic field inhibits synovitis via enhancing the efferocytosis of macrophages. Biomed Res Int. 2020;2020:4307385. https://doi.org/10.1155/2020/4307385.
72. Mathiessen A, Conaghan PG. Synovitis in osteoarthritis: current understanding with therapeutic implications. Arthritis Res Ther. 2017;19(1):18. https://doi.org/10.1186/s13075-017-1229-9.
73. Cuenda A, Rousseau S. p38 MAP-kinases pathway regulation, function and role in human diseases. Biochim Biophys Acta. 2007;1773(8):1358–75. https://doi.org/10.1016/j.bbamcr.2007.03.010.
74. Walker M, Rizzuto P, Godin M, Pelling AE. Structural and mechanical remodeling of the cytoskeleton maintains tensional homeostasis in 3D microtissues under acute dynamic stretch. Sci Rep. 2020;10(1):7696. https://doi.org/10.1038/s41598-020-64725-7.
75. Mithoefer K, Peterson L, Zenobi-Wong M, Mandelbaum BR. Cartilage issues in football-today's problems and tomorrow's solutions. Br J Sports Med. 2015;49(9):590–6. https://doi.org/10.1136/bjsports-2015-094772.
76. Gobbi A, Karnatzikos G, Chaurasia S. Pulsed electromagnetic fields for the treatment of symptomatic patellofemoral cartilage lesions of the knee. In: The patellofemoral joint. Berlin Heidelberg: Springer; 2014. p. 261–6. https://doi.org/10.1007/978-3-642-54965-6_34.
77. Ackermann PW, Renström P. Tendinopathy in sport. Sports Health. 2012;4(3):193–201. https://doi.org/10.1177/1941738112440957.
78. Gehwolf R, Schwemberger B, Jessen M, et al. Global responses of IL-1β-primed 3D tendon constructs to treatment with pulsed electromagnetic fields. Cells. 2019;8(5):399. https://doi.org/10.3390/cells8050399.
79. Binder A, Parr G, Hazleman B, Fitton-Jackson S. Pulsed electromagnetic field therapy of persistent rotator cuff tendinitis. A double-blind controlled assessment. Lancet. 1984;1(8379):695–8. https://doi.org/10.1016/s0140-6736(84)92219-0.
80. Osti L, Buono AD, Maffulli N. Pulsed electromagnetic fields after rotator cuff repair: a randomized, controlled study. Orthopedics. 2015;38(3):e223–8. https://doi.org/10.3928/01477447-20150305-61.

81. Jeon HS, Kang SY, Park JH, Lee HS. Effects of pulsed electromagnetic field therapy on delayed-onset muscle soreness in biceps brachii. Phys Ther Sport. 2015;16(1):34–9. https://doi.org/10.1016/j.ptsp.2014.02.006.
82. Gessi S, Merighi S, Bencivenni S, et al. Pulsed electromagnetic field and relief of hypoxia-induced neuronal cell death: the signaling pathway. J Cell Physiol. 2019;17:15089–97. https://doi.org/10.1002/jcp.28149.
83. Cornacchione M, Pellegrini M, Fassina L, et al. β-Adrenergic response is counteracted by extremely-low-frequency pulsed electromagnetic fields in beating cardiomyocytes. J Mol Cell Cardiol. 2016;98:146–58. https://doi.org/10.1016/j.yjmcc.2016.07.004.
84. Tepper OM, Callaghan MJ, Chang EI, et al. Electromagnetic fields increase in vitro and in vivo angiogenesis through endothelial release of FGF-2. FASEB J. 2004;18(11):1231–3. https://doi.org/10.1096/fj.03-0847fje.
85. Dong JG. The role of heart rate variability in sports physiology. Exp Ther Med. 2016;11(5):1531–6. https://doi.org/10.3892/etm.2016.3104.
86. Grote V, Lackner H, Kelz C, et al. Short-term effects of pulsed electromagnetic fields after physical exercise are dependent on autonomic tone before exposure. Eur J Appl Physiol. 2007;101(4):495–502. https://doi.org/10.1007/s00421-007-0520-x.
87. Rikk J, Finn KJ, Liziczai I, Radák Z, Bori Z, Ihász F. Influence of pulsing electromagnetic field therapy on resting blood pressure in aging adults. Electromagn Biol Med. 2013;32(2):165–72. https://doi.org/10.3109/15368378.2013.776420.
88. Bragin DE, Statom GL, Hagberg S, Nemoto EM. Increases in microvascular perfusion and tissue oxygenation via pulsed electromagnetic fields in the healthy rat brain. J Neurosurg. 2015;122(5):1239–47. https://doi.org/10.3171/2014.8.JNS132083.
89. Zhang S, Clark M, Liu X, Chen D, Thomas P, Ren L. The effects of bio-inspired electromagnetic fields on healthy enhancement with case studies. Emerg Sci J. 2019;3(6):369–81. https://doi.org/10.28991/esj-2019-01199.
90. Yuan J, Xin F, Jiang W. Underlying signaling pathways and therapeutic applications of pulsed electromagnetic fields in bone repair. Cell Physiol Biochem. 2018;46(4):1581–94. https://doi.org/10.1159/000489206.
91. Cadossi R, Massari L, Racine-Avila J, Aaron RK. Pulsed electromagnetic field stimulation of bone healing and joint preservation: cellular mechanisms of skeletal response. J Am Acad Orthop Surg Glob Res Rev. 2020;4(5):e1900155. https://doi.org/10.5435/JAAOSGlobal-D-19-00155.
92. Slater G, Phillips SM. Nutrition guidelines for strength sports: sprinting, weightlifting, throwing events, and bodybuilding. J Sports Sci. 2011;29(Suppl 1):S67–77. https://doi.org/10.1080/02640414.2011.574722.
93. Stephenson MC, Krishna L, Pannir Selvan RM, et al. Magnetic field therapy enhances muscle mitochondrial bioenergetics and attenuates systemic ceramide levels following ACL reconstruction: southeast Asian randomized-controlled pilot trial. J Orthop Translat. 2022;35:99–112. https://doi.org/10.1016/j.jot.2022.09.011.
94. El-Bandrawy AM, Nashed AB, Ghareeb HO. Effect of pulsed electromagnetic therapy on bone mineral density of female athletes. Phys Ther Rehabil. 2018;5(1):7. https://doi.org/10.7243/2055-2386-5-7.
95. Trofè A, Raffi M, Muehsam D, et al. Effect of PEMF on muscle oxygenation during cycling: a single-blind controlled pilot study. Appl Sci (Basel). 2021;11(8):3624. https://doi.org/10.3390/app11083624.
96. Schweiger V, Niederseer D, Schmied C, Attenhofer-Jost C, Caselli S. Athletes and hypertension. Curr Cardiol Rep. 2021;23(12):176. https://doi.org/10.1007/s11886-021-01608-x.
97. Varga-Pintér B, Horváth P, Kneffel Z, Major Z, Osváth P, Pavlik G. Resting blood pressure values of adult athletes. Kidney Blood Press Res. 2011;34(6):387–95. https://doi.org/10.1159/000327850.

98. Guo J, Zhang X, Wang L, Guo Y, Xie M. Prevalence of metabolic syndrome and its components among Chinese professional athletes of strength sports with different body weight categories. PLoS One. 2013;8(11):e79758. https://doi.org/10.1371/journal.pone.0079758.
99. Stewart GM, Wheatley-Guy CM, Johnson BD, Shen WK, Kim CH. Impact of pulsed electromagnetic field therapy on vascular function and blood pressure in hypertensive individuals. J Clin Hypertens (Greenwich). 2020;22(6):1083–9. https://doi.org/10.1111/jch.13877.
100. Hamid HA, Sarmadi VH, Prasad V, Ramasamy R, Miskon A. Electromagnetic field exposure as a plausible approach to enhance the proliferation and differentiation of mesenchymal stem cells in clinically relevant scenarios. J Zhejiang Univ Sci B. 2022;23(1):42–57. https://doi.org/10.1631/jzus.B2100443.
101. Malling ASB, Morberg BM, Wermuth L, Gredal O, Bech P, Jensen BR. Effect of transcranial pulsed electromagnetic fields (T-PEMF) on functional rate of force development and movement speed in persons with Parkinson's disease: a randomized clinical trial. PLoS One. 2018;13(9):e0204478. https://doi.org/10.1371/journal.pone.0204478.
102. Gomes-Osman J, Field-Fote EC. Improvements in hand function in adults with chronic tetraplegia following a multiday 10-Hz repetitive transcranial magnetic stimulation intervention combined with repetitive task practice. J Neurol Phys Ther. 2015;39(1):23–30. https://doi.org/10.1097/NPT.0000000000000062.
103. Eime RM, Young JA, Harvey JT, Charity MJ, Payne WR. A systematic review of the psychological and social benefits of participation in sport for children and adolescents: informing development of a conceptual model of health through sport. Int J Behav Nutr Phys Act. 2013;10(1):98. https://doi.org/10.1186/1479-5868-10-98.
104. Rice SM, Purcell R, De Silva S, Mawren D, McGorry PD, Parker AG. The mental health of elite athletes: a narrative systematic review. Sports Med. 2016;46(9):1333–53. https://doi.org/10.1007/s40279-016-0492-2.
105. Dubin MJ, Ilieva IP, Deng ZD, et al. A double-blind pilot dosing study of low field magnetic stimulation (LFMS) for treatment-resistant depression (TRD). J Affect Disord. 2019;249:286–93. https://doi.org/10.1016/j.jad.2019.02.039.
106. Pelka RB, Jaenicke C, Gruenwald J. Impulse magnetic-field therapy for insomnia: a double-blind, placebo-controlled study. Adv Ther. 2001;18(4):174–80. https://doi.org/10.1007/bf02850111.
107. Sandyk R. Resolution of sleep paralysis by weak electromagnetic fields in a patient with multiple sclerosis. Int J Neurosci. 1997;90(3–4):145–57. https://doi.org/10.3109/00207459709000634.
108. Vitale KC, Owens R, Hopkins SR, Malhotra A. Sleep hygiene for optimizing recovery in athletes: review and recommendations. Int J Sports Med. 2019;40(8):535–43. https://doi.org/10.1055/a-0905-3103.
109. Baker-Price LA, Persinger MA. Weak, but complex pulsed magnetic fields may reduce depression following traumatic brain injury. Percept Mot Skills. 1996;83(2):491–8. https://doi.org/10.2466/pms.1996.83.2.491.
110. Cao C, Abulaban H, Baranowski R, et al. Transcranial electromagnetic treatment "rebalances" blood and brain cytokine levels in Alzheimer's patients: a new mechanism for reversal of their cognitive impairment. Front Aging Neurosci. 2022;14:829049. https://doi.org/10.3389/fnagi.2022.829049.
111. Moya Gómez A, Font LP, Brône B, Bronckaers A. Electromagnetic field as a treatment for cerebral ischemic stroke. Front Mol Biosci. 2021;8:742596. https://doi.org/10.3389/fmolb.2021.742596.
112. Chistyakov AV, Hafner H, Soustiel JF, Trubnik M, Levy G, Feinsod M. Dissociation of somatosensory and motor evoked potentials in non-comatose patients after head injury. Clin Neurophysiol. 1999;110(6):1080–9. https://doi.org/10.1016/s1388-2457(99)00029-2.
113. Reti IM, Schwarz N, Bower A, Tibbs M, Rao V. Transcranial magnetic stimulation: a potential new treatment for depression associated with traumatic brain injury. Brain Inj. 2015;29(7–8):789–97. https://doi.org/10.3109/02699052.2015.1009168.

114. Thomas AW, Drost DJ, Prato FS. Human subjects exposed to a specific pulsed (200 microT) magnetic field: effects on normal standing balance. Neurosci Lett. 2001;297(2):121–4. https://doi.org/10.1016/s0304-3940(00)01688-8.
115. Edwards DJ, Cortes M, Wortman-Jutt S, et al. Transcranial direct current stimulation and sports performance. Front Hum Neurosci. 2017;11:243. https://doi.org/10.3389/fnhum.2017.00243.
116. Forogh B, Haqiqatshenas H, Ahadi T, Ebadi S, Alishahi V, Sajadi S. Repetitive transcranial magnetic stimulation (rTMS) versus transcranial direct current stimulation (tDCS) in the management of patients with fibromyalgia: a randomized controlled trial. Neurophysiol Clin. 2021;51(4):339–47. https://doi.org/10.1016/j.neucli.2021.03.002.
117. Eating for peak athletic performance. Accessed 5 Sept 2022. https://www.uwhealth.org/health-wellness/eating-for-peak-performance/45232.
118. Buonocore D, Negro M, Arcelli E, Marzatico F. Anti-inflammatory dietary interventions and supplements to improve performance during athletic training. J Am Coll Nutr. 2015;34(Suppl 1):62–7. https://doi.org/10.1080/07315724.2015.1080548.
119. Rosado MM, Simkó M, Mattsson MO, Pioli C. Immune-modulating perspectives for low frequency electromagnetic fields in innate immunity. Front Public Health. 2018;6:323464. https://doi.org/10.3389/fpubh.2018.00085.
120. Trayhurn P. Oxygen-A critical, but overlooked, nutrient. Front Nutr. 2019;6:10. https://doi.org/10.3389/fnut.2019.00010.
121. Sohn JH, Chung SC, Jang EH. 30% oxygen inhalation enhances cognitive performance through robust activation in the brain. J Physiol Anthropol Appl Hum Sci. 2005;24(1):51–3. https://doi.org/10.2114/jpa.24.51.
122. Bostek CC. Oxygen toxicity: an introduction. AANA J. 1989;57(3):231–7. https://www.ncbi.nlm.nih.gov/pubmed/2672691
123. Forster RE, Estabrook RW. Is oxygen an essential nutrient? Annu Rev Nutr. 1993;13(1):383–403. https://doi.org/10.1146/annurev.nu.13.070193.002123.
124. Nakazawa MS, Keith B, Simon MC. Oxygen availability and metabolic adaptations. Nat Rev Cancer. 2016;16(10):663–73. https://doi.org/10.1038/nrc.2016.84.
125. Cinti S. The adipose organ: morphological perspectives of adipose tissues. Proc Nutr Soc. 2001;60(3):319–28. https://doi.org/10.1079/pns200192.
126. Cannon B, Nedergaard J. Brown adipose tissue: function and physiological significance. Physiol Rev. 2004;84(1):277–359. https://doi.org/10.1152/physrev.00015.2003.
127. Tai YK, Ng C, Purnamawati K, et al. Magnetic fields modulate metabolism and gut microbiome in correlation with Pgc-1α expression: follow-up to an in vitro magnetic mitohormetic study. FASEB J. 2020;34(8):11143–67. https://doi.org/10.1096/fj.201903005RR.
128. McKenna ZJ, Gorini Pereira F, Gillum TL, Amorim FT, Deyhle MR, Mermier CM. High-altitude exposures and intestinal barrier dysfunction. Am J Physiol Regul Integr Comp Physiol. 2022;322(3):R192–203. https://doi.org/10.1152/ajpregu.00270.2021.
129. Jiao M, Yin H, Hu J, Xu W, Zhang X, Zhang P. Effects of low-frequency pulsed electromagnetic fields on high-altitude stress ulcer healing in rats. Biomed Res Int. 2019;2019:6354054. https://doi.org/10.1155/2019/6354054.
130. Di Bartolomeo M, Cavani F, Pellacani A, et al. Pulsed electro-magnetic field (PEMF) effect on bone healing in animal models: a review of its efficacy related to different type of damage. Biology (Basel). 2022;11(3):402. https://doi.org/10.3390/biology11030402.
131. Thomas AW, Graham K, Prato FS, et al. A randomized, double-blind, placebo-controlled clinical trial using a low-frequency magnetic field in the treatment of musculoskeletal chronic pain. Pain Res Manag. 2007;12(4):249–58. https://doi.org/10.1155/2007/626072.
132. Ziegler P, Nussler AK, Wilbrand B, et al. Pulsed electromagnetic field therapy improves osseous consolidation after high tibial osteotomy in elderly patients–a randomized, placebo-controlled, double-blind trial. J Clin Med. 2019;8(11):2008. https://doi.org/10.3390/jcm8112008.

133. Elshiwi AM, Hamada HA, Mosaad D, Ragab IMA, Koura GM, Alrawaili SM. Effect of pulsed electromagnetic field on nonspecific low back pain patients: a randomized controlled trial. Braz J Phys Ther. 2019;23(3):244–9. https://doi.org/10.1016/j.bjpt.2018.08.004.
134. Giovale M, Novelli L, Persico L, et al. Low-energy pulsed electromagnetic field therapy reduces pain in fibromyalgia: a randomized single-blind controlled pilot study. Rheumatol Immunol Res. 2022;3(2):77–83. https://doi.org/10.2478/rir-2022-0013.
135. Wu Z, Ding X, Lei G, et al. Efficacy and safety of the pulsed electromagnetic field in osteoarthritis: a meta-analysis. BMJ Open. 2018;8(12):e022879. https://doi.org/10.1136/bmjopen-2018-022879.
136. Cameron IL, Sun LZ, Short N, Hardman WE, Williams CD. Daily pulsed electromagnetic Field (PEMF) therapy inhibits tumor angiogenesis via the hypoxia driven pathway: therapeutic implications. Cancer Res. 2005;65(9_Supplement):287. Accessed 9 Sept 2022. https://aacrjournals.org/cancerres/article/65/9_Supplement/287/519343/Daily-Pulsed-Electromagnetic-Field-PEMF-therapy.
137. Barbault A, Costa FP, Bottger B, et al. Amplitude-modulated electromagnetic fields for the treatment of cancer: discovery of tumor-specific frequencies and assessment of a novel therapeutic approach. J Exp Clin Cancer Res. 2009;28(1):51. https://doi.org/10.1186/1756-9966-28-51.
138. Crocetti S, Beyer C, Schade G, Egli M, Fröhlich J, Franco-Obregón A. Low intensity and frequency pulsed electromagnetic fields selectively impair breast cancer cell viability. PLoS One. 2013;8(9):e72944. https://doi.org/10.1371/journal.pone.0072944.
139. Li S, Yu B, Zhou D, He C, Zhuo Q, Hulme JM. Electromagnetic fields for treating osteoarthritis. Cochrane Database Syst Rev. 2013;12:CD003523. https://doi.org/10.1002/14651858.CD003523.pub2.
140. Miller CP, Prener M, Dissing S, Paulson OB. Transcranial low-frequency pulsating electromagnetic fields (T-PEMF) as post-concussion syndrome treatment. Acta Neurol Scand. 2020;142(6):597–604. https://doi.org/10.1111/ane.13300.

Chapter 27
Photobiomodulation: The Use of Specific Light Wavelengths to Accelerate Recovery, Repair, and Regeneration

Jennifer Novak

Introduction

Photobiomodulation: History and Safety of the Technology

The use of light therapy as a healing and recovery modality has been around for more than half a century. However, it is only in the last 15–20 years that it has gained greater awareness and popularity. The recorded "birth" of such use of low-powered laser light as a potential therapeutic agent was in 1967 at Semmelweis University in Budapest, Hungary, by Dr. Endre Mester (Hamblin et al. 2017) [19, 22, 25]. Dr. Mester was using a low-powered ruby laser to see if mice would contract cancer after being exposed to the laser's photons on their shaved backs [19]. To his surprise, the treated mice (there was a control group) did not get cancer, but what was observed was the speed with which the hair on their backs returned compared to controls [19]. Mester termed the effect "*laser bio-stimulation*" due to the stimulatory effects on hair follicle regeneration [10, 19]. With that experience, research on low level laser therapy as a means of repair, regeneration, and recovery began. Low level laser therapy, red light therapy, cold laser therapy and Mester's laser bio-stimulation are some of the names given to this method [19]. Now termed *photobiomodulation* (see box below), researchers and practitioners address the use of photons (light particles) in specific wavelengths to help bodies recover, regenerate, and heal from the stress of sport and exercise [19, 22]. Photobiomodulation (PBM) utilizes light to change or modify life forms and is currently being broadly researched. More benefits of the use of PBM are coming to "light" every day. With much of the earlier research having been successfully completed through animal tissue and in-vitro

J. Novak (✉)
PEAK Symmetry Performance Strategies, Atlanta, GA, USA
e-mail: jennifer@symmetryperformance.com

N. C. Apostolopoulos et al. (eds.), *Fundamentals of Recovery, Regeneration, and Adaptation to Exercise Stress: An Integrated Approach*,
https://doi.org/10.1007/978-3-031-44270-4_27

models, current studies with humans have also shown improvements, validating the results from earlier studies.

> **PHOTOBIOMODULATION IS:**
> **Photo** = light (particles)
> **Bio** = life forms (in this case cells, tissues, and/or fluids) and
> **Modulation** = Change or modulate

While the history of research on PBM is decades old, the first placebo-controlled trial was not conducted on athletes until 2008 (Leal et al. [56]). In this laboratory-based study, authors took a team of professional male volleyball athletes and divided them into PBM and placebo groups [56]. An initial test measuring biceps curls to fatigue was completed on all participants. Then PBM or sham was applied in four spots on the athlete's biceps brachii muscles. A second test was administered to measure time to fatigue in a biceps curl exercise, and the PBM group extended the number of curl repetitions by 8.5 reps (+/− 1.9 reps) compared to sham [56]. Thanks to this lab-controlled trial, initial evidence of PBM's positive effect on performance with regard to extending time to fatigue was established. Moving forward, several lab-based similar trials were completed. However, in 2016, Pinto et al. provided similar results in field testing, a less-controlled environment [53]. In this study, professional rugby athletes were divided into PBM and placebo groups and familiarized themselves with a Bangsbo Sprint Test consisting of seven 30-meter sprints with a change of direction after each sprint [53]. The PBM group, after familiarization with the test, were found to have reduced sprint times, as well as reduced fatigue measures both during and after the sprint test compared to placebo [53]. Thanks to these examples and to continuing study of its effects on athletes, PBM has now been well-established to provide both performance and recovery benefits in the lab as well as the field.

The research continues, and calls for consistency in reported paramters leading to correct dosing have emerged. Dosing, however, will vary for a variety of reasons, so any suggested recommendations must take these variables into account, which is part of the difficulty in determining exact effective parameters. The dosing recommendations will therefore need to be be both *device-dependent* as well as *recipient-dependent*, for the following reasons:

Both Laser and LED devices provide a total power density, measured in Watts (or mW). The power available with any given device will partially dictate the time needed to effect positive change. Discussed later in this chapter is a difference in LED technology that puts an initial demand on the total power of the device to convert electric energy to light energy that the cells can use. Cluster diode configurations will reach a broader surface area than single diodes, also changing, perhaps, the dose area covered and therefore the number of spots to address [19]. Beam shaping optics may also either shrink or expand the surface area covered. Lastly, whether or not a device has pulsing capability (shown in research to provide better therapeutic effect than continuous light) will also change the resulting cellular effects [19].

As for the recipient-dependent aspect of light therapy application, both skin tone and skin thickness will affect how many photons will reach the target tissue [19]. Darker skin tones and thicker areas of skin will decrease the number of photons reaching target tissues, so time spent applying the light may need to be extended [19]. Furthermore, consideration should be given to each individual regarding dependent variables such as inflammatory or redox state, or different hydration, nutrition, and rest status, all of which may affect the stress level of the cells and tissues, and therefore the number of photons needed to provide a therapeutic effect [19]. Lastly, Zein et al. (2018) reported that, based on results from a variety of papers, it appears that tissue containing a higher number of mitochondria may respond to lower doses of light than tissues lacking in such mitochondrial density [48].

> The initial recommendations for clinical application of **PBM** should be taken with considerations for both **DEVICE** parameters and differing aspects of **RECIPIENTS**. Being aware of these differences helps the practitioner guide light application after considering the device being used and who is receiving the therapy.

Cotler et al. [10] reported that light with wavelengths in the red to near infrared region of the spectrum between 660 and 905 nm is generally employed because these wavelengths have the ability to penetrate skin, reach deeper cells and tissues, and have been proven in clinical trials to have a positive effect on pain, inflammation, and tissue repair [10]. The above benefits, as well as other clinical uses of PBM will be discussed in this chapter. For the purposes of this text, the focus will be on photons in the broader range (from 600 to 1070 Nm) of both red (visible spectrum) and near-infrared (invisible) wavelengths [7], and their effects on recovery, repair, regeneration, and performance.

As a practitioner who combines the science of exercise physiology, neurology, and motor control with concepts from the healing arts, no technology, technique, or method has captured my attention and earned my respect as much as PBM. When appropriately applied to compromised or inflamed tissue, its researched benefits become evident very quickly. Additionally, PBM enjoys a high safety profile, rarely reporting any type of adverse events from its appropriate use. Initially acute in nature, but with targeted, extended application, PBM provides long-term positive change. Its use facilitates recovery and healing at the cellular and tissue levels (wound and tissue healing, collagen synthesis, angiogenesis, neurogenesis, etc.), yet it also offers a more systemic effect by setting in motion the electrical, chemical, and biological pathways of healing, and through stimulation/tonification of key body tissues and organs like, for example, the vagus nerve (Cranial Nerve X) [4, 30].

More recently in sports performance, focus has turned to using technology for vagus nerve stimulation. Modulation of this nerve, the longest and most complex nerve in the body, supports the autonomic nervous system through enhancing one's sympathovagal tone, or the body's ability to appropriately adapt when stressors are present as well as when they have been removed. In athletes, this response should be flexible and finely tuned if the goal is to create stronger and more resilient athletes. Machado et al. (2022) compared effects of PBM with either red, violet, or red plus violet on the vagus nerve by illuminating the ear, the common location for the external vagal branch in humans [33]. The authors discovered that using light in the red wavelength range as well as red combined with violet upregulated parasympathetic tone, while violet alone upregulated sympathetic tone [33]. Because this chapter is focused on PBM in either red or NIR wavelengths, the paper supports the ability to affect vagal tone in terms of changes to heart rate variability with red PBM application. In my humble opinion, there can be no optimizing of performance without robust and optimized recovery, and PBM stimulation of the vagus nerve may enhance recovery through improved autonomic tone [4, 30].

The human body is a dynamic, interrelated, and interdependent system. Once one part of a dynamic system is changed, the entire system is changed. Therefore, a body that achieves and maintains autonomic flexibility can dynamically enable the appropriate physiological responses to the stressors of sport and exercise. Autonomic flexibility encourages health and resilience and promotes faster recovery from the types of physiological burdens experienced through intense exercise or competition [4]. A narrative review by Ailioaie and Litscher [3] provides a graphic of a series of cytoprotective and ergogenic effects from the use of PBM [3] (Fig. 27.1). While the list is not exhaustive, it provides a quick overview of the many benefits of utilizing PBM [3].

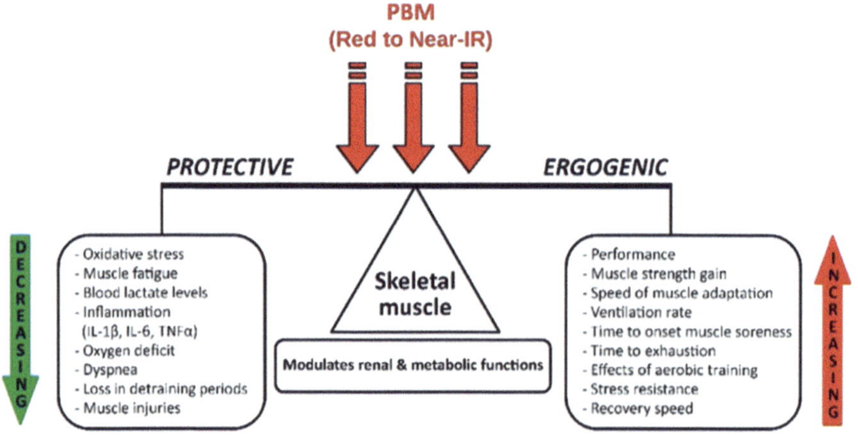

Fig. 27.1 Used with permission from a 2021 narrative review of reported PBM benefits in sports, the effects are both cytoprotective and ergogenic. (Ailioaie and Litscher [3])

Photobiomodulation: An Explanation

When considering PBM's technology, in either laser or LED form within these wavelength ranges, it is important to mention the foundational safety and efficacy of light therapy. Hamblin et al. (2021) observed that "Low-intensity light in the visible [red] and near-infrared (NIR) region has beneficial therapeutic effects. Light in this region of the spectrum lacks the carcinogenic and mutagenic properties of ultraviolet light and is capable of penetrating biological tissue and bone. Moreover, wavelengths in the NIR region are especially effective at penetrating the skull and dense muscle tissue and [have been] subsequently used in the development of NIR imaging for the noninvasive study of muscle and brain oxygenation and oxidative metabolism [19]." Current research also supports Hamblin's observations, providing strong evidence that PBM in the visible red and invisible near infrared wavelengths penetrates rather significantly into varied tissue types (soft tissue, bone, brain tissue, nerve tissue), without the deleterious properties or effects that other wavelengths on the light spectrum may produce under certain conditions [19, 28].

In another paper on the cellular effects of PBM, Hamblin explains that there seems to be a *selective* nature to the changes seen by illuminating cells and tissue [19]. The injured cells utilize the photons to evoke cellular signaling processes supporting enhanced healing and reduced apoptosis (cellular death), with normal tissue showing little or no effect after light application [19]. Hamblin provides the following example: "Illumination of normal skin or mucosa does not induce hyperplasia, and illumination of uninjured nerves does not generally induce anesthesia. This selectivity could be partially explained by the action of light on mitochondria of cells that are injured, predisposed to apoptosis or hypoxic [19]." This observation in vivo challenges the older research from in vitro models noting potential concerns of inhibitory effects or tissue damage and supports the inherent safety of utilizing PBM as an efficacious healing modality to cells and tissues in need (hypoxic, injured, or otherwise damaged through oxidative stress) while the surrounding tissue is unaffected.

> "There seems to be a selective nature to the changes seen by illuminating cells and tissue. The injured cells utilize the photons to evoke cellular signaling processes supporting enhanced healing and reduced apoptosis (cellular death), with normal tissue showing little or no effect after light application" [19]

It can be said that the following pattern is regularly established using PBM: Light therapy (red and NIR) provides a PHOTO-ELECTRIC STIMULATION that evokes PHOTO-CHEMICAL REACTIONS, resulting in a PHOTO-PHYSICAL CHANGE. Those changes indicate faster, more efficient recovery and repair, and improved cellular resilience, and by effect performance, with augmentation by this simple, yet effective method.

> Light therapy (red and NIR) provides a **PHOTO-ELECTRIC STIMULATION** that evokes **PHOTO-CHEMICAL REACTIONS**, resulting in a **PHOTO-PHYSICAL CHANGE**

Cellular energy may be equated with cellular pH. It has been well documented that during and after strenuous exercise, hydrogen ions (H+) build up around the cells, creating an acidic environment due to inhibition of the cellular redox potential [21, 26]. In an acidic environment, the pH in the localized area, as well as the blood, is temporarily lower, hampering the generation of cellular energy [21, 26, 29]. While it has been suggested that a buildup of lactate may be a key factor in the accumulation of H+ with strenuous exercise [5] and that applying PBM to the tissue post-exercise lowers lactate levels, thereby helping clear H+ ions, not all research supports this concept [19]. However, significant reductions in creatine kinase levels has regularly been noted, promoting the idea that reduced muscle damage by enhanced activation of antioxidant activity may be a key component of changing energetic status and reducing oxidative stress [19]. Therefore, the concept of PBM assisting in the change of pH suggests its use as an effective method for accelerating typical recovery and repair mechanisms [19, 21].

Exercise Recovery: A Mathematical Ratio

Recovery from the stressors of strenuous exercise and sport training can be thought of as a ratio of cellular energy to allostatic load. The greater the number or intensity of stressors present, the greater the demand for cellular energy to restore homeostasis. Once stressors are effectively mitigated, unencumbered performance can occur. These stressors can be physiological, chemical, mental/emotional, or environmental in nature. The response to any stressor, or a variety of concurrent stressors requires energy, and energy is created at the cellular level through the mitochondria [15]. Interestingly, PBM supplies benefits on both sides of this ratio: mitigating stressors on one side, while augmenting or optimizing the mitochondria's ability to create and cycle ATP on the other. PBM therefore enhances recovery and, by extrapolation, performance. The stressors of exercise, especially in athletics and competitive sports, can be varied and cumulative depending on the season, travel, and rest schedule, as well as internal, nutritional, mental/emotional, and environmental factors. By providing an all-natural, low-side effect, research-tested modality like PBM to enhance system recovery and tissue repair, PBM becomes a valuable adjunct to both healing and performance. Truly, in order to optimize performance, athletes must enjoy robust and efficient recovery and repair. In a recent review by González-Muñoz et al. (2023), the included papers supported the theory that with regard to sports performance, the well-documented recovery effects of PBM on athletes becomes the foundation of its performance benefits [53]. Some examples the authors provided were reductions in pain, reduced tissue damage and oxidative stress, improved circulation, and more [53]. They further reported that strength training and cardiovascular/endurance training both received performance benefits from the ability of PBM to improve the cellular stress response and provide the system with more ATP (energy) to either recover or train/compete [53].

Where light therapy truly "shines" in recovery and repair is in its ability to change and augment the biochemical, histological, and physiological processes of the body [19]. Current research documents positive changes in markers for varied processes such as antioxidant activity, redox potential, improved blood flow, enhanced tissue repair, and more [19]. Both the red and near infrared wavelengths have been shown to enhance recovery and repair in specific tissues locally, and the body globally [19]. The following sections will enumerate the evidence-based mechanisms of PBM as documented by utilizing red and near infrared wavelengths on both tissue and system recovery and regeneration.

I will discuss in this chapter the main takeaway of this optimization—the symbiotic relationship between recovery and performance using the mathematical ratio model of cellular energy to allostatic load. PBM offers benefits in both areas. It is this novel dual set of benefits combined with its high safety profile that sets it apart from many other modalities as a comprehensive and indispensable tool for elevating athletic recovery and performance. It is natural for the human body to both accept and utilize something as 'simple' as particles of light to promote health. The research provides ample evidence of such.

A cherished mentor once told me, "Whatever [implement, method, technique, or tool] you use, make it an extension of your intention." This statement resonated with me. Intention is a key professional foundation for one's efforts and outcomes, especially when being led by sound evidence for its use. If it is my clear intention to promote recovery and performance by providing the best, safest, and most highly effective avenues for accomplishing these goals, then PBM certainly fits the bill. By the end of this chapter, perhaps my passion for the technology will also be caught and utilized more ubiquitously by more and more practitioners.

The history of PBM spans more than 50 years from the time a Hungarian professor used a low-powered ruby laser on mouse skin to observe its effects. Moving forward, thousands of studies and case reports have been completed, continually supporting the safety and efficacy of PBM on both animal and human tissue for recovery, healing, and performance. The results are very encouraging. If one considers true performance as a mathematical ratio of cellular energy to cellular stress, PBM's ability to enhance cellular respiration and reduce oxidative stress (through various mechanisms), suggests that this safe and effective method supports athletic performance through increased energy production and decreased stress—stress that would normally consume the energy needed to return to homeostasis or recover from the rigors of sport and exercise.

Mechanisms of Photobiomodulation

Enhanced Cellular Energy: Chromophores, Calcium, and Transcription Factors

Currently, the most widely accepted theory of how PBM enhances cellular energy is based on the concept of photo-acceptors called chromophores. These mitochondrial molecules, present in all mammalian cells, respond to specific light wavelengths (red and NIR), evoking a cascade of events to support energy production through enhancing efficacy of the electron transport chain. To understand chromophores, think back to a high school biology lesson on the growth of plants. Liu and Van Iersel [30] note that chlorophyll in plant leaves is a chromophore [30]. With sunlight, the chlorophyll accepts the photons in both red and blue wavelengths from the sun, thus converting energy for the plant to grow, flower, produce an edible vegetable or fruit, etc. [30]. Another way to understand PBM is by looking at the way our skin manufactures vitamin D from sunlight. When humans are outside during daylight, the UVB rays "convert 7-dehydrocholesterol (7-DHC), a chromophore in epidermal keratinocytes and dermal fibroblasts, into previtamin D3" [38].

Chromophores, therefore, can be considered cellular "solar panels". When light particles are accepted by the chromophores, a photoelectric effect takes place—an electrical stimulus is produced at the junction between the two entities—helping the cells make more ATP through an upregulation of the mitochondrial membrane potential [19, 28]. Our cells consist of various chromophores, including hemoglobin, myoglobin, melanin, and the chromophore of interest in PBM, cytochrome c-oxidase (CCO) [19]. Each type of chromophore has an affinity for certain wavelengths within the light spectrum, and in the case of cytochrome c-oxidase, it accepts both red and near infrared wavelengths [19].

Cytochrome c-oxidase is the last enzyme in the mitochondria's electron transport chain (complex IV) and is the ideal photo-acceptor of light waves in the red and near infrared ranges [9, 19] (Fig. 27.2). As mentioned, CCO is the only chromophore that is present in every mammalian cell. By using specific wavelengths of light (typically seen in research to be between 600 and 1084 nm in length) [19], the mitochondria can generate a considerable amount of ATP, the well-known energy currency in the human body [15]. Hamblin [23] states, "Absorption of photons by molecules leads to electronically excited states and consequently can lead to acceleration of electron transfer reactions. More electron transport necessarily leads to increased production of ATP" [23]. The research reports that cells are moved, therefore, from a glycolytic state into oxidative phosphorylation, which produces more of the energy currency needed for cellular metabolism [13] (Fig. 27.2).

Hamblin [23] provided copious amounts of PBM's biochemical and bioelectric mechanisms at a symposium for optical engineers; PBM in the red and NIR areas of the light spectrum affects the mitochondria—and that is through the photo acceptor CCO, specifically, the copper centers within the enzyme (copper is an excellent conductor) [23]. CCO is located within the mitochondria of each cell and is

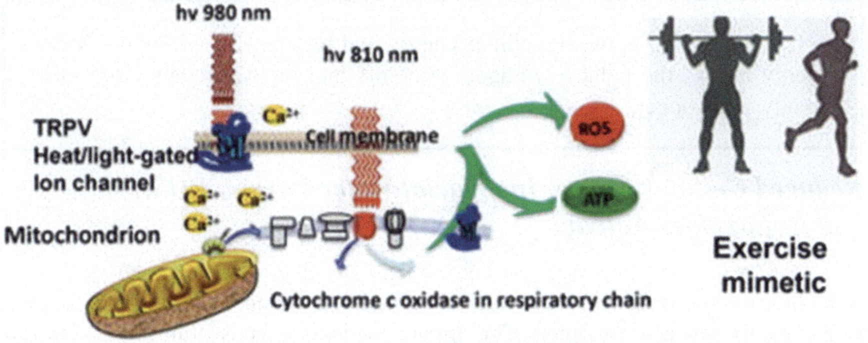

Fig. 27.2 Cytochrome c-oxidase (shown in red) in complex IV of the cellular respiratory chain, which is the chromophore thought to accept red and near infrared wavelengths and enhance ATP production in the mitochondria. (Used with permission, Hamblin et al. [19])

necessary to complete metabolic processes like producing ATP through complex IV of the electron transport chain [23]. Therefore, augmenting the mitochondrial ability to produce energy by illuminating cells with light may provide an effective means by which cells can recover, repair, and even regenerate. Later in this chapter, the relationship between CCO and Nitric Oxide (NO) with regard to cellular stress and ATP production will be described.

Along with the CCO theory of upregulated mitochondrial membrane potential, an additional benefit is the increased release of calcium ions at the light sensitive gated ion channels in the endoplasmic and/or sarcoplasmic reticulum of the cellular membrane [13, 19, 31]. As calcium is a necessary mineral for contraction of muscles and neurotransmitter release in neurons, light therapy in the red and near infrared wavelengths provides additional benefits to the communication pathways of the body (Fig. 27.2). Further restoration benefits are derived from the release of nuclear factor kappa β (Nf kβ), which is found in the mitochondria (in the intercellular membrane) but is released into the cell's nucleus as a signaling molecule to promote certain transcription factors that improve production of growth factors, cell proliferation, and migration (like satellite cells and other stem cells), and also promote cellular survival through its anti-apoptotic action [13]. In a review by Ahrabi et al. (2019) reported that PBM has profound effects (dependent on dose and wavelength) on the proliferation, migration, and differentiation to a variety mesenchymal stem cells (MSCs) [2]. Included in the review were explanations ofvarious chemical pathways stimulated upon umbilical, blood, bone, and dental pulp stem cells [2]. All MSCs seemed to experience positive effects with strategic PBM use (Fig. 27.2).

PBM's ability to provide greater availability of cellular ATP, augmented contractile capability and nerve impulse activity, as well as to promote cellular survival and repair mechanisms supports the concept that PBM is a safe and minimally invasive modality with considerable benefits. Further photo-chemical effects occur in the cells and tissues to make PBM a powerful, yet still underappreciated and perhaps underutilized tool for recovery, repair, and performance. These further effects are discussed in later sections of this chapter.

> "The relationship between cellular energy and recovery is symbiotic. More energy allows the cells to mitigate stressors and recover/repair more efficiently, and sooner."

Reduced Cellular Stress: Interfacial Water Layers, ROS, and Antioxidant Activity

In addition to the mechanisms behind mitochondrial enhancement of ATP, a more recent theory has also emerged. This theory involves areas within both the mitochondria and cell called the interfacial water layers (IWLs), a nanoscopic, two-to-three-layer deep water barrier that changes in both viscosity and density with light application (Sommer 2019) [41]. In stressed cells, these IWLs become dense and sticky. With the release of endogenous ROS, this sticky quality may cause these free radicals to remain and proliferate [13, 19, 21, 28]. Furthermore, oxidative stress creates a temporary acidic environment, thereby engaging inflammatory processes. It has been noted that pulsed light in the red and near infrared ranges enhances viscosity while decreasing the density of these water layers, providing dual benefits [41]. First, it allows for better ion exchange (perhaps the calcium ions, perhaps other ions that support either appropriate cellular stress response or other cellular processes), and second, disallowing over-proliferation or collection of reactive oxygen species (ROS) creating a kind of shield that prevents further degradation into cellular stress or even death [41]. Therefore, the more efficiently the cells release the needed chemical signals for enhancing antioxidant activity, the more resilient the tissue will be [28]. The altered IWL viscosity frees this small amount of ROS, allowing the oxidants to become the chemical 'help signal' to increase antioxidant delivery to the area [41]. Changing water viscosity, therefore, may also promote ATP production since the cells experience less oxidative stress [41]. In some cells, a pulsed frequency of merely 10 Hz with a laser diode produces such a protective, energy-promoting effect [19] (Fig. 27.3).

Continuing with the "mathematical ratio" concept of exercise recovery between cellular energy and cellular stress, perhaps the ATP-augmenting and stress-reducing mechanisms of PBM are a combined effort of chromophore acceptance of photons and the changes in viscosity of IWLs reducing the chance of over-proliferating endogenous ROS. After intense bouts of exercise or in competition, the rise in cellular energy combined with the reduction of oxidative stress enables the cells to function properly and recover more efficiently [5, 14]. A reduction in the amount of energy being dedicated to resolving the inflammatory process allows for a faster repair of injured tissue and subsequently a reduction in fatigue. Whatever the energy requirement may be—creating a reduction in cellular or tissue inflammation, repairing injured tissue, or extending an athlete's time to fatigue for longer endurance during performance—the extra ATP is there to provide [9, 11, 14, 19]. In other words, the athlete can enjoy increased endurance and enhanced performance with PBM.

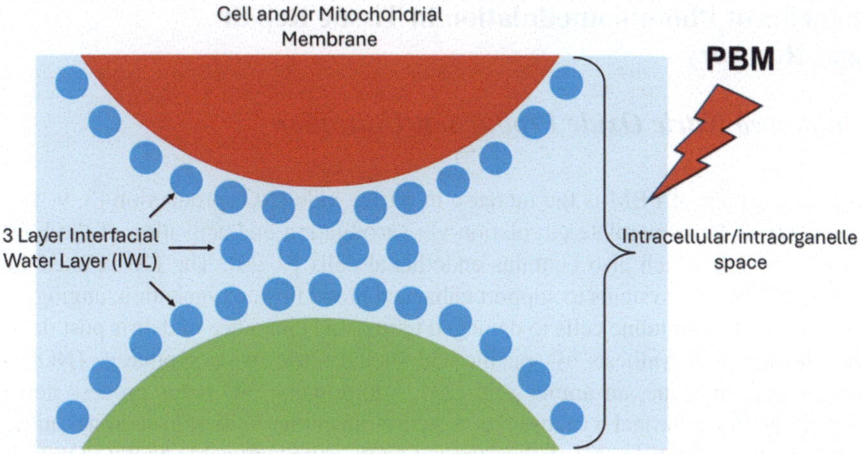

Fig. 27.3 The change in viscosity and density of the Interfacial Water Layers (IWLs) in the intracellular or intraorganelle space with PBM application enables better ion flow, and also serves as a protective mechanism to prevent over-proliferation of ROS due to hypoxic or stressed cellular conditions [41]

In addition to the increased production of ATP, there are also systemic benefits to the use of PBM that will be further discussed in this chapter. Chung et al. (2012) states, "Among its [PBM's] many effects, low-level laser therapy has been shown to cause vasodilation by triggering the relaxation of smooth muscle associated with endothelium, which is highly relevant to the treatment of joint inflammation [9]. This vasodilation increases the availability of oxygen to treated cells and allows for greater traffic of immune cells into tissue." [9] This further suggests that with more cellular energy in the form of ATP, plus the benefits of PBM at the tissue and systemic levels through inflammation-resolving, oxygenating, and tissue healing processes, allows for faster and better outcomes in repair, recovery, and performance of the athlete.

Although the explanation of the mechanisms of how PBM functions may be somewhat theoretical, the observed results suggest the need to continue investigating this safe modality. Presently, there are two main theories supported by the literature: (1) that cytochrome C-oxidase, a chromophore within the mitochondria, becomes a photo-acceptor for the light and augments ATP production; and (2) that interfacial water layers in the cell and the mitochondria alter density and viscosity to prevent over-proliferation of ROS and allow for better ion flow between membranes. Both these theories support PBM's photoelectric effect on the cells. Promoting energy production while reducing oxidative stress helps further promote the processes by which the entire human system benefits.

Benefits of Photobiomodulation in Tissue Repair and Recovery

Enhanced Nitric Oxide Production/Utilization

Another feature of PBM is the increase in nitric oxide (NO) production [3, 9, 19]. Nitric Oxide helps regulate circulation via vasodilation and activation of the lymphatic system, which also contains endothelial cells [9, 23]. The NO relaxes the walls of these two systems to support enhanced blood flow, oxygenation, angiogenesis, delivery of immune cells to damaged tissue, and clearance of debris post damage through NO synthesis by specific endothelial nitric oxide synthase's (NOS3) action on L-arginine, an amino acid [23]. Additionally, NO helps process nerve signals across neuronal synapses as a neurotransmitter utilizing neuronal nitric oxide synthase (NOS1) [23]. A third type of NOS, called inducible NOS (NOS2), is produced by macrophages, the cellular equivalent of first responders, during early stages of tissue or cellular injury. Accumulation of NO at injured tissue is known to help kill bacteria [23]. However, NO within the cells and/or mitochondria indicates cellular stress [23].

Hamblin explains that NO is produced when cells are either stressed or in a hypoxic environment [23]. When present in the mitochondria, NO is *inhibitory* to cellular respiration due to its molecular similarity to O_2 [23]. In effect, NO 'squats' on the cytochrome c-oxidase (CCO), where O_2 should adhere to promote cellular respiration and ATP production. Under certain negative circumstances, excess NO binding in the cell may contribute to cell death [9, 23]. With PBM application, the NO is dissociated from cytochrome c-oxidase (CCO), allowing for better utilization of O_2 by the cell during ATP production [23]. Chung et al. [9] reports the same concept in the description of the CCO/NO pathway: "Cellular respiration is downregulated by the production of NO by mitochondrial NO synthase (mtNOS, a NOS isoform specific to mitochondria), that binds to CCO and inhibits it. The NO displaces oxygen from CCO, inhibiting cellular respiration and thus decreasing the production of ATP. By dissociating NO from CCO, low-level laser therapy [PBM] prevents this process from taking place and results in increased ATP production [9]". In other words, without NO attaching to CCO, O_2 is free to attach, promoting cellular respiration and ATP production.

The CCO/NO activity is enhanced by both a hypoxic environment (think of O_2 being depleted by ATP production during intense or prolonged exercise as well as endogenous NO bound to CCO) and by low cellular pH [16, 20, 26]. This happens with intense exercise stressors through the presence of a hypoxic isoform, but can also be accomplished in a more normoxic environment as well, depending on the ratio of ADP:ATP [9, 19, 22]. The research suggests that light therapy at wavelengths between 509 and 691 nm (red range) have a stimulatory effect on the CCO/NO process, but wavelengths above 821 nm (NIR range) are found to be inhibitory, an important consideration when choosing dose and light wavelength (in nanometers) [19].

Nitric oxide, once released from binding to CCO through illumination in certain red and NIR wavelengths, promotes vasodilation and may enhance expression of vascular endothelial growth factor (VEGF) for angiogenesis [19]. As mentioned, Hamblin [23] explains that there also exists a process whereby NO promotes proliferation of *lymphatic* endothelial cells. The NO/sGC$\alpha 1\beta 1$/cGMP process relaxes lymphatic vessels through the expression of the $\alpha 1\beta 1$ isoform of soluble guanylate cyclase (sGC) by NO. The sGC then acts upon the cyclic guanosine monophosphate (cGMP), a secondary messenger chemical also responsible for endothelial dilation, but in lymphatic cells as opposed to vascular cells [23]. This process activates lymphatic drainage for reduction of edema and inflammation [23].

Making room for O_2 in complex IV of the electron transport chain by displacing and repurposing NO for its role in vasodilation helps protect cells from NO-induced apoptosis (cell death) [23]. Again, by NO 'squatting' on the cytochrome c-oxidase, the attachment of O_2 is inhibited at the mitochondrial level. PBM 'photo-dissociates' the NO, allowing it to be sent into the system to promote vascular wall relaxation. Consequently, more oxygen is available to feed the electron transport chain, augmenting the ATP production cycle. It is also thought that the extra NO and the resulting blood flow improvements may help skeletal muscle tissue prolong time to fatigue [12, 13, 39]. Muscles that fatigue slower may provide a performance edge, particularly for endurance athletes or athletes who, by necessity, play long segments in matches. This subset would include football (soccer), basketball, lacrosse, and more.

Balanced Cellular pH

After strenuous exercise, both muscle tissue and the blood drop in pH through accumulation of hydrogen ions [5, 26]. Vigorous and/or prolonged exercise stress creates a hypoxic and acidic environment, at least temporarily. The human body prefers a pH of approximately 7.4 [26]. During post maximal or strenuous exercise, however, pH drops an average of 0.35 (tissue) and 0.6 (serum) points, placing the body in a temporary acidic state [26]. In recent studies, use of PBM after strenuous exercise appears to accelerate the clearance of acid-producing ions through redox signaling mechanisms [21]. Because redox potential is measured in volts, we can make the correlation that PBM, through its enhancement of redox potential, is capable of raising cellular voltage/energy. Tissue healing/recovery relies on sufficient cellular voltage, or cellular energy, to be both initiated and completed.

Enhanced Tissue Redox Potential

The word "redox" is a shortened version of the oxidation-reduction potential, or the capability of cells, tissue, or other substances surrounding the tissues to either reduce (accept electrons) or oxidize (lose electrons). The balance between these two processes is used to determine cellular balance, equilibrium, and therefore resilience. When light therapy is being applied to tissue, as in strenuous exercise, a small amount of endogenous reactive oxygen species (ROS) is released in a short burst [1, 3, 21]. While over-proliferation of these species has been associated with cellular stress that leads to damage and/or death, it is now thought that perhaps, in certain amounts, they can be used as regulators to help cells become more adaptive, or resilient, to changing circumstances in their environment [21]. In other words, small amounts of endogenous ROS, freed into the system through enhanced IWL viscosity and reduced density provide the chemical signal necessary for hormetic response and adaptation to stress. This portion of PBM mechanics creates a protective shield against collections of ROS causing further hypoxic conditions, increasing the odds that the cell recovers easier from training and competition stressors. In this manner, ROS serve as the chemical signal for the body, upregulating or increasing the antioxidant activity in the area [10]. Once antioxidants arrive in the area, the tissue may improve the extracellular pH, or energetic status, providing a better environment for healing, recovery, and performance [21]. By doing so, PBM creates cellular resilience, which may logically lead to improved system resilience, promoting a more resilient athlete.

The primary ROS found in both cases is superoxide anion, which becomes the chemical signal to send superoxide dismutase to reduce oxidative stress [13, 19, 21]. Other research has shown an increased presence of glutathione, another powerful antioxidant [9]. The resulting increase in antioxidant activity from the above process provides superior ability to mitigate oxidative stress, again leading to enhanced and likely faster recovery.

Improved Collagen Synthesis & Fracture Healing

According to Luo et al. (2022) [31, 32], PBM applied to injured skeletal muscle tissue enhances expression of transforming growth factor (TGF-β1), an integral part of matrix remodeling and collagen synthesis after insult or injury to tissue [31]. TGF-β1 improves the transformation of initial outlays of type III collagen to type I [31]. A 2022 review on tendon repair (Lyu et al., 2022) reported that during the proliferation phase, PBM enhances fibroblast proliferation to the area promoting an increase in Type III collagen, a necessary repair component [32]. Additionally, PBM enhances vascular endothelial growth factor (VEGF) to increase microvascularization to the damaged area [32]. PBM, when used in conjunction with appropriate reloading, can accelerate this process, perhaps minimizing the production of scar tissue and promoting tissue regrowth [31]. Luo et al. further reported that improvement in TGF-β1 expression along with a promotion of ROS mitigation encourages cellular and tissue resilience by balancing the ability to engage in efficient electron exchange for healthy cellular processes [31]. Additionally, Berni et.al. (2023) reported accelerated fracture healing through the promotion of angiogenesis, callus formation, and stem cell proliferation and differentiation from a systematic review on PBM technology [27].

Improved Cellular Resilience

Putting together the benefits of (a) increased ATP production; (b) repurposed NO promoting blood flow and lymphatic drainage; (c) increased oxygen delivery by providing more available binding sites for O_2 to the cytochrome c-oxidase to those tissues; and (d) increasing antioxidant activity by using cellular signaling to increase the presence of superoxide dismutase and glutathione, illuminated cells may become more resilient to stress, or at least less likely to succumb to free radicals and fatigue that can predispose tissue to injury. In the current literature, little to no negative side effects are presented with appropriate use of PBM, either with laser diodes or the LED cousins. While some literature does not report a positive or modulating effect, this may be due to a lack of therapeutic dosing [19]. The determination of "ideal" parameters for each individual condition has been difficult due to a lack of complete or standardized parameters included in published papers, as well as the concept of fluence (discussed later) [19]. However, a range of beneficial doses seems to exist [19]. Dosing parameters may likely vary based on the individual, which seems to preclude the determination of precise guidelines, but *ranges* of positive effect can likely be determined due to the selectivity of photon absorption and utilization that cells in vivo seem to exhibit.

In addition, research has also shown an increase in myogenin post-PBM treatment, a key transcription factor one must have for muscle repair and regeneration [13]. Adding this benefit to the cache of mechanisms already reported provides evidence for a quite comprehensive muscle recovery and regeneration process, benefitting the athlete. According to the Handbook of Low-Level Laser Therapy (Hamblin et al. 2021), the regeneration of muscle tissue involves a complex process that includes specific growth factors in combination with activation of satellite cells proliferating, differentiating, and migrating to the target area [19]. In a separate study, Hamblin et al. (2021) reported that research on rat gastrocnemius muscles (in 2013 by Luo et al.) provided evidence that after a contusion, both tissue regeneration and reduction in scar tissue formation was observed by the modulation of two growth factors, Insulin-like growth factor (IGF-1), important for lean muscle production and transforming growth factor beta-1 (TGF-β1), known for collagen synthesis as well as matrix remodeling, with the use of PBM [19, 22, 31].

The use of PBM may increase the following in support of recovery, repair, and regeneration after the stressors of sport and exercise: Nitric Oxide production/utilization, cellular and extracellular pH, tissue redox potential, collagen synthesis, muscle repair, and cellular resilience. Enhanced cellular resilience promotes faster healing and recovery from insult, injury, or stress.

More Benefits of Photobiomodulation

Reduced Pain: Acute, Mid-Range, and Long-Term Stages

In a systematic review of the literature, Naterstad et al. [37] reported that while no adverse effects were mentioned in any included study, PBM showed a clear ability to significantly reduce pain both acutely and for longer durations, lasting anywhere from 4 to 12 weeks post therapeutic dose [37]. Included in the review were several randomized placebo-controlled trials, comparisons with other interventions, and as an additional intervention (e.g., controlled exercise alone vs. with PBM) [37]. Some research suggests that PBM is effective at reducing pain locally first by illuminating the peripheral nerves, while other research suggests the mechanism works by creating a light-induced analgesic effect to the afferent nerve bodies at the dorsal root ganglion (DRG) [10]. In other words, PBM works by stimulating the pain inhibition center at the DRG. A wide-ranging spectrum of painful conditions are reported to have responded to PBM [10]. The list of conditions responding to PBM for pain reduction includes, but is not limited to, "sprains, strains, post-surgical pain, a whiplash injury, muscular back pain, cervical or lumbar radiculopathy, tendinitis, and chronic conditions such as osteoarthritis, rheumatoid arthritis, frozen shoulder, neck and back pain, epicondylitis, carpal tunnel syndrome, tendinopathy, fibromyalgia, plantar fasciitis, post tibial fracture surgery, and chronic regional pain syndrome [10]."

The above-mentioned dorsal horn explanation of the analgesic properties of PBM may include inhibitory features at the spinal cord level [10, 19]. Some research reports that cutaneous PBM illumination at a peripheral nociceptive nerve ending may have a "cascading effect'. The effect appears to start at the epidermal nerve network, but is then transferred deeper, moving next to subcutaneous nerves, then further into sympathetic ganglia, and finally into the neuromuscular junctions for short-term pain relief [10]. However, some may choose to stimulate the nerve root itself, directly activating the pain inhibition center located there. Perhaps similar to an analgesic agent, a sufficient dose inhibits ATP synthesis in the neural mitochondria at the level of the dorsal root ganglion (DRG) after a sufficient dose (Fig. 27.4) [10]. It is thought that within the dorsal root ganglia (where the nociceptive nerve bodies reside), a dense gathering of mitochondria is formed, temporarily decreasing membrane potential, and inhibiting the respiratory chain from providing energy for pain signal transport to cortical pain matrix areas [10]. In other words, PBM in this case causes ATP production to diminish, and becomes an integral step in inhibiting both the sensitization of peripheral nerves and afferent pain/noxious stimuli being blocked. According to the researchers, the reason this happens with PBM is yet unclear, but it is not a thermal reaction [10]. Similar to utilization of local anesthetics, PBM may activate the DRG pain inhibition center at the spinal cord level, thus dampening pain signals to the brain via the spinothalamic tract.

PBM also supports *medium-term* pain remediation due to the reduction of edema and inflammation by blocking of inflammatory marker expression such as Substance P (a neurotransmitter involved in pain perception), and by mitigating oxidative

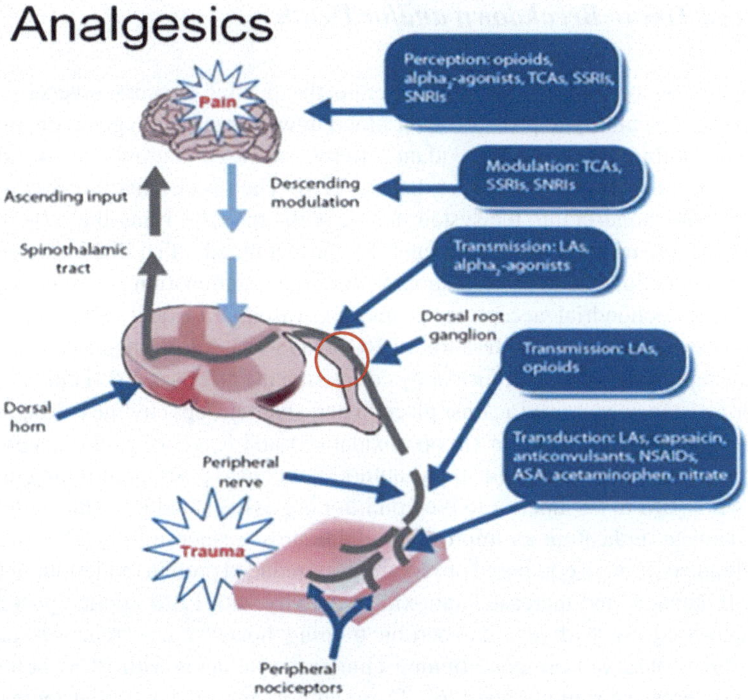

Fig. 27.4 PBM serves to activate a pain inhibition center in the dorsal root ganglion (see red circle within the illustration), keeping pain signals from reaching the central nervous system. PBM may be a valuable, non-pharmacological adjunct for pain relief. (Used with permission, Cotler et al. [10])

stress in the tissue [10]. These benefits can reduce pain for anywhere from hours to days [10]. Finally, *long-term* pain relief is likely due to the acceleration of tissue healing times [10].

Reduced Inflammation

PBM's initial small burst of endogenous ROS (similar to that experienced with acute exercise) provides the appropriate chemical signal to the athlete's system, increasing antioxidant activity at the cellular and tissue levels [19]. The increased antioxidant activity of superoxide dismutase and glutathione help tissues experience reduced inflammation. In addition, Bjordal et al. [8] reported that active Achilles tendinitis exposed to laser diode PBM application of 904 nm (near infrared) significantly increased production of prostaglandin E2 (PGE2), an inflammatory mediator and vasodilator, potentially promoting reversal of the inflammation and tissue healing [8].

Reduced Tissue Breakdown and/or Death

PBM's ability to promote and/or accelerate the healing process reverses tissue breakdown. The ability to provide better blood flow, increased oxygenation, inflammation resolution, enhanced antioxidant activity, and vascular growth factors allows tissue to become hardier and resist damage [28]. The dissociative process of NO from the mitochondria into the system for vasodilation helps bring the cells from a hypoxic to more normoxic environment [28]. Further noted is that PBM also reduces apoptosis, or cellular death, after trauma by limiting inflammation and releasing NO from the mitochondrial acceptors, promoting vasodilation and redox potential, thereby preventing over proliferation of ROS/NOS [23, 28]. Sunemi et al. (2021) took this reported benefit a bit further by comparing the effects of PBM either before or after resistance exercise (against placebo and sham groups) by measuring levels of cell membrane degradation (lipoperoxidation) and levels of SOD (superoxide dismutase, a key antioxidant) [43]. The authors concluded post-conditioning muscle tissue with PBM to be superior to pre-conditioning tissue at reducing the amount of tissue damage, indicating an improved cellular stress response [43]. These results were obtained through compared observations of reduced protein oxidation, reduced cell wall damage, and increased antioxidant activity with PBM applied post resistance exercise [43]. With regard to aerobic training, however, researchers Tomazoni et al. (2019) observed pre-conditioning elite soccer athletes with PBM before an intense progressive running test [46]. These authors reported a similar increase in SOD, but also recorded lower creatine kinase and TBARS levels (both indicators of tissue damage) [46]. Additionally reported were the following performance measures: increased VO_2 max (relative and absolute), increased time to fatigue, and and increase in the volume at which both aerobic and anaerobic thresholds were reached [46]. These papers raise an interesting question regarding timing, indicating that different exercise modalities and intensities may fare better with strategicically timing the PBM application, whether before the exercise session, after, or even both.

Reduced Scar Tissue Formation

As mentioned earlier, research suggests that PBM improves collagen synthesis when used in conjunction with guided exercise, minimizes scar tissue formation, and allows collagen alignment to take place sooner (Barolet et al. 2009) [7, 8]. PBM also modulates matrix metalloproteinase (MMP) expression, another factor in scar formation [8]. Reducing the expression of MMP with the use of light therapy provides additional support for co-treating with appropriately loaded exercise so scar tissue formation may be reduced, and in an accelerated fashion [8].

Reduced Skeletal Muscle Fatigue

Ferraresi et al. reported in a 2012 review that the use of PBM reduces skeletal muscle fatigue [13]. Muscle tissue that has been under either mechanical and/or metabolic stress results in the endogenous release of ROS, proliferation of hydrogen ions in the extracellular matrix, and depletion of ATP [13]. PBM application in small amounts directly on the tissue activates antioxidant and anti-inflammatory substances (superoxide dismutase and glutathione), thus delaying the onset of fatiguing factors (pH reduction, overproduction of ROS without clearing, etc.) [13]. Extending time to fatigue in muscles can result in longer endurance, quicker recovery, and enhanced performance. In a review by de Brito Vieira et al. (2011), PBM was used to improve performance using a near infrared laser (808 nm) on the quadriceps muscles on one side in endurance trained women [12]. The research reported that after PBM application, total work performed increased in both the dominant and non-dominant leg, indicating both a localized and possibly a systemic performance effect [12].

The chemical cascade initiated by the light application takes a short period of time to reach its desired effect [19]. As mentioned before, like exercise, PBM evokes a small amount of endogenous ROS release in the cells in a quick burst, creating a chemical signal for the migration of antioxidants to the needed area [18, 21]. Furthermore, hydrogen ions accumulate in the extracellular matrix, lowering the pH of muscle tissue [26]. Lower pH may lead to reduced cellular energy production, thereby slowing the recovery and repair process. PBM, however, helps restore pH to more desirable levels. Furthermore, research by Assis et al. [5] discovered that providing post-endurance exercise PBM (808 nm) to the tibialis anterior of rats improved tissue recovery and growth *and* promoted a larger cross-sectional area of the soleus and tibialis anterior as well, perhaps improving performance in future bouts of exercise [5]. For the endurance athlete and athletes who compete for extended periods of time (e.g., a long tennis match, football/soccer, or lacrosse), PBM may be a performance enhancer that's irreplaceable.

Lowered LF/HF HRV Ratio

Paolillo et al. [39] provided the first evidence on the effects of heart rate variability (HRV), a standard measure of an athlete's recovery status, on Wistar rats after high-intensity resistance training [39]. The 8-week study applied NIR (780 nm) laser light therapy to the tibialis anterior of the subjects [39]. While other markers were measured, HRV was closely monitored [39]. The PBM recipients demonstrated an *increased* ratio in the low frequency band (0.04–0.15 Hz) and the high-frequency band (0.15–0.40 Hz) of the HRV spectrum [39]. A lower ratio of difference in the two bands indicates enhanced sympathovagal tone, or the ability of the athlete to toggle the autonomic nervous system between sympathetic and parasympathetic states. The ability to engage parasympathetic activity typically aids recovery and tissue healing processes.

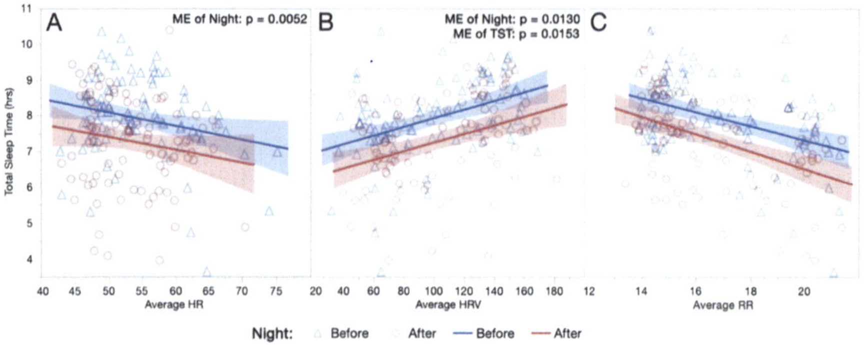

Fig. 27.5 NCAA female soccer athletes, in season, postconditioning with full-body PBM 2× per week saw heart rate and respiratory rate decrease, HRV increase, despite an average 40 min' less total sleep time. (Used with permission. Rentz et al. (2022) [40])

In a separate study, Rentz et al. (2022) [40] provided elite NCAA female soccer athletes with full body LED red and near infrared (combined) light bed therapy twice a week after training [40]. The researchers found that among participants, despite an average of 40 fewer minutes of total sleep time both heart rate and respiratory rates decreased, while HRV increased [40]. The data seems to suggest that the PBM sessions supported rest and recovery after training, without needing more sleep (Fig. 27.5) [40]. An earlier-mentioned paper on the use of PBM to the external branch of the vagus nerve (on the ear) provides further evidence using a more targeted approach to positively affect HRV through modulation of vagal tone (Machado et al., 2022) [33]. The combined research results provide promising evidence to support the use of PBM in promoting recovery after heavy training or competition loads, as well as its use as a treatment modality for overreaching/overtraining in athletes.

Shortened Return-to-Play Time After Injury or Setback

In a systematic review by Foley et al. (2016), several types of sports injury, broken down by grade/severity, received a series of PBM sessions [16]. The authors charted the days to return from initial onset, comparing each athlete's healing time to the average time for return-to-play (RTP) depending on grade or severity of the injury. Also provided was the number of light sessions (PBM) used in each study [16]. In every case, RTP time was at least slightly reduced, and in some cases, the time to return was reduced by as much as 40–60% [16]. For example, an ankle sprain (graded in the study as very severe), had an RTP time of 12 days, compared to the average of 28–42 days without PBM [16]. A hip pointer injury (bruise to the ilium) graded moderate in severity only took 4 days for RTP compared to an average of 7–10 days (Table 27.1) [16]. While recovery benefits are many, it seems that even if

Table 27.1 Injuries, treatment numbers and actual RTP *vs* historically anticipated range of RTP in 65 consecutively-treated and fully followed -up subjects (Foley et al. [16])

Damage assessment	Condition	Examined cases (N = 65)	Average no of treatments	Follow Up (F/U) (months)	Average return to play (RTP) (days)	Expected RTP range (days)
1°	Hamstring strain	19	4	2	7	10–14
1°	Knee sprain (LCL, MCL)	14	3	1	5.75	7–10
1°	Ankle sprain	11	3	2	3.6	7–10
1°	Costochondral sprain	3	4	2	5.5	7–14
1°	Hip pointer	1	3	2	4	7–10
1^+	Ankle sprain	3	3	3	7.3	10–14
1^+	Hamstring strain	3	6	3	19	14–28
1^+	Knee sprain	1	4	4	10	12–14
1^{++}	Knee sprain (LCL, MCL)	1	4	4	11	14–28
2°	Knee sprain (LCL, MCL)	6	6	4	16	28–42
2°	Costochondral sprain	1	5	4	17	28–42
2°	Hip pointer	1	5	4	7	21–24
2^+	Ankle sprain	1	5	4	12	28–42
				Means ± SEM	9.62 ± 1.41[a]	19.23 ± 2.91

1° = mild; 1^+ = moderate; 1^{++} = moderate/severe; 2° = severe; 2^+ = very severe
LCL lateral collateral ligament, *MCL* medial collateral ligament
[a]Statistically significant: P = 0.0066 (Two-tailed Student's t-test, 95% confidence interval)

tissue succumbs to injury, the benefits of red and NIR therapy promote faster healing when comparing similar injury and severity grade.

A paper by Stergiulas et al. (2008) provides evidence for combining PBM with eccentric exercise as an RTP strategy for chronic Achilles tendinopathy [42]. By utilizing PBM with eccentric exercise, RTP was shortened to only 4 weeks compared to 12 weeks for the placebo group [42]. The participants received twelve total doses of NIR therapy (820 nm, with a total dose of 5.4 J/cm^2) over 8 weeks' time, along with targeted loading of the Achilles with eccentric exercise [42]. The authors noted that "The biostimulatory effects of low-level laser therapy [PBM] have been shown to reduce cell apoptosis and promote collagen fiber synthesis" with relatively small doses [42]. Results from these papers suggest that PBM is an effective adjunctive therapy for improving healing times and promoting a faster return to sport.

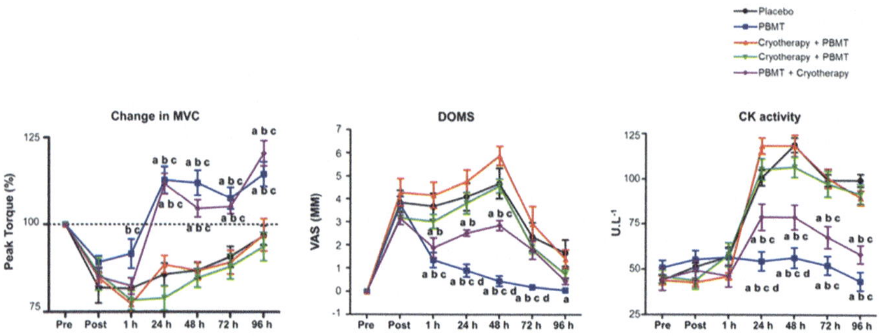

Fig. 27.6 When comparing the effects of PBM and/or cryotherapy to parameters like peak torque, DOMS, and creatine kinase (CK) levels, PBM (blue line) was the clear winner, particularly in reducing CK levels to baseline by 24 h. Cryotherapy blunted the effects of PBM, but if PBM is administered first, it overcomes the recovery-inhibiting properties of icing (violet line). (Used with permission. De Paiva et al. [50])

Additional Recovery Benefits of PBM

De Paiva et al. [50] conducted an interesting placebo-controlled study comparing the effects of PBM to cryotherapy, a common and ubiquitous modality utilized in the industry to promote recovery [50]. The authors had five groups: PBM alone, Cryotherapy alone, PBM +Cryotherapy (PBM first), Cryotherapy + PBM (PBM last), and Placebo [50]. Subjects participated in heavy eccentric training of the quadriceps muscles. The three parameters measured were MVC, DOMS, and creatine kinase levels before and directly following therapeutic administration [50]. The researchers also took measurements at 1 h, 24 h, 48 h, 72 h, and 96 h. Administering PBM alone seemed to provide the most significant benefit in all three categories—in particular it is noted that creatine kinase levels had returned to baseline at only 24 h (Fig. 27.6) [50]. Cryotherapy, on the other hand, continued to allow for elevated CK levels even 96 h post intervention. Also noted was the observation that administering PBM before cryotherapy seemed to overcome somewhat the blunting effects of ice; however, PBM after cryotherapy didn't seem to be enough to overcome the recovery-blunting effects of using ice [50]. Potential conclusions of this study provide evidence that PBM may be a better solution for recovery than traditional cryotherapy.

Photobiomodulation as a Prophylactic Injury Prevention Strategy?

Dornelles et al. [51] conducted an interesting study on football (soccer) athletes [51]. Hamstring injury is common in the sport, and two of the markers for injury susceptibility are low eccentric hamstring strength and a quadriceps:hamstring strength ratio below normal. The researchers measured eccentric hamstring torque,

Table 27.2 Utilizing PBM prophylactically before a simulated match revealed the ability to mitigate strength and power losses from competition by preconditioning the muscles before stressing the tissue, perhaps reducing vulnerability to injury due to fatigue (Dornelles et al. [51])

	PERCENT REDUCTION POST MATCH	
Parameter	PBM (%)	Placebo group (%)
Hamstring eccentric peak torque (loss)	−4.85	−8.72
Hamstring quadriceps torque ratio (loss)	−3.60	−7.75
CMJ height (loss)	−1.77	−5.47

quadriceps:hamstring strength ratio, and even countermovement jump (CMJ) height [51]. They then divided the team into PBM and placebo. They offered a warmup consisting of 5 min' dynamic stretching and 10 min light jogging, Either PBM or placebo was applied to the hamstrings, then 5 min' rest was allotted before the athletes played a full simulated match [51]. Afterwards, all three performance measures were reestablished to find what were the loss in the various parameters between the two groups [51]. Figure 27.6 shows the results, providing evidence that preconditioning muscles with PBM can decrease stress and fatiguability, mitigating the loss in strength and power. One could extrapolate that PBM may indeed reduce injury vulnerability in athletes by providing augmentation to the energy process within the tissue (Table 27.2).

The use of PBM has been shown in research to reduce the following allostatic responses to the stressors of exercise: pain, inflammation, tissue breakdown/death, scar tissue formation, skeletal muscle fatigue, LF/HF HRV ratio, and return-to-play time after injury. In addition, preconditioning with PBM may provide prophylactic benefits to reduce strength and power losses from competition. These results provide promising support for utilizing PBM to accelerate healing and recovery.

Photobiomodulation to Support Recovery Post-concussion/TBI

The research is growing significantly regarding the utilization of transcranial PBM in support of recovery after traumatic brain injury (TBI). Concussion, a ubiquitous occurrence in certain contact sports, can result in devastating effects on athletes depending on severity and number of insults experienced [45]. According to a review by Thunshelle and Hamblin (2016), it was reported in several studies that targeted PBM reduces inflammation and cellular excitotoxicity by improving antioxidant activity and reducing excess glutamate, an excitatory neurotransmitter that, in excess amounts, can cause neuronal damage post-TBI [45]. Further benefits include expression of brain-derived neurotrophic factor (BDNF), an important molecule used in

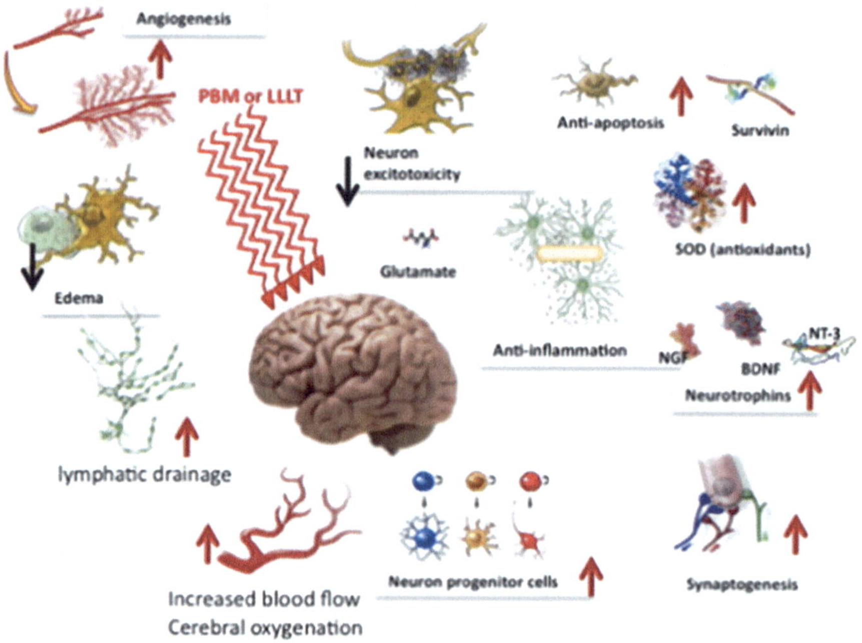

Fig. 27.7 There are numerous benefits to transcranial PBM, included reduced edema, excitotoxicity, and inflammation with increased blood flow, lymphatic drainage, neurotrophins, neurogenesis, and synaptogenesis. (Used with permission, Hamblin 2016) [56]

building memory and learning centers in the brain; improved blood flow and lymphatic drainage due to NO and vascular endothelial growth factor (VEGF) expression; and improved synaptogenesis (making new synaptic connections) (Fig. 27.7) [45].

> The time frame away from acute (traumatic brain) injury… does not appear to be a factor in providing improvement in key quantitative neurological or qualitative quality of life metrics using transcranial PBM.

For example, one study involved concussion patients aged 26–62 years who had experienced the injury ranging anywhere from 11 months to 8 years before the study's commencement, and participants showed marked improvements in quality of life (Naeser et al. [36]). The results were quite significant, especially given the length of time some participants were away from the original insult. The study reported continued improvements in cognition up to 4 months' post-PBM treatment protocol, as well as improvements in depressive symptoms [36]. Family members of participants reported improved sleep patterns, socialization, and daily living skills, and participants who also had post-traumatic stress (PTS) symptoms reported a reduction of symptoms [36].

Martin et al. [34] conducted another study on Gulf War illness (from the early 1990s) and post-traumatic stress (PTS) [34]. Participants reported reduced depression and anxiety scores as well as improved cognition [34]. The study included a sham group—interestingly, during the first week both groups improved, perhaps because there was attention given to their condition. At 1 month, however, only the PBM treatment group continued to improve, while the sham group regressed toward baseline measures [34]. In both the Naeser et al. [36] and in this case, the time distance away from the acute injury was not a factor in providing improvements in key metrics using transcranial PBM [34, 36].

In conditions such as post-concussive syndrome, post-traumatic stress, strokes, and other types of traumatic brain injuries, PBM has been proving itself as a safe and effective modality for supporting repair and recovery of brain tissue. Strategic use of PBM on the brain may provide help for other types of neurodegenerative disease conditions like Parkinson's, Alzheimer's, Multiple Sclerosis, and more. Research is starting to blossom in neurorehabilitation circles.

I've had the opportunity to consult with a laser device company that is utilizing the concept of creating research-derived custom pulse rates for different tissues and conditions. As a professional supporting the sports medicine industry, I collaborated with the clinical director of this organization to develop 120 settings specific to sports injury, some of which are dedicated to post-concussive rehabilitation. Based on the phases of recovery after concussion, one of the settings is designed to be used in the excitotoxic phase of a head injury to bring down inflammation, reduce glutamate, increase mitochondrial function, and avoid the continuation of what's called "activated glial cells", a condition that may happen when the glial or glymphatic system, ordinarily used to provide neural 'clean-up' activity to clear debris from injured or insulted tissue, becomes overactive and may begin to attack normal brain tissue as well. In fact, activated glial cells are part of the etiology of CTE (Chronic Traumatic Encephalopathy), a condition that's received a good amount of press in recent years.

However, I crossed paths with a beautiful young man who suffers with neurological complexities sometimes associated with the autistic spectrum called PANDAS (Pediatric Autoimmune Neuropsychiatric Disorders Associated with Streptococcal infections). If he gets exposed to the strep bacteria, it creates neuroinflammation that robs him of sleep and causes him to become hyperkinetic, anxious, and inappropriately vocal. They were seeing a practitioner who uses the laser device with these sports medicine settings (among others), and I suggested they try the excitotoxicity setting from the concussion section on his brain (to include cerebellum, and temporal and frontal lobes) to see if it could help attenuate this flare-up. The good news is that a short dose of light in this custom pulse rate did seem to reduce his stress, bringing him into a calmer state, enabling sleep, functionality at his school, and the ability to sit and feed himself (which had been unattainable since the onset of symptoms). While the setting was meant for helping head injured patients/athletes, it was also sufficient to calm this autoimmune-based neuroinflammatory process and provide a solution in a different way. Knowing PBM's mechanisms as discussed in this chapter provided a pathway previously unavailable to this precious family.

*Jennifer Novak MS, CSCS*D*

Important PBM Device Component Overview/Comparison

Laser Diode vs. LED

In recent years there has been a plethora of light therapy devices and tools flooding the market as popularity of the technology rises and the research continues to provide wide-ranging benefits. Devices currently on the market are being used for everything from pain relief to hair growth. Most of the commercially available products are constructed using LED lights, while some of the higher-end devices utilize true laser diodes in their products. The broad variety of light sources and power levels begs the question: Don't all photons get the job done? Does it matter which ones are used?

The answer to the question, "A photon is just a photon, right?" is not as simple as it seems. A brief explanation of the coherency of the light source and what that means in terms of efficacy is needed as clarification. An LED light offers a *non-coherent* light wave, meaning the photons cross over themselves on their way to the target tissue [19]. While a non-coherent light wave will still reach the tissue, power, number of photons reaching the desired destination, and therefore efficacy, will likely diminish [19]. A laser diode provides the user with a *coherent* wave, providing the most direct and potent form of therapeutic light doses [19]. Additionally, a laser diode provides *monochromatic* wavelength application, while LEDs may not (more than just the light wavelength of interest in the spectrum exist in LEDs than due to materials used in its construction); therefore, LED total power will likely be lower than a laser diode (in a similar wattage comparison) due to the dilution of specific wavelength being applied [19]. Additionally, LEDs utilize 20–30% of their power in first converting their electrical energy (measured in watts) into light energy (measured in joules), and this suboptimal conversion rate will generate some heat, therefore caution should be used with contact application for longer periods (Heiskanen and Hamblin 2018) [22, 24]. This conversion inefficiency may also lead to a reduction in the viable lifetime of an LED device [22]. Many of these commercial LED PBM devices have automatic shutoff mechanisms at 10–20 min, indicating the device's 'safe' range of use. A device with true laser diodes, however, may provide therapeutic effects anywhere from 10 s to 2 min' application, further suggesting enhanced efficacy of the laser diode [19].

There are several differences here that should be considered when selecting both the type of device you wish to utilize and the dose. Many of the published reviews and meta-analyses on PBM suggest more research is needed because the authors have failed to provide consistency in information regarding the power, fluence, and dose of their study parameters [3]. Another comment suggests the dosing is recorded over wider ranges than can be discerned for optimal dosing [3]. A primary reason for this issue may be due to the wide variety of power (wattage) levels, fluence levels, and overall work done in any given session (measured in joules or joules/cm^2) [19]. Some of the discrepancy may be, however, due to the use of LED vs true laser diodes.

As mentioned earlier, LEDs emit an incoherent wave that is potentially not monochromatic [10, 19]. A laser diode is monochromatic, meaning there are no other light wavelengths emitted other than the one specified for that device [10, 19]. Laser diodes offer a coherent, more direct wave than that of an incoherent LED wave [10, 19]. These features of LEDs likely reduce total delivery of photons and hence the efficiency of their application. While both laser diodes and LEDs have their place in recovery, differences in depth of penetration, device longevity and precision (in coherence and monochromaticity) and pulsing capability along with the requirements of the condition or tissue in front of the practitioner are worth further investigation – what's appropriate for each situation may vary.

The monochromaticity (spectral width) of diode lasers are typically ±0.5 nm, measured in a term called full width half minimum (FWHM) [19]. This is the measurement of the spectral width when the device is running at peak power levels. The FWHMs of LEDs range between about ±5 nm to ±20 nm, an area anywhere from 10 to 40 times larger than laser, which may translate to a reduction in total power or strength of the light source [19]. Because the chromophore of interest in PBM for tissue and cells is cytochrome c-oxidase, with its absorption spectra being rather wide, researchers suggest at least 10 nm, within the range of both cluster laser diodes as well as LED versions [19]. Therefore, LEDs may only be slightly less effective at the *superficial* tissue level, depending on need [19]. LED light sources' ability to reach deeper tissue levels may not be as effective as the laser diode due to these factors. One might induce that a laser diode, with its monochromaticity of wavelength and its coherent wave features (both unique to laser diodes), may provide advantages in efficacy for more targeted or deeper tissue applications.

Additionally, both skin tone and skin thickness should be considered when applying PBM. Darker skin may absorb some of the photons and translate that sensation as heat because melanin is a chromophore [19]. Even though PBM is a non-thermal process, the small amount of surface heat generated by LEDs conversion of electric energy to light energy can become uncomfortable in some [18]. Therefore, care should be taken to increase the distance between the athlete's skin and the diodes, and then perhaps judiciously extend the dosing time accordingly, due to possible absorption and/or scattering of photons before reaching target tissue areas. Therefore, skin tone should be especially kept in mind if the device being used contains *surface* diodes (laser or LED) as opposed to *recessed* diodes [19]. Also, partial scattering of photons may occur with varying skin thickness over the targeted tissue as well, meaning fewer photons may reach the target tissue [19]. Below is a photograph of a small sampling of the shapes, sizes, and components of some PBM devices (Fig. 27.8).

Single- vs. Multiple-Diode Devices

When single-diode devices were used, the effects mentioned in this chapter were not observed, likely because the surface area being affected by the single diode was too small to effect positive change (Abedi et al. [1]). Multiple diodes provide broader coverage, multiplying the number of photons available for tissue use [1]. Therefore, selecting a device with multiple diodes covers a broader surface area,

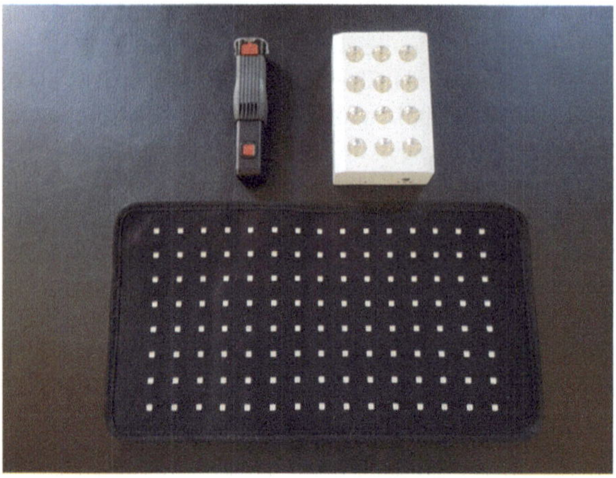

Fig. 27.8 A sampling of different types of photobiomodulation devices. From top/left: Avant Wellness LZ30 Z laser, 4 true laser diodes (two red @ 635 nm, two near infrared @ 808 nm), maximum 1400 mW power NIR, 1000 mW red. Adjustable power density, 300+ preset pulse rates according to tissue type/condition. Joovv mini, 12 LED diodes, continuous beam, available in either red (660 nm) or near infrared (850 nm), 100 mW power. Bottom, Novaalab Light Pad, 360 LED light "chips", up to 200 mW power. Includes 120 red (660 nm) and 240 near infrared (850 nm) that can be combined, or either just red, or just near infrared, and offers either continuous or a single pulsed option. These parameters are provided by manufacturers' report

enhancing the ability to provide a therapeutic dose. In addition, beam shaping optics may be a consideration, as certain configurations may either limit or broaden the surface area reached by the device.

Continuous vs. Pulsed PBM Application

Research suggests that a continuous wave is not as efficient and does not provide the level of desired positive effect as a pulsed diode cluster [1, 24]. Pulsed lasers (with on-and-off pulse durations of about 100–150 nanoseconds) appear to be more effective than continuous ones (light application without the pulsing micro-pauses) [19, 35]. In fact, research suggests that pulsed laser dosing is up to *ten times faster* than in continuous dosing [35]. Pulses are measured in 4 ways and are calculated with a simple equation to understand average power in a given cycle of PBM [19]. First, there is peak power (expressed in *Watts*); then pulse frequency or pulse repetition rate per second (expressed in *Hertz*); third is pulse width, or how long the pulse is on (expressed in *seconds*); and fourth is the duty cycle, which is the time ratio between the 'on' cycle and the 'off' cycle in the pulse (expressed as a *percentage* (%) or a *ratio* (e.g., 50:50) [19]. Then the average power is calculated as follows:

Peak power in Watts (W) multiplied by pulse width in seconds (s) multiplied by pulse frequency in Hertz (Hz) = Average power of your PBM cycle [21]

Some device manufacturers have been augmenting the concept of pulsed light through an area of scientific study called mechanobiology. In this case, custom percentages or frequencies of pulse rate are created based on instrumentation that measures and reports tissue oscillation qualities (in Hertz), applies certain custom pulse rates to the tissue, and then records the change, if any, in the tissue. This may be an important measure of specificity in supporting the restoration of cells and tissue faced with various insults or pathologies.

Scanning vs. Contact PBM Application

Research suggests that PBM is more effective when positioned and held in contact with the skin, particularly if the device has a recessed diode placement [19, 22, 35]. This method, with the proper device type, allows for slight tissue compression to assist in reaching deeper tissues without placing the actual diodes in contact with the person's skin [19]. For larger tissue areas, a device may be used to scan or paint the area, however. If using a device in this manner, the proper calculation for doing so would be to multiply the total area of tissue covered by the time of PBM application divided by the power of the device to determine the proper dose [19].

Dosing

The word "dose" in PBM refers to the amount of 'work' accomplished by the device, not solely the amount of time a device is utilized on a given recipient. Many researchers describe their dosing parameters in terms of power and peak power, defined in Joules and Joules/cm^2, respectively. While these parameters certainly play a role in the dose, or time needed to achieve the desired effect, researchers should also consider *fluence*, which is the actual interface of the light once it crosses through the skin, where tissue type, thickness, redox state, photon scatter, photon absorption, and more may alter the length of time to achieve the desired effect [19]. In The Handbook of Low-Level Laser Therapy edited by Hamblin et al. (2021), the authors state, "[Different from] Radiant exposure [which] measures the light energy that arrives at a surface (for instance, at the skin in an low-level laser therapy procedure) coming from the source; fluence, at the same point, *includes the energy of the light that backscatters from the interior of the illuminated tissue and from the surface*….fluence characterizes the *interaction* of the light beam with the illuminated material (tissue) [italics added] [19]." Due to the wide variety of tissue types and qualities in the human body, fluence is, by nature, difficult to truly measure. Some tissues may absorb, reflect, or scatter the photons as they enter through the skin, therefore, true fluence is *estimated* and measured in J/cm^2 (like peak power) to provide a consistent measurement parameter for researchers and practitioners alike [19].

An important factor to note is that the dose response *in vitro* has been observed to be "bi-phasic"—i.e., depending on the density of the light exposure, the effect can be either an inhibitory action on the cells and in effect, tissue, or excitatory (stimulating), or both [18, 19, 22]. However, the same concept has not been clearly identified *in vivo*, but more of a selectivity process takes place. Cells/tissues in need of modulation due to inflammation, oxidative stress, or damage accept the photons, but cells/tissues not in need of PBM do not [18]. Neither low-level lasers nor their LED versions (see above-mentioned possible surface diode/skin tone exceptions) evoke a significant thermal effect, suggesting a low likelihood of any tissue damage from the technology [19, 22].

The general parameters for dosing PBM, according to Moskvin [35], includes some ranges, depending on type of use [35]. For specific tissue modulatory activity, exposure time is suggested to range between 1.5- and 5-min' total (100–300 s) with laser diodes [35]. For stimulating acupuncture points (with light, not by piercing the skin with needles), the exposure is only recommended to be about 20–40 s long [35]. More 'global' applications like blood illumination, a method of transcutaneous light delivery at large blood vessels like the femoral, brachial, or carotid arteries or the vena cava at the umbilicus, suggested dosing can last up to 20 min [35]. This is designed to allow blood to flow past the light source repeatedly, absorbing the light gradually for more systemic effects on global inflammatory processes, oxygenation of tissues and organs, etc. [35] Perhaps this particular application provides substantiation for the presence of free-floating mitochondria in the blood stream that has respiratory capability but has thus far defied a clear explanation for their existence [54]. Other scientists have suggested that shorter doses, depending on the wavelength of the light source, total power, etc. may fail to be therapeutic and reach the desired effect or outcome. Called a "dose-rate effect", a clear consensus has yet to be reached due to reasons previously discussed (lack of conformity in reporting parameters in manuscripts and the concept of fluence) [19]. Some authors have arrived at a minimal range between 0.15 and 10 mW/cm^2 for targeted tissue, but again, wavelength and type of light source, the type and depth of tissue targeted, the type of effect desired by the application, and even the cells' or tissues' redox status can change these parameters [17, 31]. Redox status is the tissue's balance between the number of oxidants (reactive oxygen species) and antioxidants, indicating the level of inflammation. Higher levels of inflammation suggest the need for administering a higher dose of photons to the target tissue [20, 31].

Contraindications to Photobiomodulation

A few known contraindications to using PBM exist, in part because research has yet to be accomplished, and those lie in the following conditions: pregnancy, cancer, epilepsy at certain pulse frequencies, and directly into the eyes [19]. Further research, however, may uncover different results, but for now, avoidance of these few situations is prudent. No research has yet been completed to determine if any type of red or NIR therapy has a deleterious effect when shined directly on a developing fetus, therefore PBM is currently contraindicated during pregnancy [19].

The second, cancer, is an area of burgeoning research in terms of the benefits of using PBM to mitigate the *side effects* of cancer treatment modalities [19]. According to Tam et al. [44], positive effects have been cited on adjunct conditions such as lymphedema, radiation dermatitis, and associated pain and inflammation from standard cancer treatment interventions [44]. A more recent study measured the effects of PBM on glioblastoma, a malignant brain tumor. Excitingly, researchers measured cancer cell death up to 30% at a dose of 24.56 joules/cm^2, up to 6 h after treatment [6]. While more research is certainly needed, for now, using PBM directly on cancerous tumors is contraindicated.

Additionally, pulsed light at a power of less than 30 Hz may evoke an epileptic seizure to photosensitive epileptics [10, 19]. Finally, PBM should not be administered directly into the eyes, as it has the potential to do retinal damage [10, 19].

Also of note, the ingestion of any pharmaceutical that causes photo-sensitivity may affect the light sensitivity of the skin, where light may be applied, on such an individual [19]. Additionally, since there exists the potential for immune-stimulating properties with PBM, if a person is taking immunosuppressive medications for any autoimmune condition, the action of those medications may be dampened [19]. The above statements may not constitute an exhaustive listing of medications which might contraindicate use of PBM. As with any additional health concern, checking with a medical professional for any potential contraindications may be recommended before utilizing PBM.

Finally, upon systematic review of the literature, Cotler et al. [10] report that any adverse events involving PBM use did not differ from adverse events in placebo groups, suggesting a high safety profile [10]. Outside of specific contraindications, PBM may be used for a wide variety of conditions and tissue repair and recovery states. Therefore, keeping in mind the few contraindications to utilizing PBM, the safety profile of the technology should bring a sense of comfort to the practitioner and the recipient of its use.

Suggestions for Future Photobiomodulation Research

Literally thousands of research trials and studies have been completed on PBM, spanning over half a century and on many continents. Previous attempts at conducting reviews of the literature include inconsistency of reporting parameters used in the study; whether there was a true sham, placebo, or control group included in the study; and the difficulty in defining fluence due to subject and tissue differences in tone, thickness, base inflammation level, redox status, hydration status, and more. If research organizations could provide a set lexicon of device parameters for research going forward, more comparable effects and limitations may be observable. Practitioners can then be evidence-led and use this knowledge to think of new ways to apply PBM as a safe and effective means to promote and accelerate healing and performance, when they use the available research and apply their experience and knowledge with patients to further advance what this technology can do.

For example, most of the research applies PBM either before (preconditioning) or after (post-conditioning) a given task and measures affect the next time the task or event is attempted. I like to call this either 'preconditioning' or 'postconditioning' the cells and tissues because light application sets a cellular and systemic process in motion, aimed at either performance, or recovery, or both. However, little has been completed with co-activation of the target tissue while completing a task. One example applied PBM during a 6-week strength training program [55]. It was not exactly clear when the PBM was applied, but indications are that it was during the activities, which included both sprints and squats, with PBM or placebo applied to the quadriceps muscles [55]. No significant difference was indicated from placebo or controls, however, dosing may or may not have been adequate to achieve a therapeutic effect [55]. Regardless, if the task is metabolically stressful, such as loaded exercise to fatigue, perhaps the addition of PBM against the concurrent stress to the tissue would largely produce a zero-sum game. In other words, the properties and effects of the light waves cannot simultaneously overcome the stress of the task. However, coactivation of supporting systems may be of value in terms of providing 'energetic boosting' to those supporting physiological systems.

An example might be to stimulate associated nerves, nerve roots, or brain pathways to assist in rebuilding a communication loop between the CNS and the periphery after insult or injury. The same considerations might be interesting when applied in conjunction with vestibular, visual, proprioceptive, or motor skill rehabilitation or training. Rehabilitation specialists use targeted stimulation to rebuild and/or modulate areas of the patient's system in need of augmentation. If illumination with PBM could occur while the stimulation was being administered, would there be a compounded neurological or sensorimotor positive effect? Would PBM's tissue healing effects further augment and accelerate brain rehabilitation if used strategically and simultaneously during rehabilitative efforts? Again, these concepts might be able to further the growing research on PBM for improving symptoms after TBI, and with neurodegenerative conditions such as Parkinson's, Alzheimer's, Multiple Sclerosis, and more.

Perhaps the same concept would benefit tissue healing with musculoskeletal injury. After the acute phases, when rebuilding is taking place, it might be advantageous to add light to the physical rehabilitation process after a soft tissue or orthopedic injury *during* rehabilitation work. An additional thought is to help determine the best timing for repair, recovery, or performance applications by conducting more trials of interventions completed before, during, or after training or competition (with designated time frames), or even whether morning or evening would be more appropriate given the effects of different light wavelengths on the circadian rhythm, blue light being known to inhibit melatonin, yet red perhaps encouraging its expression [30].

Other research might move into the realm of providing the positive effects of PBM in *patterns* that focus on the supporting systems involved in a certain healing process. In other words, if a certain organ or gland is affected and/or needed in the healing process of an insult or injury, or if those body tissues are integral in making a process happen (such as digestion, or detoxification, or lymphatic, to name a few) would small doses of PBM "activate" or support the process at large? The human system is both dynamic and interdependent. Perhaps future research would recognize this concept and apply light to more areas than just the immediate affected area to support holistic recovery and resilience.

As a direct example of this concept, one of the settings we developed in the sports medicine version of the laser (to which I contributed researched input) were *theoretical* pulse rates associated with the pathophysiology of rhabdomyolysis combined with the known cellular and tissue benefits of PBM on muscle tissue and the kidney. No studies are currently available on the use of PBM to specifically support recovery from this condition, and no IRB would likely approve such a study for obvious reasons.

Interestingly, I received a phone call from an athletic trainer reporting that he was about to work with four athletes who'd contracted rhabdomyolysis in football practice. He owns one of the sports medicine versions of the laser mentioned earlier, so we discussed the possibilities of how and where to apply the rhabdomyolysis set of pulse rates to evoke the desired change in the athletes' health status. Given the damage to skeletal muscle tissue and stress on the kidneys as they filter residual myoglobin and other byproducts into the body's waste system, he and I developed a strategy for its application, also preceded by small amounts of light on the organs associated with their detoxification and lymphatic systems. At the writing of this chapter, I'm awaiting the data from this athletic trainer to write up a case study in hopes of being published. However, from a qualitative standpoint, all four athletes experienced tremendous improvements. After receiving merely two applications of the rhabdomyolysis setting and stimulating the detox and lymphatic systems with light during their sessions on consecutive days with the athletic trainer, their muscle soreness scales went from excruciating to almost gone, they reported experiencing better mental clarity and energy, and their urine samples changed from the well-known cola color back to yellow in 1 day, all clear signs they were all recovering. Once again, by knowing the positive benefits and/or the mechanisms of PBM, its safety profile, and its ability to change redox state, enhance cellular energy production, and accelerate healing, new horizons in treating many types of conditions can be reached every day.

Additional help might be provided if a consistent measure for quantifying fluence (as defined by what happens to the photons once they penetrate the skin) could be determined so researchers could better know what the effect might be at a given area, accounting for skin thickness, presence of other tissue types, tissue water content, inflammatory status, and more. The most important consideration is that athletes are individuals, so exact doses may not exist given these considerations. PBM's adverse effects are reported to be next to none, yet the benefits toward performance, recovery, and injury management are many, and growing as research continues. The practitioner can therefore be evidence-led, using proper standards of application, and likely safely advance recovery, healing, and performance measures, making for optimized, resilient athletes.

There are some important PBM concepts to note when determining how to use PBM in support of repair, regeneration, and recovery from the stressors of exercise. These include the type of diode, the number of diodes, whether they are surface or recessed diodes, and whether they provide a continuous or pulsed light stream. In addition, one should consider skin tone and thickness when determining dosing as melanin is also a photo-accepting chromophore and skin thickness may absorb some of the photons before reaching target tissue. Finally, there are a few contraindications noted, and this may not be an exhaustive list. However, PBM safety is high due to any reports of adverse events matching those experienced by placebo/control group participants. Please consult a medical professional with any question regarding a unique situation before utilizing PBM.

Further research might be easier to analyze if a consistent lexicon parameters of PBM use in each cohort or trial is decided upon and used. Additionally, the concept of co-activating tissue (soft, skeletal, brain etc.,) during rehabilitation therapy (below metabolic threshold or more maximal efforts) could provide insights into whether rehabilitation could be augmented or accelerated by providing the energy-producing and stress-reducing effects of PBM. Finally, consideration for utilizing the mechanisms of PBM to provide "energetic boosting" to supporting physiological systems to further advance the technology's capabilities within the dynamic and interdependent human system, promoting more holistic recovery and resilience.

References

1. Abedi Yekta AH, Tabeii S, Salehi M-R, et al. Effect of low-level laser therapy on muscle strength and endurance and post-exercise recovery of Young adult: a double-blind, placebo-controlled, randomized clinical trial. Ann Appl Sports Sci. 2021;9(2)
2. Ahrabi B, Rezaei Tavirani M, Khoramgah MS, et al. The effect of photobiomodulation therapy on the differentiation, proliferation, and migration of the mesenchymal stem cell: a review. J Lasers Med Sci. 2019;10(Suppl 1):S96–S103. https://doi.org/10.15171/jlms.2019.S17.
3. Ailioaie LM, Litscher G. Photobiomodulation and sports: results of a narrative review. Life (Basel). 2021;11(12):1339. Published 2021 Dec 3. https://doi.org/10.3390/life11121339.
4. Ali MK, Saha S, Milkova N, et al. Modulation of the autonomic nervous system by one session of spinal low-level laser therapy in patients with chronic colonic motility dysfunction. Front Neurosci. 2022;16(882602):1–12. https://doi.org/10.3389/fnins.2022.882602.
5. Assis L, Yamashita F, Magri AM, Fernandes KR, Yamauchi L, Renno AC. Effect of low-level laser therapy (808 nm) on skeletal muscle after endurance exercise training in rats. Braz J Phys Ther. 2015;19(6):457–65. https://doi.org/10.1590/bjpt-rbf.2014.0113.
6. Atluri R, Afkhami P, Choi T-Y, Simmons DP. Investigation of the effect of LLLT on glioblastoma multiforme [abstract]. In: Proceedings of the American Association for Cancer Research Annual Meeting 2021; 2021 Apr 10–15 and May 17–21. Philadelphia: AACR; Cancer Res 2021;81(13_Suppl):Abstract nr 1948.
7. Barolet D, Roberge CJ, Auger FA, Boucher A, Germain L. Regulation of skin collagen metabolism in vitro using a pulsed 660 nm LED light source: clinical correlation with a single-blinded study. J Invest Dermatol. 2009;129(12):2751–9. https://doi.org/10.1038/jid.2009.186.
8. Bjordal J, Lopes-Martins R, Iversen V. A randomised, placebo-controlled trial of low level laser therapy for activated Achilles tendinitis with microdialysis measurement of peritendinous prostaglandin E2 concentrations * commentary. Br J Sports Med. 2006;40(1):76–80. https://doi.org/10.1136/bjsm.2005.020842.
9. Chung H, Dai T, Sharma SK, Huang YY, Carroll JD, Hamblin MR. The nuts and bolts of low-level laser (light) therapy. Ann Biomed Eng. 2012;40(2):516–33. https://doi.org/10.1007/s10439-011-0454-7.
10. Cotler HB, Chow RT, Hamblin MR, Carroll J. The use of low level laser therapy (LLLT) for musculoskeletal pain. MOJ Orthop Rheumatol. 2015;2(5):00068. https://doi.org/10.15406/mojor.2015.02.00068.
11. de Almeida P, Lopes-Martins RA, De Marchi T, et al. Red (660 nm) and infrared (830 nm) low-level laser therapy in skeletal muscle fatigue in humans: what is better? Lasers Med Sci. 2012;27(2):453–8. https://doi.org/10.1007/s10103-011-0957-3.
12. de Brito Vieira W, Ferraresi C, de Andrade Perez S, Baldissera V, Parizotto N. Effects of low-level laser therapy (808 nm) on isokinetic muscle performance of young women submitted to endurance training: a randomized controlled clinical trial. Lasers Med Sci [online]. 2011;27(2):497–504. Available at: https://link.springer.com/article/10.1007/s10103-011-0984-0. Accessed 24 Aug 2022.
13. Ferraresi C, Hamblin M, Parizotto N. Low-level laser (light) therapy (LLLT) on muscle tissue: performance, fatigue and repair benefited by the power of light: Low-Level-Laser (Licht)-Therapie an Muskelgewebe–Möglichkeiten zur Verbesserung der Leistungsfähigkeit und zur Behandlung von Muskelermüdung und Muskelverletzungen. Photonics Lasers Med. 2012;1(4):267–86.
14. Ferraresi C, Huang Y, Hamblin M. Photobiomodulation in human muscle tissue: an advantage in sports performance? J Biophotonics. https://pubmed.ncbi.nlm.nih.gov/27874264/. Accessed 24 Aug 2022.
15. Ferraresi C, Kaippert B, Avci P, Huang Y, de Sousa M, Bagnato V, Parizotto N, Hamblin M. Low-level laser (light) therapy increases mitochondrial membrane potential and ATP synthesis in C2C12 Myotubes with a peak response at 3-6 h. Photochem Photobiol, [online].

2014;91(2):411–6. Available at: https://onlinelibrary.wiley.com/doi/10.1111/php.12397. Accessed 24 Aug 2022.
16. Foley J, Vasily D, Bradle J, Rudio C, Calderhead R. 830 nm light-emitting diode (led) phototherapy significantly reduced return-to-play in injured university athletes: a pilot study. Laser Ther. 2016;25(1):35–42. https://doi.org/10.5978/islsm.16-or-03.
17. Forsey JD. The effects of acute photobiomodulation on anaerobic exercise performance. West Virginia University; 2020.
18. Hamblin MR. Dr. Michael R Hamblin discusses photobiomodulation treatment of traumatic brain injury OSA webinar. https://youtu.be/_XA9rXDjRe0. October 2020. Accessed 17 May 2023.
19. Hamblin MR, Pires de Souza MV, Agrawal T. Handbook of low level laser therapy. Pan Stanford Publishing Pte. Ltd.; 2017.
20. Hamblin MR. Mechanisms and applications of the anti-inflammatory effects of photobiomodulation. AIMS Biophys. 2017;4(3):337–61. https://doi.org/10.3934/biophy.2017.3.337.
21. Hamblin MR. Mechanisms and mitochondrial redox signaling in photobiomodulation. Photochem Photobiol. 2018;94(2):199–212. https://doi.org/10.1111/php.12864.
22. Hamblin MR. Photobiomodulation or low-level laser therapy. J Biophotonics. 2016;9(11–12):1122–4. https://doi.org/10.1002/jbio.201670113.
23. Hamblin MR. The role of nitric oxide in low level light therapy. In: The International Society for Optical Enginneering proceedings of SPIE; 2008. https://doi.org/10.1117/12.764918.
24. Heiskanen V, Hamblin MR. Photobiomodulation: lasers vs. light emitting diodes? [published correction appears in Photochem Photobiol Sci. 2018 Oct 31;18(1):259–259]. Photochem Photobiol Sci. 2018;17(8):1003–17. https://doi.org/10.1039/c8pp90049c.
25. Huang YY, Chen AC, Carroll JD, Hamblin MR. Biphasic dose response in low level light therapy. Dose Response. 2009;7(4):358–83. Published 2009 Sep 1. https://doi.org/10.2203/dose-response.09-027.Hamblin.
26. Hermansen L, Osnes JB. Blood and muscle pH after maximal exercise in man. J Appl Physiol. 1972;32(3):304–8. https://doi.org/10.1152/jappl.1972.32.3.304.
27. Berni M, Brancato AM, Torriani C, et al. The Role of Low-Level Laser Therapy in Bone Healing: Systematic Review. Int J Mol Sci. 2023;24(8):7094. Published 2023 Apr 12. https://doi.org/10.3390/ijms24087094.
28. Karu TI. Mitochondrial signaling in mammalian cells activated by red and near-IR radiation. Photochem Photobiol. 2008;84(5):1091–9. https://doi.org/10.1111/j.1751-1097.2008.00394.x.
29. Leal E, Lopes-Martins R, Frigo L, et al. Effects of low-level laser therapy (LLLT) in the development of exercise-induced skeletal muscle fatigue and changes in biochemical markers related to Postexercise recovery. J Orthop Sports Phys Ther. 2010;40(8):524–32. https://doi.org/10.2519/jospt.2010.3294.
30. Liu J, van Iersel M. Photosynthetic physiology of blue, green, and red light: light intensity effects and underlying mechanisms. https://www.frontiersin.org/articles/10.3389/fpls.2021.619987/full. Published 2022. Accessed 2 Sept 2022.
31. Luo L, Sun Z, Zhang L, Li X, Dong Y, Liu TC. Effects of low-level laser therapy on ROS homeostasis and expression of IGF-1 and TGF-β1 in skeletal muscle during the repair process. Lasers Med Sci. 2013;28(3):725–34. https://doi.org/10.1007/s10103-012-1133-0.
32. Lyu K, Liu X, Jiang L, et al. The functions and mechanisms of low-level laser therapy in tendon repair (review). Front Physiol. 2022;13:808374. Published 2022 Feb 15. https://doi.org/10.3389/fphys.2022.808374.
33. Machado C, Machado Y, Chinchilla M, Machado Y, Foyaca-Sibat H. Assessing the autonomic effect of vagal nerve stimulation with low level lasers by heart rate variability. Ispub.com. https://ispub.com/IJN/21/1/54164. Published 2022. Accessed 24 Aug 2022.
34. Martin PI, Chao L, Krengel MH, et al. Transcranial photobiomodulation to improve cognition in gulf war illness. Front Neurol. 2021;11:574386. Published 2021 Jan 21. https://doi.org/10.3389/fneur.2020.574386.

35. Moskvin SV. Low-level laser therapy in Russia: history, science and practice. J Lasers Med Sci. 2017;8(2):56–65. https://doi.org/10.15171/jlms.2017.11.
36. Naeser MA, Zafonte R, Krengel MH, et al. Significant improvements in cognitive performance post-transcranial, red/near-infrared light-emitting diode treatments in chronic, mild traumatic brain injury: open-protocol study. J Neurotrauma. 2014;31(11):1008–17. https://doi.org/10.1089/neu.2013.3244.
37. Naterstad IF, Joensen J, Bjordal JM, Couppé C, Lopes-Martins RAB, Stausholm MB. Efficacy of low-level laser therapy in patients with lower extremity tendinopathy or plantar fasciitis: systematic review and meta-analysis of randomised controlled trials. BMJ Open. 2022;12(9):e059479. Published 2022 Sep 28. https://doi.org/10.1136/bmjopen-2021-059479.
38. Neville JJ, Palmieri T, Young AR. Physical determinants of vitamin D photosynthesis: a review. JBMR Plus. 2021;5(1):e10460. Published 2021 Jan 19. https://doi.org/10.1002/jbm4.10460.
39. Paolillo FR, Arena R, Dutra DB, et al. Low-level laser therapy associated with high intensity resistance training on cardiac autonomic control of heart rate and skeletal muscle remodeling in wistar rats. Lasers Surg Med. 2014;46(10):796–803. https://doi.org/10.1002/lsm.22298.
40. Rentz LE, Bryner RW, Ramadan J, Rezai A, Galster SM. Full-body photobiomodulation therapy is associated with reduced sleep durations and augmented cardiorespiratory indicators of recovery. Sports (Basel). 2022;10(8):119. Published 2022 Jul 31. https://doi.org/10.3390/sports10080119.
41. Sommer AP. Mitochondrial cytochrome c oxidase is not the primary acceptor for near infrared light-it is mitochondrial bound water: the principles of low-level light therapy. Ann Transl Med. 2019;7(Suppl 1):S13. https://doi.org/10.21037/atm.2019.01.43.
42. Stergioulas A, Stergioula M, Aarskog R, Lopes-Martins RA, Bjordal JM. Effects of low-level laser therapy and eccentric exercises in the treatment of recreational athletes with chronic achilles tendinopathy. Am J Sports Med. 2008;36(5):881–7. https://doi.org/10.1177/0363546507312165.
43. Sunemi SM, Teixeira ILA, Mansano BSDM, et al. Post-resistance exercise photobiomodulation therapy has a more effective antioxidant effect than pre-application on muscle oxidative stress. Photochem Photobiol Sci. 2021;20(4):585–95. https://doi.org/10.1007/s43630-021-00042-w.
44. Tam SY, Tam VCW, Ramkumar S, Khaw ML, Law HKW, Lee SWY. Review on the cellular mechanisms of low-level laser therapy use in oncology. Front Oncologia. 2020;10:1255. Published 2020 Jul 24. https://doi.org/10.3389/fonc.2020.01255.
45. Thunshelle C, Hamblin MR. Transcranial low-level laser (light) therapy for brain injury. Photomed Laser Surg. 2016;34(12):587–98. https://doi.org/10.1089/pho.2015.4051.
46. Tomazoni SS, Machado CDSM, De Marchi T, et al. Infrared low-level laser therapy (photobiomodulation therapy) before intense progressive running test of high-level soccer players: effects on functional, muscle damage, inflammatory, and oxidative stress markers-a randomized controlled trial. Oxid Med Cell Longev. 2019;2019:6239058. Published 2019 Nov 16. https://doi.org/10.1155/2019/6239058.
47. Zhevago NA, Samoilova KA. Pro- and anti-inflammatory cytokine content in human peripheral blood after its transcutaneous (in vivo) and direct (in vitro) irradiation with polychromatic visible and infrared light. Photomed Laser Surg. 2006;24(2):129–39. https://doi.org/10.1089/pho.2006.24.129.
48. Zein R, Selting W, Hamblin MR. Review of light parameters and photobiomodulation efficacy: dive into complexity. J Biomed Opt. 2018;23(12):–120901.
49. González-Muñoz A, Perez-Montilla JJ, Cuevas-Cervera M, Aguilar-García M, Aguilar-Nuñez D, Hamed-Hamed D, Pruimboom L, Navarro-Ledesma S. Effects of photobiomodulation in sports performance: a literature review. Appl Sci. 2023;13(5):3147. https://doi.org/10.3390/app13053147.
50. de Paiva PRV, Tomazoni SS, Johnson DS, et al. Photobiomodulation therapy (PBMT) and/or cryotherapy in skeletal muscle restitution, what is better? A randomized, double-blinded, placebo-controlled clinical trial. Lasers Med Sci. 2016;31:1925–33. https://doi.org/10.1007/s10103-016-2071-z.

51. Dornelles MP, Fritsch CG, Sonda FC, et al. Photobiomodulation therapy as a tool to prevent hamstring strain injuries by reducing soccer-induced fatigue on hamstring muscles. Lasers Med Sci. 2019;34(6):1177–84. https://doi.org/10.1007/s10103-018-02709-w.
52. Leal Junior EC, Lopes-Martins RA, Dalan F, et al. Effect of 655-nm low-level laser therapy on exercise-induced skeletal muscle fatigue in humans. Photomed Laser Surg. 2008;26(5):419–24. https://doi.org/10.1089/pho.2007.2160.
53. Pinto HD, Vanin AA, Miranda EF, et al. Photobiomodulation therapy improves performance and accelerates recovery of high-level Rugby players in field test: a randomized, crossover, double-blind, placebo-controlled clinical study. J Strength Cond Res. 2016;30(12):3329–38. https://doi.org/10.1519/JSC.0000000000001439.
54. Stier A. Human blood contains circulating cell-free mitochondria, but are they really functional? Am J Physiol Endocrinol Metab. 2021;320(5):E859–63. https://doi.org/10.1152/ajpendo.00054.2021.
55. Machado AF, Leal-Junior ECP, Batista NP, et al. Photobiomodulation therapy applied during an exercise-training program does not promote additional effects in trained individuals: a randomized placebo-controlled trial. Braz J Phys Ther. 2022;26(1):100388. https://doi.org/10.1016/j.bjpt.2021.100388.
56. Hamblin MR. Shining light on the head: Photobiomodulation for brain disorders. BBA Clin. 2016;6:113–124. https://doi.org/10.1016/j.bbacli.2016.09.002

Chapter 28
Cryotherapy: Extreme Cold-Based Physiological Stimulation to Improve Recovery and Enhance Adaptation to Exercise

Giovanni Lombardi, Paolo Capodaglio, and Ewa Ziemann

Introduction

> Among strategies developed to enhance/accelerate recovery, cold based therapies, and particularly cryotherapy, have been reported to have positive impact on different aspects of recovery, as well as on muscle regeneration.

Different strategies have been developed to enhance recovery from physical exercise (training and/or competition) with some of them being effective on one or more variables involved in recovery (e.g., muscle damage, soreness, fatigue, inflammation) [1]. Among the pro-regenerative and post-exercise recovery techniques, whole-body cryotherapy (WBC) has attracted great interests, in the last two decades.

G. Lombardi (✉)
Laboratory of Experimental Biochemistry and Advanced Diagnostics, IRCCS Ospedale Galeazzi-Sant'Ambrogio, Milan, Italy

Department of Athletics, Strength and Conditioning, Poznań University of Physical Education, Poznań, Poland
e-mail: giovanni.lombardi@grupposandonato.it; lombardi@awf.poznan.pl

P. Capodaglio
Musculoskeletal and Metabolic Rehabilitation Unit, IRCCS Istituto Auxologico Italiano, Milano, Italy

Department of Biomedical, Surgical and Odontoiatric Sciences, University of Milano, Milano, Italy

E. Ziemann
Department of Athletics, Strength and Conditioning, Poznań University of Physical Education, Poznań, Poland

N. C. Apostolopoulos et al. (eds.), *Fundamentals of Recovery, Regeneration, and Adaptation to Exercise Stress: An Integrated Approach*,
https://doi.org/10.1007/978-3-031-44270-4_28

Although an objective definition of its effectiveness is still debated, several reports have observed a positive impact of WBC on recovery [2].

Local or systemic exposure to cold (i.e., cold-based therapies) has been used since the ancient times to improve wellness and to relieve pain and symptoms of inflammation. Nowadays, the term cryotherapy is used to define all those treatments that extract heat from the body via either conduction (i.e., application of a cold material) or convection (i.e., exposure to a cold medium). More specifically, WBC consists of brief (generally from 2.50 to 4 min) total-body exposure to extremely cold air (from −60 to −195 °C), within a special device.

The great interest attracted by this technology in sports medicine has pushed the research towards the evaluation of its effectiveness on specific aspects of recovery from exercise including period of training, time of application or amount and frequency of sessions. In this chapter, the reader will be informed about the history of cold therapies, the available technologies, and the principles they are based on, and the physiological effects of cryotherapy in the field of recovery.

Local and systemic cold-based therapies include all the approaches based on the extraction of heat from the body in order to improve painful symptoms of inflammatory conditions.

Cryotherapy and Cryostimulation

Cold-based therapies have been known and used for a long period of time, from the Egyptians, Ancient Greeks, and Romans, as well as its recent applications in rheumatology and sports medicine. The aim of applying cold has always been the same: to reduce pain, regardless of the underlying condition.

Current cryotherapy methods consist of applying cold to the whole-body (WBC) or partial body (PBC, head-out). Besides the technical differences that may affect the final result, indications, and contraindications, for these two kind of treatments, are overlapping and cover the entire spectrum of inflammatory and rheumatological conditions, including trauma and injury, and the accelerated recovery from excessive exercise and the prevention of overreaching.

Terminology

Cryotherapy is a general term often used to describe therapeutic strategies involving cold temperatures all based on the principal purpose of withdrawing heat [3] and, therefore, by reducing core and tissue temperatures [4, 5] and altering blood flow [6, 7].

The term cryotherapy refers to all those treatments that use cold to obtain a physiological response. However, this term, nowadays, is mainly used to name two different treatments: the above-described WBC, which consists in the total-body exposure to cold air, and partial-body cryotherapy (PBC), often erroneously associated to WBC, which consists in the exposure to liquid nitrogen vapours from the feet to the base of the neck. A detailed description of the two technology follows.

Another specification needed is the distinction between the terms cryotherapy and cryostimulation. Although the term "cryotherapy" is often used generally to name WBC (and PBC) it might refer to an actual therapy to treat painful symptoms related to inflammatory conditions, like injuries, trauma, rheumatological and autoimmune diseases. On the other side, the term cryostimulation might refer to an adjuvant treatment aimed at improving metabolic/inflammatory responses and to enhance recovery from exercise and injuries [2]. In light of these differences, since the main aim of this chapter is to discuss the potential effects of cold treatment in athletes, here we refer to WBC as cryostimulation.

On the Cold-Based Treatments: Historical Notes

The use of cryotherapy for benefits to health, treatment, and recovery dates back centuries. Over time, the way in which cold temperature has been applied to the human body has remained largely unchanged, with the application of ice, cold-air, and cold-water being popular forms. The evolution of practice seemingly focused on the cold dose applied through strict control of the temperature and duration of the cooling stimulus.

The beneficial effects of cold-water immersion (CWI) were already reported in the Edwin Smith Papyrus, dating as far back as 3500 B.C., that listed several therapeutic purposes. Ancient Greeks used CWI for therapy, for relaxation and socialisation with Hippocrates, in the fourth century B.C., documenting the use of cold, in his "*On airs, waters and places*", for medicinal purposes and analgesic benefits. He suggested that water therapy could 'allay lassitude' in reducing the depletion of energy or strength, mentioning ice and snow as remedies for oedema [8]. Cold-water immersion was traditionally used to treat fever, specifically tertian fever, by the Roman physician Claudius Galen. Further, Roman warriors were known to take quick dips in icy-cold rivers following battles [9]. Hundreds of years later, the Scottish physiologist James Currie, used cold-water to treat fever, even on his own rheumatic fever. Currie investigated the impact of cold-water on human physiology (e.g., body temperature, pulse, breathing) by documenting the first records of human temperatures in health, disease, and experimental conditions to prove the value of hydrotherapy [10]. In the twentieth century, several studies started to describe reactions to CWI such as the cold allergy [11], the blood pressure response in course of hypertension [12] and the vasomotor reactions to selective sympathectomy [13].

In the '60s CWI was investigated for benefits to post-exercise recovery [14, 15], with studies investigating different exercise modalities, subject cohorts, and/or cooling doses (i.e., duration, temperature, depth). Current recommendations for the use of CWI, derived from meta-analyses, suggest a protocol of 10–15 min at 10–15 °C [16] to promote effective recovery, whilst a dose of 1.1 (i.e., 11 min at 10 °C) is required to significantly reduce muscle tissue temperature [5]. Furthermore, immediate immersion has been demonstrated preferable to delayed immersion [17] while the depth of immersion does not appear to affect the results [18].

The use of ice is recognised as the most traditional mode of cryotherapy. Napoleon's surgeon to the Grand Army, Baron Dominique Larrey, recommended ice and snow to assist in painless amputations and operations on soldiers [19]. Later, in 1840, this concept was extended by James Arnott who used local application of salt solution containing crushed ice to freeze cancerous tumours, unwittingly developing cryosurgery [20]. However, even with the use of ice, the recommendation for therapeutic treatment of musculoskeletal injury emerged in the 1960s [21, 22]. Besides several seminal studies demonstrating the decrease in skin [23–25] and muscle [26, 27] temperatures by ice packs or ice baths, in relation to skinfold thickness and adiposity [25], the investigation about its analgesic effects began in the 1970s.

Bugaj demonstrated that ice massage acutely abolished the sensation of pain in response to a surface pin prick when skin temperature was reduced to 13.6 °C [28]. It was later demonstrated that this benefit resulted from a reduction in nerve conduction velocity in sensory nerves [29–31]. In 1978, Mirkin introduced the widely used RICE acronym (rest, ice, compression, and elevation) to guide acute sport injury management. The effectiveness of the use of ice in the management of sport injuries gained limited evidence, despite its wide use. At this purpose, Guillot et al. demonstrated the positive effect of ice application on the inflammatory cytokines profile such as IL-6, IL-17A and IL-1β in arthritic rats [32] and IL-6, IL-1β, VEGF, NF-κB-p65, and PGE2 synovial levels, especially in the microcrystal-induced arthritis subgroup, while only phosphorylated NF-κB-p65 in rheumatoid arthritis and spondyloarthritis patients [33].

The application and use of extreme cold air temperatures (<−100 °C) is a relatively new technique in sport. The idea behind the development of WBC derived from Dr. Yamaguchi who observed that rheumatoid arthritis patients improved clinically when they vacationed in the mountains during winter holidays. He postulated the idea that a rapid decrease in skin temperature caused a release of endorphins and the subsequent reduction in subjective pain. This observation led to the development of the first cryosauna in 1978, in Japan. With the use of cryosauna, approximately 80% of Dr. Yamauchi's patients were completely relieved of their symptoms and chronic pain issues. The data was presented at the Rheumatology Congress in 1979 [34] whereupon an encounter with the German rheumatologist, Prof. Fricke, resulted in collaborative work culminating in the development of WBC treatments in Europe. The meeting was instrumental in the in the development of a cryotherapeutic module in 1984, focused on the treatment of hundreds of patients suffering from multiple sclerosis and arthritic conditions. However, it was not until the twenty-first century, that WBC began to be adopted in the sporting world for exercise recovery [35].

General Principle and Applications

Although ice (and cold) is routinely used in the management of soft tissue injuries, from sprains to strains and contusions, the science behind this indication is mostly ignored. A survey conducted with Irish emergency physicians reported that only 73% of the consultants "always or frequently" advocated the use of ice, while 7% "never" used any form of cryotherapy; with a 30% of the sample being "unsure" about ice therapy benefits. Interestingly, in 47% of the surveyed, the use of ice was guided by experience, in 27% by common sense with only 17% using cryotherapy based on scientific reasoning. Further, less than a fourth of the sample (23%) had read literature supporting the possibility to improve the clinical outcomes following use of ice, with 17% having never read any supporting literature while 60% were unsure [36].

The cryotherapy treatment consists of the patient being minimally dressed, with bathing suit, socks, clogs, headband, and surgical mask to avoid direct exhalation of humid air.

Types of Cryotherapy

Whole-Body Cryotherapy (WBC)

Whole Body Cryotherapy is conducted in a cryochamber with the body totally exposed to cold air. The cryochamber consists of either three chambers, each with specific set temperatures (−10, −60, and −110 to −160 °C) or two (−60 and −110 to −160 °C) with the most recent being a single chamber (−110 to −160 °C). The difference between the one chamber and the other two is the number of participants being treated at the same time, from one up to six. The chambers are cooled via a cooling liquid, generally nitrogen, flowing in pipes within the compartment walls. Two modes of behaviour are used: standing still or slowly walking around. All chambers are equipped with alarm buttons and safety openings, besides a window ensuring visual contact with the personnel.

With the chamber and cooling supply being standardized; an adequate space and size are required for safety issue. Besides fixed cryotherapy units, mobile ones exist on the market as well, which can be transported and used by teams on their grounds who cannot afford a full installed chamber [37].

Partial-Body Cryotherapy (PBC)

Partial Body Cryotherapy is performed in a cryosauna, an open tank with the body of the individual exposed to cold, but not the head and neck. This device was designed by Dr. Yamauchi [34] and further developed by Prof. Fricke in Germany [38] and Zagrobelny in Poland [39].

The participant enters the tank, through a lateral opening, with the height being adjusted so the shoulders are positioned at the upper extremity of the device. Safely positioned within the cryosauna, liquid nitrogen vapours are sprayed directly onto the body of the patient. It is important to stress the patient's head needs to be out of the tank to prevent breathing nitrogen.

The temperature within the cryosauna is between −110 and −195 °C, depending on the technical features of the system. Furthermore, the temperature within the tank ranges widely from the nozzle to the top, and from the tank wall to the centre, were the patients stays. Based on a previous study that had placed a mannequin in the cryosauna, the temperature at the nozzle was <−150 °C but increased to −60 °C in the middle of the sauna. After a 3 min of exposure, the temperature at the top was −100 and −140 °C at the lowest part. However, with a living subject the mean temperatures were between −20 °C for the chest and −40 °C for the shank. This difference in temperatures between the mannequin and the living participant, is probably due to the heat convection generated by the patient [40].

From a cost perspective, PBC are less expensive than WBC and are portable and can be transported to sports competitions such as Tour de France, the Vuelta, and the European Basketball Championship [37].

WBC Versus PBC

Partial Body Cryotherapy appears to have the most advantages in the sport recovery domain, especially when the treatment must be performed in the field, thanks to the portability of the device, the required space and the easy turn-on and turn-off procedures. Nonetheless, it is less standardized and more variable, in terms of temperature compared to WBC. Whole body cryotherapy induces greater pre-to-post variations in mean skin and core temperature and allows for the treatment of many patients per time.

However, it should be stressed that changes of body temperatures during exposure in both devices remain unknown and further and precise studies are needed.

Finally, as already mentioned, there are safety concerns with PBC related to exposure of the patient to nitrogen vapours [37].

Indications and Contraindications

The clinical application of cryotherapy has been historically driven by its analgesic efficacy and, therefore, much attention has been given to the treatment of primary and secondary tissue damage and the ensuing inflammatory response.

Based on the available evidence (i.e., systematic reviews and meta-analysis, randomized controlled trials) cryotherapy has been successful in the symptomatic treatment of inflammatory diseases such as rheumatoid arthritis [41, 42], ankylosing spondylitis [43], fibromyalgia [44, 45], and mood disturbances [46]. In addition, the literature further suggests the potential effectiveness (and safety) of cryotherapy with regards to the entire spectrum of inflammatory, autoinflammatory, immune-mediated, and autoimmune diseases, with the strongest evidence coming from the treatment of fibromyalgia [47] and rheumatoid arthritis [48], as well as atopic dermatitis [49], multiple sclerosis [50–52], chronic low back pain [53], depression and anxiety syndromes [54], tension and fatigue [55], and sleep quality [37]. By integrating the results obtained from studies investigating the efficacy of cryotherapy (mainly WBC) in improving the metabolic homeostasis in healthy and overweight-to-obese subjects, it has been hypothesized that WBC could be successfully applied as an adjuvant therapy in the treatment of dysmetabolic conditions, such as glucose intolerance, metabolic syndrome, type 2 diabetes, hypertriglyceridemia/hypercholesterolemia, and obesity [56].

Current medical indications for WBC are illustrated in Table 28.1.

Although the introduction of cryotherapy in sports medicine was delayed, compared to its use in pathology, its adoption was expedited, quickly becoming a main form of application. As often happens, besides the reported effectiveness in different domains (e.g., recovery, reduction and prevention of exercise-induced muscle damage and soreness) a huge reason for its acceptance in sports was mainly due to the media reporting on its use with famous sportsmen [2]. Table 28.2 reports the current indications for WBC in sports medicine.

Currently, with WBC being considered as a medical treatment, it needs to be performed in specialized medical centres, based on specific medical protocols and indications of use. Known contraindications reference precautions derived from the observation of adverse events in single or few cases. Some absolute contraindications (according to [57]) are: untreated high blood pressure, heart attack within six months, decompensated cardiovascular and respiratory diseases, unstable angina, pacemaker, peripheral artery occlusive disease (Fontaine stages 3rd and 4th), history of deep-vein thrombosis, acute respiratory infections, acute renal and urinary disorders, severe anaemia, cold allergy/intolerance (e.g., Raynaud's sign, microvasculature disorders), severe wasting diseases, seizure disorders, large-area bacterial and viral skin infections, wound-healing problems, alcohol and drug influence. In addition, cryoglobulinaemia, hypothyroidism, hypothermia, SNS neuropathies, cancer, and conditions that hinder the cooperation of the patient during the treatment (e.g., claustrophobia).

When performed adequately, WBC is safe, however some safety concerns do exist such as the risk of skin burns due to unwanted touching of the chamber walls, the humidity level in the room (since water vapour condensation onto the skin can cause burns at such low temperatures), as well as inappropriate behaviours from the patient increasing the risk of adverse effects. For instance, the application of hydrating creams just before treatment, access to the chamber with wet skin, not appropriately covered, the non-removal of metallic accessories (e.g., piercings), all increase the risk of localized skin burns. In addition, the case of previously unknown cold intolerances increases potential risk as represented by cold urticaria.

Table 28.1 Current medical indications, molecular targets (where known) and clinical effects (where investigated) for WBC

Disease	Molecular effects	Clinical effects
Degenerative articular pain		
Back pain from degenerative disc disease or arthritis		↓ pain, DAS; ↑ functional score index, spinal mobility parameters (AS) ↓ pain, disability index, well-being (LBP)
Skeletal muscle inflammation		
Joint, tendinous or muscle overload		↑ general well-being, mood (PJD)
Psoriasis and psoriatic arthritis		
Myositis and fibromyositis		
Chronic polyarthritis		
Autoimmune diseases (RA, MS, SEL)	↓ TNFα (RA) ↓ catalase, ↑ uric acid, TAS, SOD (MS)	↓ inflammation, fatigue, DAS, morning stiffness; ↑ walking, physician global assessment, quality of life (RA) ↓ fatigue; ↑ functional abilities (MS)
Painful osteoporosis		
Paresis and spastic contractures		
Fibromyalgia		↓ pain, fatigue, DAS; ↑ quality of life
Cellulitis		
Atopic dermatitis		↓ total water loss, ↑ barrier function
Anxious-depressive syndrome		↓ depressive-anxiety symptoms
Sleep disturbances		

Abbreviations: *AS* ankylosing spondylitis, *DAS* disease activity score, *LBP* low back pain, *MS* multiple sclerosis, *PJD* peripheral joint disease, *RA* rheumatoid arthritis, *SEL* systemic erythematous lupus, *SOD* superoxide dismutase, *TAS* total antioxidant status, *TNFα* tumour necrosis factor alpha

Besides reports of cases of urticaria [41], cold panniculitis in a patient with autoimmune neutropenia [58], transient global amnesia [59] in a patient with history of diseases and medication (comment to [59]; November 8th, 2017), intracerebral haemorrhage in a woman with a long history of ocular migraine [60], moyamoya angiopathy in a young woman with a negative medical anamnesis [61], abdominal aortic dissection in a male patient with history of hypercholesterolemia and hypertension [62], along with headache, dizziness, hypertension, and long-lasting shivering [41, 63] have been some reported cases of adverse effects occurring after the use of WBC.

On the contrary, WBC has been shown to be safe for lung [64] and heart functions [65] and, contrary to what has been previously described [66], both single and multiple WBC sessions reduce systolic and diastolic pressures slightly (unpublished data).

Table 28.2 Current indications in sports medicine, molecular targets (where known) and clinical effects (where investigated) for WBC

<table>
<tr><th>Condition</th><th>Molecular effects</th><th>Clinical effects</th></tr>
<tr><td>Muscle and soft tissue injuries, painful or fatigue syndromes</td><td rowspan="6">↓ TNFα, IL-1α, IL-1β, IL-2, IL-8, CRP, sICAM, PGE$_2$, lactate, lipid peroxidation products, muscle damage markers, GPx, TBARS
↑ IL-1ra, IL-10, IL-6, ↑ growth factors, TAS, catecholamines</td><td rowspan="6">↓ muscle damage, pain
↑ muscle strength, maximal voluntary contraction, well-being, workout power, autonomic response, sleep quality</td></tr>
<tr><td>Skeletal muscle overload</td></tr>
<tr><td>Acute and overload-induced tendinopathies</td></tr>
<tr><td>EIMD and DOMS</td></tr>
<tr><td>Post-exercise recovery</td></tr>
<tr><td>Power and resistance</td></tr>
</table>

Abbreviations: *CRP* C-reactive protein, *DOMS* delayed-onset muscle damage, *EIMD* exercise-induced muscle damage, *GPx* glutathione peroxidase, *IL* interleukin, *IL-1ra* interleukin-1 receptor antagonist, *PGE$_2$* prostaglandin E$_2$, *sICAM* soluble intracellular adhesion molecule, *TAS* total anti-oxidant status, *TBARS* thiobarbituric acid-reactive substances, *TNFα* tumour necrosis factor α

As already mentioned above, in case of partial body cryotherapy the specific safety concern is the risk of asphyxia due to the unwanted inhalation of liquid nitrogen vapours during treatment [37].

> When performed adequately, in specialized (and possibly medical) centers, cryotherapy/cryostimulation is a safe procedure. Its effectiveness has been observed for several pathological conditions, although the level of evidence is low to draw definitive conclusions. On the other hand, although there are known contraindications to treatment, further, adverse effects, although not common and still not definitively associated with the treatment, may take place. An accurate anamnestic data collection before the treatment and careful control during treatment are needed by skilled personnel.

Physiological Response to Extreme Cold Exposure

> The cold stimulus induces several responses that involve different tissues and cells, with a key involvement of the sympathetic nervous system of the skin. All cold-induced responses converge onto the inflammatory and pain signalling whose activation is limited or even abolished. Furthermore, the reaction to the cessation of the cold stimulus is relevant for determining the net result of the treatment.

Cold stimulation is responsible for the activation of several responses, with the majority being mediated by the sympathetic nervous system (SNS) through the effects on cold receptors expressed by the sympathetic afferent nerves. The consequent activation of these nerve terminals result in the release of noradrenaline (NA) and acetylcholine (ACh). These neurotransmitters target, amongst others, the macrophages via β2 and α7 receptors, with both pathways converging on nuclear factor κB (NF-κB), the master mediator of the inflammatory response, thereby inhibiting its activity. As a consequence, the following mediators are directly or indirectly expressed concerned with the establishment, progression and chronicity of inflammation: adhesion molecules [intracellular adhesion molecule (ICAM)-1], proinflammatory cytokines [interleukin (IL)-1β, IL-6, tumour necrosis factor (TNF) α], oxidative stress mediators [inducible nitric oxide synthetase (iNOS), myeloperoxidase (MPO)], and angiogenic factors [vascular endothelial growth factor (VEGF)]. Further, SNS activation results in the inhibition of expression and release in mast cells of histamine, the main mediator of allergic and itchy reactions. Noradrenaline directly affects the microvasculature system by the activation of α-adrenoreceptors, causing a vasoconstriction limiting the blood flow to the site of injury decreasing bleeding and oedema. This reduces the frequency of leukocytes diapedesis and their recruitment to the site of damage limiting inflammation. Finally, the cold stimulus directly down-regulates or inhibits enzymatic activities with a greater effect on those up-regulated due to the inflammatory process (e.g., metalloproteinases, collagenases) and subsequently affecting the synthesis and release of mediators of pain [prostaglandin (PG)E_2] [42].

Based on the evidence, cold therapies, and in particularly WBC, are considered to act like a non-steroidal anti-inflammatory drug (NSAID). Administered systemically, it acts on the cells and mechanisms activated at the time of application (i.e., inflammatory cells and flogistic pathways). However, although, the cold may limit/inhibit the classic inflammatory pathways (i.e., downregulation of enzymatic activities, pro-inflammatory mediators, cell metabolism) its primary anti-inflammatory effect is believed to be on the nervous system and its regulation of pain, since pain is dependent on neural activity (i.e., activation of the gate control, increased excitability threshold of nociceptors, reduced nerve conduction rate). In addition, the neural system also influences the muscle (i.e., reduced muscle spasms), and vasculature (i.e., decreased bleeding and oedema, thanks to the vasoconstriction and the reduced vascular permeability) [67].

Thermoregulation, is a complex homeostatic response that is still not entirely understood, especially how the core temperature is maintained at about 37 °C [68]. This response sustains the liberation of total body energy in the form of heat [69]. Skin and cutaneous circulation are a key system in cold sensation, with dermal layers containing a ten-fold higher density of cold to heat receptors [70]. Extreme cold exposure (e.g., WBC, PBC) causes peripheral vasoconstriction thereby reducing blood flow to the muscles, and some visceral organs [71], as well as a decrease in cell permeability, extracellular fluid flow with consequent reduction in oedema and inflammation [72]. After physical exercise, extreme exposure to cold induces a decrease in muscle temperature affecting enzyme activities, cell metabolism, inflammatory response, and limiting ischemia-reperfusion and hypoxia-dependent degradation. Altogether, the thermoregulatory response may help in recovery. In addition, the release of noradrenalin by the SNS in response to cold stimulation influences the sensation to pain similar to endorphins, which have been observed to be released during WBC/PBC treatment [73].

Within 4 min after the cessation of the cold stimulus, vasodilation occurs, with the peripheral blood flow increasing as high as a four-fold to normal values Although skin temperature recovers within 14 min, the changes to circulation can last several hours aiding in the discharge of catabolic waste products that have accumulated after exercise, and may be responsible for post-exercise pain, fatigue and decreased strength [73].

A similar long-lasting effect with the use of cold has been observed with the parasympathetic control of the heart [74] influencing the quality of sleep [75]. These changes influence the psyche and mood of the person, because of a decrease in the sensation of fatigue, and an increase in the quality of sleep [54]. However, it should be pointed out that this response is highly variable, depending on intrinsic and behavioural features such as body size, fat mass and distribution, gender, and fitness level [76].

> Cold activates the sympathetic nervous system that via release of noradrenaline, acetylcholine, and endorphins, activates several responses limiting the inflammatory activation of immune cells, reduced activation of nociceptive fibers, vasoconstriction and reduced vascular permeability. The net results represents a reduction of pain sensation, edema, and bleeding. Further, after the cessation of the cold-stimulus, the reactive vasodilation removes the waste products, derived the inflammatory-dependent activation of catabolism.
>
> Beside the physiology, individual features represent important variables that may affect the effectiveness of the treatment.

Cryostimulation in Recovery from Overuse and Trauma (Table 28.3)

> The domains of post-exercise recovery that warrant cryostimulatory treatments are exercise-induced muscle damage (EIMD), delayed-onset muscle soreness (DOMS), adaptation, performance, and sleep.

Exercise-Induced Muscle Damage (EIMD)

Main effects of cryostimulation are anti-inflammatory and analgesic. However, although invariably described, due to the huge inter-individual variability there is still no consensus on the reliability of the outcomes [2, 37].

Several authors tried to describe the physiological responses underlying the WBC-associated improvement. Although the participants reported a positive

Table 28.3 Five main domains of post-exercise recovery that warrant the use of cryostimulation

Five domains of post-recovery
Exercise-Induced Muscle Damage (EIMD)
Delayed-Onset Muscle Soreness (DOMS)
Adaptation
Performance
Aerobic and anaerobic
Strength
Flexibility
Sleep

outcome, the results did not correlate with changes in pro-inflammatory mediators associated with muscle injury. In a randomized controlled trial (RCT), EIMD was induced by a simulated run-in well-trained runners. The athletes completed three runs each followed by three sessions of either passive recovery, infrared therapy, or WBC (at temperatures of −10, −60, and −110 °C) at 0, 24, and 48 h. With the muscle damage marker creatine phosphokinase (CK) behaving similarly following the treatments, perception of pain, tiredness and well-being improved with WBC [77]. Similarly, in another RCT, 1, 2 or 3 min of a single WBC session were not associated with any change in IL-6 levels [78].

Other studies identified changes in soluble factors associated with cryostimulation. Tennis players, performing two sessions of WBC per day for five consecutive days experienced a change in TNFα (60% reduced) and IL-6 (30% increased) compared to the control group who did not receive WBC treatment [79]. Professional male volleyball players experienced a decrease in both IL-1 β and IL-6 with use of WBC before a submaximal exercise compared to athletes not exposed to WBC [80]. Noteworthy, while changes in classical pro-inflammatory cytokines (e.g., IL-1, TNFα) are unequivocally associated with either a pro- (upregulated/increased) or anti-inflammatory response (down-regulated/decreased), the biological meaning of IL-6 changes should be further examined, since it can behave as either a pro- or anti-inflammatory cytokine depending on the tissue source and the duration of this change. Chronically, even slightly, elevated IL-6 blood concentrations, as those associated with sedentary lifestyle and adiposity, are derived from liver, and has a pro-inflammatory character. In this case, IL-6 is considered a cytokine. On the contrary, during exercise-induced increases in IL-6, because this increase is due to the contracting muscle, IL-6 exhibits an anti-inflammatory character and is thus considered to be a myokine [81]. Contrasting findings about IL-6 levels, in response to WBC, may be due to differences in the study cohort, in the training protocol associated with WBC and the delay between the intervention and the blood sampling. Jaworska et al. demonstrated that use of WBC (10 sessions) by volleyball players, within a 2-week training program, showed improvements in muscle regeneration and a limited deterioration in physical performance as measured by an increase in circulating levels of brain-derived neurotrophic factor (BDNF) and insulin-like growth factor (IGF)-1 [82]. This data partially contrasts with those obtained in a

RCT where two daily sessions of WBC, for seven days, in elite athletes, compared to no treatment, observed a reduction in hydrogen peroxide (H_2O_2) and nitric oxide (NO), IL-1β and C-reactive protein (CRP), but also IGF-1, BDNF, hepatocyte growth factor (HGF), platelet-derived growth factor (PDGF)-BB, and vascular endothelial growth factor (VEGF) [83].

Results in leukocyte counts are not in agreement, with several studies observing no effect in rugby players [84] and kayakers [85], undergoing two sessions per day of WBC during training camps, contrasting with an increase in leukocytes (except for neutrophils and lymphocytes), but always within the physiological range, after twenty WBC sessions in military recruits (with a return to baseline after thirty sessions) [86] and tennis players [79]. More uniform was, instead the alteration of platelets the resulted mostly unaffected by WBC [84, 86, 87], contrary to what was observed in other kind of cold exposure, as in the case of winter swimming [88].

The differences observed in these studies may be a result of the different training protocols and/or sampling times, as well as the frequency and length of the WBC treatment. As demonstrated by Lubkowska et al. the number of sessions influences the results [88].

In the study consisting of forty-five healthy males, although levels of IL-1β, IL-12 and TNFα remain unchanged, IL-10 increased by 30% after five consecutive sessions (one per day) and by 14% after ten and twenty sessions, with IL-1α decreasing by 17% and IL-6 increasing by 10% after ten sessions. In addition, although the changes after five and ten sessions were lost within 2 weeks, observed changes after twenty sessions lasted longer [89].

Delayed-Onset Muscle Soreness (DOMS)

The effectiveness of WBC in reducing DOMS was confirmed by a meta-analysis [1].

WBC was more effective than passive recovery in attenuating the increase in C-reactive protein and inducing an increase in IL-1ra, with no observed effects on TNFα, IL-6, and IL-10, in runners who underwent a 45-min trail run to specifically induce muscle damage [90]. Similarly, the DOMS-associated inflammatory response induced by eccentric work was reduced (i.e., decreased IL-1 and IL-6, increased IL-10) by two sessions per day of WBC, over five days, in physically active men [87]. A similar protocol was able to reduce the circulating levels of myoglobin and IL-15, suggesting that WBC could negatively affect muscle strength, but it also decreased myostatin levels, thus suggesting that WBC could stimulate muscle hypertrophy [91].

It has been suggested that extreme temperatures of WBC might magnify the effect of cold therapies (for instance, in comparison to CWI), potentially shortening recovery time [92]. A preference was demonstrated amongst athletes about the use of 'dry' cold therapies (i.e., WBC) rather than prolonged immersion in CWI to treat DOMS (e.g., feeling and recovery time), in some studies [93, 94], but not in others

[95]. Furthermore, it is worth emphasising that amelioration of DOMS might delay or compensate adaptive changes induced by regular training, therefore suggesting the consideration of a precise time/period for use of WBC [96].

Adaptation

Several studies have warned that exposure to cold can negatively affect the adaptation to training consequently affecting physical performance. Most of these studies were performed with CWI in conditions contrasting the current indications [16].

According to a meta-analysis by Malta et al. repeated exposure to cold could affect strength performance but not aerobic performance [97] with a study suggesting that the periodization of cold therapies could possibly improve overall performance [98]. However, as reviewed by Bouzigon et al. most of the studies did not identify any real negative effect on muscle adaptation [73].

During periods of intense training load, with short recovery time, cold immersion could reduce the risks of maladaptation to training [99]. A study with Judo athletes observed an improvement in judo abilities and muscle function with the use of WBC with no effect on their performance [100]. In combination with resistance training, it was also observed that isokinetic muscle strength improved in the young active subjects, possibly as a result of the decrease in circulating myostatin and IL-15 [91].

Performance

Although hypothesized [2, 101], performance enhancement with use of WBC has not been unequivocally demonstrated with current evidence suggesting that the results are limited and contradictory. Psychological well-being and hormonal changes (i.e., decrease in cortisol and increase in testosterone) have been taken in support to the beneficial effects of WBC on performance [92, 102, 103].

Only a few studies directly investigated the effect of WBC on performance indices. In elite tennis players, two daily sessions of WBC, for five days, improved drill execution and stroke effectiveness [79]. A greater overreaching supercompensation was found in triathletes who were exposed to WBC for over 1 week [104]. Similarly, use of WBC reduced the 400 m speed in elite synchronized swimmers tested before and after the overreaching period [105].

Aerobic and Anaerobic

In a study by Klimek et al., use of WBC before and after two trials on a cycle ergometer observed an increase in maximal anaerobic power and capacity after 10 WBC sessions in the male participants, with no effects on aerobic capacity being observed in both male and female participants [106]. In another study by Klimek et al., it was observed that a single WBC session shortened the time to obtain the maximal anaerobic power, in both males and females [107].

A study observed improvement in recovery with use of WBC vs rest, from high-intensity intermittent exercise, consisting of two ramp-test protocols to exhaustion, interspersed with a high intensity running protocol (5 × 5 min at 90% of maximum velocity with 4 min of active recovery between the intervals). Further, participants exposed to WBC expressed lower values for rated perceived exertion (RPE), oxygenation of the *vastus lateralis*, heart rate, and peak oxygen uptake (VO_2 peak) at submaximal intensities [108]. In contrast to the previous study, no change compared to passive recovery was observed in performance (graded exercise test, time-to-exhaustion test, 20-km time trial, and 120-min submaximal test) with the use of WBC after each cycling high-intensity interval exercise session, over 4 weeks [109].

Strength

The neuromuscular performances, i.e., drop-jump exercise and a maximal voluntary contraction force of the wrist flexors, were tested before and after a single 2-min WBC session at the beginning and at the end of the 3-month (three WBC sessions per week) study period. Flight time decreased after a single session but only a tendency was found after repeated WBC, and a decreased co-contraction of lower leg muscles during the drop-jump while the maximal force level did not change significantly, either after a single or after repeated WBC sessions [110]. Similarly, according to a cross-over study investigating the effects of either WBC (3 min, −110 °C) or control condition (3 min, 15 °C) on maximal voluntary isometric contraction (MVIC) of the right knee extensors, no differences were found in a cohort of thirty six subjects [111].

An increase of maximal average power in the knee isokinetic extension strength test was observed by Jaworska et al. in twenty-five young participants who completed a 4-week protocol including twelve sessions of 50-min resistance training focused on the lower limbs, followed either by WBC exposure (3 min, −120 °C) or passive recovery. Biochemically, this change was associated with a drop in myostatin and IL-15 circulating levels [91].

Studies exposing participants to a single session of partial body cryotherapy did not observe a difference, compared with the control group, regarding peak torque, average power, total work, and muscle activity in thirteen participants after an elbow flexor neuromuscular activity [112] or detected an improvement in the concentric hand-grip strength in 200 healthy adults after a handgrip test [113].

Flexibility

After one or multiple cryostimulation sessions, an improvement in range of motion (ROM) was reported. According to Ma et al., after 24 WBC sessions, the ROM in flexion, abduction, internal, and external rotation of the shoulder improved in individuals suffering from adhesive capsulitis of the shoulder, compared to the use of physiotherapy alone [114]. Similarly, spine mobility and chest expandability was improved with use of WBC, at both −60 and −110 °C, compared to exercise therapy, over 8 days, in ankylosing spondylitis patients [115]. Further, patients suffering from low back pain increased the mobility of their lumbar-thoracic with use of WBC five times per week versus exercise therapy [116].

Studies examining the effectiveness of partial body cryotherapy also support these findings. A single session of PBC saw an improvement in the sit-and-reach test in men and women [117] and knee extension (standing hamstring, stretch test, sit-and-reach test, and active knee extension test), in young women [118], compared to no treatment.

Sleep

Sleep is one of the most important phases in the recovery of an athlete with sleep debt and problems being very common in elite athletes. A bad sleep hygiene is related to the increased risk of injury and prevalence of overtraining [119, 120].

WBC improved sleep perception in high-level basketball players [121]. Similarly, repeated cryostimulation exposure during intense training improved the perception of sleep in female synchronized swimmers [105] with the use of WBC post-exercise being associated with a reduced amount of movement during the sleep [75, 122].

> With EIMD, WBC improves the pro-inflammatory cytokine profile and affects the growth factor profile, however other studies have shown no effect on leukocyte behavior.
>
> With DOMS, WBC effectiveness has demonstrated changes in pro-inflammatory mediators supported by the findings of different meta-analysis.
>
> With adaptation, WBC has been associated with having negative effects however, recent studies suggest that it may increase isokinetic strength.
>
> With performance WBC is ineffective for aerobic performance but is effective for enhancing strength and flexibility.
>
> With regards to sleep studies report an extreme benefit with use of WBC.

Cryostimulation in Athletes: Other Considerations

According to a recently published consensus [73], a medical consultation is suggested before the athletes first exposure to WBC, in order to check for any contraindication, both absolute and relative, and to identify any potential risk. Further, a trial session should be performed with the expressed purpose of gathering information on how the user responds to the cryostimulation. Any feedback and observations during a trial session, ensures the best application of cryostimulation, tailoring to the needs of the athlete throughout the treatment cycle.

The following recommendations might be considered:

1. WBC is a safe treatment but, it may have health risks in case of specific conditions: absolute and relative contraindications must be excluded and any suspicious response that may take place after a WBC session must be carefully considered.
2. Expert personnel must carefully explain methods and safety warnings before each session.
3. The cold dose should be defined based on the type, severity, and stage of the symptoms; the more acute and severe the symptoms the greater should be the drop in skin temperature drops and the number of exposures, with too little not producing any beneficial effects, and too many being potentially harmful.
4. Protocol customization is needed; a one-size-fits-all approach may be less effective and, potentially, less safe.
5. Multiple sessions might be suggested; with the minimum number of sessions depending on the condition to be treated. For instance, for performance recovery, one-to-five sessions, with at least two sessions within 24 h after intense activity may be recommended, while twenty consecutive sessions, one session per day for 4 weeks, may be prescribed to treat rheumatoid arthritis.
6. Habituation throughout a cycle is possible and should be acknowledged: increasing the treatment dose over time helps to prevent it.
7. Real-time monitoring of the physiological and thermal responses combined with personal feedback would further optimize the complete WBC approach.
8. WBC must fit within an existing therapeutic regime.

Conclusion and Future Perspectives

Scientific literature, as evidenced in more or less recent reviews [1, 2, 37, 101], confirms the global growing interest around the use of cryostimulation and cryotherapy with a particular focus in sports. However, the full justification and understanding of the benefits of WBC (and even more for PBC) should come from additional and well-addressed researches [123].

An important issue that should be considered concerns the comparison of the different studies. When comparing one needs to scrutinize the heterogeneity of cold apparatuses used, the differences in the WBC protocols and the different circumstances in which data have been collected. Furthermore, an index of how much heat has been extracted from the body, from either the localized lesion or one or more areas representative of the whole-body, is lacking. Research should attempt to identify a marker of cooling whose change should represent the goal to be reached demonstrating a cooling efficacy. Such an index could be intended as a part of a standardized procedure that would allow the comparison among different studies. The working group "Whole-body cryotherapy" of the International Institute of Refrigeration (IIR), in which the authors of this chapter play an active part, is considering this specific aspect and is working on the identification of an appropriate index to be proposed to the scientific community.

References

1. Dupuy O, Douzi W, Theurot D, Bosquet L, Dugue B. An evidence-based approach for choosing post-exercise recovery techniques to reduce markers of muscle damage, soreness, fatigue, and inflammation: a systematic review with meta-analysis. Front Physiol. 2018;9:403.
2. Lombardi G, Ziemann E, Banfi G. Whole-body cryotherapy in athletes: from therapy to stimulation. An updated review of the literature. Front Physiol. 2017;8:258.
3. Kwiecien SY, McHugh MP. The cold truth: the role of cryotherapy in the treatment of injury and recovery from exercise. Eur J Appl Physiol. 2021;121(8):2125–42.
4. Stephens JM, Sharpe K, Gore C, et al. Core temperature responses to cold-water immersion recovery: a pooled-data analysis. Int J Sports Physiol Perform. 2018;13(7):917–25.
5. Vromans BA, Thorpe RT, Viroux PJ, Tiemessen IJ. Cold water immersion settings for reducing muscle tissue temperature: a linear dose-response relationship. J Sports Med Phys Fitness. 2019;59(11):1861–9.
6. Mawhinney C, Jones H, Low DA, Green DJ, Howatson G, Gregson W. Influence of cold-water immersion on limb blood flow after resistance exercise. Eur J Sport Sci. 2017;17(5):519–29.
7. Mawhinney C, Low DA, Jones H, Green DJ, Costello JT, Gregson W. Cold water mediates greater reductions in limb blood flow than whole body cryotherapy. Med Sci Sports Exerc. 2017;49(6):1252–60.
8. Tsoucalas G. Hydrotherapy: historical landmarks of a cure all remedy. Arch Balk Med Union. 2015;50:430–2.
9. Wang H, Olivero W, Wang D, Lanzino G. Cold as a therapeutic agent. Acta Neurochir. 2006;148(5):565–70; discussion 569–570.
10. Forrester JM. The origins and fate of James Currie's cold water treatment for fever. Med Hist. 2000;44(1):57–74.
11. Lamotte G, Boes CJ, Low PA, Coon EA. The expanding role of the cold pressor test: a brief history. Clin Auton Res. 2021;31(2):153–5.
12. Hines E, Brown GE. A standard stimulus for measuring vasomotor reactions: its application in the study of hypertension. Proc Staff Meet Mayo Clin. 1932;7:332–5.
13. Hines E, Brown GE. A standard test for measuring the variability of blood pressure: its significance as an index of the prehypertensive state. Ann Intern Med. 1933;7:209–2017.
14. Clarke DH. Effect of immersion in hot and cold water upon recovery of muscular strength following fatiguing isometric exercise. Arch Phys Med Rehabil. 1963;44:565–8.

15. Clarke DH, Stelmach GE. Muscular fatigue and recovery curve parameters at various temperatures. Res Q. 1966;37(4):468–79.
16. Machado AF, Ferreira PH, Micheletti JK, et al. Can water temperature and immersion time influence the effect of cold water immersion on muscle soreness? A systematic review and meta-analysis. Sports Med. 2016;46(4):503–14.
17. Brophy-Williams N, Landers G, Wallman K. Effect of immediate and delayed cold water immersion after a high intensity exercise session on subsequent run performance. J Sports Sci Med. 2011;10(4):665–70.
18. Leeder JD, van Someren KA, Bell PG, et al. Effects of seated and standing cold water immersion on recovery from repeated sprinting. J Sports Sci. 2015;33(15):1544–52.
19. Henderson AR. Cold—man's assiduous remedy. Med Ann Dist Columbia. 1971;40(9):583–8.
20. Theodorescu D. Cancer cryotherapy: evolution and biology. Rev Urol. 2004;6(Suppl 4):S9–S19.
21. Grant AE. Massage with ice (cryokinetics) in the treatment of painful conditions of the musculoskeletal system. Arch Phys Med Rehabil. 1964;45:233–8.
22. Hayden CA. Cryokinetics in an early treatment program. Phys Ther. 1964;44:990–3.
23. Bierman W. Therapeutic use of cold. J Am Med Assoc. 1955;157(14):1189–92.
24. Meeusen R, Lievens P. The use of cryotherapy in sports injuries. Sports Med. 1986;3(6):398–414.
25. Bleakley CM, Hopkins TJ. Is it possible to achieve optimal levels of tissue cooling in cryotherapy? Phys Ther Rev. 2010;15:344–51.
26. Bing HI, Carlsten A, Christiansen J. The effect on muscular temperature produced by cooling normal and ultraviolet radiated skin. Acta Med Scand. 1945;121:577–91.
27. Waylonis GW. The physiologic effects of ice massage. Arch Phys Med Rehabil. 1967;48(1):37–42.
28. Bugaj R. The cooling, analgesic, and rewarming effects of ice massage on localized skin. Phys Ther. 1975;55(1):11–9.
29. Herrera E, Sandoval MC, Camargo DM, Salvini TF. Motor and sensory nerve conduction are affected differently by ice pack, ice massage, and cold water immersion. Phys Ther. 2010;90(4):581–91.
30. Algafly AA, George KP. The effect of cryotherapy on nerve conduction velocity, pain threshold and pain tolerance. Br J Sports Med. 2007;41(6):365–9; discussion 369.
31. Ernst E, Fialka V. Ice freezes pain? A review of the clinical effectiveness of analgesic cold therapy. J Pain Symptom Manag. 1994;9(1):56–9.
32. Guillot X, Martin H, Seguin-Py S, et al. Local cryotherapy improves adjuvant-induced arthritis through down-regulation of IL-6/IL-17 pathway but independently of TNFalpha. PLoS One. 2017;12(7):e0178668.
33. Guillot X, Tordi N, Laheurte C, et al. Local ice cryotherapy decreases synovial interleukin 6, interleukin 1beta, vascular endothelial growth factor, prostaglandin-E2, and nuclear factor kappa B p65 in human knee arthritis: a controlled study. Arthritis Res Ther. 2019;21(1):180.
34. Yamauchi T. Whole body cryotherapy is method of extreme cold −175°C treatment initially used for rheumatoid arthritis. Z Phys Med Baln Med Klin. 1989;15:311.
35. Allan R, Malone J, Alexander J, et al. Cold for centuries: a brief history of cryotherapies to improve health, injury and post-exercise recovery. Eur J Appl Physiol. 2022;122(5):1153–62.
36. Collins NC. Is ice right? Does cryotherapy improve outcome for acute soft tissue injury? Emerg Med J. 2008;25(2):65–8.
37. Bouzigon R, Grappe F, Ravier G, Dugue B. Whole-body and partial-body cyostimulation/cryotherapy: current technologies and practical applications. J Therm Biol. 2016;61:67–81.
38. Fricke R. Ganzokorpekaltetherapie in einer Kaltekammer mit Temperaturen um −110°C. Z PhysMed Baln Med Klim. 1989;18:1–10.
39. Zagrobelny Z, Halawa B, Negrusz-Kawecka M, et al. Hormonal and hemodynamic changes caused by whole body cooling in patients with rheumatoid arthritis. Pol Arch Med Wewn. 1992;87(1):34–40.

40. Savic M, Fonda B, Sarabon N. Actual temperature during and thermal response after whole-body cryotherapy in cryo-cabin. J Therm Biol. 2013;38:186–91.
41. Hirvonen HE, Mikkelsson MK, Kautiainen H, Pohjolainen TH, Leirisalo-Repo M. Effectiveness of different cryotherapies on pain and disease activity in active rheumatoid arthritis. A randomised single blinded controlled trial. Clin Exp Rheumatol. 2006;24(3):295–301.
42. Guillot X, Tordi N, Mourot L, et al. Cryotherapy in inflammatory rheumatic diseases: a systematic review. Expert Rev Clin Immunol. 2014;10(2):281–94.
43. Stanek A, Cholewka A, Wielkoszynski T, Romuk E, Sieron A. Whole-body cryotherapy decreases the levels of inflammatory, oxidative stress, and atherosclerosis plaque markers in male patients with active-phase ankylosing spondylitis in the absence of classical cardiovascular risk factors. Mediat Inflamm. 2018;2018:8592532.
44. Doets JJR, Topper M, Nugter AM. A systematic review and meta-analysis of the effect of whole body cryotherapy on mental health problems. Complement Ther Med. 2021;63:102783.
45. Vitenet M, Tubez F, Marreiro A, et al. Effect of whole body cryotherapy interventions on health-related quality of life in fibromyalgia patients: a randomized controlled trial. Complement Ther Med. 2018;36:6–8.
46. Krzystanek M, Romanczyk M, Surma S, Kozmin-Burzynska A. Whole body cryotherapy and hyperbaric oxygen treatment: new biological treatment of depression? A systematic review. Pharmaceuticals. 2021;14(6):595.
47. Bettoni L, Bonomi FG, Zani V, et al. Effects of 15 consecutive cryotherapy sessions on the clinical output of fibromyalgic patients. Clin Rheumatol. 2013;32(9):1337–45.
48. Wojtecka-Lukasik E, Ksiezopolska-Orlowska K, Gaszewska E, et al. Cryotherapy decreases histamine levels in the blood of patients with rheumatoid arthritis. Inflamm Res. 2010;59(Suppl 2):S253–5.
49. Klimenko T, Ahvenainen S, Karvonen SL. Whole-body cryotherapy in atopic dermatitis. Arch Dermatol. 2008;144(6):806–8.
50. Miller E, Mrowicka M, Malinowska K, Zolynski K, Kedziora J. Effects of the whole-body cryotherapy on a total antioxidative status and activities of some antioxidative enzymes in blood of patients with multiple sclerosis-preliminary study. J Med Invest: JMI. 2010;57(1–2):168–73.
51. Miller E, Saluk J, Morel A, Wachowicz B. Long-term effects of whole body cryostimulation on uric acid concentration in plasma of secondary progressive multiple sclerosis patients. Scand J Clin Lab Invest. 2013;73(8):635–40.
52. Miller E, Kostka J, Wlodarczyk T, Dugue B. Whole-body cryostimulation (cryotherapy) provides benefits for fatigue and functional status in multiple sclerosis patients. A case-control study. Acta Neurol Scand. 2016;134(6):420–6.
53. Nugraha B, Gunther JT, Rawert H, Siegert R, Gutenbrunner C. Effects of whole body cryo-chamber therapy on pain in patients with chronic low back pain: a prospective double blind randomised controlled trial. Eur J Phys Rehabil Med. 2015;51(2):143–8.
54. Rymaszewska J, Ramsey D, Chladzinska-Kiejna S. Whole-body cryotherapy as adjunct treatment of depressive and anxiety disorders. Arch Immunol Ther Exp. 2008;56(1):63–8.
55. Huttunen P, Kokko L, Ylijukuri V. Winter swimming improves general well-being. Int J Circumpolar Health. 2004;63(2):140–4.
56. Fontana JM, Bozgeyik S, Gobbi M, et al. Whole-body cryostimulation in obesity. A scoping review. J Therm Biol. 2022;106:103250.
57. Zimmer Medizin Systeme. Consensus declaration on Whole-Body Cryotherapy (WBCT). 2006. https://www.polarior.com/images/manager/Hauptordner/Concensus%20Declaration%20on%20WBCT.pdf
58. Greenwald E, Christman M, Penn L, Brinster N, Liebman TN. Cold panniculitis: adverse cutaneous effect of whole-body cryotherapy. JAAD Case Rep. 2018;4(4):344–5.
59. Carrard J, Lambert AC, Genne D. Transient global amnesia following a whole-body cryotherapy session. BMJ Case Rep. 2017;2017:bcr-2017.

60. Cronier R, Fardellone P, Goeb V. Cerebral bleeding during a cryotherapy session: a case report. Rev Med Interne. 2020;41(12):843–5.
61. Chen PM, Chen MM, Chiang CC, Olson S, Bolar DS, Agrawal K. Moyamoya presenting after whole body cryotherapy. Acta Neurol Taiwanica. 2020;29(2):64–6.
62. Camara-Lemarroy CR, Azpiri-Lopez JR, Vazquez-Diaz LA, Galarza-Delgado DA. Abdominal aortic dissection and cold-intolerance after whole-body cryotherapy: a case report. Clin J Sport Med. 2017;27(5):e67–8.
63. Happe S, Evers S, Thiedemann C, Bunten S, Siegert R. Whole body and local cryotherapy in restless legs syndrome: a randomized, single-blind, controlled parallel group pilot study. J Neurol Sci. 2016;370:7–12.
64. Smolander J, Westerlund T, Uusitalo A, Dugue B, Oksa J, Mikkelsson M. Lung function after acute and repeated exposures to extremely cold air (−110 degrees C) during whole-body cryotherapy. Clin Physiol Funct Imaging. 2006;26(4):232–4.
65. Banfi G, Melegati G, Barassi A, d'Eril GM. Effects of the whole-body cryotherapy on NTproBNP, hsCRP and troponin I in athletes. J Sci Med Sport/Sports Med Aust. 2009;12(6):609–10.
66. Lubkowska A, Szygula Z. Changes in blood pressure with compensatory heart rate decrease and in the level of aerobic capacity in response to repeated whole-body cryostimulation in normotensive, young and physically active men. Int J Occup Med Environ Health. 2010;23(4):367–75.
67. Demoulin C, Vanderthommen M. Cryotherapy in rheumatic diseases. Joint Bone Spine: Rev Rhum. 2012;79(2):117–8.
68. Moellering DR, Smith DL Jr. Ambient temperature and obesity. Curr Obes Rep. 2012;1(1):26–34.
69. Stocks JM, Taylor NA, Tipton MJ, Greenleaf JE. Human physiological responses to cold exposure. Aviat Space Environ Med. 2004;75(5):444–57.
70. Kellogg DL Jr. In vivo mechanisms of cutaneous vasodilation and vasoconstriction in humans during thermoregulatory challenges. J Appl Physiol. 2006;100(5):1709–18.
71. Charkoudian N. Skin blood flow in adult human thermoregulation: how it works, when it does not, and why. Mayo Clin Proc. 2003;78(5):603–12.
72. Banfi G, Melegati G, Barassi A, et al. Effects of whole-body cryotherapy on serum mediators of inflammation and serum muscle enzymes in athletes. J Therm Biol. 2009;34(2):55–9.
73. Bouzigon R, Dupuy O, Tiemessen I, et al. Cryostimulation for post-exercise recovery in athletes: a consensus and position paper. Front Sports Active Living. 2021;3:688828.
74. Louis J, Theurot D, Filliard JR, Volondat M, Dugue B, Dupuy O. The use of whole-body cryotherapy: time- and dose-response investigation on circulating blood catecholamines and heart rate variability. Eur J Appl Physiol. 2020;120(8):1733–43.
75. Douzi W, Dupuy O, Theurot D, Boucard G, Dugue B. Partial-body cryostimulation after training improves sleep quality in professional soccer players. BMC Res Notes. 2019;12(1):141.
76. Miller E, Markiewicz L, Saluk J, Majsterek I. Effect of short-term cryostimulation on antioxidative status and its clinical applications in humans. Eur J Appl Physiol. 2012;112(5):1645–52.
77. Hausswirth C, Louis J, Bieuzen F, et al. Effects of whole-body cryotherapy vs. far-infrared vs. passive modalities on recovery from exercise-induced muscle damage in highly-trained runners. PLoS One. 2011;6(12):e27749.
78. Selfe J, Alexander J, Costello JT, et al. The effect of three different (−135 degrees C) whole body cryotherapy exposure durations on elite rugby league players. PLoS One. 2014;9(1):e86420.
79. Ziemann E, Olek RA, Kujach S, et al. Five-day whole-body cryostimulation, blood inflammatory markers, and performance in high-ranking professional tennis players. J Athl Train. 2012;47(6):664–72.
80. Mila-Kierzenkowska C, Jurecka A, Wozniak A, Szpinda M, Augustynska B, Wozniak B. The effect of submaximal exercise preceded by single whole-body cryotherapy on the markers of

oxidative stress and inflammation in blood of volleyball players. Oxidative Med Cell Longev. 2013;2013:409567.
81. Lombardi G, Sanchis-Gomar F, Perego S, Sansoni V, Banfi G. Implications of exercise-induced adipo-myokines in bone metabolism. Endocrine. 2016;54(2):284–305.
82. Jaworska J, Micielska K, Kozłowska M, et al. A 2-week specific volleyball training supported by the whole body cryostimulation protocol induced an increase of growth factors and counteracted deterioration of physical performance. Front Physiol. 2018;9:1711.
83. Zembron-Lacny A, Morawin B, Wawrzyniak-Gramacka E, et al. Multiple cryotherapy attenuates oxi-inflammatory response following skeletal muscle injury. Int J Environ Res Public Health. 2020;17(21):7855.
84. Lombardi G, Lanteri P, Porcelli S, et al. Hematological profile and martial status in rugby players during whole body cryostimulation. PLoS One. 2013;8(2):e55803.
85. Sutkowy P, Augustynska B, Wozniak A, Rakowski A. Physical exercise combined with whole-body cryotherapy in evaluating the level of lipid peroxidation products and other oxidant stress indicators in kayakers. Oxidative Med Cell Longev. 2014;2014:402631.
86. Szygula Z, Lubkowska A, Giemza C, Skrzek A, Bryczkowska I, Dolegowska B. Hematological parameters, and hematopoietic growth factors: EPO and IL-3 in response to whole-body cryostimulation (WBC) in military academy students. PLoS One. 2014;9(4):e93096.
87. Ziemann E, Olek RA, Grzywacz T, et al. Whole-body cryostimulation as an effective way of reducing exercise-induced inflammation and blood cholesterol in young men. Eur Cytokine Netw. 2014;25(1):14–23.
88. Lombardi G, Ricci C, Banfi G. Effect of winter swimming on haematological parameters. Biochem Med. 2011;21(1):71–8.
89. Lubkowska A, Szygula Z, Chlubek D, Banfi G. The effect of prolonged whole-body cryostimulation treatment with different amounts of sessions on chosen pro- and anti-inflammatory cytokines levels in healthy men. Scand J Clin Lab Invest. 2011;71(5):419–25.
90. Pournot H, Bieuzen F, Louis J, et al. Time-course of changes in inflammatory response after whole-body cryotherapy multi exposures following severe exercise. PLoS One. 2011;6(7):e22748.
91. Jaworska J, Rodziewicz-Flis E, Kortas J, et al. Short-term resistance training supported by whole-body cryostimulation induced a decrease in myostatin concentration and an increase in isokinetic muscle strength. Int J Environ Res Public Health. 2020;17(15):5496.
92. Rose C, Edwards KM, Siegler J, Graham K, Caillaud C. Whole-body cryotherapy as a recovery technique after exercise: a review of the literature. Int J Sports Med. 2017;38(14):1049–60.
93. Wilson LJ, Dimitriou L, Hills FA, Gondek MB, Cockburn E. Whole body cryotherapy, cold water immersion, or a placebo following resistance exercise: a case of mind over matter? Eur J Appl Physiol. 2019;119(1):135–47.
94. Poignard M, Guilhem G, de Larochelambert Q, Montalvan B, Bieuzen F. The impact of recovery practices adopted by professional tennis players on fatigue markers according to training type clusters. Front Sports Active Living. 2020;2:109.
95. Abaidia AE, Lamblin J, Delecroix B, et al. Recovery from exercise-induced muscle damage: cold-water immersion versus whole-body cryotherapy. Int J Sports Physiol Perform. 2017;12(3):402–9.
96. Hody S, Croisier JL, Bury T, Rogister B, Leprince P. Eccentric muscle contractions: risks and benefits. Front Physiol. 2019;10:536.
97. Malta ES, Dutra YM, Broatch JR, Bishop DJ, Zagatto AM. The effects of regular cold-water immersion use on training-induced changes in strength and endurance performance: a systematic review with meta-analysis. Sports Med. 2021;51(1):161–74.
98. Ihsan M, Abbiss CR, Allan R. Adaptations to post-exercise cold water immersion: friend, foe, or futile? Front Sports Active Living. 2021;3:714148.
99. Tavares F, Beaven M, Teles J, et al. Effects of chronic cold-water immersion in elite rugby players. Int J Sports Physiol Perform. 2019;14(2):156–62.

100. Jaworska J, Laskowski R, Ziemann E, et al. The specific judo training program combined with the whole body cryostimulation induced an increase of serum concentrations of growth factors and changes in amino acid profile in professional judokas. Front Physiol. 2021;12:627657.
101. Banfi G, Lombardi G, Colombini A, Melegati G. Whole-body cryotherapy in athletes. Sports Med. 2010;40(6):509–17.
102. Grasso D, Lanteri P, Di Bernardo C, et al. Salivary steroid hormones response to whole-body cryotherapy in elite rugby players. J Biol Regul Homeost Agents. 2014;28(2):291–300.
103. Hornery DJ, Papalia S, Mujika I, Hahn A. Physiological and performance benefits of halftime cooling. J Sci Med Sport/Sports Med Aust. 2005;8(1):15–25.
104. Le Meur Y, Louis J, Schaal K, et al. Whole-body cryotherapy multi exposures speed up performance supercompensation during the taper in functionally overreached endurance athletes. Sport Perform Sci Rep. 2017;1:1–2.
105. Schaal K, Lem Y, Louis J, et al. Whole-body cryostimulation limits overreaching in elite synchronized swimmers. Med Sci Sports Exerc. 2015;47(7):1416–25.
106. Klimek AT, Lubkowska A, Szygula Z, Chudecka M, Fraczek B. Influence of the ten sessions of the whole body cryostimulation on aerobic and anaerobic capacity. Int J Occup Med Environ Health. 2010;23(2):181–9.
107. Klimek AT, Lubkowska A, Szygula Z, Fraczek B, Chudecka M. The influence of single whole body cryostimulation treatment on the dynamics and the level of maximal anaerobic power. Int J Occup Med Environ Health. 2011;24(2):184–91.
108. Kruger M, de Marees M, Dittmar KH, Sperlich B, Mester J. Whole-body cryotherapy's enhancement of acute recovery of running performance in well-trained athletes. Int J Sports Physiol Perform. 2015;10(5):605–12.
109. Broatch JR, Poignard M, Hausswirth C, Bishop DJ, Bieuzen F. Whole-body cryotherapy does not augment adaptations to high-intensity interval training. Sci Rep. 2019;9(1):12013.
110. Westerlund T, Oksa J, Smolander J, Mikkelsson M. Neuromuscular adaptation after repeated WBC (−110°). J Therm Biol. 2009;34:226–31.
111. Costello JT, Algar LA, Donnelly AE. Effects of whole-body cryotherapy (−110 degrees C) on proprioception and indices of muscle damage. Scand J Med Sci Sports. 2012;22(2):190–8.
112. Ferreira-Junior JB, Vieira CA, Soares SR, et al. Effects of a single whole body cryotherapy (−110 degrees C) bout on neuromuscular performance of the elbow flexors during isokinetic exercise. Int J Sports Med. 2014;35(14):1179–83.
113. De Nardi M, Pizzigalli L, Benis R, Caffaro F, Micheletti Cremasco M. Acute effects of partial-body cryotherapy on isometric strength: maximum handgrip strength evaluation. J Strength Cond Res/Natl Strength Cond Assoc. 2017;31(12):3497–502.
114. Ma SY, Je HD, Jeong JH, Kim HY, Kim HD. Effects of whole-body cryotherapy in the management of adhesive capsulitis of the shoulder. Arch Phys Med Rehabil. 2013;94(1):9–16.
115. Romanowski MW, Straburzynska-Lupa A. Is the whole-body cryotherapy a beneficial supplement to exercise therapy for patients with ankylosing spondylitis? J Back Musculoskelet Rehabil. 2020;33(2):185–92.
116. Giemza C, Matczak-Giemza M, De Nardi M, Ostrowska B, Czech P. Effect of frequent WBC treatments on the back pain therapy in elderly men. Aging Male. 2015;18(3):135–42.
117. De Nardi M, La Torre A, Benis R, Sarabon N, Fonda B. Acute effects of whole-body cryotherapy on sit-and-reach amplitude in women and men. Cryobiology. 2015;71(3):511–3.
118. De Nardi M, Facheris C, Ruggeri P, La Torre A, Codella R. High-impact routines to ameliorate trunk and lower limbs flexibility in women. Int J Sports Med. 2020;41(14):1039–46.
119. Lastella M, Vincent GE, Duffield R, et al. Can sleep be used as an indicator of overreaching and overtraining in athletes? Front Physiol. 2018;9:436.
120. Sargent C, Lastella M, Halson SL, Roach GD. How much sleep does an elite athlete need? Int J Sports Physiol Perform. 2021;16(12):1746–57.

121. Bouzigon R, Ravier G, Dugué B, Grappe F. The use of whole-body cryostimulation to improve the quality of sleep in athletes during high level standard competitions. Br J Sports Med. 2014;48:572–57572.
122. Douzi W, Dupuy O, Tanneau M, Boucard G, Bouzigon R, Dugue B. 3-min whole body cryotherapy/cryostimulation after training in the evening improves sleep quality in physically active men. Eur J Sport Sci. 2019;19(6):860–7.
123. Dugué B, Bernard JP, Bouzigon R, et al. Whole body cryotherapy/cryostimulation, 39th informatory note on refrigeration technologies. International Institute of Refrigeration. 2020;2020:1–9.

Chapter 29
Musculoskeletal Extracorporeal Shockwave Therapy in Sports Medicine

Nikolaos Malliaropoulos and Amir Pakravan

Introduction

Extracorporeal shock wave therapy (ESWT) is a non-invasive therapeutic modality that uses high-energy transient pulses of sound waves produced through a pneumatic, electrohydraulic, electromagnetic, or piezoelectric mechanism and delivered over a short duration of approximately 10 μs. It is commonly used in sports medicine to stimulate healing and improve function in treatment of a variety of conditions including tendinopathies, muscle injuries, calcifications, and bone injuries. Extracorporeal shockwave therapy is believed to work by promoting the healing process through increasing blood flow and stimulating neovascularisation and regenerative effects at a cellular level. The procedure is generally well-tolerated, and apart from slight discomfort and other minor side effects has a very good safety profile. There is increasing evidence for effectiveness of ESWT in sports and musculoskeletal medicine; however, results may vary depending on the condition being treated (Fig. 29.1).

N. Malliaropoulos (✉)
Centre for Sports and Exercise Medicine, William Harvey Research Institute, School of Medicine and Dentistry, Queen Mary University of London, London, UK

Sports Clinic, Rheumatology Department, Mile End Hospital, London, UK

Sports and Exercise Medicine Clinic, Thessaloniki, Greece
e-mail: contact@sportsmed.gr

A. Pakravan
Centre for Sports and Exercise Medicine, William Harvey Research Institute, School of Medicine and Dentistry, Queen Mary University of London, London, UK

School of Allied Health and Science, University of Suffolk, Ipswich, UK

Super League Basketball, Canon Medical Arena, Sheffield, UK
e-mail: amir_pak@yahoo.com

N. C. Apostolopoulos et al. (eds.), *Fundamentals of Recovery, Regeneration, and Adaptation to Exercise Stress: An Integrated Approach*,
https://doi.org/10.1007/978-3-031-44270-4_29

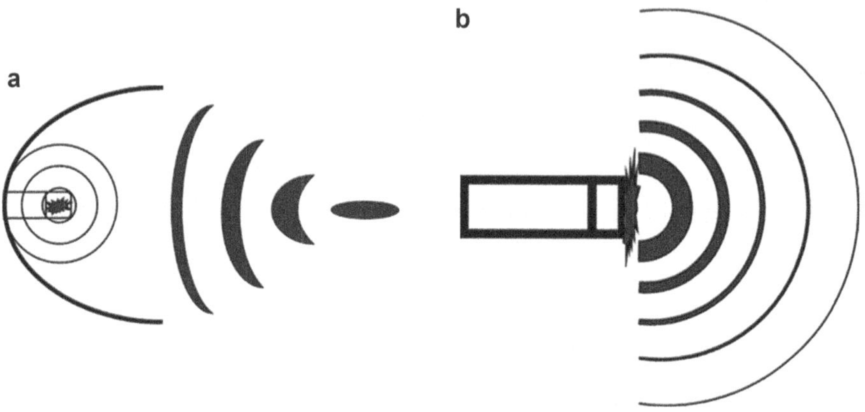

Fig. 29.1 (**a**) Pressure field of a focused shockwave device (EH-generated by means of spark gap). (**b**) Pressure filed of a radial shockwave device

Extracorporeal Shockwave Therapy (ESWT) Mechanism

The exact mechanism of action of ESWT may not yet be fully understood, but as a regenerative mechanotherapy modality it is believed to work by a combination of biological and mechanical effects commonly known as mechanotransduction. This is the process where transduction of the mechanical force generated by high-energy sound waves causes a cellular and biological response in the target tissue, leading to gene expression and production of growth factors [1, 2].

Growth factors such as platelet-derived growth factor (PDGF), insulin-like growth factor-1 (IGF-1) and vascular endothelial growth factor (VEGF) which are produced as a result of ESWT mechanotransduction, amongst other effects, lead to angiogenesis. Angiogenesis promotes the healing process through increased blood flow and subsequent improvements in tissue oxygenation, nutrients availability and removal of waste products [1–3].

A study of macrophages exposed to ESWT demonstrated its effect on reducing pro-inflammatory or M1 macrophage profile, whilst activating anti-inflammatory or M2 type macrophage [4]. This is potentially another biological mechanism through which ESWT can reduce tissue inflammation.

Up-regulation of angiogenic and osteogenic factors resulting from ESWT mechanotransduction typically starts 1 week after the treatment, peaks at about 4 weeks and continues for 12 weeks before slowly returning to baseline levels [5]. There is ample evidence for effective use of ESWT in bone healing and regeneration linked with neovascularisation and osteogenesis growth factors as well as a dose-dependent effect on the metabolism and proliferation ability of mesenchymal stem cells [2, 6, 7].

Extracorporeal shockwave therapy can mechanically break up calcifications and through activating cell membrane's mechanosensitive channels trigger release of intracellular signalling molecules that further promote healing and repair.

Release of pain-relieving substances such as endorphins following ESWT can provide effective pain management and improved function [8]. The more immediate effect on pain reduction, however, is likely as a result of high frequency stimulation of nociceptors and subsequent blocking of nerve impulses as explained by gate-control theory [1, 9]. This is one of few fundamental pain theories that since its inception in mid 60 s has undergone modifications but in essence can explain how hyperstimulation of sensory fibres, similar to what happens during ESWT, can shut the pain transmission gate [9, 10].

Proposed effects of ESWT:

- Angiogenesis
- Osteogenesis
- Chondrogenesis
- Collagen Synthesis
- Calcification breaking
- Improved oxygenation
- Pain management

Clinical Applications of ESWT in Sports Medicine

Shock Wave Therapy has been widely used as a safe non-intrusive treatment option in clinical practice in Sports and Musculoskeletal Medicine over the past 30 years and there is growing evidence of its efficacy in tissue healing and regeneration as well as pain management across a range of different presentations.

One of the most common applications is in management of fascia and tendinopathies such of Achilles, patellar tendon, gluteal tendon, elbow's common extensor origin or lateral epicondylitis, calcific and more recently non-calcific rotator cuff tendinopathy, and plantar fasciitis [1, 11–13].

There is also increasing interest in using ESWT for bone pathologies such as Medial Tibial Stress Syndrome, stress fractures, non or delayed union, and even osteonecrosis of head of femur [2, 7, 14, 15].

Studies have shown effect of ESWT on osteochondrogenesis in knee Osteoarthritis (OA) associated with osteochondral defects [1, 2, 16–18]. A recent meta-analysis of 32 studies on use of ESWT in Osteoarthritis concluded that ESWT is a viable and effective alternative to other non-surgical treatment options available for OA with meaningful improvements in pain and function [18].

Types of ESWT

There are two types of ESWT: focused (F-ESWT) and radial (R-ESWT) (See Table 29.1 below). Despite the differences in physical mechanism of shock wave production and energy delivery, there appear to be only marginal differences in clinical outcomes. Both types have proven to be successful in the management of a range of musculoskeletal disorders and both are considered as safe effective options.

Radial extracorporeal shock wave therapy (R-ESWT) is mainly generated by the more affordable pneumatic mechanism and uses a radial applicator which delivers the pressure waves perpendicular to the skin surface. This type of ESWT is generally more superficial and used in treating broader areas, such as a large muscle or joint and where the presenting condition has a broader distribution such as tendinopathies in the shoulder or the knee [1, 16, 19–21].

Focused Extracorporeal shock wave therapy (F-ESWT) uses a focus applicator, which delivers the shock waves at a specific adjustable focal point, or "focus" at specified depth. The focus applicator allows for a more precise delivery of the shock waves, which can be beneficial for treating certain conditions such as chronic heel pain, plantar fasciitis, calcifications, or management of trigger points [1, 17, 19, 22].

In practice, R-ESWT devices are smaller, more affordable, and easier to use. The choice of applicator and treatment protocol will depend on the condition being treated, the stage of the condition, and the patient's response to therapy.

ESWT Treatment Protocols

The treatment protocol for ESWT can vary depending on the type of the ESWT, the condition being treated, the stage of the condition, the patient's response to therapy, and the clinician's expertise and experience to design an individualised treatment protocol.

Choosing the correct site for treatment is a crucial part of the process which can be guided by physical examination, biofeedback, or imaging studies such as

Table 29.1 Fundamental differences between Focused-ESWT and Radial-ESWT

Focused-ESWT	Radial-ESWT
Various mechanisms (e.g. electrohydraulic)	Primarily pneumatic mechanism
Classic shockwave characteristics	Pressure waves
Deeper target tissues	More superficial target tissue
Focal point of delivery	Larger target area
Larger more expensive units	Smaller more affordable units

radiographs or ultrasound [1, 19]. The stage of the condition being treated can also play an important role in the outcome. There is evidence for better clinical results in later stage degenerative tendinopathy as opposed to early reactive phase tendinopathy [23, 24].

There are various adjustable parameters that include the energy level or intended pressure, the frequency and total number of shocks and in F-ESWT the focus depth. These variables can be accordingly adjusted during or in between sessions depending on patient response and their pain tolerance. High frequencies are likely to lead to lower overall pressure and are not generally advisable [1, 19].

Extracorporeal shockwave therapy is usually given in a series of treatment sessions, with the number of sessions varying depending on the condition being treated. For example, tendinopathies may require 6–8 treatment sessions, while plantar fasciitis depending on the chronicity of the symptoms may need more sessions. The frequency of treatment sessions is usually one every week but can be adjusted to longer intervals depending on progress and type of machine.

It is important to note that in the early stages after ESWT treatment, tendon is not able to adjust to higher loads and a short period of tendon rest following each treatment session is essential. There is evidence to suggest actively competing athletes do not respond to ESWT [24].

ESWT Contraindications

Despite an excellent safety profile, there are certain contraindications to ESWT that should be considered before offering the treatment. Some of the most common contraindications include [25–27]:

- Pregnancy: ESWT should not be performed on pregnant women as the effects on the developing foetus are not known.
- Blood clotting disorders: ESWT should not be performed on patients with blood clotting disorders as it may increase the risk of bleeding or hematoma formation.
- Active infection: ESWT should not be performed on patients with an active infection in the area to be treated as it may spread the infection.
- Malignancy: ESWT should not be performed on patients with malignant tumours in the area to be treated as it may stimulate the growth of the tumour.
- Severe osteoporosis: ESWT should not be performed on patients with severe osteoporosis as it may increase the risk of fractures.
- Pacemaker or other implanted electronic devices: ESWT is not recommended in patients with pacemakers or other implanted electronic devices as it may interfere with the functioning of these devices. However, there is still a need for more specific evidence-based guidance on this, and use of ESWT in urology for lithotripsy has been shown to be safe in this group provided the device is out of the shockwave path [28].

ESWT for Adolescents

Based on current studies in children and adolescents, ESWT is considered safe and effective for treating appropriate musculoskeletal conditions. However, further research is needed to determine the potential risks and benefits of using this therapy in this population. The decision to use it should be made on a case-by-case basis and after careful evaluation of the patient's condition [19].

The growth plates, also known as epiphyseal plates, are areas of developing cartilage located at the ends of long bones in children and adolescents. These areas are responsible for the longitudinal growth of bones and any damage to them can affect the normal growth and development of bones. Therefore, it is important to consider the potential effects of ESWT on the developing musculoskeletal system before deciding to use it in adolescents.

Additionally, the indications, dosage and energy levels used for ESWT in adolescents may be different from those used in adults and should be adjusted accordingly [19, 29].

Conclusion

Shock Wave Therapy is a well-established evidence-based treatment modality successfully used in a variety of musculoskeletal presentations. Extracorporeal shockwave therapy is considered as a regenerative procedure particularly in management of soft tissue and bone overload and degenerative injuries such as various tendinopathies, bony stress injuries and osteoarthritis commonly seen in active population. The procedure is non-invasive and safe with very few known side effects and contraindications. The evidence for effectiveness of ESWT is strong and still growing; however, the direction of future research should now be towards a more individualised approach to treatment trying to establish customised protocols for various presentations and different stages of each pathology [19].

1. How can we personalize ESWT treatment protocols for specific indications based on individual characteristics of patients for optimum effects with minimal impact on their sporting or physical activities?
2. Are there any correlations between imaging findings and ESWT treatment protocols, patient clinical response, and return to play decision making?
3. Investigate application of ESWT in acute tendon, muscle and ligament injuries as opposed to overload injuries.
4. Investigate differences between F-ESWT and R-ESWT in clinical outcomes.
5. Investigate optimum time from injury to start ESWT and whether different stages of injuries require different protocols.
6. What is the optimum rest period after ESWT session and at what stage should tissue targeted exercise program start?

References

1. van der Worp H, van den Akker-Scheek I, van Schie H, Zwerver J. ESWT for tendinopathy: technology and clinical implications. Knee Surg Sports Traumatol Arthrosc. 2013;21(6):1451–8.
2. Cheng JH, Wang CJ. Biological mechanism of shockwave in bone. Int J Surg. 2015;24(Pt B):143–6.
3. Alshihri A, Kämmerer PW, Heimes D, Niu W, Alnassar T, Spector M. Extracorporeal shock wave stimulates angiogenesis and collagen production in facial soft tissue. J Surg Res. 2020;245:483–91.
4. Sukubo NG, Tibalt E, Respizzi S, Locati M, d'Agostino MC. Effect of shock waves on macrophages: a possible role in tissue regeneration and remodeling. Int J Surg. 2015;24(Pt B):124–30.
5. Wang CJ, Sun YC, Wong T, Hsu SL, Chou WY, Chang HW. Extracorporeal shockwave therapy shows time-dependent chondroprotective effects in osteoarthritis of the knee in rats. J Surg Res. 2012;178(1):196–205.
6. Yang YM, Zhang H, Liao WX, Su XZ, Li J, Wang KT, Liu YL, Mao N, Zhang Y, Zhang N, Zhang GQ, Zhu H, Li ZL. Effects of shock wave on the proliferation and osteogenic differentiation of human bone marrow mesenchymal stem cells. Zhongguo Shi Yan Xue Ye Xue Za Zhi. 2017;25(1):209–13.
7. Zhai L, Sun N, Zhang B, Liu ST, Zhao Z, Jin HC, Ma XL, Xing GY. Effects of focused extracorporeal shock waves on bone marrow mesenchymal stem cells in patients with avascular necrosis of the femoral head. Ultrasound Med Biol. 2016;42(3):753–62.
8. Wu T, Wang D, Zhang X, Li J, Yuan B. Comparison of pain relief and limb function improvement after extracorporeal shock wave therapy and thermomagnetic therapy in the treatment of low back pain. Pak J Med Sci. 2023;39(1):268–73.
9. Melzack R, Wall PD. Pain mechanisms: a new theory. Science. 1965;150(3699):971–9.
10. Mendell LM. Constructing and deconstructing the gate theory of pain. Pain. 2014;155(2):210–6.
11. Korakakis V, Whiteley R, Tzavara A, Malliaropoulos N. The effectiveness of extracorporeal shockwave therapy in common lower limb conditions: a systematic review including quantification of patient-rated pain reduction. Br J Sports Med. 2018;52(6):387–407.
12. Malliaropoulos N, Crate G, Meke M, Korakakis V, Nauck T, Lohrer H, Padhiar N. Success and recurrence rate after radial extracorporeal shock wave therapy for plantar fasciopathy: a retrospective study. Biomed Res Int. 2016;2016:9415827.
13. Malliaropoulos N, Jury R, Pyne D, Padhiar N, Turner J, Korakakis V, Meke M, Lohrer H. Radial extracorporeal shockwave therapy for the treatment of finger tenosynovitis (trigger digit). Open Access J Sports Med. 2016;7:143–51.
14. Thompson D, Malliaropoulos N, Padhiar N. Sesamoid osteonecrosis treated with radial extracorporeal shock wave therapy. BMJ Case Rep. 2017;2017:bcr2017219191.
15. Wang CJ, Wang FS, Ko JY, Huang HY, Chen CJ, Sun YC, Yang YJ. Extracorporeal shockwave therapy shows regeneration in hip necrosis. Rheumatology. 2008;47:542e546.
16. Zhang YF, Liu Y, Chou SW, Weng H. Dose-related effects of radial extracorporeal shock wave therapy for knee osteoarthritis: a randomized controlled trial. J Rehabil Med. 2021;53(1):jrm00144.
17. Ko NY, Chang CN, Cheng CH, Yu HK, Hu GC. Comparative effectiveness of focused extracorporeal versus radial extracorporeal shockwave therapy for knee osteoarthritis-randomized controlled study. Int J Environ Res Public Health. 2022;19(15):9001.
18. Chen L, Ye L, Liu H, Yang P, Yang B. Extracorporeal shock wave therapy for the treatment of osteoarthritis: a systematic review and meta-analysis. Biomed Res Int. 2020;2020:1907821.
19. Lohrer H, Nauck T, Korakakis V, Malliaropoulos N. Historical ESWT paradigms are overcome: a narrative review. Biomed Res Int. 2016;2016:3850461.
20. Malliaropoulos N, Thompson D, Meke M, Pyne D, Alaseirlis D, Atkinson H, Korakakis V, Lohrer H. Individualised radial extracorporeal shock wave therapy (rESWT) for symptomatic

calcific shoulder tendinopathy: a retrospective clinical study. BMC Musculoskelet Disord. 2017;18(1):513.
21. Leung R, Malliaropoulos N, Korakakis V, Padhiar N. What are patients' knowledge, expectation and experience of radial extracorporeal shockwave therapy for the treatment of their tendinopathies? A qualitative study. J Foot Ankle Res. 2018;11:11.
22. Li C, Li Z, Shi L, Wang P, Gao F, Sun W. Effectiveness of focused shockwave therapy versus radial shockwave therapy for noncalcific rotator cuff tendinopathies: a randomized clinical trial. Biomed Res Int. 2021;2021:6687094.
23. Wang CJ, Ko JY, Chan YS, Weng LH, Hsu SL. Extracorporeal shockwave for chronic patellar tendinopathy. Am J Sports Med. 2007;35:972–8.
24. Zwerver J, Hartgens F, Verhagen E, van der Worp H, van den Akker-Scheek I, Diercks RL. No effect of extracorporeal shockwave therapy on patellar tendinopathy in jumping athletes during the competitive season: a randomized clinical trial. Am J Sports Med. 2011;39:1191–9.
25. Auersperg V, Trieb K. Extracorporeal shock wave therapy: an update. EFORT Open Rev. 2020;5(10):584–92. https://doi.org/10.1302/2058-5241.5.190067.
26. Crevenna R, Mickel M, Keilani M. Extracorporeal shock wave therapy in the supportive care and rehabilitation of cancer patients. Support Care Cancer. 2019;27(11):4039–41.
27. Kiessling MC, Milz S, Frank HG, Korbel R, Schmitz C. Radial extracorporeal shock wave treatment harms developing chicken embryos. Sci Rep. 2015;5:8281.
28. Ubee SS, Kasi VS, Bello D, Manikandan R. Implications of pacemakers and implantable cardioverter defibrillators in urological practice. J Urol. 2011;186(4):1198–205.
29. Lohrer H, Nauck T, Schöll J, Zwerver J, Malliaropoulos N. Einsatz der extrakorporalen Stoßwellentherapie bei therapieresistentem M. Schlatter. Extracorporeal shock wave therapy for patients suffering from recalcitrant Osgood-Schlatter disease. Sportverletz Sportschaden. 2012;26(4):218–22.

Chapter 30
Hyperbaric Oxygen Therapy

Navid Moghadam, Ali Ahmadzadeh Amiri, and Amir Ahmadzadeh Amiri

Introduction

Definition

Hyperbaric oxygen therapy (HBOT) is a type of therapeutic intervention that involves inhaling pure oxygen in a pressurized space. Candidates for this treatment breathe 100% oxygen in a chamber with pressures higher than ambient air pressure at sea level [1], meaning that the patient is exposed to a pressure of more than 1 absolute atmosphere or 760 mmHg. However, if an individual is exposed to an atmospheric pressure of 2 or 3 absolute atmosphere this is approximately equivalent to being 13–19 m below the surface of the open oceans. HBOT sessions can last between 45 and 300 minutes, depending on the indication and the diagnosis.

In our body, the need for oxygen delivery varies between different tissues under different conditions. However, on average, healthy resting tissues need about 60 ml of oxygen per liter to maintain a normal metabolism which is normally met considering the efficacy of hemoglobin in the transfer of oxygen. On the other hand, under a pressure of 3 absolute atmosphere, this level of dissolved oxygen in plasma can be achieved, even without taking into account the oxygen bound to hemoglobin [2]. As

N. Moghadam (✉)
Sports Medicine Research Center, Neuroscience Institute, Tehran University of Medical Sciences, Tehran, Iran

Spine Center of Excellence, Yas Hospital, Tehran University of Medical Sciences, Tehran, Iran
e-mail: navid.mgd@gmail.com

A. Ahmadzadeh Amiri · A. Ahmadzadeh Amiri
Sports Medicine Research Center, Neuroscience Institute, Tehran University of Medical Sciences, Tehran, Iran

N. C. Apostolopoulos et al. (eds.), *Fundamentals of Recovery, Regeneration, and Adaptation to Exercise Stress: An Integrated Approach*,
https://doi.org/10.1007/978-3-031-44270-4_30

a result, under conditions such as carbon monoxide poisoning or severe anemia, HBOT can help deliver oxygen without the need for hemoglobin transfusion [3].

Considering the above-mentioned points, HBOT is currently under investigation as a possible complementary and effective intervention for the treatment of various tissue injuries such as wounds and muscular damage.

History

Contrary to popular belief, HBOT is not a novel treatment. In fact, it has been used since the mid-seventeenth century when in 1662, the British physician Dr. Henshaw built the first pressure chamber to manage lung and gastrointestinal diseases. In 1834, under the direction of Dr. Junod of France, the first hyperbaric tank was developed and was used for the successful treatment of a variety of medical conditions [4]. The genesis for the idea can be traced back to the comparison of the resolution of the wound healing process between people living in high altitudes to those residing in submarine habitats (deep-sea divers), with the former having a slower rate of healing [5]. We have summarized the key moments in the history of HBOT in Table 30.1. All said, HBOT can be considered a relatively new application of an old technology contributing to medical science, as seen by its increased popularity within medical centers [6].

> While preparing for the 2012 Olympics in London, Michael Phelps started using HBOT as he was looking for ways to boost his recovery. In an interview, he shared that the idea came after spending a lot of time in Colorado Springs, where the altitude had made a positive impact on his ability to bounce back after training sessions.
>
> *Recovery is something that is so important to me now being older. I don't recover as fast as I used to.*

HBOT and Tissue Survival

Pathophysiology

The importance of oxygenation for tissue survival during damage and regeneration has been observed by many studies [7–9]. The rupture of the blood vessels and cells is responsible for hypoxia, accumulation of lactate, platelets, and collagen, and release of hormonal mediators, such as macrophages, aimed at clearing the site and repairing the damaged tissue [10]. This disruption of effective circulation injuries leads to tissue hypoxia, to which the tissue responds with vasodilation in endothelial cells and smooth muscles of the injured area and an increase in extracellular fluid. This process in turn could contribute to more severe hypoxia, as tissue damage can

Table 30.1 Notable historical moments in the history and development of HBOT

Year	Notable individuals	Summary
1662	J. Henshaw	Utilized compressed air for the treatment of a variety of diseases.
1775	J. Priestley	Discovered oxygen.
1796	T. Beddoes and J. Watt	Wrote the first book on the medical application of oxygen.
1834	Dr. Junod	Built the first real hyperbaric tank.
1837	C.G. Pravaz	Constructed the largest hyperbaric chamber at the time to treat a variety of maladies.
1860	A Canadian enterprise	Constructed the first hyperbaric chamber in North America, Canada.
1895	J.S. Haldane	Showed that in a jar containing oxygen at 2.0 absolute atmosphere, a mouse failed to develop CO poisoning.
1928	O.J. Cunningham	Built the largest chamber in the world in Cleveland.
1937	A.R. Benke and L.A. Shaw	Used HBOT for decompression sickness for the first time.
1956	I. Boerema (father of modern hyperbaric medicine)	Performed cardiac surgery in a hyperbaric chamber.
1963	–	Holding of the first international congress on hyperbaric medicine in Amsterdam.
1967	–	Formation of Undersea Medical Society in the USA. (Now known as the Undersea and Hyperbaric Medical Society).
1970s	–	Expansion of hyperbaric facilities in Japan and the USSR.
1983	–	Formation of the American College of Hyperbaric Medicine.
1987	K.K. Jain	Showed HBOT mixed with physical therapy.
1988	–	Formation of the International Society of Hyperbaric Medicine.

lead to edema, circulatory insufficiency, tissue ischemia, and eventually, tissue death. This potentially harmful local response to hypoxia could interfere with the host's healing processes.

Furthermore, oxygen depletion in tissue renders the cells more susceptible to infection by reducing the efficiency of leukocytes in killing the invading organism [11]. Eventually, the accumulation of lactate, hypoxia, and the production of cytokines in tissue stimulate growth, which leads to angiogenesis and collagen production. The damaged tissue starts to heal as a result of this cascade of events [7].

Due to the limited oxygen distribution within tissues, adequate oxygen delivery through an extensive capillary network is essential for tissue survival [12], as oxygen acts both as a catalyst and a source of energy for the maintenance, metabolism, and repair of damaged tissue [13].

General Effects

The effects of HBOT rely on the hyperoxygenation of damaged tissues by delivering oxygen in increased partial pressure and reduced volume, resulting in more effective oxygen delivery to and in the tissues [14].

Scientific evidence is in favor of the effectiveness of short-term HBOT in increasing oxygen delivery efficiency and compensation for hypoxia [15, 16]. Furthermore, short-term HBOT demonstrates anti-inflammatory potentials, as well as the capacity to reduce ischemia-reperfusion injury which is defined as the paradoxical exacerbation of cellular dysfunction or death, following the restoration of blood flow to previously ischemic tissues and also the risk of infection. Additionally, in long-term HBOT and with repetition, the promotion of neovascularization (formation of the new blood vessels) through angiogenesis (formation of new blood vessels from preexisting blood vessels), in addition to stimulation of collagen production by fibroblasts has been observed [15]. These studies demonstrate potential in the promotion of the rehabilitation of damaged tissue in different stages of healing.

> The effects of HBOT are mainly due to the hyperoxygenation of damaged tissues by delivering oxygen in increased partial pressure and reduced volume, resulting in more effective oxygen delivery to the tissues.

Potential Mechanisms

HBOT offers an effective method of increasing both blood and tissue oxygen levels, independently from the hemoglobulin transportation pathway. Its therapeutic properties can be studied from three different perspectives: Physical (hyperbaric 100% oxygen), physiological (hyperoxia and hyperoxemia), and cellular/molecular. These three aspects provide mechanisms that contribute to the potency of HBOT in combatting the consequences of hypoxia-derived conditions and hypoxemia. Furthermore, HBOT has demonstrated direct effects on modulating a wide variety of immune cells, cellular signaling pathways, cytokine production, and healing processes such as angiogenesis in damaged tissue [17].

Hyperoxygenation

The effect of hyperoxygenation is based on Henry's and Dalton's laws. According to Henry's law, as the oxygen pressure increases during compression, the amount of oxygen directly dissolved in the plasma is also increased, as it is directly proportional to the oxygen pressure of the atmosphere. Thus, with a slight increase in oxygen pressure in the air, a significant amount of excess oxygen is dissolved in the plasma [18].In addition, Dalton's law states that the air is a mixture of gases, and the total pressure applied is equal to the sum of the partial pressures of each of the gases in the mixture [19]. At sea level, 0.32 ml of oxygen per 100 ml of whole blood is dissolved in it. However, when breathing 100% oxygen, each extra absolute atmosphere produces 2.3% of the volume of extra oxygen in the plasma. Thus, at 2.0 absolute atmosphere, the blood oxygen content increases by 2.3% while the plasma and tissue oxygen density increase by 1000%. As a result, enough oxygen is physically dissolved in the plasma to keep the tissues alive, despite the inability of the hemoglobin-bound oxygen to reach the affected area [20]. This high oxygen pressure increases the oxygen diffusion gradient and improves oxygen delivery to damaged tissues under partial ischemia. At this pressure, enough oxygen can be dissolved in the plasma to sustain life temporarily without any red blood cells [19].

Dr. Scott Sherr, a physician certified in HBOT and Health Optimization Medicine had an interview with Passion Struck:

"When you go into a hyperbaric environment under pressure, you can get up to 1200% more oxygen in circulation" he added, *"The more oxygen you get in a blood vessel, the more that's going to diffuse out of that blood vessel and get further into tissue. So that's extremely important in acute trauma, an acute infection, or in many acute situations".*

Vasoconstriction

The vascular contraction observed in HBOT offers the potential for the reduction of harmful edema at the site of injury, leading to decreased tissue perfusion without oxygen depletion. As previously elaborated, sufficient amounts of oxygen dissolved in the blood plasma under HBOT offer enough oxygenation to effectively compensate for the reduction in blood flow to the injured area. This effective phenomenon can be seen in a subsequent 20% reduction in post-traumatic vasogenic edema which occurs due to a disruption in the Brain brain-blood barrier, resulting in the leaking of fluid and proteins such as albumin from vessels into the brain tissues evoking an increase of brain volume and intracranial pressure [21].

Collagen and Protein Production

The essential role of oxygen as a source of energy in the process of protein production is evident [22]. In the early stages of wound healing, due to the excess need for collagen production, the matrix necessary for wound healing, fibroblasts begin to migrate and replicate. Consequently, excess oxygen is needed both as a source of energy and in the process of hydroxylation of the two amino acids, proline, and lysine, which are necessary for the effective and sufficient synthesis of collagen by fibroblasts.

Studies have demonstrated that increasing oxygen levels up to 150% above the normal physiological range can result in up to a seven-fold increase in the rate of collagen production by fibroblasts [23]. Furthermore, increased oxygenation in tissues has resulted in increasing the RNA-to-DNA ratio, indicating a higher protein production [4]. In conclusion, more effective oxygen delivery can contribute to the sufficient reproduction of fibroblasts and the production of collagen.

Gas Volume Reduction

An increase in pressure in HBOT is associated with a decrease in the volume of gas, resulting in a reduction of the size of the gas bubbles in circulation. This physiological effect can be explained by Boyle's law, that "the volume of a gas is inversely proportional to its pressure at a constant temperature." In other words, the increased pressure in HBOT reduces the volume of the gas, decreasing the size of the gas bubbles that impede blood circulation in various conditions, such as embodies. Therefore, HBOT can be used as part of the primary therapy in air embolism and decompression sickness [24].

Polymorphonuclear Function

For polymorphonuclear cells to function properly, they need oxygen to eliminate target organisms [25]. These cells function by producing substances such as superoxide and hydrogen peroxide through the respiratory burst, a rapid release of the reactive oxygen species during the phagocytosis of microbes, which starts with the introduction of polymorphonuclear cells to an appropriate stimulus, aimed at eliminating the target [25].

An increased amount of oxygen provided for the polymorphonuclear cells during the respiratory burst helps protect tissues against free radicals by facilitating a detoxifying process, including the conversion of free radicals to hydrogen peroxide and the use of superoxide dismutase, catalase, and glutathione enzymes [25, 26]. Thus, the performance of polymorphonuclear cells in eliminating bacteria is dependent on oxygen aiding in the protection against free radicals [25].

Free Radical Production

Another substantial effect of HBOT is the production of free radicals in the remedial range for cellular signaling. HBOT induces oxidative stress by controlled production of both reactive oxygen and nitrogen activating cellular pathways ameliorating not only neovascularization but also post-ischemic tissue survival [27, 28].

Improved neovascularization can be achieved by increasing wound growth factor syntheses such as vascular endothelial growth factor and mobilization of bone marrow-derived stem/progenitor cells in the peripheral sites. On the other hand, the reduction of neutrophil adhesion by impairing β2 integrin function alongside diminished inflammatory response via lowering monocyte chemokine synthesis leads to an enhanced post-ischemic tissue survival [29–31].

In conclusion, beyond a more effective oxygen delivery, HBOT can provide damaged tissues with adequate amounts of free radicals to promote reperfusion and tissue regeneration.

- How does HBOT influence epigenetic modifications in injured tissues during the recovery phases?
- What are the long-term effects of repeated HBOT sessions on longevity (telomere length and cellular senescence)?

Application

Clinical Indications

Over the past decades, the effectiveness of HBOT has been widely recognized among physicians and its clinical application has been used as the primary treatment for several conditions [32]. Due to the various features of HBOT, the potential applications of this procedure are numerous. It may be used as an urgent treatment for acute pathologies in an emergency setting, in support of chronic diseases to enhance wound healing, trigger angiogenesis, or be applied in a manner stimulating antimicrobial effects [32, 33].

Currently, clinical indications of HBOT include decompression sickness, air embolism, and carbon monoxide poisoning [32]. This method is also useful as a supplementary treatment in the promotion of wound healing in crush injuries, compartment syndrome, acute thermal burn injuries, traumatic ischemia, radiation-induced tissue damage, and skin graft and flap injuries. HBOT also possesses potential antimicrobial activity as an adjunct therapy in the treatment of refractory wounds and chronic osteomyelitis, necrotizing soft tissue infections (NSTIs), clostridial myositis and myonecrosis (gas gangrene), and intracranial abscess. Moreover, in some medical emergencies, cases such as severe blood loss and anemia, central retinal artery occlusion, and idiopathic sudden sensorineural hearing loss have benefited from the rapid and effective response of tissue to HBOT under some severe or life-threatening conditions [32].

Unlike in the past, HBOT is no longer just a cure for decompression sickness in divers. Research on the application of HBOT is ongoing in all areas of medicine with HBOT on the verge of becoming a common therapeutic option for a wide variety of diseases [34, 35].

Due to the various potentials of HBOT, it is being used as a treatment for pathologies in an acute emergency setting, as a supplementary treatment for chronic diseases to enhance healing, and in a manner stimulating antimicrobial effects.

- What reliable biomarkers can be used to predict and monitor individual patients' responses to HBOT?

Clinical Contraindications

Like any other procedure, there are certain circumstances in which HBOT is either absolutely or relatively contraindicated. The only absolute contraindication of HBOT is untreated pneumothorax since it could potentially be life-threatening [36]. Nevertheless, the remaining contraindications are relative, meaning that HBOT is acceptable if, after enough caution, the physician decides the benefits outweigh the risks.

The following conditions are worth paying attention to when considering the administration of HBOT. In patients with a pacemaker or other implantable devices, it is necessary to seek cardiological consultation to verify its harmlessness under increased pressure or pure concentrations of oxygen [37]. Since HBOT could induce severe hemolysis, conditions such as hereditary spherocytosis may also be contraindicated [38]. Another potential contraindication is pregnancy, except in emergency cases of carbon monoxide poisoning [39]. Doses of HBOT must be regulated cautiously in diabetic patients to prevent hypoglycemic episodes during the procedure [40]. Other contraindications may include underlying respiratory pathologies since HBOT could increase the risk of hypercapnia and pulmonary barotrauma [32]. Patients with uncontrolled hypertension must be advised against receiving HBOT as it can cause increased blood pressure [41]. Furthermore, HBOT may induce pulmonary edema and cardiovascular distress in patients with underlying cardiac failure [42]. Finally, additional cautions are required in patients with a history of epilepsy, hypoglycemia, hyperthyroidism, current fever (due to infections), and the consumption of certain drugs such as penicillin and disulfiram since they are reported to decrease the seizure threshold during HBOT [43].

Due to its life-threatening nature, the only absolute contraindication of HBOT is untreated pneumothorax. Nevertheless, the remaining contraindications are relative, meaning that it is acceptable if the benefits outweigh the risks.

HBOT in Sports

As athletes strive to be the best in their field, they, along with coaches, seek ways to continue to perform at their peak. With high-intensity sports, training and/or competition can often lead to fatigue, mental exhaustion, and anxiety in athletes, negatively influencing their sleep and overall mood. In addition, the intense nature of the competitions develops the potential for a higher incidence of injuries in athletes. Severe injuries including broken bones, damaged muscles, torn tendons, and ligaments may be due to acute impact forces in contact sports or even the aggressiveness of training sessions. With sports medicine playing a crucial role in the rehabilitation and recovery of athletes at all levels, sports physicians are frequently challenged by teams to provide effective therapeutic methods to speed up the recovery process. The administration of HBOT for the rehabilitation of the athletes could be another method aiding in their return from injury due to training and competition [44]. Below are discussed certain applications of HBOT within sports medicine.

Rashad Jennings is a former NFL player to incorporate HBOT into his recovery regimen. During his time playing for the New York Giants, Jennings acknowledged it in an interview with the New York Post.

I use it to rejuvenate and energize... even on a cognitive level, it helps out.

Delayed Onset Muscle Soreness

Delayed onset muscle soreness (DOMS) is a condition caused by an exercise, which is predominantly eccentric, resulting in skeletal muscle strain. DOMS limits the athlete's ability to train and compete for at least several days and is believed to be correlated with the damage and inflammation of the sarcolemma, the plasma membrane of the muscle cell that is surrounded by the basement membrane and endomysial connective tissue [45]. A rise in serum C-reactive protein, a protein released during inflammation and the inflammatory response, is also associated with DOMS, suggesting the initiation of the healing process [45]. With DOMS restricting an athlete's participation in training sessions and competitions, it is substantial for trainers, coaches, and athletes to be provided with novel and effective methods expediting the recovery period [45].

Despite studies suggesting the effectiveness of HBOT in reducing the recovery time [46] and pain intensity in athletes suffering from DOMS [47, 48], some studies imply inconsistent findings [49–51]. The conflicting results might be due to the different methodological approaches of each study. In certain studies, although, the placebo protocol in the control group included less gas compression than the intervention group, instead of pure oxygen, ambient air was applied which not only influences the validity of the comparison but also raises concerns about the safety of the protocol due to the effects of compressed air or gases other than oxygen on the participants [52]. Therefore, conducting standard and reliable trials is still a challenge facing scientists struggling to identify the effectiveness of HBOT on DOMS.

> Rafael Soriano is a former MLB pitcher who played for 5 different teams. He learned to perfect his recovery regimen over his 13-year career, adding HBOT in 2012.
> *I've seen the results because I don't feel as tired or worn out when I use it... It helps.*

Muscle Injury

The effectiveness of HBOT on muscular injuries has been studied, demonstrating a significant decrease in muscle enzyme levels and a relief in pain intensity, which are advantageous for professional athletes in a highly competitive setting [44]. It also has been observed that HBOT can mitigate the amount of lactate production caused by heavy physical activity and exercise [53].

As mentioned, edema created in the acute phase of post-muscular injury increases the oxygen diffusion distance, with an increase in extracellular pressure reducing perfusion. These simultaneous occurrences could cause significant hypoxia and possibly necrosis [54, 55]. HBOT delivers oxygen without causing vascular dilation and permeability potentially reducing the edema and consequently hypoxia.

Bone marrow-derived and circulating monocytes invade the injured tissue in response to muscle injury. These monocytes then differentiate into pro-inflammatory macrophages (M1) during the acute phase of inflammation. Furthermore, pro-inflammatory M1 macrophages then differentiate into anti-inflammatory macrophages (M2) which contribute to reducing the local inflammation and promoting muscle regeneration through myoblast fusion and myofiber regeneration [56]. HBOT accelerates the transition from the inflammatory state to the proliferative state of healing by increasing the number of anti-inflammatory M2 macrophages compared to the pro-inflammatory M1 macrophages [57]. Moreover, it has been demonstrated that in the initial phase of the healing process, HBOT boosts the differentiation of migrated M1 to M2 macrophages [57].

In addition, satellite cells, which are essential for muscle rehabilitation in the proliferative phase, are transformed into myoblasts initiating muscle regeneration. Previous studies have demonstrated that using HBOT leads to an increase in the number of proliferating, differentiating, and quiescent satellite cells reserved for subsequent injuries [57, 58]. Moreover, the myogenic activity of satellite cells can be negatively influenced in a hypoxic environment [58, 59]. Therefore, it is suggestive that HBOT can enhance the regeneration processes after muscular injury [1].

Ligament and Tendon Injuries

Although damage to ligaments and tendons are some of the most common injuries in competitive sports, there are few studies investigating the effectiveness of HBOT on these tissues, with most studies still in the pre-clinical phases. In an animal study it was reported that after inducing a ligament tear and corrective surgery, a higher destructive force was required to fail the repaired medial collateral ligament (MCL) of rats that underwent HBOT compared with a control group. However, it is notable that the observed difference lasted only for 1 month, and after 1.5 months, no differences were found [60]. Furthermore, another study on rats showed a dose-response relationship between the effectiveness of HBOT on the healing process of ligaments and the concentration and pressure of oxygen. It appeared that HBOT at 2 absolute atmosphere was the most beneficial for enhancing collagen synthesis [61].

The expression of matrix metalloproteinases (MMP) and type I procollagen are suggestive of ligament healing [62]. Studies with rats, whose anterior and medial cruciate ligaments (ACLs and MCLs, respectively) were lacerated, observed a significant increase in gene expression of type I procollagen after exposure to HBOT [61, 63]. However, although macroscopic healing was observed in MCL scar formation, none of the injured ACLs physically healed after the administration of HBOT, possibly due to its intra-articular and avascular nature [62]. Gene expression of MMP was found to be significantly higher post HBOT, with the expression of tissue inhibitors of metalloproteinases (TIMPs) being higher in ACLs in the HBOT group compared to the control group [62, 63]. In a recent study, with HBOT being used as an adjunct therapy to ACL tendon graft, an increase in neovascularization in the tissue and an enhanced cohesion between tendon grafts and bone was observed, with tendon grafts showing a greater maximal pull-out strength with the use of HBOT post-surgical repair [64].

To date, limited human trials exist on the effectiveness of HBOT in ligament injuries with the results being inconclusive [65, 66]. However, in a non-randomized study, it was revealed that HBOT may reduce both the pain and accelerate the return to play in athletes with MCL injuries [67]. As discussed, basic and animal studies have justified the potential use of HBOT for treating injured ligaments and tendons, therefore designing human studies to precisely investigate its clinical effectiveness should be considered in further studies.

Fractures

Despite the contradictory findings, clinical investigations on the application of HBOT in fractures have been promising, especially in more complicated cases such as non-union fractures and fractures with a high risk of contamination and infection [68]. Discoveries such as an increment in bone mineralization, formation of hematoma and callus, callus nitrogen content, capacity for protein synthesis, alteration of the homeostatic environment, collagen formation, capillary budding, fibroblastic activity and proliferation, and osteogenesis highlight the potential role of HBOT in accelerating the recovery phase in fractures [23, 69].

Sport-Related Traumatic Brain Injury

With the incidence of sport-related traumatic brain injury (TBI) increasing globally, more scientists and physicians are paying attention to this potentially disabling issue [70]. Although in most cases spontaneous recovery from symptoms occurs within the initial days or weeks following the hit, evidence suggests that approximately one in four individuals with TBI experience chronic symptoms [71]. Symptoms include cognitive impairment, balance difficulties, headaches, fatigue, and dizziness as well as emotional disturbance and sleeping issues [72]. Risk of long-term health and cognitive difficulties, particularly in cases of repetitive injuries, has also been implicated [71, 72]. Although prevention is still the best method for reducing the rate of injury, finding effective therapeutic measures is also vital [73]. It has been suggested that HBOT may be a beneficial method in the improvement of TBI regardless of the severity and chronicity of the injury, with the most significant measurable improvements observed being in memory retention, attention, and executive function areas [70].

Apart from tissue oxygenation, other mechanisms, either cellular or vascular, have been associated with TBI treatment by HBOT. Such mechanisms include improvement in mitochondrial function, enhanced cellular metabolism, the reinforcing of the blood-brain-barrier, regulation of inflammatory reactions, reduction of apoptosis, relieving oxidative stress, increased levels of neurotrophins and nitric oxide, and the upregulation of axonal guidance factors [74]. Further, HBOT may promote neurogenesis of endogenous neural stem cells [75, 76]. Since all of the above-mentioned mechanisms are oxygen-dependent, it can be argued that HBOT could effectively facilitate metabolic interactions by supplying the energy or compensating for the lack of oxygen required for the regeneration pathways.

Yuki Okamoto, a certified athletic trainer with a degree in Sports Performance, is a San Diego based provider and the president of Oxybarix—San Diego's concierge HBOT.

The brain consumes 20% of the body's oxygen levels. If the primary mechanism in HBOT is hyperoxia, the brain is going to directly benefit from this therapy. Traumatic brain injuries, strokes, etc.

Physical Performance Enhancement

An increase in maximal aerobic fitness is associated with an increase in skeletal muscle mitochondrial quality and quantity [77, 78]. The intermittent use of HBOT has been observed to boost both the oxidative and glycolytic capacities and also the expression of proteins involved in the synthesis of mitochondria in skeletal muscles, thus, suggesting that HBOT may be effective in enhancing endurance performance [79].

A randomized controlled study observed that the use of HBOT improved aerobic fitness in athletes [77]. Therefore, the intermittent use of HBOT with an exercise training regimen may result in an increase in physical fitness compared to training exercise alone [80].

Sleep Improvement

Sleep is both a substantial element of health and an integral part of the recovery process. Accumulating evidence suggests that by increasing the duration and quality of sleep, significant improvements are made in performance resulting in greater achievements by athletes [81]. In addition, improved sleep may decrease the risk of injury, potentially reducing the periods of absence from training sessions. By improving sleep, athletes can overcome numerous challenges such as compressed training and competition schedules, travel, and mental stress [82].

Despite a limited amount of evidence, a recent study reported that exposure to high concentrations of oxygen persuades the brain to remain in deep sleep [83]. With the use of animal models, researchers first induced natural sleep, and then administered high levels of oxygen, measuring the brain activity. They observed that with a greater concentration of oxygen, the brain remains in a deactivated, slow-wave state [83]. Slow-wave sleep is the deepest stage of sleep, during which the brain waves fluctuate at a slow rhythm (mostly delta wave activity) and is reported to be the most desirable state for the proper recovery of the brain and the body [84]. Therefore, the use of HBOT could be a potential intervention for inducing slow-wave states aiding in the recovery of the athletes [83].

The administration of HBOT for the rehabilitation of the athletes could be a promising method aiding in their return from injury due to DOMS, muscle and ligament damages, fractures, and even traumatic brain injury.

- What is the role of HBOT in modulating the gut-brain axis and its potential impact on athletic performance and recovery?

Adverse Effects

Historically, HBOT has been reported to be safe, with complications being either limited, mild, or reversible [85]. The side effects of HBOT are due to the physiologic response to either the compressed and oxygen-rich setting or the patient's experience of being in the enclosed treatment chamber [86]. Below, the most common and adverse effects of HBOT are discussed with a brief description of each.

Barotrauma

Barotrauma is a term referring to an injury caused by the failure to balance the pressure between the air-compressed environment and any gas-filled cavities in the body. The most commonly affected area is the middle ear. The damage begins with the hyperemia of the tympanic membrane which can eventually lead to a tympanic membrane rupture. Sinus obstruction caused by a polyp or inflammation can also lead to increased susceptibility to barotrauma by trapping air in the sinuses. The presence of any air pocket in teeth due to dental decay can cause severe pain for the patients. Moreover, a rare yet substantial complication is barotrauma to the lungs during depressurization which requires close attention at the end of each treatment session. Patients who hold their breath or have an existing pulmonary condition are at a greater risk since it could lead to pneumomediastinum, pneumothorax, or gas embolism [86, 87].

Oxygen Toxicity

Despite its beneficial effects, large amounts of oxygen can be toxic and even life-threatening. Oxygen toxicity is one of the complications of HBOT resulting from breathing oxygen at increased partial pressures. As mentioned, certain beneficial effects of HBOT are gained via controlled oxidative stress. Even though the hyperoxic state created by HBOT aids in increasing the antioxidant defenses, the harmful results of high levels of oxygen radicals are still possible [88].

Central nervous system oxygen toxicity may be caused by acute exposure to an oxidative environment affecting neurological function. Clinical manifestations include visual disturbances, nausea, hearing changes, anxiety, confusion, and dizziness. Furthermore, during HBOT, central nervous system oxygen toxicity can present with an oxygen toxicity seizure [86]. It is noteworthy that central nervous system associated grand mal seizures have been documented at a pressure of 3.0 absolute atmosphere [89].

It has been argued that one of the most effective solutions to prevent oxygen toxicity is to avoid long sessions of treatment and take periodic breaks, inspiring the 21%-oxygen-normal air [86].

Claustrophobia

Claustrophobia, defined as an irrational fear of confined spaces, affects between 15 and 37% of people worldwide, with 5–7% presenting severe reactions to claustrophobia. Psychological effects of claustrophobia can be in response to either the tubelike cylindrical mono-place chambers or even under hoods and face masks in the multi-place chambers causing some degree of confinement anxiety. Mild symptoms are easily managed with pre-intervention sedation allowing individuals to continue receiving daily HBOT. However, in more severe cases, referrals to a psychiatrist or a psychologist might be required to prescribe cognitive-behavioral therapy, relaxation exercises, or even long-term medication therapy [90].

By taking a proper history, providing detailed education, and the reassurance of the safety of HBOT, physicians may lessen the response to claustrophobia. These aforementioned steps are the most effective methods for anticipating episodes of claustrophobia helping to manage any consequences before receiving HBOT [91].

Ophthalmologic Complications

Although rare, two of the most common ophthalmologic side effects of HBOT are hyperoxic myopia and cataracts [92]. Myopia, a condition in which individuals can clearly see only objects near them, is a known side effect of prolonged and repetitive therapeutic sessions of HBOT. At 2.0 absolute atmosphere or greater, myopia was observed after several weeks. Moreover, the development of cataracts has been reported after prolonged treatment periods, usually about 150 sessions or more [92].

Although the mechanism for ocular changes has not been fully discovered, prolonged exposure to HBOT, followed by oxidative damage to the lens proteins, may account for these changes. Several proposed hypotheses include local hyperoxia as a direct effect of high partial pressures of oxygen in contact with the eye resulting in oxygen toxicity, as well as oxidative damage to the crystalline structure of the lens proteins resulting in a decrease in the optical density of the lens [86]. In addition, Glutathione (an antioxidant responsible for maintaining the transparency of lens' crystalline proteins), when oxidized can promote abnormal protein cross-linking, increasing the synthesis of insoluble proteins, resulting in an abnormal colorization leading to scatterings of the nuclear light [93, 94].

Pre-existing ophthalmologic or systemic conditions may also contribute to the formation of cataracts after HBOT. Thus, in patients who are likely to undergo HBOT in the long term, it is beneficial to consider a thorough examination by an ophthalmologist documenting the acuity, refraction, status of the lens, and most importantly, ophthalmoscopy. Again, taking periods of normal-air breathing breaks every 20–25 minutes can decrease the risk of oxygen toxicity. It has also been reported that using an oronasal mask instead of a hood to deliver oxygen in a multiplace chamber can reduce the amount of oxygen contacting the eye, resulting in a decrease in ocular complications of HBOT [95, 96]. However, it should be noted that HBOT-induced myopic shift is reversible after discontinuation of the treatment with a return to baseline within 3–6 weeks or as long as a year [95].

Blood Pressure

According to several reports, HBOT leads to a rise in both systolic and diastolic blood pressure affecting both hypertensive and non-hypertensive patients [86]. However, hypertension caused by HBOT is often mild. To date, there have been no reports of patients experiencing hypertensive urgency or emergency. The effects can be controlled with additional protective treatments. Among medications, calcium channel blockers and beta-blockers have been reported to have a worsening effect, while angiotensin-converting enzyme inhibitors have a protective effect on rising blood pressure [97]. Discussed mechanisms consist of increased systemic vascular resistance due to alpha receptor-induced and endothelin-1-induced vasoconstriction [41, 98].

Hypoglycemia

Studies revealed that HBOT stimulates the secretion of residual insulin in patients with diabetes and can also increase glucose consumption in the brain, potentially leading to hypoglycemia in diabetic patients. As a consequence, HBOT centers set minimum serum glucose levels before the initiation of the session, ranging between as low as 100 and as high as 150 mg/dL [99].

Since HBOT is mostly an adjunct therapy to routine care, it is important to pay attention to the glycemic control of these patients. Nevertheless, the decision to elevate the patient's blood glucose level before HBOT should be taken after considering the patient's previous documented levels of serum glucose on the day of treatment and consulting with the specialists in more complicated cases.

> Complications of HBOT are mostly limited, mild, or reversible including barotrauma, oxygen toxicity, blood pressure or blood glucose complications, etc.

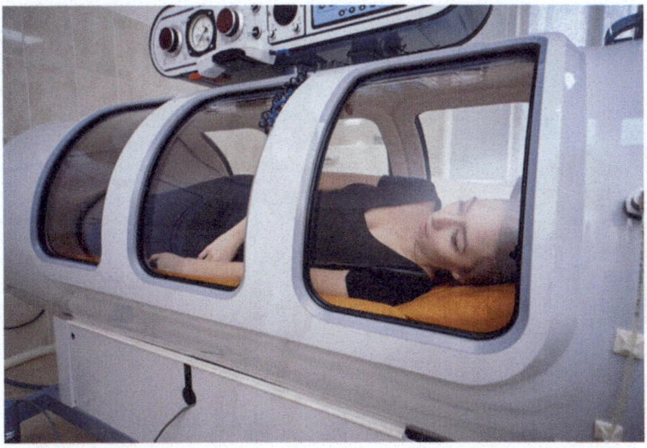

Fig. 30.1 Hard-shell Mono-place hyperbaric oxygen chamber

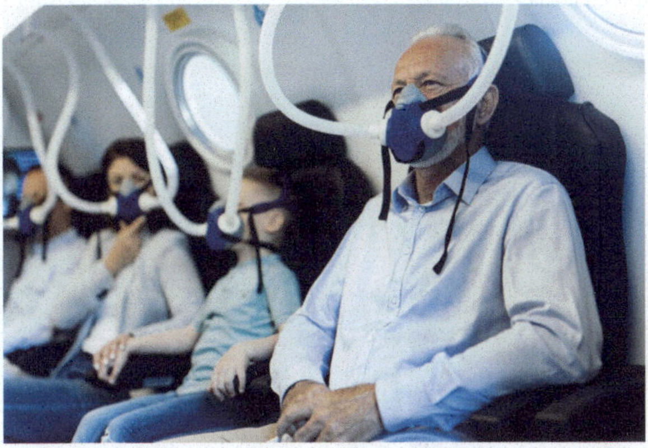

Fig. 30.2 Hard-shell multi-place hyperbaric oxygen chamber—Interior view with masks

Types of Hyperbaric Oxygen Chambers

Hard-Shell Hyperbaric Oxygen Chambers

Hard-shell chambers are steel-enclosed chambers that deliver 100% oxygen at atmospheric pressures above 2.0 absolute atmosphere. Installation of these chambers in a medical facility requires meeting various building regulations. Notably, these chambers are FDA-approved medical devices for the treatment of aforementioned conditions.

Currently, two types of hard-shell chambers are being used, the mono-place, and multi-place. The former, which is a cylinder tube, houses only one patient who slips into it (Fig. 30.1). Unlike the mono-place chamber, the latter is a room accommodating two or more individuals at once (Figs. 30.2, 30.3, and 30.4). Although the treatment protocol is mostly similar, treatment in the mono-place begins by slowly pressurizing the chamber with 100% oxygen once the patient has settled in.

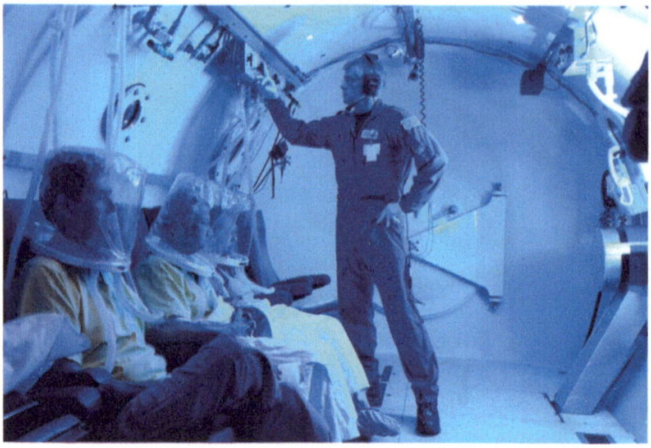

Fig. 30.3 Hard-shell multi-place hyperbaric oxygen chamber—Interior view with hoods—This chamber is located in the Institute for Exercise and Environmental Medicine (IEEM), Texas Health Presbyterian Hospital Dallas

Fig. 30.4 Hard-shell multi-place hyperbaric oxygen chamber—Exterior view—This unique chamber has 3 separate compartments which can be used for clinical or research purposes. Near side compartment is dedicated to clinical operations (see Fig. 30.3); far side compartment with white insulation is primarily for research; middle compartment is an independent lock allowing access to either compartment. This chamber is located in the Institute for Exercise and Environmental Medicine (IEEM), Texas Health Presbyterian Hospital Dallas

However, with the multi-place chamber patients breathe pure oxygen through masks or hoods.

Soft-Sided Hyperbaric Chambers

Soft-sided chambers, also known as mild HBOT which attracted public attention in the past few years, are portable chambers that seal with a zipper. They can reach pressure levels around 1.3 absolute atmosphere and typically deliver 30% oxygen levels for about 60 minutes. Although soft-sided chambers are not yet FDA-approved as a therapeutic device and they were originally designed only for divers and mountain climbers suffering from altitude sickness to use until they could be transported to a medical facility [100], some institutions are using mild HBOT as a common approach for athletic post-injury recovery.

Obviously, the notion behind HBOT is based on exposing the body to higher levels of oxygen. However, soft-sided chambers fail to offer much more than naturally aspirated air. Moreover, one of the applications of HBOT is for non-healing wounds as well as infections. Researches indicate that only at levels above 1.5 absolute atmosphere, oxygen can suppress bacterial and fungal growth. Unfortunately, with average levels of 1.3 absolute atmosphere, soft-sided chambers cannot suppress infection.

To our knowledge, no valid human studies have investigated the benefit of mild HBOT. Only scarce animal studies indicate that mild HBOT could improve oxidative metabolism in tissues without HBOT complications such as barotrauma and excessive production of reactive oxygen radicals. It showed that in rats, metabolic syndromes were inhibited by exposure to mile HBOT. It also showed improvement in the oxidative capacity of the skeletal muscles in age/injury-related processes. These findings might suggest that mile HOBT could be suitable for elderly and physically disabled patients, as well as injured athletes [101, 102]. Further studies are required to fully comprehend its efficacy and define a beneficial protocol for exposure to mild HBOT.

> Soft-sided chambers are portable chambers that seal with a zipper and can reach pressure levels around 1.3 absolute atmosphere delivering 30% oxygen levels for about an hour. Although they are not yet FDA-approved as a therapeutic device some institutions are using it as a common approach for athletic post-injury recovery.

Future Areas

Application

As discussed, due to its mechanism of action, the application of HBOT will continue to grow further, encompassing a broader spectrum of pathologies, from infections and inflammatory diseases to tissue survival and vascular complications. Presently, HBOT is also being used as an adjunctive remedy with other approved treatments like chemotherapy. The inclusion of HBOT in future clinical research could support the clinical management of multiple pathologies such as immunomodulatory properties and cancer. Recent researchers have studied and demonstrated significant improvements in nanomedicine approaches combined with HBOT [103]. Moreover, further research is underway examining its application for multiple sclerosis and its benefits in improving the quality of life in children with cerebral palsy [104, 105].

Sports Medicine

Despite considerable advances in the application of HBOT for various medical conditions, the studies investigating its use for sports medicine are mostly at the animal model stages. The design of standard human trials aimed at evaluating the effectiveness of HBOT on recovering from conditions such as muscle strains and contusions, tendon and ligament injuries, and sport-related concussions would advance the knowledge and application of HBOT on humans, thereby, lessening the reliance on animal studies. Therefore, we encourage sports physicians and scientists to conduct reliable trials and pave the way for achieving the standard protocols for the application of HBOT in sports injury rehabilitation.

Safety

Finally, apart from finding novel applications, it is equally important to shed light on the potential adverse effects derived from HBOT, as well as potential contraindications against this therapy. Future research should be designed to develop accurate protocols to precisely determine the benefits and harms for individuals before receiving HBOT. We cherish the notion of extending the possible applications of HBOT in further studies while always considering the benefits and risks it brings for the patients.

- What is the potential of Artificial Intelligence in developing adaptive HBOT protocols that adjust in real-time based on physiological feedback from wearable sensors during treatment?
- Can models be developed to analyse athlete-reported outcomes and experiences with HBOT to improve treatment protocols?

References

1. Moghadam N, Hieda M, Ramey L, Levine BD, Guilliod R. Hyperbaric oxygen therapy in sports musculoskeletal injuries. Med Sci Sports Exerc. 2020;52(6):1420–6. https://doi.org/10.1249/MSS.0000000000002257.
2. Leach RM, Rees PJ, Wilmshurst P. Hyperbaric oxygen therapy. BMJ. 1998;317(7166):1140–3.
3. Davis JC. Hyperbaric oxygen therapy. J Intensive Care Med. 1989;4(2):55–7.

4. Kumar MA, Radhika B, Gollamudi N, Reddy SP, Yaga US. Hyperbaric oxygen therapy-a novel treatment modality in oral submucous fibrosis: a review. J Clin Diagn Res. 2015;9(5):ZE01–ZE4. https://doi.org/10.7860/JCDR/2015/11500.5905.
5. Staples J, Clement D. Hyperbaric oxygen chambers and the treatment of sports injuries. Sports Med. 1996;22:219–27.
6. Brown RB, Sands M. Infectious disease indications for hyperbaric oxygen therapy. Compr Ther. 1995;21:663–77.
7. Reinke JM, Sorg H. Wound repair and regeneration. Eur Surg Res. 2012;49(1):35–43. https://doi.org/10.1159/000339613.
8. Schreml S, Szeimies RM, Prantl L, Karrer S, Landthaler M, Babilas P. Oxygen in acute and chronic wound healing. Br J Dermatol. 2010;163(2):257–68. https://doi.org/10.1111/j.1365-2133.2010.09804.x.
9. Hunt TK, Aslam R, Hussain Z, Beckert S. Lactate, with oxygen, incites angiogenesis. Adv Exp Med Biol. 2008;614:73–80. https://doi.org/10.1007/978-0-387-74911-2_9.
10. Arroyo AG, Iruela-Arispe ML. Extracellular matrix, inflammation, and the angiogenic response. Cardiovasc Res. 2010;86(2):226–35. https://doi.org/10.1093/cvr/cvq049.
11. Van Damme N, Van Hecke A, Remue E, et al. Physiological processes of inflammation and edema initiated by sustained mechanical loading in subcutaneous tissues: a scoping review. Wound Repair Regen. 2020;28(2):242–65. https://doi.org/10.1111/wrr.12777.
12. Pittman RN. Oxygen transport in the microcirculation and its regulation. Microcirculation. 2013;20(2):117–37. https://doi.org/10.1111/micc.12017.
13. Knighton DR, Fiegel VD, Halverson T, et al. Oxygen as an antibiotic. Arch Surg. 1990;12:97–100.
14. Fitzpatrick DT. Hyperbaric oxygen therapy. Mo Med. 1994;91:684–9.
15. Thom SR. Hyperbaric oxygen: its mechanisms and efficacy. Plast Reconstr Surg. 2011;127(Suppl 1):131S–41S.
16. LaVan FB, Hunt TK. Oxygen and wound healing. Clin Plast Surg. 1990;17(3):463–72.
17. Choudhury R. Hypoxia and hyperbaric oxygen therapy: a review. Int J Gen Med. 2018;11:431–42. Published 2018 Nov 20. https://doi.org/10.2147/IJGM.S172460.
18. Hajhosseini B, Kuehlmann BA, Bonham CA, Kamperman KJ, Gurtner GC. Hyperbaric oxygen therapy: descriptive review of the technology and current application in chronic wounds. Plast Reconstr Surg Glob Open. 2020;8(9):e3136. Published 2020 Sep 25. https://doi.org/10.1097/GOX.0000000000003136.
19. Bird AD, Telfer AB. Effect of hyperbaric oxygenation on limb circulation. Lancet. 1965;I:355–6.
20. Van Meter KW. The effect of hyperbaric oxygen on severe anemia. Undersea Hyperb Med. 2012;39(5):937–42.
21. Unfirer S, Kibel A, Drenjancevic-Peric I. The effect of hyperbaric oxygen therapy on blood vessel function in diabetes mellitus. Med Hypotheses. 2008;71(5):776–80. https://doi.org/10.1016/j.mehy.2008.06.016.
22. Lee P, Chandel NS, Simon MC. Cellular adaptation to hypoxia through hypoxia inducible factors and beyond. Nat Rev Mol Cell Biol. 2020;21(5):268–83. https://doi.org/10.1038/s41580-020-0227-y.
23. Růžička J, Grajciarová M, Vištejnová L, et al. Hyperbaric oxygen enhances collagen III formation in wound of ZDF rat. Physiol Res. 2021;70(5):787–98. https://doi.org/10.33549/physiolres.934684.
24. Camporesi EM, Bosco G. Mechanisms of action of hyperbaric oxygen therapy. Undersea Hyperb Med. 2014;41(3):247–52.
25. Baboir BM. Oxygen dependent microbial killing by phagocytes. N Engl J Med. 1974;298(659–68):721–6.
26. Iyer G, Islam M, Quastel J. Biochemical aspects of phagocytosis. Nature. 1961;192:535–41. https://doi.org/10.1038/192535a0.

27. Sureda A, Batle JM, Martorell M, et al. Antioxidant response of chronic wounds to hyperbaric oxygen therapy. PLoS One. 2016;11(9):e0163371.
28. Guilliod RR, Pompeo MQ. Discussion: an update on the appropriate role for hyperbaric oxygen: indications and evidence. Plast Reconstr Surg. 2016;138(3 Suppl):117S–8S.
29. Sunkari VG, Lind F, Botusan IR, et al. Hyperbaric oxygen therapy activates hypoxiainducible factor 1 (HIF-1), which contributes to improved wound healing in diabetic mice. Wound Repair Regen. 2015;23(1):98–103.
30. Benincasa JC, de Freitas Filho LH, Carneiro GD, et al. Hyperbaric oxygen affects endothelial progenitor cells proliferation in vitro. Cell Biol Int. 2019;43(2):136–46.
31. Yang Y, Wei H, Zhou X, Zhang F, Wang C. Hyperbaric oxygen promotes neural stem cell proliferation by activating vascular endothelial growth factor/extracellular signal regulated kinase signaling after traumatic brain injury. Neuroreport. 2017;28(18):1232–8.
32. Ortega MA, Fraile-Martinez O, García-Montero C, et al. A general overview on the hyperbaric oxygen therapy: applications, mechanisms and translational opportunities. Medicina (Kaunas). 2021;57(9):864. Published 2021 Aug 24. https://doi.org/10.3390/medicina57090864.
33. Memar MY, Yekani M, Alizadeh N, Baghi HB. Hyperbaric oxygen therapy: antimicrobial mechanisms and clinical application for infections. Biomed Pharmacother. 2019;109:440–7. https://doi.org/10.1016/j.biopha.2018.10.142.
34. Resanović I, Zarić B, Radovanović J, et al. Hyperbaric oxygen therapy and vascular complications in diabetes mellitus. Angiology. 2020;71(10):876–85. https://doi.org/10.1177/0003319720936925.
35. Somaa F. A review of the application of hyperbaric oxygen therapy in Alzheimer's disease. J Alzheimers Dis. 2021;81(4):1361–7. https://doi.org/10.3233/JAD-210157.
36. Oliaei S, Seyed Alinaghi S, Mehrtak M, et al. The effects of hyperbaric oxygen therapy (HBOT) on coronavirus disease-2019 (COVID-19): a systematic review. Eur J Med Res. 2021;26:96. https://doi.org/10.1186/s40001-021-00570-2.
37. DuBose KJ, Cooper JS. Hyperbaric patient selection [Updated 2022 Feb 7]. In: StatPearls [Internet]. Treasure Island: StatPearls Publishing; 2022.
38. Gawdi R, Cooper JS. Hyperbaric contraindications; StatPearls publishing: Treasure Island. USA: FL; 2020.
39. Arslan A. Hyperbaric oxygen therapy in carbon monoxide poisoning in pregnancy: maternal and Fetal outcome. Am J Emerg Med. 2021;43:41–5.
40. Stevens SL, Narr AJ, Claus PL, et al. The incidence of hypoglycemia during HBO2 therapy: a retrospective review. Undersea Hyperb Med. 2015;42:191–6.
41. Al-Waili NS, Butler GJ, Beale J, et al. Influences of hyperbaric oxygen on blood pressure, heart rate and blood glucose levels in patients with diabetes mellitus and hypertension. Arch Med Res. 2006;37:991–7.
42. Weaver LK, Churchill S. Pulmonary edema associated with hyperbaric oxygen therapy. Chest. 2001;120:1407–9.
43. Kahle AC, Cooper JS. Hyperbaric physiological and pharmacological effects gases. Treasure Island: StatPearls Publishing; 2019.
44. Chen CY, Chou WY, Ko JY, Lee MS, Wu RW. Early recovery of exercise-related muscular injury by HBOT. Biomed Res Int. 2019;2019:6289380. Published 2019 May 29. https://doi.org/10.1155/2019/6289380.
45. Yu JG, Liu JX, Carlsson L, Thornell LE, Stål PS. Re-evaluation of sarcolemma injury and muscle swelling in human skeletal muscles after eccentric exercise. PLoS One. 2013;8(4):e62056. Published 2013 Apr 15. https://doi.org/10.1371/journal.pone.0062056.
46. Staples JR, Clement DB, Taunton JE, McKenzie DC. Effects of hyperbaric oxygen on a human model of injury. Am J Sports Med. 1999;27(5):600–5.
47. Webster AL, Syrotuik DG, Bell GJ, Jones RL, Hanstock CC. Effects of hyperbaric oxygen on recovery from exercise-induced muscle damage in humans. Clin J Sport Med. 2002;12(3):139–50.

48. Mekjavic IB, Exner JA, Tesch PA, Eiken O. Hyperbaric oxygen therapy does not affect recovery from delayed onset muscle soreness. Med Sci Sports Exerc. 2000;32(3):558–63.
49. Bennett M, Best TM, Babul S, Taunton J, Lepawsky M. Hyperbaric oxygen therapy for delayed onset muscle soreness and closed soft tissue injury. Cochrane Database Syst Rev. 2005;(4):CD004713.
50. Germain G, Delaney J, Moore G, Lee P, Lacroix V, Montgomery D. Effect of hyperbaric oxygen therapy on exercise-induced muscle soreness. Undersea Hyperb Med. 2003;30(2):135–45.
51. Babul S, Rhodes EC, Taunton JE, Lepawsky M. Effects of intermittent exposure to hyperbaric oxygen for the treatment of an acute soft tissue injury. Clin J Sport Med. 2003;13(3):138–47.
52. Bennett MH. Hyperbaric medicine and the placebo effect. Diving Hyperb Med. 2014;44(4):235–40.
53. Sueblinvong T, Egtasaeng N, Sanguangrangsirikul S. Hyperbaric oxygenation and blood lactate clearance: study in sixty male naval cadets. J Med Assoc Thail. 2004;87(Suppl 2):S218–22.
54. Brancaccio P, Lippi G, Maffulli N. Biochemical markers of muscular damage. Clin Chem Lab Med. 2010;48(6):757–67.
55. Bayer ML, Magnusson SP, Kjaer M, Tendon Research Group B. Early versus delayed rehabilitation after acute muscle injury. N Engl J Med. 2017;377(13):1300–1.
56. Arnold L, Henry A, Poron F, et al. Inflammatory monocytes recruited after skeletal muscle injury switch into antiinflammatory macrophages to support myogenesis. J Exp Med. 2007;204(5):1057–69. https://doi.org/10.1084/jem.20070075.
57. Oyaizu T, Enomoto M, Yamamoto N, et al. Hyperbaric oxygen reduces inflammation, oxygenates injured muscle, and regenerates skeletal muscle via macrophage and satellite cell activation. Sci Rep. 2018;8(1):1288.
58. Chaillou T, Lanner JT. Regulation of myogenesis and skeletal muscle regeneration: effects of oxygen levels on satellite cell activity. FASEB J. 2016;30(12):3929–41.
59. Horie M, Enomoto M, Shimoda M, Okawa A, Miyakawa S, Yagishita K. Enhancement of satellite cell differentiation and functional recovery in injured skeletal muscle by hyperbaric oxygen treatment. J Appl Physiol (1985). 2014;116(2):149–55.
60. Horn PC, Webster DA, Amin HM, Mascia MF, Werner FW, Fortino MD. The effect of hyperbaric oxygen on medial collateral ligament healing in a rat model. Clin Orthop Relat Res. 1999;360:238–42.
61. Ishii Y, Ushida T, Tateishi T, Shimojo H, Miyanaga Y. Effects of different exposures of hyperbaric oxygen on ligament healing in rats. J Orthop Res. 2002;20(2):353–6.
62. Takeyama N, Sakai H, Ohtake H, Mashitori H, Tamai K, Saotome K. Effects of hyperbaric oxygen on gene expressions of procollagen, matrix metalloproteinase and tissue inhibitor of metalloproteinase in injured medial collateral ligament and anterior cruciate ligament. Knee Surg Sports Traumatol Arthrosc. 2007;15(4):443–52.
63. Mashitori H, Sakai H, Koibuchi N, et al. Effect of hyperbaric oxygen on the ligament healing process in rats. Clin Orthop Relat Res. 2004;423:268–74.
64. Yeh WL, Lin SS, Yuan LJ, Lee KF, Lee MY, Ueng SW. Effects of hyperbaric oxygen treatment on tendon graft and tendon-bone integration in bone tunnel: biochemical and histological analysis in rabbits. J Orthop Res. 2007;25(5):636–45.
65. Soolsma SJ. The effect of intermittent hyperbaric oxygen on short term recovery from grade II medial collateral ligament injuries [dissertation]. Vancouver, CA: University of British Columbia; 1996. p. 125.
66. Borromeo CN, Ryan JL, Marchetto PA, Peterson R, Bove AA. Hyperbaric oxygen therapy for acute ankle sprains. Am J Sports Med. 1997;25(5):619–25.
67. Yagishita K, Enomoto M, Takazawa Y, Fukuda J, Koga H. Effects of hyperbaric oxygen therapy on recovery acceleration in Japanese professional or semi-professional rugby players with grade 2 medial collateral ligament injury of the knee: a comparative non-randomized study. Undersea Hyperb Med. 2019;46(5):647–54.

68. Bennett MH, Stanford RE, Turner R. Hyperbaric oxygen therapy for promoting fracture healing and treating fracture non-union. Cochrane Database Syst Rev. 2012;11(11):CD004712. Published 2012 Nov 14. https://doi.org/10.1002/14651858.CD004712.pub4.
69. Al Hadi H, Smerdon GR, Fox SW. Hyperbaric oxygen therapy accelerates osteoblast differentiation and promotes bone formation. J Dent. 2015;43(3):382–8. https://doi.org/10.1016/j.jdent.2014.10.006.
70. Theadom A, Mahon S, Hume P, et al. Incidence of sports-related traumatic brain injury of all severities: a systematic review. Neuroepidemiology. 2020;54(2):192–9. https://doi.org/10.1159/000505424.
71. Boussi-Gross R, Golan H, Fishlev G, et al. Hyperbaric oxygen therapy can improve post concussion syndrome years after mild traumatic brain injury – randomized prospective trial. PLoS One. 2013;8(11):e79995. Published 2013 Nov 15. https://doi.org/10.1371/journal.pone.0079995.
72. McCrory P, Meeuwisse W, Aubry M, Cantu B, Dvořák J, Echemendia R, et al. Consensus statement on concussion in sport—the 4th international conference on concussion in sport held in Zurich, November 2012. J Sci Med Sport. 2013;16(3):178–89.
73. Manley G, Gardner AJ, Schneider KJ, Guskiewicz KM, Bailes J, Cantu RC, et al. A systematic review of potential long-term effects of sport-related concussion. Br J Sports Med. 2017;51(12):969–77.
74. Ahmadi F, Khalatbary AR. A review on the neuroprotective effects of hyperbaric oxygen therapy. Med Gas Res. 2021;11(2):72–82. https://doi.org/10.4103/2045-9912.311498.
75. Tal S, Hadanny A, Sasson E, Suzin G, Efrati S. Hyperbaric oxygen therapy can induce angiogenesis and regeneration of nerve fibers in traumatic brain injury patients. Front Hum Neurosci. 2017;11:508. Published 2017 Oct 19. https://doi.org/10.3389/fnhum.2017.00508.
76. Yang YJ, Wang XL, Yu XH, Wang X, Xie M, Liu CT. Hyperbaric oxygen induces endogenous neural stem cells to proliferate and differentiate in hypoxic-ischemic brain damage in neonatal rats. Undersea Hyperb Med. 2008;35(2):113–29.
77. Hadanny A, Hachmo Y, Rozali D, et al. Effects of hyperbaric oxygen therapy on mitochondrial respiration and physical performance in middle-aged athletes: a blinded, randomized controlled trial. Sports Med Open. 2022;8(1):22. Published 2022 Feb 8. https://doi.org/10.1186/s40798-021-00403-w.
78. Hoppeler H, Howald H, Conley K, Lindstedt SL, Claassen H, Vock P, et al. Endurance training in humans: aerobic capacity and structure of skeletal muscle. J Appl Physiol. 1985;59(2):320–7. https://doi.org/10.1152/jappl.1985.59.2.320.
79. Suzuki J. Endurance performance is enhanced by intermittent hyperbaric exposure via up-regulation of proteins involved in mitochondrial biogenesis in mice. Physiol Rep. 2017;5(15):e13349.
80. DeCato TW, Bradley SM, Wilson EL, Harlan NP, Villela MA, Weaver LK, et al. Effects of sprint interval training on cardiorespiratory fitness while in a hyperbaric oxygen environment. Undersea Hyperb Med. 2019;46(2):117–24. https://doi.org/10.22462/04.06.2019.4.
81. Hamlin MJ, Deuchrass RW, Olsen PD, et al. The effect of sleep quality and quantity on athlete's health and perceived training quality. Front Sports Act Living. 2021;3:705650. Published 2021 Sep 10. https://doi.org/10.3389/fspor.2021.705650.
82. Watson AM. Sleep and athletic performance. Curr Sports Med Rep. 2017;16(6):413–8. https://doi.org/10.1249/JSR.0000000000000418.
83. Hauer BE, Negash B, Chan K, et al. Hyperoxia enhances slow-wave forebrain states in urethane-anesthetized and naturally sleeping rats. J Neurophysiol. 2018;120(4):1505–15. https://doi.org/10.1152/jn.00373.2018.
84. Walsh JK. Enhancement of slow wave sleep: implications for insomnia. J Clin Sleep Med. 2009;5(2 Suppl):S27–32.
85. Opasanon S, Pongsapich W, Taweepraditpol S, Suktitipat B, Chuangsuwanich A. Clinical effectiveness of hyperbaric oxygen therapy in complex wounds. J Am Coll Clin Wound Spec. 2015;6(1–2):9–13. Published 2015 Apr 9. https://doi.org/10.1016/j.jccw.2015.03.003.

86. Heyboer M 3rd, Sharma D, Santiago W, McCulloch N. Hyperbaric oxygen therapy: side effects defined and quantified. Adv Wound Care (New Rochelle). 2017;6(6):210–24. https://doi.org/10.1089/wound.2016.0718.
87. Nasole E, Zanon V, Marcolin P, Bosco G. Middle ear barotrauma during hyperbaric oxygen therapy; a review of occurrences in 5,962 patients. Undersea Hyperb Med. 2019;46(2):101–6.
88. Chawla A, Lavania AK. Oxygen toxicity. Med J Armed Forces India. 2001;57(2):131–3.
89. Ciarlone GE, Hinojo CM, Stavitzski NM, Dean JB. CNS function and dysfunction during exposure to hyperbaric oxygen in operational and clinical settings. Redox Biol. 2019;27:101159. https://doi.org/10.1016/j.redox.2019.101159.
90. Liu WW, Liu K, Mao D, Xu WG, Tao HY. Claustrophobia during routine hyperbaric oxygen treatment. Undersea Hyperb Med. 2012;39(1):619–20.
91. Camporesi EM. Side effects of hyperbaric oxygen therapy. Undersea Hyperb Med. 2014;41:253–7.
92. Gengel KC, Hendriksen S, Cooper JS. Hyperbaric related myopia and cataract formation. In: StatPearls. Treasure Island: StatPearls Publishing; 2022.
93. Gesell LB, Trott A. De novo cataract development following a standard course of hyperbaric oxygen therapy. Undersea Hyperb Med. 2007;34:389–92.
94. Weikel KA, Garber C, Baburins A, Taylor A. Nutritional modulation of cataract. Nutr Rev. 2014;72:30–47.
95. Bennett MH, Hui CF, See HG, Au-Yeung KL, Tan C, Watson S. The myopic shift associated with hyperbaric oxygen administration is reduced when using a mask delivery system compared to a hood – a randomised controlled trial. Diving Hyperb Med. 2019;49(4):245–52. https://doi.org/10.28920/dhm49.4.245-252.
96. Evanger K, Haugen OH, Irgens A, Aanderud L, Thorsen E. Ocular refractive changes in patients receiving hyperbaric oxygen administered by oronasal mask or hood. Acta Ophthalmol Scand. 2004;82:449–53.
97. Heyboer M III, Smith G, Santiago W, Wojcik SM. Effect of hyperbaric oxygen therapy on blood pressure in patients undergoing treatment. Undersea Hyperb Med. 2017;44(2):93–9.
98. Abel FL, Mcnamee JE, Cone DL, Clarke D, Tao J. Effects of hyperbaric oxygen on ventricular performance, pulmonary blood volume, and systemic and pulmonary vascular resistance. Undersea Hyperb Med. 2000;27:67–73.
99. Stevens SL, Narr AJ, Claus PL, et al. The incidence of hypoglycemia during HBO therapy: a retrospective review. Undersea Hyperb Med. 2015;42:191–6.
100. Burman F. Low-pressure fabric hyperbaric chambers. S Afr Med J. 2019;109(4):12574. Published 2019 Mar 29. https://doi.org/10.7196/SAMJ.2019.v109i4.13524.
101. Ishihara A. Mild hyperbaric oxygen: mechanisms and effects. J Physiol Sci. 2019;69:573–80. https://doi.org/10.1007/s12576-019-00678-5.
102. Park SH, Park SJ, Shin MS, et al. The effects of low-pressure hyperbaric oxygen treatment before and after maximal exercise on lactate concentration, heart rate recovery, and antioxidant capacity. J Exerc Rehabil. 2018;14:980–4.
103. Wang X, Li S, Liu X, Wu X, Ye N, Yang X, Li Z. Boosting nanomedicine efficacy with hyperbaric oxygen therapy. In: Advances in experimental medicine and biology, vol. 1295. Berlin/Heidelberg: Springer; 2021. p. 77–95.
104. Bennett M, Heard R. Hyperbaric oxygen therapy for multiple sclerosis. CNS Neurosci Ther. 2010;16(2):115–24. https://doi.org/10.1111/j.1755-5949.2009.00129.x.
105. Zhang Y, Wu J, Xiao N, Li B. Hyperbaric oxygen therapy is beneficial for the improvement of clinical symptoms of cerebral palsy: a systematic review and meta-analysis. Complement Med Res. 2022;29(2):158–71. https://doi.org/10.1159/000518785.

Last Word: Challenges and Opportunities Ahead

> *You cannot hope to build a better world without improving the individuals. To that end, each of us must work for his own improvement and, at the same time, share a general responsibility for all humanity, our particular duty being to aid those to whom we think we can be most useful.*
>
> **–Marie Curie, Autobiographical notes, 1923, pp. 167–168**

In keeping with the concept described by Marie Curie in the above quote, the editors of this book had an ongoing exchange of ideas and thoughts during the pandemic, about muscle development and adaptation concerning various paradigms of exercise training, the role played by different types and patterns of recovery and regeneration, and the effect of different modalities and approaches to injury management. The discussions advanced to topics concerning training methods, the roles of nutrition and supplementation, mindfulness, sleep, muscle injury, treatment, and recovery and regeneration.

The conversations were far-reaching and sometimes contentious as our backgrounds and experience were widely varied. However, it was clear that the goal was to understand as much as possible about physical training, the limits to adaptation, the critical role of rest in the recovery and adaptation processes, the effect of past and current injuries on training, and the application and efficacy of various treatments and modalities on recovery from injury. Fundamental to the discussion was an examination of the latest published developments in basic science in muscle biology, and their importance to coaching and therapy practice. In our interchange concerning the significance of recent basic science research on the sequelae of various injuries on the short and long term recovery of muscle function to injury, it became clear that there were gaps in our collective knowledge, which hampered our ability to translate these findings into application.

N. C. Apostolopoulos et al. (eds.), *Fundamentals of Recovery, Regeneration, and Adaptation to Exercise Stress: An Integrated Approach*,
https://doi.org/10.1007/978-3-031-44270-4

Therefore, the major goals for developing this book were three-fold: (a) to provide an update on the latest developments in several areas of basic and applied exercise science that would be relevant to coaches, therapists, and practitioners involved in training athletes and promoting health through activity, and in treating individuals with activity related injuries and other movement problems; (b) to present technical and descriptive information concerning various therapy modalities and protocols currently in use, and (c) to develop an ongoing discourse and exchange of views and ideas between researchers, therapists, and other sport and fitness practitioners and professionals.

Chapters 2, 3, 4, 5, 6, 7, 8, 9, 10, 11, 12, 13, 14, 15, 16, and 17 of this book are devoted to presenting and exploring current research pertaining to muscle function, adaptation to exercise training, and information on the aetiology of muscle injury and the sequelae to significant muscle damage. Chapters 18, 19, 20, 21, 22, 23, 24, 25, 26, 27, 28, 29, and 30 provide the latest information on various therapeutic modalities and treatment protocols in terms of their use in promoting recovery and regeneration.

However, while the various chapters outline the current "state of the art" in various areas, research continues, and new thoughts on basic science, training, and therapeutic techniques and protocols are being advanced on a daily basis. For example, a search with Web of Science for the first 6 months of 2024, using the search term, "muscle damage" yielded 2697 publications, and another 2343 for the search term, "muscle injury and therapy", i.e., nearly 200 new articles per week! Obviously, this represents a considerable volume of material to read, decipher, absorb, and implement, whether it be aimed at new research, training for performance, or developing therapeutic techniques and protocols. And this material represents the output for just 6 months! Clearly, the ongoing task of keeping up with, digesting, and implementing the information from the new developments in research—whether they be basic or applied in nature, is a daunting task.

As stated in the Introduction, effective intervention strategies aimed at improving current practice and crafting future developments in addressing an injury or movement issue are best developed through an effective interaction between basic scientists and therapy practitioners via an iterative process in which therapists ask critical questions concerning their observations pertaining to treatment protocols, equipment, and injury responses to treatment.

The other "arm" of the circular communication process described in the Introduction (see Fig. 1.1, p. 7) highlights the importance of investigators distilling and communicating the critical knowledge concerning the meaning of their research in terms of relevance to therapists' issues and questions, and their potential practical application and usefulness in developing new techniques, protocols, and equipment for therapeutic uses. The distillation of research findings is integral, not only to advancing our understanding of how muscle functions and adapts to various exercise stimuli and perturbation of the cellular environment, but also how muscle responds to minor insults and more significant injuries in terms of recovery and regeneration.

The sheer volume of publications, and the issues presented by delays in research findings being "translated" for application to current practice or development of new advancements in technological innovation, underlines the need for ongoing, regular interaction between researchers and practitioners. We have attempted to begin the exchange process by requesting that authors of chapters on "basic" and "applied" muscle/exercise research include in their chapters at strategic points within their chapters, clear statements of "***Key Findings***" from the research (in light blue colour), and secondly, to develop "***Take-home Messages***" (in light green colour) in the form of boxes meant to explain the research in terms of potential application to therapy.

Similarly, and to complete the exchange, we requested that the authors of chapters on the various therapy modalities, techniques, and protocols to provide boxes that outlined the significant "***Technical/Practical Issues***" that practitioners have with the technique, modality, or protocol (in light blue colour), and in addition, boxes that present "***Research Questions***" that the authors felt would be beneficial to them in their use/development of the modality, technique, or protocol (in light green colour).

This strategy represents a start, but only a start, to creating an ongoing dialogue. However, it does show the value and potential of such regular exchanges of information, views, and ideas, but more importantly, it accentuates the possibilities and rewards that could develop for both groups—if some form of regular communications strategy was to be developed. Unfortunately, it also highlights the challenges ahead, and how much is yet to be done. What follows is some potential strategies that would be a start to the development opportunities for knowledge exchange between researchers and practitioners:

- organize a forum or symposium at conferences, or better yet, a conference devoted to the exchange of thoughts, ideas, and observations between scientists and practitioners.
- encourage journals that publish therapy and practitioner oriented papers to explore adding a section to their journal that includes the practical implications and potential application of basic and exercise science papers for assimilation by practitioners.
- utilize individuals with backgrounds in basic and exercise science to act as "translators" for practice oriented journals and/or develop publications that distill the essence of basic and exercise research for future incorporation in practice.

The opportunities and the potential rewards are there for those who seek to build the necessary bridges, and persist in creating vehicles that promote these conversations.

> *One of the chief inhibitions to human progress arises because of the extreme slowness with which the advances in knowledge become translated into action for the benefit of society as a whole. There is no step more important for the removal of that inhibition than that of providing for intimate contact between the leaders in the fields of pure and applied science.*
>
> **–Robert A. Millikan, The Autobiography of Robert Millikan, 1950**

Index

A

Active hyperaemia, 133, 134, 136
Acupressure, 524, 534
Acupuncture, 4, 534, 555–566, 704
Adaptation, 1, 20, 42, 76, 105, 165, 197, 216, 241, 304, 318, 352, 407, 429, 446, 479, 523, 566, 599, 637, 655, 688, 726
Adenosine triphosphate (ATP), 113–115, 117, 118, 132–135, 285, 319, 326, 327, 336, 418, 640, 650, 658, 663, 680, 682–687, 689, 690
Aerobic-anaerobic, 17, 118, 119, 319, 333, 350, 353, 692, 724, 726–727
Aging, 109, 242, 261
Analgesic, 3, 427–439, 556, 565, 657, 690, 715, 716, 718, 723
Anti-inflammatory, 3, 19–22, 28–30, 215–217, 219, 221–223, 227, 413, 427–439, 465, 649, 656, 664, 693, 722–724, 738, 748, 754
Athletic performance, 203–205, 295, 304, 317, 353–355, 368, 407, 416, 429, 430, 436, 438, 448, 455–457, 461, 529, 530, 534, 536, 566, 582, 583, 585, 586, 605, 638, 650
Attention, 43, 45, 134, 187, 204, 289, 296, 312, 365, 369, 373, 375–383, 414, 482, 618, 619, 664, 677, 699, 718, 752, 756, 758, 760, 762
Autonomic nervous system (ANS), 42, 80, 116, 117, 303–305, 308, 558, 564, 601, 602, 611, 678, 693

B

Balance, 20, 45, 52, 54–57, 61–63, 75, 82, 95, 114, 167–169, 227, 290, 303, 307, 318, 324, 334, 392, 395, 410, 412, 459, 497, 504, 507, 509, 527, 534, 536, 537, 584, 624, 626, 640, 661, 662, 687, 688, 704, 756, 758
Barotrauma, 758, 763
Blood flow, 23, 24, 123–152, 336, 337, 351, 465, 502, 528–530, 533, 564, 565, 604, 605, 681, 686, 687, 689, 692, 698, 722, 723, 737, 738, 748, 749
Bony stress injury, 742

C

Capillary, 23, 24, 30, 124–129, 133, 135–141, 144, 146–152, 190, 192, 225, 466, 502, 747, 756
Cardiovascular system, 303, 657, 658
Cerebral palsy, 268, 269, 271
Chiropractic, 571–573, 580, 582–586, 627
Circadian rhythm, 353–355, 661, 707
Cognitive practices, 369
Cold-based therapy, 714
Cold laser, 675
Communication, 3, 5–6, 80, 132, 135, 136, 138, 150, 152, 371, 393, 394, 398, 578, 585, 614, 683, 706
Concussion, 377, 661–662, 697, 698, 764
Connective tissue, 17, 76, 86, 91, 141, 164, 166, 167, 169, 170, 192, 205, 267, 409, 431, 446, 450, 451, 454, 455,

N. C. Apostolopoulos et al. (eds.), *Fundamentals of Recovery, Regeneration, and Adaptation to Exercise Stress: An Integrated Approach*,
https://doi.org/10.1007/978-3-031-44270-4

465, 466, 481, 495, 496, 502, 503, 524, 528, 531, 561, 562, 601, 604, 606–608, 611, 617–626, 656
Contraction-induced injury, 142
Cortical, 41–45, 48–51, 57–61, 108, 197, 198, 289, 580, 690
Cryochamber, 717
Cryosauna, 716–718
Cupping, 524, 533, 534, 597

D

Damage, 2, 13, 42, 140, 188, 214, 254, 327, 392, 408, 431, 446, 481, 530, 560, 580, 605, 679, 713, 742, 746
Destruction-induced damage, 144
Dietary intervention, 395
Dry needling, 525, 526, 557–559

E

Eccentric contraction, 14–17, 21, 27, 30, 147, 186, 188, 191, 193–200, 203, 205, 217, 253, 254, 257, 266, 267, 271, 327, 447, 462, 492
Eccentric exercise, 16–18, 20–23, 26–28, 140–147, 185–206, 213, 217–222, 254, 267, 432, 433, 454, 460, 462, 624, 695
Electrical stimulation modalities, 4, 635–642
Electroacupuncture, 561–563, 566
Electromagnetic stimulation, 647
Electron transport chain, 115, 682, 687
Energetics, 113–117, 326, 327, 334, 653, 680, 688, 706
Energy expenditure, 113, 288
Exercise performance, 29, 347–357, 411, 414, 429–431
Extracorporeal shock wave therapy (ESWT), 737–742
Extraspinal manipulative therapy (EMT), 572, 582–586

F

Fascia, 17, 449, 450, 453–455, 465, 506, 527, 528, 531, 533, 544, 560, 597, 601, 602, 604, 606, 610, 615, 618–622, 624, 739
Fascicle length, 241, 261–266, 269–271, 450, 451, 453, 454, 456, 465, 467
Fatigue, 16, 17, 48, 51, 53–55, 57–61, 63, 85, 94, 95, 189, 201, 225, 283, 285, 288–292, 294–297, 305, 306, 308–310, 312, 322–324, 326–327, 330–332, 335–337, 340, 351, 353, 374, 381, 392, 394, 396, 399–400, 402, 407, 415, 418, 419, 430, 446, 448, 459–461, 523, 527, 530, 532, 533, 545, 584, 603, 605, 623–625, 627, 637, 640, 641, 650, 665, 676, 684, 687, 689, 693, 697, 706, 713, 720, 721, 723, 753, 756
Flexibility, 117, 194, 431, 445, 448, 449, 451, 453, 456–459, 461, 463, 523, 524, 528, 530, 534–536, 544, 583, 618, 626, 656, 678, 724, 728
Flow, 94, 123, 130, 131, 133, 134, 137, 138, 143, 144, 147, 149, 152, 367, 368, 370, 378, 545, 555, 606, 662, 704, 722
Force, 1, 13, 50, 76, 106, 124, 164, 185, 214, 236, 319, 354, 447, 481, 527, 560, 573, 600, 636, 660, 727, 738, 753

G

Golgi tendon organ, 50, 57, 77, 78, 84–85, 87–95, 448, 487–489, 491, 497, 560, 577

H

Homeostasis, 2, 15, 17, 20, 24, 26, 76, 115, 141, 168–169, 216, 284, 288, 306, 324, 326, 335, 351, 354, 479, 481, 490, 495, 503, 533, 544, 545, 555, 656, 658, 659, 680, 719
Hormones, 26–31, 106–108, 110, 111, 113, 117, 165, 328, 331, 338, 412, 605, 637
Hyperbaric oxygenation, 745–764
Hyperbaric oxygen therapy (HBOT), 4, 745–764

I

Immobilization, 62, 187, 199, 221, 242–253, 258, 260, 269, 453, 454, 622
Individualization, 318, 332–333
Inflammation, 3, 18–20, 25, 26, 28–30, 51, 52, 60, 80, 141, 145, 189–190, 192, 201, 204, 213–227, 288, 329, 331, 337, 353, 372, 411, 427, 429, 431–434, 446, 450, 454, 462, 465, 481, 489, 494–497, 500, 503, 507,

523, 524, 527, 546, 560, 563–565, 601, 602, 606, 614, 618, 649, 656, 659, 661, 662, 665, 677, 684, 685, 690–692, 697, 698, 704–706, 713, 714, 720, 722, 738, 753, 754, 758
Injury risk, 203, 268, 417, 448, 457, 458, 467, 536, 655
Interdisciplinary approaches, 392–394, 398–399, 402
Intracellular signaling, 164, 168–169, 174, 178, 417

J
Joint receptors, 84, 95

L
Light therapy, 675, 677, 679, 681, 683, 686, 688, 692, 693, 700
Load monitoring, 326
Longitudinal muscle growth, 238, 465
Low-level laser therapy (LLLT), 675, 685, 686, 689, 695, 703

M
Maladaptation, 283–298, 312, 324, 726
Manual therapy, 77, 463, 523, 524, 531, 533, 535, 571–573, 582, 600, 607, 611, 612, 619, 627
Massage, 77, 395, 396, 449, 461, 463, 464, 523, 524, 532–535, 600, 602, 604–607, 609–612, 626, 627, 716
Mechanical deformation, 76, 169–171, 178, 573
Mechanical loading, 165, 167–169, 171, 172, 174, 175, 178, 284, 431, 453, 454, 465
Mechanosensors, 76, 77, 81–87, 95, 164, 168–169, 174, 178, 284
Mechanotransduction, 76, 131, 163–178, 221, 267, 453, 489, 576–578, 586, 738
Mental health, 377–382, 660
Mesenchymal cells, 24
MicroStretching®, 4, 450, 451, 461, 479–512, 535
Microvascular, 23, 123–152, 466
Mindfulness, 338, 365–384
Mitochondria, 142, 395, 648, 649, 663, 677, 679, 680, 682–684, 686, 690, 692, 704, 757
Mobilization, 108, 115, 215, 218, 223, 463, 524, 528, 531, 532, 536–538, 572–575, 582, 585, 751
Molecular and cellular responses, 178
Monitoring, 91, 289, 291–295, 304–308, 313, 324, 329, 331, 369, 375, 396, 397, 400, 432, 729
Motor control, 42, 43, 52, 53, 55, 85, 289, 503, 608, 615, 677
Motor points, 556–561, 566, 639, 640
Movement, 1, 2, 13, 25, 41–63, 81–85, 93–95, 106, 113, 129, 132, 133, 165, 167, 185, 186, 197, 201, 202, 286, 319, 329, 332, 348, 355, 447, 448, 450, 452, 456, 458, 459, 466, 479, 480, 482, 483, 487, 491, 492, 498, 503, 524, 527, 530, 531, 533, 534, 536–539, 543–545, 556, 560, 561, 565, 566, 574–581, 585, 599, 604, 606, 608–610, 614, 615, 618, 621, 624–626, 635, 636, 640, 648, 656, 728
Multi-disciplinary, 7, 598, 627
Muscle, 1, 13, 42, 75, 112, 123, 164, 185, 213, 235, 312, 319, 347, 394, 409, 430, 445, 479, 523, 560, 574, 601, 635, 650, 676, 716, 737, 746
Muscle adaptation, 165, 178, 204, 205, 221–226, 286, 419, 434, 439, 726
Muscle architecture, 235, 236, 270, 451, 467
Muscle damage, 13, 60, 141, 188, 213, 254, 329, 352, 431, 448, 497, 605, 680, 719
Muscle hypertrophy, 61, 175–177, 204, 205, 296, 297, 352, 454
Muscle lengthening, 128, 141, 213
Muscle spindles, 51, 53, 56, 77, 84–88, 91–95, 487–489, 497–501, 577, 601
Muscle stiffness, 85, 188, 192, 197, 312, 452, 458–460, 464, 466
Musculoskeletal, 2, 4, 76, 271, 332, 445, 458, 463–466, 481, 487, 492, 494, 502, 526, 531, 534, 535, 546, 559, 561, 563, 566, 581, 635, 641, 651, 656, 657, 707, 716, 737–742
Myofascial therapy, 527

N
Naps, 306, 355, 356
Neural system, 78, 722
Neuromuscular activation, 61–63

Neuromuscular electrical stimulation (NMES), 635–642
Nonsteroidal anti-inflammatory drugs (NSAIDs), 427–438, 722
Nutrient deficiency, 413, 420
Nutrition, 165, 333, 334, 354, 355, 373, 391–402, 407, 409–411, 417, 420, 659, 662–665, 677

O

Overload, 216, 221, 283, 284, 290, 305, 321, 418, 523, 524, 577, 720, 721, 742
Overtraining, 105, 256, 290–295, 298, 312, 323, 324, 329, 335, 354, 694

P

Pain, 2, 27, 50, 81, 189, 370, 400, 427, 446, 481, 523, 555, 597, 635, 649, 677, 714, 739, 753
Pain management, 3, 380–382, 436, 530, 542, 641, 651, 739
Parasympathetic nervous system (PNS), 372, 374, 382, 605, 611
Performance, 45, 93, 105, 123, 200, 216, 251, 305, 318, 349, 367, 391, 407, 429, 445, 491, 523, 565, 572, 600, 638, 649, 676, 724
Periodization, 62, 318, 321–323, 325, 334, 335, 354, 726
Pharmacological, 216, 431, 433, 436, 496
Photobiomodulation (PBM), 4, 675–709
Physical preparation, 323, 334
Physical therapy, 4, 381, 523–546, 627, 655–657, 747
Pneumothorax, 752, 758
Post-exercise recovery techniques, 713
Proactive planning, 295
Proprioception, 45, 47, 52, 56–58, 63, 75–95, 333, 479, 481, 488, 489, 532, 537, 542, 545, 580, 604, 662
Pulsed electromagnetic field therapy (PEMF), 4, 647–665

R

Range of motion/flexibility, 25, 54–56, 192, 251, 264, 269, 271, 333, 432, 445–461, 464, 466, 491–493, 495, 502, 507, 523, 524, 527–529, 531, 534–538, 542, 560, 574, 583, 584, 600, 603, 608, 612, 614, 618, 638, 728
Recommended dietary intake, 409, 419
Recovery, 14, 44, 76, 111, 141, 195, 215, 244, 306, 317, 347, 372, 391, 407, 429, 445, 481, 523, 565, 586, 599, 635, 675, 713, 753
Recovery strategies, 295, 297, 298, 306, 336, 391–395, 399–402, 657
Red light therapy, 675
Regeneration, 2–8, 13–31, 76, 85, 141–143, 146, 149, 168–169, 175, 177, 178, 215–217, 219–223, 288, 297, 317, 347, 349, 350, 352, 395, 407, 411–419, 433, 446, 450, 451, 465–467, 486, 489, 495, 502, 507, 524, 527, 535, 546, 555–566, 586, 597–628, 648, 650, 651, 655–657, 664, 665, 675, 677, 681, 689, 724, 738, 739, 746, 751, 754, 756
Rehabilitation, 7, 42, 44, 48, 52, 55, 56, 61–63, 199, 365–384, 433, 434, 446, 447, 463–465, 483, 492, 524, 528, 529, 532, 535, 539, 540, 543, 556, 557, 566, 584, 600, 604, 605, 635, 637, 638, 651, 655, 657, 665, 706, 707, 748, 753, 754, 764
Remodelling, 3, 21, 23, 427, 446, 454, 656
Repair, 3, 4, 14, 15, 18–21, 23–25, 28, 30, 141–143, 145, 152, 215, 219, 222, 337, 354, 433, 437, 455, 465, 580, 602, 605, 650, 675–708, 739, 747
Repeated bout effect, 17, 195–200, 204, 216, 217, 329
Research, 3, 22, 59, 76, 150, 198, 218, 268, 307, 322, 347, 367, 393, 418, 430, 448, 481, 528, 556, 571, 637, 648, 675, 714, 742, 751

S

Sarcomere, 13–17, 91, 113, 128, 164, 166, 172–174, 197, 213, 214, 235–271, 450, 454, 458, 465, 489, 500, 535
Sarcomerogenesis, 236–241, 254–258, 260, 261, 266–269, 271, 453, 535
Satellite cells, 14, 15, 19–25, 28–30, 141, 142, 168–169, 174, 177, 178, 215, 216, 218–220, 222, 267, 433, 434, 683, 689, 754
Segmental acupuncture, 561
Self-regulation skills, 396, 397
Sex differences, 15, 26, 194
Skeletal muscle, 1, 13, 75, 107, 123, 164, 185, 215, 236, 319, 352, 395, 416, 439,

451, 497, 524, 560, 602, 637, 658, 720, 753
Skeletal muscle damage, 145–150, 352, 416, 433
Sleep, 2, 58, 306, 318, 347, 374, 395, 508, 524, 613, 660, 694, 719, 753
Sleep hygiene, 356, 728
Soft tissue, 4, 325, 429, 436–437, 448–450, 461, 493, 523–536, 544, 575, 583, 597–628, 650, 656, 679, 707, 717, 721, 742, 751
Somatic practices, 77, 480
Spinal manipulation, 572, 578–580, 582, 583
Spinal manipulative therapy (SMT), 571–574, 576–586
Spinal mobilizations, 574
Spine pain, 572, 581, 586
Sport massage, 607
Sports, 82, 105, 140, 185, 217, 264, 305, 318, 353, 365, 391, 407, 427, 447, 482, 523, 566, 582, 603, 637, 660, 718, 737, 753
Sports medicine, 4, 214, 394, 400, 420, 445, 483, 530, 535, 557, 648, 655, 663, 665, 707, 714, 719, 721, 737–742, 753, 764
Sports performance, 291, 298, 320, 321, 325, 330, 368–370, 375, 457, 546, 582, 600, 607, 657–659, 665, 678
Stress, 2, 13, 51, 105, 131, 201, 214, 307, 317, 365, 396, 407, 430, 454, 479, 523, 566, 577, 605, 650, 675, 718, 739, 750
Stretching, 4, 51, 61, 77, 165, 166, 171, 174, 247, 257, 260, 261, 269, 270, 445–467, 479–511, 525–528, 535, 536, 538, 575, 585, 604, 622, 626, 656, 697
Stretch tolerance, 445, 450, 451, 453, 491, 493, 535
Supplements, 216, 217, 355, 392, 396, 408, 411–415, 420
Sympathetic nervous system (SNS), 106, 107, 124, 133–134, 138, 374, 383, 501, 533, 602, 604, 719, 721, 722

T
Tapping, 48
Tendinopathy, 618, 623–625, 657, 690, 721, 737, 739–742
Therapy, 4, 28, 200, 377, 464, 504, 524, 558, 573, 600, 648, 694, 715, 737, 751
Tissue remodelling, 427, 446
Training, 1, 13, 42, 105, 127, 165, 187, 217, 241, 304, 317, 350, 365, 391, 407, 429, 445, 483, 535, 557, 573, 598, 638, 658, 680, 714, 753
Training load, 290, 294, 297, 305–311, 318, 321–326, 334, 351, 431, 726
Transcutaneous electrical nerve stimulation (TENS), 641, 642
Traumatic brain injury (TBI), 651, 661, 697–699, 756–757
Trigger points, 449, 464, 523–525, 527, 556–559, 563, 616, 740

The manufacturer's authorised representative in the EU is Springer Nature Customer Service Centre GmbH, Europaplatz 3, 69115 Heidelberg, Germany. If you have any concerns regarding our products, please contact ProductSafety@springernature.com

Printed and bound by CPI Group (UK) Ltd, Croydon, CR0 4YY
30/09/2026
02225590-0019